TREATISE ON ANALYTICAL CHEMISTRY

PART I
THEORY AND PRACTICE
SECOND EDITION

TREATISE ON ANALYTICAL CHEMISTRY

PART I

THEORY AND PRACTICE

SECOND EDITION

VOLUME 8

Edited by PHILIP J. ELVING

Department of Chemistry, University of Michigan

Associated Editor: EDWARD J. MEEHAN

Department of Chemistry, University of Minnesota

Editor Emeritus: I. M. KOLTHOFF

School of Chemistry, University of Minnesota

AN INTERSCIENCE® PUBLICATION

JOHN WILEY & SONS
New York—Chichester—Brisbane—Toronto—Singapore

Professor Philip J. Elving died on March 16th, 1984. The Publisher gratefully acknowledges Professor Elving's many contributions to the successful publication of the TREATISE; he will be missed.

An Interscience® Publication

Library of Congress Cataloging in Publication Data:

(Revised for pt. 1, v. 8)

Kolthoff, I. M. (Izaak Maurits), 1894–
  Treatise on analytical chemistry.

  "An Interscience publication."
  Includes bibliographies and indexes.
  Pt. 1, v. 8–   : edited by Philip J. Elving:
associate editor, Edward J. Meehan.
  Contents: pt. 1. Theory and practice.
  1. Chemistry, Analytic. I. Elving, Philip Juliber,
1913–   . I. Title.

QD75.2.K64 1978   543   78-1707
ISBN 0-471-07995-2 (pt. 1, v. 8)

Printed in the United States of America

10 9 8 7 6 5 4 3 2 1

TREATISE ON ANALYTICAL CHEMISTRY

PART I
THEORY AND PRACTICE

VOLUME 8: SECTION H
Optical Methods of Analysis   *Chapters 9–17*

AUTHORS OF VOLUME 8

FRED W. BILLMEYER, JR.
ROBERT K. BOHN
GEORGE G. COCKS
R. W. CHRISMAN
JOHN G. DELLY
C. E. FIORI
R. O. KAGEL
H. A. LIEBHAFSKY
S. H. LIN
LUCY B. McCRONE

WALTER C. McCRONE
GIORGIO MARGARITONDO
E. A. MEYERS
GEORGE D. J. PHILLIES
J. A. ROPER III
JOHN E. ROWE
E. A. SCHWEIKERT
D. D. SHIEH
C. R. SWYT
NANCY S. TRUE

Authors of Volume 8

Dr. Robert K. Bohn

Department of Chemistry
University of Connecticut
Storrs, Connecticut 06268
Chapter 11

Dr. R. W. Chrisman

Dow Chemical U.S.A.
2030 Willard H. Dow Center
Midland, Michigan 48640
Chapter 10

Dr. George G. Cocks

E55-4, MS J979
Los Alamos National Laboratory
P.O. Box 1663
Los Alamos, New Mexico 87545
Chapter 16

Dr. John G. Delly

Walter C. McCrone Associates, Inc.
2820 South Michigan Avenue
Chicago, Illinois 60606
Chapter 15

Dr. Ronald O. Kagel

Dow Chemical U.S.A.
2030 Willard H. Dow Center
Midland, Michigan 48640
Chapter 10

Dr. H. A. Liebhafsky *is deceased*

A copy of the volume (and royalties)
go to his widow
Sybil Small Liefhafsky
1116 Hedgewood Lane
Schenectady, N.Y. 12309
Chapter 13

Dr. S. H. Lin

Department of Chemistry
Arisona State University
Tempe, Arizona 85281
Chapter 12

Dr. Giorgio Margaritondo

Department of Physics
University of Wisconsin
1150 University Avenue
2531 Sterling Hall
Madison, Wisconsin 53706
Chapter 17

Dr. Lucy B. McCrone

Walter C. McCrone Associates, Inc.
2820 South Michigan Avenue
Chicago, Illinois 60606
Chapter 15

Dr. Walter C. McCrone

Walter C. McCrone Associates, Inc.
2820 South Michigan Avenue
Chicago, Illinois 60606
Chapter 15

Dr. E. A. Meyers

Department of Chemistry
Texas A & M University
College Station, Texas 77843
Chapter 13

Dr. George D. J. Phillies

Department of Chemistry
University of Michigan
Ann Arbor, Michigan 48109
Chapter 9

Dr. J. A. Roper

Dow Chemical U.S.A.
2030 Willard H. Dow Center
Midland, Michigan 48640
Chapter 10

Dr. John E. Rowe

Bell Laboratories
IC-318
600 Mountain Avenue
Murray Hill, NJ 07974
Chapter 17

Dr. E. A. Schweikert

Department of Chemistry
Texas A & M University
College Station, Texas 77843
Chapter 13

Dr. Nancy S. True

Department of Chemistry
University of Connecticut
Storrs, Connecticut 06268
Chapter 11

Authors Not Shown

Dr. Fred W. Billmeyer, Jr.

Department of Chemistry
Rensselaer Polytechnic Institute
Troy, New York 12181
Chapter 9

Dr. C. E. Fiori

Division of Research Services
Biomedical Engineering and
Instrumentation Branch
National Institutes of Health
Bethesda, Maryland 20205
Chapter 14

Dr. D. D. Shieh

Chemical Abstract Service
Columbus, Ohio
Chapter 12

Dr. C. R. Swyt

Division of Research Services
Biomedical Engineering and
Instrumentation Branch
National Institutes of Health
Bethesda, Maryland 20205
Chapter 14

Foreword

The division of chapters among Volumes 7, 8, and 9, which collectively will constitute the section "Optical Methods of Analysis," was dictated in part by unforeseen delays in the preparation of some of the individual chapters; for example, some authors who originally undertook the preparation of chapters were unable to fulfill their commitments and other experts had to be enlisted to prepare the chapters in question. In fairness to the authors whose manuscripts were ready for publication, and to the users of the Treatise, it was decided to proceed with publication of the present volume, which is the second one on optical methods.

Volume 9 will contain the remaining chapters of the section on optical methods.

Preface to the Second Edition of the Treatise

In the mid-1950s, the plan ripened to edit a "Treatise on Analytical Chemistry" with the objective of presenting a comprehensive treatment of the theoretical fundamentals of analytical chemistry and their implementation (Part I) as well as of the practice of inorganic and organic analysis (Part II); an introduction to the utilization of analytical chemistry in industry (Part III) was also considered. Before starting this ambitious undertaking, the editors discussed it with many colleagues who were experts in the theory and/or practice of analytical chemistry. The uniform reaction was most skeptical; it was not thought possible to do justice to the many facets of analytical chemistry. Over several years, the editors spent days and weeks in discussion in order to define not only the aims and objectives of the Treatise but, more specifically, the order of presentation of the many topics in the form of a table of contents and the tentative scope of each chapter. In 1959, Volume 1 of Part I was published. The reviews of this volume and of the many other volumes of Part 1 as well as of those of Parts II and III have been uniformly favorable, and the first edition has become recognized as a contribution of classical value.

Even though analytical chemistry still has the same objectives as in the 1950s or even a century ago, the practice of analytical chemistry has been greatly expanded. Classically, qualitative and quantitative analysis have been practiced mainly as "solution chemistry." Since the 1950s, "solution analysis" has involved to an ever increasing extent physicochemical and physical methods of analysis, and automated analysis is finding more and more application, for example, its extensive utilization in clinical analysis and production control. The accomplishments resulting from automation are recognized even by laymen, who marvel at the knowledge gained by automated instruments in the analysis of the surfaces of the moon and of Mars. The computer is playing an ever increasing role in analysis and particularly in analytical research. This revolutionary development of analytical methodology is catalyzed by the demands made on analytical chemists, not only industrially and academically but also by society. Analytical chemistry has always played an important role in the development of inorganic, organic, and physical chemistry and biochemistry, as well as in that of other areas of the natural sciences such as mineralogy and geochemistry. In recent years, analytical chemistry—often of a rather sophisticated nature—has become increasingly important in the medical and biological sciences, as well as in the solving of such social problems as environmental pollution, the tracing of toxins, and the dating of art and archaeological objects, to mention only a few. In the area of atmospheric science, ozone reactivity and persistence in the stratosphere is presently a topic of great priority; extensive analysis is required both for monitoring atmospheric constituents and for investigating model systems.

One example of the increasing demands being made on analytical chemists is the growing need for speciation in characterizing chemical species. For example, in reporting that lake water contains dissolved mercury, it is necessary to report in which oxidation state it is present, whether as an inorganic salt or complex, or in an organic form and in which form.

As a result of the more or less revolutionary developments in analytical chemistry, portions of the first edition of the Treatise are becoming—and, to some extent, have become—out-of-date, and a revised, more up-to-date edition must take its place. In recognition of the extensive development and because of the increased specialization of analytical chemists, the editors have fortunately secured for the new edition the cooperation of experts as coeditors for various specific fields.

In essence, it is the objective of the second edition of the Treatise, as it was of the first edition (whose preface follows this one), to do justice to the theory and practice of contemporary analytical chemistry. It is a revision of Part I, which mirrors the development of analytical chemistry. Like the first edition, the second edition is not an extensive textbook; it attempts to present a thorough introduction to the methods of analytical chemistry and to provide the background for detailed evaluation of each topic.

Minneapolis, Minnesota   I. M. KOLTHOFF
Ann Arbor, Michigan   P. J. ELVING

Preface to the First Edition of the Treatise

The aims and objectives of this Treatise are to present a concise, critical, comprehensive, and systematic, but not exhaustive, treatment of all aspects of classical and modern analytical chemistry. The Treatise is designed to be a valuable source of information to all analytical chemists, to stimulate fundamental research in pure and applied analytical chemistry, and to illustrate the close relationship between academic and industrial analytical chemistry.

The general level sought in the Treatise is such that, while it may be profitably read by the chemist with the background equivalent to a bachelor's degree, it will at the same time be a guide to the advanced and experienced chemist—be he in industry or university—in the solution of his problems in analytical chemistry, whether of a routine or of a research character.

The progress and development of analytical chemistry during most of the first half of this century has generally been satisfactorily covered in modern textbooks and monographs. However, during the last fifteen or twenty years, there has been a tremendous expansion of analytical chemistry. Many new nuclear, subatomic, atomic, and molecular properties have been discovered, several of which have already found analytical application. In the development of techniques for measuring these and also the more classical properties, the revolutionary progress in the field of instrumentation has played a tremendous role.

It has been difficult, if not impossible, for anyone to digest this expansion of analytical chemistry. One of the objectives of the present Treatise is not only to describe these new properties, their measurement, and their analytical applicability, but also to classify them within the framework of the older classifications of analytical chemistry.

Theory and practice of analytical chemistry are closely interwoven. In solving an analytical chemistry problem, a thorough understanding of the theory of analytical chemistry and of the fundamentals of its techniques, combined with a knowledge of and practical experience with chemical and physical methods, is essential. The Treatise as a whole is intended to be a unified, critical, and stimulating treatment of the theory of analytical chemistry, of our knowledge of analytically useful properties, of the theoretical and practical fundamentals of the techniques for their measurement, and of the ways in which they are applied to solving specific analytical problems. To achieve this purpose, the Treatise is divided into three parts: I, analytical chemistry and its methods; II, analytical chemistry of the elements; and III, the analytical chemistry of industrial materials.

Each chapter in Part I of the Treatise illustrates how analytical chemistry draws on the fundamentals of chemistry as well as on those of other sciences;

it stresses for its particular topic the fundamental theoretical basis insofar as it affects the analytical approach, the methodology and practical fundamentals used both for the development of analytical methods and for their implementation for analytical service, and the critical factors in their application to both organic and inorganic materials. In general, the practical discussion is confined to fundamentals and to the analytical interpretation of the results obtained. Obviously then, the Treatise does not intend to take the place of the great number of existing and exhaustive monographs on specific subjects, but its intent is to serve as an introduction and guide to the efficient utilization of these specialized monographs. The emphasis is on the analytical significance of properties and of their measurement. In order to accomplish the above aims, the editors have invited authors who are not only recognized experts for the particular topics, but who are also personally acquainted with and vitally interested in the analytical applications. Only in this way can the Treatise attain the analytical flavor which is one of its principal objectives.

Part II is intended to be very specific and to review critically the analytical chemistry of the elements. Each chapter, written by experts in the field, contains in addition to a critical and concise treatment of its subject, critically selected procedures for the determination of the element in its various forms. The same critical treatment is contemplated for Part III. Enough information is presented to enable the analyst both to analyze and to evaluate a product.

The response in connection with the preparation of the Treatise from all colleagues has been most enthusiastic and gratifying to the editors. It is obvious that it would have been impossible to accomplish the aims and objectives cited in the Preface without the wholehearted cooperation of the large number of distinguished authors whose work appears in this and future volumes of the Treatise. To them and to our many friends who have encouraged us we express our sincere appreciation and gratitude. In particular, considering that the Treatise aims to cover all of the aspects of analytical chemistry, the editors have found it desirable to solicit the advice of some colleagues in the preparation of certain sections of the various parts of the Treatise. They would like at this time to acknowledge their indebtedness to Professor Ernest B. Sandell of the University of Minnesota for his interest and active cooperation in the organizing and detailed planning of the Treatise.

Minneapolis, Minnesota   I. M. KOLTHOFF
Ann Arbor, Michigan   P. J. ELVING

Acknowledgment

In view of the wide scope of the Treatise, it has been considered essential to have the advice and aid of experts in various areas of analytical chemistry. For the section "Optical Methods of Analysis," the editor was fortunate to have the cooperation of Dr. Edward J. Meehan of the University of Minnesota as Associate Editor; his collaboration is acknowledged with gratitude.

P. J. ELVING

PART I.   THEORY AND PRACTICE

CONTENTS   VOLUME 8

SECTION H.   Optical Methods of Analysis

13.   The Nature of X-rays, Spectrochemical Analysis by Conventional X-ray Methods and Neutron Diffraction and Absorption.

17.   Electron Spectroscopy.

Part I
Section H

Chapter 9

ELASTIC AND QUASIELASTIC LIGHT SCATTERING BY SOLUTIONS AND SUSPENSIONS

By George D. J. Phillies, *Worcester Polytechnic Institute, Worcester, Massachusetts*

With the cooperation of

Fred W. Billmeyer, Jr.,* *Consultant, Schenectady, New York*

Contents

* Author of Section II.D.

I. INTRODUCTION

A. FUNDAMENTAL PHENOMENA

The scattering of light is a phenomenon of widespread occurrence, interest, and utility. Light scattering can be produced for scientific purposes in gases, liquids, solids, and plasmas; it is the source of such natural phenomena as the opacity of fog and the color of some smokes. All matter can induce light scattering, with the limitation that a sample whose index of refraction is momentarily homogeneous over light-wavelength distances does not at that moment scatter light from its bulk.

The scattering of light by gases is observed in air (81) (where it accounts for the luminosity of the daytime sky); it is also seen in pure liquids and in solids. Scientific studies have emphasized scattering in solutions of large molecules and scattering by suspended particles. Instances of the latter phenomenon include scattering by liquids or solids suspended in gases or liquids (examples include aerosols, ice clouds, and macroparticle solutions) and scattering by interstellar dusts.

Light scattering is a process in which a substance removes energy from a beam of light and re-emits the energy, still in the form of light. In the classical description of elastic scattering, whenever a beam of light encounters a particle, the electrically charged electrons and nuclei in the material are induced to vibrate at the frequency of the incident light. The oscillating charges absorb the incident light, and then re-emit it in almost all directions. Classically, it was asserted that elastically scattered light has precisely the same frequency as the light in the exciting beam. This statement is not exactly correct. With the development of quasielastic-scattering methods, it has become possible to measure the extremely small frequency shifts ($\delta v/v \approx 10^{-12 \pm 3}$) resulting from the molecular motions within a sample.

Scattering phenomena are traditionally described as being elastic, inelastic, or quasielastic. "Inelastic scattering" usually refers to processes (e.g., Raman scattering, fluorescence, and phosphorescence) in which along with the change in the energy of the light due to the scattering process, there is a concomitant well-defined change in the quantum-mechanical energy level of the sample. For example, in Raman spectroscopy, there is a measurable change in the frequency of the light and a simultaneous change in the vibrational or rotational energy level of a molecule in the sample. Inelastic-scattering methods are treated in Chapter 10 and the chapters dealing with fluorescence and phosphorescence.

In contrast, "elastic scattering" experiments investigate processes in which the quantity of interest is the intensity of light that was scattered without an appreciable change in frequency, if changes in frequency due to molecular motions are neglected. Classical light-scattering photometry is the prime example of an elastic-scattering method.

Finally, quasielastic scattering treats those processes in which the frequency of the light is changed slightly by the molecular motions, but the change in the energy of the sample cannot be identified with particular internal energy levels. Quasielastic light scattering from solutions thus includes Brillouin scattering (from thermally excited sound waves) and the so-called Rayleigh scattering from concentration and energy-density ("temperature") fluctuations. In Brillouin and Rayleigh scattering, there is a measurable change in the frequency of the light due to the scattering. Since energy is conserved, the energy gained or lost by the light must have been exchanged with the sample. However, this energy cannot in general be associated with well-defined energy levels of particular molecules.

Elastic light scattering, properly speaking, is a special case of quasielastic light scattering. If one illuminates a sample and filters the scattered light to eliminate fluorescent and Raman-shifted photons, the remaining photons are due entirely to Brillouin, Rayleigh, and other quasielastic processes. If one measures only the intensity of the filtered light, one is studying elastic scattering. If one also measures the very small shifts in the frequency of the re-emitted light, one is studying quasielastic scattering.

Interest in light scattering has recently increased, due in no small part to the noninvasive nature of this technique. In most cases, one may study a sample without significantly perturbing its physical properties and without altering the course of ongoing physical or chemical processes. This chapter treats primarily light scattering by solutions and suspensions of large molecules and polymers. Important analytical applications of such studies include determination of the molecular weights, sizes, and diffusion coefficients of proteins, oligonucleotides, synthetic macromolecules, and colloidal particles.

The determination of the solids content of a suspension by light-scattering and absorption measurements was traditionally carried out by means of nephelometry and turbidimetry. Scattering and absorption by concentrated suspensions of pigments, which accounts for the color of many paints and inks, is treated in part in this chapter and in part in the chapter on the specification and designation of color.

B. SCOPE OF THIS CHAPTER

This chapter treats analytical methods that involve the scattering of light. Applications of light scattering in physical chemistry, such as in studies of critical phenomena, are emphasized here only to the extent that they appear to have significant potential for analytical application. Central to this discussion are the concepts of the "intensity" and the "spectrum" of the scattered light. Studies of the total intensity of the scattered light are the concern of elastic light scattering. Measurement of the spectrum of the scattered light is the domain of quasielastic light scattering. The spectrum in question is the frequency spectrum of the light, that is, the amount of light present at each frequency in a narrow range of frequencies around the frequency of the incident light. (As discussed in Section III, spectra may also be represented in the time domain as correlation functions.) The line width, lineshape, and side-band splittings contain the spectral information; in quasielastic light scattering the frequency of the spectral line center is generally not significant. The intensity of the scattered light is the integrated area under the spectrum when the spectrum is plotted as power versus frequency (hence the term "integrated intensity").

Sections I.C–I.E of this chapter discuss the fundamental form of a scattering experiment, interesting properties of light, and general references on scattering. Section II deals with elastic light scattering. Sections II.A–II.C describe the fundamental physical bases of the Rayleigh, Rayleigh–Gans–Debye (or Rayleigh–Debye), and generalized Mie scattering theories. The Rayleigh theory applies to extremely dilute dispersions in which each particle is small compared with the wavelength of light. The Rayleigh–Gans–Debye theory applies to transparent, optically isotropic particles, which may be comparable in size to a light wavelength. The applications of the Rayleigh–Gans–Debye theory to liquids, to particles that are small compared with a light wavelength, to large transparent particles, to polymers, and to Zimm plots are discussed. Many analytical uses of light scattering, such as determination of the molecular weights, virial coefficients, and molecular sizes of macromolecules, involve Rayleigh–Gans–Debye concepts. Mie's original theory (79) treated uniform spheres; the generalized Mie theory treats particles of arbitrary index of refraction and size. Theories of turbid media are developed in Section II.D. Section II.E deals with experimental methods in light scattering, including apparatus, sample preparation, absolute calibration methods, and applications to polymer solutions, critical phenomena, and streaming suspensions.

Section III treats quasielastic light scattering. Section III.A is an introduction. Section III.B presents fundamental concepts, including correlation functions, spectra, and the relationship of spectra to molecular motions.

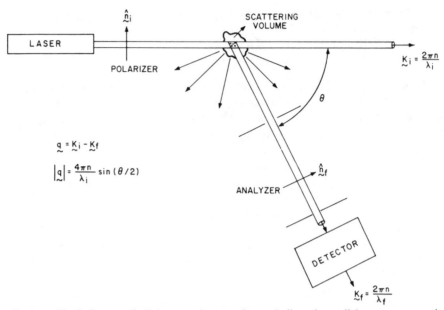

Fig. 9.1.   Block diagram of a light-scattering experiment. Indicated are a light source, a sample that scatters the light, and a detector. The light is characterized by incident and scattered polarizations ($\hat{\mathbf{n}}_i$, $\hat{\mathbf{n}}_f$) and by incident and final wavevectors ($\mathbf{K}_i$, $\mathbf{K}_f$). (From reference 10.)

Section III.C treats concepts that are fundamental to the practice of light-scattering spectroscopy, such as coherence areas, photon statistics, homodyne and heterodyne detection, correlators, methods of data analysis, and the meaning of the diffusion coefficient. Section III.D deals with experimental equipment, including lasers, correlators, detectors, and sample cells. Section III.E gives a workable procedure for such experiments, and Section III.F discusses the information that can be gained from the study of Brownian particles and small molecules. Section III.G introduces special methods for multiple-scattering suppression, for separation within the scattering cell, for light-scattering electrophoresis, and for micro-scale samples.

C.   SCATTERING EXPERIMENTS

Almost all scattering experiments are described by the same block diagram (Fig. 9.1), consisting of a source, a sample, and a detector; their relative positions determine the scattering angle θ. The three points occupied by the source, the sample, and the detector determine the scattering plane. Regardless of its actual orientation, the scattering plane is by custom defined to be horizontal. In a few specialized cases, one uses several sources, samples, or detectors simultaneously, as discussed in Section III.G.4 (46,89).

The source controls the properties (most notably the power, frequency, polarization, and amplitude stability) of the incident light. In the techniques discussed here, the incident light is monochromatic and of constant intensity.

In quasielastic scattering, the amplitude stability of the source—the absence of high-frequency irregularities in the incident power—is at least as important as the source's average power.

The sample may be a solid, liquid, or gas; in general, any transparent object may be studied by light scattering. The essential optical requirement is that the sample must scatter enough light to give a perceptible signal. One usually requires that multiple scattering of light by the sample not be significant. The study of turbid samples is treated in Sections II.D and III.G.4.

The fundamental detectable properties of the scattered radiation are the intensity and the spectrum, both of which may depend on the polarization of the light and on the scattering angle. The measured intensity of the scattered light depends in part on experimental conditions, including the intensity of the incident light and the volume from which scattered light is actually collected. If one corrects for these experimental factors, one obtains the intrinsic scattering cross section of the material being studied. The intensity of the scattered light contains information about the size, shape, and relative positions of the scattering particles. In particular, from measurements of the intensity at different concentrations and scattering angles, one may infer the molecular weight, second virial coefficient, and radius of gyration of a polymer or other molecule in solution.

The spectrum of the scattered light is determined directly by the exchange of energy between the radiation and the sample, and indirectly by the time-dependent properties of the sample. In an equilibrium system, local mechanical properties (density, energy, and momentum) fluctuate around their thermodynamic average values, so that at thermal equilibrium a material object still exhibits time-dependent behavior around a time-independent average. From the spectrum, one can derive information about transport and response coefficients. In particular cases, one can infer from the spectrum the sample's translational or rotational diffusion coefficients, thermal conductivity, viscosity, or specific heat, or the speed of sound in the sample.

Quasielastic light scattering is known by a variety of names, including photon correlation spectroscopy (PCS), dynamic light scattering (DLS), laser light scattering, inelastic light scattering (ILS), Rayleigh line broadening, light beating spectroscopy, optical mixing spectroscopy, and intensity fluctuation spectroscopy (IFS). (Only those abbreviations that the author has seen in the literature are noted here.) This plethora of synonyms has interesting effects on computer-based literature searches.

D.   LIGHT

The theory of light is treated in standard monographs (e.g., reference 59); we summarize here those results that are central to our discussion. We first treat terms for describing the amount of light; we then discuss the physical properties of light.

To avoid confusion, we begin by discussing the concepts of the power, intensity, and brightness of a beam of light. There is a significant difference between the way these and related terms are used in physics (e.g., see reference 59) and in radiometry (18,47). In this chapter, we will use the physics terminology, which is more likely to be familiar to most readers. We will note here how these two sets of terms are related, and in the following will refer to the appropriate radiometric terminology when appropriate.

The power in a beam of light is the total energy per unit time delivered by that beam; power is expressed in such units as watts. The intensity of a beam of light at a point (cf. equation 2, below) is the power per unit area at that point, and is expressed in such units as watts per square centimeter. If the intensity in question is that of a beam of light incident on a sample, in radiometric terms the intensity is properly known as the radiance. If the intensity in question is that of a beam of light that has been scattered from a sample, in radiometric terms the intensity is known as the irradiance. In radiometry, the irradiance is assigned the units of power per unit area per steradian.

The intensity and the irradiance are assigned different units, namely power per area and power per area per steradian, respectively. These units are not inconsistent. To see why, recall the difference between the units of a quantity (e.g., Joules per second, liters) and the dimensions of a quantity. There are a large number of named units; however, all physical quantities can be reduced to products of powers of three fundamental dimensions: length, time, and mass. (A pure number has the dimension of unity.) In particular, power has dimensions of (mass) $\cdot$ (length)^2/(time)^3, and areas have dimensions of (length)^2. An angle is the ratio of a part of the circumference of a circle (a length) to the radius of that circle (another length). Angles (and steradians) thus have dimension unity. The scattered intensity (in physics) and the irradiance thus have the same dimensions: (mass)/(time)^3.

It is worthwhile to explain why radiometric convention assigns irradiance units of power per area per steradian, and why these units are numerically and operationally the same as the units assigned to the intensity. Suppose one has a uniform source, that is, a source that emits light at the same rate in every direction. The total rate at which light is emitted by the source, if expressed in energy units, is the power P of the light source. For many purposes, one wishes to discuss the power sent into some range of solid angle Ω. To describe this quantity, one uses the radiant intensity $dP/d\Omega$, which has units of power per steradian; in our special case, the radiant intensity is $P/4\pi$. One may also wish to treat the power received from the source by a detector located at a distance R from the source. This requires introducing the irradiance, which is obtained by dividing the radiant intensity by R^2. The irradiance thus has units of power per steradian per (length)^2; in our special case the irradiance of the source is $P/4\pi R^2$. On the other hand, if one follows, for example, reference 59, the power received by a detector is to be calculated from the intensity—the power per unit area—at the detector. Since the area of a sphere is $4\pi R^2$, in our example the intensity incident on a sphere centered at the source is $P/4\pi R^2$. The irradiance is thus exactly identical, numerically and operationally, to the intensity, as this term is used here. However, in the former case $4\pi R^2$ is assigned the units area · steradians, whereas in the latter case $4\pi R^2$ is assigned the units of area. Two different names have been given to the same measurable quantity.

Light is an electromagnetic wave. Its wavelength in a vacuum is here denoted by λ. While light is passing through a substance, its wavelength is reduced to $\lambda' = \lambda/n$, where n is the index of refraction of the medium. The frequency of light does not change when it passes from a vacuum into a transparent object. To describe the motion of light, it is convenient to introduce the wavevector $\mathbf{k}$ of the light. The wavevector points in the direction of the light's propagation and has a magnitude of $2\pi n/\lambda$.

The equation for light as a wave can be written in the equivalent forms

$$E(\mathbf{r}, t) = a \cos(\mathbf{k} \cdot \mathbf{r} - \omega t) + a' \sin (\mathbf{k} \cdot \mathbf{r} - \omega t) \tag{1a}$$

$$E(\mathbf{r}, t) = b \exp(i\mathbf{k} \cdot \mathbf{r} - i\omega t) \tag{1b}$$

Here $E(\mathbf{r}, t)$ is the amplitude of the wave's electric field at the point $(\mathbf{r}, t)$; a, a', and b are constants; and ω is the frequency of light in radians per second. In general, b has both real and imaginary components; Chu (25) provides an interpretation of complex electric field amplitudes. The complex exponential form, although less familiar than the sine–cosine form, is computationally convenient, because a product of complex exponentials is itself a complex exponential.

Light is a transverse wave; the oscillations of its electric and magnetic fields are confined to the plane perpendicular to the wave's direction of propagation. Because the electric field vector E of a light beam is confined to a plane, it has two components. The intensity I of light is related to its electric field E by the equation

$$I(\mathbf{r}, t) = \frac{c}{8\pi} \sqrt{\frac{\varepsilon}{\mu}} \, | E(\mathbf{r}, t) |^2 \tag{2}$$

with all values in cgs–esu units. A constant other than $(c/8\pi)\sqrt{\varepsilon/\mu}$ must be used if different units are used for E. The intensity has dimensions of power per area or energy per area per time. Since E is a vector with two independent components, the measurable intensity of a beam of light also has two independent components ("polarizations"), which can vary independently of each other. We will speak primarily of horizontally and vertically polarized light. The choice of components is not unique; for example, instead of horizontally and vertically polarized light, one could speak of left- and right-circularly polarized light. At any time, we will be interested in only one polarization, and will therefore refer to "the" electric field E (rather than to a component E_x of E) of the light.

One can only detect slow (relative to the fundamental optical frequencies of 10^{15} Hz) variations in the light intensity. If we write

$$E(\mathbf{r}, t) = a(t) \cos(\mathbf{k} \cdot \mathbf{r} - \omega t) \tag{3}$$

where $a(t)$ changes on a time scale of milliseconds or microseconds, then the time dependence of I is

$$I(t) = \tfrac{1}{2} \mid a(t) \mid^2 [1 + \cos(2\omega t)] \tag{4}$$

The function $a(t)$ is sometimes referred to as an "envelope function," and ω is the "carrier frequency." No conventional detector can respond at 10^{15} Hz, so the modulation of the intensity at the frequency ω is not apparent in intensity measurements.

If the light described by equation 3 were passed through a classical spectrometer of infinite resolving power, one would find the spectrum of the light. Equation 3 appears to contain only the single frequency ω, suggesting that the spectrum is composed of a sharp spike at a single frequency. However, it can be shown that if one correctly writes the equation for a spectrum as a sum of components, with one component at each frequency, the coefficients of the different components are independent of time. The time-dependent function $a(t)$ therefore is not the amplitude of the spectrum at the frequency ω. Instead, the frequencies implicit in $a(t)$ combine with ω in such a way that the true spectrum corresponding to equation 3 has a center at ω and a width typical of the frequencies that describe the time variations in $a(t)$.

As an argument, whose equivalence to the above will become apparent momentarily, consider a ray of light containing two different frequencies:

$$E(\mathbf{r}, t) = a_1 \cos(\mathbf{k} \cdot \mathbf{r} - \omega_1 t) + a_2 \cos(\mathbf{k} \cdot \mathbf{r} - \omega_2 t) \tag{5}$$

If the intensity of the light is measured at $\mathbf{r} = 0$, one obtains

$$\begin{aligned}
I = \ & \tfrac{1}{2}[\mid a_1 \mid^2 + \mid a_2 \mid^2] \\
& + \tfrac{1}{2}[\mid a_1 \mid^2 \cos(2\omega_1 t) + \mid a_2 \mid^2 \cos(2\omega_2 t) + 2a_1 a_2 \cos(\omega_1 + \omega_2)t \\
& + 2a_1 a_2 \cos[(\omega_1 - \omega_2)t]
\end{aligned} \tag{6}$$

by applying standard trigonometric identities. The first line of this equation is the (time-independent) average intensity of the light. The terms on the second line all oscillate at very high (10^{15} Hz!) frequencies; these will not be seen with a conventional detector. The term on the third line is a "beat note," an intensity modulation at the difference frequency $\omega_1 - \omega_2$. The measurable intensity of the light provides no information on the carrier frequency ω; however, the presence of components of slightly different frequency is revealed by the appearance of a low-frequency beat note superposed on the average intensity. By measuring the time-dependent fluctuations in the intensity of nominally monochromatic light, we can determine the spectral line shape very close to the carrier frequency. Conversely, a beam of nominally purely monochromatic light whose intensity fluctuates over time necessarily contains light at a series of slightly different frequencies; we can measure the spectral line shape of light whose line width is in the kilohertz region by looking at fluctuations in the intensity.

E. GENERAL REFERENCES

The most extensive reference book on elastic light scattering by synthetic polymers (including information on instrumentation, experimental techniques, data analysis, and the interpretation of results in specific systems) is that edited by Huglin (58). ASTM Standard D4001-81, *Standard Practice for Determination of Weight-Average Molecular Weights of Polymers by Light Scattering* (3), is also of practical value. Kerker (64) provides a thorough treatment of the calculation of scattering by particles of specified form (spheres, cylinders, and stratified bodies). Extensive discussions and tables of particle scattering are provided by van de Hulst (114) and Wickramasinghe (122). Elastic and quasielastic light scattering by liquids and solutions is more extensively discussed by Fabelinskii (41). Hermans (55) has assembled an annotated collection of classic papers on light scattering by synthetic polymers.

Chu (25), especially in his later chapters, gives a clear introduction to quasielastic light scattering. Berne and Pecora (10) emphasize the statistical-mechanical relations between light-scattering spectra, transport coefficients, and the fundamental theory of molecular motion in liquids. Crosignani et al. (28) treat photon statistics, the detailed information on a sample that may be obtained by timing the arrival of individual photons. The NATO Advanced Study Institute volumes (24,29,30) and several conference proceedings (31,34) contain extensive fundamental discussions of photon correlation techniques. Recent reviews have treated biophysical hydrodynamics (17), electrophoretic light scattering (117,119), and applications to biopolymers (99).

We are primarily concerned with studies of equilibrium solutions and suspensions. The difficulties encountered in studying polymer glasses have been carefully analyzed by Patterson and coworkers (87); light scattering in crystals is discussed by Hayes and Loudon (52). Light-scattering spectroscopy can also be used to measure fluid flow velocities in a noninvasive manner (1). Laser anemometry is treated by Drain (38), whose volume also incorporates a variety of useful discussions of general optical problems. Light scattering from surfaces and films has recently been reviewed (116).

II. ELASTIC SCATTERING: PRINCIPLES AND TECHNIQUES

Our treatment of experimental methods useful in studying light-scattering intensities is preceded by a treatment of the theories that can be used to interpret measured intensities in terms of the microscopic properties (e.g., size, molecular weight) of a sample. The Mie theory (79) [and its generalizations to nonspherical and nonuniform particles (64)] correctly treats elastic scattering by particles of arbitrary size and refractive index. However, this theory is sufficiently complicated that the Rayleigh and Rayleigh–Gans–Debye approximations are preferable to it when they are adequate. For particles very small in comparison with a light wavelength, the Rayleigh theory is entirely adequate. For particles of arbitrary size that do not absorb light and whose index of refraction is close to that of their solvent, the Rayleigh–Gans–Debye theory serves. Much practical work involves systems in which the Rayleigh–Gans–Debye theory is adequate. We treat these theories in ascending order of complexity, beginning with Rayleigh's original work.

A. RAYLEIGH (POINT) SCATTERING

In 1881 Rayleigh (96) applied classical electromagnetic theory to the problem of light scattering by a gas. He showed that if light is incident on a transparent, optically isotropic particle whose dimensions are small compared with the wavelength of the light, the electric field of the light induces in the particle an oscillating electric dipole. Oscillating electric dipoles emit electromagnetic radiation at their oscillation frequency, so the particle acts as a source of light, absorbing power from the incident light and reradiating it in all directions. The intensity (power per unit area) I of the light scattered from the sample to a distant detector is trivially proportional to the intensity I_0 of the incident light and the volume v of the scattering region, and is inversely proportional to the square of the distance r from the sample to the observer. Rayleigh's calculations led to the equation

$$I_\theta = \left(\frac{I_0 v}{4\pi r^2}\right) R_\theta = \frac{I_0 v}{4\pi r^2} G(\theta) \frac{2\pi^2 (n - 1)^2}{\lambda^4 \rho} \qquad (7)$$

for the scattering of monochromatic light by a gas composed of identical particles. Here R_θ is the Rayleigh ratio [which has units of 1/(length)], n is the refractive index of the gas, ρ is the number density of the scattering particles, and λ is the wavelength of light in a vacuum. R_θ and I_θ are wavelength dependent because the ability of a dipole to radiate light depends on its oscillation frequency.

In radiometric terms, I_0 is the irradiance of the incident light, and has dimensions of power per area. Similarly, I_θ is the radiance of the scattered light, which by radiometric convention is assigned units of power per area per steradian. If one follows radiometric conventions, R_θ is assigned units of (length)$^{-1}$ (steradian)$^{-1}$.

The Rayleigh ratio is often replaced by the turbidity, τ, which is the total scattering integrated over all angles:

$$\tau = \int_{4\pi} d\Omega \, R_\theta \tag{8}$$

In the absence of absorption, τ is related to the intensities of the incident beam before and after passing through the sample. If the incident intensity is reduced to an intensity I_1 after passing through a length l of sample, then

$$\frac{I_1}{I_0} = \exp[-\tau l] \tag{9}$$

For particles that are small compared with λ, $\tau = 8\pi R_{\theta=0}/3$, whence

$$\tau = \frac{32\pi^3(n-1)^2}{3\lambda^4 \rho} \tag{10}$$

Equation 10 presumes that the space between the scattering particles is a vacuum. In a solution, n is to be interpreted as the ratio between the indices of refraction of the particle and the solvent.

The factor $G(\theta)$ depends on the polarization of the incident light and the polarizations of light that can reach the detector. $G(\theta)$ appears in equation 7 because the incident light excites dipoles that lie parallel to the incident electric field, whereas the intensity of scattering in a given direction (in radiometric terms, the irradiance) is proportional to the square of the projection of the excited dipole into the plane perpendicular to the scattering direction. For pointlike, optically isotropic particles, if the incident light is vertically polarized, the scattered light will be vertically polarized with $G(\theta) = 1$. If the incident light is horizontally polarized, the scattered light will also be horizontally polarized with $G(\theta) = \cos^2\theta$. If the incident light is nonpolarized and initially contains an even mixture of both polarizations, the scattered light will also be nonpolarized, with the ratio of the intensities of the two polarizations depending on θ, so that $G(\theta) = 1 + \cos^2\theta$. (A number of secondary references quote incorrect forms for $G(\theta)$, usually by associating a correct $G(\theta)$ with the wrong polarization.) Optically anisotropic particles and nonspherical particles cause depolarized scattering, in which vertically polarized incident light produces horizontally polarized scattered light, and vice versa, as treated in Section II.B.3.b. The reader should recall that the source-sample-detector plane is *defined* to be horizontal.

B. RAYLEIGH–GANS–DEBYE SCATTERING (THE FIRST-ORDER BORN APPROXIMATION)

1. General

Rayleigh's calculation of the light scattered by a gas, as treated above, relies on two implicit assumptions. First, the oscillating electric field incident on each particle is assumed to be the field of the incident light, and not the total field (which includes not only the incident field but also the fields set up by all the other induced dipoles in the sample.) Second, each particle is assumed to contribute independently to the scattered intensity; that is, the light scattered from different particles is assumed not to interfere.

The first of these assumptions is the Rayleigh–Gans–Debye (or Rayleigh–Debye) or first-order Born approximation. Kerker (64) gives the history of these names. The essence of the approximation is that the scattering process removes very little power from the incident beam, so that the total field set up by all the dipoles in the sample is only a small fraction of the original incident field. This turns out to be a reasonable approximation if the sample is nearly transparent. (If the sample is turbid, then one must take into account multiple scattering, treated in Sections II.D and III.G.4. If the individual scattering particles interact strongly with the incident light, one must take into account the local distortion of the incident field due to the shape and dielectric constant of the particles. This latter correction includes, for example, Mie scattering, treated in Section II.C.]

The second of these assumptions is physically incorrect, although it gives the right answer to Rayleigh's question. If one illuminates a sample with monochromatic light, there is always interference between the fields scattered by different particles. If the particles are those of a dilute gas, such interferences have no effect on the average intensity of the scattered light. Interference effects do, however, cause time-dependent fluctuations in the scattering intensity. The study of these fluctuations by quasielastic-light-scattering spectroscopy is treated in Section III.

2. Interference

In this section, the scattering of light by a set of particles is discussed from the standpoint of the Rayleigh–Gans–Debye approximation; that is, the field incident on each particle is assumed to be the same as the field that would be present if the scattering particles were absent. The light scattered by each particle is assumed to interfere with the light scattered by each of the other particles in the system. An elegant exposition on interference is provided by Benedek (9). We present here a simplified form, based on Fig. 9.2, that allows calculation of the interference between the scattering from two particles separated by a distance r_{12}. The particles are illuminated by plane-wave monochromatic light of wavevector k_I. The scattered light is observed by a detector D. The distance between the particles is much less

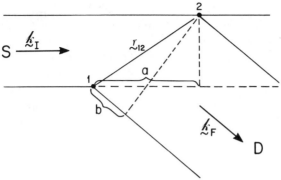

Fig. 9.2. Sketch for the calculation of the relative phase $\Delta\phi$ of the light scattered by two particles 1 and 2 separated by a distance $\mathbf{r}_{12}$.

than the distance between the particles and the detector, so light that passes from either particle to the detector has the same wavevector $\mathbf{k}_F$. Interference is due to the relative phase of the fields scattered from the two particles, so the scattering from particle 1 may be assigned a phase of zero. As seen in the figure, the second particle is farther from the source than the first particle is, so the scattering from it lags behind scattering from the first particle by a phase $\mathbf{k}_I \cdot \mathbf{r}_{12}$. However, the second particle is also closer to the detector, so its scattered field leads the field of the first particle by a phase $-\mathbf{k}_F \cdot \mathbf{r}_{12}$. The phase difference is

$$\Delta\phi = \mathbf{k} \cdot \mathbf{r}_{12} \tag{11}$$

where we have introduced the scattering vector $\mathbf{k} = \mathbf{k}_I - \mathbf{k}_F$. The scattering vector $\mathbf{k}$, sometimes denoted $\mathbf{q}$, is the fundamental quantity in all scattering experiments, since it determines the length scale to which the experiment is sensitive. Any particular value of $|\mathbf{k}|$ may be obtained by using a variety of choices of incident wavelengths and scattering angles. Elastic-scattering experiments characterized by the same value of $\mathbf{k}$ provide the same information about the sample. Since the scattering process does not change the frequency of the scattered light significantly,

$$k = |\mathbf{k}| = \frac{4\pi n}{\lambda} \sin\left(\frac{\theta}{2}\right) \tag{12}$$

where n is the index of refraction of the solution and θ is the scattering angle.
    The scattered field due to a suspension of N particles is

$$E(\mathbf{r}, t) = E_0 \sum_{i=1}^{N} \alpha_i \exp(i\mathbf{k} \cdot \mathbf{r}_i - i\omega t) \tag{13}$$

where E_0 is the incident field, $\mathbf{r}_i$ is the location of particle i, α_i is the scattering efficiency of particle i in isolation, and $\mathbf{r}$ is the location of the detector. For pointlike particles, α_i^2 is the Rayleigh cross section of a single particle, as treated in Section II.A. The right-hand side of equation 13 is, up to a multiplicative constant, the $\mathbf{k}$th spatial Fourier component of the density of scattering particles. In a multicomponent system in which only some components scatter light, the density of scattering particles is given by the concentration of scatterers. The scattering intensity for a suspension of N particles is therefore

$$I(\mathbf{r}, t) = |E_0|^2 \left\{ \sum_{i=1}^{N} \alpha_i^2 + \sum_{\substack{i=1 \\ i \neq j}}^{N} \sum_{j=1}^{N} \alpha_i \alpha_j \exp[i\mathbf{k} \cdot (\mathbf{r}_i - \mathbf{r}_j)] \right\} \qquad (14)$$

up to a dimensional constant (see equation 2). On the right, the double sum has been divided into two parts: a "self" part due to single-particle scattering, and a "distinct" part that describes interference between scattering by the pair i,j.

In Rayleigh scattering, the particles are assumed not to interact with each other. Each pair i,j of molecules is then equally likely to be found at any distance apart, and the phase of $\exp[i\mathbf{k} \cdot (\mathbf{r}_i - \mathbf{r}_j)]$ is equally likely to have any of its possible values. When one averages over all particles, the pairs of particles for which $\exp[i\mathbf{k} \cdot (\mathbf{r}_i - \mathbf{r}_j)] > 0$ (and that hence have constructive interference in their scattering) are exactly matched by the pairs of particles for which $\exp[i\mathbf{k} \cdot (\mathbf{r}_i - \mathbf{r}_j)] < 0$ (for which there is destructive interference between scattered fields). The constructive and destructive scattering effects thus cancel on the average, so for a dilute gas the distinct term of equation 14 vanishes, leaving on the average only the self term—the sum of the intensities scattered by each of the individual particles. This cancellation of constructive and destructive interferences has the same consequence for the calculation of the average intensity of the scattered light as the assumption that there is absolutely no interference between scattering by different particles.

For contrast, consider a model system in which the particles are very rigidly packed, so that the number of particles in a volume, each dimension of which much less than a light wavelength, is the same as the number of particles in every other volume of the same size. To a first approximation, a rigid crystal has this property; so do the molecules in a normal glass or liquid. In this ideal system, the number of particles scattering with each phase $\mathbf{k} \cdot \mathbf{r}_i$ is the same as the number of particles scattering with any other phase. Inspection of equation 13 shows that in this case the scattered field vanishes. Equivalently, in this model system the self and distinct terms of equation 14 cancel each other out. A system that is homogeneous over light-wavelength distances therefore does not scatter light. This explains why irregularly packed systems, such as a molecular liquid, a glass, or the lens of the eye, can be transparent (80).

3. Real Systems

a. FLUIDS

In a real substance, thermal fluctuations in the particle positions create microscopic inhomogeneities in the density. Although the density of a macroscopic volume of a substance, averaged over long times, is a constant, on a short time and distance scale the density fluctuates. In a liquid or solid, the strong repulsive interactions between the molecules greatly suppress the density fluctuations. A liquid (away from its critical points, at which its compressibility diverges) scatters vastly less light than would a gas with the same number density of particles. The intensity of scattering by a liquid was first treated by Smoluchowski (106) and Einstein (40), who used statistical mechanics to estimate the size of the density fluctuations. The fluctuations in the density ρ can be related to the compressibility κ of the liquid by comparing the thermal energy $k_B T$ with the amount of energy needed to achieve a bulk compression of the fluid. In terms of experimentally observed quantities,

$$\tau = \frac{32\pi^3}{3\lambda^4} \left[n\rho \left(\frac{\partial n}{\partial \rho} \right)_T \right]^2 k_B T \kappa \tag{15a}$$

or

$$\tau = \left(\frac{8\pi^3}{3\lambda^4} \right) (n^2 - 1)^2 k_B T \kappa \tag{15b}$$

b. PARTICLES SMALL COMPARED WITH A LIGHT WAVELENGTH

The above results apply to a fluid; a similar approach can be used to treat scattering by solutions of pointlike particles. In multicomponent systems, fluctuations in the density of scattering particles may be described as concentration fluctuations at constant total pressure. Debye (33) calculated the effects of these fluctuations on light scattering, relating them to the change in the concentration c:

$$\tau = \frac{32\pi^3}{3\lambda^4} \frac{RTc}{\left(\dfrac{\partial\pi}{\partial c} \right)_{P,T}} \left[n \left(\frac{\partial n}{\partial c} \right) \right]^2 \tag{16}$$

Here the size of the fluctuations is determined in part by the osmotic compressibility $(\partial\pi/\partial c)_{P,T}$ of the system. A nonideal osmotic compressibility [i.e., $(\partial\pi/\partial c)_{P,T} \neq RT$] results if interactions between scatterers are important, that is, if the distinct term of equation 14 is nonvanishing. Inserting in eqn.

16 the relation between osmotic pressure and the molecular weight M yields the Debye equation

$$\frac{Kc}{R_{90}} = \frac{Hc}{\tau} = \frac{1}{M} + 2A_2c \tag{17}$$

where

$$K = \frac{2\pi^2 n^2}{N\lambda^4} \left(\frac{\partial n}{\partial c}\right)^2 \tag{18a}$$

and

$$H = \frac{32\pi^3 n^2}{3N\lambda^4} \left(\frac{\partial n}{\partial c}\right)^2 \tag{18b}$$

Equation 17 has been extensively applied to solutions of moderate molecular weight polymers. Here A_2 is the second virial coefficient of the solution, and R_{90} is the Rayleigh ratio at 90° for unpolarized incident light. If a solution contains a single type of polymer whose weight distribution is not monodisperse, M is the weight-average molecular weight.

Equation 17 forms the basis for the determination of the molecular weights of small polymer molecules by light scattering. Aside from R_{90} or τ, only the refractive index of the solution and the specific refractive increment $\partial n/\partial c$ of the polymer need be measured. The latter quantity is a constant for a given polymer, solvent, and temperature, and is measured by the techniques discussed in the chapter on refractometry of the first edition of the *Treatise*.

Equation 17 is applicable to solutions of nonelectrolytes. It may also be applied to solutions of polyelectrolytes if the solvent includes a sufficiently high concentration of added salt. For 10 g/liter isoelectric bovine serum albumin in water, more than 0.10 M NaCl is required. As seen in the work of Timasheff et al. (110), in the absence of added salt the interaction term A_2c of equation 17 is replaced by a term $A_2'c^{1/2}$; in some salt-free polyelectrolyte systems A_2 is k-dependent.

In mixed solvents, allowance must be made for the fact that fluctuations in the concentrations of different species may be correlated. If, for example, a local increase in the concentration of a strongly scattering species is on the average accompanied by a local increase in the concentration of a weakly scattering species, then the scattering due to fluctuations in the concentration of the former species may be weaker than would be expected from that species' index of refraction. An example of this effect, which may be used to determine some of the thermodynamic interaction parameters of the components of a ternary solution, is found in Kirkwood and Goldberg (66).

The above description is applicable to isotropic particles. If the scattering particles are optically anisotropic (and collisional effects ensure that this is true even of gases of spherical molecules), the scattered light is partially depolarized; that is, some horizontally polarized light is observed at a scattering angle of 90°. If one illuminates the sample with nonpolarized light and uses a detector that responds equally to horizontally and to vertically polarized light, the right-hand side of equation 17 has to be multiplied by the Cabannes factor $(6 - 7\sigma)/(6 + 3\sigma)$ (see reference 21). Here σ is the ratio of the intensities of the horizontally and vertically polarized light, measured at 90° using unpolarized incident light. For scattering at angles other than 90°, a more complex correction factor is required. If the scattering particles have a definite shape (e.g., if they are rods or discs), measurement of the depolarization may yield information about shape.

c. DILUTE SOLUTIONS OF LARGE PARTICLES

When the size of a scattering particle exceeds about $\lambda/20$, at any moment different parts of it are exposed to light rays of significantly different phase. There can then be not only interference between scattering by different objects, but also interference between the light scattered by different parts of the same object. The general solution describing the distribution of light scattered by large particles is complicated, and has been obtained only for simple particle shapes such as spheres, rods, and stratified spheres (64; see also Section II.C). Fortunately, the Rayleigh–Debye approximation for the interference is often adequate.

For nearly transparent particles, a modification of the method discussed in Sections II.B.1–II.B.3 gives a good estimate of the scattering function. This approximation requires that the particle be sufficiently small that

$$ka(n_r - 1) \ll 1 \tag{19}$$

where a is the largest dimension of the particle and n is the ratio between the indices of refraction of the particle and the solvent. With this assumption, the incident fields inside and outside the particle are nearly the same. Comparison of the approximate and exact calculations indicates that this approximation is satisfactory to within <10% for nonabsorbing particles of relative refractive index (particle/medium) less than about 1.2. In this approximation, the construction of Fig. 9.2 is used to calculate the interference between scattering from different parts of a single particle, with points 1 and 2 being taken to represent different parts of the same particle rather than a pair of pointlike particles. If the particle interacted strongly with the light, equations 11 and 14 would not apply, because the phase of the incident field at point 2 would not be given by the construction of Fig. 9.2.

In normal liquids, each particle is independently equally likely to be found with each possible orientation. One then sees in scattering measurements only a spherical average over all possible particle orientations. If one rewrites equation 14 for the scattered intensity, replacing the sum over discrete particles with integrals over a continuous range of positions within each particle, and if the solution is dilute enough that interference between scattering from different particles cancels on the average, one obtains

$$P(\theta) \equiv S(k) = \frac{4\pi}{k} \int_0^\infty [rg(r) \sin(kr)] \, dr \tag{20}$$

where the scattering function $P(\theta)$ [also sometimes denoted by $S(k)$] has been normalized to unity at $\theta = 0$. Here $g(r)$ is the radial distribution function: the density of pairs of scattering points (r_i, r_j) separated by a distance $|\mathbf{r}|$. It is not necessary that either $\mathbf{r}_i$ or $\mathbf{r}_j$ be at the center of an object. Thus, for a sphere of radius a, $g(r)$ is not zero for $r > a$, but is zero for $r > 2a$. It has here been assumed that all points within the particle have the same elastic-scattering cross section. [This approximation is easy to relax by redefining $g(r)$ slightly.]

For a uniform sphere of radius a, one has

$$P(\theta) = \left[\frac{3}{u^3} (\sin u - u \cos u) \right]^2 \tag{21}$$

where

$$u = ka = \frac{4\pi na}{\lambda} \sin\left(\frac{\theta}{2}\right) \tag{22}$$

For randomly oriented rods of length L, one finds (64) that

$$P(\theta) = \frac{1}{x} \int_0^{2x} \frac{\sin \omega}{\omega} \, d\omega - \left(\frac{\sin x}{x}\right)^2 \tag{23}$$

where

$$x = \frac{kL}{2} = \frac{2\pi nL}{\lambda} \sin\left(\frac{\theta}{2}\right) \tag{24}$$

For a thin disc of radius B (6,10,69),

$$P(\theta) = \frac{2}{u^2}\left[1 - \frac{2}{u}J_1(u)\right] \approx 1 - \frac{u^2}{6} \tag{25}$$

where J_1 is the first-order Bessel function and

$$u = kB \tag{26}$$

The functions $P(\theta)$ are plotted in Fig. 9.3. One sees that the scattering approaches a maximum as u or x approaches zero (i.e., the scattering is

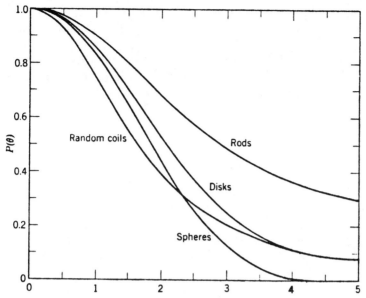

Fig. 9.3. Variation of the particle scattering function $P(\theta)$ with the size parameter ka (spheres), kR (monodisperse random coils) kL (rods), or kB (discs).

largest in the forward direction). The scattering falls off smoothly with increasing scattering angle except for very large particles.

If one also has other information about the size of the particle, such as its hydrodynamic radius, one can attempt to use the measured scattering intensity to deduce whether the particle is more likely to be a sphere, an oblate ellipsoid, or a prolate ellipsoid. In dealing with rodlike particles, the concentration must be low enough that the average center-to-center distance is much greater than the length of the rod. A corresponding requirement applies for discs. If the concentration is too high, the orientations and positions of neighboring particles will be correlated. The special-case formulas given above will thus be inapplicable.

d. Polymer Solutions, Zimm's Method, and Zimm Plots

Equation 17 was obtained on the assumption that the scattering particles are small in comparison with λ. Zimm (127–129) has shown that if this is not the case, equation 17 becomes

$$\frac{Kc}{R_\theta}\frac{1}{MP(\theta)} + 2A_2c \approx \frac{1}{M}\left(1 + \frac{k^2R_g^2}{3}\right) + 2A_2c \qquad (27)$$

The approximation for $1/P(\theta)$ is valid if kR_g is not too large. Zimm's derivation of equation 27 assumes that the likelihood that a polymer molecule has a given conformation is not changed while that molecule is colliding with one of its neighbors. This theoretical conjecture has been studied less extensively than appears desirable. Fortunately, in the limit of low polymer concentration the assumption is not important.

If one averages over possible configurations and orientations of the polymer molecules in a monodisperse highly dilute solution of random coils (33,129), one obtains

$$P(\theta) = \frac{2}{v^2} [\exp(-v) - 1 + v] \approx 1 - \frac{k^2 R_g^2}{3} + \cdots \cdot \qquad (28a)$$

where

$$v = (kR_g)^2 \qquad (28b)$$

and R_g is the radius of gyration of the polymer. For polydisperse systems, Zimm has shown that M is the weight-average molecular weight M_w, and that the radius of gyration obtained from $P(\theta)$ is a z-averaged dimension. By an average of a quantity Q over the molecules in a solution, we mean

$$\langle Q \rangle = \frac{\sum\limits_i Q_i W_i}{\sum\limits_i W_i} \qquad (29)$$

where Q_i is the value of Q for species i, the sum over i is taken over all N species in solution, and W_i is the statistical weight assigned to species i in the average. In a number average, W_i is the number of particles N_i of species i. In a weight average, W_i is the mass $N_i m_i$ of species i in solution, m_i being the molecular weight of species i. In a weight-squared average, W_i is $N_i m_i^2$. In a z-average, W_i is the amount of light scattered by species i; for particles that are much smaller than a light wavelength, the z-average is nearly equal to the weight-squared average.

Equation 27 is commonly applied by Zimm's method, in which the left-hand side of the equation is plotted against $\sin^2(\theta/2) + jc$, where j is an arbitrary constant. In a successful experiment on a conventional system, c and θ will be sufficiently small that a rectilinear grid is obtained in the plot (Fig. 9.4, open circles). If one obtains a rectilinear grid (at least for small θ and c), one may extrapolate to $\theta = 0$ at fixed c and to $c = 0$ at fixed θ (Fig. 9.4, filled circles), and then extrapolate the extrapolated points to obtain the joint limit $\theta = 0$, $c = 0$. This limiting point should be found on the vertical axis of the plot. From equation 27, the joint intercept is the inverse of the weight-average molecular weight of the scattering particles. The slope of the lines at a fixed angle is A_2, the second virial coefficient. [In some polyelectrolyte systems, the lead term of the virial expansion is proportional to $\sqrt{c}$ rather than c (110).] The limiting slope (in the limit of zero concentration)

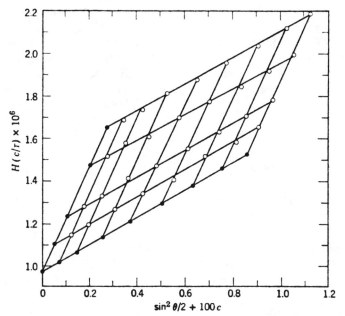

Fig. 9.4. Zimm plot showing the scattering from a sample of polystyrene in butanone (13). Molecular parameters are $M_w = 1{,}030{,}000$; $A_2 = 1.29 \times 10^{-4}$ ml·mol/gram^2, $(s_z^2)^{1/2} = 460$ Å, and $z = 1.44$.

of the lines at fixed concentration determines the radius of gyration R_g, since the slope is $k^2 R_g^2/3$.

To a first approximation sufficient for most polymer solutions, the second virial coefficient is independent of the scattering angle. Detailed theoretical considerations indicate that the second virial coefficient depends on the angle of observation if the range of the intermolecular interactions is comparable to the inverse of the scattering vector. This condition is rarely satisfied. With most systems, careful experiments show only that A_2 tends toward zero with increasing scattering angle. A decrease in A_2 with decreasing angle of observation is usually indicative of contamination of the sample with extraneous, highly scattering material; molecular weights obtained from such systems should be accepted only with considerable reservations.

Under modern conditions, the task of producing a Zimm plot and measuring slopes and limits may be replaced by a computer fit. The usual aversion to fitting data to more than two parameters can be set aside; the parameters in equation 27 are sufficiently independent that an error in one parameter will not be correlated with errors in other parameters. To obtain a limiting slope at $c = 0$ or $k = 0$, one should include quadratic fitting parameters (i.e., c^2 and k^2 terms) to absorb any curvature or noise in the data. (The quadratic parameters will not be meaningful unless cubic parameters are also included in the fit.)

C. MIE SCATTERING

For systems in which the indices of refraction of solute and solvent are very different (so that criterion 19 is not satisfied), or in which the index of refraction of the solute is complex (i.e., in which the sample absorbs light), the Rayleigh and Rayleigh–Debye scattering theories are not adequate. Recourse must then be made to Maxwell's equations for dielectric and conducting media. Mie carried out the relevant calculations for scattering by homogeneous dielectric spheres of arbitrary size and index of refraction (79). For small or nearly transparent spheres, Mie's theory gives essentially the same results as the Rayleigh and Rayleigh–Debye methods. For very large spheres, the Mie theory gives essentially the same results as classical geometric and physical optics. For spheres of intermediate size, the Mie theory is the only way to calculate the scattering functions with reasonable accuracy.

Maxwell's equations may also be used to calculate scattering by nonspherical particles, such as cylinders, discs, and stratified spheres. Results for various particle shapes are discussed by Kerker (64). If the particles are sufficiently close together, their inductive fields lead to mutual polarization, which changes the scattering intensity (74).

The Mie equations for spheres, and the corresponding equations for particles of other shapes, are extremely complicated. With modern digital computation techniques, this complexity is no longer a substantial obstacle to their use. However, it is still of considerable interest to examine approximate solutions to the Mie equations.

Figure 9.5a shows in a qualitative way the general relations to be expected between the intensity of the scattered light and the angle of observation for particles of different size. As may be seen from equations 21–26, I is always a function of the product ka, where a is the characteristic particle dimension. With visible light and accessible scattering angles, k has only a limited range of values. As the particle shape is changed, different parts of the scattering function become experimentally accessible. For large particles (>0.1 μ or so) the drop in scattering intensity with angle is very rapid, although (if one has a monodisperse preparation) the scattering intensity at large angles shows a series of maxima and minima (Fig. 9.5a, "Large" curve). As the particle size is reduced (to 0.1–0.01 μ or so), one can study only a more limited range of ka values. The scattering function (Fig. 9.5a, "Medium" curve) shows a primary maximum and perhaps an indication of a secondary maximum. For very small particles ($\ll 0.01$ μ), the accessible range of ka values is small. Except for the polarization factor $G(\theta)$, the scattering function is either entirely independent of angle (true Rayleigh scattering) or shows a slight diminution in intensity at large k.

Tables of Mie-scattering intensities as functions of particle size, shape, and structure and of real and imaginary components of the particle's index of refraction would fill many volumes (64,122). Figures 9.5b and 9.5c present

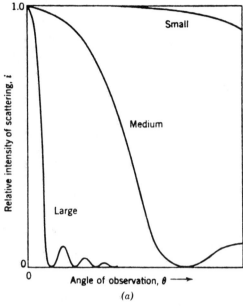

Fig. 9.5. (a) Relative intensity of light scattered as a function of angle of observation for large, medium, and small particles (105). Intensities are adjusted to be equal at $\theta = 0$. (b, c) Scattering intensities as a function of particle size for (b) perfectly reflecting spheres and a 90° scattering angle and (c) spheres with $n/n_0 = 1.20$ and a 90° scattering angle. G_1 and G_2 are normalized scattering intensities for VV and HH scattering, respectively, the normalization factor being 4/(ka).[2] Particle size is in units of $\alpha = ka$, where a is the particle size, $k = 2\pi/\lambda$ is the light wavelength in the surrounding medium, and the incident intensity is unity. (From reference 64.)

examples (64) of scattering at 90° as a function of particle size a for perfectly reflecting (i.e., metallic) spheres and for spheres with a purely real index of refraction which is 20% greater than the index of refraction of the surrounding medium. Here G_1 and G_2 correspond to VV (vertically polarized light incident, vertically polarized light detected) and HH (horizontally polarized light incident and detected) scattering powers for unit incident power. The normalizations of G_1 and G_2 are given in the figure caption, and the particle size is normalized by $k = 2\pi/\lambda$, the wavevector of the incident radiation.

In Fig. 9.5b, note that the scattering power of a reflecting sphere does not reduce, in the limit of very small sphere size, to the Rayleigh form. This difference is found because in the Rayleigh form only scattering due to induced electric dipoles is important, whereas scattering from perfectly reflecting spheres also includes a substantial contribution due to the induced magnetic dipoles. The spacing and relative amplitudes of maxima and minima in G_1 and G_2 are seen to be quite different for reflecting and for dielectric spheres.

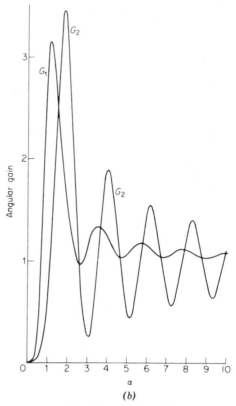

Fig. 9.5. (*continued*)

If a dielectric sphere is illuminated with white light, the scattering maxima will appear at different angles for light of different colors. This effect, which is a form of higher-order Tyndall scattering, allows one to determine particle size. Essentially, one views the scattered light at a series of angles and notes the locations at which bands of a particular color are seen; these angles are the points at which the scattering function for that color and for its additive optical complement have the largest ratios (60).

Scattering functions for particles of fixed size (1.0 μ) and different shape are plotted in Fig. 9.6. The shape of the particle has a substantial effect on the positions and heights of the secondary maxima. However, the position of the major fall in the normalized intensity (from $i = 1$ to $i = 0.1$) occurs in all cases at roughly the same value of ka. Thus, by examining the angle at which the intensity drops steeply, one can estimate the size of the particles without knowing their exact shape. In fact, most substances are not monodisperse with respect to size, so the secondary maxima of Fig. 9.6 are not always experimentally detectable. In many practical cases, it is more useful to observe the initial decrease in scattered intensity with increasing angle rather than to search for maxima and minima.

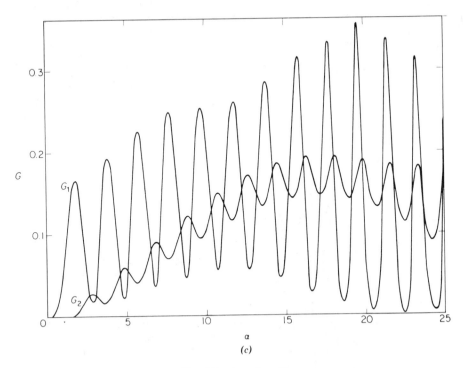

Fig. 9.5.   (*continued*)

D.   MULTIPLE SCATTERING (TURBID-MEDIUM THEORY)*

1.   Development of Phenomenological Turbid-Medium Theories

In this section we consider the interactions of radiation with matter in the form of a turbid medium that both scatters and absorbs the radiation; in particular, we are concerned with a medium in which there is a substantial amount of multiple scattering. Such a material can be considered *optically thin* if most of the scattered radiation that is observed has been scattered only once and much unscattered radiation emerges from the sample. The material has *intermediate* behavior if most of the scattered radiation has been multiply scattered but some unscattered radiation still emerges from the sample, and is designated *optically thick* if all the radiation has been multiply scattered. All three cases are industrially important; examples of each are a hazy film material (optically thin), a diffusing shade through which objects can still be seen by virtue of the unscattered light (intermediate),

* This section was written by Fred W. Billmeyer, Jr.

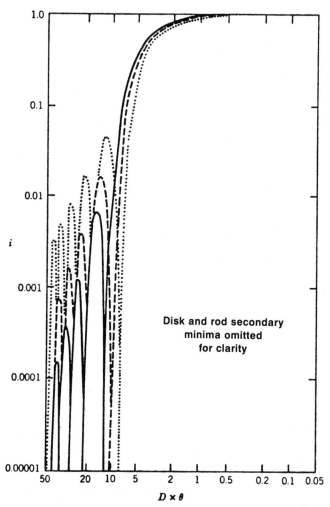

Fig. 9.6. Scattered light intensity as a function of size and angle for several approximate solutions of the Rayleigh–Gans–Debye equations. Key: ——, spheres; – – –, discs; and ······, rods.

and all remaining translucent and opaque objects (optically thick). Note that optical thickness and physical thickness are not necessarily related: Interstellar space is optically thin, whereas a paint film is optically thick.

The study of the interaction of radiation with media of the types just described, which is known as radiative-transfer theory and was developed in large part by 1983 Nobel Prize winner S. Chandrasekhar (23), is quite complex. The theories developed in this section can be considered first and second approximations to the full treatment, and are adequate for most practical calculations.

a. EARLY WORK

The first major turbid-medium theory was developed by the astronomer Schuster (100), who sought to explain the escape of light from the self-luminous foggy atmosphere of a star, and why some stars show bright-line spectra and others dark-line spectra. He introduced a concept of importance for the problem we wish to consider, namely the idea that the passage of light through a turbid medium may be represented by two diffuse fluxes, or moving beams of light, going in opposite directions, with each losing some of its light to the other by scattering in the backward direction. Another part of the scattered light, in the forward direction, would of course stay in the same flux. He assumed that the scattered light was divided equally between the forward and backward directions. No assumptions were made about the nature or size of the scattering particles, as this was not a molecular but a phenomenological theory, relating the amounts of scattering and absorption (and emission, in Schuster's case) of radiation to phenomena associated with the model, such as transmittance, reflectance, and the like.

In 1927 L. Silberstein (104) considered the behavior of a layer of turbid material with a collimated beam incident on its front surface. Inside, there were both a collimated beam and the two diffuse fluxes resulting from forward and backward scattering, respectively. He also removed the restriction that the scattered light be divided equally between the two directions. However, he made two other assumptions that have had to be re-examined in later work: first, that the same fraction of the light was removed by scattering from the collimated and from the diffuse beam while they traveled the same distance through the material; and second, that there was no optical interface—that is, no change of refractive index, with an accompanying reflection of radiation—at the surface of the sample. Silberstein's model was based on the infinite-sheet approximation, in which the medium is assumed to be bounded by parallel planes and to extend over a region very large compared with its thickness. The assumptions made by Schuster and Silberstein have been adopted in essentially all subsequent turbid-medium theories.

b. TECHNIQUES

In the most common applications of turbid-medium theory, to which we direct our attention because of their usefulness in analytical chemistry, the phenomena to be measured are the transmitted and reflected radiation from the sample. The geometric arrangement must be such that all of the radiation is measured, and typically involves incorporating an integrating sphere with a highly reflective diffusing interior surface, as indicated in Fig. 9.7, into a spectrophotometer covering the wavelength range of interest. If the sample is placed at the transmission port of the sphere (Fig. 9.7a), the measured quantity is the transmittance, the ratio of the transmitted flux to the flux incident on the sample (usually measured as the flux with no sample present). If the sample is placed at the reflection port (Fig. 9.7b), the measured quan-

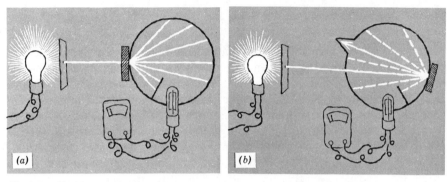

Fig. 9.7. Sketches of the geometric arrangements for the measurement of (*a*) the transmittance and (*b*) the reflectance factor of a turbid medium using an integrating sphere (16).

tity is the reflectance factor, the ratio of the reflected flux to the flux reflected by a perfect (100%) diffusing reflector identically irradiated. For the application of turbid-medium theory, the reflection function of interest is not the reflectance factor but instead the reflectance, defined as the ratio of the reflected flux to the incident flux. However, the two are numerically equivalent and in this section we will henceforth refer to the reflectance.

The integrating sphere usually includes a gloss trap (a hornlike protuberance on the sphere), which for the application of turbid-medium theories is made equivalent in reflectance to the remainder of the interior of the sphere so that all the reflected flux is collected. The reverse of the above geometries, with the radiation source and detector interchanged, gives equivalent results. Not shown in the figure is the location of the monochromator, which is usually placed before the detector.

If the sample is translucent, its reflectance will depend on the nature of a material placed behind it, commonly known as the background. The reflectance of the background, measured separately, appears in the turbid-medium theory equations for such samples. For opaque samples, the background by definition has no influence and the transmittance is of course zero.

c. Kubelka–Munk Equations for Opaque Samples

In 1931, P. Kubelka and F. Munk published (71–73) what has since become the most widely known and frequently used treatment of turbid-medium theory. Their work is very similar to that of Schuster, except that all of the light is considered to reach the sample from outside rather than being emitted within the material. The Kubelka–Munk (K-M) analysis has found widespread practical application because it yields results in terms of quantities familiar in the colorant-using industries, particularly the paper, paint, plastics, and textile industries.

The K-M theory is, like that of Schuster, a two-flux theory, and applies only to optically thick samples. The K-M model, illustrated in Fig. 9.8, is

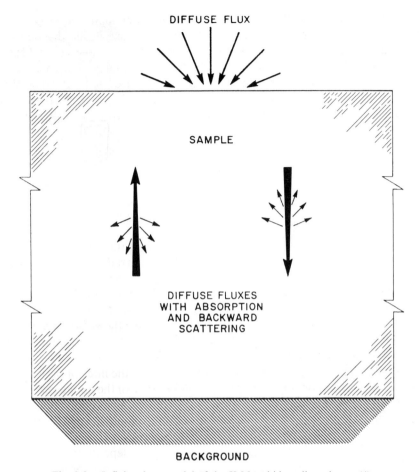

Fig. 9.8. Infinite-sheet model of the K-M turbid-medium theory (4).

based on the infinite-sheet concept in order to eliminate edge effects. Other assumptions of the theory are: (1) there is no change in refractive index at the boundaries of the sheet; (2) the differential volume element within the sheet is large compared with the sizes of the scattering and absorbing elements, so that microscopic details can be ignored; (3) the sample is homogeneous and isotropic, implying that the scattering particles are randomly oriented, so that angular dependences can be ignored; (4) polarization effects are negligible; (5) no radiant energy is generated within the sample, as by incandescence; (6) the particles scatter independently; (7) the surrounding medium has no effect on the scattering or absorption; (8) the scattering and absorption coefficients are constant (meaning, practically, that the radiation is confined to a sufficiently narrow wavelength range); and (9) both the illumination and the viewing are perfectly diffuse. The latter assumption is not met in practice when the usual instrument geometries described above are used, but for optically thick materials there is enough scattering that a collimated incident beam very quickly becomes diffuse inside the sample.

Consider a differential layer of thickness dx (where x is the vertical space coordinate and is positive upward) within the sample and parallel to its faces. Within the layer, the diffuse flux traveling downward, ϕ_i, is decreased by absorption and by scattering with K-M absorption and scattering coefficients K and S, respectively, and is increased by scattering from the upward-traveling diffuse flux, ϕ_j:

$$-\frac{d\phi_i}{dx} = -(K + S)\phi_i + S\phi_j \tag{30}$$

Similarly, for ϕ_j,

$$\frac{d\phi_j}{dx} = -(K + S)\phi_j + S\phi_i \tag{31}$$

The change in sign results from the reversal of the direction of travel. These two equations can be integrated in a straightforward manner and appropriate boundary conditions applied. At this time we consider only the case of a completely opaque sample. The sample thickness therefore does not appear in the equations. The reflectance R, evaluated from the boundary conditions according to its definition, is independent of the thickness and is usually designated as the reflectance at infinite thickness, R_∞. The simple equation that results,

$$\frac{K}{S} = \frac{(1 - R_\infty)^2}{2R_\infty} \tag{32}$$

forms the basis of virtually all industrial applications of turbid-medium theory. The K-M optical constants K and S are usually considered to be empirical and unrelated to more fundamental absorption and scattering quantities (but see Section II.D.3).

d. MIXING LAW

Application of the K-M equation requires some assumptions about the relations among the amounts of the scattering and absorbing constituents of the sample; their individual K-M optical constants, here designated with primes (11); and the optical constants of the entire sample, K and S. The usual assumption is that each component acts independently in proportion to its amount, specified as some sort of concentration. The mixing law, sometimes attributed to Duncan (39), is

$$K = \sum_i C_i K_i', \quad S = \sum_i C_i S_i' \tag{33}$$

the surfaces of the sample. Usually there is such a change, with accompanying reflection losses. If the sample has glossy surfaces, these reflection losses can be calculated by application of Fresnel's equations if the angular distributions of flux inside and outside the sample are known. The derivation of the relation between measured reflectance R' and corrected reflectance R (which would be measured just inside a glossy surface) has been published in several places (63,68) and is often attributed erroneously (see, e.g., reference 68) to Saunderson. This relation is

$$R = \frac{R' - k_1}{1 - k_1 - k_2(1 - R')} \tag{40}$$

or the reverse

$$R' = k_1 + \frac{(1 - k_1)(1 - k_2)R}{1 - k_2 R} \tag{41}$$

For transmittances, the equation is (103)

$$T' = \frac{(1 - k_1)(1 - k_2)T}{1 - k_2^2 T^2} \tag{42}$$

The reverse equation is complex but should not be needed (103). In these equations, k_1 is the Fresnel reflection coefficient for collimated light, and k_2 is the Fresnel reflection coefficient for diffuse light incident on the surface from the medium of higher refractive index; the latter coefficient is tabulated in the literature (12).

c. THEORY APPLICABLE TO ALL OPTICAL THICKNESSES

Despite its adequacy and widespread use, K-M theory is limited in its applicability to optically thick materials. As the need arose to apply turbid-medium theory to intermediate and optically thin materials, various types of approach were followed. Those familiar with K-M theory, primarily workers in the dye- and pigment-using industries, extended the K-M model, first to include the collimated fluxes that are present in intermediate-type materials, and subsequently to include still more fluxes to obtain a second approximation to the underlying radiative-transfer theory by including the angular dependence of flux. These developments are considered in this subsection. Other workers, mostly in different fields and not as familiar with K-M theory, tried a variety of quite different approaches, described in the following subsection.

Workers following the lead of Silberstein developed in the 1960s independently and nearly simultaneously two theories (5,115) in which two collimated fluxes traveling in opposite directions were added to the diffuse

For pigmented systems, volume concentrations are often used. When only opaque samples, for which only the ratio K/S is important, are considered, the mixing law can be further simplified in some practical cases (see the chapter on color specification), and the concentrations can be specified as fractions of the total pigment concentration independent of the quantity of medium present.

2. Extensions of Kubelka–Munk Theory

a. TRANSLUCENT SAMPLES

By the application of different boundary conditions in the derivation of the K-M theory, a large number of equations can be obtained relating reflection and transmission quantities to K and S. Extensive tabulations of these equations have been published, (61,63,68); only a few key ones will be given here. The equations can be derived in either exponential or hyperbolic form, with the latter being more compact and more commonly used. In the equations that follow, R is the reflectance of the translucent sample measured over a background of reflectance R_g, R_0 is its reflectance measured over a background of zero reflectance, and T is its transmittance:

$$R = \frac{1 - R_g(a - b \text{ ctgh } bSX)}{a - R_g + b \text{ ctgh } bSX} \tag{34}$$

$$R_0 = (a + b \text{ ctgh } bSX)^{-1} \tag{35}$$

$$T = \frac{b}{a \sinh bSX + b \cosh bSX} \tag{36}$$

$$R_\infty = a - b \tag{37}$$

where

$$a = \frac{K + S}{S} = \frac{1}{2}\left(R + \frac{R_0 - R + R_g}{R_0 R_g}\right) \tag{38}$$

$$b = (a^2 - 1)^{1/2} \tag{39}$$

and X is sample thickness. The abbreviations sinh, cosh, and ctgh represent the hyperbolic sine, cosine, and cotangent, respectively. Many other useful relationships can be derived from these. Because of the relative complexity of the equations, of the measurements, and of making practical use of the results, these equations are not widely used (again, further comments may be found in the chapter on color specification).

b. SURFACE-REFLECTION CORRECTIONS

The reflectances and transmittances discussed so far have been the quantities that would be observed if there were no change in refractive index at

TABLE 9.I

Applicabilities of the Various Calculation Methods in the Three
Regions of Optical Thickness

Method	Optical thickness		
	Thin	Intermediate	Thick
Doubling	Yes	Yes	Yes[a]
Monte Carlo	Yes	Yes[a]	No
Hartel	Yes	Yes[a]	No
Diffusion	No	No	Yes
Discrete Ordinate			
Two-flux	No	No	Yes
Four-flux	No	Yes[b]	No
Six- or many-flux	Yes	Yes	Yes

[a] Rather large amounts of computer time can be required.

[b] Only if internal reflection coefficients for diffuse flux are
known.

fluxes of the K-M theory. These four-flux theories were cumbersome and
involved a large number of adjustable constants requiring much calibration
to determine, but were adequate for the description of some systems of
practical importance, such as translucent plastics, showing optically thin or
intermediate behavior. The main source of error in using these theories was
lack of detailed knowledge of the internal surface-reflection coefficient, since
by definition the flux within the sample is no longer perfectly diffuse in these
cases.

In the 1940s Wick (121) and Chandrasekhar (23) developed a generalization of the two-flux method in which the radiation flux can be calculated
in as many different directions (sometimes termed discrete ordinates) as
desired. This method, here called the "many-flux" method, was applied to
pigment optics by Richards and coworkers (15,83–85,97). With the aid of
matrix-inversion computer programs, now readily available, the equations
can be solved for as many as 20–30 fluxes. Richards has shown however,
that unless the angular dependence of the reflected and transmitted fluxes
is extremely complex, however, useful results require no more than six
fluxes; the six-flux calculation is powerful enough to allow the determination
of the internal reflection coefficients as well.

d. OTHER APPROACHES

Several other theories have been used to describe the properties of turbid
media of various optical thicknesses. They are described briefly below, and
their applicabilities to optically thin, intermediate, and thick regions are summarized, together with those of the discrete-ordinate methods, in Table 9.I.
It can be seen that only the many-flux treatment is useful over the entire
range of optical thicknesses without restriction.

(1) Doubling Method

This method, sometimes called invariant embedding (2), is derived from the fact that if R and T are known for two separate layers of a turbid medium, it is relatively easy to determine the properties of the medium obtained by combining them. By first calculating the properties of a very thin layer, and then those of layers 2, 4, 8, . . . times as thick, one can extend the method to samples intermediate in optical thickness.

(2) Monte Carlo Methods

This powerful calculation method assigns a probability to each event that could happen to a photon in entering and passing through a turbid medium. Repeated calculations are then made, selecting actual events from the probabilities by means of random-number generation. The method is applicable only to optically thin media because of the large number of calculations required (95).

(3) Hartel's Method

Hártel (51) proposed a method in which each order of scattering is considered separately. First the light flux in the medium that has not been scattered is calculated. Then it is possible to calculate the once-scattered flux, then the twice-scattered flux, and so on. As with the Monte Carlo method, Hartel's is useful only for optically thin media, for which high orders of scattering need not be considered.

(4) Diffusion Method

For cases in which the amount of absorption is small compared with that of scattering, it is possible to describe the angular distribution of flux by an equation that has the same form as Fick's law of diffusion. Widely used to describe neutron scattering in nuclear reactors, this method has also been applied to light-scattering problems (44).

3. Relation of Phenomenological to Molecular Theories

It was long believed that phenomenological turbid-medium theories, in particular the K-M theory, had no firm theoretical basis, and that their optical constants, specifically K and S of the K-M theory, bore no direct relationship to the fundamental structure and properties of the medium. Both of these conjectures have been disproved by the theoretical work of L. W. Richards and its experimental confirmation by D. G. Phillips (92).

The relation between the K-M constant K and the single-scattering absorption coefficient k has been understood for some time. Kubelka (71) pointed out that when equal amounts of light are traveling in all directions within a turbid medium (isotropically diffuse light), the path length for absorption is twice that for a collimated beam of light, and $K = 2k$. This has been confirmed experimentally (44,83), and a general relationship between

K and k applicable when the light is not isotropically diffuse has been derived (84,85).

Because of the changes in the direction of travel of light, scattering represents a more complicated situation than absorption. Kubelka thought that $S = 2s$, where s is the single-scattering scattering coefficient; this was analogous to his result that $K = 2k$. Because s is the total amount of light scattered in single scattering, whereas S represents the backward-scattered light only in multiple scattering, this is clearly a great oversimplification and of doubtful validity, yet it still occasionally appears in the literature.

Since S is said to represent isotropic scattering of isotropic fluxes, some workers have suggested that $S = s/2$. Others have stated that S is the fraction of light scattered in the backward direction, implying that $S = \int R_\theta d\theta$ from $\theta = 90°$ to $\theta = 180°$ only, where R_θ is the Rayleigh ratio. Since s represents single scattering and S multiple scattering, it is unlikely that such simple definitions are correct.

Several workers have derived empirical relations between S and s, and as late as 1970 it was believed that these two could not be related theoretically. In 1971, however, Richards (84) stated that for an isotropic distribution of light within a turbid material,

$$\frac{S}{s} = \tfrac{3}{4}(1 - \overline{\cos \theta}) \tag{43}$$

where $\overline{\cos \theta}$ is the asymmetry factor given by

$$\overline{\cos \theta} = \frac{\displaystyle\int_{-1}^{1} R_\theta \cos \theta \, d(\cos \theta)}{\displaystyle\int_{-1}^{1} R_\theta \, d(\cos \theta)} \tag{44}$$

Richards later (84,85) derived a completely general relation between S and s holding regardless of the distribution of the light flux within the sample. Both of the relationships were confirmed experimentally by Phillips (92), and it can now be stated that the K-M constants are firmly related to corresponding single-scattering constants.

4. Applications

By far the most widespread application of turbid-medium theory is the use of K-M theory to describe scattering and absorption in dyed or pigmented media. The information thus obtained is applied in the industrial formulation of colored paints, plastics, textiles, papers, and the like, which is discussed in the chapter on color specification.

Other applications of turbid-medium theory not as closely related to analytical chemistry occur in the study of astrophysics, atmospheric physics, remote sensing, photography, heat transfer, neutron scattering, and the behavior of certain translucent natural materials such as foods and plant leaves. A partial list of references up to 1973 is given in reference 15.

E.   TECHNIQUES

1.   Apparatus

A variety of commercial light-scattering photometers that provide accurate measurement of the light-scattering intensity as a function of angle are currently available. Some of these instruments simultaneously measure the spectrum of the light as well. Of particular interest in the design of these devices are their light sources, optical trains, and photodetectors.

For measurements of the static light-scattering intensity at large angles, a laser offers little advantage over a classical source, and has several minor disadvantages. [At small angles ($<10°$), the linearity of the laser beam gives lasers a significant advantage over conventional light sources.] For one thing, the specific refractive increment is wavelength dependent; standard tables of specific refractive increments of polymer–solvent mixtures are reported for mercury arc wavelengths, not laser wavelengths. Furthermore, because of the extreme monochromaticity of laser light, very small optical imperfections in the optical train, or within the laser itself, lead to laser speckling: If one projects the scattered light onto a screen, one sees small, motionless, bright and dark patches. These have nothing to do with the properties of the sample, but are an obstacle to light-scattering photometry. Speckle effects arising from within the laser may be eliminated by mounting a spatial filter on the laser.

No comparison of different light sources should neglect their operating wavelengths and the consequences of these for the signal-to-noise ratio. The Rayleigh scattering law indicates that the light-scattering intensity—the irradiance—increases as λ^{-4}. Thus, for the study of small particles ($ka \ll 1$), light sources that operate in the blue or violet range are preferred. Furthermore, many detectors have better sensitivity at shorter wavelengths, so replacing a mercury arc (blue–green region) with a helium–neon (red region) laser of the same power may substantially reduce the signal.

The optical train between the light source and the sample is generally of either a focusing or a colinearizing type (112). Focusing optical trains use converging lenses to bring the illuminating beam to a focus at some point within the sample cell. The increased light intensity within the sample cell substantially improves the signal-to-noise ratio. Absolute calibration of instruments with focusing optics is difficult to accomplish, but the use of secondary standards to calibrate a photometer is now well understood. One should recognize that focusing-type optics lead to difficulties in determining the depolarization ratio.

In contrast to focusing optics, colinearizing optics are designed to illuminate the sample with parallel rays of light; a laser without ancillary lenses provides such illumination. Detailed optical calculations are much easier with colinearizing optics, so most absolute light-scattering photometers incorporate them.

The optical train between sample and detector generally includes a series of slits, irises, and lenses so placed that the detector registers only light scattered by the sample through a limited range of angles and not light scattered from the incident beam by a wall of the scattering cell. Difficulties with stray light, including scattering and diffraction by the edges of pinholes and irises, are reduced if one increases the distance between sample and detector and enlarges other optical components, such as pinholes, in an appropriate way. The optical train may also include polarizers, such as Nicol or Glan–Thompson prisms, arranged so that the sample and detector are exposed only to vertically or horizontally polarized light.

A modern instrument will be equipped with photon-counting electronics, which record the stream of photons incident on the detector as a series of digital pulses rather than as an analog current. Photon-counting electronics appreciably improve measurement accuracy with very dilute, weakly scattering samples. Digital counting methods are entirely linear over an extremely wide range of light intensities, so by using photon counting one nearly eliminates the need to change the intensity of the illuminating beam during the course of an experiment. One then does not have to calibrate a set of attenuating filters, eliminating a source of experimental error. However, good analog circuits and measurement of the photocurrent are sufficient for many light-scattering intensity studies.

2. Sample Preparation

The proper preparation of the sample prior to measurements is of decisive importance for the obtaining of reliable data. Light scattering is a nonselective technique that determines the total scattering power of everything (solvent, solute, dust, lint, etc.) in the sample cell. Typical dust particles have far larger scattering cross sections than do small polymer molecules, so the presence of small amounts of contaminant may have an extremely adverse effect on scattering measurements, especially at low scattering angles.

If one is interested only in molecular-weight measurements, one may be able to improve the accuracy of one's results by a proper choice of solvent. One should choose a solvent in which the specific refractive increment of the polymer is large, as this will increase the scattering intensity. [For very large polymer molecules, one should keep in mind that one does not wish to have multiple scattering or to exceed the criterion (equation 19) for the validity of the Rayleigh–Gans–Debye scattering theory.] In most cases, the index of refraction increment of the polymer should be at least 0.1 cm^3/gram. It is convenient to choose a solvent that is not itself a strong scatterer. The proper choice of solvent may also reduce the second virial coefficient. The use of a Flory theta solvent will reduce the second virial coefficient to zero (125). As a general rule, it is easier to clean solutions in nonpolar solvents than it is to clean solutions in polar solvents.

The most general method of removing dust from a polymer solution is ultrafiltration using microporous filters (Nucleopore, Millipore) of the etched precision-bore (typically polycarbonate based) or cellulosic types. Some filters with large surface areas bind appreciable amounts of some solutes, most notably biological macromolecules, so that the solute concentration after filtration may differ appreciably from the concentration before filtration. Teflon-based filters are necessary for the processing of certain solvents. The traditional fritted-glass filters remain entirely acceptable for most systems, but their high cost, and the difficulty of cleaning them, discourage their use.

For small molecules, vacuum distillation is a useful alternative to filtration. This process is best carried out in a small-volume sealed unit containing the sample cell and a sample reservoir. One repeatedly uses vacuum distillation to move the sample into the sample cell. After each distillation the sample is poured back into its original container, transferring with it, one hopes, any dust originally found in the sample cell. Since the apparatus is sealed, the pouring process admits no new dust into the system. It is critically important to avoid boiling the fluid during the distillation step, because boiling transfers dust (as an aerosol) back into the sample cell. With liquid mixtures, one must be aware during the final distillation of the need to effect quantitative transfer of all components of the mixture.

It does little good to clean a sample unless the cell is also clean. Most cell-cleaning methods are similar: One repeatedly adds to the sample vessel a volume of dustfree solvent, shakes the cell or subjects it to an ultrasonic bath, and removes the solvent from the cell. The process is more effective if the solvent flow through the cell is laminar; one can sometimes inject filtered solvent into one end of a cell and simultaneously remove solvent from the other end. After the cell is clean (examination under side illumination, using a black background material, is often an effective check), the cell may be dried by passing filtered nitrogen or air through it. Many utility compressed-air lines are badly contaminated with pump oil and metal dust; tanked nitrogen or nitrogen boil-off from liquid nitrogen are better sources of pressurized gas. The use of detergents is sometimes necessary in early stages of the cleaning process; however, one must then rinse the cell sufficiently thoroughly that the detergent does not contaminate the sample cell. The modern understanding is that the traditional "chromic acid" bath, besides being hazardous and difficult of disposal, is not good for glass surfaces. Sanders and Cannell (98) discuss more sophisticated methods for handling biopolymers.

With some caution, it is possible to centrifuge certain types of scattering cells after they have been filled with sample. This procedure removes gas bubbles and particulate matter that differs in density from the solvent. To balance the hydrostatic pressure head during centrifugation, the centrifuge bucket should be filled with solvent to a height above the top of the sample cell. In using this method, recall that a sample's properties may change when it is purged of macroparticulate components, and that a macroparticulate component may be viewed either as a contaminant or as an interesting part of the sample. The author has seen centrifugation used successfully with small-molecule solvents and a low-speed centrifuge. (An alternative method, in which polymer solutions are cleaned by prolonged ultracentrifugation prior to being placed in the sample cell, has fallen into disuse, in part because the method can lead to an artifactual fractionation of polydisperse substances.)

3. Operation and Calibration

After one's instrument and samples have been readied, the first step in a light-scattering study is the measurement of the scattering intensity of the pure solvent, including any low molecular weight components (added salts, buffers, and sucrose) found in the sample. The light scattering due to the solvent has to be subtracted from the scattering due to the solutions to obtain the scattering power of the solute of interest. Any background signal due to dark noise or anode leakage in the photodetector is removed by this subtraction. The light scattering of the solution is then measured at a series of different angles and solute concentrations. A standard procedure is to fill the sample cell partially with solvent, measure the scattering intensity, and then add to the sample cell a series of aliquots of concentrated solute, measuring the intensity after each addition of solute. Some photometers permit ready measurement on a working standard during a series of angular measurements.

To interpret one's measurements, one must convert measured intensity into a Rayleigh ratio. This requires subtraction of the solvent scattering, correction for geometric factors as described below, and conversion of the corrected intensity into absolute units by means of an empirical calibration constant. An effective procedure for obtaining this calibration constant is to measure the scattering intensity of a known substance. Although a variety of standards are described in the literature, 12-tungstosilicic acid ($H_4W_{12}SiO_{40}$, mol. wt. = 2879) in 3.0 M aqueous NaCl, at concentrations in the range 0.05–0.5 grams/ml, is preferred. The Rayleigh ratio of 12-tungstosilicic acid may be calculated correctly using the simple Rayleigh theory (equations 16–18) and the measured indices of refraction of the calibration solutions. Before comparing the theoretical value of R with the measured intensity, the solvent scattering intensity must be subtracted and all geometric corrections made. After the geometric corrections are made, the scattering intensity of 12-tungstosilicic acid should be independent of angle. Instead of 12-tungstosilicic acid, one may use as a standard a small-molecule neat liquid. For safety reasons, toluene is to be preferred to the more traditional benzene. The use of a polymer as a standard is not presently a recommended practice.

To convert measured intensities into absolute intensities, one must apply geometric and other correction factors, as discussed by Utiyama (112) and by Billmeyer and coworkers (13). The following factors must be used as corrections in all measurements, including those on the intensity standard:

1.   The source power and detector efficiency may fluctuate with time. The source power can be monitored directly. The use during measurements of a working standard of intensity, such as an opal glass diffuser plate, allows one to compensate for slow drifts in detector efficiency and background count levels.

2.   The sensitivity of the detector may depend on the polarization of the scattered light. This may be tested by studying a fully depolarizing sample.

3.   The effective scattering volume—the region within the scattering cell in which light scattering can take place and from which scattered light can reach the detector—depends on the scattering angle and the index of refraction of the sample. The dependence of the scattering volume on the angle of observation is determined by the detector geometry. If at all scattering angles the detector views the entire length of the scattering volume, there is no correction for scattering angle. This geometry is rarely used, because of the background due to scattering from the sample cell's walls. In the more common detector geometry, only light scattered by the sample can reach the detector. The slits that define the size of the scattering volume are left fixed when the scattering angle is changed, so the apparent length of the scattering volume, as projected onto a plane perpendicular to the sample–detector axis, is a constant. However, the physical volume of the sample is determined by the length of the scattering volume as measured along the illuminating beam (and by a depth and height fixed by the illuminating beam). This length decreases as the scattering angle approaches 90°. Simple geometry shows that the scattering volume therefore depends on the scattering angle as $1/\sin \theta$. This correction factor does not include effects caused by the nonuniform brightness of the illuminating beam.

The effective scattering volume depends also on the index of refraction of the sample, because the path of the scattered light is bent as it passes from the sample into the air. This phenomenon has the consequence that light that might otherwise have reached the detector is deflected out of the detector's acceptance angle. This correction has the magnitude $(n_s/n_{air})^2$, n_s being the index of refraction of the solvent and n_{air} being the index of refraction of the medium between the sample cell and the detector. Note that calibration standards will not include this correction in a correct way unless the sample and the standards have the same index of refraction. Since light passing from the geometric center of a cylindrical cell is not deflected when it passes through the cell wall, it was at one time believed that for cylindrical cells the correction for this effect is $(n_s/n_{air})^1$. However, with a cylindrical cell one observes scattering over a considerable distance to each side of the cell's center. Light emanating from points other than the geometric center of a cylindrical cell is deflected by the effect. The corrections for cylindrical and rectangular scattering cells are in fact the same (13).

4.   The transmission of any attenuating filters placed in the illuminating beam must be determined. In general, one cannot assume that the attenuation of a series of neutral density filters is simply the product of the attenuation percentages of the individual filters.

5.   Internal reflections within the scattering cell present substantial difficulties. When the incident beam passes out of the scattering cell, some of it is reflected back through the scattering volume. The illuminating intensity within the scattering volume is therefore several percent larger than one might expect. Furthermore, some of the light scattered directly away from the detector is reflected by the rear cell wall back toward the detector, and vice versa. This second scattering effect is generally much weaker than the first. Its magnitude, which depends in detail on the type of collecting optics, is seldom substantial unless the light-scattering intensity has a very pronounced angular dependence. By using the Fresnel formulas, one can estimate the amount of backwards scattering, as seen in equations 45 and 46 below. Careful tests have indicated that the correction formulas for this effect are not entirely reliable. Several schemes of differing effectiveness exist for eliminating backscattering. Coating cell surfaces with nonreflective paint is usually neither adequate nor reproducible in its results. The Rayleigh horn may be used to eliminate reflection of the incident beam. Heller and Witeczek (54) have described a scattering cell with two Rayleigh horns that will eliminate backscattering terms, at least for a range of angles around 90°.

If one combines all of these effects, the true Rayleigh ratio R_θ is predicted to be related to the apparent intensity i_θ by the equation

$$R_\theta = \left(\frac{R_g}{K_g}\right) \frac{\sin \theta}{G(\theta)} \left(\frac{n_s}{n_m}\right)^2 \frac{1}{t_i^2(1 - 4f_i^2)} (i_\theta - 2f_i i_{\pi - \theta})\rho_\mu \qquad (45)$$

where R_g is the calculated Rayleigh factor of the calibrating standard, K_g is the measured scattering intensity of the standard at 90°, i_θ and $i_{\pi - \theta}$ are the measured scattering intensities of the sample at the complementary angles θ and $\pi - \theta$, ρ_μ is the Cabannes factor, $G(\theta)$ is as defined in Section II.A, and (for perpendicular incidence of the light on all surfaces)

$$f_i = f_1 + (1 - f_1)^2 f_2 \qquad (46a)$$

$$t_i = 1 - f_i \qquad (46b)$$

$$f_1 = \frac{(n_s - n_g)^2}{(n_s + n_g)^2} \qquad (46c)$$

$$f_2 = \frac{(n_g - n_m)^2}{(n_s + n_m)^2} \qquad (46d)$$

where f_1 is the fraction of the light reflected at the sample–glass surface, f_2 is the fraction of the light reflected at the glass–outside interface, and n_s, n_g, and n_m are the indices of refraction of the sample, the cell wall, and the bath outside the cell, respectively. Utiyama (112) notes that effects 3 and 5 may be measured by filling the sample cell with a fluorescent dye and determining the fluorescence intensity as a function of angle.

4.   Applications of Light Scattering

a. POLYMER SOLUTIONS

The determination of the absolute value of the weight-average molecular weight by light scattering has major applications in the study of synthetic polymers and biological macromolecules. In the case of synthetic polymers, the absolute value of the molecular weight is most often required for comparison with the results of other methods. For example, the breadth of the molecular weight distribution of a polymer preparation can be estimated by measuring the number-average molecular weight, as with osmometry, and the weight-average molecular weight with light scattering. By means of Zimm's method, as treated in Section II.B.3.d, the weight-average molecular weight, second virial coefficient, and radius of gyration of a polymer molecule may be obtained.

A widespread use of molecular weights derived from light scattering is in the establishment of empirical correlations between the intrinsic viscosities and the molecular weights of linear polymers. Ever since Staudinger's conjecture that the intrinsic viscosity of a polymer should be linearly proportional to its molecular weight (108), the solution viscosity has provided a simple and useful technique for estimating the molecular weight of high molecular weight polymers. For a linear polymer in a given solvent, the intrinsic viscosity [η] and the molecular weight are empirically found to be related by an equation of the form

$$[\eta] = \lim_{c \to 0} \left(\frac{\eta - \eta_0}{\eta_0 c} \right) = K' M^a \tag{47}$$

where η is the solution viscosity and η_0 is the solvent viscosity. In this equation K' and a are usually found to be constant, within experimental error, over very broad ranges of the polymer molecular weight M.

For polydisperse systems, it can be shown that the molecular weight in equation 47 is properly interpreted as the viscosity-average molecular weight, which is closer to a weight-average molecular weight than to any of the other averages (19). For unfractionated polymers, equation 47 is often applied in a more empirical way. One measures the intrinsic viscosity and weight-average molecular weight of a series of polymer preparations, determines empirical values for K' and a, and then uses measurements of [η] to estimate the weight-average molecular weights of other polymer preparations. For examples of this procedure, see references 35 and 42.

b. CRITICAL PHENOMENA

It has been known for many years that the light scattering of a fluid becomes very great near the gas–liquid critical point. The phenomenon of critical opalescence is also observed in binary, ternary, and higher mixtures near their consolute points: the equilibrium concentrations, temperatures, and pressures at which two or three separate liquid phases become indistinguishable. Similar phenomena are observed in other types of material near their critical points. The light scattering can be related to the distance over which concentration fluctuations are correlated. Applications of light-scattering photometry and spectroscopy to critical-phenomenon studies have been extensively reviewed (57,101,102).

c. STREAMING SUSPENSIONS AND SOLUTIONS

Light scattering by oriented particles is anisotropic: The scattered intensity depends not only on the angle of observation (which fixes the magnitude of the scattering vector), but also on the angle between the scattering vector and the direction of flow. Measurement of light scattering in streaming systems may be used to obtain information on particle shape (53).

III.   QUASIELASTIC-LIGHT-SCATTERING SPECTROSCOPY

A.   INTRODUCTION

This section deals with quasielastic-light-scattering spectroscopy and its analytical applications. Quasielastic light scattering yields information about the motions of small molecules and macromolecules in solution. The motion may be the Brownian motion of the individual particles; one may also study the motion of particles due to their interactions, due to an applied electric field (light-scattering electrophoresis), or due to solvent flow (laser Doppler velocimetry).

The concepts underlying quasielastic light scattering have not yet found a place in the undergraduate curriculum. Although none of these ideas is particularly difficult, they may be unfamiliar. We therefore first present an observational description of light-scattering spectroscopy dealing with phenomena that—under favorable circumstances—are visible to the unaided eye. We then refine this picture.

Suppose one prepares a monodisperse suspension of macroparticles, such as virus particles or polystyrene latex spheres. The scattered light may be observed by placing the sample in a laser beam and allowing the scattered light to fall onto a piece of white card. If the macroparticles are large, the card will be brightly illuminated by scattering in the forwards direction, and more faintly illuminated at 45° or 90°. Before the sample reaches thermal equilibrium, the scattering intensity will shimmer wildly. Once the sample achieves a state of quiescence, the scattering pattern will superficially appear constant. On close examination, the scattered light will be seen not to be locally uniform, but instead to have a mottled pattern of bright and dark patches. The individual patches will fluctuate and flicker, being brightly lit at some moments, but only dimly lit in others. The flicker rate of a patch depends on the scattering angle; at larger angles, the patches are smaller and fluctuate more rapidly. If one examines the scattering in a series of particle suspensions, the flicker rate will be seen to be more rapid for the suspensions containing smaller particles.

A quasielastic-light-scattering spectrometer works by measuring the temporal evolution of these fluctuations in the intensity of the scattered light. The most important single piece of information obtained by a spectrometer is the correlation time τ_c, the average period for which a patch of scattered light remains bright or dim. From τ_c, the diffusion coefficient of the particles can be calculated. The measurement is absolute, being calibrated by the light wavelength, the apparatus geometry, and an external clock; the use of reference samples or calibration standards is not required. In the following sections, we refine the qualitative description given above.

Fundamental concepts in light-scattering spectroscopy are treated in Section III.B. Section III.C discusses concepts of importance to the practice of light-scattering spectroscopy; the components of a spectrometer are described in Section III.D. Section III.E presents the experimental methods used to obtain a spectrum. Section III.F treats standard applications of quasielastic-light-scattering spectroscopy to solutions of small molecules and macromolecules, and Section III.G covers special methods for micro-scale samples, turbid solutions, and resolved mixtures.

B.   FUNDAMENTAL CONCEPTS

1.   Correlation Functions

a. DEFINITION

The previous section proposed characterizing fluctuations in the scattering intensity $I(\mathbf{r}, t)$. An appropriate way to characterize $I(\mathbf{r}, t)$ is by means of the intensity–intensity autocorrelation function $G^{(2)}(t)$. $G^{(2)}(t)$ is defined below; its use is then motivated.

The basic observation noted above was that the scattering intensity fluctuates with time. From the discussion in Section I.D, it is evident that if the light scattering is time dependent, the scattered light must contain a range of frequencies. If the incident light had the single frequency ω_0 and the intensity variations take place over times on the order of τ, then the scattered light must have as its frequency spectrum a spectral line spread out over the frequency range $\omega_0 \pm (2\pi\tau)^{-1}$. [The coefficient 2π appears because frequencies are quoted in hertz, whereas the "natural" unit for this discussion is radians per second. In quasielastic-light-scattering spectroscopy, it is not actually necessary that the incident light be truly monochromatic, or that the line width of the incident light be less than the line broadening due to the molecular motions. (See Pusey's chapter of reference 30.)]

An obvious way to observe the frequency spectrum of visible light is to pass the light into an optical spectrometer, such as a grating monochromator, and measure the amount of light present at each of a series of frequencies. This approach is not effective here. The frequency of visible light is perhaps $5 \cdot 10^{14}$ Hz, whereas the intensity fluctuations take place during intervals of 10^0 to 10^{-8} sec. The scattered light therefore has a spectral line width of 0.2–$0.2 \cdot 10^7$ Hz. To measure this line width, one would need a spectrometer of resolution $3 \cdot 10^7$–$3 \cdot 10^{15}$. The smaller of these numbers lies within the range accessible to a good Fabrey–Perot or other type of interferometer; the larger of these numbers is entirely beyond the range of any classical instrument. Classical optical spectroscopy is therefore not useful for studying light-scattering intensity fluctuations.

In favorable cases, intensity fluctuations are visible to the naked eye, suggesting that a direct analysis of the time dependence of the intensity of the light is appropriate. An adequate procedure is to record the intensity as a function of time for a period T and then calculate

$$G^{(2)}(\tau) \equiv \langle I(0)I(\tau) \rangle = \int_0^T dt\, I(t)I(t + \tau) \tag{48}$$

In this equation, $G^{(2)}(\tau)$ and $\langle I(0)I(\tau) \rangle$ are both symbols for the intensity–intensity correlation function. (See McQuarrie (77) or Berne and Pecora (10) for further discussion of the ensemble-average "$\langle \ldots \rangle$" notation.) $I(t)$ is the intensity recorded at time t. Within the integral, the product of the intensities at the pair of times t and $t + \tau$ is taken. For equilibrium systems, $G^{(2)}(\tau)$ is independent of the absolute time, and depends only on the interval τ separating t and $t + \tau$. As shown in Section III.B.4, the diffusion coefficient is obtained from the functional dependence of $G^{(2)}$ on τ.

b. HEURISTIC ARGUMENT FOR CORRELATION FUNCTIONS

Correlation functions are not commonly encountered in undergraduate courses. The next few paragraphs suggest a motivation for using correlation functions. (Readers familiar with correlation functions may skip to III.B.2.)

If one wants to characterize $I(t)$, one can measure its average $\bar{I}$. This is the basis of light-scattering photometry. As a next step, one can record $I(t)$ and study the recorded dependence of I on time. There are many ways to do this. One interesting method is to begin by sorting out, from the record of $I(t)$ measurements, all segments in which the initial intensity $I(t)$ has a specified value I_1, where $I_1 > \bar{I}$, as shown in Fig. 9.9. At the beginning of each segment, the intensity has the value I_1; each segment is thereafter different. (As seen in Fig. 9.9, these segments of the record may overlap,

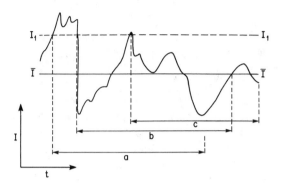

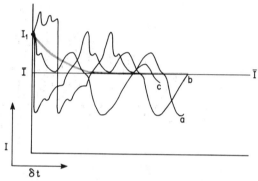

Fig. 9.9.   Construction of the correlation function $\langle I(0)I(t)\rangle$ (lower figure, shaded line) from a series of segments (a, b, c, upper and lower figures) of the recorded intensity versus time curve (upper figure). Segments were chosen having the value I_1 for their initial intensities. The average scattering intensity is $\bar{I}$.

because the intensity I_1 may repeat several times within a few moments.) Very close to the start of a segment, the intensity is necessarily almost equal to I_1. Far from the beginning of a segment, the system will have forgotten its initial intensity I_1; all values of the intensity may be encountered. An average of all segments with initial intensity I_1 yields a decaying exponential. If s is the time since the beginning of the segment, one finds that on average

$$I_1(s) = (I_1 - \bar{I}) \exp(-\Gamma s) + \bar{I} \qquad (49)$$

Equation 49 really represents an infinite set of identical equations, one for each value of I_1. Equation 49 indicates that (on the average) if the initial intensity I_1 satisfies $I_1 > \bar{I}$, the intensity decays exponentially down toward $\bar{I}$. If one repeats this analysis, sorting out from the record all sections with initial intensity $I_1' < \bar{I}$ and calculates the same average, one again obtains equation 49; the exponential now decays upward toward $\bar{I}$ with the same decay constant Γ.

Because only a single decay constant Γ is encountered, it is parsimonious to combine one's results on all decays beginning from I_1, all decays beginning from I_1', and so forth. Combining these results by averaging together the decay curves is not satisfactory. Some of the exponentials decay upward, whereas others decay downward; the average of the averaged curves is the constant $\bar{I}$. However, if one multiplies each curve i by $I_i - \bar{I}$, where I_i is the initial intensity for that curve, and averages the decay curves, the cancellation does not occur. The curves that decay downward are multiplied by a number $I_i - \bar{I} > 0$, making them larger or smaller. However, the curves that decay upward are multiplied by negative numbers, which effectively flips them upside-down. The average of a set of exponentials with identical decay coefficients, all of which decay downward, is a decaying exponential.

Mathematically, the weighted average of equation 49 over all initial intensities I_i is

$$\langle [I_i(s)][I_i - \bar{I}] \rangle = \langle (I_i - \bar{I})^2 \rangle \exp[-\Gamma s] + \langle \bar{I}(I_i - \bar{I}) \rangle \tag{50}$$

where the brackets "$\langle . . . \rangle$" denote the average over the possible values of I_i.

Equation 50 contains the decay constant Γ, which characterizes the temporal behavior of $I(t)$. The described method of obtaining this constant is, to say the least, tedious. One must first work through the record of $I(t)$, finding all segments of the record in which the initial intensity is I_1; these segments are averaged together. This process is then repeated for each initial value I_i. Finally, a weighted average of the averaged segments is constructed. Fortunately, there is a more direct way to obtain equation 50, namely, by computing the correlation function, equation 48.

This result may be seen as follows. If one writes

$$I(t) = \bar{I} + \delta I(t) \tag{51}$$

one has

$$\langle I(t)I(t + \tau) \rangle = \int_0^T dt [\bar{I}^2 + \bar{I}\delta I(t) + \bar{I}\delta I(t + \tau)$$
$$+ \delta I(t)\delta I(t + \tau)] \tag{52}$$

The first term on the right-hand side of equation 52 is a constant. The second and third terms are proportional to the average value of $I(t) - \bar{I}$; by the definition of $\bar{I}$, these terms vanish. The final term of this equation is the time-dependent part.

Equations 50 and 52 have the same numerical content. The only difference between these equations is in the methods of their computation. In equation

50, all segments of the record with initial intensity I_i are first averaged together (to yield equation 49); the averages over segments for each initial value I_i are then averaged again to yield equation 50. In contrast, in equation 52 each moment in time is taken to mark the beginning of a segment of the record, and the segments are averaged together with weighting factors I_i − $\bar{I}$. (As shown in Fig. 9.9, there is nothing to prevent the segments from overlapping.) The segments used in equation 52 each have some starting intensity, so each of these segments is included in one or another of the equations 50. Thus, each segment of the record used to compute equation 52 is also used in generating equation 50, and vice versa. Because equations 50 and 52 are generated by averaging the same data with the same weighting factors, they are equal.

From the standpoint of this heuristic treatment, the intensity–intensity correlation function allows one to answer the question: If the scattered light is especially bright at one moment, how bright is the light, on the average, at slightly later moments?

2.   The Correlation Function and the Spectrum: Theoretical Relations

This section treats the connection between $G^{(2)}(t)$ and the spectrum of the scattered light; they contain the same information in different forms. This section also summarizes a variety of theoretical terms often used in the primary literature. These terms refer to fundamental properties of scattered light, its random statistical fluctuations, and the connection between the correlation functions and the particle motions. These terms are not necessary if one wishes merely to use light-scattering spectroscopy, but are important if one wishes to understand what one is doing. The original statistical ideas are found in Wax (120); the connection between these ideas and light scattering is treated in the major references 10, 26, and 28–31. If one is willing to accept a few results without any demonstration, most of this part can be skipped.

The first necessary result from this section is that the intensity–intensity correlation function $G^{(2)}(t)$ is related to the averaged particle motions by the equation

$$G^{(2)}(\tau) = [G^{(1)}(\tau)]^2 + \bar{I}^2 \tag{53}$$

where

$$G^{(1)}(\tau) = \left\langle \sum_{i,j=1}^{N} \alpha_i \alpha_j \mid E_0 \mid^2 \exp[i\mathbf{k} \cdot (\mathbf{r}_i(t) - \mathbf{r}_j(t + \tau))] \right\rangle \tag{54}$$

is the electric field correlation function. This result describes the intensity of the light that has been scattered in a single direction. It is valid if the number of particles in the illuminated volume is large (say, $>10^6$) and if the interactions between the scattering particles are effective only over distances that are small in comparison with the linear extent of the illuminated volume. In conventional experiments on polymers or biopolymers, these assumptions are nearly always met. These assumptions may fail if N is very small (as with extremely dilute solutions or very small scattering volumes) or if an external field is present.

The second necessary result from this section is, as treated in reference 77, Chapter 22, that the correlation function $G^{(2)}(t)$ and the optical power spectrum $S^I(\omega)$ are a Fourier transform pair, namely

$$G^{(2)}(t) = \frac{1}{2\pi} \int_{-\infty}^{\infty} d\omega S^I(\omega) \exp[2\pi i \omega t] \tag{55}$$

and

$$S^I(\omega) = \int_{-\infty}^{\infty} G^{(2)}(t) \exp[-2\pi i \omega t] \tag{56}$$

Equations 55 and 56 comprise the Wiener–Khintchine theorem. The factors of 2π in the exponentials refer to ω in units of hertz rather than radians per second. In particular, if $G^{(2)}(t)$ is exponential,

$$G^{(2)}(t) = A \exp(-2Dk^2t) + B \tag{57}$$

then the spectrum is a Lorentzian:

$$S^I(\omega) = \frac{1}{\pi} \left(\frac{4Dk^2A}{\omega^2 + (2Dk^2)^2} \right), \qquad \omega > 0 \tag{58}$$

The members of a Fourier transform pair contain the same information, so $G^{(2)}(t)$ and $S^I(\omega)$ are both called the "spectrum" of the scattered light. Reference is also sometimes made to $2Dk^2$ as the "linewidth" of the spectrum; equation 58 shows that $2Dk^2$ is the half width at half maximum of the spectrum $S^I(\omega)$. Deviations of $G^{(2)}(t)$ from a single exponential form are often referred to as changes in the "line shape"; a nonexponential correlation function implies that $S^I(\omega)$ is not a Lorentzian.

The electric field $E(\mathbf{r}, t)$ of the scattered light is given by equation 13. The positions of the individual particles are random functions of time, so $E(\mathbf{r}, t)$ is also a random function of time. Suppose N is large, and suppose that the particles are sufficiently far apart that the position $\mathbf{r}_i(t)$ and the displacement $\mathbf{r}_i(t) - \mathbf{r}_i(t + \tau)$ of each particle are both statistically independent of the positions and displacements of most of the other particles in the system. The central limit theorem then predicts the behavior of the total scattered field $E(\mathbf{r}, t)$.

The central limit theorem (32) states that a variable S that is the sum of a large number M of independent, identical, random quantities is itself a random variable; furthermore, in the limit $M \to \infty$, S has a Gaussian probability distribution. In equation 13 the functions $\exp(i\mathbf{k} \cdot$

$r_i(t)$) are all identical and random; the suppositions assure us that most of the $\exp(i\mathbf{k} \cdot \mathbf{r}_i)$ are independent of one another. The central limit theorem therefore indicates that $E(\mathbf{r}, t)$ and $E(\mathbf{r}, t')$ are jointly Gaussian random variables; that is, the probability P of finding that $E(\mathbf{r}, t)$ and $E(\mathbf{r}, t')$ have a particular pair of values is given by a bivariate Gaussian distribution. The central limit theorem does not require that the individual $\exp(i\mathbf{k} \cdot \mathbf{r}_i)$ have Gaussian distributions.

The central limit theorem is the basis for references to the "Gaussian" or "non-Gaussian" nature of a particular experiment. A "Gaussian" light-scattering experiment is one in which the scattered field at two times has a joint Gaussian probability distribution. A "non-Gaussian" experiment is one in which N is effectively very small, so that the scattered field does not have a Gaussian probability distribution. Equation 53 is at best applicable only to Gaussian experiments. In non-Gaussian experiments, $G^{(2)}(\tau)$ and $G^{(1)}(\tau)$ may contain different information about the system. (Strong short-range interactions between the scattering particles do not inherently lead to non-Gaussian spectra. If each particle in the system does not interact with most of the other particles in the system, the central limit theorem is applicable.)

Variables that are jointly Gaussian have an important property: The average value of a higher-order moment of a Gaussian distribution can be written as a sum of products of the second-order moments of the distribution. In particular,

$$\langle I(t)I(t + \tau)\rangle = \langle E(t)E^*(t)E(t + \tau)E^*(t + \tau)\rangle \tag{59a}$$

or

$$\langle I(t)I(\tau + \tau)\rangle = \langle E(t)E^*(t + \tau)\rangle \langle E^*(t)E(t + \tau)\rangle + \langle|E(t)|^2\rangle\langle|E(t + \tau)|^2\rangle \tag{59b}$$

where only the nonzero terms of the second equation have been written. The average intensity $\bar{I}$ is $\langle E(t)E^*(t)\rangle$. Equation 59b implies equation 53.

Equation 54 gives $G^{(1)}(\tau)$ in terms of pairs of particle positions. If a solution is very dilute, the positions of different particles are independent of one another. For a dilute solution, on the average

$$\exp(i\mathbf{k} \cdot [\mathbf{r}_i(t) - \mathbf{r}_j(t + \tau)]) = 0 \qquad \text{for } i \neq j \tag{60}$$

in which case $G^{(1)}(\tau)$ may be written

$$G^{(1)}(\tau) = \left\langle \sum_{i=1}^{N} | E_0 |^2\alpha_i^2 \exp(i\mathbf{k} \cdot [\mathbf{r}_i(t) - \mathbf{r}_i(t + \tau)])\right\rangle \tag{61}$$

Equation 61 embodies a much stronger assumption than the particle-independence assumption required by the central limit theorem. The central limit theorem requires that each particle's position be independent of most of the other particles in the system, where "most" can exclude the immediate neighbors of the particle of interest. In contrast, equation 61 requires that particle i move independently of every other particle in the system, including particle i's near neighbors.

In dilute solutions, the motion of each particle is a "random walk." The displacement of a particle in one time interval gives no information about how the particle moves in any other time interval, nor about how any other particle moves at any time. Doob's theorem (37) states that if $r_i(t)$ is a random walk (technically, a Markov process) the right-hand side of equation 61 is

$$G^{(1)}(t) = \bar{I} \exp(-Dk^2t) \tag{62}$$

so

$$G^{(2)}(t) = \bar{I}^2 \exp(-2Dk^2t) \qquad (63)$$

where D is the translational diffusion constant of the individual particles.

The above results relate the spectrum to the translational diffusion coefficient of the particles, at least in an extremely dilute suspension. Theoretical treatments of concentrated solutions are based on other mathematical approaches. The calculation of $G^{(2)}(t)$ in Section III.B.4 is more physical in nature.

3. Particle Motions That Lead to Intensity Fluctuations

Intensity fluctuations arise from the positions and motions of the scattering particles. The discussion here is based on the Rayleigh–Gans–Debye theory (Section II.B), and thus does not apply without slight corrections to particles that interact strongly with light. The theory has been used successfully to interpret such phenomena as the motion of biological macromolecules in water and of synthetic polymers in a variety of solvents.

At any moment, the intensity of the scattered light is described by equations 13 and 14, namely

$$E(t) = E_0 \sum_{i=1}^{N} \alpha_i \exp[i\mathbf{k} \cdot \mathbf{r}_i - i\omega t] \qquad (13')$$

and, up to a dimensional constant,

$$I(t) = |E_0|^2 \left\{ \sum_{i=1}^{N} \alpha_i^2 + \sum_{\substack{i=1 \\ i \neq j}}^{N} \sum_{j=1}^{N} \alpha_i \alpha_j \exp[i\mathbf{k} \cdot (\mathbf{r}_i - \mathbf{r}_j)] \right\} \qquad (14')$$

These equations, and hence the intensity of the scattered light, depend on time through three physical factors, given here in ascending order of importance:

1. N may change because particles move into and out of the illuminating beam of light. Under normal operating conditions, in which the dimensions L of the illuminated volume and the value of N are substantial (say, $L > 10$–$100\ \mu$ and $N \gg 10^3$), changes in N are too small and too slow to affect $I(t)$ significantly.

2. The scattering amplitudes α_i may depend on time. This affect is prominent in the "depolarized" or "VH" spectrum, in which the incident light is vertically polarized and the horizontal component of the scattered light is observed. The depolarization is strongest with nonspherical macromolecules, though the intrinsic optical anisotropy of solute and solvent is also important. The depolarization is time dependent because it depends on the orientation of the molecules with respect to source, optical polarizations,

and detector. As the molecules rotate, the intensity of their depolarized light scattering changes. The VH spectrum gives information on rotational diffusion.

3.   Because of diffusion, the scattering centers move with respect to each other, so that $r_i - r_j$ is time dependent. Included in this effect are both the relative motion of pairs of macromolecules and the internal motions of scattering centers within a single large macromolecule. To compute $G^{(1)}(t)$, one needs a theory for the motions of the particles.

4.   Spectrum of the Scattered Light

a. CONTINUUM DESCRIPTION

This section presents a calculation of the correlation function $G^{(1)}(t)$, based on a continuum description of the scattering solution, Fick's macroscopic diffusion equations, and the Onsager regression hypothesis. This approach does not incorporate molecular rotation or internal molecular motions.

The mathematical description of light scattering by a solution is given in equation 13, which shows that $E(r, t)$ is proportional to the kth spatial Fourier component $a_k(t)$ of the concentration of scattering particles. If one writes

$$a_k(t) = \sum_{i=1}^{N} \alpha_i \exp[i k \cdot r_i(t)] \tag{64}$$

the correlation function of the scattered field is

$$G_1(\tau) = |E_0|^2 \langle a_k(t) a_{-k}(t + \tau) \rangle \tag{65}$$

As suggested by the shaded dots in Fig. 9.10, a spatial Fourier component of the concentration is simply a sinusoidal concentration fluctuation. This description of light scattering by a solution is mathematically the same as the standard description of Bragg scattering by planes of atoms in a crystal. The crystalline atomic layers correspond to time-independent Bragg planes, while the concentration fluctuations behave as evanescent Bragg planes: They are created by the random motions of the scattering molecules and are rapidly destroyed by diffusion.

The Onsager regression hypothesis states that on the average, a random fluctuation in a variable obeys the macroscopic relaxation equation for the decay of nonequilibrium values of the same variable. This hypothesis is based on the plausible assumption that a molecule has a position and a momentum, but does not "know" whether its present position was reached by random thermal motion or by an external intervention. For example, the nonuniform (and hence "nonequilibrium") concentration fluctuation shown

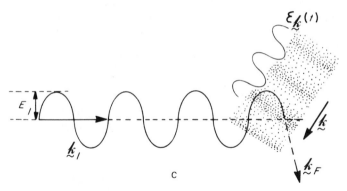

Fig. 9.10. Light scattering from a representative spatial Fourier component $\varepsilon_k(t)$ (shaded area, $\varepsilon_k(t)$ curve) of the polarizability of a particle suspension. For purposes of clarity, the solution has been drawn as having an extremely large fluctuation in a single spatial Fourier component of the concentration, and only minimal fluctuations in other spatial Fourier components. (From reference 31.)

in Fig. 9.10 might have been produced by random macromolecular motions. One could also in principle produce this concentration distribution by layering solutions of different concentration on top of one another, as is done to make a pousse-café. However, once the molecules are at the positions indicated in the figure, they are always subject to the same laws of motion. The relaxation of a macroscopic concentration gradient is described by Fick's law; the Onsager regression hypothesis indicates that the microscopic concentration fluctuations produced by Brownian motion also decay in accordance with Fick's law.

From Fick's law of diffusion

$$\mathbf{J} = -D\nabla c \qquad (66)$$

and the conservation equation

$$\frac{\partial c}{\partial t} = -\nabla \cdot \mathbf{J} \qquad (67)$$

an initially sinusoidal concentration distribution

$$c(\mathbf{r}, t) = c_0 + a_k(t)\cos(\mathbf{k} \cdot \mathbf{r}) \qquad (68)$$

decays as

$$\frac{\partial c(\mathbf{r}, t)}{\partial t} = -D_0 k^2 a_k(t)\cos(\mathbf{k} \cdot \mathbf{r}) \qquad (69)$$

This equation has the solution

$$a_\mathbf{k}(t) = a_\mathbf{k}(0) \exp(-Dk^2 t) \tag{70}$$

Inserting this result into equation 65 gives

$$G^{(1)}(t) = \bar{I} \exp(-Dk^2 t) \tag{71}$$

where $I = |E_0|^2 \langle |a_\mathbf{k}(0)|^2 \rangle$. Equivalently,

$$G^{(2)}(t) = \bar{I}^2 \exp(-2Dk^2 t) + B \tag{72}$$

where in the ideal case $B = \bar{I}^2$.

From equations 71 and 72, it can be seen that the intensity–intensity correlation function of light scattered by diffusing macromolecules is a decaying exponential. The amplitude of the correlation function is proportional to the mean-square $\langle |a_\mathbf{k}(t)|^2 \rangle$ concentration fluctuations. The diffusion coefficient is usually obtained from the logarithmic derivative of $G^{(2)}(t)$,

$$\lim_{t \to 0} \frac{d}{dt} \{\ln[G^{(2)}(t) - B]\} = -2Dk^2 \tag{73}$$

To use equation 73, one must know the base line B to which the spectrum decays. Experimental methods for determining the base line are treated in Section III.E.

b. DOPPLER-SHIFT DESCRIPTION

The purpose of this section is to show that quasielastic-light-scattering spectroscopy may properly be said to study Doppler line broadening. While clumsy for detailed calculations, this is a useful description of light-scattering spectroscopy for presentations to nonspecialists.

Consider the scattered field of the light. The displacement of a particle between t and $t + \tau$ can be written in terms of the particle velocity $\mathbf{v}_i$ as

$$\mathbf{r}_i(t + \tau) - \mathbf{r}_i(t) = \int_t^{t+\tau} ds \, \mathbf{v}_i(s) \tag{74}$$

so that

$$E(t + \tau) = E_0 \sum_{i=1}^{N} \alpha_i \exp(i\mathbf{k} \cdot [\mathbf{r}_i(t) + \int_t^{t+\tau} \mathbf{v}_i(s)ds] - i\omega t) \tag{75}$$

Suppose all particles move with a constant velocity $\mathbf{V}$. Equation 75 becomes

$$E(t + \tau) = E_0 \sum_{i=1}^{N} \alpha_i \exp[i\mathbf{k} \cdot \mathbf{r}_i(t)] \exp[-i(\omega - \mathbf{k} \cdot \mathbf{V})t] \tag{76}$$

The scattered field has been shifted from the frequency ω to the frequency $\omega - \mathbf{k} \cdot \mathbf{V}$.

In solutions, $\mathbf{k} \cdot \mathbf{v}_i(t)$ is different for each particle and is not a constant; its sign is equally likely to be positive or negative. Thus, instead of shifting the scattered field to a single new frequency, particle motion acts by taking the incident light, initially at a single frequency ω, and speading the light to higher and lower frequencies over a symmetric band centered on ω. The frequency range $\Delta\omega$ over which the scattered light is dispersed depends on the diffusion coefficient D. Specifically, $\Delta\omega$ is inversely proportional to the time required for a typical particle to diffuse through a distance of one light wavelength.

One might ask why light-scattering spectroscopy determines D rather than the Maxwell–Boltzmann thermal velocity of the particles; that is, why does one see the diffusion of the particles over light-wavelength distances rather than the thermal velocity of the particles? To measure a frequency ω, $E(\mathbf{r},t)$ must pass through zero twice, which requires a time $\tau \approx (2\pi\omega)^{-1}$. Over this time period, particle motion is diffusive, not ballistic. In terms of equations 74–76, the change in frequency $\omega - \mathbf{k} \cdot \mathbf{V}$ might better be written

$$\left[\omega\tau - \mathbf{k} \cdot \int_t^{t+\tau} \mathbf{v}_i(s)\, ds \right] / \tau.$$

Over the interval τ, the motion of particles in a fluid is a random walk, so one sees low-frequency diffusive motion, and not the larger Maxwellian thermal velocity.

C. FUNDAMENTAL CONSIDERATIONS IN THE PRACTICE OF LIGHT-SCATTERING SPECTROSCOPY

We now discuss relations between the properties of scattered light and the practice of light-scattering spectroscopy. Sections III.C.1–III.C.7 treat coherence areas, photon statistics, the wavelength dependence of the signal-to-noise ratio, homodyne and heterodyne detection, digital correlators, data analysis, and diffusion coefficients.

1. Coherence Areas

A real detector (Fig. 9.11) is not a mathematical point. If the intensity fluctuations at D and D^1 are not the same, they will tend to cancel, so with a detector of nonzero size the time-dependent part of a measured intensity–

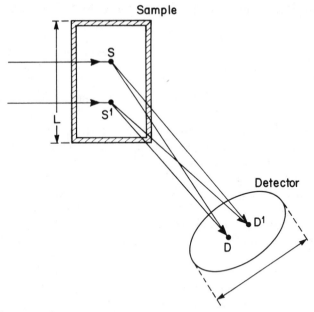

Fig. 9.11.   Construction for estimating the size of a coherence area.

intensity correlation function might be smaller than implied by equation 53. If the distance DD^1 is very small, the intensities at D and D^1 are nearly the same; if DD^1 is large, intensity fluctuations at D and D^1 are independent. The region over which the intensity is nearly constant is a "coherence area." In terms of the picture of Section III.A, a coherence area is one of the flickering patches. The allowable area of a detector, expressed as a number of coherence areas, depends on the experimental method, as discussed in Section III.E.

Figure 9.11 shows a construction for estimating the size of a coherence area. The interference between scattering from the points S and S^1 is determined by the path lengths from the source through S or S^1 to the detector. The source, S, and S^1 have fixed positions, so to calculate the relative amounts of interference at D and at D^1, only the path lengths SD, SD^1, S^1D, and S^1D^1, as the differences $d = SD - S^1D$ and $d' = SD^1 - S^1D^1$, need to be taken into account. If $d - d'$ is much less than a light wavelength, the degrees of interference at D and at D^1 will be roughly the same. However, if $d - d'$ approaches a light wavelength, there may be constructive interference at D and destructive interference at D^1, or vice versa. The area over which the interference, and hence the intensity, is roughly constant is a coherence area. A coherence area has linear dimension of order (25)

$$\ell = \frac{R\lambda}{L} \tag{77}$$

where R is the distance between the sample cell and the detector and L is the extent of the illuminated volume. One may readily focus and collimate in such a way that $L \approx 100$ μ, so that at 1 m from the scattering cell coherence areas are of order 0.5 mm across.

The discussion of light-scattering photometry did not mention coherence areas. Increasing the acceptance angle of the detector in a light-scattering photometer simply increases the total signal, thereby improving the signal-to-noise ratio without changing variables other than the angular resolution of the photometer.

2.   Photon Statistics

Section III.B described light in terms of a continuous field $E(\mathbf{r},t)$. Light is actually composed of discrete particles, namely photons. A measurement of the light intensity carried out with ultimate limiting precision consists of counting how many photons of each frequency are received during a specific time interval. $E(\mathbf{r},t)$ and $I(\mathbf{r},t)$ determine the probability of counting a given number of photons. However, even with a fixed light intensity, the actual number of detected photons fluctuates randomly from one intensity measurement to the next. For light-scattering spectroscopy, we need to consider only monochromatic light. With monochromatic light, all photons carry the same amount of energy, so the measured intensity I (energy flux per area) and the measured photon flux $n(\mathbf{r},t)$ (in units of number of photons per unit time per unit area) contain exactly the same information.

The light-scattering intensity in a quasielastic-light-scattering experiment is therefore a doubly random function of time. First, $I(\mathbf{r},t)$ is random because it is determined by the random Brownian motions of the scattering molecules. Second, the number of photons $n(\mathbf{r},t)$ counted in a given time interval is random; for a fixed $I(\mathbf{r},t)$, $n(\mathbf{r},t)$ has a statistical distribution whose average is $I(\mathbf{r},t)$ and whose width is determined by the quantized nature of light.

Much theoretical work has been done on $n(\mathbf{r},t)$ and $I(\mathbf{r},t)$ under different experimental conditions (28). A major important result is that the signal-to-noise ratio in a quasielastic-light-scattering experiment is determined by the scattering intensity, the fundamental quantity being N_r, the number of photons detected per coherence area per correlation time $[Dk^2]^{-1}$. (In contrast, in a classical intensity measurement, the signal-to-noise ratio depends only on the total number of photons detected, not on the count rate.) A reasonable standard is that one needs $N_r > 1$ for a practical spectroscopic experiment. If $N_r \gg 1$, quantum effects are negligible; calculations based on a continuum $I(\mathbf{r},t)$ are usually sufficient.

3.   Wavelength Dependence of the Signal-to-Noise Ratio

The signal-to-noise ratio in a quasielastic-light-scattering spectroscopy experiment is determined by the number of detected photons per coherence area per correlation time. For the study of small particles, this criterion implies that the signal-to-noise ratio depends only weakly on the laser wavelength.

The Rayleigh equation 7 shows that the scattered intensity is proportional to λ^{-4}; by shifting to shorter-wavelength illumination, one may greatly increase the signal intensity. However, a change in λ also changes other factors. First, from equation 77 the size of a coherence area is proportional to λ^{-2}, so shifting to shorter wavelengths shrinks the coherence areas. Second, since the correlation time is proportional to Dk^2, changing to shorter wavelengths reduces the correlation time by a factor of λ^{-2}. If the coherence areas and correlation times are made smaller, N_r will be reduced. Finally, the Rayleigh formula gives the scattered intensity (power per area). In quasi-elastic light scattering, one is interested in the photon flux (number per area per time). The energy of a photon is hc/λ, so at shorter wavelengths a given scattering intensity corresponds to fewer photons. Conversion from intensity to photon flux thus involves another factor, λ^{-1}. The ideal signal-to-noise ratio therefore depends on λ^{+1} rather than on λ^{-4}; that is, in principle the efficiency of a light-scattering spectrometer is higher if the incident beam contains red rather than blue or green light.

The quantum efficiency of most real detectors decreases with increasing wavelength. Furthermore, the standard commercially available red lasers (primarily He–Ne) have less power than a typical Ar^+ laser offers in the blue and green, so replacing a He–Ne (5–30 mW, 632.8 nm) laser with an Ar^+ laser (100 mW–1 W, 514.5 nm) will usually increase the sensitivity of a spectrometer. The increase in sensitivity is not markedly different from the change in laser power.

4.   Homodyne and Heterodyne Detection

Thus far, we have discussed only direct observation of the scattered intensity. Direct observation is properly referred to as "homodyne" detection of the scattered light. In contrast, in "heterodyne" detection the detector is simultaneously illuminated by light scattered from the sample and by light deflected, without a shift in frequency, from the original laser beam; the total intensity is measured. Because of interference, the total intensity is not the sum of the intensities of the scattered light and the deflected light. The source of the unshifted light is the "local oscillator"; if the light scattered by the sample and the light deflected by the local oscillator do not follow the same path, the light from the local oscillator travels along a "reference beam."

In elastic light scattering, heterodyne detection would contribute to the scattering intensity only an extra background term, which would have to be subtracted to obtain the scattering due to the sample. However, heterodyne detection is a reasonable procedure to use in light-scattering spectroscopy, because it allows the detection of frequency shifts as well as spectral line broadenings. For example, suppose light was scattered from a set of objects all having the same velocity $\mathbf{V}$. From equations 53 and 61, the correlation function for this system, as observed with homodyne detection, is

$$G^{(2)}(t) = \left| \left\langle \sum_{i,j=1} \exp[i\mathbf{k} \cdot (\mathbf{r}_i(0) - \mathbf{r}_j(0) - \mathbf{V}t)] \right\rangle \right|^2 + B \qquad (78a)$$

$$= \left| \left\langle \sum_{i,j=1}^{N} \exp[i\mathbf{k} \cdot (\mathbf{r}_i(0) - \mathbf{r}_j(0))] \right\rangle \right|^2 + B \qquad (78b)$$

which has no time dependence. Although each particle moves, the phase differences $\mathbf{k} \cdot (\mathbf{r}_i - \mathbf{r}_j)$ are not changed by a uniform displacement of all particles. On the other hand, with heterodyne detection the total field incident on the detector is

$$E = E_x \exp(-i\omega t) + E_0 \sum_{i=1}^{N} \alpha_i \exp[i\mathbf{k} \cdot \mathbf{r}_i(t) - i\mathbf{k} \cdot \mathbf{V}t - i\omega t] \qquad (79)$$

where E_x is the field due to the local oscillator. The correlation function in a heterodyne experiment is

$$\langle I(t)I(t + \tau) \rangle = I_{\mathrm{LO}}^2$$

$$+ \xi I_{\mathrm{LO}} \bar{I} \left\{ \left\langle \sum_{i,j=1}^{N} 2\alpha_i\alpha_j \cos (\mathbf{k} \cdot [\mathbf{r}_i(t) - \mathbf{r}_j(t + \tau) + \mathbf{k} \cdot \mathbf{V}\tau]) \right\rangle \right\}$$

$$+ \bar{I}^2 \left\{ 1 + \left| \left\langle \sum_{i,j=1}^{N} \alpha_i\alpha_j \exp(i\mathbf{k} \cdot [\mathbf{r}_i(t) - \mathbf{r}_j(t + \tau)] \right\rangle \right|^2 \right\} \qquad (80)$$

where $I_{\mathrm{LO}} = |E_x|^2$ is, up to a dimensional constant, the average intensity of the local oscillator, and ξ is the heterodyning efficiency, which is determined by the alignment of the scattered and reference beams. ξ has an ideal value of unity. This equation depends on $\mathbf{k} \cdot \mathbf{V}$. Thus, with heterodyne detection, one can measure the uniform motion of scattering particles, as is done in laser Doppler velocimetry (38) and light-scattering electrophoresis (117–119).

The light-scattering spectrum of a solution, as observed with heterodyne detection, is

$$\langle I(t)I(t + \tau) \rangle = I_{\mathrm{LO}}^2 + 2\xi I_{\mathrm{LO}}\bar{I}G^{(1)}(\tau) + \bar{I}^2(1 + [G^{(1)}(\tau)]^2) \qquad (81)$$

The experiment is typically run with $I_{\mathrm{LO}} \gg \bar{I}$, so the cross term $I_{\mathrm{LO}}\bar{I}G^{(1)}(t)$ dominates the homodyne spectral term $\bar{I}^2(1 + [G^{(1)}(\tau)]^2)$. From equations 53 and 81, homodyne and heterodyne detection give the same information about suspensions of diffusing macromolecules.

The signal-to-noise ratio obtained with homodyne detection is nominally half of the signal-to-noise ratio obtained under the same conditions with heterodyne detection. This is only an idealization. Fluctuations in the intensity of the reference beam and in the heterodyning efficiency are extra sources of noise for a heterodyne detector; because of these effects, the signal-to-noise ratio is typically nearly independent of the type of detection.

There are two general ways to perform heterodyne detection. One can use a beam splitter to deflect some of the light from the incident beam, and then use another beam splitter to merge the deflected light into the path of the light scattered by the sample. Alternatively, one may place a local oscillator—a stationary object such as a glass capillary or metal needle—inside the scattering volume. With this latter arrangement, the deflected light reaches the detector along the path used by the scattered light, making it easier to obtain a large heterodyning efficiency.

An intense scatterer that is virtually immobile, for example, a particle whose diffusion coefficient is much less than that of the sample, acts as a local oscillator. Indeed, any process that scatters light and has a very long relaxation time is effective as a local oscillator. This effect can complicate the interpretation of spectra that include series of very different relaxation times. Because the method is not reproducible, it is not an acceptable practice deliberately to use dust in a sample to cause heterodyning. There are samples, such as gels or highly viscous fluids, that cannot readily be freed of dust. The introduction of additional local oscillator strength to drive the spectrum toward its pure heterodyne form may permit interpretation of mixed homodyne–heterodyne spectra.

Discussions of heterodyne and homodyne detection are substantially hindered by geographic nonuniformity among the names used for these methods. In British usage, heterodyne detection is often referred to as "homodyne detection," and homodyne detection is referred to as "photon correlation spectroscopy" or "light beating spectroscopy."

5.   Digital Correlators

A digital correlator is a device that computes the correlation function of a fluctuating signal. Modern correlators are digital rather than analog instruments, and so describe the intensity of the scattered light by counting photons.

A correlator functions by dividing time into contiguous intervals, the length of an interval being the "sample time." The number of photons $n(t_i)$ received during each sample time t_i is stored electronically; the correlator then forms the correlation function as a sum of cross products,

$$\langle I(0)I(aS) \rangle = \sum_{i=1}^{M} n(t_i)n(t_i - aS) \tag{82}$$

where S is the sample time, M is the total number of sample times in the experiment, and a is an integer, $a = 1, 2, 3, \ldots$. The $n(t_i)$ are stored in a series of accumulator circuits or "channels." The sample time is sometimes called the "channel time" or the "bin width." Although aS refers to the time *interval* between a pair of measurements, it is sometimes referred to as the "time" or the "delay time."

The memory storage of a correlator is not arbitrarily large. The correlator retains each $n(t_i)$ until a fixed number of sample times has elapsed, and then discards its record of $n(t_i)$. The memory registers that hold the individual $n(t_i)$ are typically limited to recording numbers in a restricted range, such as 0–15. If $n(t_i)$ can exceed 15, that is, if the scattering is sufficiently intense, a procedure is needed to reduce the actual $n(t_i)$ to a number that can be accommodated in the memory. Instead of lowering the laser power, the number of counts entered as $n(t_i)$ can be manipulated with signal clipping or random scaling.

In signal clipping, the number of photons in each sample time is counted and converted to a clipped count

$$n_c(t_i) = \begin{cases} 1 & \text{if } n(t_i) \geqslant N \\ 0 & \text{if } n(t_i) < N \end{cases} \tag{83}$$

where N is the clipping level. This process appears very harsh. However, for conventional samples and a detector that views no more than one coherence area, the single-clipped correlation function

$$\langle I(0)I_c(aS) = \sum_{i=1}^{M} n(t_i)n_c(t_i - aS) \tag{84}$$

has the same functional form as the full spectrum $\langle I(0)I(aS)\rangle$. The double-clipped correlation function $\langle I_c(t)I(t_c - aS)\rangle$ is distorted by the clipping in a complex way.

In random scaling, one places a scaling counter between the detector and the correlator memory. If one randomly scales by k, then only every kth photocount is added to the memory. The "random" refers to the fact that the scaling counter is not reset at the start of each sample time. Instead, the scaling counter retains counts from sample time to sample time, and is reset to zero only when it accumulates k counts. Random scaling preserves the functional form of the correlation function. Some "scaled" counters are scaled non-randomly: the scaling counter is reset to zero at the beginning of each sample time. Non-random scaling is similar, in its effects on the correlation function, to clipping.

Proper adjustment of the scaling or clipping level is necessary for taking spectra efficiently. If the clipping level is too low or too high, $n_c(t_i)$ will be largely a string of ones or zeros; thus the correlation function will accumulate very slowly. If the scaling level k is too high, the memory registers will be loaded primarily with zeros, and the correlation function will not accumulate. If k is too low, the memory registers will overflow, which (details depend on the particular instrument) will distort the measured correlation function.

The cross product $\langle I(0)I(aS)\rangle$ is referred to as the "value" of the correlation function at the time aS. This is a slight misnomer. A photon contributes to $n(t_i)$ whether it is received at the beginning or the end of a sample time, so $\langle I(0)I(aS)\rangle$ actually includes cross products between photons that were detected as little as $(a - 1)S$ apart in time or as much as $(a + 1)S$ apart in time. $\langle I(0)I(aS)\rangle$ is therefore the average of the true correlation function and a wedge-shaped weighting function. If the slope of the correlation function is nearly constant between $(a - 1)S$ and $(a + 1)S$, this averaging does not distort the spectrum appreciably. For an exponentially decaying correlation function, the time required for the correlation function to decay to $1/e$ of its initial value (the "$1/e$ time") should be at least 8 or 10 times the sample time. If the $1/e$ time is less than this, the measured correlation function may be appreciably distorted by the wedge averaging.

As explained in Section III.E, it is desirable to know $\langle I(0)I(t)\rangle$ for very large values T of t. Rather than computing $\langle I(0)I(aS)\rangle$ for each value of aS between 0 and T, one may store, say, the 128 most recent values of $n(t_i)$, but compute $\langle I(0)I(aS)\rangle$ only for $a = 1$ to $a = 56$ and for $a = 121$ to $a = 128$. The correlator channels corresponding to a in the range 121–128 are the "delay channels." The use of delay channels provides the desired values of $\langle I(0)I(t)\rangle$ while reducing the instrumental complexity. (The use of delay channels does not reduce the number of values of $n(t_i)$ that must be stored by the correlator.)

6.   Data Analysis

From equation 72, the light-scattering spectrum of a dilute suspension of monodisperse particles has the form

$$G^{(2)}(t) = N_1^2\alpha_1^4 \mid E_0 \mid^4 \exp[-2\Gamma_1 t] + B \tag{85}$$

The spectrum of a dilute mixture of particles is

$$G^{(2)}(t) = \left(\mid E_0 \mid^2 \sum_{i=1}^{m} N_i\alpha_i^2 \exp(-\Gamma_i t) \right)^2 + B \tag{86}$$

where the sum is taken over all m species of Brownian macromolecules in the solution; $\Gamma_i = D_i k^2$; and N_i, α_i^2, and D_i are the number of particles, the scattering cross section per particle, and the diffusion coefficient, respectively, of species i.

The summation over i may formally be replaced with a continuous integral:

$$[G^{(2)}(t) - B]^{1/2} = \int dDa(D) \exp(-Dk^2 t) \qquad (87)$$

In this equation, $a(D)$ is the diffusion-coefficient distribution: the contribution to the intensity by all species with diffusion coefficient D. Equation 87 has the form of a Laplace transform.

If the system is monodisperse, equation 85 shows that the slope obtained from a least-mean-squares fit of $\ln[G^{(2)}(t) - B]$ to a straight line is proportional to the diffusion coefficient. If the solution is known to contain precisely two diffusing species, and if their diffusion coefficients are sufficiently different, the spectrum may be fit to the sum of two exponentials using the diffusion coefficients and scattering intensities of the two species as unknown parameters. If the two diffusion coefficients are very different, then one exponential will decay nearly to zero before the other exponential decays substantially; thus the multiexponential fitting procedure will be well behaved. If the two diffusion coefficients differ by less than a factor of 3 or so, a two-exponential fit to real data is difficult to carry out with any accuracy. The underlying problem is that a sum of exponentials that have very similar decay constants looks a great deal like a single exponential; the noise present in real data makes it impossible to distinguish between different ways of writing a nearly exponential curve as a sum of exponentials.

Unless the system is known to contain more than two species, multiexponential fitting procedures are generally not helpful. Although one can obtain a best fit of n exponentials to the spectrum, the calculated diffusion coefficients and amplitudes are often not very reproducible. Furthermore, if the spectrum of a solution that contains m species is incorrectly assumed to be represented by n exponentials, where $n \neq m$, the calculated and real diffusion coefficients will not be related to each other in any simple way. If noise is present, the correct number of exponentials cannot easily be deduced by comparing the statistical goodness of a series of fits. In considering the statistical validity of a particular fit, it is important to remember that there is random noise in the spectrum. A series of spectra of the same sample, fit to the same functional form, will show a random distribution in their root-mean-square (rms) fitting errors. Furthermore, if a single spectrum is fit to two different mathematical forms, and the rms errors in the fits to the two forms are both comparable with the random noise in the spectrum, the fact that one form gives a smaller rms error than the other form does not mean that either form is statistically preferable. All functional forms that fit a spectrum to within the random noise in that spectrum, after one has allowed for the number of independent parameters, have comparable statistical validity.

For interpreting a nonexponential spectrum, the most successful method in general use is Koppel's method of cumulants (67), in which the spectrum is expanded as

$$[G^{(2)}(t) - B]^{1/2} = \exp\left[\sum_{i=0}^{N} \frac{(-t)^i K_i}{i!} \right] \tag{88}$$

Here the K_i are the cumulants (or central moments) of the diffusion-coefficient distribution; the true expansion has $N = \infty$. The central moments are related to $a(D)$ by the equations

$$K_1 = \overline{D}k^2 = k^2 \left(\frac{\int_0^\infty dDa(D) \cdot D}{\int_0^\infty dDa(D)} \right) \tag{89a}$$

$$K_2 = [Dk^2 - \overline{D}k^2]^2 = k^4 \left(\frac{\int_0^\infty dDa(D) \cdot D^2}{\int_0^\infty dDa(D)} - \overline{D}^2 \right) \tag{89b}$$

The first cumulant, which is the slope of $S(k, t)$ as t approaches 0, is determined by the light-scattering-intensity-weighted average diffusion coefficient D. If the scattering particles are much smaller than a light wavelength, D is the molecular-weight-square- or z-average diffusion coefficient. If $a(D)$ is a Gaussian, K_2 gives the width of the diffusion-coefficient distribution. In general, K_2 contains information about the range of diffusion coefficients present. K_2 is often expressed as the variance V:

$$V = 100 \, \text{sgn}(K_2) \frac{\sqrt{|K_2|}}{K_1} \tag{90}$$

where $\text{sgn}(K_2)$ is the sign function, so that V and K_2 have the same sign. There are no common physical effects that lead to a physically interesting negative value for V. The finding $K_2 < 0$ generally implies that experimental artifacts are present. The most common errors that lead to negative values for K_2 are:

1.   Use of too many cumulants in the expansion.
2.   An incorrect (too high) value for the base line B, typically due to an insufficient time delay between the nondelayed and the delay channels.

In the author's experience, with a signal-to noise ratio of 500 or so, and spectra that cover three to four decay ($1/e$) times, one can obtain $\overline{D}$ and V with reasonable reproducibility (<0.5% and ±10 percentage points, respectively). To apply the cumulant fitting procedure, one needs a method of deciding how many cumulants to fit to the data. A reasonable procedure is to fit the spectrum to the cumulant expansion with $N = 1, 2, 3, \ldots$ and then choose the best value of N. Criteria for choosing the best value of N include the rms difference R between the calculated and measured spectrum and the so-called quality parameter Q. If $C(t_i)$ is the calculated value of the spectrum for the ith channel of a T-channel correlator, then

$$Q = \sum_{i=1}^{T-1} [G^{(2)}(t_i) - C(t_i)][G^{(2)}(t_{i+1}) - C(t_{i+1})] \qquad (91)$$

If there is no systematic difference between the calculated and the measured spectrum, the measured data will be distributed randomly on both sides of the theoretical curve, and the magnitude of Q will approach zero. In the author's experience, R and Q show either minima or saddle points near the best fit, the best fit being the fit that requires fewer parameters than the other fits near the minima in R and Q. Substantially negative values for K_2 are unacceptable. In a series of measurements on the same preparation of the same material, more reproducible results are typically obtained if one consistently uses the same-order fit for interpreting all spectra, even though small differences in R and Q may indicate that fits of different order are to be preferred for analyzing different spectra in a series.

Some complex forms for the diffusion-coefficient distribution are characterized by two or three coefficients. If $a(D)$ is known to have one of these specific forms, light-scattering spectroscopy may be able to obtain the characteristic coefficients. The fact that a numerical analysis of the spectrum yields a set of coefficients does not prove that $a(D)$ has the assumed functional form.

Equation 87 suggests that the diffusion-coefficient distribution $a(D)$ may be obtained from an inverse Laplace transform of the spectrum. This idea is not correct. The inverse Laplace transform is mathematically ill-posed, so a small error in measuring the spectrum can have a large effect on the calculated diffusion-coefficient distribution.

There has recently been increased interest in obtaining $a(D)$ from the spectrum. The most important results appear to be those due to McWhirter and Pike (78), who in effect have obtained a set of orthogonal polynomials suitable for estimating the inverse of equation 87. As an example of the difficulty of the inversion problem, Pike et al. (93) have shown, for a specified set of circumstances, that if one exponential gives a satisfactory description, with a given signal-to-noise ratio, for a specific range of τ, then in order to fit the same range of τ with equal validity to two, three, four, or more exponentials, one must increase the integrating times by factors of 10^2, 10^4, 10^6, and so on, respectively. Many other authors have approached this problem (48,94). The practical value of much of this work is still under active investigation. Extreme caution is necessary before believing results from a method that has not been tested experimentally against a series of samples with known diffusion-coefficient distributions. Imposing special conditions on a fitting function, for example that it be the "smoothest" function which is consistent with the data, may improve the mathematical quality of the fit, but only by suppressing the physical reliability of the results. Parameters obtained by fitting spectra in the presence of constraints (such as "smoothness") may be of great value in process control or industrial testing, even though the parameters may have little relation to the actual properties of substance being tested. Other, direct, methods of measuring a diffusion-coefficient distribution are noted in Section III.G.1.

7.   Diffusion Coefficients

For a dilute suspension of spherical particles, the diffusion coefficient is given by the Stokes–Einstein equation

$$D = \frac{k_B T}{6\pi\eta r} \tag{92}$$

where $6\pi\eta r$ is the Stokes' law drag coefficient for spheres with stick boundary conditions. For particles of known size, equation 92 predicts with good accuracy the diffusion coefficient of large molecules or macroparticles dissolved in small-molecule solvents. The measured diffusion coefficient is almost always slightly smaller than the value predicted on the basis of other physical measurements, probably because the D value from quasielastic light scattering is a z-weighted average.

For ellipsoids of revolution, the formulas of Perrin supply replacements for the Stokes' law drag coefficient. Calculation of the light-scattering spectra of large nonspherical particles is complicated by the correlation between the anisotropic scattering factors and anisotropic diffusion coefficients, as discussed by Wilcoxon and Schurr (123). For particles of more complex shape, the methods of de la Torre and coworkers (36) appear to be the best available ways of estimating D. Since the diffusion coefficient is affected by the fine structure of the particle surface and by the hydrodynamic boundary conditions (i.e., by the details of the solute–solvent forces), it is probably wise to avoid interpreting diffusion coefficients in terms of elaborate models unless substantial supporting evidence is available.

It has long been known that D is modified by interactions between diffusing molecules, so that D depends on concentration. If careful attention is given to the effect of intermolecular interactions, it is possible to gain much interesting information about, for example, molecular aggregation effects in concentrated solutions (124). On the other hand, there is a substantial body of published work that is simply wrong because concentration effects were treated improperly or ignored. We therefore briefly survey how D is modified by concentration effects.

The fact that D appears to be independent of concentration does not mean that intermolecular interactions are unimportant. For neutral globular particles, intermolecular interactions may be presumed to be significant if the volume fraction of particles in solution is greater than 0.01. Interactions between rods are certainly important if the average distance between them is less than the length of the rods, even if the volume fraction of the rods is very low. For polyelectrolytes, a dependence of D on pH or ionic strength may signal that intermolecular interactions are important, though internal conformational change can also affect D.

Two physically distinct diffusion coefficients characterize the random translational motions of macromolecules in solution. The mutual diffusion coefficient D_m describes the relaxation of a gradient in the total concentration. The self (or tracer) diffusion coefficient D_s characterizes the motion of labeled solute molecules through a solution that also contains unlabeled solute molecules. To measure D_s, one sets up countervailing gradients in the concentrations of the labeled and unlabeled solute molecules in such a way that the total solute concentration is everywhere the same. Fick's law (equation 66), with D replaced by D_s, predicts the motion of the labeled molecules.

In a two-component (solvent + solute) system, quasielastic light scattering determines the mutual diffusion coefficient. Common literature errors include the conclusion that light-scattering spectroscopy measures D_s, or that solute–solute interactions do not affect the diffusion coefficient measured by light scattering. An extensive, rigorous treatment of light-scattering spectroscopy and its relation to D_s and D_m shows that (88,91)

$$D_s = \frac{K_B T}{f_s} \tag{93a}$$

$$D_m = D_s(1 - \phi)\frac{\left(\frac{\partial \pi}{\partial c}\right)_{P,T}}{k_B T} \tag{93b}$$

where k_B is Boltzmann's constant, T is the absolute temperature, f_s is the translational drag coefficient of the solute diffusion at the appropriate concentration, π is the osmotic pressure, c is the solute concentration, and the volume fraction $\phi = c\bar{v}$, where $\bar{v}$ is the partial volume of the diffusing species. The $(1 - \phi)$ term is the reference-frame correction of Kirkwood et al. (65). Equations 93 have been verified to give results of good accuracy for globular proteins at concentrations as high as 350 grams/liter (49,50).

In systems with long-range intermolecular interactions, D_m may depend on the scattering vector. However, D_m may be independent of k even if intermolecular forces are strong. A k dependence of D_m has possible causes other than intermolecular interactions.

D_s and D_m are sometimes called the "single-particle" and "pair" diffusion coefficients, respectively, because D_s describes the displacement of each molecule relative to its own initial position, and D_m describes the displacement of a molecule relative to the initial positions of all of the diffusing molecules (including itself) in the system.

The "cooperative diffusion coefficient" is operationally synonymous with D_m. The term "cooperative" refers to a particular theoretical model for the relaxation of a concentration gradient. The term "interdiffusion" refers to the requirement in a two-component solute–solvent system that the diffusion of solvent and of solute occur simultaneously. The relaxation of a concentration gradient is due to antiparallel motion (hence the term "interdiffusion") of solute and solvent.

For nonelectrolytes at low concentration, D_s and D_m are equal to within experimental error. The concentration dependences of D_s and D_m are not usually the same. If D_s and D_m are unequal, intermolecular interactions must be important. For macromolecules in simple solvents, D_s falls substantially with increasing concentration. The interactions that modify D_m sometimes tend to cancel, so D_m may be nearly the same at high and at low concentrations. Interactions may change D_m by a factor of 2 or more from its zero-concentration limit. The sign of dD_m/dc depends on the system. The dD_m/dc value of a given protein may change sign as the pH is varied, even if the protein undergoes no conformational changes in the pH range in question (91). The concentration dependence of D depends on the chemical identities of the diffusing substance and its solvent. There is therefore no universal curve for interpreting the D_m value of a concentrated solution of a previously unstudied material in terms of the molecular size of that material.

C. EXPERIMENTAL CONSIDERATIONS

This section discusses components of a quasielastic-light-scattering spectrometer. Although one can purchase a complete instrument, many workers prefer to build their own. Furthermore, a detailed understanding of the apparatus can increase one's ability to obtain reliable data.

Figure 9.12 shows a block diagram of a typical light-scattering spectrometer. Light proceeds from the laser (a) through the focusing optics (b) to a sample cell (c). The scattered light is collected (d) and allowed to fall on a photomultiplier tube (e). The signal is amplified and sent to a digital correlator (f). Finally, the observed spectrum (g) is analyzed.

1. Lasers

A light-scattering spectrometer requires a source of intense monochromatic light. All practical instruments use lasers to obtain the needed intensity within the sample cell. The major factors influencing the choice of a laser are the power output, the amplitude stability, and the laser medium, the last of which determines the possible wavelengths of the incident beam. Most spectrometers described in the literature incorporate helium–neon (He–Ne) or argon-ion (Ar^+) gas lasers with incident-beam powers of 5–500 mW. A tunable dye laser can be used in a quasielastic-light-scattering instrument (62).

For a protein of molecular weight 65,000 at a concentration of 10 grams/liter, good spectra may (with patience on the part of the investigator) be obtained with a 20-mW He–Ne laser. A 5-mW source is adequate for the study of very large molecules and macroparticles. If lower molecular weight

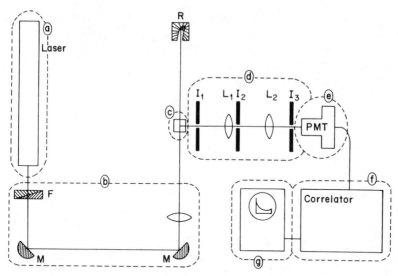

Fig. 9.12. Block diagram of a light-scattering spectrometer, including (a) laser, (b) focusing optics, (c) scattering cell, (R) Rayleigh horn, (d) collecting optics, (e) photomultiplier tube (PMT), (f) digital correlator, and (g) data analyzer. Also indicated are a filter F, mirrors M; lenses L_1 and L_2; and irises I_1, I_2, and I_3.

materials or lower concentrations are to be used, or more rapid data acquisition is needed, a more powerful incident beam may be desired. Conservative design practice for a general-purpose instrument calls for a minimum power of 10–15 mW. Many workers would prefer to have several hundred milliwatts available. Because the fundamental parameter is the intensity (power per area) of the illuminating beam (i.e., the radiance) within the sample cell and not the total power, improvements in focusing optics can compensate for a lack of laser power.

Artifacts due to sample heating limit the total incident power. With colored samples, the dependence of the measured diffusion coefficient on the incident power should be examined; one can sometimes extrapolate to zero incident power. The use of a laser line that the sample absorbs weakly may be helpful. Sanders and Cannell (98) present a thorough discussion of experimental methods of value with colored samples.

As shown in Section III.C.3, the signal-to-noise ratio in light-scattering spectroscopy is almost independent of the incident wavelength. For colorless samples, the exact choice of laser wavelength is therefore not very important.

Less obvious in importance than the power of a laser is its amplitude stability, usually quoted as the rms noise in the laser power integrated over some range of frequencies and given as a percentage of the total laser power. One needs high-frequency amplitude stability because random fluctuations in the beam intensity are indistinguishable from intensity fluctuations due to the sample. Amplitude stability is especially important with heterodyne detection. Over the frequency range of interest (usually 10 Hz–1 MHz), one can readily obtain lasers with no more than a few tenths of a percent of noise. Above 30 mW, water-cooled argon-ion lasers frequently have much less noise than helium–neon lasers.

The rms integrated noise is not a uniquely important datum. A light-scattering spectrometer is exquisitely sensitive to periodic modulation of the source brightness. If the laser noise occurs at harmonics of 60 Hz, or otherwise appears at a few discrete frequencies, the laser is unusable. A simple control experiment tests for laser instability. If the laser is allowed to illuminate a static scatterer, such as a piece of lens paper, the scattered light should have a flat, time-independent correlation function. Laser artifacts typically appear as "ringing"—a cosinusoidal modulation of the flat spectrum—or as "spikes"—a "picket fence" profile superposed on the correlation function. There is no reliable postdetection processing method for subtracting these artifacts. They must be eliminated at the source. Ringing and spiking may also be due to faults in the photomultiplier tube or correlator; such faults persist when the laser is replaced with a flashlight.

Almost all gas lasers are manufactured with Brewster-angle windows, the purpose of which is to give the laser light a vertical polarization. The emission from a laser without Brewster-angle windows may be linearly polarized, but the polarization orientation, and hence the relative intensity of VV and HV scattering, changes unpredictably from moment to moment. Lasers without Brewster-angle windows are not suitable for light-scattering spectroscopy. A polarizer in the focusing optics cannot correct for the lack of Brewster-angle windows.

2. Focusing Optics

The focusing optics direct the laser beam into the scattering cell. Within this optical train may be found a polarizer, filters, beam directors, and a focusing lens.

a. FILTERS

A sample may scatter so much light that the photodetector is overloaded. The preferred solution to this problem is to reduce the acceptance angle of the collection optics. An alternative is to dim the incident beam, as with neutral density filters. A filter may substantially deflect or displace the incident beam. Nondeflecting, continuously tunable neutral density filters may be made by overlapping a pair of glass wedges, as in some commercial products.

b. MIRRORS

It is easier to position a laser beam by rotating mirrors than by moving the laser. To position a laser beam in an arbitrary way requires two mirrors, because there are four degrees of freedom: two coordinates to fix a point through which the beam passes, and two angles to fix the beam direction at that point.

c. LENS

One can usefully increase the intensity of the laser beam by focusing it. A laser beam is slightly divergent, so the maximum intensity is inversely proportional to the focal length of the condensing lens. The signal quality can therefore be improved by reducing the focal length of the condensing lens. A lens with a very short focal length may impair the angular resolution of the spectrometer, since within the scattering volume the incident light will move on convergent rather than parallel paths.

3. Sample Cells

Samples prepared for light-scattering spectroscopic studies must meet the same standards of cleanliness as samples used for light-scattering intensity measurements. For cell-cleaning methods, see Section II.E.2.

Sample cells for light-scattering spectrometers vary in design. Many workers use rectangular cells (typically fluorimeter cuvettes with four sides polished), but thin-walled cylindrical cells are also popular. The major significant difference between the cell shapes is in the ease of determining the scattering angle. For general work, square cells are preferable to round cells.

The major practical limit on one's knowledge of the scattering angle is one's ignorance of the direction of the incident and scattered beams inside the scattering cell. As a ray of light passes from one medium to another, the ray is deflected (except in the special case of a ray perpendicular to the interface). The true scattering angle—the angle measured inside the cell—may thus be different from the apparent scattering angle measured at the laboratory bench. An error in determining the scattering angle leads to an error in calculating $|\mathbf{k}|$.

With rectangular cells, perpendicular incidence of the incident beam on the cell is assured if the entrance window reflects the incident beam back along the beam's original path. A circular cell functions as a lens. An effective, though not very accurate, way to place the incident beam at the optical center of a cylindrical cell is to mark the path of the beam in the absence of the cell, insert the cell, and translate the scattering cell until the transmitted beam is not deflected by the cell. This method is particularly unsatisfactory for low scattering angles.

Scattered light is also deflected at the sample–air interface. For a cylindrical cell, this deflection is zero only if the collecting optics view the optical center of the cell, which is difficult to guarantee with high accuracy. If the collecting optics view a noncentral part of a cylindrical cell, the light is deflected to an indeterminable extent. For a rectangular cell, the deflection is given by Snell's law,

$$\frac{n_s}{n_a} = \frac{\cos(\theta_a)}{\cos(\theta_s)} \tag{94}$$

where n_s and n_a are the indexes of refraction, and θ_s and θ_a the angles of incidence, in the sample and in air, respectively. The corrections for measurements done through the side window of a cell and for measurements done through the front or rear window are not the same.

A diffusion apparatus should include a reliable temperature control. Mounting the cell in a massive thermostatted metal block is sufficient for routine work. For high-accuracy methods, consult the references on critical phenomena (101,102).

4. Collecting Optics

As shown in Fig. 9.12, the collecting optics are a series of lenses and irises whose purpose it is to collect the light scattered from a fixed part of the sample (the "scattering volume") into a fixed range of directions (the "acceptance angle") and bring the light to the detector.

To do this, one needs one lens and two irises (the lens is not absolutely necessary). Figure 9.12 indicates positions for three irises, because there is more than one rational way to arrange the optical train. The lens collects the light; with two lenses, the photomultiplier tube can be left at a fixed distance from the sample. The irises serve two purposes. One iris acts as a mask, limiting the region within the cell that is seen by the photomultiplier tube. The second iris limits the range of angles through which light can be scattered and reach the detector. The limitation on the range of scattering angles is required because the spectrum of the scattered light depends on the scattering angle. A description of the acceptance angle in terms of coherence areas is given in Section III.C.1.

One standard arrangement uses irises at I2 and I3; the lens serves to create at I3 a real image of the scattering volume. The iris at I2 limits the acceptance angle. The iris at I3 determines how much of the scattering volume illuminates the detector. Window flare due to scattering of the incident beam by the entrance and exit windows may readily be seen in the real image at I3 and masked from the detector.

An alternative arrangement places a lens at its focal length from I3, so that rays of light emitted by the scattering volume in parallel directions come ot a point at I3. Light emitted in different directions is focused at different points on I3, so the acceptance angle is determined by the opening of the iris at I3. The other iris is placed at I1, where it masks parts of the scattering cell.

The collecting lenses cause the light to diverge strongly beyond I3. If the detector is close enough to I3, this divergence does no harm. If the phototube has a very small photocathode, the divergence may cause the light to strike optically inactive parts of the detector. An additional lens placed between I3 and the photomultiplier tube can focus light onto the photocathode; such a lens can also align the scattered beam with the photoactive section of the photomultiplier tube.

It is convenient to mount a periscope between I3 and the detector. Multiprism periscopes are expensive. An adequate substitute can be constructed by positioning a mirror at a place where it may be rotated into or out of the path of the scattered light.

5. Detectors

Quasielastic light scattering is a photon-counting technique in which a photomultiplier tube, preamplifier, and pulse shaper are used to transform incident photons into electronic pulses. A variety of photomultiplier tubes have been used successfully, including large photocathode tubes (typified by the RCA 7265 with S-20 photosurface) and low-dark-count (see below), small photocathode tubes (typified by the ITT FW-130).

A photomultiplier tube contains a light-sensitive surface, the photocathode, that is likely to emit one or several electrons when it absorbs a photon. These photoelectrons pass through an intense electric field (due to the high voltage applied between the photocathode and anode) and scatter repeatedly off targets ("dynodes"); in this process, the targets emit additional electrons. The initial photon is thereby converted to a group (a "photopulse") of approximately $1 \cdot 10^6$ electrons, spread out in time over 5–50 nsec. After leaving the tube, the pulse passes through a preamplifier–discriminator circuit, which separates pulses from background noise and converts the pulses into electrical signals compatible with TTL, ECL, or other digital circuitry. In the absence of light, photomultiplier tubes still emit a few pulses each second. This background pulse rate is the "dark noise," and is composed of "dark counts."

The quality of a detector is determined by its maximum allowable photocurrent, quantum efficiency, and dark-noise level. The maximum allowable photocurrent determines the maximum rate at which photons can be detected without signal distortion or (in some cases) permanent damage to the photomultiplier tube. The signal-to-noise ratio is determined by the photon-counting rate; a detector with a small limiting photocurrent will give poor spectra. For typical applications, photomultiplier tubes limited to fewer than $5 \cdot 10^4$–$1 \cdot 10^5$ counts/sec are not desirable.

Most preamplifier–discriminator circuits assume that the photomultiplier tube has a gain of $1 \cdot 10^6$, that is, that a single photon leads to a current pulse containing $1 \cdot 10^6$ electrons. The manufacturer's specifications indicate what voltage must be applied to the photomultiplier tube to obtain this gain. The dark noise increases nearly exponentially with applied voltage; increasing the voltage beyond the level needed for a gain of $1 \cdot 10^6$ will only increase the dark noise.

Many photomultiplier tubes show a very high dark current when high voltage is first applied to them. This current largely disappears if the tube is left in the dark with the high voltage on for several days. The dark current from some photomultiplier tubes is reduced by cooling them to 0° or −20°C, which can be done with a refrigerated photomultiplier tube mount. The advantage of cooling should not be overstated. A cooled FW-130 tube may have less than 1 count/sec of dark noise, but the dark noise in an uncooled large cathode tube is typically under 500 counts/sec, which level is already much less than a typical signal. If the base line is measured directly (see Section III.E), the presence of a weak dark current is irrelevant. (VH scattering can be quite weak; thus a cooled low-dark-count photomultiplier tube is essential for depolarized scattering.)

To observe very short correlation times, one must protect against afterpulsing and related effects. The detection of a photon disturbs the internal voltages within a photomultiplier tube, so for a few hundred nanoseconds or a few microseconds after detection of a photon the sensitivity of the phototube is changed. To test for afterpulsing, measure the spectrum of an unchanging source, such as an illuminated piece of tissue paper. The correlation function should be completely flat. At very short sample times, afterpulsing causes the correlation function to oscillate.

The traditional method of avoiding photomultiplier tube afterpulsing is to use a specially selected detector. Measurements on rapidly decaying correlation functions are limited by the count rate as well as by afterpulsing. To acquire a signal, one must detect two photons within the same correlation time. If the correlation time is very short, an intense photocurrent is needed or there will be little signal. Unfortunately, many detectors with good afterpulsing qualities have low limiting photocurrents. Bendjaballah (8) and Bursteyn et al. (20) report an alternative method, effective with any photomultiplier tube, for suppressing afterpulsing effects. The scattered beam is divided with a beam splitter; cross correlations between the divided beams are then studied with two detectors.

6. Correlators

Features found in a well-designed correlator include:

1.   Direct counting of received photocounts and total run time, permitting calculation of the base line from equation 53.

2.   Base-line monitor channels that permit simultaneous measurement of the correlation function and the level to which it decays at long delay times.

3.   Continuous, as opposed to batch, processing of the received photocounts. Batch processing restricts one to viewing a fraction of a coherence area.

4. A means of modifying input signals that exceed the capacity of the correlator shift registers. This may include clipping, scaling, or random scaling.

5. A method of transferring correlation functions, when they are measured, directly to a computer or other mass storage device, as by means of an RS-232C or IEEE-488 output port.

6. An analog output (or its equivalent) permitting one to see the shape of the correlation function as it is accumulated.

Optional features in a correlator include:

1. The number of channels. A good lower bound is 64 channels, perhaps including base-line monitoring channels. There appears to be little advantage to instruments with many more than 64 channels unless one is interested in substantially nonexponential spectra or in velocimetry.

2. Nonuniform channel spacing and width. A correlator whose channels increase in width at larger times will give better signal-to-noise ratios at those times without losing resolution at smaller times. Nonuniform channel widths increase the effectiveness of each channel without increasing the number of channels.

3. Internal analysis of spectra during data acquisition. With the advent of inexpensive microprocessors, it is possible to compute the apparent diffusion coefficient while the correlator is still operating. This can save time by allowing one to identify flawed samples.

7. Tables

For experiments using homodyne detection, any solidly built table with optical rails for mounting the lenses and mirrors will usually give adequate stability. A light-scattering spectrometer that uses heterodyne detection is a multibeam interferometer. The optical components must be held in position to within a small fraction of a light wavelength, as is possible with vibration-isolation tables and rigid supports. Typically, the paths followed by the laser beams must be enclosed to protect against drafts.

E. EXPERIMENTAL PROCEDURE

This section discusses the practice of measuring and interpreting quasi-elastic-light-scattering spectra.

Light-scattering spectroscopy is absolutely calibrated in terms of the wavelength of light and the external time standard that fixes the sample time. It is therefore not necessary to calibrate light-scattering spectra against known standards. Nonetheless, especially with a newly assembled spectrometer, it is good practice to first study materials with known diffusion coefficients. One should obtain good agreement between the measured and the nominal diffusion coefficients of one's standards. If the measurements do not give the desired result, something is wrong with the apparatus, the data analysis, or the "standard" samples. The deficiency must be identified and corrected before proceeding further. It is not proper to use measurements on a known standard to establish a calibration coefficient for one's equipment. Errors of precisely a factor of 2 typically result from unanticipated heterodyne detection due to stray laser light.

The best standards presently available are polystyrene latex sphere suspensions. The sphere suspensions are of two sorts: surface modified (generally with carboxylic acid groups) and unmodified. The surface-modified spheres are stable in distilled water; the "unmodified" spheres are stabilized with surfactant and precipitate slowly in very pure distilled water or salt solutions. Addition of surfactant can stabilize the unmodified spheres. Because of the z-averaging of D, the apparent hydrodynamic radii of these preparations will typically be found to be 4–7% larger than the nominal radii given by their manufacturers. The quality of commercial polystyrene latex suspensions varies substantially from batch to batch. Whereas some preparations have no detectable polydispersity, in other cases spectra will be found to be severely nonexponential, indicating appreciable aggregation or other difficulties with the manufacture or storage of the material. In the author's experience, very small (<0.05 μ) and very large (>0.8 μ) latex particles have greater degrees of polydispersity.

In making a measurement, one should begin with the irises in the collecting optics at their minimum diameters. One should then bring the laser power up toward its maximum value while monitoring the intensity of the scattered light to be sure that the limiting current of the photomultiplier tube is not exceeded. After maximum laser power has been reached, the irises may be widened; the limiting iris width is decided by one's method of measuring the base line (see below).

In observing an unknown, one should at first make the correlator sample time too long, so that the measured correlation function decays completely in the first few channels. The sample time should then be reduced to meet the criterion of Section III.C.5: $(2Dk^2)^{-1}$ should cover at least 8–10 channels. Beginning with a sample time that is too short, and then increasing the sample time until the criterion appears to be met, readily leads to errors.

There are two correct methods for determining the base line to which the correlation function decays. These methods have different implications for other aspects of the measurements.

Method 1: Direct measurement of the base line by means of delay channels on the correlator. To minimize noise with this method, the spectrum should be acquired in a single long measurement. If the correlator does not do batch processing of $I(t)$, the signal-to-noise ratio in the experiment will be largely independent of the number of coherence areas viewed by the detector, so scattering into several coherence areas may safely be observed simultaneously. Increasing the acceptance angle will suppress some types of background noise, such as photomultiplier tube dark counts, but will not improve the intrinsic signal-to-noise ratio. In a system with an extremely wide range of decay coefficients, heterodyne detection (as discussed in Section III.C.4) may be of value.

Method 2: Deduction of the base line from the average count rate.   In this method, the total number of counts (or clipped or scaled counts) and the total running time are used to determine the average counting rate $\bar{I}$. The base line is then $\bar{I}^2$, as shown in equation 53. The acceptance angle of the detector should be less than one coherence area. If a long integration time is required, the spectrum should be measured repeatedly in short (1–10 sec) spectrometer runs. Each spectrometer run should be separately normalized by its measured $\bar{I}$; the normalized runs should then be combined.

Either of these methods appears to give satisfactory results.

If the spectrum is a single exponential, the base line level may alternatively be determined from a nonlinear least-mean-squares fit of the data to equation 57. Because of the nonlinear nature of the fit, this procedure is generally not as accurate as Methods 1 and 2.

A similar, but incorrect, procedure for locating the base line treats B as a fitting parameter and simultaneously fits the spectrum to a cumulant expansion. Since errors in the base line can be compensated for with errors in the upper cumulants, most reasonable values of the base line will give statistically sound fits to the spectrum. One could, for example, choose B by minimizing the second or higher cumulants. However, there is no *a priori* reason to suppose that an unknown sample does not have a large value for K_2 or K_3 so this fitting procedure gives meaningless results.

F.   APPLICATIONS OF QUASIELASTIC-LIGHT-SCATTERING SPECTROSCOPY

The quantity directly measured in quasielastic-light-scattering spectroscopy is the spectrum $\langle I(0)I(t)\rangle$ of the scattered light. For particular systems, $\langle I(0)I(t)\rangle$ can be interpreted in terms of microscopic parameters to obtain information about the motions and relative positions of molecules in solution. The discussion of such applications is conveniently divided between studies on Brownian macroparticles and studies on solutions of small molecules.

1.   Measurements on Brownian Macroparticles

The primary datum obtainable from the spectrum of a solution of Brownian molecules is the mutual diffusion coefficient D_m. As shown in eqn. 86, if the solution is monodisperse, the spectrum is a single exponential. The cumulant expansion (equation 88) may be used to interpret the spectrum of a polydisperse solution; the first cumulant is the initial slope of $S(k, t)$. K_1 determines a diffusion coefficient:

$$\bar{D} = \frac{K_1}{k^2} \tag{95}$$

$\overline{D}$ is the average diffusion coefficient of all scattering species, weighted by their light-scattering intensities. For molecules much smaller than a light wavelength, $\overline{D}$ is the z-weighted average diffusion coefficient (cf. equation 29). For large molecules, scattering intensities are given in Sections II.B and II.C. Since large particles scatter preferentially in the forward direction, D may depend on the scattering angle; furthermore, dust contamination is generally more of a problem in measurements made at low scattering angles.

The diffusion coefficient is sometimes of direct interest in transport problems. For dilute solutions, the Swedberg equation (22)

$$M = \frac{sRT}{\overline{D}(1 - \overline{v}\rho_s)} \tag{96}$$

allows one to combine D with the sedimentation coefficient S, the partial volume $\overline{v}$, and the solution density ρ_s to calculate the molecular weight M of a macromolecule. Although one can obtain D from a careful analysis of sedimentation profiles, direct measurement of D with quasielastic light scattering is substantially more accurate. Equation 96 assumes that the solution is sufficiently dilute that $(d\pi/dc) = RT$.

If the scattering macromolecules are known to be spheres, D and the Stokes–Einstein equation (equation 92) give the spheres' hydrodynamic radius. For nonspherical particles, equation 92 can still be used to calculate an "apparent hydrodynamic radius," but this quantity contains no information other than that given by the diffusion coefficient. If the system being studied is polydisperse, equation 96 and other equations of its type suffer from the difficulty that the D value obtained from light-scattering spectroscopy is a z-averaged quantity, whereas s is a weight-averaged quantity. The ratio of these two numbers yields a value of M that in general cannot be expressed as a simple average over the molecular weight distribution.

Conformational changes in a macromolecule may result in changes in its translational drag coefficient, so that one can monitor, for example, protein denaturation by observing changes in the diffusion coefficient (86). In dilute solutions, changes in D are directly connected to changes in the effective hydrodynamic radius of the macromolecule. In concentrated solutions, changes in D indicate changes in the molecular conformation. However, conformational changes affect both the intermolecular interactions and the molecular drag coefficients, so that in concentrated solution one can not infer, from the observation that D has increased, whether the size of the macroparticles has increased or decreased, and vice versa (88).

The irreversible aggregation of biopolymers at low concentration may be observed with light-scattering spectroscopy. For example, by measuring the spectrum at 1-sec intervals, Benbasat and Bloomfield (7) determined kinetic rate constants for the aggregation of T7 phage particles. This technique is limited by the speed at which single light-scattering spectra can be acquired and recorded; with present instrumentation, this rate does not exceed a few spectra per second.

Efforts have been made to apply quasielastic-light-scattering spectroscopy to the analysis of micelle solutions. These studies are greatly complicated by the interactions between micelles in solution (27); the apparent dependence of the sizes and shapes of micelles on the pH, ionic strength, and amphiphile concentration of the solutions (126); and the destruction of micelles on dilution of their solutions. Unambiguous results are difficult to obtain (70,90).

2. Solutions of Small Molecules

In pure liquids and liquid mixtures, the Brownian motions of individual particles are replaced by the collective motions of the liquid. The VV (vertically polarized light incident, vertically polarized light detected) spectrum of a pure liquid has the Rayleigh–Brillouin form. The spectrum $S'(\omega)$ (equation 56) has a central Lorentzian line due to energy fluctuations and two side bands—spectral lines of nearly Lorentzian form whose centers are not at zero frequency—due to scattering from thermally excited sound waves. In many cases, line widths and separations occur at such large frequencies that the spectrum is readily observed with a Fabrey–Perot interferometer.

The detailed form of the VV spectrum of a pure liquid is treated in references 10 and 25. From the spectrum of the scattered light, one can infer the thermal diffusivity, the isothermal compressibility, the adiabatic sound velocity, and the classical sound-attenuation coefficient of a pure fluid. For a liquid mixture, the mutual diffusion coefficient and the osmotic compressibility are accessible. Although these constants can be obtained in other ways, light-scattering spectroscopy has the advantage that it is entirely noninvasive. It is therefore particularly well suited to the observation of systems that are easy to perturb, such as liquids and mixtures near their critical points (101,102).

3. Depolarized Light-Scattering Spectra of Small and Large Molecules

The VH spectrum (vertically polarized incident light, horizontally polarized scattered light) of a solution of an optically anisotropic molecule such as benzene contains information about molecular rotation. For small molecules, the molecular rotation time is so short that interferometric rather than photon correlation techniques must be used to measure it. In solvents of low viscosity, the line width of the spectrum is

$$\Gamma_R = (2\pi\tau_R)^{-1} = \frac{k_B T}{f_R} \tag{97}$$

where f_R is a drag coefficient for rotational motion.

One may attempt to use continuum hydrodynamics to compute f_R and τ_R for reorienting particles of known shape. For ellipsoids of revolution, Perrin's formulas are complete. Experiments indicate that continuum hydrodynamics is not adequate, except at a relatively crude level, for either small or large molecules. For small molecules, the difficulty may be said to lie in choosing the appropriate hydrodynamic boundary conditions. Slip boundary conditions give better results than stick boundary conditions. For small molecules in solutions of high viscosity (10), one observes non-Lorentzian line shapes, including Brillouin-like side bands and central dips in the spectral line, due to coupling between the orientational motions of the molecules and damped high-frequency shear waves in the liquid. Spectra of small molecules may also show features caused by internal molecular relaxations (82).

The depolarized scattering spectra of large particles such as tobacco mosaic virus can be obtained with quasielastic light scattering. Few experimental results have been obtained on depolarized scattering by large molecules of known shape. The use of Perrin-type formulas to interpret experimental data on translational and rotational diffusion coefficients does not appear to give self-consistent results (123). Hence, the inference of molecular shape from the VV and VH spectra is difficult at best.

G. SPECIAL METHODS

1. Separations *In Situ*

The light-scattering spectrum does not give detailed information on the diffusion-coefficient distribution. This is a fundamental mathematical limit not readily overcome with better data. Instead of trying to interpret nonexponential spectra, it may be more effective to separate a sample into its components and study each component. Loewenstein and Birnboim (76) and Lim et al. (75) present methods for separating the macrocomponents of a mixture by ultracentrifugation or free (Tiselius, unsupported) electrophoresis. Elastic and quasielastic scattering studies are then made on the resolved species within the centrifuge or electrophoresis cell. The diffusion coefficients and concentration profiles may thereby be determined. As in conventional centrifugation or electrophoresis, components that are in rapid equilibrium with each other cannot necessarily be resolved.

Loewenstein and Birnboim (76) centrifuged a solution in a capillary tube and determined the scattering intensity and diffusion coefficient at a series of points. Intensity measurements revealed a sedimenting boundary; the sedimentation coefficient inferred from the boundary position agreed with conventional values. D was obtained from the spectrum, allowing use of the Swedberg equation for calculation of a molecular weight.

Lim et al. (75) performed Tiselius (unsupported) electrophoresis of their sample, turned off the applied field, and used light-scattering intensity measurements to locate the migrating bands. The diffusion coefficients of the resolved species were also accessible. Theoretical analysis indicates that this technique compares most favorably with electrophoretic light scattering (see next section) for the study of small macromolecules and less favorably for the study of large macromolecules.

2. Light-Scattering Electrophoresis

If an electric field is applied to a solution of charged molecules, the molecules acquire a constant uniform motion. The heterodyne (and not homodyne!) spectrum of such a solution shows cosinusoidal modulations superimposed on the usual exponential decay. As was first shown by Ware and Flygare (118), the electrophoretic mobility of molecules can be determined with light-scattering spectroscopy.

The need to apply an electric field to the scattering volume imposes special design requirements on the sample cell. These have been reviewed in detail (113,117,119). How this method works is better understood in terms of the frequency spectrum, which, for a mixture, is composed of a series of Lorentzian lines each with a different center frequency

$$\omega = \mathbf{k} \cdot \mathbf{U} = \mathbf{k} \cdot \mu \mathbf{E} \tag{98}$$

where $\mathbf{U}$ is the electrophoretic velocity.

The electrophoretic mobility is given by Henry's law. For a particle of radius r, charge q, and surface charge density σ in a solvent in which the Debye length is K, (119)

$$\mu = \begin{cases} \dfrac{q}{6\pi\eta r(1 + Ka)}, & Kr \ll 1 \\[4mm] \dfrac{\sigma}{\eta K}, & Kr \gg 1 \end{cases} \tag{99}$$

It is very easy to identify a series of nonoverlapping spectral lines. Under favorable conditions, the half width at half height of a line is $2D_m k^2$, allowing simultaneous determination of the electrophoretic mobilities and diffusion coefficients. The D_m value obtained from light-scattering electrophoresis is less reliable than the D_m value obtained for separated components in the absence of an applied field. Relative to other techniques, light-scattering spectroscopy is most useful for determining the electrophoretic mobilities of particles larger than 100 Å, such as intact biological cells and their fragments.

3. Micro-Scale Samples

High-quality light-scattering spectra may be obtained of samples in melting point capillaries, as reported by Foord et al. (43), who obtained spectra of 15-μl samples. With simple optics the scattering from the capillary tube will reach the detector and act as a local oscillator; in this method, unless very turbid samples are encountered, one should assume that one is performing heterodyne detection. In the author's laboratory, a simple holder for capillary tubes was prepared from a standard fluorimeter cuvette polished on four sides by milling a vertical hole in the stopper to serve as a capillary mount and filling the sample holder with dustfree water. To compensate for the index matching of the capillary walls by the water, and to ensure pure heterodyne detection of the scattered light, it is sometimes useful to roughen the capillary with cleaning tissue. Capillaries, being round, function as cylindrical scattering cells (see Section III.D.3).

4. Multiple Scattering

The Doppler-shift description of light-scattering spectra (Section III.B.4.b) suggests that each time light is scattered by the sample it is Doppler-shifted again. The diffusion coefficient inferred from the spectrum of double-scattered light is therefore substantially larger than the true diffusion coefficient. Multiple scattering and polydispersity have the same effects on the line shape, so concentration-induced aggregation, nonexponential line shapes due to particle interactions, and multiple-scattering effects mimic each other. A standard method of avoiding multiple scattering is to use very thin sample cells. If one studies the same sample with a series of cells of different thickness, and extrapolates D to the limit of zero cell thickness, one can avoid many multiple-scattering effects. Alternatively, one may use a cross-correlation instrument containing two incident beams and two detectors to obtain spectra that are not perturbed by the presence of multiply scattered photons (89).

ACKNOWLEDGMENTS

One of us (G.D.J.P.) gratefully acknowledges the partial support of this work by the National Science Foundation under Grants CHE79-20389 and 82-13941. The comments and criticisms of P. J. Elving and F. W. Billmeyer, Jr. on my portion of the chapter were greatly appreciated.

REFERENCES

1. Abiss J. B., in ref. 30.

2. Ambarzumian, V. A., *Theoretical Astrophysics*, Pergamon, New York, 1958.

3. *Standard Practice for Determination of Weight-Average Molecular Weights of Polymers by Light Scattering*, ASTM-ANSI Designation D 4001, American Society for Testing and Materials, Philadelphia, 1981.

4. Atkins, J. T., and F. W. Billmeyer, Jr., *Color Eng.*, **6**, (3) 40 (1968).

5. Beasley, J. K., J. T. Atkins, and F. W. Billmeyer, Jr., in R. L. Rowell and R. S. Stein, Eds., *Electromagnetic Scattering*, Gordon & Breach, New York, 1967, pp 765–785.

6. Becher, P., *J. Phys. Chem.*, **63**, 1213 (1959).

7. Benbasat, J. A., and V. A. Bloomfield, *J. Mol. Biol.*, **95**, 335 (1975).

8. Bendjaballah, C., *Opt. Commun.*, **9**, 279 (1973).

9. Benedek, G. B., in M. Chretien, E. P. Gross, and S. Deser, Eds., in *Brandeis Summer Institute in Theoretical Physics, 1966*, Vol. 2, Gordon & Breach, New York, 1968, pp. 1–98.

10. Berne, B. J., and R. Pecora, *Dynamic Light Scattering with Applications to Chemistry, Biology, and Physics*, Wiley-Interscience, New York, 1976.

11. Billmeyer, F. W. Jr., and R. L. Abrams, *J. Paint Technol.*, **45**, (579) 23 (1973).

12. Billmeyer, F. W. Jr., and D. H. Alman, *J. Color Appear.*, **2**, (1) 36 (1973).

13. Billmeyer, F. W. Jr., and C. B. de Than, *J. Am. Chem. Soc.*, **77**, 4763 (1955).

14. Billmeyer, F. W., H. I. Levine, and P. J. Livesey, *J. Coll. Int. Sci.*, **35**, 204 (1971).

15. Billmeyer, F. W. Jr., and L. W. Richards, *J. Color Appear.*, **2**, (2) 4 (1973).

16. Billmeyer, F. W. Jr., and M. Saltzman, *Principles of Color Technology*, 2nd ed., Wiley-Interscience, New York, 1981.

17. Bloomfield, V. A., *Annu. Rev. Phys. Chem.*, **28**, 233 (1977).

18. Boyd, Robert W., *Radiometry and the Detection of Optical Radiation*, Wiley, New York, 1983, Chapter 2.

19. Bryant, W. M. D., F. W. Billmeyer, Jr., L. T. Muus, J. T. Atkins, and J. E. Eldridge, *J. Am. Chem. Soc.*, **81**, 3219 (1959).

20. Bursteyn, H. C., R. F. Chang, and J. V. Sengers, *Phys. Rev. Lett.*, **44**, 410 (1980).

21. Cabannes, J., and Y. Rocard, *La Diffusion Moleculaire de la Lumiere*, Les Presses Universitaires des France, Paris, 1929.

22. Cantor, C. R., and P. R. Schimmel, *Biophysical Chemistry*, Vol. 2, W. H. Freeman, San Francisco, 1980.

23. Chandrasekhar, S., *Radiative Transfer*, Clarendon Press, Oxford, 1950 (reprinted by Dover, New York, 1960.

24. Chen, S.-H., B. Chu, and R. Nossal, *Scattering Techniques Applied to Supramolecular and Non-Equilibrium Systems*, Plenum, New York, 1981.

25. Chu, B., *Laser Light Scattering*, Academic, New York, 1974.

26. Chu, B., Es. Gulari, and Er. Gulari, *Phys. Scripta (Sweden)*, **19**, 476 (1979).

27. Corti, M., and V. DeGiorgio, *J. Phys. Chem.*, **85**, 711 (1981).

28. Crosignani, B., P. Di Porto, and M. Bertolotti, *Statistical Properties of Scattered Light*, Academic, New York, 1975.

29. Cummins, H. Z., and E. R. Pike, Eds., *Photon Correlation and Light Beating Spectroscopy*, Plenum, New York, 1974.

30. Cummins, H. Z., and E. R. Pike, Eds., *Photon Correlation Spectroscopy and Velocimetry*, Plenum, New York, 1977.

31. Dahneke, B., Ed., *Measurement of Suspended Particles by Quasi-Elastic Light Scattering*, Wiley, New York, 1983.

32. Davenport, W. B., and W. L. Root, *An Introduction to the Theory of Random Signals and Noise*, McGraw-Hill, New York, 1958.

33. Debye, P., *J. Appl. Phys.*, **15**, 338 (1944); *ibid.*, *J. Phys. Coll. Chem.*, **51**, 18 (1947).

34. DeGiorgio, V., M. Corti, and M. Giglio, Eds., *Light Scattering in Liquids and Macromolecular Solutions*, Plenum, New York, 1980.

35. De la Cuesta, M. O., and F. W. Billmeyer, Jr., *J. Polymer Sci.*, **A1**, 1721 (1963).

36. de la Torre, J. G., and V. A. Bloomfield, *Biopolymers*, **17**, 1605 (1978).

37. Doob, J. L., *Ann. Math.*, **43**, 351 (1942).

38. Drain, L. E., *The Laser Doppler Technique*, Wiley, New York, 1980.
39. Duncan, D. R., *Proc. Phys. Soc. (London)*, **52**, 390 (1940).
40. Einstein, A., *Ann. Phys.*, **33**, 1275 (1910).
41. I. L. Fabelinskii, *Molecular Scattering of Light*, Plenum, New York, 1957.
42. Flory, P. J., *J. Am. Chem. Soc.*, **65**, 372 (1943).
43. Foord, R., E. Jakeman, C. J. Oliver, E. R. Pike, R. J. Blagrove, E. Wood, and A. R. Peacocke, *Nature*, **227**, 242 (1970).
44. Gate, L. F., *J. Phys. D, Appl. Phys.*, **4**, 1049 (1971); **5**, 837 (1972).
45. Giercke, T. D., and G. R. Alms, *J. Chem. Phys.*, **76**, 4809 (1982).
46. Griffin, W. G., and P. N. Pusey, *Phys. Rev. Lett.*, **43**, 1100 (1979).
47. Grum, F., and R. A. Becherer, *Radiometry*, Academic, New York, 1979, Chapter 2.
48. Gulari, Es., Er. Gulari, Y. Tsunashima, and B. Chu, *J. Chem. Phys.*, **70**, 3965 (1979).
49. Hall, R. S., and C. S. Johnson, Jr., *J. Chem. Phys.*, **72**, 4251 (1980).
50. Hall, R. S., Y. S. Oh, and C. S. Johnson, Jr., *J. Phys. Chem.*, **84**, 756 (1980).
51. Hartel, W., *Licht*, **10**, 141 (1940); 165 (1940); 190 (1940); 214 (1940); 232 (1940).
52. Hayes, W., and R. Loudon, *Scattering of Light by Crystals*, Wiley, New York, 1978.
53. Heller, W., *Rev. Mod. Phys.*, **21**, 1072 (1959).
54. Heller, W., and J. Witeczek, *J. Phys. Chem.*, **74**, 4241 (1970).
55. Hermans, J. J., Ed., *Polymer Solution Properties, Part II: Hydrodynamics and Light Scattering*, Dowden, Hutchinson & Ross, Stroudsburg, Pennsylvania, 1978.
56. Hermans, J. J., and S. Levinson, *J. Opt. Soc. Am.*, **41**, 360 (1951).
57. Hohenburg, P. C., in P. Dupuy and A. J. Dianoux, Eds., *Microscopic Structure and Dynamics of Liquids*, Plenum, New York, 1978, 333–366.
58. Huglin, M. B., Ed., *Light Scattering from Polymer Solutions*, Academic, New York, 1972.
59. Jackson, J. D., *Classical Electrodynamics*, Wiley, New York, 1962.
60. Johnson, I., and V. K. La Mer. *J. Am. Chem. Soc.*, **69**, 1184 (1947).
61. Johnston, R. M., in T. C. Patton, Ed., *Pigment Handbook*, Wiley-Interscience, New York, 1973, Chapter IIIDb.
62. Jones, C. R., and C. S. Johnson, Jr., *J. Chem. Phys.*, **65**, 2020 (1976).
63. Judd, D. B., and G. Wyszecki, *Color in Business, Science, and Industry*, 3rd ed., Wiley, New York, 1975.
64. Kerker, M., *The Scattering of Light and Other Electromagnetic Radiation*, Academic, New York, 1969.
65. Kirkwood, J. G., R. L. Baldwin, P. J. Dunlop, L. J. Gosting, and G. Kegeles, *J. Chem. Phys.*, **33**, 1505 (1960).
66. Kirkwood, J. G., and R. J. Goldberg, *J. Chem. Phys.*, **18**, 54 (1950).
67. Koppel, D. E., *J. Chem. Phys.*, **57**, 414 (1972).
68. Kortüm, G., *Reflectance Spectroscopy*, Springer-Verlag, New York, 1969.
69. Kratky, O., and G. Porod, *J. Colloid Sci.*, **4**, 35 (1949).
70. Kratohvil, J. P., *J. Colloid Interf. Sci.*, **75**. 271 (1980).
71. Kubelka, P., *J. Opt. Soc. Am.*, **38**, 448 (1948).
72. Kubelka, P., *J. Opt. Soc. Am.*, **44**, 330 (1954).
73. Kubelka, P., and F. Munk, *Z. Tech. Phys.*, **12**, 593 (1931).
74. Levine, S., and G. O. Olaofe, *J. Colloid Interf. Sci.*, **27**, 442 (1968).
75. Lim, T. K., J. G. Baran, and V. A. Bloomfield, *Biopolymers*, **16**, 1473 (1977).

76. Loewenstein, M. A., and M. H. Birnboim, *Biopolymers*, **14**, 419 (1975).

77. McQuarrie, D. A., *Statistical Mechanics*, Harper & Row, New York, 1976.

78. McWhirter, J. G., and E. R. Pike, *J. Phys.*, **A11**, 1729 (1982).

79. Mie, G., *Ann. Phys.*, **25**, 377 (1908).

80. Miller, D., and G. B. Benedek, *Intraocular Light Scattering: Theory and Clinical Applications*, Thomas, Springfield, Illinois, 1973.

81. Minnaert, M., *The Nature of Light and Color in the Open Air*, Dover, New York, 1954.

82. Mountain, R. D., *J. Res. Natl. Bur. Stds.*, **70A**, 207 (1966).

83. Mudgett, P. S., and L. W. Richards, *Appl. Opt.*, **10**, 1485 (1971).

84. Mudgett, P. S., and L. W. Richards, *J. Colloid Interf. Sci.*, **39**, 551 (1972).

85. Mudgett, P. S., and L. W. Richards, *J. Paint Technol.*, **45**, (586) 43 (1973).

86. Nicoli, D. F., and G. B. Benedek, *Biopolymers*, **15**, 2421 (1976).

87. Patterson, G. D., *Adv. Polymer Sci.*, **48**. 125 (1983).

88. Phillies, G. D. J., *J. Chem. Phys.*, **60**, 974 (1974); 983 (1974).

89. Phillies, G. D. J., *Phys. Rev. A.*, **24**, 1939 (1981).

90. Phillies, G. D. J., *J. Colloid Interf. Sci.*, **86**, 226 (1982).

91. Phillies, G. D. J., G. B. Benedek, and N. A. Mazer, *J. Chem. Phys.*, **65**, 1883 (1976).

92. Phillips, D. G., and F. W. Billmeyer, Jr., *J. Coatings Technol.*, **48** (616), 30 (1976).

93. Pike, E. R., D. Watson, and F. McNeil-Watson, in B. E. Dahneke, Ed., *Measurement of Suspended Particles by Quasi-Elastic Light Scattering Spectroscopy*, Wiley, New York, 1983, pp. 107–128.

94. Provencher, S. W., *Makromol. Chem.*, **180**, 201 (1979).

95. Querfeld, C. W., M. Kerker, and J. P. Kratohvil, *J. Colloid Interf. Sci.*, **39**, 568 (1972).

96. Rayleigh, Lord (Strutt, J. W.), *Philos. Mag.*, (IV) **41**, 107 (1871); 224 (1871); 447 (1871); (V) **12**, 81 (1881).

97. Richards, L. W., *J. Paint Technol.*, **42**, 276 (1970).

98. Sanders, A. H., and D. S. Cannell, in ref. 34, pp.

99. Schurr, J. M., *C.R.C. Crit. Rev. Bioeng.*, **4**, 371 (1977).

100. Schuster, A., *Astrophys J.*, **21**, 1 (1905); reprinted in D. H. Menzel, Ed., *Selected Papers on the Transfer of Radiation*, Dover, New York, 1966, pp. 3–24.

101. Sengers, J. V., in M. Levy, J.-C. LeGuillou, and J. Zinn-Justin, Eds., *Phase Transitions: Cargese 1980*, Plenum, New York, 1982, pp. 95–135.

102. Sengers, J. V., and J. M. H. Levelt-Sengers, in C. A. Croxton, Ed., *Progress in Liquid Physics*, Wiley, New York, 1978, pp. 103–174.

103. Shah, H. S., and F. W. Billmeyer, Jr., *Color Res. Appl.*, **26–31** (1984).

104. Silberstein, L., *Philos. Mag.*, **4**, 1291 (1927).

105. Sloan, C. K., *J. Phys. Chem.*, **59**, 834 (1955).

106. Smoluchowski, M. S., *Ann. Phys.*, **25**, 205 (1908); *Philos. Mag.*, **23**, 165 (1912).

107. Stacey, K. A., Ed., *Light Scattering in Physical Chemistry*, Butterworths, London, 1956.

108. Staudinger, H., and W. Heuer, *Ber.*, **63**, 222 (1930); Staudinger, H., and R. Nodzu, *Ber.*, **63**, 721 (1930).

109. Tanford, C., *The Physical Chemistry of Macromolecules*, Wiley, New York, 1961.

110. Timasheff, S. N., H. M. Dintzis, J. G. Kirkwood, and B. D. Coleman, *Proc. Natl. Acad. Sci. USA*, **41**,710 (1955).

111. Tirado, M. M., and J. G. de la Torre, *J. Chem. Phys.*, **73**, 1986 (1980).

112. Utiyama, H., in ref. 58, Chapter 4.

113. Uzgiris, E. E., *Adv. Colloid Interf. Sci.*, **10**, 53 (1981).

114. van de Hulst, H. C., *Light Scattering by Small Particles*, Wiley, New York, 1957.

115. Volz, H. G., *Congress FATIPEC*, **VI,** 98 (1962); **VII,** 194 (1964).

116. Vrij, A., J. G. H. Joosten, and H. M. Fijnaut, *Adv. Chem. Phys.*, **48,** 329 (1981).

117. Ware, B. R., *Chem. Biochem. Appl. Lasers*, **2,** 199 (1977).

118. Ware, B. R., and W. H. Flygare, *Chem. Phys. Lett.*, **12,** 81 (1971).

119. Ware, B. R., and D. D. Haas, in R. I. Sha-Afi and S. N. Fernandez, Eds., *Fast Methods in Physical Biochemistry and Cell Biology*, Elsevier, New York, 1983.

120. Wax, N., *Selected Papers on Noise and Stochastic Processes*, Dover, New York, 1954.

121. Wick, G. C., *Z. Phys.*, **121,** 702 (1943).

122. Wickramasinghe, N. C., *Light Scattering Functions for Small Particles*, Wiley, New York, 1973.

123. Wilcoxon, J., and J. M. Schurr, *Biopolymers*, **22,** 849 (1983).

124. Wills, P. R., and Y. Georgalis, *J. Phys. Chem.*, **85,** 3978 (1981).

125. Yamakawa, Y., *Modern Theory of Polymer Solutions*, Harper & Rowe, New York, 1971.

126. Young, C. Y., P. J. Missel, N. A. Mazer, and G. B. Benedek, *J. Phys. Chem.*, **87,** 308 (1983).

127. Zimm, B. H., *J. Chem. Phys.*, **16,** 1093 (1948); 1099 (1948).

128. Zimm, B. H., and R. W. Kilb, *J. Polymer Sci.*, **37,** 19 (1959).

129. Zimm, B. H., R. S. Stein, and P. Doty, *Polymer Bull.*, **1,**90 (1945).

RAMAN SPECTROSCOPY

By R. O. Kagel, *Environmental Quality, Dow Chemical U.S.A., Midland, Michigan*

R. W. Chrisman, *Analytical Laboratories, Michigan Division, Dow Chemical U.S.A., Midland, Michigan*

and J. A. Roper III, *Designed Latexes and Resins, Dow Chemical U.S.A., Midland, Michigan*

Contents

I. INTRODUCTION

Raman spectroscopy provides vibrational and rotational structural information and is often viewed as an alternative technique to infrared spectroscopy. However, the Raman effect is an optical scattering phenomenon and operates by a mechanism different from those of the more traditional optical absorption and emission spectroscopies. Thus, Raman spectroscopy has its own set of advantages and disadvantages, as well as unique applications.

Among the real and potential advantages of Raman spectroscopy are:

1. The ease of sample preparation, including the use of glass sample cells and water as a solvent.

2. The capability of doing *in situ* analysis of chemical systems under high-temperature or -pressure conditions.
3. The ability to achieve high spatial resolution (through Raman micro-probe techniques).
4. The possibility of obtaining unique information from species adsorbed to surfaces.

The chief disadvantage of the Raman technique is the weakness of the Raman effect and the consequent requirements for a powerful light source, good monochromator, and an efficient light-collection and -detection system. In addition, competing optical processes, such as fluorescence, often swamp out the Raman spectrum and thus pose severe experimental problems.

In the early 1970s, optimism about the possibility of routine analytical application of Raman spectroscopy ran high. The laser, as a source for Raman spectroscopy, had been developed from a few milliwatts to several hundred milliwatts in power. Mixed-gas laser systems offered a choice of lines for excitation. Fluorescence in one region of the spectrum was to be minimized by changing the exciting frequency. No less than a dozen instrument companies offered low-price table-top models aimed at the analytical market. The late Professor E. R. Lippincott conducted an extremely successful summer course in Raman spectroscopy at the University of Maryland. H. R. Morgan attempted to organize a summer course specifically for analytical spectroscopists using the table-top model instruments at Fisk University in 1973. The *Raman Newsletter* flourished. Dr. T. Herschfeld, addressing the Fourth International Conference on Raman Spectroscopy in 1974 (24), pointed out that analytical chemists far outnumbered spectroscopists and made a strong plea to the spectroscopists to orient their own work toward analytical-type applications, for this was the future of Raman spectroscopy.

While researchers pressed on in the study of new phenomena (the resonance Raman effect, coherent anti-Stokes Raman spectroscopy, etc.), several papers and publications appeared during this time extolling the virtues of routine applications of conventional Raman spectrometry in industrial laboratories. Academia also contributed to this growing body of literature. But, again, the nemesis of the Raman effect—fluorescence—took its toll. One by one, the instrument companies dropped out of the small, practical instrument market. Summer courses stressing analytical-type applications of Raman spectroscopy disappeared. Except in a few industrial and academic laboratories, practical analytical Raman spectroscopy remained a curiosity.

Now the efforts of the research community are beginning to be seen again. New analytical applications of more sophisticated techniques are beginning to emerge. During the last decade, several other Raman effects, such as the electronic Raman effect, the stimulated Raman effect, the hyper-Raman effect, the inverse Raman effect, resonance Raman spectroscopy (RRS), coherent anti-Stokes Raman spectroscopy (CARS), and surface-enhanced

Raman spectroscopy (SERS), have been extensively investigated; these effects are the subjects of a large volume of the recent theoretical and experimental literature (see, e.g., reference 55). The development of these Raman techniques has greatly improved the versatility, sensitivity, and selectivity of Raman spectroscopy. However, few practical applications have been reported and further examination of these effects is beyond the scope of this work. The present discussion is concerned with the theory, instrumentation, and analytical applications of conventional Raman spectroscopy, with a focus on the use of Raman spectroscopy as a practical analytical tool in the industrial laboratory.

II. THEORETICAL CONSIDERATIONS

The Raman effect is a weak second-order scattering phenomenon first reported by Raman (56) in 1928, only a few months before Landsberg and Mandelstam (47). Rayleigh and Tyndall scattering are examples of first-order scattering phenomena. Theoretical aspects of the effect, however, date back to 1923, when Smekal (59) predicted that the interaction of light with molecules would result in discrete line spectra. Although the classical treatment of Smekal is absolutely valid, Raman scattering is better understood in terms of quantum theory.

Light is regarded as having both wavelike and corpuscular characteristics. That is, although light has associated with it a frequency of wave motion, v_0, it is also considered to be composed of discrete energy packets, or photons. According to Planck (54), the energy of these photons is related to the frequency of the wave motion by the equation

$$E = hv_0 \tag{1}$$

where h is Planck's constant. The significance of Planck's theory is that energy is not continuous, but discrete. Bohr (10) advanced the principle that electrons orbiting about a nucleus may only occupy discrete orbitals corresponding to these energy levels.

When radiation is incident upon a transparent medium, most of it passes through the medium without undergoing collisions with the molecules making up the transparent medium. For the small fraction of photons that do collide with the medium's molecules, the majority of these collisions are elastic; that is, the energy of the scattered photons is identical to the energy of the incident radiation. This process is called Rayleigh scattering. A small number of photon–molecule collisions are inelastic; that is, an energy transfer occurs between the photon and the molecule. The frequency of the scattered radiation is shifted relative to the frequency of the incident radiation for inelastic collisions. This process is called Raman scattering and the frequency shift corresponds to a vibrational or rotational energy transition within the molecule. Similarly, acoustical modes in crystalline systems give rise to weak low-frequency-shifted Brillouin scattering. Brillouin scattering is not a commonplace analytical tool, but is used more in basic studies in physics.

If the molecule involved in the photon–molecule collision is in its ground vibrational state, it can borrow from the photon sufficient energy to cause a transition between the ground state and the first excited vibrational state. This amount of energy is given by

$$E = h\nu_j \tag{2}$$

where ν_j is the vibrational frequency of the transition. The energy of the scattered photon is thus

$$E = h(\nu_0 - \nu_j) \tag{3}$$

The frequency shift $\nu_0 - \nu_j$ is called a Raman shift and the resulting lower-frequency-shifted line is called a Stokes line.

If the molecule involved in the photon–molecule collision is already in a vibrationally excited state, it can undergo a transition to the ground state by giving its excess vibrational energy to the photon. The energy of the scattered photon is thus

$$E = h(\nu_0 + \nu_j) \tag{4}$$

The higher-frequency-shifted line is called an anti-Stokes line.

The entire process is depicted in Fig. 10.1. Raman shifts resulting from rotational transitions (gas phase) are explained in an analogous manner. The number of Rayleigh-scattered photons far exceeds the number of Raman-scattered photons. About 1 out of every 10^3–10^4 incident photons is Rayleigh scattered, whereas only 1 out of every 10^6–10^8 photons is Raman scattered. The intensity of a Raleigh line is therefore 10^3–10^4 times greater than the intensity of a Raman line. Anti-Stokes lines are less intense than Stokes lines because, by Boltzmann statistics, the population of excited states is usually lower than that of the ground state. Anti-Stokes lines are not generally used in interpretive analytical Raman spectroscopy. A typical Raman (Stokes) line spectrum is shown in Fig. 10.2. The Rayleigh-scattered component, ν_0, is set at 0 cm^{-1} on the abscissa of Raman-shifted frequencies.

The Raman effect can therefore serve as an alternative to infrared absorption spectroscopy for examining vibrational and rotational molecular motions. However, these two phenomena are distinctly different. Although a detailed discussion of Raman line intensities is beyond the scope of this text, it is important to note that the intensity of a Raman vibrational line is

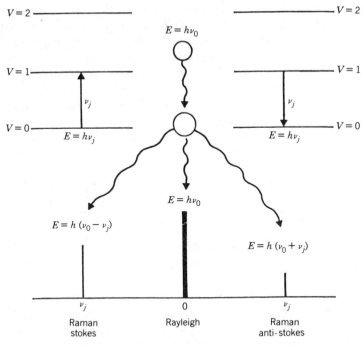

Fig. 10.1. Schematic of Raman scattering process.

related to the rate of change of α, the polarizability of the molecule with respect to the change in r, the bond length:

$$I \propto \frac{\partial \alpha}{\partial r} \tag{5}$$

Correspondingly, the intensity of an infrared absorption band is related to the rate of change of the dipole moment μ with respect to bond length r:

$$I \propto \frac{\partial \mu}{\partial r} \tag{6}$$

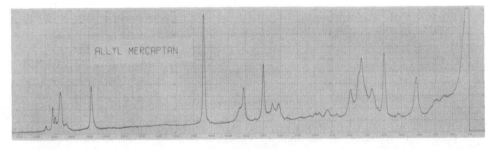

Fig. 10.2. Representative Raman spectrum.

Quite simply, this means the selection rules for the two phenomena are different. The Raman effect measures vibrations centered in highly polarizable bonds, whereas infrared absorption measures vibrations centered in polarized bonds. The Raman spectrum is dominated by vibrations that are symmetric; antisymmetric vibrations are less intense in the Raman spectrum.

In many cases, infrared and Raman spectra yield overlapping data when the molecule has low symmetry (e.g., oxylchlorofluoride, which has C_s symmetry). In cases where the molecule has high symmetry (e.g., sulfur hexafluoride, which has O_h symmetry), the two phenomena are mutually exclusive and completely complementary. In the real world, where most complex molecules have little or no symmetry, the fundamental vibrations of the molecules are both infrared and Raman active. Because of their complementary nature, it is advantageous to use both infrared and Raman data whenever possible. However, quite often Raman data stand on their own merits.

A very distinctive feature of Raman-scattered radiation (and one that is also associated with the polarizability of the particular vibrational transition) is the state of its polarization relative to the state of polarization of the exciting radiation. For totally symmetric vibrations, the corresponding Raman-scattered radiation is polarized in the same direction as the exciting radiation. For antisymmetric vibrations, the scattered radiation is polarized in a direction different from that of the incident radiation. The extent to which the polarization of the incident radiation is affected by the symmetry of the vibration is expressed by the depolarization ratio, ρ. The depolarization ratio is expressed in terms of the intensity of scattered radiation polarized perpendicular to the plane containing the direction of incident radiation, $I_\perp$, relative to the intensity of scattered radiation parallel to this plane, $I_\parallel$:

$$\rho = \frac{I_\perp}{I_\parallel}$$

For plane-polarized light interacting with an antisymmetric vibration, ρ has a maximum value of $\frac{3}{4}$. For interaction with symmetric vibrations, ρ varies between 0 and $\frac{3}{4}$. The depolarization ratio is a measure of the anisotropy of the vibration. Detailed derivations of the depolarization ratio are given in several excellent classical texts (39,65).

III. INSTRUMENTATION

The typical Raman system (Fig. 10.3) is composed of a group of independent subsystems supplied by different manufacturers. This allows the researcher to tailor the system to his individual needs. Unfortunately, it can also make the purchase of a Raman instrument a rather involved process

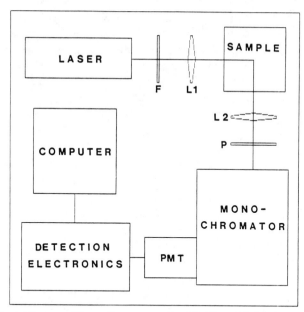

Fig. 10.3. Block Diagram of Raman instrumentation. F, Filter; L1, L2, lenses; P, polarization analyzer, PMT, photomultiplier tube.

compared with the purchase of an infrared system. The problem is not too complicated, however, since most monochromator manufacturers are willing to acquire the individual subsystems and assemble them into a complete package for a rather modest additional cost. The purpose of this section is to describe each subsystem and give generalized specifications for the design of a versatile high-quality Raman instrument for analytical measurements.

A. LIGHT SOURCE

The continuous-wave (CW) laser is now the standard light source for Raman spectroscopy. The laser has replaced the mercury arc lamp because it is stable, monochromatic, powerful, polarized, and highly directional. Although individual applications may require enhanced capabilities in one of the above-listed areas, the laser, as it is sold today, is an almost ideal light source for Raman spectroscopy. Lack of complete wavelength tunability may be the one weakness of the laser, and even in this area the frequency range available is suitable for most nonresonance Raman applications. However, to avoid some fluorescence problems and in work with colored samples, it is desirable to have laser lines at both ends of the visible spectrum. Thus, it is not uncommon to see two lasers, and sometimes many lasers, in a well-equipped Raman laboratory. However, the capital investment required to achieve versatility in the frequency range can be substantial.

Currently, the best general-purpose laser is the 3-W, all-line, argon-ion laser. This laser has proven to be very reliable and easy to operate. The more than 1 W of average power that this laser can deliver, in either of two lines, is more than enough power, except for high-sensitivity gas-phase work. Actually, very few samples can endure even 1 W of power without decomposing. Also, the 514.5-nm of 488.0-nm lines produced by the argon-ion laser eliminate many fluorescence problems while staying in the high-efficiency region of most gratings. Moreover, the 3-W argon-ion laser can be expected to operate for more than 5000 hr in the normal, quarter- to half-power range, and when the laser tube does fail, it can often be changed in the field. In addition, the power supply operates for a very long time, with only very rare failure of power transistors, which can normally be found and corrected by an operator with only a cursory knowledge of electronics. In general, the argon-ion laser is a very reliable light source. However, the reliability seems to go down as the output of the laser goes up.

For operation in the red end of the spectrum, the helium–neon (HeNe) laser was the first and most common source. This laser is also very reliable, with an expected lifetime even greater than that of the argon-ion laser. However, it is a relatively low-powered laser; 50 mW of average power is about the largest size available. It is now being replaced with the krypton-ion laser, which has considerably more power in the red. Until fairly recently, the krypton-ion laser was plagued with reliability problems and was not widely used for routine applications. The new krypton-ion lasers with pressure control appear to have solved many of these problems and are now the lasers of choice for operation in the red. The argon–krypton mixed-gas-ion lasers would appear to be the best possible choice, since they have the best line of both gases, but they have a reputation for being very unreliable in comparison with the argon-ion laser. The hope is that the changes that seem to have solved the reliability problems of the krypton laser will work for the mixed-gas-ion lasers.

Another way to cover the visible spectrum is with a high-powered krypton-ion laser. These lasers are capable of providing over 1 W of power at the blue end of the spectrum, but as is the case for the high-powered argon-ion lasers, the reliability again tends to go down as output goes up. In addition, the mirrors need to be changed if one is to operate at both ends of the visible spectrum. While mirror changing is not a major problem, it can provide enough of a barrier that if a sample cannot be run with one line, the operator will be unlikely to switch over to another line. The lowered reliability of this laser, coupled with its inconvenience of operation and a price equal to that of a 3-W argon-ion laser and a small pressure-controlled krypton-ion laser, would argue in favor of using the two lasers in place of the one big laser. The big krypton-ion lasers, and for that matter the big argon-ion lasers, do offer the potential for operation in the uv, but this is not normally of great value unless resonance Raman studies are to be undertaken. If this capability is desired, it is still advisable to have a small argon- or krypton-ion laser to serve as a workhorse laser for normal samples.

One final way to cover the visible spectrum (without going to much more expensive light sources) is to use an ion laser–pumped dye laser system. In general, a dye laser system is more trouble than it is worth to an analytical lab. In fact, currently there are almost no nonresonance Raman applications that cannot be done with the two ion lasers described earlier. However, if an analytical laboratory wants to work at the forefront of research in Raman spectroscopy, a dye laser is essential. A dye laser system should be purchased not as the primary light source for Raman spectroscopy, but rather as the second, or preferably the third, source after the purchase of an argon-ion and a krypton-ion laser. A useful discussion on the relative merits of different dye laser systems is given in a recent review article (14).

If a dye laser is chosen, a much more powerful ion laser should be used as a pump, though a 4-W all-line argon-ion laser will provide reasonable dye laser output throughout the visible spectrum. However, if one wishes to use frequency doubling into the uv or to operate in the near ir, more powerful ion pump lasers are recommended. Often, the use of these more complex dye laser systems or the more expensive glass laser systems must be justified in terms of laser spectroscopy and not solely for what they can do for Raman spectroscopy. In general, combined Raman-spectroscopy and laser-spectroscopy groups in an analytical laboratory can be very useful for progress in two areas that may have trouble growing independently.

B. BEAM-TRANSFER COMPONENTS

Since power is usually no problem, almost any type of mirror can be used to direct the beam into the sample compartment. The other optical components needed before the sample compartment are the polarization rotator, a single-frequency filter, a beam stop, possibly neutral density filters, and a beam-power monitor. The types of components needed are somewhat dependent on the laser system used.

A broadband polarization rotator is the best choice if more than one laser is to be routinely used, and a must if a dye laser system is used. If only one laser is used, less expensive half-wave plates can be purchased for the laser lines of choice. Either method for changing the polarization of the incoming laser radiation will allow the depolarization ratios to be determined simply by changing the plane of polarization of the incoming beam by 90°. The rotator may be aligned by placing it in the beam and adjusting the rotation to give the maximum signal for the 459 cm^{-1} line of CCl_4. A rotation of the polarization by 90° should give almost no signal for the band that arises from the totally symmetric vibration of the molecule. This method of measuring depolarization ratios gives a value of $\frac{6}{7}$ for bands that arise from non-totally symmetric vibrations of the molecule. The other method of measuring depolarization ratios, by placement of a polarization analyzer between the sample and the monochromator, gives a value of $\frac{3}{4}$ for bands that arise from non-totally symmetric vibrations of the molecule.

A single-frequency filter is needed because almost all lasers put out light at frequencies other than the lasing frequency. The intensity of this light is low, but of the same order of magnitude as that of the Raman lines. This light can be reflected into the monochromator by any glass surfaces in the field of view of the collection lens. Small glass capillaries and powders give the most problems. For single-line operation, spike filters are the easiest to use. Spike filters often transmit only 50% of the incoming light. They also sometimes age, with a resultant shift in band pass. The transmission can often be maximized by rotating the filter slightly about an axis perpendicular to the beam. If a dye laser is used, a small monochromator will be needed to screen out the background radiation from the laser.

With the ion lasers, it is often difficult to reduce the power to a level that will not damage the sample under investigation. It is also handy to be able to align the laser with less than full power on the sample. For these situations, a series of neutral density filters can be very useful. The filters are also useful when the laser line is scanned to calibrate the monochromator. Usually a set of filters of optical density 0.3, 1, and 2 will do the job.

Two types of power monitors are needed for the instrument. One is needed to measure total beam power along the beam path. The head of this monitor should be as small as possible, so that it can be placed in the beam at various spots without disturbing the setup. It should be able to read from 1 mW to the maximum power of the laser. The other monitor is needed to follow the laser-power drift with time. Most modern lasers can be bought with automatic light-intensity stabilization; this is usually the best approach to ensure constant power at the sample. If such a device is not available, a beam splitter can be placed in the beam to pick off a small portion of the beam and direct it to a silicon photodiode. The output from the diode must then be used to correct the Raman signal for laser-power variations. One can also monitor laser power by mounting a detector behind the 100% back-reflector of the laser to look at the small amount of light that passes through this mirror, or by mounting a detector above one of the Brewster-angle windows of the laser. If a monitor is placed above the Brewster-angle windows, a piece of Polaroid sheeting must be used to select the proper orientation. Only one orientation is linear with laser power.

C. SAMPLE HANDLING AND COMPARTMENT

One very useful feature of Raman spectroscopy is the flexibility of sample presentation. Solids of many shapes and sizes can often be looked at with no preparation. Liquids and gases can be contained in anything that is transparent and colorless in the region of the spectrum that is defined by the laser line and the appropriate Raman shift. Glass is usually the best sample container, but salt crystals, diamonds, plastics, or anything that meets the above requirements can be used. It must be remembered, however, that if the laser hits the container material in the field of view of the collection lens, the

Raman spectrum of the container may appear as interference in the spectrum. This ease of sampling allows samples in clear nuclear magnetic resonance (NMR) tubes to be run in the NMR instrument and then the Raman instrument with no additional sample handling. In addition, samples prepared in a vacuum line can be placed in the instrument in the same container in which they were isolated.

Many specialized sample-handling conditions can be accommodated in a Raman instrument. The use of unusual sampling systems is aided by the great variation allowed in the angle between excitation to collection. Although 90° is the usual angle of choice for reducing stray light and facilitating polarization studies, any angle from 0° to 180° can be used. This feature, coupled with the ease of directing a laser beam and the wide choice of construction materials, allows, among other things, great variations in both temperature and pressure to be handled fairly easily.

The microscope can be added to most Raman instruments to allow the collection of spectra from 1 micron sized particles. A normal microscope with bright field or dark field reflecting optics can be used for opaque samples. A trinocular eyepiece works well for the experiment. One eyepiece can be used to project the field of view onto a white surface for alignment. The internal mirror can then be switched to direct the light to the camera port which can be fitted with an aperture to limit the field of view to the point to be analyzed. The light coming through the aperture can then be directed through an appropriate lens and then onto the entrance slit of the monochromator. The normal beam splitter used for bright field reflecting work is a 50% splitter, which means half of the Raman scattered light is lost. While bright field systems are easy to use, the laser beam passing through the same lens, which is used to collect the light, can cause background problems in addition to the reduced intensity. Dark field operation is much more difficult to align but does not have the background problems. Transmission can be used if the condensing lens is replaced with an objective lens for tighter focus. This method is very easy to use and gives better quality spectra for small isolated particles. Great care must be taken never to look into any of the eyepieces when using a microscope. The eyepieces must be blocked when doing transmission work since the laser beam may come out of them at full intensity (5). References to Raman systems designed exclusively for microwork are given later in the chapter.

Bulk samples that are sensitive to beam damage can be handled in several ways. Liquids can be placed in a circular cell that can be rotated at high speed. Alternately, a liquid can be rapidly pumped through a fixed tubular cell. Solids can be packed onto a disk with a circular channel about its circumference that can be rotated at high speeds. Cooling can often reduce damage problems; these usually result from absorption of the laser beam coupled with poor heat transfer.

Gases can often be viewed at very high laser powers (>1 W) to obtain increased sensitivity. Although many types of containers can be used, a 5-

cm uv cell with two ports works very well. The gas is flushed through the cell, which is then sealed. The laser enters through one flat, exits through the other, and then is reflected back through the cell; this provides a convenient double-pass system. The Raman scattering is collected from the side. The cell is large enough that the flats are out of the field of view of the collection lens, which means the background will be low. In general, Rayleigh-scattered light will be trapped in the glass cell by total internal reflection due to refractive-index differences, and will only escape to create background problems at the edges. Unstable gas-phase species can often be studied using matrix-isolation methods. Matrix isolation of species in an argon matrix will often sharpen bands to the point that bands arising from different natural-abundance isotope vibrations will be resolved and can be studied and assigned. Matrix-isolation interpretation and experimentation often demands such specialized knowledge and equipment that its general use as an analytical tool is precluded.

Liquids can be placed in a melting-point capillary or a shortened clear NMR tube. They can be drawn into a long-tipped disposable pipet, and the tip with the liquid can be broken off and sealed with a torch or clay. Liquids can be placed in various uv cells. Walrafen (63) actually took advantage of internal reflection, mentioned above, and filled a hollow fiber with a liquid to create a cell with a very long path length. For the method to work, the refractive index of the liquid must be greater than that of the fibers; if this condition is met, very weak bands can be observed.

Various problems can develop with liquids. Dust particles in the liquid can cause background problems. This problem can be overcome by filtering the liquids through filters with 250-nm pores. This method is very easy: The filters are held in a filter body that fits on the end of a syringe, thus making it possible to filter the liquid directly into the cell. Gas bubbles can also cause background problems. They arise when low-boiling-point solvents boil in the beam. Bubbles also arise from solutions that have been aerated during attempts to filter out dust. Monochromators with horizontal slits tend to have fewer problems with bubbles, because the bubbles tend to move out of the beam. Unfortunately, this does not always happen, since sometimes pinpoint bubbles will form on the glass where the beam enters the cell and not move. In addition, vertical-slit instruments must have the cells either filled to the top or slightly inclined so that the beam does not strike the meniscus, since it, like bubbles, will defract the beam path. Fortunately, these problems, while inconvenient, can be overcome using various techniques of sample preparation.

The major problem with liquids and solids is fluorescence. The Raman instrument is a very sensitive fluorimeter that is capable of detecting strong fluorophors down to concentrations of less than 1 part per trillion. This means that even very minor impurities can cause major background problems. Background fluorescence, however, very seldom makes it impossible to obtain the spectrum of a compound as long as the compound is not a strong fluorophor. Fluorescence can be reduced by choosing a laser line that is as far as possible from the absorption band of the fluorophor. Because fluorescence is often due to impurities, various sample-cleaning procedures, such as distillation, sublimation, recrystallization, and chromatography, will often solve the problem. The laser can also be used to cause a photolytic reaction in the impurity and thus destroy the impurity. If the signal-to-noise ratio is not drastically reduced by the additional intensity due to the fluorescence, various digital methods can be used to remove the broad fluorescence background signal from the spectrum and improve the spectral appearance. Finally, it has been shown that the Raman and fluorescence signals can be resolved in the time domain (35). Time resolution, though interesting and potentially very useful, is currently an expensive and complex technique, and one beyond the scope of this discussion.

In a Raman experiment, one should use the neat solid sample, if at all possible, because of sensitivity problems. However, if polarization information is desired, the solid can be run either as a melt or in solution. If a single crystal is available, it can be oriented in the beam with respect to the various crystal axes to obtain polarization information, although this is a more involved method. A 10% solution will normally give good-quality spectra, although the more concentrated the solution, the better. Many solvents can be used, but the small-molecule solvents such as CCl_4 and H_2O are best, since they cause less spectral interference. In fact, water is a very good solvent for Raman spectroscopy, which is one key advantage of Raman over ir spectroscopy. Water has a continuous series of weak, overlapping bands until almost 6000 cm^{-1} (64), but fortunately they overlap so well as to give an almost flat base line throughout much of the spectrum. Tobias et al. (60), in an article with 571 references, describe many applications of Raman spectroscopy to inorganic chemistry. A substantial portion of this work deals with methods of measuring species concentration and structure in aqueous solution.

Powders are normally run in melting-point capillaries, which can be sealed with a torch and kept almost indefinitely. If large crystals are available that do not fit into a capillary, a larger tube should be used, since grinding them to small particles will often degrade the quality of the spectrum by increasing the scattering of the light. Larger solids, such as geological samples, are best mounted directly in the beam. Soft polymers can be mounted on a pin or small nail. Molding clay can often be used to position unusually shaped objects in the beam, though care must be taken, since many clays contain added dyes that are fluorescent. Cross-linked polymers, which are difficult to prepare for ir studies, can be placed directly in the beam of a Raman instrument with no sample preparation. It is possible to study polymer suspensions if they are placed in a small-diameter tube, such as a long-tipped disposable pipet, whose tip can be broken off and placed in the beam. Although scattering from both fine powders and suspensions can be a problem, the effects of scattering on the quality of the spectrum are greatly reduced by the new monochromators.

Finally, a very useful sampling method is Raman difference spectroscopy (RADS). This method, like all difference methods, allows very precise solvent subtractions and the determination of subtle band shifts due to interactions in solution. Tobias and coworkers (1,16,61) have described the most useful technique so far, judging by the number of papers published using the technique. Digilab has supplied a commercial sample system based on their description. It uses the cell-in, cell-out method to obtain the spectrum. The advantage of this setup for an analytical laboratory is that, as it is described (61), it can also be used as an automatic multisample cell holder for running up to eight different samples automatically.

Current Raman instruments have versatile sample compartments; these are usually large and easily disassembled to accommodate many of the types of sampling system described above. An enclosed compartment is needed to keep out room-light interferences, especially from fluorescent lights. The stage the sample rests on should be able to make X-, Y-, and Z-translations. The sample compartment should have some type of mirror to back-reflect the beam through the sample. The collection lens should have a low f-number such as $f/1.8$ or below, to collect as much light as possible. This lens should have a fine-focus adjustment to focus the image on the slits. It must also be matched to the monochromator so that it fills the slits without overfilling the gratings. The manufacturers usually do a very good job of specifying the proper collection lens. A fine adjustment is also needed to translate the lens in a direction perpendicular to the slit so that one may sweep the sample image across the slit for final fine tuning. The distance between the lens face and the point where the laser beam crosses the sample compartment should be several centimeters to allow room for the Dewar vessels and other assorted sampling systems that may be used. Finally, if at all possible, the system should be in a room that can be completely darkened. A dark room helps one to see the beam for alignment, and it is also useful if the sample compartment must be left open for some unusual sampling system.

Usually very few optical components are placed between the sample compartment and the monochromator, in order to reduce light losses. A polarization analyzer, a polarization scrambler, and possibly a neutral density filter are usually all that are used. The neutral density filters are used only when the laser line is scanned for calibration purposes. The polarization analyzer is a piece of Polaroid glass that can be rotated 90°. Polarization ratios measured in this manner will have a maximum value of $\frac{3}{4}$. The depolarizer or polarization scrambler is the last optical component before the slits. The scrambler is a quartz wedge that is used to scramble polarizations, since grafting efficiency varies with polarization.

A recent development is the use of a fiber-optic probe for Raman spectroscopy (49). Such an approach has several advantages. It can eliminate the need for a sample compartment. Also, fiber-optic probes do not require special sample positions or realignment between samples. In addition, samples can be studied in a wide range of environments remote from the spectrometer. Future developments in the use of fiber-optic Raman spectroscopy are expected.

D. MONOCHROMATOR

Almost all current Raman instruments use monochromators rather than the spectrographs that were formerly used. However, for some special applications, spectrographs are making a comeback, due to the increasing sensitivity of the multichannel detectors. The advantages of a spectrograph–multichannel detector system are the greatly increased speed of data acquisition (for example, 1 min versus 40 min for recording a spectrum) and improved signal-to-noise ratio due to longer signal-accumulation times. These capabilities allow one to observe monolayer coverages on catalytic surfaces without resonance enhancement (34) and also facilitate the use of time-resolved Raman spectroscopy. Although the potential of spectrographic systems is great when considered in terms of increased speed or increased signal-to-noise ratio, double monochromators are the basis of almost all Raman instruments being used today, due to the higher sensitivity of photomultipliers. However, because multichannel detectors are undergoing rapid development, the change to spectrographs could come very rapidly. Certainly, for a large Raman–laser spectroscopy research group, a multichannel detector system is a worthwhile, albeit expensive, option to consider.

Current monochromators are of two types: those with vertical and those with horizontal slits. Although there are advantages to each orientation with respect to ease of sample placement, liquids are more easily handled in the horizontal-slit systems. Also, beam-steering optics tend to stay cleaner in the horizontal-slit systems, where they are held in a vertical position that minimizes the likelihood of powders and liquids being spilled on them. Historically, the vertical-slit systems were more commonly available, and as a result, almost all types of sampling problems have been dealt with for these systems. Thus, almost any type of sampling system can be handled by either slit system.

The double monochromator used today for high-quality Raman work has holographic gratings. This means the stray-light rejection is very good, usually about 10^{-14} at 20 cm^{-1} from the exiting line. The holographic gratings have also eliminated grating ghosts, which used to give spurious bands in the spectra. For most applications, holographic gratings have eliminated the need for triple monochromators.

In a Raman experiment, it is desirable to have the best stray-light rejection possible while maintaining the light throughput of the system. Throughput is determined by the efficiency of the gratings, the reflectivity of the mirrors in the system, the f-number, and the dispersion. Holographic gratings are generally about 70% efficient through much of the visible region of the spectrum. The mirrors all are highly reflective, so only the number of mirrors is important, and even this is not critical. Large gratings are used in the 1-m focal-length systems to give an f-number of 8 or less. The focal length and number of grooves per millimeter determine the dispersion. The current holographic gratings have between 1800 and 2000 grooves/mm and allow less stray light than does the ruled, 1200 grooves/mm grating. In general, the throughput goes up as the square of the slit width. This means the smaller the dispersion in nanometers per millimeter is, the higher the throughput is for a given resolution.

The resolution of the current systems is better than 0.5 cm^{-1} in the visible portion of the spectrum, which is much better than is needed for most analytical applications. Actually, the 1-m focal-length systems have even better resolution than this, but because the throughput falls off as the square of the slit width, relatively few samples scatter well enough to use the narrower than 50μ slits needed for higher resolution.

One other type of monochromator, based on a design developed by Delhaye (18) maintains spatial resolution of the imaged sample through the monochromator. Although this system can be used for classical Raman spectroscopy, it was developed for mapping the surface of a heterogeneous sample. Thus, only those features in the field of view that emit light at a characteristic wavelength will be visible.

Control systems for monochromators are beginning to be microprocessor based, which gives them much greater flexibility. The minimum controls needed are those for variable scan speed, automatic variable stop point, laser-line protection, variable chart recorder coupling, externally controllable functions, timing pulses to synchronize external devices, and a step-count, step-mode of operation. Variable scan speed is required so that one may observe the old rule of thumb that it should take a minimum of four time constants to scan the optical slit width. Thus, at 1 cm^{-1} band pass, with a time constant of 1 sec on the analog output signal, the scan speed should be $1 \text{ cm}^{-1}/4$ sec or $15 \text{ cm}^{-1}/\text{min}$ to allow accurate tracking of the spectrum. A variable automatic stop point is needed so that the instrument will scan only the region of interest, leaving the operator free to do other things while the instrument is running. The laser-line protection is needed to protect the photomultiplier from inadvertent scanning of the laser line. The variable chart recorder coupling, when used with a digital chart recorder, allows the wavenumber scale to be synchronized with the chart paper scale for convenient unattended operation. The ability to control all instrument functions externally is needed because most future systems will have some type of computer system associated with them. Controllable functions allow computer-controlled systems to be designed so that only regions of interest need be scanned. External timing pulses allow the computer to act as a data logger. For computer operation, the step-count-step operation is probably the best mode of operation. In this mode of scanning, the monochromator rapidly takes some preset step size, such as one wavenumber, and then collects the signal for a preset time. Since the monochromator is stopped, it is easy to do such things as automatically changing samples in the beam or automatically changing polarizations and maintain perfect wavenumber registration. With data-logging systems, this mode of operation is more efficient, since the counter can dump its data while the monochromator is going on to the next point. However, for normal computer-logged survey scans, there is not much difference between changing the slew speed with continuous counting or changing the dwell time in the stop-and-count mode. The point is that with several options available, it is usually easy to find some type of simple computer-compatible hookup.

E. DETECTION SYSTEM

Although multichannel detectors are getting better, they have not improved to the point that they have begun to replace the single-channel photomultiplier systems in normal laboratory applications. Photomultipliers are used instead of multichannel detectors because they are sensitive, cheap, and easy to use; have a large dynamic range; and have a broad spectral response. Although monochromator manufacturers usually recommend the best types of photomultiplier to use with their systems, there are some general features to consider. The spectral response curve (in milliamperes per watt) should be as high, as broad, and as flat as possible. This is required because the Raman process is weak, several laser lines may be used, and bands at either end of the spectrum may be compared. The response of GaAs photocathodes is usually very flat out to 800 nm when measured in incident watts on the photocathode. For Raman work, interest is in quantum efficiency, or the efficiency with which photoelectrons are ejected when photons hit the photocathode. The GaAs photocathodes are still the best, but the spectral response curve is not as flat when plotted on the quantum-efficiency scale. The quantum efficiency of the GaAs photocathodes is greater than 25% at 400 nm and falls only to half that number at 850 nm. A large photocathode (23 × 7 mm) is used to collect as much light as possible with little focusing. The number and type of dynodes determines the current amplifications and pulse rise time. The tubes should have a gain of greater than 5×10^5 with a rise time of less than 3 nsec to give a pulse that is easily detected and counted. Finally, the tubes should be selected for pulse counting and for low dark-count rates. The dark-count rate of a cooled tube should be less than 20 counts/sec. As long as the tube has the desired features, the operating voltage and whether it is an end-on or side-on tube are fairly unimportant.

The large red-sensitive photocathode requires that the tube be cooled to achieve a low dark-count rate. Thermoelectric coolers (water cooled) are used to reach the −30°C needed for low-dark-count operation. The housings also provide radio frequency (RF) and magnetic shielding for the tubes. They have a double-pane window insulator that should be specified as quartz. The housings come with bases that are prewired for the photomultiplier that is to be used.

The detection electronics currently used are all of the pulse-counting type. The equipment should be able to handle 10 million counts/sec; some instruments can handle 100 million counts/sec. Although the count rate should never get that high, the maximum count rate determines the maximum linear count rate, which is about $\frac{1}{10}$ of the maximum count rate. Since tubes have a maximum anode current, a picoammeter should be used to determine the maximum safe count rate.

The electronics needed for a Raman system are a high-voltage power supply, an amplifier and discriminator, a counter or rate meter, and a recorder. The power supply should be able to supply a minimum of -2000 V at 1 mA with less than 10 mV peak-to-peak ripple. Since the maximum voltage rate of side-on tubes is only -1200 V, a smaller supply can be used with these tubes. The amplifier–discriminator system is usually a single box that can be mounted near the tube. The length of the cable between the tube base and the amplifier–discriminator should be no longer than 6 in., to reduce interferences with the low-level signals. Some photomultiplier housing suppliers have built the amplifier–discriminator unit into the tube base.

It is useful to have a variable discriminator setting. If a variable discriminator is available, it can be set by first setting the tube voltage to give a gain of 10^6, as determined from the tube literature, and then by plotting the discriminator setting versus the count rate for the dark count and also versus the count rate for medium-level signal. It should be set where the difference between count rates is the largest. If the discriminator cannot be set, one can do a similar experiment by plotting the tube voltage versus count rate for a medium signal, taking care not to go over the medium voltage or current ratings. In this case, the voltage should be set a few volts above the beginning of the plateau region, which is that region where the signal count rate does not much increase as the voltage is turned up.

The counter or rate meter should have an analog output to drive a recorder. The counter or rate meter should be compatible with the discriminator, which is not always the case with the fast discriminators. If a counter is used, its recorder output should not be autoranging or have rollover if usable Raman spectra are to be recorded. Since Raman spectra often have some fluorescence background, zero suppression is a very useful control. The more suppression available, the better.

The recorder should have a digital drive that can be driven by the monochromator and will stop when the monochromator stops; this allows unattended operation. It is useful if the monochromator output driving the recorder has a scale change at 2000 cm^{-1}, so that ir-compatible plots can be produced. Finally, the recorder input voltage range should be compatible with the counter or rate meter output signal.

A computer is not required for the detection system; however, it is a very useful addition. Much of the future of Raman spectroscopy will be dominated by the improvement in data presentation and manipulation made possible by the addition of data processing to the system. Although very complex computer-controlled systems have been designed (16,61), most analytical labs will benefit by the addition of even a simple data logger to the system. These data-logging systems will make possible the easy use of multicomponent analysis methods to provide rather straightforward quantitative analyses. Loy (48) has described, in a rather cryptic note, an ir method for the multicomponent analysis of samples with indeterminant path length. Since many quantitative methods are designed for use on well-characterized samples, Loy's method applies directly to the quantitative analysis of complex, but well-defined, mixtures by Raman spectroscopy. Just as the computer systems currently being introduced are greatly improving the use of ir spectroscopy in quantitative analysis, so they will also make possible rapid quantitative analysis by Raman spectroscopy. In general, all the features in ir spectroscopy that have been improved by the addition of a computer will be improved in the Raman method as well. Thus, if a computer is used in a Raman system, it should have all the same features sought in a computer used for an ir system.

IV. ANALYTICAL APPLICATIONS

Raman spectroscopy is usually considered the complement of ir spectroscopy for qualitative structural analysis. However, prior to the 1940s and the advent of the automated-recording ir spectrometers, Raman spectroscopy was widely applied both in academia and industry for solving molecular structure problems. But the use of Raman spectroscopy as an analytical tool was limited for the most part to only the most persevering laboratories, due to experimental difficulties and practical considerations. In contrast to ir spectroscopy, which is a strong first-order effect, the Raman effect is, of course, a weak second-order effect that depends on light scattering.

In early Raman work, the sample requirements were very stringent. The sample had to be hygienically free of dust and other suspended material that could give rise to first-order scattering phenomena. The sample could not be one that fluoresced. Since it is a first-order effect, even moderate fluorescence obscured the weak Raman effect. The sample had to be colorless; otherwise the exciting radiation was absorbed, resulting in no spectrum, an unusable weak spectrum, sample decomposition, and the like. For the most part, studies were restricted to the liquid phase. Gas-phase and solid-phase studies could be obtained, but only with great difficulty. These studies were of questionable analytic usefulness.

With the advent of the laser as a source of excitation, many of these experimental difficulties disappeared, or means were devised to minimize their contributions in recording Raman spectra. The field experienced a re-surgence of activity and has since gone through several phases of activity focused on optimizing and maximizing those applications peculiar to the Raman effect.

Numerous applications of Raman spectroscopy in vibrational-assignment studies have been reported. These range from studies on inorganic com-pounds, either as single crystals or as ions in aqueous solution, to polymeric conformation and structure studies of complex organic molecules. The present discussion is concerned only with qualitative and quantitative ana-lytical applications of Raman spectroscopy.

A.   QUALITATIVE ANALYSIS

Spectrum–structure correlations were established in the early works of Kohlrauch (46), Hibben (40), Glockler (31), and others. Several recent works have dealt with Raman group-frequency studies. Nakamoto (50) discussed group frequencies for inorganic and coordination compounds. Anderson (2) surveyed the Raman group frequencies of organosilicon compounds. Group-frequency studies of organic compounds were treated by Dollish et al. (20) from a theoretical point of view. Freeman (25) used an empirical but effective approach to describe group frequencies. The latter work (25) was directed toward the chemist who dabbles in ir spectroscopy, but would like to know more about Raman spectroscopy, whereas the former (20) was directed to-ward an audience composed primarily of spectroscopists. Nyquist and Kagel (52) more recently reported practical applications of spectrum–structure cor-relations and group-frequency studies in the industrial environment.

The reader is referred to several review articles that focus on group-frequency studies of interest to the analytical chemist (27,32,33). The de-velopments in analytical applications of Raman spectroscopy from 1977 to 1981 have been reviewed by Gardiner (28,29).

Qualitative applications of Raman spectroscopy take advantage of the unique properties of the laser as a source of excitation. Many of these ap-plications, some of which could be considered spectroscopic tricks, are of interest to the analytical chemist.

Much emphasis was placed on microsampling techniques in the late 1960s and early 1970s (3,26,43). Microsampling techniques took advantage of the coherent source characteristics of the laser beam, namely that the laser beam could be focused to a spot about 50 μ in diameter. Spectra of nonvolatile liquids were obtained from small-diameter (about 100 μ) capillary sample tubes using axial illumination and transverse viewing (relative to the cell axis). Working with a sealed capillary, Bailey et al. (3) obtained spectra of as little as 0.25 nl. The sealed-capillary concept has been modified for use with a gas chromatograph (12,42). In this technique, gas-chromatography fractions are trapped in the capillaries by using liquid nitrogen and the cap-illaries are sealed. The obvious advantage of using gas-chromatographic techniques is that fluorescing impurities, scattering particulate matter, and so forth are excluded from the sample.

A very different approach to microsampling was introduced by Delhaye (18) and Rosasco (57,58). Both workers demonstrated the feasibility of obtaining not only the Raman spectra of microsamples, but also topological information about their sizes, shapes, and positions. The Raman microprobe involves analysis of individual particles with diameters in the micrometer range. This technique allows precise chemical definition of micrometer-size particles.

Delhaye and coworkers developed laser mapping and imaging techniques using both single-channel and multichannel detection (18). Different substances in an inhomogeneous medium are separately identified by mapping the surface of the medium. The system has been used for inorganic and organic materials, geological samples, polymers, and biological samples (19).

Rosasco and coworkers have studied a variety of environmental samples, including urban dust. The spectra of these samples are relatively simple and usually result from only one or two inorganic compounds. Spurious bands in the spectrum of urban particulate matter was attributed to carbon in a form analogous to polycrystalline graphite. The polycrystalline graphite forms by conversion of graphite soot on the particles or decomposition of organic impurities attached to the host particulate by laser radiation (9). Blahar and Rosasco (8) have reported Raman microprobe spectra of individual microcrystals and fibers of talc, tremolite and related silicate minerals, including low-iron anthrophyllite and actinolite. Spectra of organic polymers and of biological systems mounted on different inorganic substrates have also been reported (23). In these cases, significantly reduced irradiation levels were used, because organic materials tend to absorb the excitation radiation. The Raman microprobe technique was successfully used for a 9-μ polystyrene microsphere, a 2.8- $\times$ 4.0-μ polyvinyl chloride particle, and a 3.2- $\times$ 3.6-μ cholesterol particle. This recent work indicates that the Raman microprobe holds great promise for the characterization and identification of molecular constituents at the cellular or subcellular level in biological specimens.

The use of Raman spectroscopy for the characterization of species adsorbed on surfaces was first reported in the early 1960s in the European literature (44,45). This early work, which used mercury arc illumination, was focused on catalyst surfaces, in much the same way that ir spectroscopy had been since the 1950s. Raman spectroscopy offers the advantage over ir spectroscopy that the M—O stretching frequency of metal oxides commonly used as catalysts yields very weak Raman scattering, in contrast to the situation with ir spectroscopy, where the M—O stretching frequency is quite intense. Large regions of the ir spectra are obscured because of this. Fluorescence and scattering effects due to particle size are obvious disadvantages of the Raman technique.

Hendra and coworkers in the United Kingdom first reported laser Raman results on adsorbed species in late 1967 and 1968 (37,38). Shortly after this, a paper on Raman spectroscopy of 2-chloropyridine adsorbed on silica gel appeared in the American literature (41). Both physisorbed and chemisorbed 2-chloropyridine were studied. The sample cell was designed to be part of a vacuum system. Particular attention has been paid to Raman spectra of species adsorbed on zeolites (22,51). Numerous reports of species adsorbed on catalysts have appeared in the literature; these have been summarized in several excellent reviews and books (15,30,53).

B. QUANTITATIVE ANALYSIS

Few practical quantitative applications of the classical Raman effect have been reported in the recent literature. One reason for this is that the current trend is toward trace-level analysis, and Raman spectroscopy is not the preferred tool for trace-level studies. Even in aqueous solutions, where Raman spectroscopy offers certain advantages over ir spectroscopy, both techniques fail to produce acceptable results in the sub-part-per-million range, where matrix effects overshadow the behavior of the material of interest. A few attempts to quantify organic and inorganic anions (4,11) in parts-per-million level concentrations have been reported in the literature. However, the preferred techniques for quantification of trace organic components in aqueous solutions, particularly waste waters, are still gas chromatography and gas chromatography–mass spectrometry.

The Raman effect is most suitable for quantitative applications when the components of interest are present at concentrations greater than a few tenths of a percent. Unlike in ir spectroscopy, where concentration varies exponentially with peak height, in the Raman effect concentration varies linearly with peak height.

It is necessary to establish a calibration curve at the beginning of a quantitative Raman experiment, and the standards should be checked regularly during the course of the experiment. The calibration curve is prepared using standard solutions whose concentrations bracket the range of concentration in the unknown solution. Band peak heights are plotted against concentration to establish the calibration curve. Standard solutions should be sealed in capillary tubes. They can be used repeatedly for checking the calibration curve or re-establishing the curve for day-to-day operation. Each point on the calibration curve should represent the average of values from at least four runs. Under normal operating conditions, the scan time is set equal to at least 4 times the time it takes to scan over the half-band width of the band. The spectral slit width is chosen so that it is less than the half-band width. The period is adjusted so that the pen exhibits a rapid response to small signals. Once the instrument has been set up in this manner, it is a relatively simple matter to determine the peak height of an unknown and read the concentration from the calibration curve.

The system should remain stable over a period of hours. Most systems do not possess long-term stability; for this reason, it is recommended the standard solutions be run periodically to check the calibration curve during the experiment. Certainly, new calibration curves should be constructed on successive days.

Using this method, Nyquist and Kagel (52) have routinely analyzed aqueous solutions of acrylonitrile in the concentration range 500–5000 ppm. The C≡N stretching vibration at 2240 cm^{-1} was selected over the C=C stretching vibration at 1616 cm^{-1} because there are fewer interferences in the 2200 cm^{-1} region. The lower limits of detection and quantification of acrylonitrile in water depend on laser power. Nyquist and Kagel estimated that with a 1-W laser, the limit is about 125 ppm.

Loy et al. (48) described a rapid Raman method for determining the percentage conversion of polyacrylamide (PAC) to poly-N-dimethylaminomethylacrylamide (PDMAMAC) via the Mannich reaction. The method uses peak heights of the complex C—N stretching modes at 1212 and 1112 cm^{-1} for PAC and PDMAMAC, respectively. The precision of the method was determined by analyzing commercial samples 10 times on 2 consecutive days, with a resultant rms deviation of 2.1%.

V. EMERGING RAMAN ANALYTICAL TECHNIQUES

In Section I, it was noted that a whole panoply of Raman techniques had been developed that improve the sensitivity of the Raman process. Several of these techniques have emerged as useful analytical tools for specialized application. Brief mention will be made here of the three techniques with the greatest potential for practical application.

The first technique is coherent anti-Stokes Raman spectroscopy (CARS), which has been applied to a few analytical systems, with some success (6). A CARS experiment requires the use of two laser sources, one of which must be tunable, so that stimulated coherent anti-Stokes emission can be produced. The efficiency of the CARS process is very high and discrimination against background fluorescence is several orders of magnitude higher than in conventional Raman scattering. However, the experimental arrangement for CARS is not simple, and the instrumental costs are high.

The development of CARS of necessity parallels the development of tunable laser sources, which has been slow. The efficiency of the CARS process drops rapidly with concentration, and therefore the process is of little use for trace analysis. It is not apparent that use of CARS will become commonplace in the analytical laboratory in the near future. However, for certain applications, such as studies in combustion chemistry, *in situ* flame diagnostics, and photochemistry, CARS provides unique information and is often the best optical analytical technique (21,34).

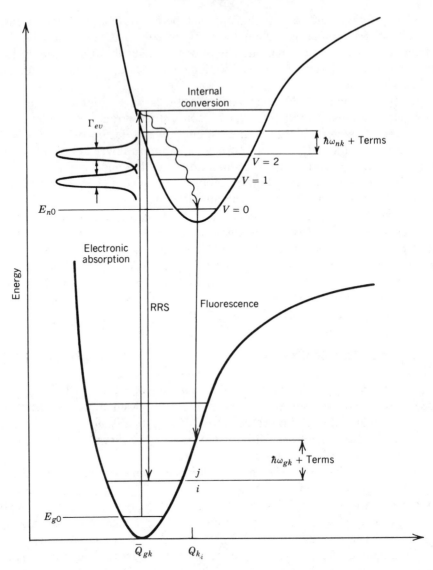

Fig. 10.4.   Resonance Raman scattering, shown with other electronic transition phenomena.

The second area in which rapid progress is being made is resonance Raman spectroscopy (RRS). The resonance Raman effect is depicted in Fig. 10.4, which shows that the laser excitation matches an electronic transition of the molecule of interest. By coupling with the strong first-order electronic transition, a great ($\sim10^6$) increase in the Raman scattering intensity is achieved. In addition, by selective tuning of the laser frequency to match an absorption band of the molecule of interest, the Raman scattering of that species can be preferentially enhanced. The chief shortcoming of RRS is that given the present state of laser technology, it is limited to species having an electronic absorption band in the visible region of light. Thus, RRS has been most widely applied to the study of dye molecules and biological molecules having suitable chromophores. Particularly in the latter case, the high selectivity and sensitivity of RRS offer several real advantages (13). Often, complex biomolecules are difficult to obtain in high concentrations and isolated from other components, and it is in this area that the potential analytical application of RRS appears to be the greatest. Another area that takes advantage of the enhanced intensity of RRS is the study of species adsorbed on solid surfaces (66). With continuing developments in laser technology, the range of applications for RRS is expected to increase.

From an experimental point of view, RRS does not differ much from conventional Raman spectroscopy. The primary requirement of RRS is that the laser excitation frequency fall close to an electronic transition. Thus, for flexibility in the experiment, a tunable dye laser source is required. With the addition of a dye laser system and its necessary accessories, such as a monochromator filter and dye laser beam-transfer components, the cost and complexity of instrumentation for the experiment is increased. Therefore, for practical applications, the use of RRS is usually not warranted unless one is studying trace amounts of a dye or a suitable biomolecule.

The third technique to be considered is surface-enhanced Raman scattering (SERS). The SERS phenomenon is a greatly increased intensity of Raman scattering for molecules near suitably prepared metal surfaces. The increase of five to six orders of magnitude in Raman scattering intensity allows monolayer coverage of molecules adsorbed to metal surfaces to be observed. The SERS effect is a powerful tool for investigating interfacial phenomena and appears to be general for the types of molecules and interfaces studied. However, SERS is highly dependent on the metal surface, both in its composition and in its morphology. To date, only copper, silver, and gold have been confirmed as SERS substrates. The bulk of the work has been done on silver surfaces and in electrochemical environments. This, of course, limits the potential usefulness of SERS for studying catalytic surfaces. In addition, the mechanism of the SERS effect is not yet fully understood; this is an area of very active research. SERS is a relatively young field in which rapid progress is still being made. The developments since its discovery in the early 1970s are covered in recent reviews (7,62).

Although SERS is not limited to electrochemical environments, from a practical point of view, the immediate application of SERS will probably involve the electrode–electrolyte interface. Three broad areas of potential analytical application can be identified. The first is studies of electrochemical reactions at the electrode surface. The second is studies of inorganic complexes in solution. The third area, which looks especially promising, is investigations of biologically important molecules. Complex biomolecules such as myoglobin and cytochrome *c* can be adsorbed onto silver electrode surfaces from dilute solutions without the protein structure being radically affected (17). Combined RRS and SERS leads to a multiplicative increase in sensitivity, opening the possibility of qualitative trace analysis on suitable species. In addition, adsorption of species on silver electrodes often leads to quenching of fluorescence, permitting Raman spectra to be obtained from highly fluorescent species such as laser dye molecules. Finally, unlike RRS, SERS does not need compounds with a visible absorption band to produce high Raman signal intensities; this may make it possible to look at colorless biological molecules under very dilute conditions. These are merely demonstrated capabilities of SERS. No routine analytical applications of SERS have yet been developed, although the method continues to hold promise.

In terms of instrumentation, no major modifications of a conventional Raman setup are required for SERS investigations. The heart of the SERS experiment lies in the sample preparation. Thus, the experimenter needs to construct a suitable spectroelectrochemical cell and acquire a silver electrode assembly, polishing equipment, and a potentiostat with a waveform generator. The electrode surface needs to be roughened by an oxidation–reduction cycle. Unfortunately, the preparation of the electrode surface is still an art rather than a science. Thus, to determine the optimum conditions for observing a given molecule on a given electrode surface requires experimentation, and the operator has to be skilled at doing SERS. So although the additional equipment costs for SERS are modest, the experimenter needs to invest time to learn how to make the experiment work. In addition, because of the dependence of SERS intensity on surface-preparation conditions, SERS provides qualitative, not quantitative, information. However, as noted above, this is a rapidly developing field, and over time this barrier to the routine use of SERS as an analytical tool may be removed.

In summary, for specialized applications, the techniques of CARS, RRS, and SERS offer advantages over conventional Raman spectroscopy and often provide unique and useful information. It is up to the individual experimenter to decide whether the areas of special application for each of these techniques are of sufficient importance to warrant the additional cost and complexity.

VI.   SUMMARY

The advantage of Raman spectroscopy as an analytical tool centers around the unique inherent properties of the Raman effect itself and the unique properties of the laser as a source of excitation.

For reasons relating to symmetry, the Raman spectrum is usually simpler than the ir spectrum. Raman line intensities are dependent on changes in the polarizability of the molecule, rather than changes in the dipole moment as in ir absorption. Therefore, selection rules for Raman scattering differ significantly from selection rules for ir absorption. The well-known principle of mutual exclusion, which states that when a molecule contains a center of symmetry, those vibrations that are ir active will be Raman inactive, and vice versa, can be a definite aid in routine identification work. Under other symmetry conditions, ir absorption and Raman scattering often complement one another.

One very distinct feature of Raman-scattered light is its state of depolarization. Depolarization ratios are useful for distinguishing symmetric vibrations from antisymmetric vibrations, and hence aid spectrum–structure correlation. Experimental determination of depolarization ratios in a relatively straightforward matter.

A disadvantage of Raman spectroscopy for spectrum–structure correlation work is the limited number of standard spectra available. A vast, ever increasing body of ir standard spectra has been accumulated over many years, making possible a vast number of spectrum–structure correlations. Standard Raman spectra have begun to emerge only over the past few years. However, today there is a growing body of Raman standard spectra and spectrum–structure correlations covering many organic, inorganic, and organometallic compounds.

A most attractive feature of Raman spectroscopy is that it offers an opportunity for studying systems in aqueous solution. In fact, quantification in aqueous solution is relatively straightforward at higher concentrations. With careful control, experimental errors of less than 5% (at 2 σ) can be achieved in more complex systems.

Trace-level quantification in solution by means of Raman spectroscopy remains a problem. The method lacks sensitivity. Raman spectroscopy is not at all suitable for the determination of substances present at trace concentrations of parts per billion and parts per trillion, the concentration levels of greatest interest to advocates of ecological chemophobia.

The coherent source characteristics of the laser beam make Raman spectroscopy particularly suitable for handling extremely small samples. Spectra of micro-scale (nanoliter) samples are easily obtained without specialized instrumentation. Specialized instruments with sample-positioning stages allow spectra of individual particles with diameters in the micrometer range to be obtained. This technique holds great promise for the characterization and identification of molecular constituents at the cellular or subcellular level in biological specimens.

Fluorescence remains the most formidable experimental problem in Raman spectroscopy. Advances in laser technology may eventually alleviate this situation. Sample color and clarity are no longer major problems, although applications of Raman spectroscopy are still somewhat limited by these sample properties. Problems due to these properties are generally minimized by use of multiple laser sources or lasers with multiple excitation frequencies. Unfortunately, not all laboratories are well enough equipped to eliminate completely all problems due to color and clarity.

Finally, sophisticated Raman techniques such as CARS, RRS, and SERS have already proved useful for specialized applications. Continued developments in technology and academic research may lead eventually to these methods becoming much more general analytical techniques.

REFERENCES

1. Amy, J. W., R. W. Chrisman, J. W. Lundeen, T. I. Ridley, J. C. Sproules, and R. S. Tobias, *Appl. Spectrosc.*, **28**, 262 (1974).

2. Anderson, D. R., "Infrared, Raman and Ultraviolet Spectroscopy," in A. L. Smith, *Analysis of Silicones*, Wiley, New York, 1974, pp. 247–86.

3. Bailey, G. F., S. Kint, and J. R. Sherer, *Anal. Chem.*, **39**, 1040 (1967).

4. Baldwin, S. F., and C. W. Brown, *Water Res.*, **6**, 1601 (1972).

5. Bartz, A. M., C. B. Pratt, and R. O. Kagel, *Appl. Spectrosc.*, **25**, 474 (1971).

6. Begley, R. F., A. B. Harvey, R. L. Byer, and B. S. Hudson, *Am. Lab.*, **16**, 11 (1974).

7. Birke, R. L., J. R. Lombardi, and L. A. Sanchez, "Surface Enhanced Raman Spectroscopy," in K. M. Kadish, Ed., *Electrochemical and Spectrochemical Studies of Biological Redox Compounds* (Advances in Chemistry Series, Vol. 201), American Chemical Society, Washington, D.C., 1982, pp. 69–74.

8. Blahar, J. J., and G. J. Rosasco, *Anal. Chem.*, **50**, 892 (1978).

9. Blahar, J. J., G. J. Rosasco, and E. S. Etz, *Appl. Spectrosc.*, **32**, 292 (1978).

10. Bohr, N., *Philos. Mag.*, **26**, 476, 857 (1913).

11. Bradley, F. B., and C. A. Frenzel, *Water Res.*, **4**, 125 (1970).

12. Bulkin, B. J., K. Dill, and J. J. Dannenberg, *Anal. Chem.*, **43**, 974 (1971).

13. Bunow, J. R., "Raman Spectroscopy," in C. Marton, Ed., *Methods of Experimental Physics*, Vol. 20, Academic, New York, 1982, pp. 123–161.

14. Compaan, A., *Appl. Spect. Rev.*, **13**, 295 (1977).

15. Cooney, R. P., G. J. Chrthroys, and T. T. Nguyen, *Adv. Catal.*, **24**, 293 (1975).

16. Crisman, R. W., S. C. English, and R. S. Tobias, *Appl. Spectrosc.*, **30**, 168 (1976).

17. Cotten, T. M., S. G. Schultz, and R. P. Van Duyne, *J. Am. Chem. Soc.*, **102**, 7960 (1981).

18. Delhaye, M., and P. Dhamelincourt, *J. Raman Spectrosc.*, **3**, 33 (1975).

19. Delhaye, M., E. DaSilva, and G. S. Hayat, *Am. Lab.*, **9**, 83 (1977).

20. Dollish, F. R., W. G. Fateley, and F. F. Bently, Eds., *Characteristic Raman Frequencies of Organic Compounds*, Wiley-Interscience, New York, 1973.

21. Eckbreth, A. C., P. A. Bonczyk, and J. F. Verdieck, *Appl. Spect. Rev.*, **13**, 15 (1977).

22. Egerton, T. A., A. H. Hardin, and N. Sheppard, *Can. J. Chem.*, **54**, 586 (1976).

23. Etz, E. S., and G. J. Rosasco, *Res. Devel.*, **528**, 20 (1977).

24. *Fourth International Conference on Raman Spectroscopy*, Bowdoin College, Brunswick, Maine, August 25–29, 1974.

25. Freeman, S. K., *Applications of Laser Raman Spectroscopy*, Wiley-Interscience, New York, 1974.

26. Freeman, S. K., P. R. Reed, and D. O. Landon, *Mikrochem. Acta*, **3**, 288 (1972).

27. Gardiner, D. J., *Anal. Chem.*, **50**, 131R (1978).

28. Gardiner, D. J., *Anal. Chem.*, **52**, 96R (1980).

29. Gardiner, D. J., *Anal. Chem.*, **54**, 165R (1982).

30. Gilson, T. R., and P. J. Hendra, *Laser Raman Spectroscopy*, Wiley-Interscience, London, 1970.
31. Glockler, G., *Rev. Mod. Phys.*, **15**, 112 (1943).
32. Grossman, W. E. L., *Anal. Chem.*, **46**, 345R (1974).
33. Grossman, W. E. L., *Anal. Chem.*, **48**, 261R (1976).
34. Haller, G. L., *Catal. Rev.*, **23**, 447 (1981).
35. Harris, J. M., R. W. Crisman, F. E. Lytle, and R. S. Tobias, *Anal. Chem.*, **48**, 1937 (1976).
36. Harvey, A. B., and J. W. Nibler, *Appl. Spect. Rev.*, **14**, 101 (1978).
37. Hendra, P. J., and E. J. Loader, *Nature*, **216**, 789 (1967).
38. Hendra, P. J., and E. J. Loader, *Nature*, **217**, 637 (1968).
39. Herzberg, G., *Molecular Spectra and Molecular Structure*, Part II, Van Nostrand-Reinhold, Princeton, New Jersey, 1945.
40. Hibben, J. H., *The Raman Effect and Its Chemical Applications*, Reinhold, New York, 1939.
41. Kagel, R. O., *J. Phys. Chem.*, **74**, 4518 (1970).
42. Kagel, R. O., Industrial applications of Raman Spectroscopy in ref. 24.
43. Kagel, R. O., and R. A. Nyquist, Paper 284, in *Abstracts of Pittsburgh Conference on Analytical Chemistry and Applied Spectroscopy*, Cleveland, 1972.
44. Karagounis, G., and R. Issa, *Nature*, **195**, 1196 (1962).
45. Karagounis, G., and R. Issa, *Z. Electrochem.*, **66**, 874 (1962).
46. Kohlrauch, K. W. F., "Ramanspektren," in A. Euken and L. L. Wolf, Eds., *Hand- und Jahrbuck der Chemischen Physik*, Academische Verlidsgellschaft Becker and Erler, Leipzig, 1943, pp. 1–469.
47. Landsberg, G., and L. Mandelstam, *Naturwissenschaften*, **16**, 57 (1928).
48. Loy, B. R., R. W. Chrisman, R. A. Nyquist, and C. R. Putzig, *Appl. Spectrosc.*, **33**, 638 (1979).
49. McCreery, R. L., M. Fleischman, and P. Hendra, *Anal. Chem.*, **55**, 148 (1983).
50. Nakamoto, K., *Infrared Spectra of Inorganic and Coordination Compounds*, Wiley, New York (1970).
51. Nguyen, T. T., R. P. Cooney, and G. J. Curthoys, *J. Chem. Soc., Faraday Trans. I*, **72**, 2577, 2592, 2598 (1976).
52. Nyquist, R. A., and R. O. Kagel, "Organic Materials," in E. G. Brame and J. G. Grasselli, Eds., *Handbook of Practical Spectroscopy*, Vol. 1, Dekker, New York, 1977, Chapter 6.
53. Paul, R. L., and P. J., Hendra, *Miner. Sci., Eng.*, **8**, 171 (1976).
54. Planck, M., *Verh. Dtsch. Phys. Ges.*, **2**, 237 (1900).
55. *Proceedings of 6th International Conference on Raman Spectroscopy*, Bangalore, India, September 4–9, 1978.
56. Raman, C. V., *Ind. J. Phys.*, **2**, 387 (1928).
57. Rosasco, G. J., and E. S., Etz, Investigation of Raman spectra of individual micro-sized particles in ref. 24.
58. Rosasco, G. J., E. S., Etz, and W. A., Cassatt, *Appl. Spectrosc.*, **29**, 396 (1975).
59. Smekal, A., *Naturwissenschaften*, **11**, 873 (1923).
60. Tobias. R. S., "Applications to Inorganic Chemistry" in Anderson, A., *Raman Effect*, Dekker, New York, 1974, pp. 405–518.
61. Tobias, R. S., T. H., Bushaev, and J. C., English, *Ind. J. Pure Appl. Phys.*, **16**, 401 (1978).
62. Van Duyne, R. P., "Laser Excitation of Raman Scattering from Absorbed Molecules on Electrode Surfaces," in C. B. Moore, Ed., *Chemical and Biochemical Application of Lasers*, Vol. 4, Academic, New York, 1979, pp. 101–185.

63. Walrafen, G. E., *Phys. Bl.*, **30,** 540 (1974).

64. Walrafen, G. E., and L. A., Blatz, *J. Chem. Phys.*, **59,** 2646 (1973).

65. Wilson, E. B., J. C., Decius, and P. C., Cross, *Molecular Vibrations*, McGraw-Hill, New York, 1955.

66. Yamada, H., *Appl. Spect. Rev.*, **17,** 227 (1981).

MICROWAVE SPECTROMETRY

By ROBERT K. BOHN, *University of Connecticut, Storrs, Connecticut*

AND NANCY S. TRUE, *University of California, Davis, California*

Contents

I. INTRODUCTION

In gases the rotational energies of molecules are distributed among quantized states and are characterized well by quantum mechanics. The transitions among these states generally occur in the microwave region of the spectrum; most studies have observed the 8 to 40-GHz frequency or 4- to 0.7-cm wavelength range. For low-pressure (~ 10 μ Hg) gases, the ratio of line width to absolute frequency is so small, $\sim 10^{-5}$, that a single transition often can identify a component and specify its concentration. Resolution is so great that isotopic as well as vibrationally excited species are generally resolvable from the normal ground-state species. In spite of this complexity and the large number of transitions for heavy, asymmetric molecules, the overlap problem in microwave spectrometry is less severe than in any other spectral technique.

The first microwave experiments were carried out by Cleeton and Williams (13) at the University of Michigan in 1934. They observed a broad absorption in NH_3 at ~ 0.8 cm^{-1}, which indicates the splitting of the ground vibrational state by inversion, that is, by the passage of the nitrogen atom through the plane of the hydrogen atoms. Ten years passed before further microwave experiments were done. The research effort spurred by World War II produced microwave sources, detectors, and circuitry developed for radar applications. Expanding uses of microwaves in radar, weapon guidance, and communications have continued to encourage the development of microwave technology.

The primary microwave source developed during World War II was the klystron, a tunable intense monochromatic source with a line width narrower than most absorption lines. Tunable monochromatic sources of ultraviolet, visible, and infrared radiation in the form of lasers have only recently been developed, and yet they are already being used in a myriad of applications. Although similar capabilities in the microwave spectral region have been available for decades, the application of microwaves to spectroscopic problems has remained concentrated in a relatively small number of laboratories scattered about the world. Microwave spectroscopy has not been widely applied to analytical problems, partly because of outdated historical attitudes that remain in force.

The high resolution and accuracy of microwave studies already available in the 1940s provided the most demanding tests of the validity of the quantum theory, and the exquisite agreement between microwave experiments and quantum theory remains the best evidence of that theory's validity. Such research was the major focus of the pioneering laboratories of Wilson at Harvard, Gordy at Duke, and Townes at Columbia. And since most of the world's microwave spectroscopists emerged from these centers, that focus has remained important. Microwave spectroscopists have contributed a wealth of information on molecular geometries, dipole moments, magnetic moments, vibrational and internal rotation potential functions, centrifugal distortion constants, molecular quadrupole and magnetic quadrupole moments, and much more. However, applications of microwave techniques to analytical problems have not been extensive except in radioastronomy.

The microwave spectra of low-pressure gases consist of very sharp absorption lines whose frequencies are highly distinctive of the absorbing species and from which much information can be derived. A corollary of this fact is that *mere* identification and quantification of absorbing species has not been seriously exploited in most of the world's microwave laboratories. The dominant theme in microwave spectroscopy has been the extraction of molecular information from microwave spectra.

Commercial microwave spectrometers were produced by the Hewlett-Packard Corporation and Cambridge Scientific Instruments during the late 1960s and early '70s, but both firms have since stopped production of the complete spectrometers; many of the components, however, are still available. As a result, application of microwave-spectroscopic techniques to analytical problems is not convenient for nonspecialists and other techniques are usually chosen for analytical problems.

It is our purpose to demonstrate the usefulness of microwave spectroscopy for analytical problems, to provide enough background and resource information to allow the reader to evaluate microwave spectrometry for solution of analysis problems, and to encourage the microwave industry to apply its considerable technological expertise to problems of chemical analysis. A recurrent theme of this and other reviews of analytical applications of microwave spectroscopy (14,25,48,65,66,88,124) is that the potential for application exists but has not been exploited.

A number of excellent general references on microwave spectroscopy have been published, including recent monographs by Gordy and Cook (34), Kroto (60), Wollrab (133), and Sugden and Kenney (100) as well as the classic texts by Gordy, Smith, and Trambarulo (35) and Townes and Schawlow (109).

II. THEORY OF MICROWAVE SPECTROSCOPY

Since microwave transitions span differences between rotational energy levels of a molecule, they have two essential characteristics. For a transition between states to have an observable intensity, *the molecule must be polar,* since the basic interaction is between the molecule's electric dipole and the electric field of the radiation. Also, the lifetimes of the states must be long enough (according to Heisenberg's Uncertainty Principle) that transitions between them are defined accurately enough to enable one to take advantage of the precision of experimental microwave technology. The lifetimes are generally limited by molecular collisions. This means that *low-pressure* (a few μ *Hg) gases* are required to insure that collisions will be infrequent enough not to dominate the absorption line width. Microwave spectra with useful resolution can be obtained at pressures up to ~100 μ Hg.

A. LINE FREQUENCIES

The theory of rotational energy states is simplest for rotors of highest symmetry and becomes more complex for less symmetric rotors. We will

TABLE 11.I
Rotational Properties of Molecules

Type of molecule	Moments of inertia	Examples
Linear	$I_a = 0, I_b = I_c$	NO, OCS
Spherical[a]	$I_a = I_b = I_c$	CH_4, SF_6
Symmetric		
Prolate rotor	$I_a < I_b = I_c$	SiH_3Cl, CH_3CCH
Oblate rotor	$I_a = I_b < I_c$	NF_3, $CHCl_3$
Asymmetric	$I_a \neq I_b \neq I_c$	H_2O, CH_3OH

[a] Spherical molecules are nonpolar and have no microwave spectra except when subjected to perturbations that destroy the spherical symmetry.

discuss only the outlines of the theory, and refer the readers to any of the general references for details.

All rotating objects have three mutually perpendicular principal axes; for objects as small as molecules, rotational states must be treated quantum mechanically. For molecules, rotational spectra of the three types (not counting that for a spherical molecule) described in Table 11.I are important. The simplest type is that of the *linear molecule,* which has $I_a = 0$ and $I_b = I_c$, where the I are moments of inertia about each principal axis. The a-axis is the molecular axis. The energy levels of a linear rigid molecule are given by

$$E_J = hBJ(J + 1) \tag{1}$$

where B, the rotational constant, is given by $B = h/8\pi^2 I_b$, and J is the total angular momentum quantum number ($J = 0, 1, 2, 3, \ldots$). The square of the total angular momentum is given by

$$\langle P^2 \rangle = \frac{BJ(J + 1)h^2}{8\pi^2}. \tag{2}$$

If the molecule is polar, the frequencies of the allowed transitions are

$$\nu = 2B(J + 1) \tag{3}$$

A *symmetric molecule* also has two equal moments of inertia (a threefold or higher axis of symmetry insures that a molecule is symmetric), but the third moment is not equal to zero. If the unique moment of inertia is the smallest, the molecule is a prolate symmetric top; if the largest, the molecule is an oblate symmetric top. The energy levels of a rigid prolate symmetric top are

$$E_{JK} = BhJ(J + 1) + (A - B)hK^2 \tag{4}$$

Here K, the projection of J along the symmetry axis, is introduced. K is an integer and takes the value $K = 0, \pm 1, \pm 2, \ldots, \pm J$. The rotational constant, A, is given by $A = h/8\pi^2 I_a$. Small centrifugal distortion terms may be added for nonrigid molecules. The frequencies of the transitions allowed (selection rules $\Delta J = 0, \pm 1; \Delta K = 0$) for a prolate symmetric top are

$$\nu = 2B(J + 1) \tag{5}$$

Thus, spectral absorptions occur at integral values of $2B$. Recall that the energies of the rotational levels depend on K as well as J, but the K dependence drops out in the difference because K does not change. In the symmetric-rotor absorption spectrum, each absorption line is composed of $2J + 1$ (the number of allowed K values) degenerate lines. If one imagines that the molecule's geometry distorts from symmetric-top symmetry, that is, that I_b becomes different from I_c, one can also imagine that the simple equation 5 for the observed frequencies breaks down and that the $2J + 1$ lines that are degenerate in the case of a symmetric top begin to shift and are no longer degenerate. If the geometry is still close to that of a symmetric top, the lines will still be nearly degenerate and one expects a piling up of lines around $\nu = 2B(J + 1)$, or, what is more relevant to the nearly symmetric top,

$$\nu \approx (B + C)(J + 1) \tag{6}$$

where $(B + C)$ is twice the average of the two nearly equal rotational constants. This is the fundamental expression used in low-resolution microwave spectroscopy (96). It is important, because many molecules happen to be nearly symmetric rotors.

For the oblate symmetric top, one interchanges A with C in equations 4–6.

Finally, the rotational energy levels of an *asymmetric rotor* may be expressed as

$$E_{J_\tau} = \frac{h}{2}(A + C)J(J + 1) + \frac{h}{2}(A - C)E(\kappa) \tag{7}$$

where the asymmetry is expressed in terms of Ray's parameter,

$$\kappa = \frac{2B - A - C}{A - C} \tag{8}$$

which varies between -1 for a symmetric prolate rotor and $+1$ for an oblate rotor. Details of the energy-level expression $E(\kappa)$ may be found in King et al. (53). The subscripts for E are J, which has been defined, and τ, which is given by

$$\tau = K_{-1} - K_{+1} \tag{9}$$

K_{-1} and K_{+1} are the K values in the prolate and oblate symmetric limits, respectively. The selection rules (53) vary according to the direction of the molecule's dipole moment relative to its principal axes. If the molecule has an a-component of dipole moment, there are μ_a-type spectra, and similarly for b- and c-components. The selection rules derived are as follows:

1. μ_a-type: ee ↔ eo and oo ↔ oe,
2. μ_b-type: ee ↔ oo and eo ↔ oe,
3. μ_c-type: ee ↔ oe and oo ↔ eo,

where the letters represent the evenness or oddness of K_{-1} and K_{+1}.

Line frequencies can be strictly related to the effective rotational constants. However, because molecules are nonrigid, the effective rotational constants are affected by molecular motions. To equations 1, 3, 4, 5, and 7 centrifugal distortion terms may be added. These are small for low J values and nearly rigid molecules but become more important at high J values. Also, the lines in rotational spectra are usually narrow. Thus, spectra of vibrationally excited species are often distinguishable from those of the ground-state species. Spectra of pure molecules are complicated by resolvable spectra from excited vibrational states. Also, overall angular momentum is not completely uncoupled from internal angular momentum arising from molecular vibrations (40). The interactions are largest for internal rotation; this motion has been analyzed for many systems involving symmetrical internal rotors, especially methyl groups (73,130) and systems that undergo low-frequency flapping motions (11,86).

The frequencies of lines may also be shifted through interactions with external electric fields (Stark effect) or magnetic fields (Zeeman effect). These effects are used for detection of spectra, assignment of transitions, and determination of molecular moments. Details of these interactions are explained in the general references.

B.   INTENSITY AND LINE SHAPE

The power absorbed by the gas in the cell is

$$\Delta P = \alpha_\nu l P_0 \tag{10}$$

where P_0 is the power with no sample present, l is the length of the cell, and α_ν is the absorption coefficient given in the pressure broadened limit by the Lorentzian line shape

$$\alpha_\nu = \frac{8\pi^2 N F_{J\tau} \nu^2 \mu_g^2 \lambda_g (J, \tau, J', \tau')}{3ckT(2J + 1)} \frac{\Delta\nu}{(\nu_0 - \nu)^2 + \Delta\nu^2} \tag{11}$$

where N is the number of molecules per unit volume, $F_{J\tau}$ is the fraction of molecules in the lower state of the transition, ν is the radiation frequency, ν_0 is the resonant frequency, $\Delta\nu$ is the half width of the lines, μ_g is the dipole moment component for the transition under consideration, $\lambda_g(J. \tau, J', \tau')$ is the line strength, c is the speed of light, k is Boltzmann's constant, and T is the temperature. Line strengths, λ_g, have been tabulated as a function of κ (125). For linear and symmetric molecules, F and λ can be expressed simply. Several features of α should be noted. Typical values of α range from 10^{-4} to 10^{-9} cm^{-1}. The absorption coefficient increases as frequency squared and as dipole moment squared. Also, α_ν increases with decreasing T because $F_{J\tau}$ depends exponentially on T. At very low pressures, up to ca. 1 μ Hg, the narrow absorption line width is determined mostly by Doppler and wall-collision broadening. At higher pressures, pressure broadening dominates line shape, and above ca. 10 μ Hg the peak absorption does not change but the line width increases. A transition's width can also be increased if the microwave power is sufficient to upset the thermal equilibrium between the two states. Saturation effects can be eliminated by reducing the radiation power level. Sometimes the resultant signals are too weak to be useful, however. A special method that uses power saturation for quantitative analysis is discussed in Section V.B (36).

III. EXPERIMENTAL TECHNIQUES

Commercial microwave spectrometers were available from 1965 to 1974 from Hewlett-Packard and Cambridge Scientific Instruments, Limited. Unfortunately, demand was insufficient and production of complete spectrometers was discontinued; however, many components are still available. A variety of microwave spectrometers are described by Carrington (10). Basically, a microwave spectrometer is composed of a source, a sample cell, and a detection system. We will survey each of these elements.

A. SOURCES

Traditionally, microwave generation and amplification are accomplished by means of electron-beam devices such as magnetrons, klystrons, and traveling-wave tubes. In recent years, paramagnetic amplifiers and solid-state devices have been used increasingly to perform these functions, although application of these devices to conventional microwave spectroscopy is not extensive. We will discuss the more common electron-beam devices, the reflex klystron and backward-wave oscillator, and their application as microwave sources in spectroscopy, and will also briefly mention the solid-state devices.

Reflex klystrons, which are widely used for microwave spectroscopy, typically have outputs ranging from milliwatts to hundreds of milliwatts. They are essentially valves in which a beam of electrons is accelerated through a cavity and reflected back through the cavity by a high-voltage electrode. Electromagnetic oscillations build up in the cavity and modulate the electron beam, causing the electrons to bunch together. The bunching occurs because fast electrons are retarded by the resonant field, whereas the slower ones are accelerated. If the dimensions and voltages are propitious, the oscillations of the bunches can deliver power to the cavity, which can be coupled into the wave guide. Electrons generated at the cathode are accelerated toward the gap by the beam voltage. In the gap, the electron beam is alternately accelerated and decelerated by a radio-frequency (rf) electronic field. In the drift space between the cavity and the reflector, the fast electrons drift from the slow ones and produce bunches of electrons. The reflector turns the beam around and returns it to the rf gap. In the gap the electrons become debunched and give up energy to the cavity, which is coupled to a transmission line. The rf beam voltage and reflector voltages must be adjusted for proper phase. The resonant frequency can be tuned over a wide range by mechanical deformation of the cavity and simultaneous adjustment of the reflector potential to maintain resonant oscillations. The frequency can also be tuned over a small range by variation of the reflector potential alone. The electronic tuning range may be only a few tens of megahertz. For short-term stability (1 part in 10^7) careful temperature control of the klystron is adequate and phase locking is unnecessary (48). Operation of reflex klystrons is discussed by Poole (80), and a more theoretical presentation is given by Soohо (93).

Backward-wave oscillators are related to klystrons, but have the advantage that they can be tuned electronically over a frequency range of several gigahertz. In these devices, the beam of electrons passes through a series of consecutive holes in a spiral cavity in which the radiation is generated. The frequency generated is not as stable as that from a klystron and generally requires phase locking.

Microwave triodes (space-charge-limited tubes) are employed as generators in the S-band (1–4 GHz). These tubes typically supply power outputs ranging from 1 W at 1 GHz to 0.1 W at 4 GHz. Both klystrons and triodes are available in this frequency region. The advantage of a triode source is that it needs a lower supply voltage and tends to be more stable in frequency. It is difficult to achieve wide band tuning with triode sources.

The principal application of magnetrons is to systems in which the generation of very high microwave power outputs is needed, particularly if the power is pulsed rather than continuous. They are available over a large frequency range and are capable of furnishing up to 1 MW of peak power output during the duration of the pulse. In a typical case the pulse is 0.1 or 1 μsec long, with a 10^{-3} duty cycle. Theoretically, the sensitivity of a spectrometer increases with power. Factors such as molecular saturation and the inability of detectors to handle high power usually prevent realization of the advantages deriving from the higher power available from magnetrons.

Solid-state devices generating microwaves include IMPATT (*imp*act ionization *a*valanche-*t*riggered *t*ransit) devices and Gunn-effect oscillators (101). These devices require only low-voltage power supplies and lend themselves to monitoring applications. Both devices are broad banded.

B. SAMPLE CELLS

Two fundamental facts are relevant to sample-cell design: (1) Absorption coefficients are small (values of $\alpha = \Delta P / l P_0 = 10^{-4} - 10^{-9}$ cm^{-1} are typical, and samples with 10^{-12} cm^{-1} absorption coefficients have been observed in special cases); and (2) the line width is collision broadened above ~10 mtorr. Therefore, one usually uses a partly evacuated cell, which is either a length of wave guide, a Stark cell, a parallel-plate cell, or a free-space cell. Alternatively, one can use a high-Q [Q (quality factor) = $2\pi \cdot$ (average energy stored)/(energy lost per cycle)] cavity. The wave guide cell is used for high-frequency ($\geqslant$50 GHz) work with some type of source modulation (see below). A Stark cell has an insulated septum running the length (~3 m) of the wave guide cell. A square-wave electric field (<100 kHz) is applied to the septum to Stark-modulate the molecular energy levels. Wave guide and Stark cells are usually made of copper. Parallel-plate cells do not suffer from low-frequency cutoffs at specific wavelengths and are useful for special applications such as high-temperature measurements, since the metallic plates are simply mounted within larger glass vacuum envelopes. Free-space cells are coupled to the source and detector by microwave horns and are especially useful for samples undergoing metal-catalyzed decomposition. All the other cells are constructed of metal. Cavity cells can provide large effective path lengths even though they are only about 1 cm across, but are inherently narrow-banded devices. For monitoring specific frequencies, cavity-type cells are the most useful (76).

C. DETECTION SYSTEMS

Since absorption of power in the microwave region is small (typically 0.1% or less), detection of the transmitted microwave is usually coupled with a modulation scheme to enhance the signals. We will first discuss detectors, and then modulation schemes.

The most widely used detectors are point-contact diodes and Shottky barrier diodes. We will also briefly mention bolometers, acoustical detectors, and low-noise microwave amplifiers. Noise in detectors can be characterized as thermal or shot noise, which is essentially frequency independent, and "flicker" noise, which depends on the value of $1/f$, where f is the modulating frequency. Certain types of source modulation (107) can be carried out at megahertz frequencies, minimizing flicker noise. The more common Stark, Zeeman, or source modulation is carried out below 100 kHz, however. Bolometers are sometimes used as detectors and are essentially low-temperature thermometers (79). Krupnov (61) has used acoustical detection up to

1.2 THz (1 THz = 10^{12} Hz). Borchert (8) and Gillies and collaborators (32) have developed acoustical detectors for the 10 to 40-GHz region.

Diode detection is usually coupled with Stark modulation (44), in which the molecular energy levels are shifted by an electric field applied as a square wave. Phase-sensitive detection is used at the Stark modulating frequency. Frequencies can be measured accurately, and the frequencies of the Stark lobes are used to determine dipole moments.

Zeeman modulation (26) can be employed similarly to Stark modulation, but shifts are generally smaller unless the compound is paramagnetic. With Zeeman modulation, frequency shifts can be used to determine magnetic moments.

Ekkers and Flygare (21) reported construction of a Fourier-transform microwave spectrometer that utilizes a superheterodyne detection system and is capable of a band width of 50 MHz around the carrier frequency. Since the detection of the signal takes place in the absence of any microwave power from the master oscillator, the need for a balanced bridge is eliminated and the spectrometer signal-to-noise ratio is unaffected by any source noise. Fast-switching microwave diodes are used to generate the high-power microwave pulses. The spectrometer signal-to-noise ratio is dramatically increased over those of conventional spectrometers, and resolution is increased due to the total absence of any power and modulation broadening. Studies of the van der Waals molecules KrHCl, ArHCl, KrHBr, and ArHBr using this device were reported by Flygare and collaborators (3,52).

IV. QUALITATIVE ANALYSIS

The combination of high-resolution spectrometers and narrow line widths in microwave spectroscopy allows greater discrimination among gaseous compounds than is provided by almost any other spectroscopic method. Ultra-high-resolution mass spectroscopy may rival microwave spectroscopy in resolution, but with the former technique, for a single peak, all isomers are degenerate. In microwave spectroscopy all isomers, as well as several vibrationally excited species, are often resolved. Of course, microwave spectroscopy requires gaseous, polar samples, but millitorr of pressure and dipole moments as small as that of propane (0.083 D) are sufficient to yield useful spectra. Some aspects of microwave spectroscopy's discriminatory ability, spectrometer sensitivity, sample requirements, and speed of analysis; examples of analyses; and specialized methods such as low-resolution techniques and applications to ions, unstable species, and molecules of astrophysical interest will be discussed in this section. Quantitative aspects will be discussed in Section V.

A. HIGH-RESOLUTION MICROWAVE SPECTROSCOPY

1. Discrimination

Rotational transition frequencies can be measured with high-resolution microwave spectroscopy with kilohertz accuracy, often to seven or eight

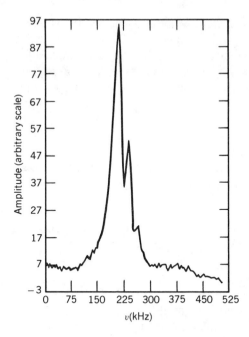

Fig. 11.1. Fourier-transformed spectrum of 0.2-mtorr CD_2O averaged over 10^4 pulses. The markers indicate frequency difference from the 6083 MHz carrier frequency. (From reference 21.)

significant figures. Resolution is limited by the natural line width of the transition and not by the source radiation, which is essentially monochromatic. A molecule's rotational spectrum is a function of its unique rotational constants and observation of this spectrum allows an unambiguous identification to be made. Also, isotopic species, vibrationally excited species, and conformational isomers of a compound usually have resolvable microwave spectra.

The sensitivity of a microwave analysis to the amount of substance depends on the sensitivity of the spectrometer and molecular characteristics. Spectrometer sensitivity may be determined by calibration with a molecule with lines of known absorption coefficients. Spectrometers can generally observe transitions with absorption coefficients as small as 10^{-9} cm^{-1}, or even 10^{-12} cm^{-1} in special circumstances. Computer averaging and Fourier-transform techniques can improve spectral signal-to-noise ratios. Frequency resolution may also be improved, since it is possible to use lower sample pressures. An example of the sensitivity obtained with Fourier-transform methods is shown in Fig. 11.1, which shows the $1_{11}-1_{10}$ transition of CD_2O. The pressure was 0.2 mtorr and the temperature $-77°C$. The spectrum is the Fourier transform of the sum of 10^4 transients. The splitting of the line is due to nuclear quadrupolar interactions.

Detection sensitivity depends also on molecular characteristics. The strength of a microwave transition is affected by (see equation 11) the following: (1) the square of the transition's dipole moment matrix element (this is a function of the quantum numbers involved and the dipole moment of the molecule); (2) the square of the transition frequency; (3) a line shape term, $\Delta \nu / [(\nu - \nu_0)^2 + \Delta \nu^2]$, where $\Delta \nu$ is the natural line width, ν_0 is its center, and ν is the frequency of observation; and (4) the factor Nf/T, where N is the number density of absorbing molecules, f is the fraction of molecules in the lower energy state of the transition, and T is the absolute temperature.

Values of the peak absorption coefficient may range as high as 10^{-5}/cm. An absorption coefficient of 10^{-6}/cm corresponds to a strong line in a microwave spectrum.

For a molecule with known rotational constants and dipole moments, line strengths can be calculated with available computer programs. Knowledge of the spectrometer sensitivity allows calculation of the minimum pressure at which the compound is detectable. In general, small, rigid, polar molecules can be detected at lower concentrations than large, relatively nonpolar molecules.

Jones and Beers (50) determined by computer simulation the extent to which spectral lines in mixtures of gases interfere with each other. They considered a mixture of 33 gases of moderate complexity by combining lists of measured line frequencies. In this mixture, measurement of a single line with 0.2-MHz accuracy resulted in overlaps in 45% of the cases. Measurement of two lines with 0.2-MHz accuracy resulted in overlap in only 2% of the cases. This study used published tables of lines, but for most of the molecules included the actual spectra contain more lines. In any case, it is necessary to measure only a few lines, perhaps five, to identify unambiguously a compound with a known spectrum, even in a complex mixture. For a molecule with strong absorption coefficients that is present in high concentration, relatively crude measurements of a few strong lines establish its identity. This requires only a few minutes of experimental time. For species with weak spectra in low concentration, analysis may require much slower spectral scanning rates and signal averaging.

2. Sample Requirements

Microwave spectroscopy is limited to gas-phase samples and spectra are usually obtained at 1 to 100-mtorr pressure. Therefore, many solids and liquids with normal boiling points below 250°C have sufficient vapor pressure to produce useful microwave spectra at room temperature. For a typical wave guide absorption cell of 2-liter volume, about 1 μmol of sample is required. For cavity cells, ~10 nmol is sufficient. Microwave spectroscopy is nondestructive, and the sample can be recovered, although adsorption on the cell walls and cell-wall degassing may make this impractical. For studies of reaction intermediates, ions, radicals and other unstable species, it is necessary to generate a sufficient concentration in a flow system. In these cases much larger samples are required. An example is the study of HCO^+ by Anderson et al. (1).

The coefficient of absorption for a rotational transition depends on the square of the permanent electric dipole moment. Molecules with dipole moments of >0.1 D generally produce useful microwave spectra. Hirota's group has observed pure rotational transitions of $^{16}O^{12}C^{18}O$, which has a dipole moment of ~7 × 10^{-4} D (22).

3. Examples

a. OZONOLYSIS

The mechanism of the reaction of ozone with unsaturated compounds has fascinated chemists for a long time (16). Several microwave laboratories are actively investigating this complex reaction and have made major contributions to our understanding of it. Gillies, Kuczkowski, and their collaborators (31,33) have studied ozonolyses of several simple olefins and have discovered new products as well as refined the mechanism. Günthard, Bauder, and their collaborators have made significant contributions to our understanding of the mechanism of ozonolysis of ethylene. In addition to microwave studies, they have carried out infrared studies of matrix-isolated products (62). In studying the ozonolysis of ethylene at low temperatures, Suenram and Lovas (98) discovered and characterized the unusual three-membered ring compound dioxirane ($\overline{CH_2-O-O}$).

b. PYROLYSIS

Microwave analysis has been used extensively in pyrolysis studies. Saito has identified and characterized the pyrolysis products of ethylene episulfoxide (81) and N-sulfinylaniline (84). The microwave group at the National Bureau of Standards has characterized pyrolyses with microwave spectroscopy, notably the pyrolyses of sulfur compounds (49) and of amines (78). They have been especially interested in generating possible interstellar molecules by pyrolysis, and have also been successful in elucidating pyrolysis pathways. Bak and collaborators in Copenhagen have studied the pyrolyses of many organic compounds, among them NCCNO (2).

Hrubesh and collaborators have greatly advanced the microwave technique as a tool for analyzing complex mixtures; their analysis of toxic components in oil-shale retort residues is an example (see pp. 139 ff. of reference 124). They searched for 120 compounds within 3 min. Quantitative analysis required longer times. The major drawback of the technique is that it is less sensitive to higher molecular weight species, due to their low volatilities.

4. Catalogs of Spectral Lines

Microwave spectral lines have been tabulated in several sources. The National Bureau of Standards' Monograph 70 (77) is a large collection. White (128) has also compiled an extensive spectral catalog. The Lawrence Berkeley Laboratory (56,57) has tabulated lines relevant to atmospheric pollutants. Extensive listings on molecules of astrophysical interest have been published recently (4–7,19,30,38,47,55,63,64,67–71,105,106,129,132). The group at the Jet Propulsion Laboratory (51) has compiled an extensive spectral catalog that can be obtained on magnetic tape.

B. LOW-RESOLUTION MICROWAVE SPECTROSCOPY

The microwave spectra of nearly symmetric rotors are very similar to those of strictly symmetric rotors (see equations 5 and 6). If such spectra are scanned under low-resolution conditions, that is, a fast frequency sweep coupled with a long detector time constant, the nearly degenerate K components of each $J + 1 \leftarrow J$ transition form bands with frequencies given by (96)

$$\nu \approx (B + C)(J + 1) \tag{6}$$

where $(B + C)$ very nearly equals $(B_0 + C_0)$ for the molecule's ground state (24).

Large and heavy molecules have so many low-energy rotational states that the intensity of a transition from a particular state is weak, since the fraction of molecules in the initial state is small. Under low-resolution conditions, however, the cumulative intensity of many nearby lines often compensates for the intensity problem. The experimental requirement is that one must be able to sweep over a pile-up of lines, typically 100 MHz wide, with a base line visible on either side of the pile-ups. Such broad-banded sweeps are extremely difficult with klystron sources, but simple with backward-wave oscillator sources. Luckily, many common molecules are nearly symmetric rotors and have the required spectral characteristics. The authors have observed such spectra from ethers (110), esters (111–118), thiolesters (89–91), aldehydes and acid halides (103,104), alkyl nitrites (120,121), and a variety of substituted benzenes (12). Other groups have observed microwave band spectra in several other classes of compounds. The low-resolution microwave spectrum of *tert*-butyl formate is shown in Figure 11.2. This spectrum demonstrates that only one conformer is present (a single series of equally spaced bands) and that the *tert*-butyl group is oriented *syn* to the carbonyl group [from a comparison of the observed $(B + C)$ value, 3950 MHz, with calculated values, 3950 (*syn*) and 3230 MHz (*anti*)].

A molecule will have a low-resolution microwave band spectrum if it is volatile, polar, and nearly symmetric. Extra band series have been observed in the band spectra of all the classes of compounds listed above except ethers. These bands are due to unresolved torsionally excited species lying above a low internal rotation barrier (120). Thus, low-resolution microwave spectroscopy can be used to characterize low internal rotation barriers involving asymmetric groups.

C. UNSTABLE SPECIES AND IONS

Many species with limited stability have been studied by microwave spectroscopy. Rare and reactive compounds have been studied extensively by

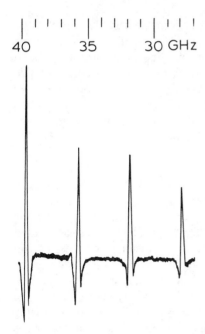

Fig. 11.2. Low-resolution microwave R-band (26.5–40 GHz) spectrum of *tert*-butyl formate. (From reference 118.)

Saito (82), Kroto and his collaborators (54), and Johnson and collaborators at the National Bureau of Standards (NBS) (46). The Berlin group has extensively studied diatomic molecules of low volatility (108), and the NBS group has carried out beautiful studies of glycine (98). Hydrogen-bonded dimers have been extensively studied by Millen and collaborators (29) and by Muenter and collaborators (20) using molecular beams. Several groups have studied van der Waals molecules prepared in molecular beams (18,52,95).

Transient species having lifetimes of the order of milliseconds can be detected by present microwave-spectroscopic techniques. Observations of transient species often requires spectrometers modified to incorporate, for example, external or internal discharges, flow systems, and Zeeman or source instead of Stark modulation. Relatively long lived ($\sim$100 msec) transient species can be produced externally by a discharge, pyrolysis, or chemical reaction and then flowed through the sample cell. For example, HO_2 radicals can be produced by reacting allyl alcohol at 0.02 torr with the products of a microwave discharge in 0.02–0.04 torr oxygen (82). The radicals then flow through the cell. Use of free-space or parallel-plate cells can minimize decomposition. For standard metal Stark cells, coatings such as Corning K can minimize reactions catalyzed by the metal walls. In all of the microwave studies of free radicals reported, radical concentrations have been sufficiently large that signal averaging was unnecessasry (see, e.g., reference 10).

Strong interactions between electron spin and molecular rotation generally complicate the microwave spectra of free radicals. Hyperfine interactions between electron and nuclear spins become extremely complicated for high-J transitions in asymmetric radicals. The magnitude of a perturbation depends on the three spin–rotation coupling constants, the Fermi contact term, and the magnetic dipole–dipole coupling constant. The size of the Fermi contact term yields information about the type of orbital occupied by the unpaired electron. Only radicals having four or fewer atoms have been studied by microwave spectroscopy. Low-resolution microwave studies of larger radicals of more general chemical interest would probably be fruitful.

Ions are much more difficult to detect. They are usually generated by a discharge within the sample cell, which precludes the use of Stark modulation for detection, owing to the charged plasma. Signal averaging is essential and the plasma containing the sample seriously attenuates the probing microwave radiation and generates noise. These problems are exacerbated by increased free-electron concentration and the closeness of the electron collision frequency in the plasma to the microwave radiation frequency. The problems are minimized at low pressure and low degrees of ionization. Woods and his collaborators (1,85,134) have successfully studied ions with microwaves by maintaining a glow discharge within the sample cell and using detection by Zeeman modulation. With their technique, signal averaging is necessary and the entire large cell is cooled with liquid nitrogen to increase the signal-to-noise ratio.

Application of Fourier-transform microwave spectroscopy (21) should greatly expand the number of experimentally accessible transient species. This technique has been applied very successfully in studies of van der Waals molecules using a Fabry–Perot cavity Fourier-transform spectrometer with a pulsed supersonic nozzle as the molecular source (52).

D. RADIOASTRONOMY

Rotational spectroscopy is a powerful tool for studying molecules in interstellar space (28). In the last 20 years, more than 50 molecules, ranging in complexity from diatomics such as the hydroxyl radical to exotic 11-atom molecules such as cyanotetra-acetylene (HC_9N), have been identified in dense interstellar clouds using radioastronomy. It is convenient to divide interstellar space into two regions: clouds, consisting mainly of neutral gases, and intercloud regions of partly ionized gases. Intercloud regions have densities of <0.1 particle/cm^3 and are hot ($\sim 10^3$–10^5 K). Clouds, which occupy <10% of the interstellar volume, are dense (10–10^6 particles/cm^3) and cool ($\lesssim$100 K). Low-density clouds ($\lesssim$10^2 particles/cm^3) attenuate starlight and can be observed optically. Denser clouds are optically opaque and can be observed only with radiotelescopes. In dense clouds with higher collisional frequencies, most of the elements are bound into molecules. Since interstellar-cloud temperatures are typically <100 K, their thermal energies are comparable to rotational energy level spacings. Molecules are usually detected by observation of rotational emission spectra in the ground electronic and vibrational states. The dominant mechanism of microwave emission in a dense star cloud is through collisions with H_2 molecules.

Radioastronomy is identifying many interstellar molecules, their abundances, and the isotopic compositions, densities, temperatures, and velocities of interstellar clouds. Watson (127) and Gammon (28) have discussed how this information can be used to characterize mechanisms of interstellar molecular formation.

Usually, the laboratory microwave spectrum is determined before the astronomical search is made. Identification rests on agreement between laboratory and interstellar transition frequencies, which must be corrected for Doppler shifts due to interstellar cloud velocities. Lines are detected with radiotelescopes, and considerable signal averaging is usually necessary. For example, detection of the 6_{06}–5_{15} transition of dimethyl ether with a 36-ft radiotelescope required 3 days of data accumulation before a signal-to-noise ratio of 2 was achieved (92). The group at the NBS has organized the compilation of microwave data on molecules of astrophysical interest. These compilations include data on H_2CO, $HCONH_2$, and H_2CS (47); CH_2NH (55); CH_3OH (64); H_2S (38); H_2O (19); OCS and HCN (71); CO, SiO, and CS (68); SO (105); diatomics (70); CH_3CHO (6); HNC (132); SiS (106); OH (7); NC_3H (63); triatomics (67); $NCCHCH_2$ (30); CH_3CCH (5); $HCOOCH_3$ (4); $(CH_3)_2O$ (69); HCOOH (129); and CH_3CN (9). Another extensive listing of relevant data is the *JPL Submillimeter, Millimeter and Microwave Spectral Line Catalogue* (51).

Interstellar radicals and ions have also been detected. For many of these, laboratory spectra were not known and initial assignments were based on spectra calculated from assumed molecular geometries. The radicals CN, C_3N, C_2H, and C_4H and the ions HCO^+ and N_2H^+ were identified in this way. The abundance of neutral nonpolar interstellar molecules unobservable by radioastronomy may be inferred from observations of their protonated ions (39). For example, observed N_2H^+ column densities have been used to obtain molecular nitrogen abundances in dense interstellar clouds.

Molecular isotopic fractionation in interstellar clouds can also be studied by radioastronomy. A zero-point energy difference among isotopic forms of a molecule can cause significant isotopic fractionation if the isotopic exchange reactions approach equilibrium. For example, observations of the DCN/HCN (reference 45) and DCO^+/HCO^+ (reference 42) concentration ratios indicate enhancements of 10^2 and 10^4, respectively, over the D/H ratio in the interstellar medium. Wilson et al. (131) observed carbon isotope fractionation in interstellar formaldehyde. In 10 molecular clouds the ratio between [^{12}C]- and [^{13}C]formaldehyde was found to be 46 ± 18, somewhat lower than that in our solar system (131).

V. QUANTITATIVE ANALYSIS

In quantitative analysis, spectral intensities are converted into concentrations. Techniques for quantitative analysis can be classified according to the microwave power level used in the experiment. Low-power methods are performed at radiation intensities too low to disturb the Boltzmann distribution of energy level populations significantly. Measurements can also be performed at higher power levels at which the populations of the two states involved in the transition become equal (saturated conditions). This method eliminates effects due to variable line widths associated with different sample compositions and pressures.

A. LOW-POWER METHODS

Low-power (typically 1 mW radiation power) methods utilize either peak-intensity or line-area measurements. The former are easier and quicker, but probably less accurate. Peak-intensity measurements are most useful for spectral lines of similar shape. The integrated area is independent of these factors.

1. Peak Intensities

Peak intensities are generally obtained by slowly scanning over a few hundred kilohertz around the absorption maximum. About 10 min/line is required. A base-line scan with the sample pumped out is useful. The selection of appropriate lines is easy for small molecules and simple mixtures, but more difficult for large and asymmetric molecules or for complex mixtures, due to the large number of spectral lines which often overlap.

Under optimum conditions, peak-intensity measurements can yield concentrations to about 2% accuracy. Esbitt and Wilson (23) have discussed factors contributing to uncertainties in peak-intensity measurements. Many potential sources of error have been minimized in recently designed microwave spectrometers. Wave guide reflections, which effectively alter the cell path length, can be a serious source of error at certain frequencies. Detector reflections can be reduced to 0.5% by insertion of a ferrite isolator before the crystal detector. The detector's response may not be linear. A possible source of error dependent on the sample is interference from Stark lobes and nearby lines. The Stark electric field should be sufficient to shift all Stark components away from the zero-field absorption line. Microwave power levels should be low enough that the state's population will not be perturbed significantly by the radiation field. Since the power level for saturation differs for different transitions and different components of a mixture, intensity ratios should be measured at a series of power levels to insure that saturation does not affect the results. Perhaps the greatest source of uncertainty is the base line. Drifting irregularities, recorder pickup at the modulation frequency, and high noise levels can all present problems.

Changes in sample composition may occur during the course of the measurements. Degassing from cell walls may dilute the sample, the component of interest may adsorb onto the walls, or the metal walls may catalyze decomposition.

Conversion of intensities to concentrations requires calibration curves or knowledge of the molecules rotational constants dipole moments along the principal axis and assignment of the levels of the transition. The peak absorption coefficient, α_{max}, is directly related to concentration:

$$\alpha_{max} = \beta N \tag{12}$$

where N is the concentration and β is everything else from equation 11. If the rotational constants and dipole-moment components are known, β can be calculated for all the possible rotational transitions. Then the concentrations of two species can be compared according to the formula

$$\frac{N_1}{N_2} = \frac{\alpha_{max1}\beta_2}{\alpha_{max2}\beta_1} \tag{13}$$

where α_{max1} and α_{max2} are the observed line strengths for species 1 and 2, respectively, and β_1 and β_2 are the respective calculated line strengths. Equation 13 assumes the line widths of the two transitions to be equal. Alternatively, calibration curves can be determined from mixtures of known composition.

Use of peak intensities for quantitative analysis may be troublesome, because the line width of a given transition may depend on sample composition. The natural line width is determined by the molecule's rotational relaxation time. Relaxation occurs primarily through collisions and depends on collisional cross sections. In general, molecules with large dipole moments have large collisional cross sections. For a sample containing high concentrations of polar molecules, rotational lifetimes will be shorter and lines broader. An extreme example, that of methanol lines broadened by water vapor, is shown in Fig. 11.3.

2. Integrated Absorption Intensities

Integrated-intensity methods require intensity measurements across the full width of the spectral line. This method requires more time, but is more accurate than the peak-intensity method, because collisional effects are not important. These measurements suffer from some of the same sources of uncertainty as do peak-height measurements: base-line inaccuracy, overlap from neighboring transitions, and incomplete Stark modulation. No assumptions regarding line width are necessary.

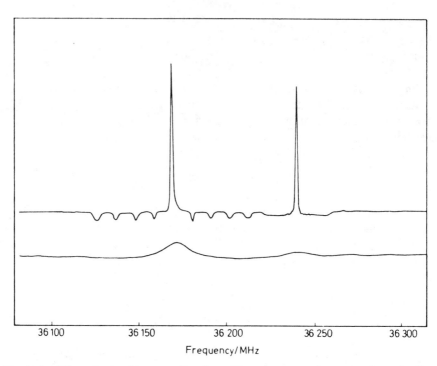

Fig. 11.3.   Effect of polar diluents on line shape. The upper curve shows two absorptive lines of gaseous methanol. The lower curve shows the spectrum of the same sample with a 49-fold excess of water vapor. (Reproduced with permission of Cambridge Scientific Instruments, Ltd.)

B.   SATURATION METHODS

The advantage of the saturation method is that intensity data can be obtained under maximum signal conditions. The method was developed by Harrington (36,37). It allows intensity measurements to be factored into a concentration factor and a line-width (relaxation time) factor. Harrington defines

$$\Gamma = \eta\phi \tag{14}$$

where

$$\eta = \frac{\alpha_0}{K^{1/2}} \tag{15}$$

α_0 is the peak absorption coefficient, K is the power saturation coefficient, and ϕ is the dimensionless power density (36). The factor η has the advantage that it depends on the concentration of the absorbing species but not the line width, and hence not on the sample composition. Γ is proportional to the signal voltage if the detector crystal current remains constant. Then Γ_{max}, the maximum signal amplitude, can be measured, along with the power level ϕ_{max} that provides the maximum signal. One can then obtain η from

$$\Gamma_{max} = \eta\phi_{max} \qquad (16)$$

The Hewlett-Packard 8460A system contains a microwave bridge for measurements at high power levels. A bridge configuration is necessary so that the power may be attenuated before hitting the detector. Power-saturation results are adversely affected by poor line shapes due to incompletely resolved fine structure, incomplete Stark modulation, and interference from other lines.

C. EXAMPLES

Mathur and Harmony (72) characterized the kinetics of an isomerization reaction by analyzing products with microwave spectroscopy. Exo-2,3-dideuterobicyclo[2.1.0]pentane thermally isomerizes into the endo isomer, and endo-2-methylbicyclo[2.1.0]pentane thermally isomerizes to the exo form. The reactions

were carried out by heating 100 ml gas at 3 torr in a variable-temperature bath. Reactions were run for different times (on the order of hours) at several temperatures (~200°C), and then quenched at room temperature. Relative intensity measurements were made using a Hewlett-Packard 8460A microwave spectrometer. For the dideutero sample, three analogous transitions from each of the endo and exo isomers were used in the analysis. Peak intensities were measured carefully at low microwave power. Reproducibility of the relative intensity measurements was 2–3%. Calibration curves for the microwave intensities were determined by analyzing one set of experiments with microwave spectroscopy and vapor-phase chromatography. For the dideutero compound, the microwave analyses over a range of temperatures yielded an activation energy of 38.5 ± 1.4 kcal/mol. For the methyl compound at 206.6°C, the rate constant determined using microwave spectroscopy was 5.07×10^4 sec^{-1}, compared with 5.28×10^4 sec^{-1} from the vapor-phase chromatography results. For these compounds, the spectral signal-to-noise ratio was quite poor, due to the small electric dipole moments (~0.03 D).

The relative abundances of deuterated propenes in hydrogen exchange reactions was determined by microwave spectroscopy by Morino and Hirota (75) and Scharpen et al. (87). Further studies on this type of reaction utilizing microwave analysis include those of Kondo et al. (58,59), Saito et al. (83), and Morino (74).

Microwave spectroscopy should be particularly useful in the analysis of products of laser-induced gaseous reactions, for which mass spectroscopy may yield ambiguous results. Isotopic enrichment can be determined by mass spectroscopy, but the site of enrichment may not be. Microwave spectroscopy can easily yield both the extent and the site of enrichment.

Curl et al. (17) developed a technique for measuring absolute intensities of rotational transitions by calibration against a reference material. They used the intensities to determine the free energies of several compounds. The errors from the microwave method are no larger than those of values obtained calorimetrically. The microwave method requires only a few days of experimental effort to determine room-temperature free energies. Determination of enthalpy and entropy values requires that these measurements be repeated over a range of temperatures.

A common use of microwave relative intensity measurements is for the determination of relative energies of conformational isomers. Consider, for example, the two conformers of ethyl cyanoformate, *syn-anti* (SA) and *syn-gauche* (SG) ethyl cyanoformate (99):

   syn-anti (SA)   *syn-gauche* (SG)

Suenram and coworkers calculated reference intensities according to equation 11. The line widths for all the transitions used were assumed equal. Observed intensities were then compared with the reference intensities, and the difference in energy, $\Delta E = E(SA) - E(SG)$, was determined according to

$$\Delta E = -RT \ln \left[\frac{\dfrac{g(SA)}{g(SG)} \cdot \dfrac{I_{obs}(SG)}{I_{ref}(SG)}}{\dfrac{I_{obs}(SA)}{I_{ref}(SA)}} \right] \tag{17}$$

TABLE 11.II

Energy Difference Between *Syn-anti* (SA) and *Syn-gauche* (SG) Conformers of Ethyl Cyanoformate

Transition	Observed[a] intensity (24°C)/ reference[b] intensity (24°C)
SA conformer	
$11_{0\ 11} \rightarrow 12_{0\ 12}$	1.258
$11_{3\ 8} \rightarrow 12_{3\ 9}$	1.317
$12_{1\ 12} \rightarrow 13_{1\ 13}$	1.240
$12_{0\ 12} \rightarrow 13_{0\ 13}$	1.344
$12_{3\ 9} \rightarrow 13_{3\ 10}$	1.332
Average:	1.300 ± 0.046
SG conformer	
$11_{3\ 8} \rightarrow 12_{3\ 9}$	2.081
$11_{2\ 10} \rightarrow 12_{2\ 11}$	1.985
$11_{1\ 10} \rightarrow 12_{1\ 11}$	1.979
$12_{2\ 10} \rightarrow 13_{2\ 11}$	1.940
$12_{1\ 11} \rightarrow 13_{1\ 12}$	1.975
Average:	1.992 ± 0.053

[a] Observed intensities corrected to -65 dB for both forms.

[b] Calculated molecular parameters of reference 99.

where the g are degeneracy factors ($g(\text{SA}) = 1$, $g(\text{SG}) = 2$) and the I are intensities. It is convenient to use this method because only intensity ratios are required and the units do not matter if all the observed intensities are referenced to the same gain setting on the spectrometer. The results are summarized in Table 11.II. ΔE is 55 ± 27 cm^{-1}. The error is derived nearly equally from the intensity errors given in Table 11.II and errors in the calculated reference intensities, since the dipole-moment components have errors of up to 0.07 D.

STARK SWEEP MW CAVITY SPECTROMETER

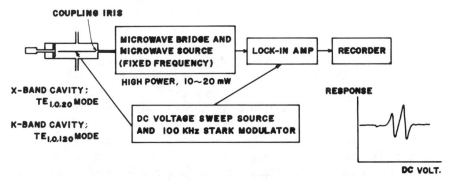

Fig. 11.4.   Stark sweep microwave cavity spectrometer. (From reference 123.)

SENSITIVITY OF THE PRESENT SPECTROMETER

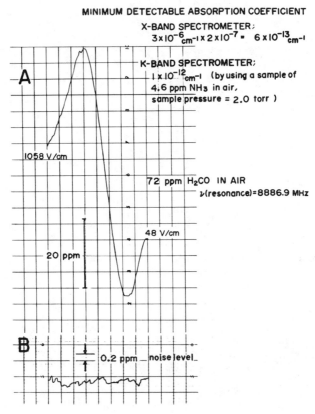

MINIMUM DETECTABLE ABSORPTION COEFFICIENT

X-BAND SPECTROMETER:
3×10^{-6} cm^{-1} $\times 2 \times 10^{-7}$ = 6×10^{-13} cm^{-1}

K-BAND SPECTROMETER:
1×10^{-12} cm^{-1} (by using a sample of 4.6 ppm NH$_3$ in air, sample pressure = 2.0 torr)

1058 V/cm

72 ppm H$_2$CO IN AIR
ν(resonance)=8886.9 MHz

48 V/cm

20 ppm

A

B

0.2 ppm — noise level

Fig. 11.5. Illustration of the sensitivity of the spectrometer shown in Fig. 11.4. See text for discussion. (From reference 122.)

If dipole moments have not been measured, conformer relative energies can be determined from the temperature dependence of spectral intensities (119). This technique works best if the conformers have nearly equal energies, because temperature uncertainties make a minor contribution to the total uncertainty.

A microwave cavity spectrometer of the Stark–dc-voltage type designed by Uehara and collaborators (123) for monitoring pollutants is shown in Fig. 11.4. The spectrometer resonates near 9 and 24 GHz. Inside the rectangular cavity a Stark electrode is placed perpendicular to the microwave electric field. A dc sweep voltage, modulated by a 100-kHz sine-wave voltage, is applied to the Stark electrode. The microwave and amplifier systems come from a commercial electron spin resonance apparatus. High sensitivity is important for monitoring pollutants. The spectrum shown in Fig. 11.5A was obtained from 72 ppm of formaldehyde in air. Figure 11.5B shows that the measurement is sensitive to about 0.2 ppm of formaldehyde. Since the absorption coefficient for the transition is 3×10^{-6} cm^{-1}, the sensitivity of this spectrometer is about 6×10^{-13}. Using the same spectrometer, Tanimoto and Uehara (102) detected acrolein in automobile exhaust. Preconcentration was required to detect 5 ppm of acrolein.

Analysis of gaseous mixtures by microwave spectroscopy has been investigated by several groups. Funkhouser et al. (27) examined mixtures of acetone, methanol, and freon in nitrogen. Analyses of many gaseous mixtures have been carried out at the Lawrence Livermore Laboratory by Hrubesh and collaborators (references 124, pp. 137 ff., and 43). Crable (15) has also studied gaseous mixtures; his studies have focused on pollutants in air. These workers conclude that the microwave method is generally applicable if parts-per-million sensitivity is adequate, but parts-per-billion sensitivity can be achieved only in special cases. Hrubesh (reference 124, pp. 154 ff.) has demonstrated that very inexpensive yet dependable microwave monitors for specific gases can be built.

Microwave technology has also been applied in meteorology. For example, Hogg and Guiraud (41) have studied absorption of microwave radiation by water vapor in the lower atmosphere, and Waters et al. (126) have measured carbon monoxide levels in the mesosphere by measuring the absorption of solar microwave radiation.

REFERENCES

1. Anderson, T. G., C. S. Gudeman, T. A. Dixon, and R. C. Woods, *J. Chem. Phys.* **72,** 1332 (1980).
2. Bak, B., H. S. Svanholt, and A. Holm, *Acta Chem. Scand.,* **A33,** 71 (1979).
3. Balle, T. J., E. J. Campbell, M. R. Keenan, and W. H. Flygare, *J. Chem. Phys.,* **72,** 922 (1980).
4. Bauder, A., *J. Phys. Chem. Ref. Data,* **8,** 583 (1979).
5. Bauder, A., F. J. Lovas, and D. R. Johnson, *J. Phys. Chem. Ref. Data,* **5,** 53 (1976).
6. Bauer, A., D. Boucher, J. Burie, J. Demaison, and A. Dubrulle, *J. Phys. Chem. Ref. Data,* **8,** 537 (1979).
7. Beaudet, R. A., and R. L. Poynter, *J. Phys. Chem. Ref. Data,* **7,** 311 (1978).
8. Borchert, S. E., Ph.D. Thesis, Dept. of Chemistry, Harvard University, Cambridge, Massachusetts, 1978.
9. Boucher, D., J. Burie, A. Bauer, A. Dubrulle, and J. Demaison, *J. Phys. Chem. Ref. Data,* **9,** 659 (1980).
10. Carrington, A., *Microwave Spectroscopy of Free Radicals,* Academic, New York, 1974.
11. Chan, S. I., J. Zinn, J. Fernandez, and W. D. Gwinn, *J. Chem. Phys.,* **33,** 1643 (1960). [See also Scharpen (86).]
12. Chieffalo, A., R. K. Bohn, M. S. Farag, M. A. MacGregor, J. Radhakrishnan, and N. S. True, *J. Phys. Chem.,* **87,** 4622 (1983); *J. Phys. Chem,* **87,** 4628 (1983).
13. Cleeton, C. E., and N. H. Williams, *Phys. Rev.,* **45,** 234 (1934).
14. Cook, R. L., and G. E. Jones, "Microwave Spectrometry," in J. H. Richardson and R. V. Peterson, Eds., *Systematic Materials Analysis,* Academic, New York, 1974, Chapter 11.

15. Crable, G. F., *Study of Air Pollutants by Microwave Spectroscopy*, Report NTIS PB212-554, National Technical Information Service, Springfield, VA 1972; G. F. Crable and J. C. Eahr, *J. Chem. Phys.*, **51**, 5181 (1969).

16. Criegee, R., *Rec. Chem. Progr.*, **18**, 11 (1957).

17. Curl, R. F., T. Ikeda, R. S. Williams, S. Leavell, and L. H. Scharpen, *J. Am. Chem. Soc.*, **95**, 6182 (1973).

18. DeLeon, R. L., and J. S. Muenter, *J. Chem. Phys.*, **72**, 6020 (1980).

19. DeLucia, F. C., P. Helminger, and W. H. Kirchhoff, *J. Phys. Chem. Ref. Data*, **3**, 211 (1974).

20. Dyke, T. R., K. M. Mack, and J. S. Muenter, *J. Chem. Phys.*, **66**, 498 (1977).

21. Ekkers, J., and W. H. Flygare, *Rev. Sci. Instrum.*, **47**, 448 (1976).

22. Endo, Y., K. Yoshida, S. Saito, and E. Hirota, *J. Chem. Phys.*, **73**, 3511 (1980).

23. Esbitt, A. S., and E. B. Wilson, *Rev. Sci. Instrum.*, **34**, 901 (1963).

24. Farag, M. S., and R. K. Bohn, *J. Chem. Phys.*, **62**, 3946 (1975).

25. Flygare, W. H., "Microwave Spectroscopy," in A. Weissberger and B. W. Rossiter, Eds., *Techniques of Chemistry, Vol. I: Physical Methods of Chemistry*, Part IIIA, Wiley-Interscience, New York, 1972, Chapter V.

26. Flygare, W. H., W. Hüttner, R. L. Shoemaker, and P. D. Foster, *J. Chem. Phys.*, **50**, 1714 (1969).

27. Funkhouser, J. T., S. Armstrong, and H. W. Harrington, *Anal. Chem.*, **40**, 22A (1968).

28. Gammon, R. H., *Chem. Eng. News*, October 2, 21 (1978).

29. Georgiou, A. S., A. C. Legon, and D. J. Millen, *J. Mol. Struct.*, **69**, 69 (1980).

30. Gerry, M. C. L., K. Yamada, and G. Winnewisser, *J. Phys. Chem. Ref. Data*, **8**, 107 (1979).

31. Gillies, C. W., R. P. Lattimer, and R. L. Kuczkowski, *J. Am. Chem. Soc.*, **96**, 1536 (1974).

32. Gillies, C. W., M. Marino, and P. A. Timpano, *J. Mol. Spectrosc.*, **78**, 185 (1979).

33. Gillies, C. W., S. P. Sponseller, and R. L. Kuczkowski, *J. Phys. Chem.*, **83**, 1545(1979).

34. Gordy, W., and R. L. Cook, *Microwave Molecular Spectra*, Wiley-Interscience, New York, 1970.

35. Gordy, W., W. V. Smith, and R. F. Trambarulo, *Microwave Spectroscopoy*, Wiley, New York, 1953 (Republished by Dover, New York, 1966).

36. Harrington, H. W., *J. Chem. Phys.*, **46**, 3698 (1967).

37. Harrington, H. W., *J. Chem. Phys.*, **49**, 3023 (1968).

38. Helminger, P., F. C. DeLucia, and W. H. Kirchhoff, *J. Phys. Chem. Ref. Data*, **2**, 215 (1973).

39. Herbst, E., S. Green, P. Thaddeus, and W. Klemperer, *Astrophys. J.*, **215**, 503 (1977).

40. Herschbach, D. R., *J. Chem. Phys.*, **31**, 91 (1959).

41. Hogg, D. C., and F. O. Guiraud, *Nature*, **279**, 408 (1979).

42. Hollis, J. M., L. E. Snyder, F. J. Lovas, and D. Buhl, *Astrophys. J. Lett.*, **209**, L83 (1976).

43. Hrubesh, L. W., *Radiat. Sci.*, **8**, 167 (1973).

44. Hughes, R. H., and E. B. Wilson, *Phys. Rev.*, **71**, 562 (1947).

45. Jefferts, K. B., A. A. Penzias, and R. W. Wilson, *Astrophys. J. Lett.*, **179**, L57 (1973); A. A. Penzias, P. G. Wannier, R. W. Wilson, and R. A. Linke, *Astrophys. J.*, **211**, 108 (1977).

46. Johnson, D. R., and F. J. Lovas, *Chem. Phys. Lett.*, **15**, 65 (1972).

47. Johnson, D. R., F. J. Lovas, and W. H. Kirchhoff, *J. Phys. Chem. Ref. Data,* **1,** 1011 (1972).

48. Johnson, D. R., and R. Pearson, Jr., "Microwave Region," in D. Williams, Ed., *Methods of Experimental Physics: Spectroscopy,* Vol. 13B, Academic, New York, 1976, pp. 102–133.

49. Johnson, D. R., and F. X. Powell, *Science,* **169,** 679 (1970).

50. Jones, G. E., and E. T. Beers, *Anal. Chem.,* **43,** 656 (1971).

51. *JPL Submillimeter, Millimeter and Microwave Spectral Line Catalogue,* 2nd ed. (1981). Write H. M. Pickett or R. L. Poynter, Jet Propulsion Laboratory, 4800 Oak Grove Dr., Pasadena, CA 91109.

52. Keenan, M. R., E. J. Campbell, T. J. Balle, L. W. Buxton, T. K. Minton, P. D. Soper, and W. H. Flygare, *J. Chem. Phys.,* **72,** 3070 (1980).

53. King, G. W., R. M. Hainer, and P. C. Cross, *J. Chem. Phys.,* **11,** 27 (1943).

54. Kirby, C., H. W. Kroto, and D. R. Walton, *J. Mol. Spectrosc.,* **83,** 261 (1980).

55. Kirchhoff, W. H., D. R. Johnson, and F. J. Lovas, *J. Phys. Chem. Ref. Data,* **2,** 1 (1973).

56. Kolbe, W. F., H. Buscher, and B. Leskovar, *Microwave Absorption Coefficients of Atmospheric Pollutants and Constituents,* UCLBL Report 4467, University of California, Lawrence Berkeley Laboratory, Berkeley, California, 1976.

57. Kolbe, W. F., B. Leskovar, and H. Buscher, *Absorption Coefficients of Sulfur Dioxide Microwave Rotational Lines,* UCLBL Report 3249, University of California, Lawrence Berkeley Laboratory, Berkeley, California, 1975.

58. Kondo, T., E. Hirota, and Y. Morino, *J. Mol. Spectrosc.,* **28,** 471 (1968).

59. Kondo, T., S. Saito, and K. Tamaru, *J. Am. Chem. Soc.,* **96,** 6857 (1974).

60. Kroto, H. W., *Molecular Rotation Spectra,* Wiley-Interscience, London, 1975.

61. Krupnov, A. F., and A. V. Burenin, "New Methods in Submillimeter Microwave Spectroscopy," in K. N. Rao, Ed., *Molecular Specrtroscopy: Modern Research,* Academic, New York, 1976, pp. 93–126.

62. Kuehne, H., S. Vaccani, A. Bauder, and Hs. H. Günthard, *Chem. Phys.,* **28,** 11 (1978).

63. Lafferty, W. J., and F. J. Lovas, *J. Phys. Chem. Ref. Data,* **7,** 441 (1978).

64. Lees, R. M., F. J. Lovas, W. H. Kirchhoff, and D. R. Johnson, *J. Phys. Chem. Ref. Data,* **2,** 205 (1973).

65. D. R. Lide, Jr., "Analytical Applications of Microwave Spectroscopy," in C. N. Reilley and F. H. McLaffery, Eds., *Advances in Analytical Chemistry and Instrumentation,* Vol. 5, Wiley-Interscience, New York, 1966, pp. 235–277.

66. D. R. Lide, Jr., in "Microwave Spectroscopy," in D. Williams, Ed., *Molecular Physics, Methods of Experimental Physics,* Vol. 3, 2nd ed., Academic, New York, 1974, Chapter 11.

67. Lovas, F. J., *J. Phys. Chem. Ref. Data,* **7,** 1445 (1978).

68. Lovas, F. J., and P. H. Krupenie, *J. Phys. Chem. Ref. Data,* **3,** 245 (1974).

69. Lovas, F. J., H. Lutz, and H. Dreizler, *J. Phys. Chem. Ref. Data,* **8,** 1051 (1979).

70. Lovas, F. J., and E. Tiemann, *J. Phys. Chem. Ref. Data,* **3,** 609 (1974).

71. Maki, A. G., *J. Phys. Chem. Ref. Data,* **3,** 221 (1974).

72. Mathur, S. N., and M. D. Harmony, *Anal. Chem.,* **48,** 1509 (1976).

73. Meyer, R., and W. Caminati, *J. Mol. Spectrosc.,* **90,** 303 (1981).

74. Morino, Y., *J. Mol. Spectrosc.,* **19,** 1 (1973).

75. Morino, Y., and E. Hirota, *Bull. Chem. Soc. Jpn.,* **85,** 535 (1964).

76. Morrison, R. L., A. Maddux, and L. Hrubesh, "A Portable Microwave Spectrometer Analyzer for Chemical Contaminants in Air a Feasibility Study" Report UCRL-51945, University of California, Lawrence Livermore Laboratory, Berkeley, California, 1975.

77. National Bureau of Standards, *Microwave Spectral Tables,* NBS Monograph 70, Vol. 1, *Diatomic Molecules;* Vol. 2, *Polyatomic Molecules with Internal Rotation;* Vol. 3, *Polyatomic Molecules Without Internal Rotation;* Vol. 5, *Spectral Line Catalogue;* U.S. Dept. of Commerce, Washington, D.C., 1964–69.

78. Pearson, R. Jr., and F. J. Lovas, *J. Chem. Phys.,* **66,** 4149 (1977).

79. Penzias, A. A., and C. A. Burrus, *Annu. Rev. Astron. Astrophys.,* **11,** 51 (1973).

80. Poole, C. P. Jr., *Electrin Spin Resonance,* Wiley-Interscience, New York, 1967.

81. Saito, S., *Tetrahedron Lett.,* **48,** 4961 (1968).

82. Saito, S., *Pure Appl. Chem.,* **50,** 1239 (1978).

83. Saito, S., T. Kondo, M. Ichikawa, and K. Tamaru, *J. Phys. Chem.,* **76,** 2184 (1972).

84. Saito, S., and C. Wentrup, *Helv. Chim. Acta,* **54,** 273 (1971).

85. Saykally, R. J., T. A. Dixon, T. A. Anderson, P. G. Szanto, and R. C. Woods, *Astrophys. J.,* **205,** L101 (1976).

86. Scharpen, L. H., *J. Chem. Phys.,* **48,** 3552 (1968).

87. Scharpen, L. H., R. F. Rauskolb, and C. A. Tolman, *Anal. Chem.,* **44,** 2010 (1972).

88. Sheridan, J., "Microwave Spectroscopy," in T. S. West, Ed., *MTP International Review of Science, Physical Chemistry Series One, Analytical Chemistry—Part 1,* Vol. 12, Butterworths, London, 1972, Chapter 8.

89. Silvia, C. J., N. S. True, and R. K. Bohn, *J. Phys. Chem.,* **82,** 483 (1978).

90. Silvia, C. J., N. S. True, and R. K. Bohn, *J. Mol. Struct.,* **51,** 163 (1979).

91. Silvia, C. J., N. S. True, and R. K. Bohn, *J. Phys. Chem.,* **85,** 1132 (1981).

92. Snyder, L. E., D. Buhl, P. R. Schwartz, F. O. Clark, D. R. Johnson, F. J. Lovas, and P. T. Giguere, *Astrophys. J.,* **191,** L79 (1974).

93. Sooho, R. F., *Microwave Electronics,* Addison-Wesley, Boston, 1971.

94. Srinivasan, T. M., *Phys. Scripta.,* **7,** 84 (1973).

95. Steed, J. M., L. S. Bernstein, T. A. Dixon, K. C. Janda, and W. Klemperer, *J. Chem. Phys.,* **71,** 4189 (1979).

96. Steinmetz, W. E., *J. Am. Chem. Soc.,* **96,** 685 (1974).

97. Suenram, R. D., and F. J. Lovas, *J. Am. Chem. Soc.,* **100,** 5117 (1978).

98. Suenram, R. D., and F. J. Lovas, *J. Am. Chem. Soc.,* **102,** 7180 (1980).

99. Suenram, R. D., N. S. True, and R. K. Bohn, *J. Mol. Spectrosc.,* **69,** 435 (1978).

100. Sugden, T. M., and C. N. Kenney, *Microwave Spectroscopy of Gases,* Van Nostrand, London, 1965.

101. Sweet, A. A., *Elec. Des. News,* **21,** 40 (1974).

102. Tanimoto, M., and H. Uehara, *Env. Sci. Tech.,* **9,** 153 (1975).

103. Thomas, L. P., N. S. True, and R. K. Bohn, *J. Phys. Chem.,* **82,** 480 (1978).

104. Thomas, L. P., N. S. True, and R. K. Bohn, *J. Phys. Chem.,* **84,** 1785 (1980).

105. Tiemann, E., *J. Phys. Chem. Ref. Data,* **3,** 259 (1974).

106. Tiemann, E., *J. Phys. Chem. Ref. Data,* **5,** 1147 (1976).

107. Törring, T., *J. Mol. Spectrosc.,* **48,** 148 (1973).

108. Törring, T., and E. Tiemann, *NBS Spec. Publ. (U.S.),* **561-1,** 695 (1979).

109. Townes, C. H., and A. L. Schawlow, *Microwave Specroscopy,* McGraw-Hill, New York, 1955.

110. True, N. S., and R. K. Bohn, *J. Chem. Phys.,* **62,** 3951 (1975).

111. True, N. S., and R. K. Bohn, *J. Am. Chem. Soc.,* **98,** 1188 (1976).

112. True, N. S., and R. K. Bohn, *J. Mol. Struct.,* **36,** 173 (1977).

113. True, N. S., and R. K. Bohn, *J. Am. Chem. Soc.,* **99,** 3575 (1977).

114. True, N. S., and R. K. Bohn, *J. Phys. Chem.,* **81,** 1667 (1977).

115. True, N. S., and R. K. Bohn, *J. Phys. Chem.,* **81,** 1671 (1977).

116. True, N. S., and R. K. Bohn, *J. Phys. Chem.,* **82,** 466 (1978).

117. True, N. S., and R. K. Bohn, *J. Phys. Chem.,* **82,** 474 (1978).

118. True, N. S., and R. K. Bohn, *J. Phys. Chem.,* **82,** 478 (1978).

119. True, N. S., and R. K. Bohn, *J. Mol. Struct.,* **50,** 205 (1978).

120. True, N. S., and R. K. Bohn, *Chem. Phys. Lett.,* **60,** 332 (1979).

121. True, N. S., and R. K. Bohn, *J. Phys. Chem.,* **86,** 2327 (1982).

122. Uehara, H., and Y. Ijuuin, *Chem. Phys. Lett.,* **28,** 597 (1974).

123. Uehara, H., M. Tanimoto, and Y. Ijuuin, *Chem. Phys. Lett.,* **26,** 578 (1974).

124. Varma, R., and L. Hrubesh, *Chemical Analysis by Microwave Rotational Spectroscopy,* Wiley-Interscience, New York, 1979.

125. Wacker, P. F., and M. R. Pratto, *Microwave Spectral Tables,* NBS Monograph 70, Vol. 2, U.S. Dept. of Commerce, Washington, D.C., 1964.

126. Waters, J. W., W. J. Wilson, and F. I. Shimabukuro, *Science,* **191,** 1174 (1976).

127. Watson, W. D., *Acc. Chem. Res.,* **10,** 221 (1977).

128. White, W. F., *NASA Technical Notes TND-8053, Microwave Spectral Line Listing; TND-7450, Microwave Spectra of Some Sulfur Compounds; TND-8002, Microwave Spectra of Some Chlorine and Fluorine Compounds; TND-7904, Microwave Spectra of Some Volatile Organic Compounds;* NASA Langley Research Center, Hampton, Virginia, 1975.

129. Willemot, E., D. Dangoisse, N. Monnanteuil, and J. Bellet, *J. Phys. Chem. Ref. Data,* **9,** 59 (1980).

130. Wilson, E. B., *Chem. Soc. Rev.,* **1,** 293 (1972).

131. Wilson, T. L., J. Bieging, D. Downes, and F. F. Gardner, *Astron. Astrophys.,* **51,** 303 (1976).

132. Winnewisser, G., W. H. Hocking, and M. C. L. Gerry, *J. Phys. Chem. Ref. Data,* **5,** 79 (1976).

133. Wollrab, J. B., *Rotational Spectra and Molecular Structure,* Academic, New York, 1967.

134. Woods, R. C., T. A. Dixon, R. J. Saykally, and P. G. Szanto, *Phys. Rev. Lett.,* **35,** 1269 (1975).

MAGNETIC CIRCULAR DICHROISM

By S. H. Lin, *Department of Chemistry, Arizona
State University, Tempe, Arizona*

and D. D. Shieh,* *Department of Anesthesiology,
University of Kansas Medical Center, Kansas City,
Kansas*

Contents

* Dr. Shieh's current address: Chemical Abstract Service, 2540 Olentangy River Road, Columbus, Ohio 43210.

I. INTRODUCTION

It was Faraday (27) who first observed that optical activity can be induced in matter by a magnetic field; in 1845 he found that the plane of polarization of light was rotated after the light passed through lead borate glass between the poles of an electromagnet. By means of this and subsequent experiments, Faraday and later workers demonstrated that the plane of polarization of linearly polarized light is rotated on passage through any substance placed in a magnetic field with a nonzero component in the direction of the light beam (58,69). In other words, a longitudinal magnetic field makes all substances appear optically active. For a weak magnetic field, the optical rotation was shown to be proportional to the strength of the magnetic field:

$$\phi = VH$$

where ϕ is the rotation of the polarization plane per unit path length, H represents the strength of the magnetic field in the direction of the light propagation, and the proportional constant V is the so-called Verdet constant, which is a function of optical frequency and a general property of matter.

The optical rotation of matter results from the inequality of its refractive indexes for right- and left-circularly polarized light. In magnetic optical rotation (MOR), this inequality is created by the magnetic field (69). In the absorption region of matter, the absorption coefficients of right- and left-circularly polarized light also differ in the presence of a magnetic field; this gives rise to magnetic circular dichroism (MCD). The discovery of magnetic optical activity was important evidence for the electromagnetic nature of light. Early research interest (in the period 1900–1920) in this area was centered on the anomalous dispersion of MOR compared with the frequency dependences predicted by different applications of the classical electron theory (58,69). In the period between 1920 and 1930, the temperature dependence of the MOR of crystalline salts of paramagnetic ions at very low temperatures supplemented the magnetic susceptibility in providing information about the interaction of these ions with the crystal lattice (112). After the successful use of natural optical rotatory dispersion and circular dichroism techniques for obtaining structural information (21) in the early 1960s, the dispersion of MOR and MCD through absorption bands was again studied, with the aim of obtaining similar information (9,90,94).

The quantum-mechanical theory of MOR outside absorption bands was developed around 1930 by Rosenfeld (76), Kramers (42), and Serber (89) and has satisfactorily accounted for experimental data in these regions. The qualitative form of MOR and MCD in regions of absorption has also long been understood in principle (14), and has been semiquantitatively verified both for the narrow line spectra of monatomic gases and rare-earth crystals and for the broad absorption bands of liquids and solutions. However, it is only recently that quantitative expressions for the dispersion have been suggested and used with experimental data to elucidate the nature of individual absorption bands (12,16,78,93,104).

Although the first observation of the Faraday effect was made over a century ago, this effect has until recently found relatively little application to chemical systems, at least in comparison with natural optical activity. The latter yielded early triumphs in both organic and inorganic chemistry, and the former has enjoyed tremendous popularity in recent years, what with the development of techniques for making measurements over a wide range of frequencies, particularly through absorption bands. Our present research activity in MOR and MCD has been greatly stimulated by the excellent work of Buckingham and Stephens and their coworkers.

In recent years there has been considerable interest in the use of circularly polarized emission (CPE) spectroscopy as a probe of molecular stereochemistry and electronic structure (25,55,59,101,103). Circularly polarized emission is the emission analog of circular dichroism (CD), and is defined as the difference between the intensities of left- and right-circularly polarized radiation in the luminescence of a chiral molecular system. Circularly polarized emission may also be observed in nonchiral luminescent systems that are placed in a magnetic field whose field direction is parallel to the direction of emission detection (74). This latter phenomenon is the emission analog of MCD and is usually referred to as magnetic circularly polarized emission (MCPE) (74). The Faraday effect in the present context refers to MOR, MCD, and MCPE.

The principal objects of this chapter are to provide a general account of the nature of magnetic optical activity, particularly in regions of absorption, and to show in some detail how measurements of MOR and MCD through absorption bands can be used as a powerful means of obtaining both qualitative and quantitative information (spectroscopic assignments, symmetry properties, angular moments, etc.) about the ground and excited states of molecules and ions. Representative experimental work is also discussed.

II. PHENOMENOLOGICAL THEORY

We shall for convenience summarize the relations that are used in treatments of optical activity, as the same relations are also used for magnetic optical activity. The two special characteristics exhibited by an optically active medium are (1) circular birefringence, manifested as a rotation of the plane of polarization of transmitted linearly polarized light, and (2) circular dichroism, the change from linear to elliptic polarization evinced by transmitted electromagnetic waves in the region of an optically active absorption band. Treating these two effects from a purely phenomenological point of view, one can show that the first can be accounted for by assuming a difference in the indexes of refraction of the medium for left- and right-circularly polarized light, and the second, by assuming a difference in the absorption coefficients. More specifically, the angle of rotation per unit length ϕ is given by (18,21,58,62)

$$\phi = \frac{\pi \nu}{c}(n_L - n_R) = \frac{\pi}{\lambda}(n_L - n_R) \tag{2-1}$$

where n_L and n_R are the indexes of refraction for left- and right-circularly polarized light, respectively.

The expression for the ellipticity per unit length is similar to equation 2-1, but before we write it down, we summarize the symbols for various spectroscopic absorption constants (18,62):

$$I = I_0 e^{-kl} \tag{2-2}$$

$$I = I_0 e^{-4\pi\chi l/\lambda} \tag{2-3}$$

$$I = I_0^{-\epsilon Cl} \tag{2-4}$$

where I_0 is the initial intensity, C represents the concentration of absorbing material in moles per unit volume enter, I is the light intensity after it has traversed l centimeters of that medium, k is the absorption coefficient, χ is the absorption index, and ϵ is the molar absorption coefficient. In terms of absorption coefficients k_L and k_R, the ellipticity θ' is defined by

$$\tan \theta' = \tanh \frac{l}{4}(k_L - k_R) \tag{2-5}$$

In practice, $(k_L - k_R)l$ is usually much smaller than unity, in which case the ellipticity per unit length θ may be expressed as

$$\theta = \frac{\theta'}{l} = \tfrac{1}{4}(k_L - k_R) = \frac{\pi}{\lambda}(\chi_L - \chi_R) \tag{2-6}$$

Finally, we may simultaneously express the birefringence and the dichroism by defining the complex rotatory power Φ,

$$\Phi = \phi - i\theta = \frac{\pi}{\lambda}(\hat{n}_L - \hat{n}_R) \tag{2-7}$$

where $\hat{n}_L = n_L - i\chi_L$, and similarly for $\hat{n}_R$.

If the cgs system of units is used, both ϕ and θ have dimensions of radians per centimeter. They must be multiplied by $1800/\pi$ to be converted to the common experimental units of degrees per decimeter. The most frequently encountered ways of expressing rotation data are in terms of the specific rotation $[\alpha]$ and the molar rotation $[\phi]$, defined below:

$$[\alpha] = \left(\frac{\phi}{C'}\right)\left(\frac{1800}{\pi}\right) \tag{2-8}$$

and

$$[\phi] = [\alpha]\left(\frac{M}{100}\right) = \phi\left(\frac{18}{\pi}\right)\left(\frac{M}{C'}\right) \tag{2-9}$$

where C' is the concentration of the optically active material in grams per cubic centimeter, and M is its molecular weight. Likewise, the molar ellipticity is given by

$$[\theta] = \theta\left(\frac{18}{\pi}\right)\left(\frac{M}{C'}\right) \tag{2-10}$$

or in terms of the molar absorption coefficients, by

$$[\theta] = 2.303\left(\frac{4500}{\pi}\right)(\epsilon_L - \epsilon_R) \tag{2-11}$$

For making comparisons among compounds, the molecular quantities $[\phi]$ and $[\theta]$ are the most suitable.

In magnetic rotation spectra (MRS), if I_0 and I are the initial and final intensities, which are polarized parallel to the polarizer and crossed analyzer, respectively, and l is the path length, then in weak fields, we have (31)

$$\frac{I}{I_0} = l^2 \mid \Phi \mid^2 \exp\left[-\omega(\chi_+ + \chi_-)\frac{l}{c}\right] \tag{2-12}$$

In MCPE, particular attention is given to the absolute magnitude and time dependence of the quantity (74)

$$g_{em} = \frac{\Delta I}{I} = \frac{I_L - I_R}{\frac{1}{2}(I_L + I_R)} \tag{2-13}$$

where I_L and I_R are the emission intensities of left- and right-circularly polarized light, respectively.

In the following, we shall briefly present the phenomenological theory of optical activity and magnetic optical activity. The phenomenological equation 2-7 of course gives no clue as to why in an optically active (or magneto-optically active) medium n_L is different from n_R (or χ_L is different from χ_R). To answer this question one might first inquire into what sort of bulk electric and magnetic properties an isotropic medium must have if n_L is not to equal n_R. Let us consider a circularly polarized wave propagating along the z-axis described by the electric field vector

$$\vec{E}_{\pm} = E_0(\hat{\imath} \pm i\hat{\jmath}) \exp\left[i\omega \left(t - \frac{\hat{n}_{\pm}z}{c} \right) \right] \tag{2-14}$$

where the plus and minus signs refer to right- and left-circular polarization, $\hat{n} = n - i\chi$ is the complex refractive index, and n and χ are the real refractive index and absorption coefficient, respectively. A plane-polarized wave is equivalent to the sum of a right- and a left-circularly polarized wave of equal amplitude. The Maxwell equations for an electromagnetic wave in a medium in the absence of macroscopic charges and currents are

$$\vec{\nabla} \cdot \vec{D} = 0, \qquad \vec{\nabla} \cdot \vec{B} = 0, \qquad \vec{\nabla} \times \vec{H}'$$

$$= \frac{1}{c} \frac{\partial \vec{D}}{\partial t}, \qquad \vec{\nabla} \times \vec{E} = -\frac{1}{c} \frac{\partial \vec{B}}{\partial t} \tag{2-15}$$

The usual constitutive properties for a homogeneous, isotropic, optically inactive medium, namely, that the electric displacement $\vec{D}$ is directly proportional to the electric field $\vec{E}$ through the dielectric constant and that the magnetic induction $\vec{B}$ is directly proportional to the magnetic field $\vec{H}'$ through the magnetic permeability, are inappropriate for our purposes, because as is well known, a solution of Maxwell's equations subject to these conditions requires merely that there be an index of refraction for the medium which is equal to the square root of the dielectric constant. As Condon (18) points out, the essential feature of a successful theory of optical activity is the modification of $\vec{D}$ and $\vec{B}$ in such a way that part of $\vec{D}$ and part of $\vec{B}$ are also proportional to $\partial \vec{H}'/\partial t$ and $\partial \vec{E}/\partial t$, respectively.

This can be accomplished by taking into account the finite extension of a molecule in space, which means that the field vectors of a light wave will vary over the finite dimensions of the molecule (74,79), and proceeding as follows. Notice that the connection between the macroscopic and the microscopic levels is provided through the statistical interpretation one can give to the polarization vector $\vec{P}$ and the magnetization vector $\vec{M}$, which relate $\vec{D}$ with $\vec{E}$ and $\vec{B}$ with $\vec{H}'$ through the relations

$$\vec{D} = \vec{E} + 4\pi\vec{P}, \qquad \vec{B} = \vec{H}' + 4\pi\vec{M} \tag{A}$$

Since $\vec{P}$ represents the induced electric dipole moment per unit volume, and $\vec{M}$ represents the induced magnetic dipole moment per unit volume, these quantities may therefore be expressed as (11,12,62)

$$\vec{P} = \sum_a N_a\vec{R}^a, \qquad \vec{B} = \vec{H}' + \sum_a N_a\vec{\mu}^a \tag{B}$$

where N_a is the number of molecules per unit volume in state a. Because of the connections between $\vec{R}^a$ and $\vec{D}$ and between $\vec{\mu}^a$ and $\vec{B}$, for a medium to exhibit optical activity, $\vec{R}^a$ should be proportional to $\partial \vec{H}'/\partial t$ as well as to $\vec{E}$, and $\vec{\mu}^a$ should be proportional to $\partial \vec{E}/\partial t$ as well as to $\vec{H}'$. In other words, for the high-order approximation, it is necessary to drop the point-molecule supposition and to allow for the fact that different parts of the molecule "see" different electromagnetic field strengths at the same instant (26).

To the first order in the electromagnetic fields, the components of the complex induced moment vectors are (where $i,j,k = x,y,z$) (11,12,18)

$$R_i = \alpha_{ij}E_j + \beta_{ij}H_j', \quad \mu_i = \gamma_{ij}E_j + \chi_{ij}H_j' \qquad (2\text{-}16)$$

where α_{ij} and χ_{ij} are the electric and magnetic polarizability tensors, and are complex functions of the static magnetic field $\vec{H}$, for example,

$$\alpha_{ij} = \alpha_{ij}^{(0)} + \alpha_{ijk}^{(1)}H_k + \cdots \qquad (2\text{-}17)$$

In an isotropic medium, $\alpha_{ij}^{(0)} = \alpha_{(0)}\delta_{ij}$ and $\alpha_{ij}^{(1)} = \alpha_{(1)}\epsilon_{ijk}$, where $\alpha_{(0)} = \frac{1}{3}\alpha_{ij}^{(0)}\delta_{ij} = \frac{1}{3}\alpha_{jj}^{(0)}$, $\alpha_{(1)} = \frac{1}{6}\alpha_{ijk}^{(1)}\epsilon_{ijk}$, $\delta_{ij} = 0$ if $i \neq j$, $\delta_{ii} = 1$, and ϵ_{ijk} is 1 or -1 if (ijk) is an even or odd permutation of (xyz), respectively, and 0 if any two suffixes are identical. Using these relations, equation 2-16 becomes (12,18,26)

$$\vec{R} = \alpha_{(0)}\vec{E} + \beta_{(0)}\vec{H}' + \alpha_{(1)}(\vec{E} \times \vec{H}) + \beta_{(1)}(\vec{H}' \times \vec{H}) \qquad (2\text{-}18)$$

$$\vec{\mu} = \gamma_{(0)}\vec{E} + \chi_{(0)}\vec{H}' + \gamma_{(1)}(\vec{E} \times \vec{H}) + \chi_{(1)}(\vec{H}' \times \vec{H}) \qquad (2\text{-}19)$$

Solving equations 2-15–2-17 and B for the complex refractive index gives

$$\hat{n}_{\pm} = 1 + \pi \sum_a N_a[(\alpha_{xx}^a + \alpha_{yy}^a) \pm i(\alpha_{xy}^a - \alpha_{yx}^a) + (\chi_{xx}^a + \chi_{yy}^a)$$

$$\pm i(\chi_{xy}^a - \chi_{yx}^a) \mp i(\beta_{xx}^a + \beta_{yy}^a) + (\beta_{xy}^a \pm \beta_{yx}^a) \pm i(\gamma_{xx}^a + \gamma_{yy}^a)$$

$$- (\gamma_{xy}^a - \gamma_{yx}^a)] \qquad (2\text{-}20)$$

where α_{xx}^a and so forth are the complex polarizability components of a molecule in state a. Using equations 2-7 and 2-20 we obtain

$$\Phi = -\frac{i\pi\omega}{c} \sum_a N_a[(\alpha_{xy}^a - \alpha_{yx}^a)$$

$$+ (\chi_{xy}^a - \chi_{yx}^a) - (\beta_{xx}^a + \beta_{yy}^a) + (\gamma_{xx}^a + \gamma_{yy}^a)] \qquad (2\text{-}21)$$

Alternatively, with equations 2-18 and 2-19, equation 2-20 can be written as

$$\hat{n}_\pm = 1 + 2\pi \sum_a N_a[\alpha^a_{(0)} + \chi^a_{(0)} \mp i\beta^a_{(0)} \pm i\gamma^a_{(0)}$$

$$+ H_z\{\pm i\alpha^a_{(1)} \pm i\chi^a_{(1)} + \beta^a_{(1)} - \gamma^a_{(1)}\}] \qquad (2\text{-}22)$$

It follows that

$$\Phi = -\frac{2i\pi\omega}{c} \sum_a N_a[-\beta^a_{(0)} + \gamma^a_{(0)} + H_z\{\alpha^a_{(1)} + \chi^a_{(1)}\}] \qquad (2\text{-}23)$$

Equations 2-18, 2-19, and 2-23 indicate that optical activity arises from the components of the induced electric and magnetic moments perpendicular to the electric and magnetic fields of the wave, respectively, and to the direction of propagation. Notice that $\beta_{(0)}$ and $\gamma_{(0)}$ determine the components of $\vec{R}$ and $\vec{M}$ perpendicular to $\vec{E}$ and $\vec{H}'$ when $\vec{H} = 0$, and that in the presence of $\vec{H}$, $\alpha_{(1)}$ and $\chi_{(1)}$ play roles analogous to those of $\beta_{(0)}$ and $\gamma_{(0)}$ and are the terms responsible for magnetic optical activity. For natural optical activity, equation 2-23 reduces to

$$\Phi = \frac{2i\pi\omega}{c} \sum_a N_a[\beta^a_{(0)} - \gamma^a_{(0)}] \qquad (2\text{-}24)$$

The real and imaginary parts of $\beta_{(0)}$ and $\gamma_{(0)}$ are responsible for CD and optical rotatory dispersion (ORD), respectively. It is useful to recognize the relative orders of magnitude of the various terms in equation 2-22. In general, $\alpha_{(0)} \gg \beta_{(0)},\chi_{(0)}$; $\alpha_{(1)} \gg \beta_{(1)},\gamma_{(1)},\chi_{(1)}$; and for reasonable field strengths, $\alpha_{(1)}H_z \approx \beta_{(0)},\gamma_{(0)}$. Thus magnetic optical activity is similar in magnitude to natural optical activity and the dominant contribution is from the term in $\alpha_{(1)}$. It is also clear that magnetic optical activity is not a perturbation of natural optical activity, but arises from entirely different "polarizabilities" that equal zero when $H_z = 0$. The natural and magnetic optical activity are additive in isotropic media.

Quantum-mechanical expressions for MOR and MCD can be obtained using equation 2-21 or equation 2-23 if one expresses the molecular polarizabilities in terms of molecular wavefunctions and introduces the effect of the applied magnetic field on energy levels and wavefunctions. This approach has been discussed by Buckingham and Stephens (12) and will not be presented here. Instead, in Section III we shall derive the quantum-mechanical expression for MCD (95,96,105) and employ the Kronig–Kramers transform to obtain the corresponding expression for MOR (107).

III. QUANTUM THEORY

Our objectives in this and subsequent theoretical sections are to present an introduction to the theory of MCD, to demonstrate the techniques used in its analysis, and to discuss the kinds of information made available by such analysis. The Kronig–Kramers relations (100,117) will be applied to MCD theory to obtain the corresponding theoretical expressions for MOR.

In the Appendix, we show that the $\Delta k(\omega)$ function appearing in the expression for ellipticity per unit length for the electronic transition $K \to N$ can be expressed as

$$\Delta k(\omega) = H[\Delta k(\omega)_A^{(1)} + \Delta k(\omega)_B^{(1)} + \Delta k(\omega)_C^{(1)}] \qquad (3\text{-}1)$$

In this chapter, we shall regard atoms as a particular case of molecules. Note that in the adiabatic approximation (52,53),

$$\vec{R}_{Kek,Nen} = \langle \oplus_{Kek}\Phi_{Ke} \mid \vec{R} \mid \oplus_{Nen}\Phi_{Ne} \rangle = \langle \oplus_{Kek} \mid \vec{R}_{KeNe} \mid \oplus_{Nen} \rangle \qquad (3\text{-}2)$$

and

$$\vec{\mu}_{Nen,Mm} = \langle \oplus_{Nen}\Phi_{Ne} \mid \vec{\mu} \mid \oplus_{Mm}\Phi_{\mu} \rangle = \langle \oplus_{Nen} \mid \vec{\mu}_{NeM} \mid \oplus_{Mm} \rangle \qquad (3\text{-}3)$$

where $\vec{R}_{KeNe}$ and $\vec{M}_{NeM}$ represent the matrix elements of the electronic dipole and magnetic dipole moments, respectively, and are in general functions of nuclear coordinates.

A commonly used approximation is $E_{Mm}^0 - E_{Nen}^0 = E_M^0 - E_{Ne}^0$, that is, the neglect of the energy of nuclear motion in comparison with the electronic energy or the replacement of $E_{Mm}^0 - E_{Nen}^0$ by its average value. In this case, the Faraday B-coefficient becomes

$$B(Kk \to Nn) = \frac{1}{g_K} \sum_{Ke} \sum_{Ne} I_m[\vec{R}_{Kek,Nen} \cdot \langle \oplus_{Nen} \mid \vec{M}_{NeKe} \mid \oplus_{Kek} \rangle]$$

$$= \frac{1}{g_K} \sum_{Ke} \sum_{Ne} I_m(\vec{R}_{Kek,Nen} \cdot \vec{M}_{Nen,Kek}) \qquad (3\text{-}4)$$

after applying the closure relation to m, where

$$\vec{M}_{NeKe} = \sum_M \frac{\vec{R}_{MKe} \times \vec{\mu}_{NeM}}{E_M^0 - E_N^0} + \sum_M \frac{\vec{R}_{NeM} \times \vec{\mu}_{MKe}}{E_M^0 - E_K^0} \qquad (3\text{-}5A)$$

and

$$\vec{M}_{Nen,Kek} = \langle \oplus_{Nen} \mid \vec{M}_{NeKe} \mid \oplus_{Kek} \rangle \qquad (3\text{-}5B)$$

According to the Herzberg–Teller theory, the nuclear coordinate dependence of the electronic matrix elements of electric and magnetic dipole moments can be expanded in power series of say, normal coordinates Q_i:

$$\vec{R}_{MM'} = \vec{R}_{MM'}^{(0)} + \sum_i \left(\frac{\partial \vec{R}_{MM'}}{\partial Q_i}\right)_0 Q_i + \cdots \tag{3-6}$$

$$\vec{\mu}_{MM'} = \vec{\mu}_{MM'}^{(0)} + \sum_i \left(\frac{\partial \vec{M}_{MM'}}{\partial Q_i}\right)_0 Q_i + \cdots \tag{3-7}$$

It follows that (97)

$$\Delta k(\omega)_A^{(1)} = -\frac{16\pi^2\omega\alpha^2 N}{3c\hbar^2}[A(K \to N)f'_{KN}(\omega) + \sum_i A_i^{(1)}(K \to N)f'_{KN}(\omega)_i^{(1)}]$$
$$\tag{3-8}$$

$$\Delta k(\omega)_B^{(1)} = -\frac{16\pi^2\omega\alpha^2 N}{3c\hbar}[B(K \to N)f_{KN}(\omega) + \sum_i B_i^{(1)}(K \to N)f_{KN}(\omega)_i^{(1)}]$$
$$\tag{3-9}$$

and

$$\Delta k(\omega)_C^{(1)} = -\frac{16\pi^2\omega\alpha^2 N}{3c\hbar}\beta[C(K \to N)f_{KN}(\omega)$$

$$+ \sum_i C_i^{(1)}(K \to N)f_{KN}(\omega)_i^{(1)}] \tag{3-10}$$

where

$$A(K \to N) = \frac{1}{2g_K}\sum_{K_e}\sum_{N_e}(\vec{\mu}_{N_eN_e}^{(0)} - \vec{\mu}_{K_eK_e}^{(0)}) \cdot I_m(\vec{R}_{K_eN_e}^{(0)} \times \vec{R}_{N_eK_e}^{(0)}) \tag{3-11}$$

$$A_i^{(1)}(K \to N) = \frac{1}{2g_K}\sum_{K_e}\sum_{N_e}(\vec{\mu}_{N_eN_e} - \vec{\mu}_{K_eK_e}) \cdot I_m\left[\left(\frac{\partial \vec{R}_{K_eN_e}}{\partial Q_i}\right)_0 \times \left(\frac{\partial \vec{R}_{N_eK_e}}{\partial Q_i}\right)_0\right]$$
$$\tag{3-12}$$

$$B(K \to N) = \frac{1}{g_K}\sum_{K_e}\sum_{N_e}I_m(\vec{R}_{K_eN_e}^{(0)} \cdot \vec{M}_{N_eK_e}^{(0)}) \tag{3-13}$$

$$B_i^{(1)}(K \to N) = \frac{1}{g_K}\sum_{K_e}\sum_{N_e}I_m\left[\left(\frac{\partial \vec{R}_{K_eN_e}}{\partial Q_i}\right)_0 \cdot \left(\frac{\partial \vec{M}_{N_eK_e}}{\partial Q_i}\right)_0\right] \tag{3-14}$$

$$C(K \to N) = \frac{1}{2g_K}\sum_{K}\sum_{N}\vec{\mu}_{K_eK_e}^{(0)} \cdot I_m[\vec{R}_{K_eN_e}^{(0)} \times \vec{R}_{N_eK_e}^{(0)}] \tag{3-15}$$

$$C_i^{(1)}(K \rightarrow N) = \frac{1}{2g_K} \sum_{K_e} \sum_{N_e} \vec{\mu}_{K_e K_e}^{(0)} \cdot I_m \left[\left(\frac{\partial \vec{R}_{K_e N_e}}{\partial Q_i} \right)_0 \times \left(\frac{\partial \vec{R}_{N_e K_e}}{\partial Q_i} \right)_0 \right] \quad (3\text{-}16)$$

$$f_{KN}(\omega) = \sum_{nk} P_{Kk}^{(0)} \, | \langle \oplus_{Kk} \, | \, \oplus_{Nn} \rangle \, |^2 \, \delta(\omega - \omega_{Nn,Kk}^0) \quad (3\text{-}17)$$

and

$$f_{KN}(\omega)_i^{(1)} = \sum_{nk} P_{Kk}^{(0)} \, | \langle \oplus_{Kk} \, | \, Q_i \, | \, \oplus_{Nn} \rangle \, |^2 \, \delta(\omega - \omega_{Nn,Kk}^0) \quad (3\text{-}18)$$

Notice that $f'_{KN}(\omega)$ and $f'_{KN}(\omega)_i^{(1)}$ denote the frequency derivatives of $f_{KN}(\omega)$ and $f_{KN}(\omega)_i^{(1)}$, respectively. In equations 3-8–3-18, we have ignored the normal coordinate dependence of $\vec{\mu}_{K_e K_e}$ and $\vec{\mu}_{N_e N_e}$. The functions $f_{KN}(\omega)$, $f_{KN}(\omega)_i^{(1)}$, $f'_{KN}(\omega)$, and $f'_{KN}(\omega)_i^{(1)}$ are the so-called band shape functions of MCD.

The electronic transition that is symmetry forbidden but vibronically allowed is defined by

$$\vec{R}_{K_e N_e}^{(0)} = 0 \text{ and } \left(\frac{\partial \vec{R}_{K_e N_e}}{\partial Q_i} \right)_0 \neq 0$$

The normal mode Q_i involved in $\partial \vec{R}_{K_e N_e} / \partial Q_i$ is usually called the promoting mode or inducing mode. The allowedness of the symmetry-forbidden transition can also be invoked by the breakdown of the adiabatic approximation (52,53,67). This effect can easily be taken into account, and will not be discussed here. For allowed transitions, the contributions from $\Delta A_i^{(1)}(K \rightarrow N)$, $\Delta B_i^{(1)}(K \rightarrow N)$, and $\Delta C_i^{(1)}(K \rightarrow N)$ are small and can usually be neglected (the Franck–Condon approximation).

To obtain the expression for MOR, we use the Kronig–Kramers relation,

$$\phi(\omega) = \frac{2}{\pi} P \int_0^\infty \frac{\omega^2 \theta(\omega') d\omega'}{\omega'(\omega'^2 - \omega^2)} \quad (3\text{-}19)$$

or

$$\theta(\omega) = -\frac{2}{\pi} P \int_0^\infty \frac{\omega^3 \phi(\omega') d\omega'}{\omega'^2(\omega'^2 - \omega^2)} \quad (3\text{-}20)$$

where P indicates the principal value of the integral. Substituting equations 3-1 and 2-8 into equation 3-19 yields

$$\phi(\omega) = \phi_A(\omega) + \phi_B(\omega) + \phi_C(\omega) \quad (3\text{-}21)$$

where if we use equations 3-8–3-10, then $\phi_A(\omega)$, $\phi_B(\omega)$, and $\phi_C(\omega)$ are given by

$$\phi_A(\omega) = -\frac{4\pi^2\omega\alpha^2 NH}{3c\hbar^2}[A(K \to N)g'_{KN}(\omega) + \sum_i A_i^{(1)}(K \to N)g'_{KN}(\omega)_i^{(1)}]$$

(3-22)

$$\phi_B(\omega) = -\frac{4\pi^2\omega\alpha^2 NH}{3c\hbar}[B(K \to N)g_{KN}(\omega) + \sum_i B_i^{(1)}(K \to N)g_{KN}(\omega)_i^{(1)}]$$

(3-23)

and

$$\phi_C(\omega) = -\frac{4\pi^2\omega\alpha^2 NH}{3c\hbar}\beta[C(K \to N)g_{KN}(\omega)$$

$$+ \sum_i C_i^{(1)}(K \to N)g_{KN}(\omega)_i^{(1)}] \quad (3\text{-}24)$$

where $g_{KN}(\omega)$, $g_{KN}(\omega)_i^{(1)}$, $g'_{KN}(\omega)$, and $g'_{KN}(\omega)_i^{(1)}$ represent the Kronig–Kramers transforms of $f_{KN}(\omega)$, $f_{KN}(\omega)_i^{(1)}$, $f'_{KN}(\omega)$, and $f'_{KN}(\omega)_i^{(1)}$, respectively, and are given by

$$g_{KN}(\omega) = \frac{2}{\pi}P\sum_{nk}P_{Kk}^{(0)}|\langle\oplus_{Kk}|\oplus_{Nn}\rangle|^2\frac{\omega^2}{\omega_{Nn,Kk}^0[(\omega_{Nn,Kk}^0)^2 - \omega^2]} \quad (3\text{-}25)$$

$$g_{KN}(\omega)_i^{(1)} = \frac{2}{\pi}P\sum_{nk}P_{Kk}^{(0)}|\langle\oplus_{Kk}|Q_i|\oplus_{Nn}\rangle|^2\frac{\omega^2}{\omega_{Nn,Kk}^0[(\omega_{Nn,Kk}^0)^2 - \omega^2]} \quad (3\text{-}26)$$

$$g'_{KN}(\omega) = \frac{2}{\pi}P\sum_{nk}P_{Kk}^{(0)}|\langle\oplus_{Kk}|\oplus_{Nn}\rangle|^2\frac{\omega^2[3(\omega_{Nn,Kk}^0)^2 - \omega^2]}{(\omega_{Nn,Kk}^0)^2[(\omega_{Nn,Kk}^0)^2 - \omega^2]^2} \quad (3\text{-}27)$$

and

$$g'_{KN}(\omega)_i^{(1)} = \frac{2}{\pi}P\sum_{nk}P_{Kk}^{(0)}|\langle\oplus_{Kk}|Q_i|\oplus_{Nn}\rangle|^2\frac{\omega^2[3(\omega_{Nn,Kk}^0)^2 - \omega^2)}{(\omega_{Nn,Kk}^0)^2[(\omega_{Nn,Kk}^0)^2 - \omega^2]^2}$$

(3-28)

These functions, $g_{Kn}(\omega)$, $g_{KN}(\omega)_i^{(1)}$, $g'_{KN}(\omega)$, and $g'_{KN}(\omega)_i^{(1)}$, are the band shape functions of MOR.

It is useful to have the expression for the absorption coefficient for comparison with MCD. This can be obtained from equations A-11–A-13 (see Appendix) as (46,96)

$$k(\omega) = \frac{8\pi^2\omega}{c\hbar}\alpha^2\sum_{nk}N_k|\hat{\epsilon}\cdot\vec{R}_n|^2\delta(\omega_n - \omega) \quad (3\text{-}29)$$

where $\hat{\epsilon}$ represents the unit vector of polarization. For an isotropic medium, equation 3-29 becomes

$$k(\omega) = \frac{8\pi^2}{3c\hbar}\alpha^2\sum_{nk}N_k|\vec{R}_{nk}|^2\delta(\omega_{nk} - \omega) \quad (3\text{-}30)$$

For molecular systems, the adiabatic approximation can be used. In this case, we find

$$k(\omega) = \frac{8\pi^2\omega N}{3c\hbar} \alpha^2 \sum_n P^{(0)}_{Kk} \mid \vec{R}_{Nn,Kk} \mid^2 \delta(\omega^0_{Nn,Kk} - \omega) \qquad (3\text{-}31)$$

For the electronic transition $K \rightarrow N$, equation 3-31 can be written, using equation 3-6, as

$$k(\omega) = \frac{8\pi^2\omega N}{3c\hbar} \alpha^2 [D(K \rightarrow N)f_{KN}(\omega) + \sum_i D^{(1)}_i(K \rightarrow N)f_{KN}(\omega)^{(1)}_i] \qquad (3\text{-}32)$$

where

$$D(K \rightarrow N) = \mid \vec{R}^{(0)}_{NK} \mid^2 \qquad (3\text{-}33)$$

and

$$D^{(1)}_i(K \rightarrow N) = \left| \left(\frac{\partial \vec{R}^0_{NK}}{\partial Q_i}\right)_0 \right|^2 \qquad (3\text{-}34)$$

The quantity $\mid \vec{R}_{NK} \mid^2$ is usually called the dipole strength. For symmetry-forbidden transitions, $\vec{R}^{(0)}_{NK} = 0$ and equation 3-32 reduces to

$$k(\omega) = \frac{8\pi^2\omega N}{3c\hbar} \alpha^2 \sum_i D^{(1)}_i(K \rightarrow N)f_{KN}(\omega)^{(1)}_i \qquad (3\text{-}35)$$

On the other hand, for electric dipole-allowed transitions, the second term in equation 3-32 is small compared with the first term and can usually be ignored (the Franck–Condon approximation).

IV. INSTRUMENTATION

Placing magnets into cell compartments of spectropolarimeters or circular dichrographs enables one to measure MOR or MCD. The directions of the magnetic field and the light beam are parallel. Permanent magnets of 7 kG, electromagnets of 15 kG, and superconducting magnets of 100 kG are commercially available. The photomultiplier and xenon arc lamp (in the Cary 60 spectropolarimeter) should be properly shielded to minimize the effect of magnetic fields (1).

A. MAGNETIC OPTICAL ROTATION (FARADAY ROTATION)

Two identical magnets and two identical cells are needed to cancel the MOR contributed by solvents and cell windows. One magnet contains a

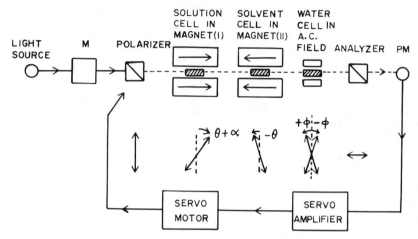

Fig. 12.1.   Instrumentation for MOR measurement (92).

solvent cell and the other magnet a solution cell. These two identical magnets are arranged to produce magnetic fields of the same strength but in opposite directions (22,92). The MOR line shapes are complicated when they have contributions from several electronic transitions. The tails from the transparent region are generally the sums of contributions of different rotations. These complications in instrumentation and difficulties in interpretation of MOR finally forced early investigators in the field to concentrate on the measurement of MCD(1,8,22).

The instrumentation for MOR is shown in Fig. 12.1 (92). This MOR spectropolarimeter was used by Shashoua in the study of biological molecules. A 150-W xenon arc lamp is used as the light source, "M" is a Cary Model 14 double-beam monochromator, "PM" is a photomultiplier, θ is the solvent MOR, α is the solute MOR, and ϕ is the modulation angle in the Faraday coil. The Faraday rotation induced in a water sample 20 cm long in the ac magnetic field is used to modulate the light beam. The angle-sensing and -recording sections of the apparatus are activated by this carrier frequency. The specific magnetic rotation $[\alpha]_{sp}$ is

$$[\alpha]_{sp} = \frac{\alpha}{lCH} \times 10^4$$

where α is the angle of rotation in degrees, l is the path length in decimeters, C is concentration in grams per milliliter, and H is the magnetic field in gauss. Two 5-kG permanent magnets are used.

B.   MAGNETIC CIRCULAR DICHROISM

1.   The Spectropolarimeter or Circular Dichrograph

Technological advances in the photomultiplier tube, in the electronics (e.g., synchronous detection), and in the light source (e.g., xenon arc lamps)

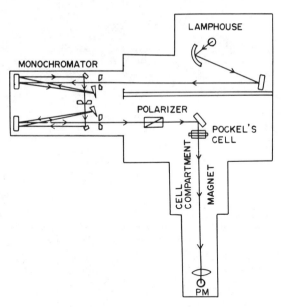

Fig. 12.2. Measurement of MCD using the optical system of a Cary Model 60 spectropolarimeter (1). A 50-kG superconducting magnet is used. "PM" is a photomultiplier.

have improved the sensitivity of the modern ORD or CD instrument. The availability of transparent optical materials extended the spectral range both toward the vacuum ultraviolet (82,83) and toward the infrared (15,75). Photoelastic (stress or piezoelastic) modulators (PEMs), which have been used in several laboratory-built instruments (75,82), are now replacing the old Pockels cell [electro-optic modulator (EOM)] in Jasco's J-40 and J-41 models. The PEM has a wide aperture and needs a very low driving voltage. Linearly polarized light is converted to circularly polarized light by a block of stressed transparent material (e.g., fused quartz or germanium) working as a quarter-wave plate. The oscillating stress is created in the fused quartz transducer (the crystal oscillator oscillates in a longitudinal mode) cemented to one end of the fused quartz block (Fig. 12.2) (15,75,82).

The detection of CD or MCD by Fourier-transform spectroscopy is particularly important in the infrared region, where the sources tend to be weak. Future developments in the instrumentation may include the use of tunable lasers as light sources and fast-response instruments for studies of the rates of processes.

Magnetic circular dichroism measured using alternating magnetic fields (2) is not additive to the natural CD. It represents only the magnetically induced CD. This method could be very useful in studies of the MCD of natural products, which often have natural CD. The measurable spectral range can be wide, since the polarization modulator is not needed.

2.   The Magnet

An electromagnet of 15 kG was used in the Jasco J-40 circular dichrograph and was shown to produce good-quality MCD spectra (5). The recent report distributed by Jasco concerning use of their 10-kG MCD-1 electromagnet for measuring the MCD spectra of heme-conjugated proteins is rather encouraging. The expense of using a superconducting magnet (due to liquid helium consumption) has limited the wide application of this technique in both research and routine analytical work.

In the Cary Model 60 ORD/CD spectropolarimeter (manufactured by Cary Instrument, Varian Associates), the CD optics of the instrument has to be relocated to provide a large sample compartment to adapt to a superconducting magnet (50 kG) between the Pockels cell and the photomultiplier (1) (Fig. 12.2). The authors (reference 1) superconducting magnets are made by the Analytical Instrument Division of Varian Associates. A power-supply unit is used to energize the superconducting solenoid to produce any field strength up to 50 kG. Jasco's J-20 ORD/CD spectropolarimeters are designed with a large cell compartment to adapt to various magnets. With minor modifications, their J-40 and J-41 models can also adapt to superconducting magnets manufactured by Oxford Instrument Company, Ltd. The Jasco instruments also include permanent magnets of 4–7 kG and electromagnets of 10–15 kG.

In the superconducting solenoid system supplied by Varian Associates, the solenoid is housed in the liquid helium Dewar vessel. This Dewar vessel is surrounded by a baffle, which is connected to a liquid nitrogen reservoir near the top of the cryostat. There are spaces between the helium Dewar vessel and the nitrogen baffle, and between these and the outer, room-temperature wall. These spaces are evacuated with a diffusion pump before the liquid nitrogen reservoir is filled. The helium Dewar vessel is cooled first by filling it with liquid nitrogen. The liquid nitrogen is then forced out with helium gas after cooling for about 30 min. Finally, liquid helium is introduced into the helium Dewar vessel. It usually takes about 5 liters of liquid helium to fill the 3.3-liter Dewar vessel in the beginning. This amount of liquid helium will maintain the solenoid in superconducting condition for about 24 hr.

3.   Cell Compartment and Sample Handling

The bore of the superconducting magnet is a 1-in. diameter room-temperature tube about 13 cm long. For liquid nitrogen- or liquid helium-temperature studies, the room-temperature tubes or the liquid nitrogen-temperature tubes may be removed. Silica windows are used for low-temperature measurements. These windows are necessary for maintaining the vacuum.

A sample cell with strainfree silica windows is used. It has a small opening in the cylindrical wall for loading samples. A piece of wax paper or Teflon tape is used to cover the hole. A brass tube with a piece of double adhesive tape to hold the cell is inserted into a stainless steel tube about 10 cm long. The whole assembly is then inserted carefully into the bore of the magnet.

4. Measurement

Calibration and operating specifications for MCD measurements are the same as those used in measuring natural CD. Several references discuss in detail ORD and CD measurements (19,116). Magnetic field strengths are determined with a Bell "110" gaussmeter and $CoSO_4$ solution. Calibration of ORD and CD was re-examined by Krueger and Pschigoda (43). The compound d-10-camphorsulfonic acid (Eastman Organic Reagent) was found to form a stable hydrate (7.29% water) after being left in a room at 50% relative humidity for 2 days. Circular dichroism experiments should be calibrated with the equilibrated hydrated d-10-camphorsulfonic acid, which gave a molar ellipticity, $[\theta]_M$, of 7775° · cm^2/dmol at 290 nm. The water content of the equilibrated sample was found to be constant.

To obtain the best results in measuring MCD, one should check the absorbance (A) of the sample. If we consider only shot noise in the system, we can express the signal-to-noise ratio (S/N) as a function of the absorbance (8,48):

$$S/N = k_s A 10^{-A/2}$$

where k_s is a proportional constant. To obtain the maximum S/N we differentiate the above equation with respect to the absorbance. The value A = 0.868 is found by setting the resulting expression to zero. By plotting S/N versus A in the above equation, we see that the best results (higher S/N) can be obtained if we keep the absorbance of the sample between 0.5 and 1.5 throughout the spectral range of interest. Data may become unreliable if the sample absorbance exceeds 2.0.

The Jasco J-40 DP data processor, provided as standard equipment in the J-41 model, can also be used in Model J-40. This is a small computer with a memory of 1 K 18 bits. It corrects the base line automatically and improves S/N by repetitive scanning.

Measured values from the chart are converted into magnetic molar ellipticity per gauss as follows:

$$[\theta]_M = \frac{\theta°}{ClH} \times 100° \cdot dl/(dm \cdot mol \cdot G)$$

where $\theta°$ is the reading from the chart in degrees, C is the concentration of the solution in moles per liter, l is the path length in centimeters, and H is the magnetic field in gauss.

V.   BAND SHAPE FUNCTIONS

In this section we will discuss the properties and calculation of band shape functions. We will be concerned only with $f_{KN}(\omega)$, $f'_{KN}(\omega)$, $f_{KN}(\omega)_i^{(1)}$, and $f'_{KN}(\omega)_i^{(1)}$, since the band shape functions of MOR can be obtained by the use of the Kronig–Kramers transform (47,61).

From the vibrational structuring of MCD spectra in condensed media, we can see that in most cases only a small number of normal modes of molecular vibration determine the intensities of the vibronic bands, and that the remaining normal modes of vibration (including the intermolecular modes) determine the shapes of these vibronic bands (48). We first consider $f_{KN}(\omega)$. It should be noted that

$$\int d\omega f_{KN}(\omega) = \sum_{nk} P_{Kk}^{(0)} |\langle \oplus_{Kk} | \oplus_{Nn} \rangle|^2 = 1 \qquad (5\text{-}1)$$

by the closure relation. Introducing the integral representation for the delta function (53),

$$\delta(\omega - \omega_{Nn,Kk}^0) = \frac{1}{2\pi} \int_{-\infty}^{\infty} dt \, \exp[-it(\omega - \omega_{Nn,Kk}^0)] \qquad (5\text{-}2)$$

into equation 3-17 yields

$$f_{KN}(\omega) = \frac{1}{2\pi} \sum_{nk} \int_{-\infty}^{\infty} dt \, \exp[it(\omega_{Nn,Kk}^0 - \omega)] P_{Kk}^{(0)} |\langle \oplus_{Kk} | \oplus_{Nn} \rangle|^2 \qquad (5\text{-}3)$$

To demonstrate the calculation of the band shape function, we shall for simplicity assume that only the vibrational motion is involved in the Franck–Condon factor $|\langle \oplus_{Kk} | \oplus_{Nn} \rangle|^2$, and that the vibration is harmonic. In other words, $\oplus_{Nn}$ and $\oplus_{Kk}$ are written as products of harmonic-oscillator wavefunctions (47,48,61):

$$\oplus_{Nn} = \prod_i X_{Nv_{i'}}(Q_i''), \qquad \oplus_{Kk} = \prod_i X_{Kv_{i'}}(Q_{i'}) \qquad (5\text{-}4)$$

Using equation 5-4, equation 5-3 becomes

$$f_{KN}(\omega) = \frac{1}{2\pi} \int_{-\infty}^{\infty} dt \, e^{it(\omega_{NK}^0 - \omega)} \prod_i G_i(t) \qquad (5\text{-}5)$$

where

$$G_i(t) = \sum_{v_{i'}} \sum_{v_{i''}} P_{Kv_{i'}}^{(0)}$$

$$\times |\langle X_{Kv_{i'}} | X_{Nv_{i''}} \rangle|^2 \exp[it(v_{i''} + \tfrac{1}{2})\omega_{i''} - it(v_{i'} + \tfrac{1}{2})\omega_{i'}] \qquad (5\text{-}6)$$

which can be simplified to (45)

$$G_i(t) = \frac{2\beta_{i'}\beta_{i''}\sinh\dfrac{\hbar\omega_{i'}}{2kT}\exp\left[-d_i^2\beta_{i'}^2\beta_{i''}^2\left(\beta_{i'}^2\coth\dfrac{\mu_{i''}}{2} + \beta_{i''}^2\coth\dfrac{\lambda_{i'}}{2}\right)^{-1}\right]}{\left[\sinh\lambda_{i'}\sinh\mu_{i''}\left(\beta_{i'}^2\coth\dfrac{\lambda_{i'}}{2} + \beta_{i''}^2\coth\dfrac{\mu_{i''}}{2}\right)\right]^{1/2}}$$

$$\times \frac{1}{\left(\beta_{i'}^2\tanh\dfrac{\lambda_{i'}}{2} + \beta_{i''}^2\tanh\dfrac{\mu_{i''}}{2}\right)^{1/2}} \quad (5\text{-}7)$$

where $\beta_{i'} = (\omega_{i'}/\hbar)^{1/2}$, $Q_{i''} - Q_{i'} = d_i$, $\mu_{i''} = -it\omega_{i''}$, and $\lambda_{i'} = it\omega_{i'} + \hbar\omega_{i'}/(kT)$. For the case that $\omega_{i'} = \omega_{i''}$ (displaced oscillators), equation 5-7 reduces to

$$G_i(t) = \exp\left[-\frac{\beta_i^2 d_i^2}{2}\left\{\coth\frac{\hbar\omega_i}{2kT} - \operatorname{csch}\frac{\hbar\omega_i}{2kT}\cosh\left(it\omega_i + \frac{\hbar\omega_i}{2kT}\right)\right\}\right]$$

$$(5\text{-}8)$$

In this case, the band shape function $f_{KN}(\omega)$ is given by

$$f_{KN}(\omega) = \frac{1}{2\pi}\int_{-\infty}^{\infty} dt\,\exp\left[it\Delta\omega - \sum_i \frac{\beta_i^2 d_i^2}{2}\left\{\coth\frac{\hbar\omega_i}{2kT}\right.\right.$$

$$\left.\left. - \operatorname{csch}\frac{\hbar\omega_i}{2kT}\cosh\left(it\omega_i + \frac{\hbar\omega_i}{2kT}\right)\right\}\right] \quad (5\text{-}9)$$

where $\Delta\omega = \omega_{NK}^0 - \omega$. If $\sum_i (\beta_i^2 d_i^2/2)\coth[\hbar\omega_i/(2kT)] > 1$, which is usually referred to as the strong coupling case, we may expand the hyperbolic function in power series of t. It follows that

$$f_{KN}(\omega) = \left[\pi\sum_i \beta_i^2 d_i^2\omega_i^2\coth\frac{\hbar\omega_i}{2kT}\right]^{1/2}\exp\left[\frac{-\left(\Delta\omega + \sum_i \dfrac{\beta_i^2 d_i^2\omega_i}{2}\right)^2}{\sum_i \beta_i^2 d_i^2\omega_i^2\coth\dfrac{\hbar\omega_i}{2kT}}\right]$$

$$(5\text{-}10)$$

In other words, in the strong coupling approximation, $f_{KN}(\omega)$ is Gaussian. The condition under which the strong coupling approximation holds has been provided by Lin et al. (54) and Knittel et al. (41).

It should be noted that the band shape function $f_{KN}(\omega)$ given by equation 5-10 takes into account only the normal coordinate displacements between the two electronic states K and N, and does not include the effect of normal frequency displacements between the two electronic states. The band shape function $f_{KN}(\omega)$ in the strong coupling case that takes into account the effect of both normal coordinate and normal frequency displacements has been derived, and can be expressed as (48)

$$f_{KN}(\omega) = \left[\frac{4 \log 2}{\pi(\Delta\omega_{KN})^2_{1/2}}\right]^{1/2} \exp\left[-\frac{(4 \log 2)(\omega_{NK}^{(max)} - \omega)^2}{(\Delta\omega_{KN})^2_{1/2}}\right] \qquad (5\text{-}11)$$

where $\omega_{NK}^{(max)}$ and $(\Delta\omega_{KN})_{1/2}$ represent the frequency maximum and half width of the Gaussian band shape function $f_{KN}(\omega)$ and are defined by

$$\omega_{NK}^{(max)} = \omega_{NK}^0 + \frac{1}{2}\sum_i \omega_{i'}\left[d_i^2\beta_i^2(1 - \rho_i)^4 - \rho_i\left(1 - \frac{\rho_i}{2}\right)\coth\frac{\hbar\omega_{i'}}{2kT}\right] \qquad (5\text{-}12)$$

and

$$(\Delta\omega_{KN})_{1/2} = (4 \log 2)\sum_i \omega_{i'}^2\left[d_i^2\beta_i^2(1 - \rho_i)^4 \coth\frac{\hbar\omega_{i'}}{2kT}\right.$$
$$\left. + \rho_i^2\left(1 - \frac{\rho_i}{2}\right)^2 \coth\frac{\hbar\omega_{i'}}{2kT}\right] \qquad (5\text{-}13)$$

The quantity ρ_i in equations 5-12 and 5-13 has been introduced for the normal frequency difference between the two electronic states, $\rho_i = (\omega_{i'} - \omega_{i''})/\omega_{i'}$. Equation 5-11 reduces to equation 5-10 if $\rho_i = 0$. If the expansion involved in equation 5-10 is carried out beyond t^2, then one gets the skew Gaussian band shape; in other words, the deviation from the Gaussian form may arise from this effect. Once $f_{KN}(\omega)$ is obtained, $f'_{KN}(\omega)$ can be found by differentiating $f_{KN}(\omega)$ with respect to ω.

Next we consider $f_{KN}(\omega)_i^{(1)}$. In contrast with $f_{KN}(\omega)$, $f_{KN}(\omega)_i^{(1)}$ is not normalized:

$$\int d\omega f_{KN}(\omega)_i^{(1)} = \sum_{v_{i'}} P_{Kv_{i'}}^{(0)} \left|\langle X_{Kv_{i'}} \mid Q_{i'}^2 \mid X_{Kv_{i'}}\rangle\right|^2 \qquad (5\text{-}14)$$

The promoting mode Q_i is usually not totally symmetric; in this case, $Q_{i'} = Q_{i''}$. If the harmonic-oscillator approximation is applied to Q_i, then equation 5-14 can be simplified to

$$\int d\omega f_{KN}(\omega)_i^{(1)} = \frac{\hbar}{2\omega_{i'}}\coth\frac{\hbar\omega_{i'}}{2kT} = (1 + 2\bar{n}_i)\frac{\hbar}{2\omega_{i'}} \qquad (5\text{-}15)$$

where $\bar{n}_i$ represents the phonon distribution $\bar{n}_i = (\hbar\omega_{i'}/e^{kT} - 1)^{-1}$. Using equation 5-2, $f_{KN}(\omega)_i^{(1)}$ can be written as

$$f_{KN}(\omega)_i^{(1)} = \frac{1}{2\pi} \int_{-\infty}^{\infty} dt\, e^{it\Delta\omega} K_i(t) \prod_j' G_j(t) \qquad (5\text{-}16)$$

where $K_i(t)$ is defined by

$$K_i(t) = \sum_{v_{i'}} \sum_{v_{i''}} P_{K v_{i'}}^{(0)} \left| \langle X_{K v_{i'}} \mid Q_i \mid X_{N v_{i''}} \rangle \right|^2 \exp[it(v_{i''}$$

$$+ \tfrac{1}{2})\omega_{i''} - it(v_{i'} + \tfrac{1}{2})\omega_{i'}] \qquad (5\text{-}17)$$

The evaluation of $K_i(t)$ is given by (45)

$$K_i(t) = K_i^{(0)}(t) G_i(t) \qquad (5\text{-}18)$$

where to the first-order approximation with respect to ρ_i, $K_i^{(0)}(t)$ can be simplified to

$$K_i^{(0)}(t) = \frac{1}{2\beta_{i'}^2} [(\bar{n}_i + 1)(1 - \rho_i \bar{n}) e^{it\omega_{i'}} + \bar{n}_i\{1 + \rho_i(1 + \bar{n}_i)\}\, e^{-it\omega_{i'}}]$$

$$(5\text{-}19)$$

Substituting equation 5-19 into equation 5-16 yields

$$f_{KN}(\omega)_i^{(1)} = \frac{1}{2\beta_{i'}^2} [(1 + \bar{n}_i)(1 - \bar{n}_i\rho_i) f_{KN}(\omega - \omega_{i'})$$

$$+ \bar{n}_i\{1 + \rho_i (1 + \bar{n}_i)\} f_{KN}(\omega + \omega_{i'})] \qquad (5\text{-}20)$$

Equation (5-20) indicates that the band shape of $f_{KN}(\omega)_i^{(1)}$ is the superposition of $f_{KN}(\omega - \omega_{i'})$ and $f_{KN}(\omega + \omega_{i'})$ weighted by the Boltzmann factors.

One can obtain $f'_{KN}(\omega)_i^{(1)}$ from $f_{KN}(\omega)_i^{(1)}$ by differentiating the latter with respect to ω.

In many cases, the vibrational bands of MCD spectra are partially resolved; the band shape functions of these vibrational bands, $f_{Kk,Nn}(\omega)$ and $f_{Kk,Nn}(\omega)_i^{(1)}$, are of the same functional forms as $f_{KN}(\omega)$ and $f_{KN}(\omega)_i^{(1)}$, but with ω_{NK}^0 replaced by $\omega_{Nn,Kk}^0$ for a particular vibrational band.

In the above discussion, we have shown that for the strong coupling case (i.e., $\sum_i (\beta_{i'}^2 d_i^2/2) \coth [\hbar\omega_{i'}/(2kT)] > 1$), $f_{KN}(\omega)$ can be expressed in the Gaussian form. For the case in which the vibration bands of MCD spectra are well resolved, we may not have the strong coupling situation; in that case, the damping effect has to be included, and we have

$$f_{Kk,Nn}(\omega) = \frac{1}{2\pi} \sum_{v_{I'}v_{I''}} \cdots$$

$$\sum_{v_{M'}v_{M''}} \int_{-\infty}^{\infty} dt (\prod_{i} P_{Kv_{i'}}^{(0)} \mid \langle X_{Kv_{i'}} \mid X_{Nv_{i'}} \rangle \mid^2) \exp[it(\omega_{Nn,Kk}^0 - \omega)$$

$$- \Gamma_{Nn(v_{i'}),Kk(v_{i'})} \mid t \mid + \sum_{i} it\{(v_{i''} + \tfrac{1}{2})\omega_{i''} - (v_{i'} + \tfrac{1}{2})\omega_{i'}\}] \quad (5\text{-}21)$$

where the relation

$$\frac{1}{\pi} \operatorname{Re} \frac{1}{i\omega + \gamma} = \frac{1}{\pi} \cdot \frac{\gamma^2}{\omega^2 + \gamma^2} = \frac{1}{2\pi} \int_{-\infty}^{\infty} dt \exp(i\omega t - \gamma \mid t \mid) \quad (5\text{-}22)$$

has been used. In other words, the delta function has been replaced by the corresponding Lorentzian. Notice that equation 5-21 indicates that in this case the overall band shape is the superimposition of progressive Lorentzians.

VI.   MOMENT RELATIONS

In this section, the moment relations of MCD spectra, defined as (29,33,50,80,105)

$$\int \omega^n \frac{\Delta k(\omega)}{\omega} d\omega = \left[\omega^n \frac{\Delta k(\omega)}{\omega} \right] \quad (6\text{-}1)$$

will be called the nth moments. The corresponding $[\omega^n f_{KN}(\omega)]$ and $[\omega^n f_{KN}(\omega)_i^{(1)}]$ appeared in equation 6-1. Only the low-order moment relations are important. For $n = 0$, we have

$$\left[\frac{\Delta k(\omega)}{\omega} \right] = H \left\{ \left[\frac{\Delta k(\omega)_A^{(1)}}{\omega} \right] + \left[\frac{\Delta k(\omega)_B^{(1)}}{\omega} \right] + \left[\frac{\Delta k(\omega)_C^{(1)}}{\omega} \right] \right\} \quad (6\text{-}2)$$

where

$$\left[\frac{\Delta k(\omega)_A^{(1)}}{\omega} \right] = 0 \quad (6\text{-}3)$$

$$\left[\frac{\Delta k(\omega)_B^{(1)}}{\omega}\right] = \frac{16\pi^2\alpha^2 N}{3c\hbar}\left[B(K \to N)\right.$$

$$\left. + \sum_i B_i^{(1)}(K \to N)(1 + 2\bar{n}_i)\frac{\hbar}{2\omega_{i'}}\right] \quad (6\text{-}4)$$

and

$$\left[\frac{\Delta k(\omega)_C^{(1)}}{\omega}\right] = -\frac{16\pi^2\alpha^2 N}{3c\hbar}\beta[C(K \to N)$$

$$+ \sum_i C_i^{(1)}(K \to N)(1 + 2\bar{n}_i)\frac{\hbar}{2\omega_{i'}}\right] \quad (6\text{-}5)$$

The summations over i in equations 6-4 and 6-5 cover all the promoting modes. In obtaining equation 6-3, the relations $[f'_{KN}(\omega)] = 0$ and $[f'_{KN}(\omega)_i^{(1)}] = 0$ have been used.

Similarly, for $n = 1$, we have

$$[\Delta k(\omega)] = H\{[\Delta k(\omega)_A^{(1)}] + [\Delta k(\omega)_B^{(1)}] + [\Delta k(\omega)_C^{(1)}]\} \quad (6\text{-}6)$$

where

$$[\Delta k(\omega)_A^{(1)}] = \frac{16\pi^2\alpha^2 N}{3c\hbar^2}\left[A(K \to N) + \sum_i A_i^{(1)}(K \to N)(1 + 2\bar{n}_i)\frac{\hbar}{2\omega_{i'}}\right] \quad (6\text{-}7)$$

$$[\Delta k(\omega)_B^{(1)}] = -\frac{16\pi^2\alpha^2 N}{3c\hbar}\{B(K \to N)[\omega f_{KN}(0)] +$$

$$\sum_i B_i^{(1)}(K \to N)[\omega f_{KN}(w)_i^{(1)}]\} \quad (6\text{-}8)$$

and

$$[\Delta k(\omega)_C^{(1)}] = -\frac{16\pi^2\alpha^2 N}{3c\hbar}\beta\{C(K \to N)[\omega f_{KN}(\omega)]$$

$$+ \sum_i C_i^{(1)}(K \to N)[\omega f_{KN}(\omega)_i^{(1)}]\} \quad (6\text{-}9)$$

Notice that for symmetry-forbidden transitions,

$$A(K \rightarrow N) = B(K \rightarrow N) = C(K \rightarrow N) = 0$$

From equations 6-8 and 6-9, we can see that to evaluate the first moment ($n = 1$), it is necessary to evaluate $[\omega f_{KN}(\omega)]$ and $[\omega f_{KN}(\omega)_i^{(1)}]$. Notice that they can be put into the forms

$$[\omega f_{KN}(\omega)] = \omega_{NK}^0 + \frac{1}{\hbar} \sum_k P_{Kk}^0 \langle \oplus_{Kk} | \hat{H}_N - \hat{H}_K | \oplus_{Kk} \rangle \quad (6\text{-}10)$$

and

$$[\omega f_{KN}(\omega)_i^{(1)}] = \omega_{NK}^0 \sum_k P_{Kk}^{(0)} | \langle \oplus_{Kk} | Q_{i''} | \oplus_{Nn} \rangle |^2$$

$$+ \frac{1}{\hbar} \sum_k P_{Kk}^{(0)} [\langle \oplus_{Kk} | Q_{i'} \hat{H}_N Q_{i'} - Q_{i'}^2 \hat{H}_K | \oplus_{Kk} \rangle] \quad (6\text{-}11)$$

where $\hat{H}_N$ and $\hat{H}_K$ are the Hamiltonians of nuclear motion of the electronic states N and K, respectively. The above equations indicate that the moment relations are invariant to the choice of the excited manifolds Nn.

To find explicit expressions for $[\omega^n f_{KN}(\omega)]$ and $[\omega^n f_{KN}(\omega)_i^{(1)}]$, a model for nuclear motion is necessary. We shall employ the harmonic-oscillator model for molecular vibrations. In this case, we find

$$[\omega f_{KN}(\omega)] = \omega_{NK}^0 + \sum_i \sum_{v_{i'}} \sum_{v_{i''}} P_{Kv_{i'}}^{(0)} \times$$

$$[(v_{i''} + \tfrac{1}{2})\omega_{i''} - (v_{i'} + \tfrac{1}{2})\omega_{i'}] | \langle X_{Kv_{i'}} | X_{Nv_{i''}} \rangle |^2 \quad (6\text{-}12)$$

which can be rewritten as (49,51)

$$[\omega f_{KN}(\omega)] = \omega_{NK}^0 + \sum_i \left[\frac{(\omega_{i''}^2 - \omega_{i'}^2)}{2\omega_{i'}} (\tfrac{1}{2} + \bar{n}_i) + \frac{\omega_{i'}^2 d_i^2}{2\hbar} \right] \quad (6\text{-}13)$$

Similarly, we have (49, 51)

$$[\omega f_{KN}(\omega)_i^{(1)}] = (\bar{n}_i + \tfrac{1}{2}) \frac{\hbar}{\omega_{i'}} [\omega f_{KN}(\omega)] + \frac{\hbar}{2} + \frac{\hbar(\omega_{i''}^2 - \omega_{i'}^2)}{\omega_{i'}^2} (\bar{n}_i + \tfrac{1}{2})^2$$

$$(6\text{-}14)$$

In other words, for $[\omega f_{KN}(\omega)_i^{(1)}]$, there is, in addition to $[\omega f_{KN}(\omega)]$, a shift due to the promoting mode.

It is important to note that in the moment relations, as n increases, the outer wings of the band contribute increasingly heavily, and that an individual moment is a single number describing an averaged property of the full dispersion curve. In fact, knowledge of all moments is required for construction of the actual spectrum. To demonstrate this point, and to present a general method for calculating the moment relation, we consider $f_{KN}(\omega)$. Equation 5-5 can be written as (51)

$$f_{KN}(\omega) = \frac{1}{2\pi} \int_{-\infty}^{\infty} dt\, e^{-it\omega} G_{KN}(t) \tag{6-15}$$

where

$$G_{KN}(t) = e^{it\omega_{KK}^0} \prod_i G_i(t) \tag{6-16}$$

By the Fourier transformation, we obtain (51)

$$G_{KN}(t) = \int_{-\infty}^{\infty} d\omega\, e^{it\omega} f_{KN}(\omega) \tag{6-17}$$

Equation 6-15 shows that to determine $f_{KN}(\omega)$, we need to determine $G_{KN}(t)$. Expanding $e^{it\omega}$ in power series of t yields

$$G_{KN}(t) = \sum_{n=0}^{\infty} \frac{(it)^n}{n!} [\omega^n f_{KN}(\omega)] \tag{6-18}$$

which shows that $G_{KN}(t)$ can in turn be determined by knowledge of the full moment relations $[\omega^n f_{KN}(\omega)]$ (51,96).

Expanding $G_{KN}(t)$ in power series of t, we find (31,62)

$$G_{KN}^{(n)}(0) = i^n [\omega^n f_{KN}(\omega)] \tag{6-19}$$

where $G_{KN}^{(n)}(0)$ represents the nth derivative of $G_{KN}(t)$ evaluated at $t = 0$. Equation 6-19 shows that the moment relations can be obtained from $G_{KN}(t)$. In other words, equation 6-19 provides a general method for calculating the moment relations. To find $G_{KN}(t)$, we use equation 6-16, and in general a model is required. Notice that the exact analytical expression for $G_i(t)$ in the harmonic-oscillator model has been obtained and is presented in Section V. For example, for the case in which $\rho_i = 0$, $G_i(t)$ is given by equation 5-8, and in this case equation 6-16 becomes

$$G_{KN}(t) = \exp\left[it\omega_{NK}^0 - \sum_i \frac{\beta_i^2 d_i^2}{2} \left\{ \coth \frac{\hbar\omega_{i'}}{2kT} \right. \right.$$
$$\left. \left. - \operatorname{csch} \frac{\hbar\omega_i'}{2kT} \cosh\left(it\omega_{i'} + \frac{\hbar\omega_i'}{2kT} \right) \right\} \right] \tag{6-20}$$

The corresponding moment relations can then be obtained by using equation 6-19.

The moment relations of absorption spectra can be defined similarly:

$$\int \omega^n \frac{k(\omega)}{\omega} d\omega = \left[\omega^n \frac{k(\omega)}{\omega} \right] \qquad (6\text{-}21)$$

They can be calculated in a manner similar to that presented above for the MCD spectra. In particular, for $n = 0$,

$$\left[\frac{k(\omega)}{\omega} \right] = \frac{8\pi^2 N}{3c\hbar} \alpha^2 \left[D(K \rightarrow N) + \sum_i D_i^{(1)}(K \rightarrow N)(1 + 2\bar{n}_i) \frac{\hbar}{2\omega_{i'}} \right] \qquad (6\text{-}22)$$

and for $n = 1$,

$$[k(\omega)] = \frac{8\pi^2 N}{3c\hbar} \alpha^2 \{ D(K \rightarrow N)[\omega f_{KN}(\omega)]$$

$$+ \sum_i D_i^{(1)}(K \rightarrow N)[\omega f_{KN}(\omega)_i^{(1)}] \} \qquad (6\text{-}23)$$

Notice that $[\omega f_{KN}(\omega)]$ and $[\omega f_{KN}(\omega)_i^{(1)}]$ are given in equations 6-13 and 6-14, respectively.

VII. DETERMINATION OF FARADAY PARAMETERS

The coefficients A, B, and C involved in the MCD expression are often called the Faraday parameters (or coefficients). In this section, we shall discuss how to determine these parameters from experimental MCD spectra.

A commonly used method for determining A, B, and C is based on the use of the moment relations (33,80,107). For the allowed electronic transition, the Franck–Condon approximation is a good approximation. In this case, the zeroth-moment relation of the MCD spectra is given by

$$\left[\frac{\Delta k(\omega)}{\omega} \right] = - \frac{16\pi^2 \alpha^2 N}{3c\hbar} H[B(K \rightarrow N) + \beta C(K \rightarrow N)] \qquad (7\text{-}1)$$

and the zeroth-moment of the absorption spectra is given by

$$\left[\frac{k(\omega)}{\omega} \right] = \frac{8\pi^2 N \alpha^2}{3c\hbar} D(K \rightarrow N) \qquad (7\text{-}2)$$

Similarly, the first-moment relations of the MCD and absorption spectra are given by

$$[\Delta k(\omega)] = \frac{16\pi^2\alpha^2 N}{3c\hbar} H \left(\frac{1}{\hbar} A(K \rightarrow N) - \{B(K \rightarrow N) \right.$$

$$\left. + \beta C(K \rightarrow N)\}[\omega f_{KN}(\omega)] \right) \quad (7\text{-}3)$$

and

$$[k(\omega)] = \frac{8\pi^2\alpha^2 N}{3c\hbar} D(K \rightarrow N)[\omega f_{KN}(\omega)] \quad (7\text{-}4)$$

respectively.

From equations 3-8–3-10, we can see qualitatively that the A term changes sign at the absorption maximum, whereas the B and C terms peak at this wavelength; that the C term is inversely proportional to T and is observable only if the ground state is degenerate; and that the A term is possible (i.e., nonzero) only if either the ground or excited state in the transition is degenerate, whereas the B term is generally present for all transitions in molecules. The magnitude of C is determined by the ground-state magnetic moment, a quantity that often either is known experimentally or is easy to calculate by using the transition moment. This can make the observation of C terms a powerful diagnostic tool for determining the symmetry of transitions. In addition, the C term is sensitive to interactions that quench the ground-state angular momentum, such as the Jahn–Teller distortion, low-symmetry crystal-field perturbation, and exchange interaction. Magnetic circular dichroism may thus be used to investigate these effects, particularly if a wide range of temperatures is available. Qualitatively, the appearance of an A term is unambiguous proof that at least one of the states involved in the transition is degenerate. This is an especially useful tool for identifying degenerate excited states when the ground state is known to be nondegenerate. Quantitative determination of the A value permits one to determine experimentally the excited-state magnetic moment, which is often used to test molecular wavefunctions.

Experimentally, the moment relations are computed numerically by integration of the MCD and absorption spectra according to equations 6-1 and 6-22. For example, if we use the molar ellipticity per unit magnetic field $[\theta]_m$ in the conventional units of natural optical activity (degrees times deciliters per decimeter per mole) and express A in units of (debye)^2 β_b and $(B + C/kT)$ in (debye)^2 β_b per centimeter (β_b is a Bohr magneton), then from equation 7-1 we obtain the zeroth moment of MCD as

$$\left[\frac{[\theta]_m}{\omega} \right] = -33.53 \left(B + \frac{C}{kT} \right) \quad (7\text{-}5)$$

Notice that in calculating $[[\theta]_m/\omega]$ from experimental MCD spectra, it is immaterial what frequency unit is used in the calculation. In equation 7-5, the solvent effect has been ignored. From equations 7-1 and 7-5, we can see that from experimental measurements of the zeroth moment of MCD at various temperatures, we can obtain the Faraday parameters B and C simultaneously.

From equations 7-2 and 7-4, we find

$$\frac{[k(\omega)]}{\left[\dfrac{k(\omega)}{\omega}\right]} = [\omega f_{KN}(\omega)] = \langle\omega\rangle \qquad (7\text{-}6)$$

which indicates that the quantity $[\omega f_{KN}(\omega)]$ appearing in $[\Delta k(\omega)]$ can be determined from the zeroth and first moments of the absorption spectra. Substituting equation 7-6 into equation 7-3 yields

$$[\Delta k(\omega)] = \frac{16\pi^2\alpha^2 N}{3c\hbar} H\left[\frac{A}{\hbar} - \langle\omega\rangle\left(B + \frac{C}{kT}\right)\right] \qquad (7\text{-}7)$$

In terms of $[\theta]_m$, equation 7-7 can be expressed in a form similar to equation 7-5 as

$$[[\theta]_m(\bar{\nu})] = 33.53\left[A - \langle\bar{\nu}\rangle\left(B + \frac{C}{kT}\right)\right] \qquad (7\text{-}8)$$

where the frequency $\bar{\nu}$ is in inverse centimeters. Thus, A, B, and C can be extracted from the measured moments of MCD and absorption spectra. The moment method is clearly not applicable if data are not available completely through a band, and the moment expressions will involve sums of Faraday parameters if several overlapping bands are involved. However, when applicable, the moment method is clearly of great utility.

Other methods for determining the Faraday parameters A, B, and C have been developed (44,80). For example, one obvious method is to fit the experimental data to equation 3-1 using a least square procedure assuming a specific (usually Gaussian or Lorentzian) band shape. The fitting method based on a Gaussian band shape is usually good, but if both A and $(B + C/kT)$ contributions exist and if one term dominates, the values obtained for the nondominant terms by this fitting method are less reliable. Nevertheless, in cases of overlapping bands and cases in which data can be obtained only part way through a band, the curve-fitting method seems to be the only practical alternative.

For symmetry-forbidden transitions, the same basic considerations apply, but it is necessary to consider the details of the intensity-borrowing mechanism to perform a complete analysis (44,106). For a detailed discussion of this topic, the original papers should be consulted.

For a given transition, the magnitudes of the A, B, and C terms in MCD are dependent on ω, kT, and Γ (the apparent damping constant if the band shape is assumed to be Lorentzian). To estimate their importance in the region of the absorption band (12,106,107), we assume $A \approx C \approx B \times 10^4$ cm^{-1}, $\Gamma \approx 10^3$ cm^{-1}, and $kT \approx 200$ cm^{-1} (room temperature). In this case, the A, B, and C terms are approximately in the ratio 20:1:50.

VIII. APPLICATIONS

Magnetic circular dichroism is a useful tool for the assignment of absorption bands and for the investigation of magnetic properties of the ground and excited states of atoms, ions, and molecules. A major accomplishment in the application of MCD has been the observation of the spectra of spin-forbidden transitions and weak transitions of paramagnetic molecules, and the resolution of the spectra of symmetry-forbidden transitions into the contributions from various promoting modes; in other words, by using MCD one can obtain information on the promoting modes involved in symmetry-forbidden transitions.

In view of the rapid growth of research activity on MCD (and MOR), it would clearly be impossible and unwise to review the literature related to magnetic optical activity in this chapter. Instead, in this section we shall present some representative examples of important applications of MCD.

A. DIATOMIC MOLECULES

The MCD and uv spectra (800–250 nm) of I_2, Br_2, and Cl_2 in solution have been investigated by Brith et al. (10). The ground-state configuration of the halogen molecules is $(ns\sigma_g)^2(ns\sigma_u^*)^2(np\sigma_g)^2(np\pi_u)^4(np\pi_g^*)^4$ $^1\Sigma_g^+$ (63,64,65). The lowest excited configuration is $(np\pi_g^*)^3(np\sigma_u^*)$, which gives rise to $^1\Pi_u$ and $^3\Pi_u$, and spin–orbit coupling splits $^3\Pi_u$ into component states 2_u, 1_u, 0_u^+, and 0_u^- (66–68). The MCD spectra have been interpreted in terms of the transitions to the states $1_u(^3\Pi)$, $0_u^+(^3\Pi)$, and $^1\Pi$. Brith et al. have resolved the observed absorption spectrum of I_2 into $0_u^+(^3\Pi)$ and $^1\Pi$ components.

The occurrence of oxygen in the atmosphere has naturally led many investigators of the solar spectrum to study its absorption spectrum. Besides the well-known Schumann–Runge bands and the atmosphere bands, Wulf (119) noticed another distinct absorption present in oxygen at moderate pressure and in the liquid state. Finkelnburg and Steiner (28) investigated this progression in more detail and found that each band consisted of three diffuse peaks and that the intensity varied with the square of the pressure. They interpreted these bands as due to a $^3\Sigma_g^- \rightarrow {}^3\Delta_u$ transition, with the triplet structure being caused by the splitting of the $^3\Delta_u$ state by spin–orbit coupling. Herzberg (34,35) has studied the absorption of oxygen in this region of the spectrum using path lengths of up to 800 m and pressures of 2 or 3 atm. He has been able to identify three separate band systems, which he assigns as transitions to the $^3\Sigma_u^+$, $^1\Sigma_u^-$, and $^3\Delta_u$ states.

The MCD spectra of atmospheric oxygen condensed on a cold window have been measured in the region 240–270 nm by Douglas et al. (23). The MCD spectra at 19 K and 48 K are dominated by two progressions of oppositely signed peaks that decrease in intensity with increasing temperature, a characteristic of MCD C terms. The positions of the MCD peaks are in very close agreement with the bands observed by Finkelnburg and Steiner. The observation of C terms of opposite sign has been shown to arise when the spin–orbit coupling occurs in the excited state and the ground state is spin degenerate. As a result, the excited state cannot be a singlet or Σ state. The lowest electron configuration of oxygen is $(\sigma_g 2s)^2(\sigma_u 2s)^2(\sigma_g 2p)^2$ $(\pi_u 2p)^4(\pi_g 2p)^2$, which gives rise to the ground states $X^3\Sigma_g^-$ and two low-lying excited states $b^1\Sigma_g^+$ and $a^1\Delta_g$. The first excited configuration, $(\sigma_g 2s)^2(\sigma_u 2s)^2(\sigma_g 2p)^2(\pi_u 2p)^3(\pi_g 2p)^3$, gives rise to a number of states, including $^3\Sigma_u^+$ (responsible for the Schumann–Runge bands) and $^3\Sigma_u^+$ (the Herzberg I-bands). Of the remaining states, only $^3\Delta_u$ has the spin–orbit coupling necessary to produce the temperature dependent C terms seen in MCD. Other states in which the spin–orbit coupling could give rise to C terms are $^3\Pi_u$ and $^3\Pi_g$ from the configurations $(\sigma_g 2s)^2(\sigma_u 2s)^2(\sigma_g 2p)^2(\pi_u 2p)^4$ $(\pi_g 2p)^1(\sigma_u 2p)^1$ and $(\sigma_g 2s)^2(\sigma_u 2s)^2(\sigma_g 2p)^1(\pi_u 2p)^4(\pi_g 2p)^3$, respectively. However, Douglas et al. showed that the transition involved is $^3\Sigma_g^- \rightarrow {}^3\Delta_u$, based on the following arguments: (1) Calculations of the spin–orbit coupling and the expected signs of the C terms for $^3\Delta_u$, $^3\Pi_u$, and $^3\Pi_g$ show that the spectrum is inconsistent with $^3\Sigma_g^- \rightarrow {}^3\Pi_g$; (2) the observed separation of the C terms agrees with that predicted for the spin–orbit splitting of the $^3\Delta_u$ state, but is twice that expected for the $^3\Pi$ states; and (3) the transition is extremely weak and could not be detected in absorption, so it is unlikely to be the electric-dipole-allowed $^3\Sigma_g^- \rightarrow {}^3\Pi_u$ transition. This example shows how sensitive MCD spectroscopy is for detecting very weak transitions of paramagnetic materials at low temperatures.

B. ORGANIC MOLECULES

1. Spin-Forbidden Transitioins

It might be expected that MCD would provide a facile means of detecting spin-forbidden transitions (e.g., singlet–triplet transitions) in organic molecules, since the triplet state should give the A term not possible in a singlet–singlet transition. This has not been found in practice until recently. We shall describe here two systems in which the MCD spectra of singlet–triplet transitions have been reported.

Linder et al. (55) demonstrated the utility of MCD spectroscopy in the location of singlet–triplet absorption bands by measuring the vapor-phase MCD spectra associated with the singlet–triplet $n \rightarrow \pi^*$ absorption band of formaldehyde and formaldehyde-d_2. Seamans et al. (86) have carried out a

TABLE 12.I
Wavefunctions of Formaldehyde

$$\psi_{1A_1} = |(\overset{+}{\sigma})(\overset{-}{\sigma})(\overset{+}{\pi})(\overset{-}{\pi})(\overset{+}{n})(\overset{-}{n})|$$

$$\psi_{3A_2}(1) = |(\overset{+}{\sigma})(\overset{-}{\sigma})(\overset{+}{\pi})(\overset{-}{\pi})(\overset{+}{n})(\overset{+}{\pi^*})|$$

$$\psi_{3B_1}(1) = |(\overset{+}{\sigma})(\overset{+}{\pi})(\overset{-}{\pi})(\overset{+}{n})(\overset{-}{n})(\overset{+}{\pi^*})|$$

$$\psi_{3B_2}(1) = |(\overset{+}{\sigma}_1)(\overset{-}{\sigma}_1)(\overset{+}{\pi})(\overset{-}{\pi})(\overset{+}{n})(\overset{+}{\sigma^*})|$$

theoretical analysis of the fine structures of these gaseous MCD spectra. Because of the importance of the carbonyl chromophore in molecular spectroscopy and photochemistry, Yoon et al. (120,121) have carried out a theoretical calculation of the magnetic rotational strength $A\ (^1A_1 \rightarrow {}^3A_2)$ of the spin-forbidden $n \rightarrow \pi^*$ transition of carbonyl compounds. The wavefunctions used in the calculation are given in Table 12.I; only the wavefunction of the highest M_s value (the z-component of the spin angular momentum) is given. Matrix elements of the spin–orbit coupling are given in Table 12.II. Magnetic moments of each spin sublevel of the $^3A_2$ state can be calculated easily. They are

$$\vec{\mu}(^3A_2(0)^3A_2(0)) = 0 \tag{8-1}$$

$$\vec{\mu}(^3A_2(1)^3A_2(1)) = \hat{k}\frac{e\hbar}{mc} = -\vec{\mu}(^3A_2(-1)^3A_2(-1)) \tag{8-2}$$

where $\hat{k}$ is the unit vector in the z-direction. The matrix elements of the electric transition moments are given in Table 12.III. For numerical calculation, the simple molecular orbitals of the carbonyl group that were adopted by Pople and Sidman (73) were used. Only one-center integrals were taken into account, which is consistent with the simple molecular orbital calculation. In the perturbation calculation, only the lower excited states n

TABLE 12.II
Matrix Elements of Spin–Orbit Coupling[a]

$$\langle\psi^{(0)}_{3B_2}(1)\mid\hat{H}'_{so}\mid\psi^{(0)}_{1A_1}\rangle = -\langle\psi_{3B_2}(-1)\mid\hat{H}'_{so}\mid\psi^{(0)}_{1A_1}\rangle = -26.8\,i$$

$$\langle\psi^{(0)}_{3B_1}(1)\mid\hat{H}'_{so}\mid\psi^{(0)}_{1A_1}\rangle = \langle\psi^{(0)}_{3B_1}(-1)\mid\hat{H}'_{so}\mid\psi^{(0)}_{1A_1}\rangle = -26.2$$

$$\langle\psi^{(0)}_{3A_2}(0)\mid\hat{H}'_{so}\mid\psi^{(0)}_{1A_1}\rangle = -41.4\,i$$

$$\langle\psi^{(0)}_{1B_2}\mid\hat{H}'_{so}\mid\psi^{(0)}_{3A_2}(1)\rangle = \langle\psi^{(0)}_{1B_2}(-1)\mid\hat{H}'_{so}\mid\psi^{(0)}_{3A_2}(-1)\rangle = 7.11$$

$$\langle\psi^{(0)}_{1B_1}\mid\hat{H}'_{so}\mid\psi^{(0)}_{3A_2}(1)\rangle = -\langle\psi^{(0)}_{1B_1}\mid\hat{H}'_{so}\mid\psi^{(0)}_{3A_2}(-1)\rangle = 25.5\,i$$

[a] Given in units of $e^2\hbar^2/12m^2c^2a_0^3$, where e is charge, m is mass, c is the velocity of light, and a_0 is Bohr radius.

TABLE 12.III

Matrix Elements of Electric Moments

$$\langle\psi_{B_1}^{(0)}(1)|\vec{R}|\psi_{A_2}^{(0)}(1)\rangle = \langle\psi_{B_1}^{(0)}(-1)|\vec{R}|\psi_{A_2}^{(0)}(-1)\rangle = -\langle(n)|er_i|(\sigma)\rangle = -0.320jK_0$$

$$\langle\psi_{B_2}^{(0)}(1)|\vec{R}|\psi_{A_2}^{(0)}(1)\rangle = \langle\psi_{B_2}^{(0)}(-1)|\vec{R}|\psi_{A_2}^{(0)}(-1)\rangle = \langle(\sigma^*)|er_i|(\pi^*)\rangle = \hat{i}(0.368\,K_c + 0.144\,K_0)$$

$$\langle\psi_{A_1}^{(0)}|\vec{R}|\psi_{B_2}^{(0)}\rangle = \sqrt{2}\,\langle(n)|e\vec{r}_i|(\sigma^*)\rangle = -0.339jK_0$$

$$\langle\psi_{A_1}^{(0)}|\vec{R}|\psi_{B_1}^{(0)}\rangle = \hat{i}(0.396\,K_c - 0.271\,K_0)$$

$\to \pi^*$, $n \to \sigma^*$, $\pi \to \pi^*$, and $\sigma \to \pi^*$ of the carbonyl group were included. Notice that using the Slater atomic orbitals, Yoon et al. obtained

$$K_c = \langle(2P_x^c)\,|\,ex\,|\,(2s^c)\rangle = 0.9124\,ea_0 \qquad (8\text{-}3)$$

$$K_0 = \langle(2P_x^0)\,|\,ex\,|\,(2s^0)\rangle = 0.6481\,ea_0 \qquad (8\text{-}4)$$

Using Tables 12.II and 12.III and equation 3-11, they calculated the magnetic rotational strength A ($^1A_1 \to {}^3A_2$) to be -0.549×10^{-7} (debyes)$^2\beta_b$. The result seems to be in qualitative agreement with the reported gas-phase MCD spectra of formaldehyde for $^1A_1 \to {}^3A_2$. It is therefore desirable to measure solution MCD spectra of $^1A_1 \to {}^3A_2$ transitions of carbonyl compounds.

Seamans et al. (87,88) reported the MCD spectrum of the $^1A_{1g} \to {}^3B_{3u}$ ($n \to \pi^*$) transition of pyrazine (1.4-diazine). Both vapor-phase and solution-phase (in iso-octane) MCD spectra were measured. Three vibronic bands are observed in the absorption and MCD spectra of pyrazine; the one at the lowest frequency is the 0–0 band. The spacings and intensities of the other two bands are consistent with assignments as the 5_0^1 and 3_0^1 bands ($\nu_5 = 621$ cm^{-1} and $\nu_3 = 1149$ cm^{-1}). The solution MCD spectrum (see Fig. 12.3) of the 0–0 band exhibits a single A term of negative sign, which indicates that line broadening has greatly reduced the contribution of the sharp feature that dominates the gas-phase spectrum, and the amplitude of the MCD band is decreased 100-fold compared with that observed in the gas-phase spectrum. The Faraday A and B coefficients, dipole strengths D, and frequencies of the band centers for different vibronic bands, along with the corresponding parameters for the entire electronic band, are presented in Table 12.IV.

2. Symmetry-Forbidden Transitions

In this section we shall discuss the MCD spectroscopy of the symmetry-forbidden transitions for two important molecules, benzene and formaldehyde.

Shieh et al. (98) have reported the MCD and absorption spectra of benzene in n-heptane at room temperature. Figure 12.4 shows the MCD spectrum of benzene in the wavelength range 230–270 nm, which corresponds to the $^1A_{1g} \to {}^1B_{2u}$ transition. In the figure, $[\theta]_M$ represents the molar ellipticity per gauss

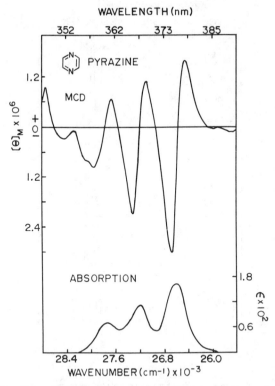

Fig. 12.3.   Uv and MCD spectra of spin-forbidden transitions of pyrazine.

measured with a modified Cary Model 60 spectrophotometer furnished with
a Varian superconducting solenoid system (at a magnetic field of 45 kG).
The absorption spectra were measured with a Cary Model 14 spectrophotometer. It is well known that this electronic transition is symmetry forbidden
but vibronically allowed. The symmetry-forbiddenness is removed by the

TABLE 12.IV

Experimental Results for Spin-Forbidden Transitions of Pyrazine ($^1A_{1g} \rightarrow {}^3B_{3u}$, $n \rightarrow \pi^*$)[a]

| Band | Frequency (cm^{-1}) | | $D \times 10^{6}$[b] | $A \times 10^{7}$[c] | $B \times 10^{17}$[d] |
	Absorption	MCD			
0–0	26,600	26,543	2.49	−1.8	2.81
5^1_0	27,233	27,178	1.60	−1.4	2.32
3^1_0	27,772	27,781	0.99	−0.9	0.62

[a] Entire electronic band (moment analysis): $\bar{\nu} = 27,040$ cm^{-1}, $A = -4.0$, $D = 4.9 \times 10^{-6}$.
[b] In units of (debyes)^2.
[c] In units of (debyes)$^2\beta_b$.
[d] In units of (debyes)$^2\beta_b$/cm^{-1}.

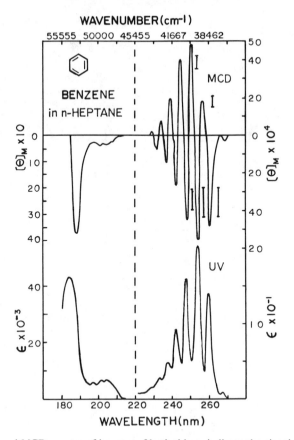

Fig. 12.4.   Uv and MCD spectra of benzene. Vertical bars indicate the signal-to-noise ratios.

vibronic coupling by the normal vibrations of e_{2g} symmetry, which mix the $^1B_{2u}$ state with a higher $^1E_{1u}$ state, allowing the transition to appear. Benzene has four fundamentals of e_{2g}, with ground-electronic-state frequencies of 608 (ν_6), 3056 (ν_7), 1596 (ν_8), and 1178 (ν_9) cm^{-1}. Besides the ν_6 mode, two of the remaining three modes also appear in the uv spectrum of benzene, the exception being the ν_8 mode, which involves mainly in-plane CH bending (13). From the spacings between successive vibrational bands in the uv and MCD spectra (see Fig. 12.4), we can see that the normal mode a_{1g} (ν_1, 992 cm^{-1}, ground electronic state) plays an important role in vibrational structuring of the absorption spectrum and MCD spectrum of benzene. It should be noted that MCD spectra of the $^1A_{1g} \rightarrow {}^1B_{2u}$ transition of benzene in solution at room temperature (77,108) and liquid nitrogen temperature (77) and in the vapor phase (24) have been reported by other workers.

    A main feature of the MCD spectrum of benzene is that it shows consecutive positive and negative peaks even in the electronically nondegenerate transition. Based on analysis of the MCD spectrum and calculation of the magnetic rotational strengths of the four promoting e_{2g} modes (44,77), it is now generally agreed that the positively and negatively signed bands in the MCD spectrum of the $^1A_{1g} \rightarrow {}^1B_{2u}$ transition of benzene result from the vibronic effects of different promoting modes: The negative bands can be attributed to a progression in the ν_1 mode vibronically induced by the ν_6 and ν_8 modes, and the positive bands can be attributed to a progression in the ν_1 mode vibronically induced by the ν_7 and ν_9 modes.

This example demonstrates an important application of MCD spectroscopy: finding the promoting modes involved in symmetry-forbidden transitions. In the following, we briefly outline the vibronic coupling calculation and the calculation of the magnetic rotational strengths of different promoting modes based on the Herzberg–Teller theory.

If we let V represent the interaction energy between electrons and nuclei,

$$V = \sum_i \sum_\sigma \frac{Z_\sigma e^2}{r_{i\sigma}} \tag{8-5}$$

then the Hamiltonian for the vibronic coupling can be expressed as

$$\hat{H}'_{vc} = \sum_\alpha \left(\frac{\partial V}{\mathrm{pt}Q_\alpha}\right)_0 Q_\alpha + \cdots$$

$$= \sum_\alpha \sum_i \sum_\sigma \left(\frac{Z_\sigma e^2}{r_{i\sigma}^3}\right)_0 \left(\vec{r}_{\sigma i} \cdot \frac{\partial \Delta \vec{r}_\sigma}{\partial Q_\alpha}\right)_0 Q_\alpha + \cdots \tag{8-6}$$

where $r_{i\sigma} = |\vec{r}_i \rightarrow \vec{r}_\sigma|$ and

$$\Delta \vec{r}_\sigma = \vec{r}_\sigma - \vec{r}_{\sigma 0} = \hat{i}\Delta x_\sigma + \hat{j}\Delta y_\sigma + \hat{k}\Delta z_\sigma \tag{8-7}$$

For numerical calculation of the vibronic coupling of benzene, the conventional ASMO functions for benzene (see Table 12.V) are used, only π electrons and single-electron excitations are considered, and the overlap of the atomic orbitals is taken into account. The normal coordinates of the e_{2g} vibrations are taken from Albrecht (3), and the values of $(\partial \Delta \vec{r}_\sigma / \partial Q_\alpha)_0$ are given in Table 12.VI. The Z_σ value of 3.18 is used for the $2p\pi$ atomic orbital and the Z_σ value of 1.00 is used for both nearest- and nonnearest-neighbor interactions. The one-center integrals in the vibronic coupling calculation for benzene vanish, and all the multicenter integrals are calculated exactly (44). The computed matrix elements of the vibronic coupling of benzene are given in Table 12.VII. Using these matrix elements, the magnetic rotation strengths and dipole strengths can be calculated; the computed results are shown in Tables 12.VIII and 12.IX along with the experimental results and the computed results of Rosenfeld et al. (75).

Seamans and Moscowitz (85) have shown that in an electric-dipole-forbidden transition such as the 290-nm $n \rightarrow \pi^*$ transition of saturated ketones,

TABLE 12.V
Wavefunctions and Molecular Orbitals of Benzene

$$\phi_1 = 0.3238(x_1 + x_2 + x_3 + x_4 + x_5 + x_6)$$
$$\phi_2 = 0.2637(2x_1 + x_2 - x_3 - 2x_4 - x_5 + x_6)$$
$$\phi_3 = 0.4567(x_2 + x_3 - x_5 - x_6)$$
$$\phi_4 = 0.5855(-x_2 + x_3 - x_5 + x_6)$$
$$\phi_5 = 0.3381(2x_1 - x_2 - x_3 + 2x_4 - x_5 - x_6)$$
$$\phi_6 = 0.5485(x_1 - x_2 + x_3 - x_4 + x_5 - x_6)$$

State function	Energy (eV)
$\psi_{B_{2u}} = \dfrac{1}{\sqrt{2}}(\psi_3^5 - \psi_2^4)$	4.9
$\psi_{B_{1u}} = \dfrac{1}{\sqrt{2}}(\psi_3^4 + \psi_2^5)$	6.2
$\psi_{E_{1ux}} = \dfrac{1}{\sqrt{2}}(\psi_3^5 + \psi_2^4)$	7.0
$\psi_{E_{1uy}} = \dfrac{1}{\sqrt{2}}(\psi_3^4 - \psi_2^5)$	7.0
$\psi_{E_{1u}(1)} = \dfrac{1}{\sqrt{2}}(\psi_{E_{1ux}} + i\psi_{E_{1uy}})$	
$\psi_{E_{1u}(2)} = \dfrac{1}{\sqrt{2}}(\psi_{E_{1ux}} - i\psi_{E_{1uy}})$	

TABLE 12.VI
Cartesian Displacement Coordinates $(\partial \Delta \vec{r}_\sigma / \partial Q_\alpha)_0 \times 10^2$

	E_{2ga}				B_{2u}
	Q_6	Q_7	Q_8	Q_9	Q_{14}
xC_1	0	0	0	0	−4.375
yC_1	−7.635	−3.011	−3.987	2.637	0
xC_2	5.442	1.139	−1.799	−0.642	2.187
yC_2	−1.790	1.039	7.102	−1.525	−3.789
xC_3	5.447	1.139	−1.799	−0.642	2.187
yC_3	1.790	−1.039	−7.102	1.525	3.789
xC_4	0	0	0	0	−4.375
yC_4	7.635	3.011	3.987	−2.637	0
xC_5	−5.442	−1.139	1.799	0.642	2.187
yC_5	1.790	−1.039	−7.102	1.525	−3.789
xC_6	−5.442	−1.139	1.799	0.642	2.187
yC_6	−1.790	1.039	7.102	−1.525	3.789

TABLE 12.VII
Vibronic Coupling Matrix Elements

Symmetry	Normal mode	ψ_A	ψ_B	$\langle \psi_A \mid \hat{H}'_{vc} \mid \psi_B \rangle$ (eV)[a]	$\langle \psi_A \mid \dfrac{\partial}{\partial Q} \mid \psi_B \rangle \times 10^{-19}$ $\left(\dfrac{1}{(grams)^{1/2} \, cm} \right)$
E_{2ga}	Q_6	B_{2u}	E_{1ux}	0.0874	1.94
		B_{1u}	E_{1uy}	0.0238	1.4
	Q_7	B_{2u}	E_{1ux}	0.0086	0.427
		B_{1u}	E_{1uy}	0.0139	1.808
	Q_8	B_{2u}	E_{1ux}	0.0164	0.589
		B_{1u}	E_{1uy}	0.1074	10.138
	Q_9	B_{2u}	E_{1ux}	0.080	0.247
		B_{1u}	E_{1uy}	0.0298	2.419
B_{2u}	Q_{14}	A_{1g}	B_{2u}	0.0829	1.155

[a] For convenience of tabulation, the matrix elements $\langle \psi_A \mid \hat{H}'_{vc} \mid \psi_B \rangle$ have been multiplied by $\overline{Q_i^2} = \sqrt{\hbar/4\pi c \bar{\nu}_i}$ at $T = 0$.

the intensity of the associated MCD curve must be second or higher order in the perturbations responsible for that intensity, and that perturbations belonging to different irreducible representations can contribute in a simple, additive manner. Complications can arise, however, in connection with the concurrence of vibrational (or vibronic coupling) and structural perturbations, especially when these perturbations belong to the same irreducible

TABLE 12.VIII
Magnetic Rotational Strengths for $^1A_{1g} \to {}^1B_{2u}$ and $^1A_{1g} \to {}^1B_{1u}$ Transitions in Benzene

Normal mode	Frequency (cm^{-1})	$\dfrac{\hbar}{2\omega_i} B_i \times 10^5 \left(\dfrac{D^2 \beta_b}{cm^{-1}} \right)^a$		
		$^1A_{1g} \to {}^1B_{2u}$		$^1A_{1g} \to {}^1B_{1u}$
		Calculated	Reference[b]	Calculated
Q_6	608	0.522	0.65	5.99
Q_7	3056	-0.051	-0.001	-0.19
Q_8	1596	0.677	0.244	12.67
Q_9	1178	-0.096	-0.003	-0.03
$Q_6 + Q_8$		1.199	0.901 $(0.407)^c$	18.66
$Q_7 + Q_9$		-0.147	-0.005 $(-0.297)^c$	-0.22

[a] The B values here are obtained at $T = 0$.
[b] From reference 75; $Z_\sigma = 0.4$.
[c] our experimental values.

TABLE 12.IX
Oscillator Strengths for $^1A_{1g} \rightarrow {}^1B_{2u}$ and $^1A_{1g} \rightarrow {}^1B_{1u}$ Transitions in Benzene

Normal mode	Frequency (cm^{-1})	$\dfrac{\hbar}{2\omega_i} f_i \times 10^2$			
		Exp.[a]	Calc.	Ref.[b]	Ref.[c]
		$^1B_{2u}$ transition			
Q_6	608	0.14	0.166	0.09	0.0498
Q_7	3056	0	0.0016	0.00	0.0008
Q_8	1596	0.001	0.0058	0.00	0.0248
Q_9	1178	0	0.0014	0.00	0.0040
Total		0.14	0.1748	0.09	0.0794
		$^1B_{1u}$ transition			
Q_6	608	1.9	0.108	1.5	1.628
Q_7	3056	0	0.036	0.0	0.066
Q_8	1596	7.5	2.184	20.0	6.816
Q_9	1178	0	0.168	1.5	0.588
Total		9.4	2.496	23.0	9.098

[a] Garforth, F. M., C. K. Ingold, and H. G. Poole, *J. Chem. Soc.*, 406 (1948) for $^1B_{2u}$ transitions; Birth, M., R. Lubart, and T. Sternberger, *J. Chem. Phys.*, **54**, 5104 (1971) for $^1B_{1u}$ transitions.

[b] Roche, M., and H. H. Jaffe, *J. Chem. Phys.*, **60**, 1193 (1974).

[c] Ziegler, L., and A. C. Albrecht, *J. Chem. Phys.*, **60**, 3558 (1974).

representation (84,96). A prime prerequisite for a fuller understanding of the effects of both vibrational and structural perturbations in MCD is a quantitative grasp of the role played by vibrations that obtain in the absence of interfering structural perturbations. To acquire such knowledge, Linder et al. (55,56) have carried out theoretical investigations of the vibronically induced MCD intensity for the $n \rightarrow \pi^*$ transition of formaldehyde. From equations 3-20 and 6-4, we obtain

$$B(K \rightarrow N) = \sum_i B_i^{(1)}(K \rightarrow N)(1 + 2\bar{n}_i) \frac{\hbar}{2\omega_i'} = \sum_i B(K \rightarrow N)_i \quad (8\text{-}8)$$

where $B(K \rightarrow N)_i$ is the integrated partial magnetic rotational strength associated with the ith promoting mode. Equation 8-8 shows that in general the magnetic rotational strength depends on temperature through the phonon distribution $\bar{n}_i$, increasing with increasing T. If T is low ($\hbar\omega_i'/kT \ll 1$), then $\bar{n}_i$ may be set to zero.

Linder et al. (56) calculated the magnetic rotational strengths associated with the non-totally symmetric vibrations in formaldehyde using CNDO/2 wavefunctions. For the planar (C_{2v}) formaldehyde molecule, there are three non-totally symmetric vibrational modes: the out-of-plane bending vibration $\nu_4(b_1)$, the asymmetric C—H stretch $\nu_5(b_2)$, and the in-plane bend $\nu_6(b_2)$.

TABLE 12.X

Experimental and Calculated Magnetic Rotational Strengths and Oscillator Strengths for the $n \rightarrow n^*$ ($^1A_1 \rightarrow {}^1A_2$) Transition in Formaldehyde[a]

	Calculated			
Property	Minimal basis set, no CI	Minimal basis set, CI included	Extended basis set, CI included	Experimental[b]
$-B_4$	-78.11	-9.36	-11.88	—
$-B_5$	2.93	0.94	0.89	—
$-B_6$	0.97	1.39	1.33	—
$B_4/(B_5 + B_6)^c$	-20.0	-4.0	-5.3	-4
f_4	23.2	0.81	1.57	—
f_5	0.13	0.006	0.006	—
f_6	0.05	0.039	0.039	—
$f_4/(f_5 + f_6)$	>100	18	35	10^d
$f_4 + f_5 + f_6$	23.4	0.86	1.61	2.40

[a] Units for magnetic rotational strengths are 10^{-7} Bohr magnetons (debyes)^2/cm^{-1}. All oscillator strengths have been multiplied by 10^4.

[b] In the experimental determination of the MCD spectrum of formaldehyde, polymerization precluded absolute measurements. Hence we could estimate only the ratios of the intensities from the b_1 and b_2 modes.

[c] The experimental determination of the ratio $B_4/(B_5 + B_6)$ is complicated by the large degree of overlap of positive and negative bands in the MCD spectrum. The value reported here, -4, is therefore only approximate.

[d] Estimated by Job, Sethuraman, and Innes (37) as 10:1 for b_2:b_2.

The B coefficient associated with the $n \rightarrow \pi^*$($^1A_1 \rightarrow {}^1A_2$) transition in formaldehyde is therefore given by

$$B(^1A_1 \rightarrow {}^1A_2) = B_4 + B_5 + B_6 \qquad (8\text{-}9)$$

In their MCD calculation, Linder et al. used the so-called geometric perturbation (GP) method. In this method, the nuclei are displaced by some distance, say $\langle 0 \mid Q_i^2 \mid 0 \rangle^{1/2}$, in the direction of a normal coordinate Q_i, and the electronic wavefunctions appropriate to the new geometry are calculated; these electronic wavefunctions are then used to calculate the partial rotational strength B_r directly from equation 3-4 and the total strength $B(n \rightarrow \pi^*)$ from equation 8-8 using the formula $\langle 0 \mid Q_i^2 \mid 0 \rangle^{1/2} = \hbar/2\omega_i'$. Only one-center matrix elements are retained, and the one-center magnetic dipole matrix elements coupling s and p orbitals are neglected. Linder et al. carried out three types of CNDO/2 calculation: (1) with the minimal basis set but without configuration interactions (CI); (2) with the minimal basis set and including CI; and (3) with the extended basis set and including CI (72). The calculated partial B_i values and the partial oscillator strengths f_i induced by each normal mode are given in Table 12.X. The experimental values of B_i and f_i are due to Seamans et al. (86) and Job et al. (37), respectively. The agreement with experiment when CI are included is not bad. From this table

TABLE 12.XI

$B(^1A_{1g} \rightarrow {}^1B_{2u})$ and q_m of Substituted Benzenes

Substituent	Position	$B(^1A_{1g} \rightarrow {}^1B_{2u})$ $(D^2\beta_b/cm^{-1} \times 10^5)$	$q_m \times 10^{10}$
CH$_3$	1	2.15	5.0
	1,2	2.35	
	1,3	1.34	
	1,4	5.72	
	1,2,4	4.88	
	1,3,5	−0.68	
	1,3,4,5	1.31	
	1,2,4,5	7.23	
	Penta	0.95	
	Hexa	−1.36	
Cl	1	3.38	6.0
	1,2	4.26	
	1,3	3.72	
	1,2,4	11.58	
	Penta	3.27	
	Hexa	−3.45	
F	1	1.16	12.5
	Hexa	−1.34	
Br	1	3.38	6.0
I	1	2.80	8.0
CN	1	−14.61	−11.0
OH	1	16.22	20.0
COOH	1	−28.57	−17.0
	Hexa	−5.29[a]	
NO$_2$	1	−34.84	
CHO	1	−36.06	−20.0
NH$_2$	1	46.55	24.0

[a] Methanol solution.

we can see that inclusion of CI is necessary if any sort of reasonable agreement with experiment is to be achieved.

3. Substituted Benzenes

Shieh et al. (98) have investigated the effects of chemical substitution in aromatic hydrocarbons on the electronic transitions associated with MCD; their goal is to find rules that will predict the sign of the MCD of substituted aromatic hydrocarbons and that will describe the changes of the MCD intensity that occur with changes in the types and number of substituents and the position of substitution. For this purpose, they measured MCD and absorption spectra of 27 substituted benzenes for the $^1A_{1g} \rightarrow {}^1B_{2u}$ transition of benzene (see Table 12.XI). All the spectra were obtained at room temperature using n-heptane as the solvent, except those of benzene hexacarboxylic acid, which was dissolved in methyl alcohol. Shieh et al. have developed a semiquantitative theory to interpret and correlate the MCD data for these compounds, which will be briefly outlined below.

For a nondegenerate electronic transition $A \rightarrow B(^1A_{1g} \rightarrow {}^1B_{2u}$ in this case), the magnetic rotational strength $B(A \rightarrow B)$ is given by

$$B(A \rightarrow B) = I_m(\vec{R}_{AB} \cdot \vec{M}_{BA}) \qquad (8\text{-}10)$$

If we let $B(A \rightarrow B)$ represent the rotational strength before substitution, then after substitution, if the changes in energies and wavefunctions are small, $\vec{R}_{AB}$ and $\vec{M}_{BA}$ are changed by the small amounts $\Delta \vec{R}_{AB}$ and $\Delta \vec{M}_{BA}$, respectively. Thus the change in the magnetic rotational strength $\Delta B(A \rightarrow B)$ caused by $\Delta \vec{R}_{AB}$ and $\Delta \vec{M}_{BA}$ is given by

$$\Delta B(A \rightarrow B) = I_m(\Delta \vec{R}_{AB} \cdot \vec{M}_{BA}) + I_m(\vec{R}_{AB} \cdot \Delta \vec{M}_{BA}) + I_m(\Delta \vec{R}_{AB} \cdot \Delta \vec{M}_{BA})$$

$$(8\text{-}11)$$

The change in electric transition moment $\Delta \vec{R}_{AB}$ caused by the introduction of substituents consists of two parts, $(\Delta \vec{R}_{AB})_q$ and $(\Delta \vec{R}_{AB})_v$, where $(\Delta \vec{R}_{AB})_q$ represents the substituent-induced component (due to inductive and resonance effects) and $(\Delta \vec{R}_{AB})_v$, the vibronically induced component. The contribution to $\Delta \vec{M}_{BA}$ may originate from the changes in electric transition moment, magnetic transition moment, and energy levels. For convenience, equation 8-11 can be rewritten as

$$\Delta B(A \rightarrow B) = \Delta B(A \rightarrow B)_q + \Delta B(A \rightarrow B)_v + \Delta B(A \rightarrow B)_m \quad (8\text{-}12)$$

where

$$\Delta B(A \rightarrow B)_v = I_m[(\Delta \vec{R}_{AB})_v \cdot (\vec{M}_{BA} + \Delta \vec{M}_{BA})] \qquad (8\text{-}13)$$

$$\Delta B(A \rightarrow B)_q = I_m[(\Delta \vec{R}_{AB})_q \cdot (\vec{M}_{BA} + \Delta \vec{M}_{BA})] \qquad (8\text{-}14)$$

and

$$\Delta B(A \rightarrow B)_m = I_m(\vec{R}_{AB} \cdot \Delta \vec{M}_{BA}) \qquad (8\text{-}15)$$

For the $^1A_{1g} \rightarrow {}^1B_{2u}$ transition of benzene, which is symmetry forbidden, both $\vec{R}_{AB}$ and $\vec{M}_{BA}$ of benzene itself vanish unless the vibronic effect is taken into consideration. From investigation of the electronic spectra of $^1A_{1g} \rightarrow {}^1B_{2u}$, it is found (70,110) that in most cases $|(\Delta \vec{R}_{AB})_v| < |(\Delta \vec{R}_{AB})_q|$, and thus we may expect that $|\Delta B(A \rightarrow B)_v| < |\Delta B(A \rightarrow B)_q|$. Shieh et al. (98,99) ignored $\Delta B(A \rightarrow B)_m$.

According to the Skar–Platt–Petruska (SPP) theory (70,110), the change in the electric transition moment $(\Delta \vec{R}_{AB})_q$ induced by substitution can be expressed in terms of the so-called spectroscopic moments q_m as

$$(\Delta \vec{R}_{AB})_q = \sum_m q_m \hat{i}_{-zm} \qquad (8\text{-}16)$$

where q_m represents the spectroscopic moment of the substituent at position m and the unit vector $\hat{t}_{-zm}$ is defined by

$$\hat{t}_{-zm} = \hat{i} \sin (2m\rho) + \hat{j} \cos (2m\rho) \tag{8-17}$$

where $\rho = \pi/3$ for benzene. The unit vectors $\hat{i}$ and $\hat{j}$ are in the x- and y-directions, respectively. For homosubstitutions, the q_m are equal and equation 8-16 reduced to

$$(\Delta \vec{R}_{AB})_q = q \sum_m \hat{t}_{-zm} \tag{8-18}$$

The experimental MCD rotational strength for the $^1A_{1g} \rightarrow {}^1B_{2u}$ transition of benzene is $B(^1A_{1g} \rightarrow {}^1B_{2u}) = 0.14 \times 10^{-5}$ $D^2\beta_b$ cm^{-1} and is at least one order of magnitude smaller than the corresponding magnetic rotational strengths of substituted benzenes (see Table 12.XI). It can easily be shown that $(\Delta \vec{R}_{AB})_q = 0$ for the 1,2,3-, 1,3,5-, and hexa-homosubstitutions. The q value is positive for *orth*- and *para*-directing substituents and negative for *meta*-directing substituents (70,110). Thus from equations 8-11 and 8-18, we expect that for a given type of homosubstituted benzene other than a 1,2,3-, 1,3,5-, or hexa-substituted benzene, the *ortho-/para*- and *meta*-directing substituents will give rise to oppositely signed MCD (see Tables 12.XI and 12.XII). From the three cases of homosubstitution, 1,2,3-, 1,3,5-, and hexa-benzenes, $\Delta B(^1A_{1g} \rightarrow {}^1B_{2u})_q = 0$. In these cases, the experimental values of $B(^1A_{1g} \rightarrow {}^1B_{2u})$ vary regularly with the number of substituents [e.g., $B(\text{hexa}) = 2B(1,3,5)$], and the signs of the rotational strengths of these homo-substitutions are always negative regardless of the types of substituents.

Other useful relations for $\Delta B(A \rightarrow B)$ have been derived by Shieh et al. (98). For example, they theoretically showed that for bisubstituted benzenes, $|B(1,4)| > |B(1,2)| > |B(1,3)|$; for trisubstituted benzenes, $|B(1,2,4)| > |B(1,2,3)| \approx |B(1,3,5)|$; and for tetrasubstituted benzenes, $B(1,2,4,5)$ is the largest B value among the three substituted benzenes (see Tables 12.XI and 12.XII). For a more detailed discussion of the theoretical method for estimating the $B(^1A_{1g} \rightarrow {}^1B_{2u})$ values of substituted benzenes developed by Shieh et al., the original paper should be consulted.

Kaito et al. (40) measured MCD spectra of some monosubstituted benzenes (aniline, anisole, phenol, styrene, benzonitrile, benzaldehyde, and nitrobenzene) and determined the magnetic rotational strengths $B(^1A_{1g} \rightarrow {}^1B_{2u})$ for these compounds. They also calculated the Faraday B terms using the Pariser–Parr–Pople (PPP) approximation, and found that the calculated B terms are in good agreement with the experimental values in both sign and magnitude. In their calculation of the Faraday B coefficients, only four or five lower-lying excited states of the monosubstituted benzenes were taken into account.

Hatano and his coworkers (38,39) have been concerned with the investigation of the Faraday A term in substituted benzenes. In such compounds,

TABLE 12.XII

$\Delta B(^1A_{1g} \rightarrow {}^1B_{2u})$ of Substituted Benzenes ($D^2\beta_b/cm^{-1} \times 10^5$)

Substituent	Position	$\Delta B(^1A_{1g} \rightarrow {}^1B_{2u})$	$\Delta B(^1A_{1g} \rightarrow {}^1B_{2u})_v$	$\Delta B(^1A_{1g} \rightarrow {}^1B_{2u})_q$
CH_3	1	2.01	-0.27	2.28
	1,2	2.21	-0.54	2.75
	1,3	1.20	-0.54	1.74
	1,4	5.58	-0.54	6.12
	1,2,4	4.74	-0.82	5.56
	1,3,5	-0.82	-0.82	0
	1,2,3,5	1.17	-1.08	2.25
	1,2,4,5	7.09	-1.08	8.17
	Penta	0.81	-1.35	2.16
	Hexa	-1.50	-1.50	0
Cl	1	3.24	-0.60	3.84
	1,2	4.12	-1.20	5.32
	1,3	3.58	-1.20	4.78
	1,2,4	11.44	-1.80	13.24
	Penta	3.13	-3.00	6.13
	Hexa	-3.59	-3.59	0
F	1	1.02	-0.25	1.27
	Hexa	-1.48	-1.48	0
Br	1	3.24		
I	1	2.66		
CN	1	-14.75		
OH	1	16.08		
COOH	1	-28.71	-0.91	-27.80
	Hexa	-5.43	-5.43	0
NO_2	1	-34.98		
CHO	1	-36.20		
NH_2	1	46.41		

the magnetic moments of the degenerate excited electronic states can be extracted from the observed Faraday A coefficient. Kaito et al. (38) tried to study the MCD of the $^1A_{1g} \rightarrow {}^1E_{1u}$ transitions of benzene and its derivatives in order to obtain the magnetic moment of the $^1E_{1u}$ state. However, the $^1A_{1g} \rightarrow {}^1E_{1u}$ transition of benzene itself lies in the wavelength region 1700–1900 Å (see Fig. 12.4), in which uv region MCD cannot be detected with a conventional spectropolarimeter including quartz plates. They thus chose to investigate hexachlorobenzene, 1,3,5-trichlorobenzene, hexabromobenzene, and 1,3,5-tribromobenzene, because the $^1A_{1g} \rightarrow {}^1E_{1u}$ transitions in these compounds would be expected to be shifted to longer-wavelength regions. All measurements were carried out at room temperature using n-heptane as a solvent. They obtained the A/D and B/D values from the observed MCD and uv spectra using the method of moments; these values are listed in Table 12.XIII. From this table we can see that the contribution of the Faraday B term cannot be neglected and that the values of the magnetic moments in the $^1E_{1u}$ states of hexachlorobenzene and hexabromobenzene

TABLE 12.XIII

Faraday Parameters for the $^1A_1 \rightarrow {}^1E$ (or $^1A_{1g} \rightarrow {}^1E_{1u}$) Transitions of Benzene Derivatives

Derivative	$\nu \times 10^{-3}$ (cm^{-1})	A/D (β_b)	$B/D \times 10^5$ (β_b/cm^{-1})
Phloroglucin	49.7	0.127	-1.29
1,3,5-trimethoxybenzene	49.2	0.262	2.49
1,3,5-tricyanobenzene	48.0	0.172	-2.35
1,3,5-trichlorobenzene	49.6	0.323	-5.79
1,3,5-tribromobenzene	48.2	0.301	-7.54
Hexachlorobenzene	46.4	0.182	-1.98
Hexabromobenzene	44.2	0.207	-4.45

are smaller than those of 1,3,5-trichlorobenzene and 1,3,5-tribromobenzene. Kaito et al. (39) have also measured the MCD and uv spectra of phloroglucin, 1,3,5-trimethoxybenzene, 1,3,5-tricyanobenzene, 1,3,5-trinitrobenzene, 1,3,5-benzenetricarboxylic acid, 1,3,5-trinitrobenzene, and 1,3,5-benzenetricarbonyl chloride at room temperature. For the $^1A_1 \rightarrow {}^1E$ transitions of phloroglucin (in water), 1,3,5-tricyanobenzene (in acetonitrile–water), and 1,3,5-trimethoxybenzene (in n-heptane), the Faraday A terms were observed and the A/D values have been reported (see Table 12.XIII). However, 1,3,5-trinitrobenzene, 1,3,5-benzenetricarboxylic acid, and 1,3,5-benzenetricarbonyl chloride showed no MCD in this spectral region. This indicates that the magnetic moments in the ^1E states of these molecules seem to be quenched by the effects of substituents and that the magnetic moments in the ^1E states of benzene derivatives are sensitive to substitution in the benzene ring.

C.   ATOMS

Research on the magnetic optical activity of atoms has been much less active than that on molecules. However, in view of the recent revival of the application of atomic spectroscopy to analytical chemistry, one may anticipate a similar change with regard to the magnetic optical activity of atoms. Early work in this area focused on MOR; this work has been reviewed by Buckingham and Stephens (12). A full discussion of the MOR of atoms has been given by Rosenfeld (70,77) for the case of Russell–Saunders coupling (113,116). It should be noted that MOR or MCD studies of atoms may provide a useful means of investigating the nature of the band shape function (or, in other words, the mechanisms of line broadening).

In this section, to show what other information is obtainable from the magnetic optical activity of atoms, we shall present the calculation of the Faraday parameters of alkali atoms for the transition $^2S \rightarrow {}^2P$. The MOR

TABLE 12.XIV
Matrix Elements of the Electric Transition Moment of Alkali Atoms[a]

(SJM_J)	$(P\frac{3}{2}\frac{3}{2})$	$(P\frac{3}{2}\frac{1}{2})$	$(P\frac{3}{2}-\frac{1}{2})$	$(P\frac{3}{2}-\frac{3}{2})$	$(P\frac{1}{2}\frac{1}{2})$	$(P\frac{1}{2}-\frac{1}{2})$
$(S\frac{1}{2}\frac{1}{2})$	$\frac{1}{\sqrt{2}}(\hat{i}+i\hat{j})d\ \hat{k}\sqrt{\frac{2}{3}}d$	$\frac{1}{\sqrt{6}}(\hat{i}-i\hat{j})d$	0		$\frac{1}{\sqrt{3}}\hat{k}d$	$\frac{1}{\sqrt{3}}(\hat{i}-i\hat{j})d$
$(S\frac{1}{2}-\frac{1}{2})$	0	$\frac{1}{\sqrt{6}}(\hat{i}+i\hat{j})d$	$\hat{k}\sqrt{\frac{2}{3}}d$	$\frac{1}{\sqrt{2}}(\hat{i}-i\hat{j})d$	$-\frac{1}{\sqrt{3}}(\hat{i}+i\hat{j})d$	$-\frac{1}{\sqrt{3}}\hat{k}d$

[a] $d = (1/\sqrt{3})\langle R_{ns}\mid er\mid R_{n'p}\rangle$.

of sodium close to the D lines has been reported by Wood (117) and Schütz (84). Due to spin–orbit coupling (26),

$$\hat{H}'_{so} = \frac{1}{2m^2c^2}\sum_i \frac{Ze^2}{r_i^3}(\vec{M}_i\cdot\vec{S}_i) \tag{8-19}$$

the ^2P term will split as $^2P_{3/2}$ and $^2P_{1/2}$ by

$$E_{so}(^2P_{3/2}) = \frac{\hbar^2}{4m^2c^2}\left\langle\frac{Ze^2}{r^3}\right\rangle \tag{8-20}$$

and

$$E_{so}(^2P_{1/2}) = -\frac{\hbar^2}{2m^2c^2}\left\langle\frac{Ze^2}{r^3}\right\rangle \tag{8-21}$$

First we consider the calculation of the A term (see equation 3-11). The matrix elements of the electric transition moment and the magnetic moments of the alkali atom are given in Tables 12.XIV and 12.XV. It follows that $A(^2S_{1/2}\to{}^2P_{1/2}) = (2\mid e\mid\hbar/9mc)d^2$ and $A(^2S_{1/2}\to{}^2P_{3/2}) = (7/18)(\mid e\mid\hbar/mc)d^2$. Notice that the corresponding dipole strengths are given by $D(^2S_{1/2}\to{}^2P_{1/2}) = d^2$ and $D(^2S_{1/2}\to{}^2P_{3/2}) = 2d^2$. In other words, from the A coefficient, we can determine the dipole strength of the absorption spectrum.

Similarly, the C term can be calculated using equation 3-15 and Tables 12.XIV and 12.XV. We find $C(^2S_{1/2}\to{}^2P_{1/2}) = -(\mid e\mid\hbar/6mc)d^2$ and $C(^2S_{1/2}\to{}^2P_{3/2}) = (\mid e\mid\hbar/6mc)d^2$. In other words, the dipole strengths can also be obtained from the C coefficients.

To calculate the B term, we notice that

$$B(K\to N) = \frac{1}{g_K}\sum_{K_e}\sum_{N_e}I_m\left[\sum_m{}'\frac{(\vec{R}_{MK_e}\times\vec{R}_{K_eN_e})\cdot\vec{\mu}_{N_eM}}{E_M^0 - E_N^0}\right.$$

$$\left. + \sum_m{}'\frac{(\vec{R}_{N_eM}\times\vec{R}_{K_eN_e})\cdot\vec{\mu}_{MK_e}}{E_M^0 - E_K^0}\right] \tag{8-22}$$

TABLE 12.XV
Magnetic Moments of Alkali Atoms

	$(P \frac{3}{2} \frac{3}{2})$	$(P \frac{3}{2} \frac{1}{2})$	$(P \frac{3}{2} - \frac{1}{2})$	$(P \frac{3}{2} - \frac{3}{2})$
$^{1/2}P_{3/2}$	$\hat{k}\,\dfrac{e\hbar}{3mc}$	$\hat{k}\,\dfrac{e\hbar}{3mc}$	$-\hat{k}\,\dfrac{e\hbar}{3mc}$	$-\hat{k}\,\dfrac{e\hbar}{mc}$

	$(P \frac{1}{2} \frac{1}{2})$	$(P \frac{1}{2} - \frac{1}{2})$		
$^{1/2}P_{1/2}$	$\hat{k}\,\dfrac{e\hbar}{6mc}$	$-\hat{k}\,\dfrac{e\hbar}{6mc}$		

	$(S \frac{1}{2} \frac{1}{2})$	$(S \frac{1}{2} - \frac{1}{2})$		
$^{1/2}S_{1/2}$	$\hat{k}\,\dfrac{e\hbar}{2mc}$	$-\hat{k}\,\dfrac{e\hbar}{2mc}$		

In view of the small energy difference between $^2P_{3/2}$ and $^2P_{1/2}$, other excited states involved in equation 8-22 can be ignored and the second term in equation 8-22 is negligible. The matrix elements of magnetic transition moment required for calculation of the B term are given in Table 12.XVI. The calculated B terms are given by

$$B(^2S_{1/2} \to {}^2P_{1/2}) = \frac{-\dfrac{|e|\hbar}{9mc}\,d^2}{E_{so}(^2P_{3/2}) - E_{so}(^2P_{1/2})} \tag{8-23}$$

and

$$B(^2S_{1/2} \to {}^2P_{3/2}) = \frac{\dfrac{|e|\hbar}{9mc}\,d^2}{E_{so}(^2P_{3/2}) - E_{so}(^2P_{1/2})} \tag{8-24}$$

TABLE 12.XVI
Matrix Elements of Magnetic Transition Moments for $^{1/2}P_{1/2} \to {}^{1/2}P_{3/2}$ (in units of $e\hbar/2mc$)

	(PJM_J)			
(PJM_J)	$(P \frac{3}{2} \frac{3}{2})$	$(P \frac{3}{2} \frac{1}{2})$	$(P \frac{3}{2} - \frac{1}{2})$	$(P \frac{3}{2} - \frac{3}{2})$
$(P \frac{1}{2} \frac{1}{2})$	$-\dfrac{1}{\sqrt{6}}(\hat{i} + i\hat{j})$	$\sqrt{\tfrac{2}{9}}\,\hat{k}$	$\dfrac{1}{\sqrt{18}}(\hat{i} - i\hat{j})$	0
$(P \frac{1}{2} - \frac{1}{2})$	0	$-\dfrac{1}{\sqrt{18}}(\hat{i} + i\hat{j})$	$\sqrt{\tfrac{2}{9}}\,\hat{k}$	$\dfrac{1}{\sqrt{6}}(\hat{i} - i\hat{j})$

In other words, if the dipole strength can be determined from the A term or C term, the B term can provide the spin–orbit coupling constant.

From the above example, we can see that measurement of the MOR or MCD of atoms can provide not only the energy levels, but also the dipole strength, spin–orbit coupling constant (if the spin–orbit interaction is significant) and nature of band shape functions.

D. INORGANIC MOLECULES

In the following we give two examples demonstrating the application of the C term and A term to the assignment of electronic transitions of inorganic molecules. These two systems are due to Schatz and his coworkers (80–82).

An excellent illustration of the use of the C term is provided by the charge-transfer transitions of octahedral d^5 complex ions (80,81), for example, $Fe(CN)_6^{-3}$ in aqueous solution. The absorptioin and MCD spectra (in the region from 20,000 to 40,000 cm^{-1}) of $Fe(CN)_6^{-3}$ have been measured by Schatz et al. (81). The three absorption bands observed are allowed charge-transfer transitions involving the transfer of a ligand $t_u(\sigma,\pi)$ electron to the vacancy in the t_{2g} shell. Thus the three absorption bands correspond to the three transitions $^2T_{2g} \rightarrow {}^2T_{2u}, {}^2T_{1u}^{(1)}, {}^2T_{1u}^{(2)}$. The question that cannot be answered on the basis of the absorption spectra alone is which of the three transitions is $^2T_{2g} \rightarrow {}^2T_{2u}$. The measurements of the MCD spectra by Schatz et al. settle this question with ease. The three observed MCD bands are due mainly to C terms, since the MCD shows $1/T$ dependence down to liquid helium temperatures. Schatz et al. also calculated C/D for each transition; the results are $C/D = 0.5 \beta_b$ for $^2T_{2g} \rightarrow {}^2T_{1u}$, and $C/D = -0.5 \beta_b$ for $^2T_{2g} \rightarrow {}^2T_{2u}$. Noting that a positive C value corresponds to a negative MCD, and vice versa, it is clear by inspection of the MCD spectrum given by Schatz et al. (80,81) that the second band must correspond to $^2T_{2g} \rightarrow {}^2T_{2u}$. The experimental C/D values are in reasonable agreement with the theoretical predictions. Other d^2 systems such as $IrCl_6^{-2}$ have also been investigated by Schatz and his coworkers.

Now we turn our attention to the example that illustrates the use of the A term. For this purpose we consider the d^{10} systems $SbCl_6^-$ and $SnCl_6^{-2}$, both of which exhibit a moderately intense uv absorption; the absorption and MCD spectra of these ions have been reported by Schatz et al. (80,82). They showed that the two most probable one-electron excitations are $t_{1u}(\pi) \rightarrow a_{1g}$ and $e_g \rightarrow t_{1u}^*$, each of which gives rise to a transition to a T_{1u} excited state from the $^1A_{1g}$ ground state. An A term calculation performed by Schatz et al. gives A/D values of $\frac{1}{2} \beta_b$ and $-\frac{1}{2} \beta_b$ for these two possibilities. Their experimental MCD spectrum showed that A/D is positive, this enabling them to choose the assignment $t_{1u}(\pi) \rightarrow a_{1g}$ by simple qualitative observation of the sign of the A term. This assignment is reinforced by the fact that the observed A/D value is quantitatively in good agreement with theory (82).

E. Biological Systems

Metalloporphyrins serve as coenzymes for biologically important proteins. The molecular symmetry of the metalloporphyrins can be designated as D_{4h}. Simpson (100) assigned the $\pi \rightarrow \pi^*$ electronic transitions, $^1A_{1g} \rightarrow {}^1E_u^{(1)}$ and $^1A_{1g} \rightarrow {}^1E_u^{(2)}$, polarized in the molecular plane. The absorption spectrum consists of a very intense peak at about 400 nm and low-intensity vibronic peaks at about 500–550 nm and 560–590 nm. The 400-nm peak is known as a Soret band and the peaks in the 500 to 550-nm and 560 to 590-nm regions are called β (or Q_1) and α (or Q_0) bands, respectively (102). Several molecular orbital (MO) studies on the electronic structure and transitions of metalloporphyrins have been reported (30,57,104). Molecular orbital analyses of MCD have been made by Stephens et al. (109,110). Treu and Hopfield (112) developed a spin–orbit coupling model to explain the MCD of deoxyhemoglobin.

The MCD spectra of metalloporphyrins show characteristic A terms associated with the absorption bands. The free porphyrin bases, which have D_{2h} symmetry, give all B terms (22). The excited states are no longer degenerate in the free base. The large A terms of MCD spectra were used in the quantitative detection of very small amounts of these substances (22).

Deoxyhemoglobin, which has an iron (II) atom in an $S = 2$ spin state, shows an MCD C term. The intensity of the Soret band is thus strongly temperature dependent. The MCD sign of the Soret band is reversed compared with that of oxyhemoglobin, which has an iron (II) of zero spin state (111). The Soret-band intensity of ferricytochrome oxidase is also temperature dependent (68). The MCD intensity of this peak is a function of the amount of low-spin hemochrome present. The amount of low-spin heme "a" in the cytochrome oxidase can be estimated by comparison with data obtained with model heme "a" compounds (68). From studying the MCD of horseradish peroxidase, Nozawa et al. (66) concluded that the Soret MCD band is more sensitive to the structures of the heme and to the spin state than are the visible Q bands.

Collman et al. (17) made comparative studies of the MCD of simple model systems and cytochrome P-450, and obtained evidence for mercaptide as the fifth ligand to iron in stages 1, 2, and 5 of the enzymatic cycle for P-450. Cytochrome P-450 is a hemoprotein of a mixed-function oxidase. Its ferrous carbonyl adduct absorbs light at 450 nm. Iterative extended Hückel (IEH) calculations (20) showed that the additional peak at 370 nm for the ferrous carbonyl adduct originated from the strong mixing of in-plane allowed charge-transfer transitions with the normal Soret ($\pi \rightarrow \pi^*$) transition of similar energy. This assignment was supported by MCD data.

The electronic state of the heme is very sensitive to the change in quaternary structure of the heme protein in solution. Both MOR and MCD were used in the study of the quaternary structure of deoxyhemoglobin (89). Treu and Hopfield (112) also made comparative MCD studies of the native tetrameric deoxyhemoglobin. They separated the α and β chains and found that the most striking difference in MCD between these chains occurs between 8° and −55°C.

MCD detection of the different cytochromes present in a suspension of liver microsomes is possible (5). The light-scattering effect makes using absorption spectroscopy difficult in this case. The high-signal A term of cytochrome P450 at 450 nm and the A term at 559 nm due to cytochrome b_5 can be used to determine the approximate quantities of these enzymes. Shashoua (91) used MOR to study the kinetics of redox reactions of cytochrome c. The intensities of tyrosine's and tryptophan's MCD are large in comparison with those of other naturally occurring amino acids. Their longest-wavelength B terms appear at different wavelengths. The MCD spectra were found to be insensitive to conformational changes. The number of tryptophan residues in intact protein can be determined by MCD spectroscopy (6,7,4,36). The absorption spectra of adenine (electronic transistors) overlap at 261 nm.

The polymer MCD bands of adenine-3′-(2′)-phosphate show two oppositely signed B terms, a minimum at 271 nm and a maximum at 252 nm. Maestre et al. (60) found that the MCD decreases in specific magnitude as the chain length of the polymer increases, due to the stacking interactions of the polymer.

Although our discussion of the applications of MCD has focused on the qualitative application of MCD (e.g., determination of molecular structure, detection of a molecular species, assignment of electronic transitions, measurement of physical properties, and nature of band shape), due to the fact that molar ellipticity is proportional to concentration, MCD can also be used for quantitative purposes (120) in the conventional way. We now present an example showing how to determine the equilibrium constant of the dimerization of benzoic acids, to resolve the MCD spectra for monomers and dimers, and to obtain the magnetic rotational strengths B of monomers and dimers (120), from the charge-transfer (CT) bands of benzoic acids.

The maxima of the CT bands of benzoic acids, which occur at 229 mμ in an $\sim 10^{-5}$ M n-heptane solution, continuously shift toward longer wavelengths with increasing concentration, finally reaching 233 mμ in a solution higher than 10^{-2} M in concentration, and the intensities of these maxima, ϵ_{max}, increase with increasing concentration.

The equilibrium constant of dimerization can be expressed using the equations

$$2M = D \tag{8-25}$$

and

$$K = \frac{X}{2C(1 - X)^2} \tag{8-26}$$

where $C = 2C_D + C_M$, $X = 2C_D/C$, and C_M and C_D are the monomer and dimer concentrations, respectively. Any physical property E that is proportional to the concentration of a particular species can be used to determine the equilibrium constant K as follows. Notice that

$$D_E = E_M C_m + E_D C_D \qquad (8\text{-}27)$$

which can be rewritten as

$$E = E_M(1 - X) + \tfrac{1}{2} E_D X \qquad (8\text{-}28)$$

where $E = D_E/C$. Eliminating X from equation 8-26 by using equation 8-28 yields

$$K^* = \frac{\Delta E_0^*}{(1 - \Delta E_0^*)^2} \qquad (8\text{-}29)$$

where $K^* = 2CK$, $\Delta E_0^* = X = \Delta E_0/\Delta E$, $\Delta E_0 = E - E_M$, and $\Delta E = \tfrac{1}{2} E_D - E_M$.

If the solution is very dilute (i.e., if $X \ll 1$ or $\Delta E_0^* = \Delta E_0/\Delta E \ll 1$), then equation 8-29 can be rewritten as

$$\Delta E_0^* = \frac{\Delta E_0}{\Delta E} = \frac{K^*}{1 + 2K^*} + O(\Delta E_0^{*2}). \qquad (8\text{-}30)$$

Equation 8-30 indicates that in the low-concentration range ($K^* < 1$), the plot of E versus C is linear and has the intercept E_M. This provides one way to determine E_M.

From equation 8-29 we can see that a universal plot for dimerization equilibrium can be obtained by plotting $\log \Delta E_0^*$ versus $\log K^*$ (Fig. 12.5). With this plot one can determine the equilibrium constant of any dimerization, provided one knows E_M, by superimposing the experimental plot of $\log \Delta E_0^*$ versus $\log C$ on the theoretical plot of $\log \Delta E_0^*$ versus $\log K^*$. The amount of the shift on the ΔE_0^* axis will give ΔE or E_D, and the amount of the shift on the $\log C$ axis will give K. After determining K and E_D, one plots $\log \Delta E_0^*$ versus $\log K^*$ using the experimental data on ΔE_0 and C, and compares this plot with the corresponding theoretical plot to evaluate the determination of K and E_D.

The above process has been adapted to MCD for measurement of the equilibrium constants of dimerization of benzoic acid and its derivatives. Equation 8-28 can be applied to the area under the MCD spectrum of an electronic transition or to the intensity of the MCD spectrum at a particular wavelength. In Fig. 12.5, we also show the experimental plot of $\log \Delta E_0^*$ versus $\log K^*$ for benzoic acid. Once the equilibrium constant is obtained, equatioin 8-28 can be used to resolve the observed absorption spectra or

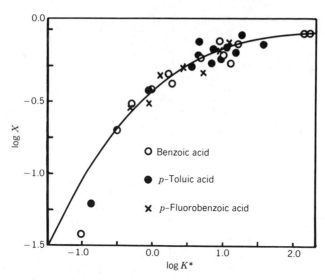

Fig. 12.5. Theoretical (line) and experimental (points) plots of $\log X$ (or $\log \Delta E_0^*$) versus $\log K^*$.

MCD spectra into the spectra of the monomers and dimers; for each wavelength one solves equation 8-28 for two different concentrations to find E_M and E_D. For each concentration C, X can be determined from Fig. 12.5, since $X = \Delta E_0^*$.

In Fig. 12.6, we demonstrate how the observed MCD spectrum of the CT band of benzoic acid is resolved into the MCD spectra of the CT bands of the monomer and dimer. From the resolved MCD spectra it is possible to obtain the magnetic rotational strengths of the intramolecular CT band for monomers and dimers by measurement of the area under the MCD spectrum of the CT band. The dimerization equilibrium constants and magnetic rotational strengths of the CT bands, B_{CT}, of monomers and dimers of benzoic acid and its derivatives are summarized in Table 12.XVII. In general, the B_{CT} value of the dimer is larger than that of the corresponding monomer. Also notice that the dimerization constants of p-toluic acid and p-fluorobenzoic acid are larger than that of benzoic acid.

TABLE 12.XVII
Experimental Results on Dimerization

Compound	$K \times 10^{-3}$	$B_{CT} \times 10^4 \ (D^2\beta_b/cm^{-1})$	
		Monomer	Dimer
Benzoic acid	8.5	1.17	3.98
p-Toluic acid	20.4	0.713	3.43
p-Fluorobenzoic acid	19.5	0.333	1.58

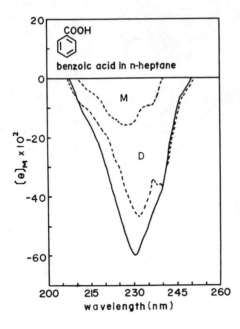

Fig. 12.6.   MCD spectra for CT bands of the monomer (M) and dimer (D) of benzoic acid: $C = 1.00 \times 10^{-4} M$, path length = 1 cm, and $X = 0.479$.

In conclusion, we can see that MCD is a powerful tool for the study of electronic structures, for obtaining the physical constants of both ground and electronic states, and for determining the concentrations of mixtures. Its applications can be expected to become far more numerous and varied in the future.

APPENDIX: DERIVATION OF THE QUANTUM-MECHANICAL EXPRESSION FOR MAGNETIC CIRCULAR DICHROISM

In this appendix, we shall derive the expression for MCD given in Section III. For this purpose we employ semiclassical radiation theory, which treats the radiation fields classically and the absorbing medium quantum mechanically. In this case the total Hamiltonian of the absorbing system plus its interaction with the radiation field can be written as (26,32)

$$\hat{H} = \hat{H}_0 + \hat{H}' \qquad \text{(A-1)}$$

where $\hat{H}_0$ and $\hat{H}'$ are the unperturbed Hamiltonian and the interaction with the radiation, respectively. The corresponding time-dependent Schrödinger equation is given by

$$\hat{H}\Psi = i\hbar \frac{\partial \Psi}{\partial t} \qquad \text{(A-2)}$$

In most cases the dipole interaction between the system and radiation field is sufficient. In that case, $\hat{H}'$ can be expressed as

$$\hat{H}' = -\vec{R} \cdot \vec{E}_{\text{eff}} \tag{A-3}$$

where $\vec{R} = \sum e\vec{r}_i$ is the electric dipole operator and $\vec{E}_{\text{eff}}$ represents the effective electric field due to the light wave. We shall assume that $\vec{E}_{\text{eff}}$ is proportional to the macroscopic field $\vec{E}$:

$$\vec{E}_{\text{eff}} = \alpha\vec{E} = \tfrac{1}{2}\alpha\vec{E}_0(e^{it\omega} + e^{-it\omega}) \tag{A-4}$$

For optical absorption equation A-2 can be integrated to yield $P_{k \to n}$, the so-called transition probability.

In terms of left- and right-circularly polarized light, we have (96)

$$P^{(\pm)}_{k \to n} = \frac{\pi}{2\hbar^2} \alpha^2 \, | \, X_{nk} \pm iY_{nk} \, |^2 \, | \, \vec{E}_0(\pm) \, |^2 \, \delta(\omega_{nk} - \omega) \tag{A-5}$$

Notice that $| \, \vec{E}_0(\pm) \, |^2$ is related to the light intensity by $| \, \vec{E}_0(\pm) \, |^2 = (8\pi/c)I^{(\pm)}$. Thus $\Delta k(\omega)$ in the expression of MCD is given by

$$\Delta k(\omega) = k^{(-)}(\omega) - k^{(+)}(\omega) = \frac{4\pi^2\omega}{c\hbar} \alpha^2 \sum_{kn} N_k \Delta P(k \to n)\delta(\omega_{nk} - \omega) \tag{A-6}$$

where $\Delta P(k \to n) = | \, X_{nk} - iY_{nk} \, |^2 - | \, X_{nk} + iY_{nk} \, |^2$. It should be noted that in the above derivation the damping effect is ignored; when it is included, the delta function $\delta(\omega_{nk} - \omega)$ should be replaced by

$$D(\omega_{nk} - \omega) = \frac{1}{\pi} \, \text{Re} \, \frac{1}{i(\omega_{nk} - \omega) + \Gamma_{nk}} \tag{A-7}$$

where Γ_{nk} represents the damping constant. When the lifetime broadening is responsible for the damping effect, $\Gamma_{nk} = \tfrac{1}{2}(\Gamma_n + \Gamma_k)$, where $1/\Gamma_n$ and $1/\Gamma_k$ are the lifetimes of the n and k states.

Now notice that the wavefunctions $\psi_n, \psi_k, \ldots$ and energies $E_n, E_k, \ldots$ are functions of the applied magnetic field H. This effect can be taken into account by the perturbation method:

$$\psi_n = \psi_n^0 - \sum_m \frac{(\hat{\mu}_z)_{mn}H}{E_m^0 - E_n^0} \psi_m^0 + \cdots \tag{A-8}$$

and

$$E_n = E_n^0 + (\hat{\mu}_z)_{nn}H + \cdots \tag{A-9}$$

where H is the static magnetic field and $\hat{\mu}$ is

$$\hat{\mu} = \sum_i \frac{|e|}{2mc} [(\vec{r}_i \times \vec{p}_i) + 2\vec{s}_i] \tag{A-10}$$

In equation A-6, the quantities N_k, $\Delta P(k \rightarrow n)$, and $\delta(\omega_{nk} - \omega)$ depend on H. For example, to the first order in H,

$$\Delta P(k \rightarrow n) = \Delta P(k \rightarrow n)^{(0)} + H\Delta P(k \rightarrow n)^{(1)} \qquad \text{(A-11)}$$

where $\Delta P(k \rightarrow n)^{(1)}$ is given by (97)

$$\Delta P(k \rightarrow n)^{(1)} = 4I_m \left[\sum_m \frac{(\hat{\mu}_z)_{mn}X^0_{nk}Y^0_{km} + (\hat{\mu}_z)_{nm}X^0_{mk}Y^0_{kn}}{E^0_m + E^0_n} \right.$$
$$\left. + \sum_m \frac{(\hat{\mu}_z)_{km}Y^0_{mn}X^0_{nk} + (\hat{\mu}_z)_{mk}X^0_{nm}Y^0_{kn}}{E^0_m - E^0_k} \right] \qquad \text{(A-12)}$$

Combining equations A-11 and A-12 with equation A-6, we obtain (97)

$$\Delta k(\omega) = \Delta k(\omega)^{(0)} + H\Delta k(\omega)^{(1)} \qquad \text{(A-13)}$$

where $\Delta k(\omega)^{(0)}$ represents the zeroth-order expression of $\Delta k(\omega)$, which vanishes for the isotropic system, and $\Delta k(\omega)^{(1)}$ is given by (105)

$$\Delta k(\omega)^{(1)} = \Delta k(\omega)^{(1)}_A + \Delta k(\omega)^{(1)}_B + \Delta k(\omega)^{(1)}_C \qquad \text{(A-14)}$$

where

$$\Delta k(\omega)^{(1)}_A = -\frac{4\pi^2\omega\alpha^2 N}{c\hbar^2} \sum_{nk} P^{(0)}_k \Delta\omega_{nk}\Delta P(k \rightarrow n)^{(0)}\delta'(\omega - \omega^0_{nk}) \qquad \text{(A-15)}$$

$$\Delta k(\omega)^{(1)}_B = \frac{4\pi^2\omega\alpha^2 N}{c\hbar} \sum_{nk} P^{(0)}_k \Delta P(k \rightarrow n)^{(1)}\delta(\omega - \omega^0_{nk}) \qquad \text{(A-16)}$$

with $\Delta\omega_{nk} = [(\hat{\mu}_z)_{nn} - (\hat{\mu}_z)_{kk}]/\hbar$, and

$$\Delta k(\omega)^{(1)}_C = -\frac{4\pi^2\omega\alpha^2 N}{c\hbar} \beta \sum_{nk} P^{(0)}_k (\hat{\mu}_z)_{kk}\Delta P(k \rightarrow n)^{(0)}\delta(\omega - \omega^0_{nk}) \qquad \text{(A-17)}$$

For randomly oriented systems, a spatial average has to be carried out over $\Delta k(\omega)$. Notice that if the k state and n state are nondegenerate, then $\hat{\mu}_{kk} = 0$, $\hat{\mu}_{nn} = 0$, $\Delta k(\omega)^{(1)}_A = 0$, and $\Delta k(\omega)^{(1)}_C = 0$.

For molecular systems, the adiabatic approximation can be used in most cases. In this case, the total wavefunction can be written as a product of nuclear and electronic wavefunctions, for example $\Phi_N(q)\oplus_{Nn}(Qq)$. Thus for the electronic transition $K \rightarrow N$, equations A-15–A-17 can be expressed as

$$\Delta k(\omega)_A^{(1)} = -\frac{16\pi^2\omega\alpha^2 N}{3c\hbar^2} \sum_{nk} P_{Kk}^{(0)} A(Kk \to Nn)\delta'(\omega - \omega_{Nn,Kk}^0) \tag{A-18}$$

$$\Delta k(\omega)_B^{(1)} = -\frac{16\pi^2\omega\alpha^2 N}{3c\hbar} \sum_{nk} P_{Kk}^{(0)} B(Kk \to Nn)\delta(\omega - \omega_{Nn,Kk}^0) \tag{A-19}$$

and

$$\Delta k(\omega)_C^{(1)} = -\frac{16\pi^2\omega\alpha^2 N}{3c\hbar} \sum_{nk} P_{Kk}^{(0)} C(Kk \to Nn)\delta(\omega - \omega_{Nn,Kk}^0) \tag{A-20}$$

where

$$A(Kk \to Nn) = -\frac{1}{2g_K} \sum_{K_e} \sum_{N_e} (\vec{\mu}_{Nen,Nen}$$

$$- \vec{\mu}_{Kek,Kek}) \cdot I_m(\vec{R}_{Nen,Kek} \times \vec{R}_{Kek,Nen}) \tag{A-21}$$

$$B(Kk \to Nn) = \frac{1}{g_K} \sum_{K_e} \sum_{N_e} I_m \left[\sum_{Mm}{}' \frac{(\vec{R}_{Kek,Nen} \times \vec{R}_{Mm,Kek}) \cdot \vec{\mu}_{Nen,Mm}}{E_{Mm}^0 - E_{Nen}^0} \right.$$

$$\left. + \sum_{Mm}{}' \frac{(\vec{R}_{Kek,Nen} \times \vec{R}_{Nen,Mm}) \cdot \vec{\mu}_{Mm,Kek}}{E_{Mm}^0 - E_{Kek}^0} \right] \tag{A-22}$$

and

$$C(Kk \to Nn) = \frac{1}{2g_K} \sum_{K_e} \sum_{N_e} \vec{\mu}_{Kek,Kek} \cdot I_m(\vec{R}_{Kek,Nen} \times \vec{R}_{Nen,Kek}) \tag{A-23}$$

In equations A-18–A-23, $P_{Kk}^{(0)}$ represents the Boltzmann factor of the nuclear motion, g_K is the degeneracy of the K electronic state, and the summations over K_e and N_e in equations A-21–A-23 cover the degenerate electronic states of the Kth and Nth manifolds.

Notice that at extremely low temperatures, where $\beta \mid (\hat{\mu}_z)_{nn} \mid H \gg 1$, the systems in this case will occupy the lowest level and $\Delta k(\omega)_C^{(1)}$ disappears.

From the above derivation, we can see that MCD arises from the effect of the magnetic field on the energy difference between the initial and final states and on the Boltzmann population (A and C terms), and from the perturbation of the wavefunctions of the initial and final states by the magnetic field (B term).

REFERENCES

1. Abu-Shumays, A., and J. J. Duffield, *Appl. Spectrosc.*, **24**, 67 (1970).
2. Abu-Shumays, A., G. E. Hooper, and J. J. Duffield, *Appl. Spectrosc.*, **25**, 238 (1971).

3. Albrecht, A. C., *J. Mol. Spectrosc.*, **5**, 236 (1960).

4. Barth, G., E. Bunnenberg, and C. Djerassi, *Anal. Biochem.*, **48**, 471 (1972).

5. Barth, G., J. H. Dawson, P. M. Dolinger, R. E. Linder, E. Bunnenberg, and C. Djerassi, *Anal. Biochem.*, **65**, 100 (1975).

6. Barth, G., R. Records, E. Bunnenberg, C. Djerassi, and W. Voelter, *J. Am. Chem. Soc.*, **93**, 2545 (1971).

7. Barth, G., W. Voelter, E. Bunnenberg, and C. Djerassi, *J. Am. Chem. Soc.*, **94**, 1293 (1972).

8. Briat, B., in F. Ciardelli and P. Salvadori, Eds., *Fundamental Aspects and Recent Developments in Optical Rotatory Dispersion and Circular Dichroism*, Heyden and Son, London, 1973, pp. 375–401.

9. Briat, B., M. Billardon, and J. Badoz, *Compt. Rend.*, **256**, 3440 (1963).

10. Brith, M., O. Schnepp, and P. J. Stephens, *Chem. Phys. Lett.*, **26**, 549 (1974).

11. Buckingham, A. D., in D. Henderson, Ed., *Physical Chemistry: An Advanced Treatise*, Vol. 4, New York, Academic Press, 1970, pp. 349–386.

12. Buckingham, A. D., and P. J. Stephens, *Annu. Rev. Phys. Chem.*, **17**, 399 (1966).

13. Callomon, J. H., T. M. Dunn, and I. M. Mills, *Philos. Trans. R. Soc. Lond.*, **A259**, 499 (1966).

14. Carroll, T., *Phys. Rev.*, **52**, 822 (1937).

15. Chabay, I., E. C. Hsu, and G. Holzwarth, *Chem. Phys. Lett.*, **15**, 211 (1972); Hsu, E. C., and G. Holzwarth, *J. Am. Chem. Soc.*, **95**, 6902 (1973); Osborne, G. A., J. C. Cheng, and P. J. Stephens, *Rev. Sci. Instrum.*, **44**, 10 (1973).

16. Clogston, A. M., *J. Phys. Radium*, **20**, 151 (1959).

17. Collman, J. P., T. N. Sorrell, J. H. Dawson, J. R. Trudell, E. Bunnenberg, and C. Djerassi, *Proc. Natl. Acad. Sci. U.S.A.*, **73**, 6 (1976).

18. Condon, E. U., *Rev. Mod. Phys.*, **9**, 432 (1937).

19. Crabbé, P., *Optical Rotatory Dispersion and Circular Dichroism: An Introduction*, Academic Press, New York, 1972.

20. Dawson, J. H., J. R. Trudell, G. Barth, R. E. Linder, E. Bunnenberg, C. Djerassi, M. Gouterman, C. R. Connell, and P. Sayer, *J. Am. Chem. Soc.*, **99**, 641 (1977).

21. Djerassi, C., *Optical Rotatory Dispersion*, McGraw-Hill, New York, 1960.

22. Djerassi, C., E. Bunnenberg, and D. L. Elder, *Pure Appl. Chem.*, **25**, 57 (1971).

23. Douglas, I. N., R. Grinter, and A. J. Thomson, *Chem. Phys. Lett.*, **28**, 192 (1974).

24. Douglas, I. N., R. Grinter, and A. J. Thomson, *Mol. Phys.*, **26**, 1257 (1973); **29**, 673 (1975).

25. Emeis, C. A., and L. J. Oosterhoff, *J. Chem. Phys.*, **54**, 4809 (1971); *Chem. Phys. Lett.* **1**, 129 (1967).

26. Eyring, H., J. Walter, and E. G. Kimball, *Quantum Chemistry*, Wiley-Interscience, New York, 1944, Chapter 17.

27. Faraday, M., *Philos. Mag.*, **28**, 294 (1846); *Philos. Trans. R. Soc.*, 1, (1846).

28. Finkelnburg, W., and W. Steiner, *Z. Phys.*, **79**, 69 (1932).

29. Gordon, R., *Adv. Mag. Reson.*, **3**, 1 (1968).

30. Gouterman, M., *J. Chem. Phys.*, **30**, 1139 (1959).

31. Hameka, H. F., *J. Chem. Phys.*, **36**, 2540 (1962).

32. Heitler, W., *Quantum Theory of Radiation*, Oxford University Press, Oxford, 1954.

33. Henry, C. H., S. E. Schnatterly, and C. P. Slichter, *Phys. Rev.*, **137**, A583 (1965).

34. Herzberg, G., *Can. J. Phys.*, **30**, 185 (1952).

35. Herzberg, G., *Can. J. Phys.*, **31**, 657 (1953).

36. Holmquist, B., and B. L. Vallee, *Biochemistry*, **12**, 4409 (1973).

37. Job, V. A., V. Sethuraman, and K. K. Innes, *J. Mol. Spectrosc.*, **30**, 365 (1969).

38. Kaito, A., A. Tajiri, and M. Hatano, *Chem. Phys. Lett.*, **25**, 548 (1974).

39. Kaito, A., A. Tajiri, and M. Hatano, *Chem. Phys. Lett.*, **28**, 197 (1974).

40. Kaito, A., A. Tajiri, and M. Hatano, *J. Am. Chem. Soc.*, **98**, 384 (1976).

41. Knittel, D., H. Raizdadeh, H. P. Lin, and S. H. Lin, *J. Chem. Soc. Faraday Trans. II*, **73**, 120 (1977).

42. Kramers, H., *Proc. Acad. Sci., Amsterdam*, **33**, 959 (1930).

43. Krueger, W. C., and L. M. Pschigoda, *Anal. Chem.*, **43**, 675 (1971).

44. Lee, S. T., Y. H. Yoon, S. H. Lin, and H. Eyring, *J. Chem. Phys.*, **66**, 4349 (1977).

45. Lin, S. H., *J. Chem. Phys.*, **44**, 3759 (1966).

46. Lin, S. H., *Theor. Chim. Acta*, **10**, 301 (1968).

47. Lin, S. H., *J. Chem. Phys.*, **55**, 3546 (1971); **54**, 1177 (1971).

48. Lin, S. H., *J. Chem., Phys.*, **59**, 4458 (1973).

49. Lin, S. H., *J. Chem. Phys.*, **62**, 4500 (1975).

50. Lin, S. H., *J. Chem. Phys.*, **65**, 1053 (1976).

51. Lin, S. H., L. Colangelo, and H. Eyring, *Proc. Natl. Acad. Sci. U.S.A.*, **68**, 2135 (1971).

52. Lin, S. H., and H. Eyring, *Proc. Natl. Acad. Sci. U.S.A.*, **71**, 3415 (1974).

53. Lin, S. H., and H. Eyring, *Proc. Natl. Acad. Sci. U.S.A.*, **71**, 3802 (1974).

54. Lin, S. H., H. P. Lin, and D. Knittel, *J. Chem. Phys.*, **64**, 441 (1976).

55. Linder, R. E., G. Barth, E. Bunnenberg, C. Djerassi, L. Seamans, and A. Moscowitz, *Chem. Phys. Lett.*, **28**, 490 (1974); **38**, 28 (1976).

56. Linder, R. E., E. Bunnenberg, L. Seamans, and A. Moscowitz, *J. Chem. Phys.*, **60**, 1943 (1974).

57. Longuet-Higgins, H. C., C. W. Rector, and J. R. Platt, *J. Chem. Phys.*, **18**, 1174 (1950).

58. Lowry, T. M., *Optical Rotatory Power*, Dover, New York, 1964, pp. 160–168.

59. Luk, C. K., and F. S. Richardson, *Chem. Phys. Lett.*, **25**, 215 (1974); Luk, C. K., *J. Am. Chem. Soc.*, **96**, 2006 (1974); Luk, C. K., and F. S. Richardson, *J. Am. Chem. Soc.*, **97**, 6666 (1975).

60. Maestre, M. F., D. M. Gray, and R. B. Cook, *Biopolymers*, **10**, 2537 (1971).

61. Moffitt, W., and A. Moscowitz, *J. Chem. Phys.*, **30**, 648 (1959).

62. Moscowitz, A., *Adv. Chem. Phys.*, **4**, 67 (1962).

63. Mulliken, R. S., *J. Chem. Phys.*, **7**, 20 (1939).

64. Mulliken, R. S., *Phys. Rev.*, **57**, 500 (1940).

65. Mulliken, R. S., *J. Chem. Phys.*, **55**, 288 (1971).

66. Nozawa, T., N. Kobayashi, and M. Hatano, *Biochim. Biophys. Acta*, **427**, 652 (1976).

67. Orlandi, G., and W. Siebrand, *J. Chem. Phys.*, **58**, 4513 (1973); Geldof, P. A., R. P. H. Rettschnick, and G. J. Hoytink, *Chem. Phys. Lett.*, **10**, 549 (1971).

68. Palmer, G., G. T. Babcock, and L. E. Vickery, *Proc. Natl. Acad. Sci. U.S.A.*, **73**, 2206 (1976).

69. Partington, J. R., *An Advanced Treatise on Physical Chemistry*, Vol. 4, Longmans-Green, London, 1953, pp. 592–632.

70. Perrin, M. H., M. Gouterman, and C. L. Perrin, *J. Chem. Phys.*, **50**, 4137 (1969).

71. Petruska, J., *J. Chem. Phys.*, **34**, 484 (1961); 1120 (1961).

72. Pople, J. A., and D. L. Beveridge, *Approximate Molecular Orbital Theory*, McGraw-Hill, New York, 1970.

73. Pople, J. A., and J. W. Sidman, *J. Chem. Phys.*, **27**, 1270 (1957).

74. Riehl, J. P., and F. S. Richardson, *J. Chem. Phys.*, **65**, 1011 (1976) (and references therein).

75. Rosenfeld, J. S., A. Moscowitz, and R. E. Linder, *J. Chem. Phys.*, **61**, 2427 (1974).

76. Rosenfeld, L., *Z. Phys.*, **52**, 161 (1928).

77. Rosenfeld, L., *Z. Phys.*, **57**, 835 (1929).

78. Russel, M. F., M. Billardon, and J. Badoz, *Appl. Opt.*, **11**, 2375 (1972); Dudley, R. J., S. F. Mason, and R. D. Peacock, *Chem. Commun.*, **25**, 1084 (1972).

79. Sage, M. L., *J. Chem. Phys.*, **35**, 969 (1961).

80. Schatz, P. N., and A. J. McCaffery, *Quart. Rev.*, **23**, 552 (1969); **24**, 324 (1970).

81. Schatz, P. N., A. J. McCaffery, W. Suetaka, G. N. Henning, A. B. Ritchie, and P. J. Stephens, *J. Chem. Phys.*, **45**, 722 (1966).

82. Schatz, P. N., P. J. Stephens, G. N. Henning, and A. J. McCaffery, *Inorg. Chem.*, **7**, 1246 (1968).

83. Schnepp, O., S. Allen, and E. F. Pearson, *Rev. Sci. Instrum.*, **41**, 1136 (1970); Johnson, W. C. Jr., *Rev. Sci. Instrum.*, **42**, 1283 (1971).

84. Schütz, W., *Magnetooptik,* Akademische Verlags-gesellschaft, Berlin, 1936.

85. Seamans, L., and A. Moscowitz, *J. Chem. Phys.*, **56**, 1099 (1972).

86. Seamans, L., A. Moscowitz, G. Barth, E. Bunnenberg, and C. Djerassi, *J. Am. Chem. Soc.*, **90**, 6464 (1972).

87. Seamans, L., A. Moscowitz, R. E. Linder, G. Barth, E. Bunnenberg, and C. Djerassi, *J. Mol. Spectrosc.*, **54**, 412 (1975); **56**, 441 (1975).

88. Seamans, L., A. Moscowitz, R. E. Linder, G. Barth, E. Bunnenberg, and C. Djerassi, *Chem. Phys. Lett.*, **13**, 135 (1976).

89. Serber, R., *Phys. Rev.*, **41**, 489 (1932).

90. Sharanov, Yu. A., N. A. Sharanov, and B. P. Atanasov, *Biochim. Biophys. Acta,* **434**, 440 (1976).

91. Shashoua, V. E., *J. Am. Chem. Soc.*, **86**, 2109 (1964).

92. Shashoua, V. E., *Arch. Biochem. Biophys.*, **111**, 550 (1965).

93. Shashoua, V. E., *J. Am. Chem. Soc.*, **87**, 4044 (1965).

94. Shen, Y. R., *Phys. Rev.*, **133**, A511 (1964).

95. Shen, Y. R., and N. Bloembergen, *Phys. Rev.*, **133**, A515 (1964).

96. Shieh, D. J., Y. C. Fu, and H. Eyring, *Proc. Natl. Acad. Sci. U.S.A.*, **71**, 209 (1974).

97. Shieh, D. J., S. H. Lin, and H. Eyring, *J. Phys. Chem.*, **76**, 1844 (1972).

98. Shieh, D. J., S. H. Lin, and H. Eyring, *Proc. Natl. Acad. Sci. U.S.A.*, **69**, 2000 (1972).

99. Shieh, D. J., S. H. Lin, and H. Eyring, *J. Phys. Chem.*, **77**, 1031 (1973); Shieh, D. J., S. T. Lee, Y. C. Yim, S. H. Lin, and H. Eyring, *Proc. Natl. Acad. Sci. U.S.A.*, **72**, 452 (1975).

100. Simpson, W. T., *J. Chem. Phys.*, **17**, 1218 (1949).

101. Slichter, C. P., *Principles of Magnetic Resonance,* Harper & Row, New York, 1963.

102. Snir, J., and J. A. Schellman, *J. Phys. Chem.*, **78**, 387 (1974).

103. Sober, H. A., Ed., *Handbook of Biochemistry,* The Chemical Rubber Co., Boca Raton, Florida, 1970, p. J-258.

104. Steinberg, I. Z., in R. F. Chen and H. Edelhoch, Eds., *Concepts in Biochemical Fluorescence,* Vol. 1, Marcel Dekker, New York, 1975, Chapter 3.

105. Stephens, P. J., *Theoretical Studies on Magneto-Optical Phenomena,* Ph.D. Thesis, Oxford University, Oxford, 1964.

106. Stephens, P. J., *J. Chem. Phys.*, **52**, 3489 (1970).

107. Stephens, P. J., *Annu. Rev. Phys. Chem.*, **25**, 201 (1974).

108. Stephens, P. J., *Adv. Chem. Phys.*, **35**, 197 (1976).

109. Stephens, P. J., P. N. Schatz, A. B. Ritchie, and A. J. McCaffery, *J. Chem. Phys.*, **48**, 132 (1968).

110. Stephens, P. J., W. Suetaka, and P. N. Schatz, *J. Chem. Phys.*, **44**, 4592 (1966).

111. Stevenson, P. E., *J. Chem. Ed.*, **41**, 434 (1964).

112. Treu, J. I., and J. J. Hopfield, *J. Chem. Phys.*, **63**, 613 (1975).

113. van den Handel, J., *Handbuch Phys.*, **15**, 1 (1956).

114. See reference 75 of Van Vleck, J. H., *The Theory of Electric and Magnetic Suscepti-bilities*, Oxford Press, Oxford, 1932.

115. Weiss, C., H. Kobayashi, and M. Gouterman, *J. Mol. Spectrosc.*, **16**, 415 (1965).

116. Wong, K. P., *J. Chem. Ed.*, **51**, A573 (1974).

117. Wood, R. W., *Philos. Mag.*, **14**, 145 (1907).

118. Wu, T. Y., and T. Ohmura, *Quantum Theory of Scattering*, Prentice-Hall, 1962.

119. Wulf, O. R., *Proc. Natl. Acad. Sci. U.S.A.*, **14**, 609 (1928).

120. Yoon, Y. H., S. T. Lee, S. H. Lin, and H. Eyring, *Proc. Natl. Acad. Sci. U.S.A.*, **73**, 2964 (1976).

121. Yoon, Y. H., S. T. Lee, D. J. Shieh, H. Eyring, and S. H. Lin, *Chem. Phys. Lett.*, **38**, 24 (1976).

Chapter 13

THE NATURE OF X-RAYS, SPECTROCHEMICAL ANALYSIS BY CONVENTIONAL X-RAY METHODS, AND NEUTRON DIFFRACTION AND ABSORPTION*

By H. A. LIEBHAFSKY,† E. A. SCHWEIKERT, AND E. A. MYERS, *Department of Chemistry, Texas A&M University, College Station, Texas*

Contents

* The authors are grateful to Wiley-Interscience for permission to use certain passages, tables, and figures from *X-rays, Electrons, and Analytical Chemistry*, by H. A. Liebhafsky, H. G. Pfeiffer, E.-H. Winslow, and P. D. Zemany, Wiley-Interscience, New York, 1972. To save space, occasional references to this work (identified by "LPWZ" and page number) will be made.
† Deceased.

I.   INTRODUCTION

Although we intend this chapter to be generally useful, we have written it with the analytical *chemist* mainly in mind—most particularly the analytical chemist whose job it is to accomplish the characterization or control of materials effectively at minimum cost. This is not a simple assignment. The embarrassment of riches in methods and instrumentation gives every excuse for uneasiness on the part of anyone who must decide in which of many competing ways the analytical work of his laboratory should be done. To what extent should "workhorse" equipment and methods be supplemented by new, costly, "exotic" instruments? If these are not installed, sooner or later there may come a problem that simply *must* be solved, but cannot be for lack of equipment the purchase of which had seemed unjustified. As concerns x-rays, and the electrons rapidly becoming interchangeable with these, we hope this chapter will give the analytical chemist any fundamental knowledge he may need, describe for him the workhorse x-ray methods, and lead him on to chapters [such as those on electron spectroscopy (Chapter 17) and microprobe technique] in which the exotic methods are treated. The brief discussion of neutrons was added at the request of the Editors of the *Treatise*.

II.   ORIGIN AND PROPERTIES OF X-RAYS

A.   THE DISCOVERY AND GENERATION OF X-RAYS

On November 8, 1895, Wilhelm Konrad Roentgen ended the era of classical physics and enlarged the eventual scope of analytical chemistry with his discovery of x-rays, so named by him because they were then new and mysterious. Roentgen had been experimenting with "highly evacuated" tubes whose glass walls had been subjected to pulsating electron bombardment. Even when such a tube was covered with black cardboard, a neigh-

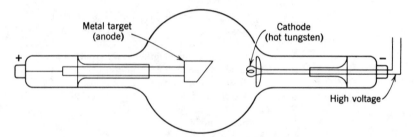

Fig. 13.1.   Schematic diagram of the Coolidge (high-vacuum) x-ray tube. Coolidge tubes are widely used because they are stable, long-lived, and permit tube current and voltage to be controlled independently. The target and cooling circuit are grounded so that the voltage is fixed by the high voltage impressed on the cathode.

boring barium platinocyanide phosphor fluoresced visibly and synchronously with the electric discharge through the tube. We know now that this chain of events was possible because x-rays have both high energy *and* high penetrating power, a combination of properties that sets them apart from electrons and from light of all wavelengths. Roentgen studied the properties of the new rays so well that he laid the foundations not only for important methods of x-ray detection (fluorescence of a phosphor, darkening of a photographic plate, 2nd ionization of a gas) and for radiography, but for the application of x-ray absorption to analytical chemistry as well; he showed, for example, that such absorption could be used to reveal the presence of lead paint on wood.

In the early type of tube designed specifically for the generation of x-rays, positive ions bombard a metal cathode to release electrons that strike an anode, or target, from which x-rays are released in a controllable fashion. The maximum energy of the rays increases with the effective potential difference, and this in turn increases with decreasing gas pressure. A constant and reproducible x-ray beam therefore requires such careful stabilization of the gas pressure as to restrict severely the usefulness of these tubes in analytical chemistry.

By contrast, in the Coolidge tube (Figs. 13.1 and 13.2) the vacuum is so good that gases or positive ions derived therefrom do not influence the x-ray beam. The tube is sealed, a great advantage. Electrons for excitation are produced by heating a tungsten filament, the heating circuit being separate from that which provides the accelerating potential. Consequently, the rate at which electrons strike the target (tube current) can be adjusted independently of their energy (tube voltage). This makes possible the independent variation of intensity and wavelength in the x-ray beam.

These characteristics make the Coolidge tube the prime x-ray source for analytical chemistry. The high temperature of the cathode (filament) leads eventually to the contamination of the tube by evaporated tungsten, which problem is most serious when the target (anode) has been made of another metal and the tube planned to give only x-rays derived from that metal.

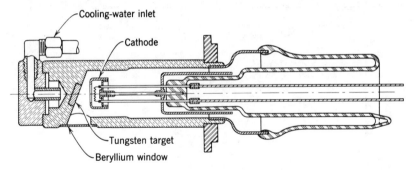

Fig. 13.2. A modern version of the Coolidge x-ray tube. The tube operates at high vacuum with a storage battery (portable models) or step-up transformer as power source. Tube voltages range from a few hundred volts up to values (a million or more) beyond the needs of ordinary analytical chemistry. Because most of the energy fed into the tube is degraded to heat, effective cooling is of first importance.

Coolidge tubes with targets of tungsten, molybdenum, platinum, chromium, copper, silver, nickel, cobalt, iron, and even carbon are now available. Tubes with targets of other materials could be made if needed. Dual-target Coolidge tubes have found wide use.

B.   EXCITATION OF X-RAYS

"Generation" or "production" of x-rays implies a broad view. If we are concerned with the process in more detail, we speak of the excitation of x-rays, which occurs whenever matter is bombarded by agents sufficiently high in energy—agents such as electrons, other x-rays, protons, and other ions. Only the first two warrant consideration in this chapter.

When the target of an x-ray tube is struck by electrons, these are retarded by the atoms of the target. The energy the electrons lose is radiated in a spectrum that ranges from the x-ray region into the infrared; we say this spectrum has been produced by electron excitation. The elementary processes involved in this energy transfer can take place also in the sample.

The Coolidge tube is the x-ray source most important in analytical chemistry. We must therefore examine the spectra that have been observed when an "infinitely thick" tungsten target in such a tube is bombarded by electrons of various energies. In Fig. 13.3, intensities (in arbitrary units) are plotted against wavelengths (in angstrom units) for these spectra. Three items are noteworthy:

1. All spectra begin abruptly at a short-wavelength limit that shifts systematically to shorter wavelengths as the electron energy increases. The short-wavelength limit shows that we are dealing with a quantum phenomenon, and from it reliable values for Planck's constant, h, can be calculated.

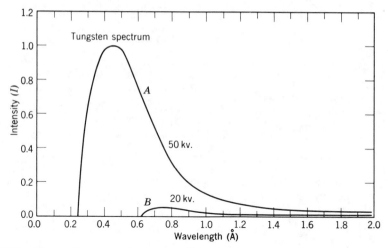

Fig. 13.3. The continuous x-ray spectrum. Note that the short-wavelength limit (equation 1) is 0.248 Å for 50 kV. *(A)* and 0.620 Å for 20 kV *(B)*. A spectrum from a rectified ac tube would have the peak displaced to the right, and for a given input energy it would have less x-ray output. [After Ulrey, C. T., *Phys. Rev.*, **11** (2), 401 (1918).]

2. Each spectrum is a continuum that extends over a wide range of wavelengths. The curves are left unfinished to call attention to the difficulty of working with long-wavelength x-rays.

3. The spectra in Fig. 13.3 are observed spectra, that is, spectra obtained experimentally. The spectra as generated at the target differ from these in several important respects (see below).

A useful relationship between the short-wavelength limit, λ_0, and V, the maximum potential difference (in kilovolts) across the tube, is

$$\lambda_0 = \frac{12.393}{V} \qquad (1)$$

The total power, or integrated intensity (I_{int}), of the x-ray beam (in watts) is the product of the (empirical) efficiency of x-ray production and cathode-ray power iV:

$$I_{int} = (1.4 \times 10^{-9})iZV^2 \qquad (2)$$

where i is the electron current in amperes, Z is the atomic number, and V is the potential difference across the tube in volts. This estimate of the beam energy becomes less reliable as V decreases, but it emphasizes the inefficiency of x-ray production by electron bombardment. In typical cases, less than 1% of the electron energy appears in the x-ray beam. Almost all the rest is degraded to head, and special provisions for cooling the target often are necessary.

When electrons strike an "infinitely thick" target, they are quickly slowed as they interact with the atoms of the target. Consequently, they do not penetrate at all deeply. Furthermore, different electrons lose different amounts of energy as they interact with individual atoms of the target, so that an electron-energy spectrum is set up even in a target being struck only by electrons of the same energy. The actual excitation of x-rays is therefore brought about by electrons differing widely in energy, with the result that the emitted x-rays show a corresponding wavelength range. Even mono-energetic electrons will generate a polychromatic x-ray beam.

The absorption of x-rays on their way out of the target also is important. Because x-rays are usually absorbed more readily the longer their wave-length, the intensity distribution is changed by absorption, with the inten-sities of the longer wavelengths being decreased. The target thus tends to filter the longer wavelengths out of the beam.

The two effects just described help to determine the x-ray distribution at the target. Before an x-ray beam strikes a sample to be analyzed, the beam is usually modified further. For example, there may be absorption (and fil-tering) by the window of the x-ray tube, by an air path between the tube and the sample, by the walls of the cell containing the sample, and finally by the sample itself. Analogous considerations govern the absorption (and, with polychromatic beams, the filtering) of the beam entering the detector from the sample.

Experiments on target materials other than tungsten show that the nature of the continuous spectrum does not change, although its integrated intensity increases, with increases in atomic number. As a means of identifying ele-ments, such spectra are virtually useless to the analytical chemist.

But the *characteristic* spectra are another story. When the energy of the electron beam is sufficiently high (above 70 kV in the case of tungsten **K** lines), the smooth spectra of Fig. 13.3 assume the spiked appearance of Fig. 13.4. The spikes are lines characteristic of the elements. These emitted lines can be used as *analytical lines* for the elements, just as the lines in the visible or the ultraviolet range, which are some thousandfold longer in wavelength, have been used in emission spectrography by analytical chemists since the days of Kirchhoff and Bunsen.

The characteristic spectra can be excited also by other x-rays sufficiently high in energy. The chief usefulness of the continuous spectrum (Fig. 13.3) to the analytical chemist is as a means of x-ray excitation for whatever analytical lines he wishes to employ. In most analyses by x-ray methods, samples are exposed to continuous x-ray beams to excite analytical lines, the wavelengths and intensities of which are determined in a spectrograph.

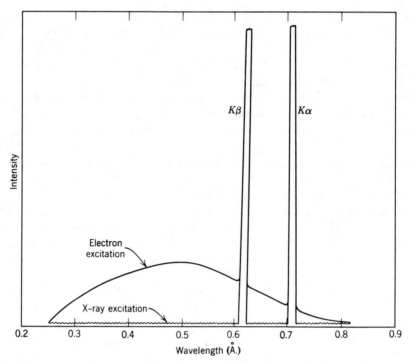

Fig. 13.4.   An idealized drawing of the molybdenum spectra excited by 35-kV electrons and by the polychromatic beam from a 35-kV x-ray tube. With x-ray excitation, most of the energy appears in the characteristic lines. With electron excitation, much of it is wasted in the continuous spectrum.

C.   INTERACTION OF X-RAYS WITH MATTER

Nowhere is the high energy (or short wavelength) of x-rays more evident than in their interaction with matter, which consequently—for the purposes of analytical chemistry—may be described as differing in kind from the interactions with matter of the longer wavelengths (infrared, visible, and ultraviolet) more generally used by the analytical chemist in the past. This interaction of x-rays with matter can be broken down into *absorption, scattering,* and *diffraction.*

1.   Absorption of X-rays

The absorption of x-rays by matter is important not only for its own sake, but also because it can lead to emission. Inasmuch as atoms (or ions derived from them) act as absorbing centers for x-rays, Beer's law might be expected to govern the absorption process. Owing to the high energies of x-rays, the electrons primarily involved in absorption will lie close to the nucleus. As a consequence, one might expect also that x-ray absorption coefficients will be more simply interrelated than are absorption coefficients for radiant energy of longer wavelengths, that is, of wavelengths long enough to affect valence electrons and hence to make the absorption process subject to chemical influences. Both expectations are realized. In the main, x-ray absorption is a physical process governed by Beer's law.

In x-ray absorptiometry, especially when polychromatic beams are used, I_0, the intensity of the beam as it strikes the sample, is often too high for direct measurement by the detector. Consequently, Beer's law is conveniently written in the form

$$\log \left(\frac{I_1}{I_2}\right) = k \, \Delta m = k_d \, \Delta d \tag{3}$$

where Δm and Δd are the differences in the masses and thicknesses of two uniform samples of the same material, and I_1 and I_2 are the emergent intensities. Equation 3 is strictly valid where the same parallel, monochromatic x-ray beam is perpendicularly incident on a unit area of each sample. The values of the proportionality constants k and k_d depend not only on the units in which the amount of sample is measured, but on the wavelength and on the elements in the sample as well. For a sample of unit area, k_d is k times the density.

The general validity of the empirical equation 3 points to a mass-absorption coefficient, μ, as being the most logical for x-ray absorption. If the absorption equation is written in terms of natural logarithms, and μ is introduced, the relationship becomes

$$\ln \left(\frac{I_1}{I_2}\right) = \mu \, \Delta m = \mu\rho \, \Delta d \tag{4}$$

where Δm is the difference between the masses m_2 and m_1 of the two samples expressed in grams per square centimeter of irradiated sample area, and ρ is the sample density expressed in grams per cubic centimeter. If cgs units are used also in equation 3, then $\mu/2.303$ is equal to k in equation 3 when Δd is the sample thickness in centimeters.

The use of equation 4 is obviously restricted to the area of the sample that is uniformly irradiated by the beam. It is also obvious that the part of the sample not in the beam and the part of the beam that does not strike the sample do not contribute to the absorption process.

Other absorption coefficients also are used, the most important of which is the linear-absorption coefficient, μ_l, used in thickness measurements. In accordance with its definition, μ_l is given by the equation

$$\ln \left(\frac{I_1}{I_2}\right) = \mu_l \, \Delta d \tag{5}$$

where Δd is the thickness (in centimeters) of the sample irradiated.

There is convincing experimental evidence that to a degree of approximation satisfactory for most analytical work, the mass-absorption coefficient of an element is independent of chemical or physical state. This means, for example, that an atom of bromine has the same chance of absorbing an x-ray quantum incident upon it in bromine vapor, whether completely or partially dissociated; in potassium bromide or sodium bromate; or in liquid or solid bromine. X-ray absorption is predominantly an atomic property. This simplicity is without parallel in absorptiometry.

Equation 4 was written for a sample containing a single element upon which monochromatic x-rays are incident. Insofar as x-ray absorption is an atomic property, the mass-absorption coefficients for other samples are additive functions of the weight fractions of the elements present, whether in free or combined form; that is,

$$\mu_s = W_A \mu_A + W_B \mu_B + \cdots + W_M \mu_M \tag{6}$$

for a sample s containing M elements each at a weight fraction W with a mass-absorption coefficient μ.

Equation 6 is rigorously valid for absorptiometry with monochromatic x-rays, and usually an excellent first approximation for absorptiometry with polychromatic x-rays as well.

The simplicity of x-ray mass-absorption coefficients is shown in Fig. 13.5. It can be fully appreciated only if one notes the large spread of atomic numbers represented by the three metals, the order in which they appear, and the enormous (200-fold) wavelength range on the abscissa.

Although simple, x-ray absorption is of two general kinds: (1) *photoelectric absorption,* in which the entire energy of an incident x-ray quantum is transformed into the kinetic energy of a photoelectron and the potential energy of an excited atom that can subsequently emit a characteristic line, or emit a second photoelectron, or have an even more complex history; and (2) *absorption leading to scattering.* Inasmuch as these two processes are affected differently by changes in wavelength and in the atomic number of the absorber, it is convenient to subdivide the (overall) mass-absorption coefficient as follows:

$$\mu = \tau + \sigma \tag{7}$$

Of the two latter coefficients, the photoelectric-absorption coefficient τ generally makes the greater contribution to the mass-absorption coefficient. Absorption leading to scattering (represented by σ) gains in relative importance as atomic number Z and wavelength λ decrease.

Between absorption edges, the photoelectric (true) mass-absorption coefficient τ can be expressed by the following approximate empirical function of Z and λ:

$$\tau = \left(\frac{CN}{A} \right) (Z^4 \lambda^3) \simeq \mu \tag{8}$$

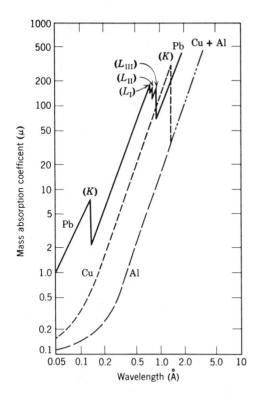

Fig. 13.5. Log–log plot showing the mass-absorption coefficient as a function of wavelength for three common metals. Note that the discontinuities locate the absorption edges, (**K**) and (**L**). [After H. A. Liebhafsky, *Ann. N. Y. Acad. Sci.*, **53**, 997 (1951).]

This equation contains a proportionality constant C and the number of atoms per gram (Avogadro's number N divided by atomic weight A), the introduction of which is necessary to change from the atomic photoelectric-absorption coefficient τ_a to the photoelectric mass-absorption coefficient τ. In light of what was said above, it is clear that the values of Z and λ determine the degree of approximation involved in substituting μ for τ in equation 8, the substitution being more nearly justified as Z and λ increase.

An examination of Fig. 13.5 will substantiate the statements just made. The following features are noteworthy: (1) Within the wavelength range shown, aluminum has no absorption edges, copper has only the **K** edge, and lead has the **K** edge and the **L** edges, of which there are three, each associated with a different characteristic spectrum; (2) the slopes between absorption edges calculated from the logarithmic values approach the value 3 as Z and λ increase; (3) the decrease in slope at short wavelengths, which occurs mainly because μ (and not τ) has been plotted (see equation 8), is most pronounced for aluminum and is imperceptible for lead.

In the discussion above, the fine structure of absorption edges has been neglected, and no mention has been made of the (small) effect of chemical state on the wavelength of an edge.

Equation 8 shows why filtering (preferential removal of the longer wavelengths) occurs when a polychromatic beam passes through a sample: τ (and hence μ between absorption edges) increases very sharply with wavelength. This important change in a polychromatic beam is best described as a decrease in its *effective wavelength* (see below).

2.   Scattering and Diffraction of X-rays

The importance of x-ray scattering is easy to underestimate, especially for analytical chemists, to whom it is generally a nuisance. Yet scattering is the basis of x-ray reflection, diffraction, and refraction, and it is the only process by which x-rays can be polarized. Scattering increases x-ray absorption, enhances the background present when analytical-line intensity is measured, and is sometimes a useful reference standard for such measurements. Scattering may be *unmodified* (no change in x-ray wavelength) or *modified* (wavelength increased by the Compton effect).

Owing to their high energies, the wavelengths of x-rays often are comparable to spacings in crystals. Consequently, a crystal can act as an x-ray grating made up of equidistant parallel planes (Bragg planes) of atoms or ions from which unmodified scattering of x-rays can occur in such a fashion that the waves from different planes are in phase and reinforce each other. When this happens, the x-rays are said to undergo *Bragg reflection* by the crystal, and a *diffraction* pattern results.

The conditions for Bragg reflection are diagrammed in Fig. 13.6, where AA' and BB' are the traces of successive Bragg planes, d is the distance between them, and σ is the glancing angle of the x-ray beam incident on these planes. Bragg reflection occurs when a wave scattered at O' can reinforce an identical wave scattered at O, these being the points at which the incident beam meets the Bragg planes.

According to the laws that govern the reflection of light, the reinforcement in question will occur if the path length of a beam specularly reflected at O' exceeds by an integral number n of wavelengths the path length of a beam similarly reflected at O. This requires for Fig. 13.6 that

$$\theta = \theta' \tag{9}$$

and

$$n\lambda = 2d \sin \theta \quad \text{(Bragg's law)} \tag{10}$$

Although only two Bragg planes have been considered, it is clear that planes below BB' also will contribute to the reflected beam, although to an exponentially decreasing extent. The relationship between λ and $\sin \theta$ evidently will be important in determining the useful wavelength range of a particular crystal.

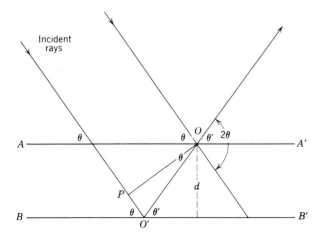

Fig. 13.6.   Simplified derivation of Bragg's law:

$$OO' = \frac{d}{\sin \theta}$$

$$O'P = OO' \cos(180° - 2\theta) = -OO' \cos 2\theta = -\frac{d \cos 2\theta}{\sin \theta}$$

$$OO' + O'P = n\lambda = d\left[\frac{1 - \cos 2\theta}{\sin \theta}\right] = d\left[\frac{1 - 1 + 2 \sin^2 \theta}{\sin \theta}\right]$$

$$n\lambda = 2d \sin \theta$$

In equation 10, the integer n gives the order of the reflection, $n\lambda$ always being the path difference in Fig. 13.6. A second-order beam, for example, of wavelength $\lambda/2$ will be reflected at the same angle as a first-order beam of wavelength λ, and a crystal cannot of itself distinguish between them. But according to equation 8, these two beams will have greatly different mass-absorption coefficients in a suitable absorber, that of the first-order beam being the larger. Insertion of such an absorber between crystal and detector will therefore attenuate the first-order beam to the greater extent, and this technique can be used to estimate the relative intensities of the two beams.

Bragg's law (equation 10) is obeyed so well that it is possible to use x-ray diffraction from crystals for highly precise determinations of either d or λ. The former type of determination is basic in establishing crystal structure; the latter, in x-ray emission spectrography.

The determination (or selection) of x-ray wavelengths by means of diffraction from crystals is highly precise ultimately because d is remarkably constant for the same Bragg planes in different crystals properly grown. When a polychromatic beam strikes such a crystal, only those wavelengths for which equations 9 and 10 are obeyed are detectable above background. By positioning the detector to intercept a reflected beam that makes an angle of 2θ with the beam incident on the crystal, one can thus measure the combined intensities of the wavelengths for which equations 9 and 10 are satisfied with a given crystal. By varying 2θ (and, if necessary, by changing crystals), the intensity–wavelength distribution of a polychromatic beam can be obtained, as in Fig. 13.3. In this way, also, a "monochromatic" beam of desired wavelength can be selected from a polychromatic beam. If an even purer beam is needed, two crystals in series (a double monochromator) can be used. Clearly, the purer the beam is, the lower will be its intensity.

D.   X-RAYS AND ATOMIC STRUCTURE

The short-wavelength limit (Fig. 13.3), the presence of absorption edges (Fig. 13.5), and the production of characteristic lines—whether by electron or by x-ray excitation (Fig. 13.4)—all point to quantum transitions and therefore to a close relationship between x-ray spectra and atomic structure. This relationship cannot be explored here, but it warrants mention as the fundamental reason for the simplicity of x-ray spectra—a matter of utmost importance to the analytical chemist. Perhaps the best-known aspect of this simplicity is the fundamental significance of the atomic number in determining the frequencies of characteristic lines. This relationship, discovered by Moseley (42), is expressible as the linear function

$$v = (0.248 \times 10^{16}) (Z - 1)^2 \tag{11}$$

where v is the frequency of the Kα line for the element of atomic number Z.

For our purposes, the absorption spectra are straight lines broken by absorption edges in plots, such as those in Fig. 13.5. The edges are called **K, L, M,** . . . edges in the order of increasing wavelength. Each edge has associated with it an emission spectrum (or series of lines) similarly identified by letter (e.g., **K** series), as are the lines (e.g., **K** lines) constituting the spectrum. The wavelength of every line in the spectrum exceeds that of the corresponding edge. Only the **K** spectra will be described here; the others, although more complex because of an increased number of participating energy levels, are governed by the same fundamental considerations. Only the **K** and **L** spectra are currently important in analytical chemistry.

Every chemist knows that atomic nuclei are surrounded by shells of electrons that, when filled, contain 2 (inner shell), 8, 8, 18, . . . electrons, this being the explanation of periodicity in chemical properties. For reasons that will be discussed, these may be called (in the same order) **K, L, M, N,** . . . electrons on the basis of conclusive x-ray evidence.

The question now is: What role do the **K, L, M,** . . . electrons play in generating the **K, L, M,** . . . series? The answer is not obviously predictable from a knowledge of visible or ultraviolet spectra. Neither hydrogen nor helium has a **K** series, although each has **K** electrons. This is because the **K** series is generated only when the **K** shell contains a hole that is filled by an electron that leaves one of the outer (**L, M, . . .**) shells; in other words, the generation of the **K** series requires (1) the *absence* of a **K** electron, and (2) the presence of an outer-shell electron whose transition to the **K** shell is permitted by the selection rules. This explains why—no matter what the method of excitation—all **K** lines for a given element have the same excitation threshold and appear together if they appear at all.

An atom with a **K** electron removed from the **K** shell is usually a singly charged positive ion, but an ion different (for the elements beyond helium) from the kind formed when one valence electron is lost. In rare cases, the **K** electron in question will lodge in an unfilled external shell and the atom will remain uncharged. If a **K** electron is ejected from a conventional ion (e.g., from Na^+), then the usual charge will increase by 1. We shall say that an atom or ion with a single hole in the **K** shell is in the **K** state no matter what its charge.

The generation of the **K** lines may be represented as follows:

$$\textbf{K state} \rightarrow \textbf{L state} + \textbf{K}\alpha$$

and

$$\textbf{K state} \rightarrow \textbf{M state} + \textbf{K}\beta$$

The **K** spectrum is thus a *series* of lines originating from a *single* initial state, and the wavelengths of the characteristic lines in each series differ because each line represents a transition to a *different final state*. Because the energy of any **M** state is less than that of any **L** state, it follows that **K**β will have a shorter wavelength than **K**α. The **L** and the **M** states are each at a higher energy than that of the neutral atom, and this is the reason why the wavelength of the **K** absorption edge is shorter than that of any **K** line by an amount set by the Einstein equivalence law. The relationship of other edges to their lines is determined according to analogous considerations.

Figure 13.7 shows, for a typical atom, the characteristic lines of possible importance in analytical chemistry. The simplicity of x-ray emission is attested to by the small number of lines, but even more eloquently by the fact that such a simple drawing can represent the x-ray excitation of analytical x-ray lines for any element with atomic number 29 (copper) or greater. The **L** spectra are useful substitutes for the **K** spectra when, as with very heavy atoms, the energies required to excite the **K** spectra are beyond the capabilities of the equipment on hand.

The electrons ejected from an excited atom (see Fig. 13.7) have become important in analytical chemistry. Electron spectroscopy for chemical anal-

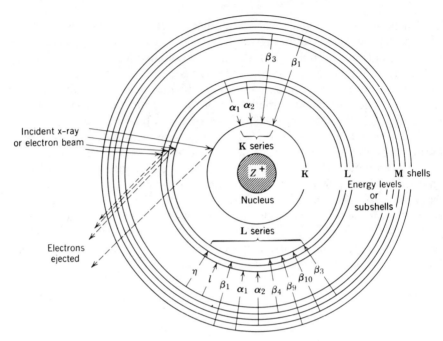

Fig. 13.7.   Schematic diagram showing x-ray or electron excitation of analytical x-ray lines for any element with atomic number 29 (copper) or greater. The **K** spectrum results when x-rays of wavelengths shorter than that of the **K** edge are absorbed to eject a **K** electron, and the vacancy is filled by an electron from an outer shell. Electrons may replace x-rays to cause ejection of the electron. **L** spectra are produced in a similar manner.

ysis, or ESCA, a modern application of photoelectron spectrography (7,20,50), is treated along with Auger electron spectrography in Chapter 17.

E.   CHEMICAL INFLUENCES IN X-RAY ABSORPTION AND EMISSION

Up to this point, our position has been that the elementary processes by which x-rays are absorbed and emitted are free of chemical influences. This simplified position suffices for most x-ray applications in analytical chemistry. However, chemical influences on both types of elementary processes have been demonstrated, but only at very high resolution, much higher resolution than the analytical chemist usually requires. These "chemical shifts" or "valence effects" are due to processes involving energy levels of valence electrons. They affect fine structure and cause shifts in wavelength, and thus change the intensity–wavelength distribution. They can provide information on the chemical state of the element studied. Examples of x-ray lines affected by valence electrons are the **Kβ** rays of elements with atomic numbers 11–17. This includes the elements sulfur, phosphorus, and chlorine; abroad range of applications of chemical shifts have arisen due to the variety of their valence states. Hurley and White (32) have illustrated such an application: the use of the **Kβ** line of sulfur to determine its chemical form in coal.

One of the attractive features of analytical methods using x-rays is that they can be applied without destroying the samples. This generalization, though valid as written, must not be taken to imply that samples necessarily come through unchanged when analyzed by such methods. For example, a glass slide can be permanently darkened when exposed for a few minutes in an x-ray spectrograph. It is well to be on the lookout for such effects, especially when the element being determined is present in a small amount and is particularly susceptible to the action of x-rays.

III. SCHEMATIC SUMMARY OF INSTRUMENTS FOR ANALYSIS

To show how x-ray instruments important to the analytical chemist are related, they are summarized schematically in Fig. 13.8 in terms of five components: the *x-ray source, sample, crystal* (Bragg reflector), *collimator*, and *detector*. A *photometer* consists of a source, sample, and detector; the sample acts as an absorber to filter and attenuate a polychromatic beam. Let us replace the sample with a crystal undergoing Bragg reflection. If the crystal is being used to produce a monochromatic beam of known wavelength, the instrument is a *spectrometer*. If in place of the crystal is a sample whose lattice spacings are being determined, we have a *diffractometer*. Let us add a sample as a fourth component of the spectrometer in the figure. We then have an *x-ray emission spectrograph* using x-ray excitation and with a crystal serving to resolve the emitted spectrum. If the sample primarily attenuates an x-ray beam, we have a *spectrophotometer*. Filtering has been ignored in Fig. 13.8.

In the spectrophotometer, the crystal may be placed between source and sample, in which case it acts as a monochromator, or it may be inserted between sample and detector, where it acts as an analyzer of the transmitted beam. Heating of the sample is negligible in the first arrangement.

For the sake of the general reader, we shall try to use simple, consistent nomenclature, but we must warn the reader to expect a shock when turning to the scientific literature for more detailed information. Consider just one aspect of the communication problem: The Bragg crystal scatters, diffracts, reflects, resolves, disperses, selects, isolates, and monochromatizes x-rays. The spectrograph system sketched in Fig. 13.8 exists to measure the intensity of analytical lines emitted by the sample. Which of the eight verbs listed above best describes what happens to the analytical line on its way to the detector? We shall speak of *wavelength resolution* via Bragg reflection. As discussed below, Bragg reflection is today often supplemented or replaced by *pulse-height selection,* an inadequate description of several highly sophisticated electronic processes. When wavelength resolution has been replaced, we shall speak of *energy resolution* (see LPWZ, p. 209). Other authors favor the term "dispersion" over "resolution."

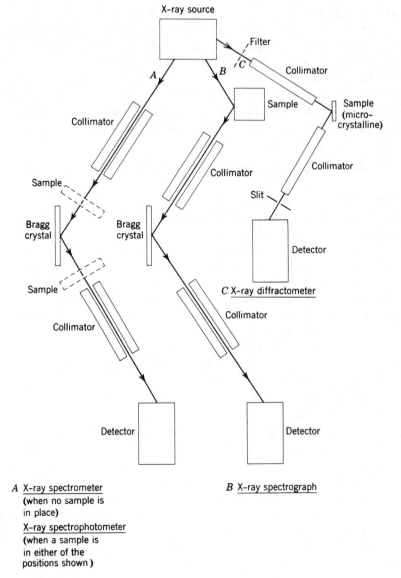

Fig. 13.8. Schematic introductory diagrams of instruments for x-ray spectral analysis.

IV. MEASUREMENT OF X-RAY INTENSITY

A. INTRODUCTION

Table 13.I places in perspective the radiant energies of five spectra for which intensities are often measured.

Table 13.I

A Crude Correlation of Various Spectra[a]

Spectrum	Spectral region	Representative $\lambda(\text{Å})$[b]	Quantum energy (ergs)	eV (eV)[c]
γ-Ray	γ-ray	0.01	2×10^{-6}	12.4×10^{5}
X-ray	X-ray	1	2×10^{-8}	12.4×10^{3}
Electronic	Ultraviolet	1000	2×10^{-11}	12.4
Vibrational	Infrared (near)	10^{4}	2×10^{-12}	1.24
Rotational	Infrared (far)	10^{6}	2×10^{-14}	0.0124

[a] It is assumed the atom is combined. Only molecules have vibrational and rotational spectra.

[b] The wavelengths are chosen for convenience. The spectral regions are poorly defined; for example, the ultraviolet region shades into the visible.

[c] One electron-volt is 1.6×10^{-12} ergs. Often an electron with energy eV electron-volts is called a V electron. Kilovolts often replace volts in this designation.

X-rays are detected by observing an effect of their interaction with matter; examples include (1) latent image formation on a photographic plate (chemical effect), (2) ionization in a gas or a solid (electrical effect), and (3) excitation of a phosphor (optical effect). The last two are often translated into electric currents, either pulsed or continuous, for the readout or recording of which complex electronic circuitry is usually needed.

We take intensity I to mean the number of x-ray quanta per second measured by a detector. An *instantaneous detector* measures I directly; that is, the detector gives a reading that is always proportional to intensity over the linear range of the detector.

At very low intensities, an instantaneous detector often cannot give sufficiently precise intensity readings. An *accumulative detector* is then needed. If, for example, the x-ray quanta reach the detector at a rate low enough for them to be counted, the counting can be continued until the accumulated total count N is large enough to ensure a sufficiently precise value of the mean intensity $\bar{I}$, which is proportional to the counting rate $N/\Delta t$, where Δt is the counting interval.

The x-ray intensity is never constant, but the fluctuations are usually negligible in a good spectrograph. An instantaneous detector would indicate these fluctuations; an accumulative detector would not. The analytical chemist usually wants these fluctuations ironed out; that is, he is generally interested in intensity integrated over time. As an accumulative detector accumulates counts, it necessarily integrates them when the counting rate is variable. Hence, accumulative detectors will be called *integrating detectors*.

An instantaneous detector is related functionally to an integrating detector as an ammeter is to a coulometer. Most detectors can operate in either way, although not necessarily with comparable efficiency or facility.

The absorption of x-rays influences their detection in two important ways. Absorption on the way to the detection medium reduces the measured in-

Table 13.II

X-ray Transmission and Absorption in Spectrographs

	0.1 Å	0.3 Å	1.0 Å	3.0 Å	10 Å
Fraction *transmitted* by 35-cm[a] beam path (STP)					
Air	0.9935	0.9863	0.896	0.074	~0
He	0.9993	0.9991	0.9985	0.9912	0.740
Fraction *transmitted* by various windows					
Be 10 mil (0.025 cm)	0.994	0.992	0.972	0.625	~0
Be 1 mil (0.0025 cm)[b]	0.9994	0.9992	0.997	0.953	0.216
Mylar 0.25 mil (0.0006 cm)[c]	0.9999	0.9998	0.9985	0.963	0.287
Mylar 0.1 mil (0.00025 cm)[c,d]	0.9999	0.9999	0.9994	0.985	0.607
Al 0.2 mil (0.0005 cm)	0.9998	0.9993	0.980	0.606	0.487
Fraction *absorbed* by detector gas (3-cm path, STP)					
Argon[e]	0.0009	0.0065	0.176	0.9852	>0.9999
Krypton	0.006	0.089	0.343	>0.9999	>0.9999
Xenon	0.025	0.320	0.831	>0.9999	>0.9999
Fraction *absorbed* by solid detectors					
NaI 0.2 cm	0.58	>0.9999	0.9999	>0.9999	>0.9999
Si 0.3 cm	0.11	0.35	0.9999	>0.9999	>0.9999
Ge 0.3 cm	0.47	>0.9999	0.9999	>0.9999	>0.9999

[a] Length of beam path in a typical spectrometer.

[b] The 1 mil Be is liable to be porous.

[c] Gases used in detector and spectrometer permeate thin polymers readily.

[d] Ultra thin polymer windows (10^{-4} cm or less) with thin, electrically conducting coatings, transmit to very long wavelengths (200 Å). They are extremely fragile and are often mounted on supporting screens, which of course reduce transmission.

[e] "Argon" used in detectors usually contains 10% CH_4. This mixture is known as P-10.

tensity and is undesirable; this includes absorption by the detector window. Conversely, absorption within the detecting medium produces the effect to be measured and is desirable.

These absorption problems may be considered under three headings: (1) attenuation along the beam path; (2) attenuation by the detector window; and (3) absorption by the detecting medium. The results of absorption calculations (Table 13.II) show the importance of these problems and suggest ways of dealing with them.

B.　FUNCTIONS OF ELECTRONIC CIRCUITRY

X-rays being quanta, it follows that the effects used in x-ray detection are quantum effects. In most detectors important in analytical chemistry, these quantum effects eventually yield electrons. Under the simplest conditions, these electrons appear as separate, well-defined pulses (bundles), one pulse for each x-ray quantum. Under these conditions, the pulses can be counted as individuals, and each pulse (hence each x-ray quantum) will register as a unit. As the intensity increases, it becomes increasingly difficult to maintain the individuality of the pulses. Discreteness may be lost within the detector, in the external electronic circuitry, or in both.

The detector output is processed through *linear* and *logic* circuits. The primary function of a linar circuit is to amplify, the most common example being increasing the *amplitude* or number of electrons in a pulse so that the energy of an absorbed quantum can be measured. The logic circuit makes decisions about pulses; for example, it decides whether the energy of a pulse, which is proportional to its amplitude, exceeds a given minimum. For such decisions to be possible, the pulses entering the circuit must have a fixed *shape*. Both linear and logic circuits are required for pulse-height selection and the counting of x-ray quanta.

The principal functions performed by electronic circuitry in the measurement of x-ray intensity are listed below. The considerations that follow are limited to the detector system itself; stable x-ray tube voltage and current, and a stable voltage across the detector, are taken for granted.

1. *Impedance matching.* The charge resulting from the absorption of an x-ray quantum must be transferred from the detector to the preamplifier smoothly, without loss, and without loss of pulse individuality. Proper matching of impedances, particularly that pertaining to the preamplifier input capacitance, is thus important.

2. *Amplification as required to give a useful signal.* Considerably more energy than that directly available from absorption of an x-ray photon in a detector ($\sim$10–15 W $\cdot$ sec) is usually needed before the detector system can produce a useful readout. Amplification is possible within the *detector* (internal) or within the *preamplifier* and the *amplifier* (external). A *Geiger detector* has enormous internal amplification, but the *ionization chamber* produces only very weak pulses and requires sophisticated amplifiers. The *proportional detector* and the *solid ionization detectors* [Si(Li), Ge(Li), intrinsic Ge], give stronger pulses, but need an amplifier with a wide linear range of gain, so that pulse heights can be satisfactorily recorded.

3. *Pulse shaping.* The pulses produced by a detector vary in amplitude and occur at random intervals because x-ray emission itself is a random process. Before these pulses can be selected and counted, they must be "clipped"—shortened in duration—to make each pulse individual, and these individuals must be properly shaped for pulse-height selection. These functions are carried out by the preamplifier and the amplifier.

4. *Pulse-height selection.* A pulse-height selector or analyzer has one or more "windows" through which pulses of selected height pass. A multichannel analyzer consists of a series of windows each set to admit a range of pulse heights successively higher in energy. The pulses from a sample pass through these windows independently and simultaneously, each pulse through its proper window, for accumulation and storage prior to readout.

5. *Counting and scaling.* These functions consist of the *accumulation* of pulses to give either a number of counts over a fixed counting interval or the time required for a fixed number of counts; either way ultimately yields an average intensity, $\bar{I}$, measured in counts per second. Scaling is necessary because the number of counts is often too large for a counter to handle. In scaling, a fixed number of counts (usually 2 or 10) is accumulated into a larger unit (i.e., is counted as a single, "higher-level" count), these units are accumulated into still larger units, and so on through as many stages as necessary. In solid-state circuits, counting and scaling are combined into a single operation. With multichannel analyzers, computers are often used to store the sorted counts.

6. *Instantaneous readout.* A detector that yields discrete pulses can be used as an instantaneous detector if the pulses can be averaged to form a continuous electric current, as in a counting-rate meter. An important point is that counting-rate meters alone cannot give energy resolution because all pulses that trigger the counting mechanism, no matter what their height, appear identically in the readout of the meter.

7. *Pulse rejection.* Pulse-height selection is achieved by rejecting pulses *lower* and *higher* in energy than those to be selected. The lower-energy pulses are rejected by a pulse-height discriminator. The higher-energy pulses are isolated by a second pulse-height discriminator and then rejected by an anticoincidence unit, which blocks the passage of pulses simultaneously surviving the two discriminators.

The precision of the x-ray intensity measurements that rely on counting is limited ultimately by statistical errors inherent in the counting. These will combine with, and may be overshadowed by, errors arising in the electronic circuitry, such as drift or fluctuations in the detector output and *dead-time losses* at high counting rates, that is, losses that occur during counting if one component of the circuitry is considerably more sluggish than the others. As in television sets, solid-state devices have replaced electron tubes in electronic circuitry for x-ray detection.

C. PHOTOELECTRIC DETECTORS

Today we use photoelectric detectors to *measure* and *record* what Roentgen *saw* when he discovered x-rays; the *photomultiplier* has replaced the human eye. We shall discuss only the *scintillation detector*, which serves in many "workhorse" x-ray emission spectrographs; for a more complete discussion, see LPWZ, pp. 70–75.

The visible light Roentgen saw when he discovered x-rays can be "seen" also by a photoelectric surface, from which the light will eject photoelectrons. The photomultiplier surpasses the human eye in that it can multiply the number of these electrons by some six powers of 10 and never feel fatigue. How it does this is shown by Fig. 13.9.

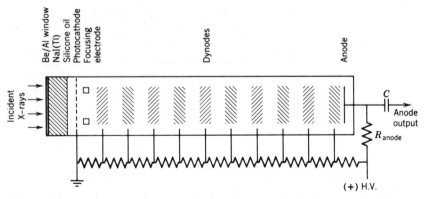

Fig. 13.9.   Schematic diagram of a scintillation detector. The window is beryllium (0.2 mm thick) coated with aluminum (about 1 μ) to give opacity. Not shown are (1) the outer glass envelope; (2) the glass seal between the phosphor and the silicone oil; (3) the crystalline MgO placed above and below the phosphor to reflect visible light; and (4) the voltage circuit for the focusing electrode. The phosphor converts the incident beam to visible light, which in turn generates electrons when it strikes the Cs–Sb photocathode in the photomultiplier. The dynodes, each acting in turn, collect and amplify the electron beam as it moves to the anode. Voltage control for the dynodes is provided within the detector assembly by the series of resistors shown. The bleeder current provides a uniformly higher voltage at each successive dynode and makes it possible for the electrode to perform its two ("dy" = "two") functions.

Thallium-activated (Tl-"doped") sodium iodide, or NaI(Tl), is a satisfactory phosphor for a photomultiplier because its mass-absorption coefficient for x-rays is high, and because its decay time (or dead time or recovery time), 250×10^{-9} sec, though longer than that of some materials (e.g., naphthalene), is generally short enough for satisfactory counting; this time is considerably longer than the electrical response time of the photomultiplier. The photomultiplier must of course respond satisfactorily to the wavelength generated by x-ray absorption; for NaI(Tl), this wavelength is near 4100 Å (blue), and a cesium–antimony photocathode is satisfactory.

We have space only for several added important facts about scintillation detectors; (1) They suffer less from *escape peaks* (the **Kα** line of iodine being the only one likely to escape at significant intensity) than do gas-filled detectors (see Section IV.D, below). (2) The two principal parts of a scintillation detector, phosphor and photomultiplier, must be *optically coupled* (as by a film of silicone oil) to minimize loss of visible quanta. (3) The detector must be hermetically sealed because NaI(Tl) is hygroscopic. (4) The detector window must transmit x-rays but not visible light. Satisfactory devices are on the market.

D.   GAS-FILLED DETECTORS

Gas-filled detectors make use of the ionization, called here the *initial ionization,* that follows the absorption of x-rays by a gas and the ejection

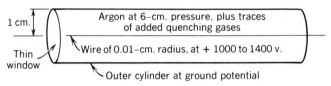

Fig. 13.10. Schematic diagram of a Geiger counter tube, showing a typical end-window arrangement. The nature of the window depends on the kind of rays to be detected.

of photoelectrons from the molecules involved. These photoelectrons subsequently ionize other molecules. The relatively large energy of the x-ray quantum thus leads to the production of a number of ion pairs, each consisting of an electron and a relatively immobile positive ion; on the average, about 30 eV is needed to produce an ion pair. If these ion pairs do not recombine, the extent of this initial ionization is determined by (and measures) the energy of the x-ray quantum.

But the generation of positive and negative charges is not enough. Information about the absorbed x-rays will be forthcoming only if the charges are separated and collected. This requires an electrical field. Such a field accelerates the light electrons much more effectively than it can accelerate the heavy positive ions. Electrons thus accelerated will collide with gas molecules, which they may ionize if the electron energy is sufficiently large. In this way, the electrons resulting from the absorption of an x-ray quantum can amplify (multiply) the initial ionization. The extent of amplification increases in a complex way with the electrical field, and it can be used to distinguish one type of gas-filled detector from another.

Three detector types are distinguishable. In the *ionization chamber,* a nineteenth-century device, there is no amplification, and the electric field need only be strong enough to keep electrons and positive ions from recombining. In the *proportional detector,* first used by Rutherford and Geiger in 1908, the amplification is moderate, as is the electric field. In the *Geiger detector,* developed soon thereafter and so greatly improved in 1928 by Geiger and Müller that it is often called by both their names, the degree of amplification is high and the field is strong; the field, in fact, is near the maximum that the gas can sustain without breaking down in the absence of x-rays.

In principle, which is as far as we shall go here, the device in Fig. 13.10 can serve as any one of the three important detectors just listed; it is necessary only to choose a potential difference between the central wire and the envelope that will give the amplification characterizing the desired detector.

1.   Ionization Chambers

In these detectors, the potential difference is minimal; it is low enough to prevent amplification and high enough to prevent a decrease in the initial ionization through recombination. Because they provide no amplification of the initial ionization, the electric currents they generate are minute. With appropriate circuitry, the ionization chamber can serve as either a current-measuring (instantaneous) or a pulse-counting (integrating) device. When used in conjunction with amplifiers now available, the current-measuring chamber deserves to be considered by the analytical chemist for measuring the intensities of strong polychromatic beams.

2. The Geiger (Geiger–Müller) Detector

The Geiger detector was widely used in x-ray emission spectrography some years ago. It has now been superseded by the proportional, scintillation, and solid ionization detectors. The virtues and faults of the Geiger detector stem from its enormous internal amplification, which is the highest possible without electrical breakdown of the filler gas. The virtues are simple ancillary circuitry, good stability, trouble-free operation, powerful output signals, and sensitivity to soft x-rays. Unfortunately, the response of the Geiger detector is the same no matter the energy of the x-ray quantum. Another limitation is its sizable dead time ("recovery time"), which limits the useful maximum counting rate of the detector to about 500 counts per second (cps).

3. The Proportional Detector

The proportional detector strikes a happy mean between the two other gas-filled detectors. Its internal amplification can be sufficient to make possible energy resolution, yet moderate enough that dead time is negligible and adequate counting rates ($\sim 10^5$ cps) are possible. Methane mixed with argon is a good filler gas. Proportional detectors have proved most valuable in x-ray emission spectrography as an adjunct to Bragg resolution. They are usable in three ways: (1) as simple detectors, in the manner of Geiger detectors, especially for high counting rates; (2) with a pulse-height analyzer, as an adequate means of resolving an x-ray beam when the wavelengths of interest have no close neighbors; and (3) with a pulse-height selector or discriminator, as a means of obtaining reduced background.

So-called *escape peaks* are mentioned here because in the third use just named, they can wrongly lower the count for an analytical line. Here is what may happen: An analytical line of the requisite energy excites the characteristic line of a filler gas. The excess energy appears as a pulse, called the escape peak, which is normally too low in energy to pass the selector window; it therefore *remains behind* and is not counted. If its accompanying characteristic line is counted, all is well. But it may not be, because all elements (including the principal filler gas) transmit their characteristic lines quite readily. To the extent that these characteristic lines escape the detector, the count will be low. Suitable replacement of the filler gas is the best remedy. Owing to the relatively high absorbance of x-rays by solid phosphors, escape peaks can usually be disregarded when scintillation detectors are used.

a. FLOW PROPORTIONAL DETECTORS

Sealed proportional detectors, the only kind so far discussed, are not adequate for x-ray emission spectrography of light elements because the absorbance of their windows for the analytical lines, which are necessarily all of long wavelength, is too high. They are replaced in this application by flow detectors, which have thin and ultrathin windows (typically 0.1–1 μ thick formvar or polypropylene), usually supported on a metal screen and (if nonconducting) coated with a thin conducting layer (aluminum or carbon) on the side facing the anode, a measure that gives a uniform electrical field. Whether thin or ultrathin, the windows are leaky; satisfactory detector operation requires continuous flow of the filler gas through the detector; hence the name. Argon containing 10% methane (so-called P-10 gas) is a favorite filler. Formvar windows generally serve for the 10–100 Å region, whereas polypropylene is used when all significant wavelengths exceed the critical absorption wavelength of the carbon **K** edge, which is near 44 Å (29).

b. NOVEL DETECTORS

New gas proportional detectors with improved performance have been developed in recent years. Cairns et al. (16) have described an end-window-type detector featuring better resolution and efficiency for soft and ultrasoft x-rays. Improved resolution is achieved by varying the anode-to-window distance, thus enabling more selective detection of such x-rays. Substantial improvements in resolution have been obtained with the novel scintillation proportional detector pioneered by Policarpo et al. (48). A relative resolution of ~36% has been reported for C**K**α with such a device.* The counter is based on the detection of the "secondary light" generated in the gas atoms (preferably noble gases) by the primary or secondary electrons gently accelerated in an electrical field. The field strength must be sufficient to allow production of light for absorption by this scintillation detector, but it must remain below the threshold for charge multiplication. Under these conditions, resolution comparable to that of semiconductor detectors is claimed for soft x-rays. These new detectors have a promising future.

E.   SEMICONDUCTOR DETECTORS

McKay (40), in 1949, seems to have been the first to think of the "solid state equivalent of an ionization chamber [detector]" as a device that might be better than gas-filled detectors. To be an improvement over gas-filled detectors, a solid ionization detector would, to begin with, have to:

---

* Henceforth we shall refer to the **K** lines of a given element by closing up the symbol for that element with name of the **K** line. Thus C**K**α is the carbon **K**α line.

1. Show adequate x-ray absorption.
2. Produce, for each quantum absorbed, a number of ion pairs adequate for satisfactory counting after amplification. There is no internal multiplication in these solid detectors as normally used.
3. Have this number of ion pairs be proportional to the energy of the absorbed x-ray quantum over a wide energy range.
4. Maintain an internal electric field or gradient suitable for charge collection without recombination.

The first requirement means that the mass-absorption coefficient must be sufficiently high and the "sensitive volume" (the volume within which x-ray absorption will give ion pairs) must be sufficiently large; a thin film or boundary layer will not do. The second requirement can be met only if the energy needed to produce an ion pair is considerably less than 30 eV. With respect to the third requirement, the solid ionization detector should be competitive with the proportional detector.

Among all the materials known, only two semiconductors, germanium and silicon, can meet all four requirements. They can do so only if they are nearly perfect crystals of high purity. Both these elements have four electrons in the valence band. They have different filled shells between these valence electrons and the atomic nucleus, but this difference is unimportant here. Electrons in the conduction band, just above the valence band, are free to move; electrons in the valence band belong to particular atoms, and hence are more tightly bound. In silicon, the energy difference ("band gap") between conduction and valence bands is 1.1 eV; in germanium, it is only 0.66 eV. This means that at a given temperature, far fewer electrons will be in the conduction band of silicon owing to thermal excitation than even the small number in the conduction band of germanium. The need to reduce this number by cooling below room temperature (to suppress electronic noise) is of course greater for germanium. The band gap is analogous to the ionization energy in a gas.

Impurities that *donate* electrons to the *conduction* band can produce an n-type region in a semiconductor; impurities that leave a *deficiency* of electrons in the *valence* band can produce a p-type region. (Here "n" means negative, and "p" means positive.) There is normally a potential difference between an n-type and a neighboring p-type region in the same semiconductor. Antimony (five valence electrons) in silicon or germanium (four valence electrons each) will tend to form n-type regions; boron (three valence electrons) will tend to form p-type. An impurity such as lithium (one valence electron, loosely held) will always tend to produce an n-type region: It is not built into the crystal lattice as are boron and antimony.

In semiconductors, the hole–electron ($\oplus$–e^-) pair is the analog of the ion pair in gases; both can be produced in the same way. But pair production requires on the average only 3.5 eV in silicon, and even less (only 2.94 eV) in germanium; in the rare gases some 30 eV is demanded. Unlike positive ions in gases, the holes have mobilities approaching those of electrons. In germanium at 78 K, the mobility of holes is 1.5×10^4 cm/sec, for a gradient of 1 V/cm; the mobility of electrons is only 3 times greater.

A difference in the conduction mechanism accounts for the difference just mentioned between gases and semiconductors. In an n-type region, the current is naturally carried almost entirely by electrons (the *majority* carriers), because they are free to move in the conduction band. Any positive holes (the *minority* carriers) exist in the valence band below. In a p-type region, any free electrons are drawn into the valence band by these positive holes. When an electron in such a valence band jumps from one atom to its neighbor, the process is conveniently regarded as the movement of a positively charged hole in the opposite direction, which makes these holes the *majority,* and the electrons the *minority,* carriers of current in the p-type regions. This movement of holes is far faster than the movement of positively charged ions in gases at appreciable pressures.

The properties of germanium and silicon thus make them uniquely good choices for solid x-ray detectors, provided they can be produced as nearly perfect single crystals of ultrahigh purity and large enough for use as detectors.

For more than a decade, neither germanium nor silicon could be made pure enough to serve as an "intrinsic" detector. By the ingenious use of lithium in a process called "lithium drifting," it proved possible to create a sensitive "intrinsic zone" in the purest material then available (see Fig. 13.11). Si(Li) and Ge(Li) detectors, affectionately called "silly" and "jelly," were the result. First made by R. N. Hall of the General Electric Company, "intrinsic" Ge has now become commercially available; "jelly" is on the way out. In welcome contrast to the lithium-drifted detectors, intrinsic Ge needs cooling only during use.

We now give a quick appraisal of Si(Li), Ge(Li), and intrinsic Ge in terms of the four requirements named earlier. With regard to requirement 1, the absorption of x-rays in a unit volume of these materials is obviously much greater than in any filler gas under usable conditions, and of course greater in germanium than in silicon. Requirement 2 is easily met; an x-ray quantum of 1 Å wavelength creates about $12,400/3.5 = 3,540$ hole–electron pairs in a Si(Li) detector. For comparison, note that the same x-ray will produce about 400 ion pairs in a gas-filled ionization detector. Requirement 3 can be met, but electronic difficulties are to be expected because the pulses for long wavelengths are very small. Requirement 4 is satisfactorily met by the potential difference applied across the $n^+ - p^+$ diode separated by a semiconductor of intrinsic resistance. The charge-carrier lifetime is long enough ($\sim 10^{-3}$ sec), and the charge-collection time short enough ($\sim 10^{-7}$ sec), to preserve individuality of pulses at fairly high counting rates. The great promise of solid ionization detectors would have come to little in analytical chemistry had it not been for the tiny field-effect transistors (FET) to which they are intimately joined, and by means of which amplification can be accom-

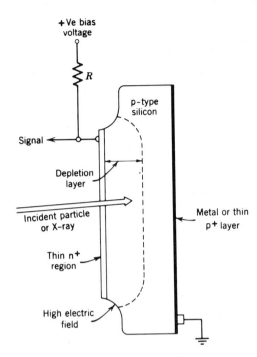

+Ve bias
voltage

R

p-type
silicon

Signal

Depletion
layer

Incident particle
or X-ray

Metal or thin
p+ layer

Thin n+
region

High electric
field

Fig. 13.11. Conventional representation of an n^+–p^+ junction in a solid ionization detector. The "depletion layer," from which charge carriers have been removed by drifting lithium in the direction of the arrow, is the sensitive zone in which the element approaches intrinsic resistivity. (After Gibson, W. M., G. L. Miller and P. F. Donovan in K. Siegbahn, Ed., *Alpha-, Beta- and Gamma-ray Spectroscopy*, Vol. 1, North-Holland, Amsterdam, 1966, Chapter 6.)

plished by use of an external load resistor. In this way, and by cooling of both detector and FET to 77 K, electronic "noise" is reduced to the point where the detectors fulfill their promise. Of course, the cryostat must have a window, which unavoidably means reduced analytical-line intensity.

For wavelengths greater than about 0.3 Å (energies below 40 keV), Si(Li) detectors are preferred. The SiK escape peaks affect analytical-line intensities less than do those of germanium. The GeK absorption edge is responsible for further difficulties. But for x-rays of wavelengths shorter than 0.3 A, germanium detectors are indispensable, because silicon, being of low atomic number, cannot absorb the rays effectively enough.

Solid ionization detector systems joined to multichannel pulse-height analyzers have revolutionized x-ray emission spectrography in recent years; Figure 13.12 shows why. The two principal advantages of this approach are instantaneous intensity readout of all analytical lines and a more favorable geometry, because the sample and the (highly efficient) detector can be close together. Consequently, the analytical lines need not be excited at high intensity, and a radioactive source may often replace the x-ray rube and its bulky auxiliary equipment. Lower capital investment should follow as a third advantage. Counting rates, linear to beyond 10^4 cps, are normally adequate, though marginal for some trace determinations. Stability, given continuous cooling, is good. Other advantages sometimes claimed vary with x-ray wavelength. The resolution attainable is an important "figure of merit" for solid

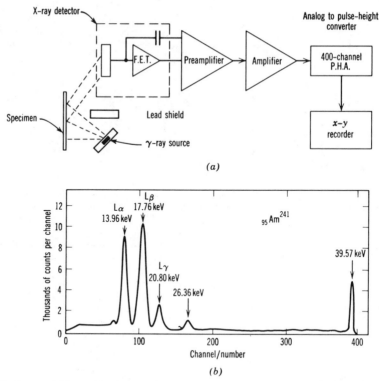

(a)

(b)

Fig. 13.12.   (a) Diagram of an x-ray emission spectrography system with unaided energy resolution. Broken lines enclose the low-temperature region. (b) Gamma- and x-ray spectra of Am^{241} obtained with this system. A 2-mm plastic cover on the radioactive source barred alpha-particles from the Si(Li) detector. [After Bowman, H. R., E. K. Hyde, S. G. Thompson, and R. G. Jared, *Science*, **151**, 562 (1966).]

ionization detector systems: they cannot generally compete in this respect with wavelength resolution via Bragg reflection aided by pulse-height selection.

F.   ENERGY RESOLUTION (OR DISPERSION)

Because solid ionization detectors, proportional detectors, and scintillation detectors all give pulses of heights proportional to the energies of absorbed x-ray quanta, they offer the possibility of replacing wavelength (Bragg) resolution (or dispersion) with *energy resolution* (or dispersion). The replacement has much to offer, notably, the simultaneous recording of the intensities of all analytical lines and the elimination of the serious intensity losses that accompany Bragg reflection. Energy resolution is accomplished by pulse-height selection in a multichannel analyzer. The *pulse-height distributions* (Fig. 13.13) for proportional and scintillation detectors are so wide as to make these detectors unpromising for quantitative determinations by *unaided* energy resolution. Fortunately, the solid ionization detector is an-

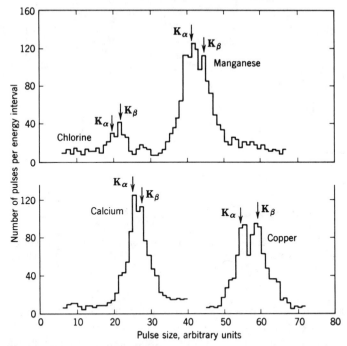

Fig. 13.13. Histograms for Kα and Kβ lines of chlorine and manganese (from manganese chloride), calcium, and copper. The abscissa may also be read as *pulse height, pulse energy,* or *x-ray-quantum energy*. [After Curran, S. C., J. Angus, and H. L. Cockroft, *Philos. Mag.,* **40**(7), 36 (1949).]

other story. With the other detectors, energy resolution combined with wavelength resolution is now the accepted way to eliminate interference by lines reflected at orders higher than the first and to reduce background, particularly in trace determinations.

All quantitative observations vary. Small variations occur more frequently than large; positive and negative variations tend to balance; and huge variations occur seldom, if at all, and are generally unpredictable. This summary of universal human experience is best represented by the well-known, bell-shaped, Gaussian distribution curve. For all Gaussian curves, the standard deviation (observed value, s) is the horizontal distance from the mean to the point of inflection, and the *full width of the curve at half maximum* (FWHM) is $2.35s$ (see Fig. 13.14). When the Gaussian represents the results of a *single, completely random* process, it has a further characteristic that makes in unique: s, the standard deviation, then equals the *square root of the mean value* of the observations.

When a *monochromatic* x-ray beam is absorbed by a detector whose *mean* output is proportional to x-ray energy, the pulses in that output vary considerably in height, so that a *pulse-height distribution* results. If the output results from a single random process (ionization and multiplication being

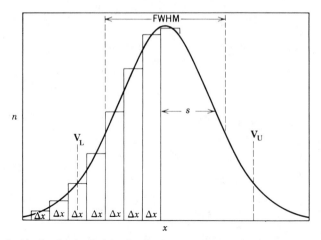

Fig. 13.14.   An idealized pulse-height distribution superimposed on a Gaussian. V_U and V_L are the upper and lower voltages defining a *window* in a pulse-height selector. Here, $s = s_i$, the *ideal* standard deviation.

regarded as one such process), then the pulse-height distribution should be Gaussian (see Fig. 13.14).

Various ordinate are used, the most fundamental being $\bar{n}$, the mean number of electrons in a pulse of energy E. These two quantities are proportional in the linear range of the detector. One may therefore use as abscissa either n or energy; in the latter case, the energy of the monochromatic x-ray is the mean, and the energies of the individual pulses are each proportional to the value of n for the pulse. Because the n electrons of a pulse may be used to charge a capacitor to voltage V when the detector operates, V is also a logical abscissa. Thus, with a linear detector and linear amplification, n, E, or V determines pulse height. Figure 13.14 is therefore a pulse-height distribution for the idealized detection of a monochromatic x-ray, the energy of which fixes the mean.

Under the simple conditions just outlined,

$$s = \sqrt{E} = s_i \tag{12}$$

$$\text{FWHM} = 2.35s = 2.35\sqrt{E} \tag{13}$$

If other components (e.g., the preamplifier or amplifier) of the detector system also introduce random fluctuations, then by the rule for the combination of errors,

$$s = \sqrt{s_i^2 + s_p^2 + s_a^2 + \ldots} \tag{14}$$

where s, the total standard deviation, is now larger than the ideal s_i (equation

12) because of contributions from the other components (designated by sub-scripts).

For the solid ionization detector, it is experimentally established that

$$s_i < \sqrt{E} \tag{15}$$

presumably because of fluctuations in the amount of energy degraded to heat in the detector. The theoretical considerations involved need not concern us. What matters here is the *experimental* observation that a correction factor—the Fano factor, *F*—needs to be put under the radical to change equation 15 into an equality, and that *F* seems to have a value between 0.10 and 0.15 for solid ionization detectors.

The relationship between pulse-height distribution and counts per second (or per any fixed counting interval) is of first importance. Monochromatic x-ray quanta absorbed in any counting interval will give rise to pulses that may have any height included in the pulse-height distribution. Each of these pulses, regardless of its height, *will register as one count* in the rate meter or in the counting–scaling system. Therefore, the x-ray intensity expressed in counts per second by the rate meter or as counts per Δt in an accumulative system will be proportional to the *area* of the pulse-height distribution. With x-ray lines of neighboring wavelength, these areas will overlap. Any system of energy resolution will have to reject many pulses in these overlapping areas—a serious disadvantage.

For the last time, we must contrast "resolution" and "dispersion." When a Bragg crystal "disperses" an x-ray spectrum, it performs a useful service *prior to* detection; movement of the crystal (or detector) then makes it ideally possible to measure the intensity of one wavelength at a time. "Energy dispersion" has been defended as a name on the basis of Fig. 13.14. Notice, however, that this "dispersion" occurs *after* detection and is, as we shall soon see, a *step backward*. The term "energy resolution" at least avoids this ambivalence.

The sharpness of energy resolution is important (1) as a *figure of merit* in rating detectors, and (2) as a criterion of what unaided energy resolution can do in analytical chemistry.

The FWHM is satisfactory as a *rough* figure of merit in rating detectors. By selecting a reference line [e.g., the MnKα line (2.10 Å, 5.9 keV) from $^{55}_{26}$Fe] and measuring for it the FWHM achieved by a detector, one obtains an experimental value that normally includes any broadening attributable to the electronic circuitry. The reference line for which the FWHM was measured should always be given. Obviously, the FWHM does not give the *shape* of the distribution curve, and such curves are often far from Gaussian.

The problem in analytical chemistry is very complex, for several reasons: (1) *Resolution* is a vague concept. Resolution for qualitative identification is more easily attained than resolution for quantitative determination. when the FWHMs for adjacent pulse-height distributions do not overlap, complete resolution can usually be achieved for major constituents from half the maximum height upward, but that is not enough for quantitative work, in which complete resolution down to the abscissa is desirable. (2) In general, the amounts of the elements in a smaple can vary over a wide range. the concomitant intensity variations complicate the resolution problem. (3) The presence of background further complicates the problem. (4) The analytical chemist needs to know the resolution achievable over the entire wavelength range in which his analytical lines are found.

In this situation, calculations employing FWHM values are unlikely to be of much use; it is usually futile to do more than see how far roughly drawn pulse-height distributions overlap. In terms of Fig. 15.14, a logical measure of energy resolution is the *relative resolution* RR, defined by

$$RR = \frac{2.35s}{\bar{x}} = \frac{FWHM}{\bar{x}} \qquad (16)$$

where $\bar{x}$ is commonly taken as E, and s as $\sqrt{E}$ (see equation 13). The RR, usually expressed as a percentage, can be used to *estimate* the way in which resolution changes with changes in wavelength or energy, as the following example will show. One manufacturer (33) gives the following sepcifications for a Si(Li) system: FWHM < 155 eV for the 5.9-keV, 210-A MnKα line from ^{55}Fe at 1000 cps for a detector area of 10 mm^2, which corresponds to

$$RR = \frac{155}{5900} = 0.026 = 2.6\%$$

As concerns light elements, a FWHM of 99 eV has been reported for the carbon Kα line with a windowless Si(Li) detector (44). This corresponds to 2 RR of 0.35, or 35%. Thus, the present resolution of Si(Li) detectors is not sufficient for quantitative emission spectrography of the light elements. Today, this field belongs to wavelength resolution via Bragg reflection aided by pulse-height selection, and is accomplished by use of suitable crystals or of the multilayered soap films ("Langmuir–Blodgett gratings") introduced by Henke.

The mathematical treatment of unresolved pulse-height distributions is a welcome and much needed complement to energy resolution. The problem is illustrated by Figure 13.15 for the simple case of idealized composite peaks resulting from equal contributions by neighboring elements (for further details, see LPWZ, pp. 406 *et seq*). The computer is invaluable for work of ths kind.

To summarize: Every analytical laboratory doing precise x-ray emission spectrography on diverse samples must employ wavelength resolution aided by pulse-height selection. Energy resolution is useful for survey work and for the less demanding quantitative determinations. Ideally, both methods should be available.

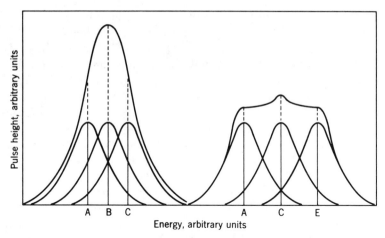

Fig. 13.15.   Idealized composite pulse-height distributions from a multichannel analyzer un-aided by wavelength resolution. The taller distribution is for neighboring elements A, B, and C; unfolded, it would give the three individual Gaussian distributions (one for each element) shown. The broader distribution is for A, C, and E, for which the atomic-number interval is 2 between neighbors; individual contributions again are shown.

V.   ANALYTICAL METHODS BASED ON X-RAY DIFFRACTION*

A.   INTRODUCTION

Roentgen, knowing nothing about x-ray wavelengths, could scarcely have been expected to discover x-ray diffraction. That discovery had to await a true flash of genius, which came to M. F. T. von Laue when he realized that if the wavelengths of x-rays were comparable to the distances between atoms and ions in crystals, and if crystals had highly ordered interiors cor-responding to the highly ordered exteriors long known in crystallography, then crystals ought to scatter x-rays *coherently* (form them into patterns). A single epochal experiment showed that both conditions were met. The Laue flash of genius was not directly translatable into an analytical method. That translation followed W. L. Bragg's realization that Laue's discovery meant that crystals could serve as x-ray gratings that (by analogy with their effect on light) ought to reflect ("scatter coherently without modification") x-rays incident upon them. Bragg's law (equation 10) resulted.

Because crystals have highly ordered internal structures, one can pass through them different *kinds* of identical planes. Each kind is distinctively populated by atoms or ions; thus each kind distinctively diffracts x-rays. To emphasize this distinction, we write Bragg's law as

$$d_{hkl} = \frac{\lambda}{2 \sin \theta_{hkl}} \tag{17}$$

* Much of this section is excerpted from LPWZ, Chapter 6.

The equation takes for granted the first-order reflection of a "monochromatic" x-ray beam, the usual situation in analytical chemistry. The subscripts, the *Miller indexes,* are a heritage from crystallography; the analytical chemist usually need regard these only as a set of numbers unique for each kind of plane in a crystal.* The *interplanar spacing* ("repeat distance") for planes of one kind is d, and is calculated from the measured 2θ value.

X-ray diffraction continues to make its greatest contributions in establishing the structures of crystals and molecules, and these contributions have grown more spectacular as the technique has come to be applied in the life sciences, revealing the structures of complex molecules such as insulin, hemoglobin, and DNA. Although such work is properly describable as the characterization of materials, we must exclude it here for the same reason that we shall not concern ourselves with work on single crystals: neither category is likely to describe the work of an analytical-chemistry laboratory.

Usually, then, the analytical chemist is concerned with x-ray diffraction measurements made on samples containing crystalline powders. If only one crystalline species is present, the diffraction pattern from a family of planes, each with its own set of Miller indexes, is a "fingerprint" of the crystal. In this fingerprint, each plane has its d value and reflects x-rays at its own intensity. When more than one crystalline species is present, the pattern will be a superposition of the individual "fingerprints." Qualitative identification can generally be made by matching diffraction patterns of standards and unknowns (the comparative method once more). Quantitative determinations require intensity measurements.

The diffraction pattern can be established *simultaneously* by using a camera with a photographic plate as detector, or *sequentially* by using a diffractometer with a proportional or a scintillation detector.

B. THE HULL–DEBYE–SCHERRER METHOD

1. Introduction

In the Hull–Debye–Scherrer method, the diffraction pattern produced when a collimated beam of monochromatic x-rays strikes a small, finely powdered sample is photographically recorded. Ideally, the sample is uniform, so that a known λ value and measured θ values from a single pattern will give *all* values of d_{hkl} (and, for higher-order reflections, all values of d_{hkl}/n) for the sample. This is possible because the finely divided sample, if it has *random* crystal orientation, properly presents *all* the kinds of crystal faces to the x-ray beam. Slow rotation (about 1 rpm) of a properly prepared sample can virtually ensure that this condition is met.

---

* Initially, crystallography was indispensable to x-ray diffraction; now, x-ray diffraction is indispensable to crystallography. For this story and much more, see Klug H. P., and L. E. Alexander, *X-ray Diffraction Procedures,* John Wiley & Sons, New York, 1954, to which we shall refer as KA.

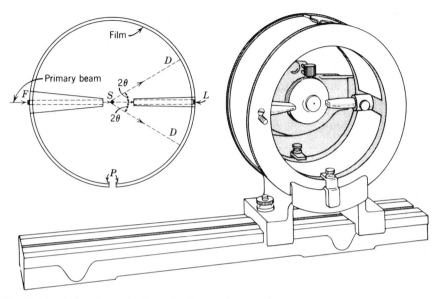

Fig. 13.16. A drawing and schematic diagram (insert) of a powder-diffraction camera. At F is a filter for the x-ray beam; S is the sample; θ is the Bragg angle; DD is the cone of diffraction that intercepts the cylindrical film; L is an absorber for the direct beam; and the film stops are at P. Better patterns are often obtained if L is eliminated and the x-ray beam allowed to escape.

From the wealth of information in KA, we shall excerpt only the more significant experimental details, and we shall restrict ourselves to one kind of camera containing film positioned according to Straumanis.

2. The Camera

A typical Debye–Scherrer-type powder camera is shown in Fig. 13.16. It is cylindrical, and there is provision for mounting the sample on the axis of the cylinder. The x-ray entrance and exit ports are extended to hold to a minimum the scattered x-rays that reach the film from the primary beam.

Commercial cameras include a means for aligning the sample in the center of the pinhole system. The entrance pinhole ought to be filled uniformly by the x-ray beam. To ensure this, the diameter of this pinhole is usually restricted to about 1.2 mm. So that the entire sample will be bathed by the entering beam, the sample diameter is slightly less than that of the pinhole. To reduce the need for fine grinding, provision is usually made for rotating the sample about its axis.

The camera must be positioned in such a ways that its optical system views the target of the x-ray source at a small angle (say, 6°), so that the target appears almost square. The camera mounting must be adjustable vertically, horizontally, and about the sample axis. The camera support must permit return of the camera to precisely the position it occupied before removal.

The film is held against the inside wall by some means such as spring clips, film stops, or beveled rings. To reduce local blackening and general scattering of x-rays, holes are often punched in the film where the x-ray beam enters and leaves the camera. In the Straumanis film position (see Fig. 13.16), the complete 2θ range is covered without a break in the film. In that figure, if SL $= r$ is the radius of the camera, then

$$\text{arc LD} = 2\pi r \times \frac{2\theta}{2\pi} = 2r\theta \qquad (18)$$

because 2θ is recorded on the film and θ (not 2θ) appears in Bragg's law. It follows that with a cylindrical camera 5.73 (or 11.46) cm in diameter, 1 mm along the film is equivalent to $1°$ (or $\frac{1}{2}°$) in θ. Mechanical devices with hand wheels and vernier scales are available for measuring distances along the film with high precision.

3. Selected Experimental Details

Examination of a sample by the Hull–Debye–Scherrer method consists of three steps: (1) sample preparation; (2) generation and interpretation of the diffraction pattern; and (3) comparison of the pattern with those of known substances.

Samples, as received, are often unsatisfactory for analysis, and the analyst must exercise considerable care to make them suitable. If the sample is not a mass with crystals of the right size range, the analyst must reduce it to a satisfactory size (325 mesh) without introducing foreign substances, segregating the components, or producing strains and distortions in the crystal lattice. The most common foreign substance to be guarded against is material from the mortar in which the samples are ground.

Powders much coarser than 325 mesh produce individual diffraction spots in the lines and contribute graininess to the pattern. Most substances can be ground quite satisfactorily, but some that are too soft or too malleable require special treatment (KA, p. 194). Care also must be exercised to ensure that the softer components of a powder are not preferentially selected.

The powder, having been reduced to the proper fineness, must be mounted in the x-ray beam without introducing a support that contributes a pattern of its own, or that scatters x-rays too strongly. A massive polycrystalline sample can be mounted directly in the x-ray beam, but a fine powder needs some support. When possible, the sample may be mixed with a binder and molded into small cylinders. Other approaches are to coat a small amorphous rod or hair with the powder, or to pack it into a glass or polymer tube. The sample diameter must in all cases be kept as small as possible to reduce absorption and scattering effects. In fact, for heavy elements it is sometimes necessary to introduce a diluent.

The best wavelength for the primary beam depends on the nature of the sample, especially on the complexity of the diffraction pattern and the in-

Table 13.III
Typical Exposure Times[a] for Different Conditions

Camera diameter (mm)	Pinhole diameter (mm)	K Filter[b]	Material		
			Metal filings	Simple salts and minerals	Organic compounds, complex inorganic compounds, and minerals
57.3	1.0	No	5 min	10 min	20 min
57.3	1.0	Yes	10 min	20 min	40 min
57.3 114.6	0.5 1.0	No	15 min	30 min	1 hr
57.3 114.6	0.5 1.0	Yes	30 min	1 hr	2 hr
114.6	0.5	No	1 hr	2 hr	4 hr
114.6	0.5	Yes	2 hr	4 hr	8 hr

SOURCE: KA, p. 210.

[a] Times given are for the K lines of molybdenum and copper; those of chromium require up to twice these times.

[b] The need to filter out $K\beta$ (see Fig. 13.16) is greatest when little is known about the sample and a complex diffraction pattern is expected. Proper filters are as follows: for $MoK\beta$, Zr; for $CuK\beta$, Ni; for $CrK\beta$, V.

terplanar spacings. For general use, $CuK\alpha$ (1.542Å) is favored. For organic compounds with low mass-absorption coefficients and high d values, $CrK\alpha$ (2.291 A) is useful, especially when the pattern can be recorded in a helium atmosphere to reduce x-ray absorption. For inorganic compounds with low d values and high absorption coefficients, $MoK\alpha$ (0.711 Å) is preferred. Exposure data appear in Table 13.III.

4. A Highly Precise Diffraction Pattern

Straumanis exposed for 8 h a cylindrical sample of aluminum 0.18 mm in diameter and 99.998% pure to $CuK\alpha$ as incident beam in a camera 57.7 mm in diameter at 23.10°C, and obtained the highly precise pattern shown in Figure 13.17.

C. DIFFRACTOMETRY

1. Introduction

The improvement of spectrometers and detectors has been advantageous for both x-ray diffraction and x-ray emission spectrography. Where a many-lined pattern is to be used for a complex qualitative determination, photographic methods are still to be preferred. For accurate quantitative determinations, parafocusing diffractometers are clearly superior. The parafo-

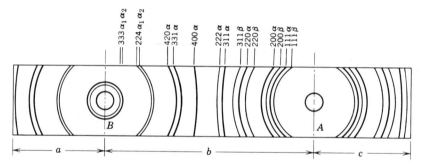

Fig. 13.17.   Diffraction pattern of aluminum by Straumanis. The unfiltered beam enters through the hole at B and exits through the hole at A. The *back-reflection* pattern (around B) gives the most precise results. Above the film are given the Miller indexes of the reflecting planes and the **K** lines reflected by them. (According to convention, 222 and 333 represent second- and third-order reflections from the 111 plane.) Note that the **K**α *doublet* has been resolved, as shown by the designation $\alpha_1\alpha_2$ (wavelength difference ~0.003 Å). Such resolution is far beyond that usually required in analytical chemistry. (After KA, p. 459.)

cusing arrangement is used because it gives the maximum intensity achievable at acceptable resolution.

Parafocusing, in Brentano's words, makes best use of "the ray-collective properties of powder arrangements." It deserves to be placed alongside true focusing, which is possible with single crystals, but not with powders. With parafocusing geometry and a flat sample, absorption effects are virtually independent of Bragg angle, because penetration of the sample is far less important in generating the diffraction pattern than it is in the photographic method; the difference is most pronounced at low θ values. With parafocusing, intensity is also increased, because a large flat sample has a large effective area (except at low θ values, where it subtends too small a part of the primary beam).

In a spectrometer, as we define it, Bragg reflection involves use of a single, known d_{hkl} for the determination of an unknown λ. In the diffractometer, as generally used in analytical chemistry, Bragg reflection of a "single" known λ involves all the d_{hkl}'s in a finely divided powder, randomly oriented. Scintillation and proportional detectors are now preferred for both instruments, mainly because they have quicker responses and longer lives than Geiger detectors.

2   Selected Experimental Details

In diffractometry, the sample should be ground considerably finer than the 325 mesh satisfactory for photographic work. Unless the sample is rotated so that graininess or preferred orientation is averaged out, large intensity deviations can result from the presence of larger crystals. (This rotation is *not* that of Fig. 13.18 which depicts operation of a parafocusing spectrometer with a *flat* sample.) The mounting of the sample also is critical,

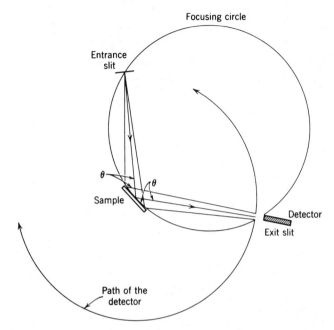

Fig. 13.18. Schematic diagram of a parafocusing spectrometer. As the flat sample rotates through an angle θ, the detector moves along the path shown. The sample-to-detector distance remains constant. The entrance slit determines the horizontal divergence of the x-ray beam; the exit-slit angular aperture determines the resolution.

because the path of the x-ray beam in the sample decreases, by a factor proportional to $1/\sin \theta$, as the Bragg angle increases. If the sample is thick enough that essentially all of the primary beam is absorbed at high values of θ, trouble from this source is eliminated. The minimum thickness (in centimeters) required to reduce the effect of the path length to 1% is given by

$$t = \frac{3.2}{\mu\rho} \qquad (19)$$

where μ is the mass-absorption coefficient of the sample, and ρ is its density, including binder and interstices. This equation is another good reason why the longer wavelengths (e.g., CrKα) are generally better for organic samples.

The sample holders is usually a piece of aluminum or glass with a circular depression sufficiently deep to provide the minimum thickness. The sample is loaded into the depression and smoothed with a spatula or other flat edge. If difficulty is experienced in holding the powder in the cavity, a small amount of binder should be added. At small Bragg angles, where the beam strikes at a glancing angle, the cavity must be long.

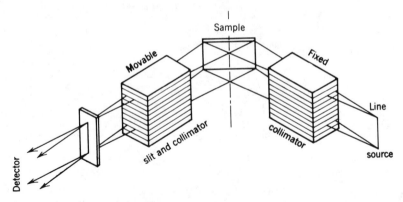

Fig. 13.19. Schematic diagram showing the use of two Soller slits to limit the vertical divergence of the x-ray beam in an x-ray spectrometer. (After KA, p. 241.)

The line resolution and the area of the sample irradiated both depend on the divergence of the x-ray beam. With a flat specimen, the vertical divergence must be limited, both between target and sample and between sample and detector. This is done (Fig. 13.19) by means of two Soller slits, in effect the equivalent of a large number of slits, arranged so that their transmitted beams add. In the horizontal plane in the figure, the x-ray tube focal spot is viewed at an angle so small that it appears as a line source. The horizontal divergence is usually kept below 2°.

Most commercial instruments have coupled drives for the sample and detector, with the detector output recorded on a strip chart. Considerable care must be taken in choosing the optimum combination of drive rates and the integrating-time constant in the detector. If the integrating-time constant of the detector is too long, the recording of the diffraction line is shifted in the direction in which the scan is going. If the time constant is too short, the statistical fluctuations make the data too erratic. A time constant of less than half the time it takes the receiving slit to travel its own width should normally be used, to avoid a large shift in the line peak and a concomitant lack of resolution. (The relation between scanning speed and slit width for different types of determinations appears on p. 310 of KA, which should be consulted to supplement the scanty experimental information given here.)

Figure 13.20 shows a simple diffractometric trace—the "signature" of powdered tungsten.

D. INTERPRETATION OF DIFFRACTION PATTERNS

1. Simple Unknowns

When an unknown has a "fingerprint" (Fig. 13.17) or a "signature" (Fig. 13.20) simple enough for visual comparison with that of a standard in a file, the analytical chemist's problem is readily solved. Visual comparison will

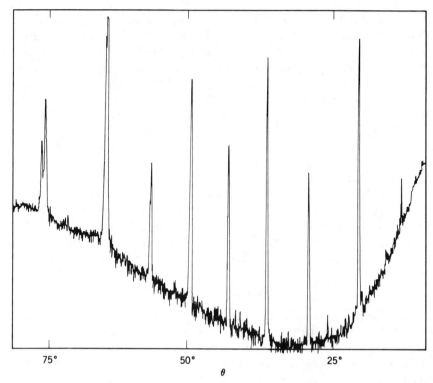

Fig. 13.20.   Spectrometer trace of a tungsten powder-diffraction pattern. The ordinate is proportional to x-ray intensity.

then usually show a matching of θ_{hkl} values (and hence of d_{hkl} values) and of intensities that is good enough to identify the unknown.

If standards for visual comparison are not available, the d_{hkl} values must be computed and the intensities measured for comparison with recorded values for standards. As we have seen, computation of the d_{hkl} values from measured θ_{hkl} values is easy.

Obtaining quantitative intensity measurements from photographic diffraction patterns requires microphotometering of the film. Microphotometering somewhat resembles x-ray absorptiometry with visible light in place of x-rays, and silver in the developed film being the absorber (see KA, pp. 368–376).

In x-ray diffractometry, as in x-ray emission spectrography, (see Section VII), we almost always deal with peak intensities. It is easy to see why. In Fig. 13.20, for example, the integrated intensity

$$I_{\text{int}} = \int_{\theta_1}^{\theta_2} I_\theta \, d\theta = N'_{\text{T}} \text{ counts} \tag{20}$$

would have to be obtained by summing the counts for the interval $\theta_2-\theta_1$, which includes virtually all the quanta of the diffracted line. The first difficulty is the proper choice of θ_1 and θ_2. The second difficulty is more serious. N_T includes N_B, the number of counts registered by the background. A glance at the figure will show that N_B over the interval $\theta_2 - \theta_1$ is difficult to establish; in addition, the fluctuations in N_B will introduce uncertainties.

The peak intensities are the differences $(N_T - N_B)$ for a given counting interval at the value of θ for the peak; when divided by Δt, these differences become counting rates. The peak intensities are thus more quickly and more easily obtained than are the integrated intensities, and this explains why peak intensities are generally preferred for determinations, whether by diffraction or by emission.

Several features of the background in powder diffractometry are of general interest. Residual $K\beta$ lines will contribute to the background if filtration is insufficient. Compton scattering also contributes. Characteristic lines excited in the diffracting crystals will appear as general background, as they do in x-ray emission spectrography without pulse-height selection.

2. Complex Unknowns

Most unknowns contain more than one crystalline species. In this case, and especially when the "fingerprints" or "signatures" are themselves complex, x-ray diffraction can use help, which x-ray emission spectrography can readily provide. Examining the unknown under a microscope is usually advisable.

In some quantitative work, internal standards are useful, as they are in x-ray emission spectrography. Comparison with external standards is obviously not possible when there is insufficient information about the unknown for preparation of a standard, which in any case requires additional work that is best avoided.

The intensity of a given diffraction line of a crystalline species present in a sample depends on the amount of the species, the absorbing characteristics of the sample, the particle size, and the perfection of the crystals of that species. The relationship between the intensity of a diffraction line and the concentration of the species is simpler than in x-ray emission spectrography, because the incident beam is monochromatic and the diffracted beam has the same wavelength. The integrated intensity of a line is given formally by the diffracting power of the unknown substance and the mass-absorption coefficient of the sample, as follows:

$$I_{CS} = \frac{K_{CS} W_{CS}}{\rho_{CS}[\mu_M(1 - W_{CS}) + \mu_{CS} W_{CS}]} \tag{21}$$

where I_{CS} is the intensity of the diffracted line (the subscript "CS" stands for "crystalline species"), K_{CS} a constant depending on the unknown substance in the sample, ρ_{CS} = the density of the unknown substance, μ_M is the mass-absorption coefficient of the matrix, and μ_{CS} is the mass-absorption coefficient of the unknown substance.

3. Information Storage and Retrieval

When comparison with external standards is ruled out, the composition of complex unknowns can be established only if recorded information is available for the crystalline species suspected of being responsible for the experimentally determined diffraction pattern of the unknown. This makes for a massive information storage and retrieval problem, one that poses a familiar dilemma: For x-ray diffraction to be used widely in analytical chemistry, a lot of stored information is needed, but until x-ray diffraction is used widely, the information will not be available for storage.

Analytical chemistry has good reason to be grateful for what the Dow Chemical Company has done to make x-ray diffraction useful. In 1938, J. D. Hanawalt, H. Rinn, and L. K. Frevel of that company published and distributed tested diffraction data for 1000 compounds. This information was the basis for a useful classification scheme that grew into the modern *Powder Diffraction File* of the Joint Committee on Powder Diffraction Standards [members of which are the American Society for Testing Materials, the American Crystallographic Association, the (British) Institute of Physics, and the National Association of Corrosion Engineers], which contained 13,000 patterns (10^7 characters) in 1967, and was then growing at the rate of 2000 patterns per year.

This classification system employs punch cards with seven regions providing information as follows:

Region A contains the d spacings and relative intensities of the most intense reflections (with the strongest given an intensity value of 100).

Region B contains the essential features of the pattern recording. The x-rays used, filter conditions, and type of instrument can all affect the intensity estimates and the appearance of spurious lines.

Regions C and D contain the essential crystallographic and physical data. The data in this part of the card can be used in connection with a polarizing microscope and hot stage to prove a doubtful identification.

Region E often is used for further sample description, details of sample preparation, and a classical analysis when available (this is particularly useful for mineral samples).

Region F contains the names and formulas of the compound.

Region G contains the complete d-value and intensity data, plus the planes responsible for the diffraction lines.

The enormous information storage and retrieval problem in x-ray diffraction reminds one of the inventory problem successfully solved by means of a computer in many large corporations. Does the inventory hold what the customer has ordered? How long to delivery? Bring on the computer!

E. COMPUTERIZATION AND AUTOMATION OF POWDER X-RAY DIFFRACTION

1. Camera Method

Frevel has succeeded in automating the measurement of powder patterns on film to produce digital output readings of d and I that can serve as input for the computerized identification of the crystalline species in an unknown. To record his patterns, he has used a more sophisticated camera than that described herein, namely, an AEG double-cylinder camera (cylinder diameter, 114.7 mm) with $CuK\alpha1$, selected by a bent-crystal monochromator, as the primary beam. This gives patterns of high quality for which the aluminum "fingerprint" serves as reference. (See LPWZ, pp. 270–273.)

2. The Dow *ZRD* Search–Match Program

Late in 1968, powder diffraction samples were being analyzed at an annual rate of nearly 3000, and x-ray emission spectrography samples at an annual rate of nearly 4200, in the Chemical Physics Research Laboratory of the Dow Chemical Company at Midland, Michigan. The second kind of determination preceded the first for many unknowns, Frevel having logically placed the horse before the cart in reversing traditional procedure. In his words:*

> In view of the observation that any dependable powder diffraction analysis of a genuine unknown is supplemented or confirmed by a semiquantitative or quantitative spectroscopic analysis, it becomes expedient to reverse the procedure and obtain the element data first to facilitate finding the appropriate diffraction standards. . . . It is thus timely to consider computer programs to carry out the entire searching and matching process by fast digital computers.

This thinking led to the creation of the Dow *ZRD* Search–Match program, which can only be outlined here.

The *ZRD* Search–Match program is carried out by a computer with an input that includes *at most* the following three kinds of data:

1. $\{Z\}$, the atomic numbers of all elements in the sample in the range $Z = 3$–93, *excluding* C, N, O, and F. Hydrogen is of course excluded also. These five ubiquitous elements are difficult to identify by spectrographic methods.
2. $\{R\}$, the encoded polyatomic groups (radicals, complex ions, ligands, and clathrates) in the sample. Inclusion of these groups is desirable because specific information about them is easily obtained by other methods, such as infrared absorption.
3. $\{D\}$, the powder diffraction data $\{v, d_v, I_v\}$, where v is the number that distinguishes a diffraction line.

---

* L. K. Frevel, letter to H. A. Liebhafsky, dated October 9, 1968.

To show what the early *ZRD* Search–Match program could accomplish, consider the following results for a Frevel sample:

1.   Principal elements detected by x-ray emission spectrography: Y, As, Ag, Pb, Se, and K.
2.   Output of program: Y_2O_3, As_2O_3, Ag_3AsO_4, KH_2AsO_4, and $PbSeO_4$ were identified as crystalline phases present.
3.   The presence of minor phases was suggested by 16 weak, unidentified lines.

Frevel and C. E. Adams have supplemented the *ZRD* Search–Match program with one based only on $\{d_v, I_v\}$. To operate at full effectiveness, these programs need certified standard patterns of greater precision than most of those that were available in 1968. Such patterns are continuing to be obtained. (see LPWZ, pp. 268–270.)

3.   The Chevron Diffractometer

Samples for the diffractometer are adapted to handling by machine; there is no need for lengthy exposures, nor any film to develop; the detector gives digital (counts per second) values of the intensities—these are among the reasons why automation and computerization of a diffractometer system are natural developments. A successful system of this kind has resulted from the joint efforts of the Chevron Research Company (La Habra, California) and the Datex Corporation (Monrovia, California).

The operating statistics are impressive. In the words of R. W. Rex:*

All operating variables are numerically controlled by the input information on a command card. Once a deck of cards is loaded into the card reader, the samples inserted into the automatic sample changer, and the system started, no further human intervention is needed until the run has been completed. Uninterrupted runs have been as long as 18 days. Samples may be added to the sample changer and cards added to the deck in the card reader during operation, as desired, to permit runs of indefinite length.

After the system has established and subtracted the background, it analyzes the net diffraction pattern to recognize peaks, shoulders, and the wing edges ("tails") of diffraction peaks. It next measures peak heights and areas, identifies phases, and calculates amounts present with reference to internal or mutual standards.

This diffractometer system can obviously control processes or materials. By August 1966 it had operated for nearly 6000 hr and it had then an availability rating of 98%. The system is said to have sufficient flexibility for use in research and for determinations on unknowns. (See LPWZ, pp. 273–275.)

As we shall see, the influence of the computer on analytical chemistry does not end here.

* Rex, R. W., *Advances in X-ray Analysis,* Vol. 10, Plenum, New York, 1967, p. 366.

F. CONCLUDING NOTES

1. Samples Uniform in Chemical Composition

X-ray diffraction is at its best when one or more crystalline species must be determined in a sample uniform in chemical composition. Such applications include detecting graphite or diamond in carbon, determining individual crystalline organic isomers in a mixture thereof, and determining the kind of silica responsible for silicosis (see KA, pp. 430–433).

2. Determination of Crystallite Size

The FWHM of a diffracted line reflects both *instrumental* and *pure* diffraction broadening. One example of the latter is the change from a sharp to a diffuse pattern as the crystallite size in a sample decreases to approach the "amorphous" state. Such broadening can be used to measure crystallite size.

Particle-size measurements have a large range of useful applications, paint pigments, carbon for electrical brushes, and metal catalysts being some of the more common materials studied. An obvious extension of the method is the determination of particle shape. If all the lines are broadened uniformly, the crystals are more or less equidimensional in all directions. Reduced broadening indicates an extension of the crystal in the direction(s) normal to the plane(s) associated with the broadening, increased broadening, the opposite. Care must be taken in interpreting line broadenings because lattice defects or strains can produce somewhat similar effects.

3. X-ray Diffraction Studies of Polymers and Fibers

Polymers consist of large molecules built up from one or more simple building blocks. The building process may produce a regular or random structure—linear, planar, or three-dimensional. A high degree of crystallinity may be associated with some of the modes of growth.

The degree of order in polymers is not great enough, in general, to provide a pattern to very high values of θ. In most cases the interpretable pattern can all be recorded on a flat plate in the forward-beam direction from the sample. Standard powder equipment is often used and gives very satisfactory results.

The diffractometer is particularly valuable in following polymerization processes. A metal or glass plate can be dipped into a reaction vessel and immediately put into the sample position. If the equipment is set up to monitor a line that changes during the process, each step can be followed rapidly.

Many linear polymers are noncrystalline until they are stretched, at which time the molecules line up in the direction of stretching. The diffraction pattern can be used to follow the stretching. Simple devices can be set up directly in the x-ray equipment to control the tension and elongation.

The interpretation of polymer diffraction patterns can be a highly specialized task (see Alexander, L. E., *X-ray Diffraction Methods in Polymer Science,* John Wiley & Sons, New York, 1969).

VI. X-RAY ABSORPTIOMETRY

A. INTRODUCTION

Immediately after discovering x-rays, Roentgen brilliantly and meticulously laid the foundation for *x-ray absorptiometry with polychromatic beams,* practice of which demands no knowledge of the x-ray wavelength; in this respect the technique resembles colorimetry with white light. Roentgen made it possible for the first time to become acquainted with the interior of unopened, opaque objects. He was probably surprised by the excitement aroused by his technique, which we today call *x-ray fluoroscopy* when a fluorescent screen serves as detector, and *radiography* when a photographic plate or film is so used. Common samples for such x-ray absorptiometry, which goes by still other names as well, are teeth and other parts of the body, castings or other metal objects suspected of having flaws, sheets or films whose thicknesses need to be measured and controlled, luggage at airports, and even the small telephone capacitors within which "bugging" devices might have been planted. All x-ray tests of this sort are *nondestructive*.

X-ray absorptiometry can serve for both the characterization and the control of materials, the two main objectives of analytical chemistry. Two imperfect generalizations provide a convenient basis for assessing the relative usefulness of mono- and polychromatic x-ray beams in absorptiometry: (1) Polychromatic beams are strong and complex; they cannot generally give specific results, but can be used with simple equipment. Monochromatic beams are weak and simple; they can sometimes give specific results, but require more elaborate equipment if their source is an x-ray tube. (2) Polychromatic beams are suited for instantaneous intensity measurements; with monochromatic beams, quanta must usually be counted and the counts accumulated over a time interval.

With monochromatic beams, there are no worries about filtering, effects of unsuspected absorption edges, or effective wavelengths, and there is far less risk of intensity fluctuations caused by equipment instabilities. With either monochromatic or polychromatic beams, suitable comparison of a standard with an unknown sample, as by commutation between them, usually makes the method more reliable.

Equations 3, 4, 6, and 8 are basic to x-ray absorptiometry. The greatest advantages of this technique derive from the simplicity and predictability of x-ray absorption spectra, and from the fact that the mass-absorption coefficients of an element are to a large extent independent of its chemical or physical state. The great weakness of all absorptiometric methods but one is their lack of specificity. Owing to this weakness, these methods continue to lose ground *relative* to x-ray emission spectrography, a specific method by nature, for the characterization of materials. We shall therefore have to give absorptiometric methods short shrift, and refer the interested reader to LPWZ, pp. 127–170, for fuller information.

B.   ABSORPTIOMETRY WITH MONOCHROMATIC BEAMS

As an illustrative example of ordinary spectrophotometry with mono-chromatic x-ray beams, we shall look quickly at the determination of sulfur in liquid hydrocarbons, which is important and interesting because it can be done with a radioactive source as well as with x-rays monochromatized by Bragg reflection. The basis of the method is equation 6 in the form

$$\mu_{sample} = \mu_H W_H + \mu_c W_c + \mu_S W_S \tag{22}$$

where the subscripts on the right denote the elements in the sample, which may be either a *standard* or an *unknown*. Obviously the reliability of the method will increase as the last term on the right increases relative to the other two. Here sulfur is the *element sought* in a *matrix* of carbon and hydrogen. If the matrix varies, the results will be affected unless a wave-length is used for which μ_H and μ_c are nearly the same. To counteract *matrix effects* and other sources of error, it is usually desirable to compare a stan-dard and an unknown, as by commutation between them, that is, to practice *comparative* absorptiometry. (The words italicized above are significant for other x-ray methods, as of course are *x-ray source, sample,* and *detector.*) Beer's law in a convenient form serves for the evaluation of results.

The one specific x-ray absorptiometric method also uses monochromatic x-rays. Figure 13.21 illustrates how it works. The attenuation of monochro-matic beams at various wavelengths is measured on both sides of an ab-sorption edge (the **K** edge is preferred for simplicity), which is shown as an inclined broken line in the figure. Extrapolation to the vertical broken line AB locates the edge and thus identifies the element, the amount thereof being fixed by the height of the "jump" from B to A. One successful use of the method has been for the determination of bromine in liquid samples at the Dow Chemical Company in Midland, Michigan, on modern equipment including a Geiger detector.

C.   X-RAY ABSORPTIOMETRY WITH POLYCHROMATIC BEAMS

Not surprisingly, the principal differences between absorptiometry with polychromatic beams and absorptiometry with monochromatic beams arise from the wide distribution of wavelengths within polychromatic beams. As regards the *control* of materials, these differences are of little importance, because control operations are almost always *comparative* operations of a kind chosen with control in view. To use absorptiometry with polychromatic

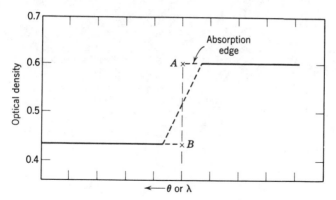

Fig. 13.21. The absorption-edge method, according to Glocker and Frohnmayer. The solid lines are obtained from photometric measurements of a photograph of the x-ray intensity as a function of the angle. The concentration is calculated from the ratio of these lines, extrapolated to the absorption edge. [After Glocker, R., and W. Frohnmayer, *Ann. Physik*, **76**, 369 (1925).]

beams for the *characterization* of materials, as in an analytical laboratory, one must have recourse to the *effective wavelength* of a polychromatic beam, a concept that poses some difficulties. The effective wavelength of such a beam is that of a monochromatic beam that behaves identically in absorbance measurements. Trouble arises when such a beam cannot be found. This unhappy situation may occur when unexpected absorption edges exist in the unknown and not in the standard with which it is compared, and it may arise also because of filtering.

The filtering that occurs in a polychromatic x-ray beam has been mentioned several times. It is an unavoidable consequence of equations 6 and 8, the latter of which shows that x-rays are inevitably going to be absorbed more strongly by elements of higher atomic number. There is nothing one can do about it. When x-ray beams pass through a sample, the emergent beam will inevitably be "harder" (of shorter mean wavelength) than the beam that entered. Suitable commutation between proper standard and unknown samples can do much to alleviate this difficulty.

To summarize: Except under special conditions, x-ray absorptiometry with polychromatic beams is more useful for the qualitative characterization of materials, and for their control during manufacture, than it is for their quantitative characterization. The three applications next discussed show why Beer's Law was given the form of equation 3.

1. A "Homemade" X-ray Photometer

Once he had been apprised of the progress in equipment, Roentgen would have felt at home with the "homemade" spectrophotometer, dating from 1947, shown in Fig. 13.22: There is no progress in principle over what his three great fundamental papers established. He of course did not know about

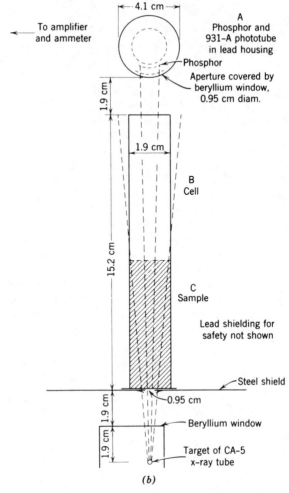

Fig. 13.22.   Schematic diagram of a simple laboratory x-ray photometer. [After Liebhafsky, H. A., H. M. Smith, H. E. Tanis, and E. H. Winslow, *Anal. Chem.*, **19**, 861 (1947).]

x-ray wavelengths, absorption edges, and filtering, but he did show what x-ray absorptiometry with polychromatic beams might do.

For the strengths and weaknesses of the photometer in Fig. 13.22 we refer the reader to LPWZ, pp. 146–160. We shall illustrate its usefulness only in Fig. 13.23, all the data in which can be obtained in *minutes* once the materials are at hand and the photometer is in operating condition. The figure shows that the photometer can accomplish both qualitative characterization (e.g., determining whether a sheet of known thickness is likely to be polystyrene or Saran wrap) and thickness gauging of a known material. The photometer can be used for solids, liquids, or gases. Equipment progress since 1947 would make it simpler today. This photometer affords the welcome oppor-

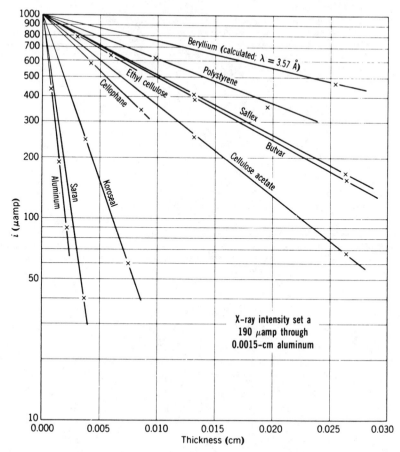

Fig. 13.23.   Attenuation of a polychromatic beam by various materials. Note the superiority of beryllium as a window material. [After Liebhafsky, H. A., H. M. Smith, H. E. Tanis, and E. H. Winslow, *Anal. Chem.*, **19**, 861 (1947).]

tunity to obtain at least preliminary information about a sample by reading on an ammeter the suitably amplified current generated by a photoelectric detector in the same category as the scintillation detector. A large laboratory might find that such a homemade photometer can answer quickly and easily occasional questions that need not be submitted to more sophisticated equipment such as an x-ray emission spectrograph.

As the two following examples will show, the modest accomplishments of x-ray absorptiometry in the *characterization* of materials are completely overshadowed by its success in their *control*.

2.   Controlling the Thickness of Tin Plate

Because tin is costly, the thickness to which it is plated on steel must be rigorously and continuously controlled. This objective was reached in the

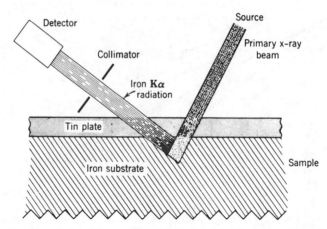

Fig. 13.24.   Diagram of Beeghly's method [H. F. Beeghly, *J. Electrochem. Soc.*, **103**, 157 (1956).] for determining the thickness of tin plates. Note that the polychromatic x-ray beam is attenuated and filtered, and that the substrate beam, being largely Fe $K\alpha$, is attenuated in the main. A manganese filter in the emergent beam will increase its spectral purity.

American steel industry before 1950 by means of the method diagrammed in Fig. 13.24. X-ray excitation converts the substrate into an x-ray source that yields a (nearly) monochromatic beam (Fe$K\alpha$ or β) for absorptiometry of the tin plating. The substrate acts as a characteristic-line generator for the thickness determination. X-rays other than the characteristic lines of the substrate are undesirable if they are incident at appreciable intensity upon the detector. Among x-rays that could distort the results are diffraction peaks and characteristic lines of the film. The detector should be positioned so that the former do not enter. The latter can be avoided by choosing an x-ray tube voltage high enough to excite the characteristic lines of the subtrate but low enough not to excite interfering lines from the film. The excitation potentials of interest are as follows; iron K spectrum, 7.11 keV; tin K spectrum, 29.2 keV; voltage used, 20 keV. The L lines of tin (wavelength 3.0 Å and longer), which are excited in the sample at 4 keV and above, are absorbed strongly enough by the air in the optical path to be practically unimportant.

Noteworthy features of a Norelco Tin Coating Weight Gauge, which records tin-plate thickness directly in *pounds per base box* (lb/bb) are as follows:*

1.   One gauge scans the upper side of the strip, while a second scans the lower. The scanning heads can traverse strips up to 48 cm wide, and their direction is reversed photoelectrically as they approach the strip edge.

2.   Each gauge contains an air-cooled tungsten-target x-ray tube, two scintillation detectors, two preamplifiers, and a built-in reference standard

* One pound per base box is about 11 g Sn/m^2, or a uniform plate about 1.8×10^{-4} cm thick.

(the comparative method once more!). An ingenious optical system compensates for fluttering of the strip.

3.   Additional electronic circuitry makes it possible to record the output of two rate meters on a calibrated strip-chart as pounds per base box for each side. The reliability (precision and accuracy) is ± 0.01 lb/bb/side over the range 0.1–0.55 lb/bb/side. The time constant of the system is 1 sec.

3.   Thickness Gauging of Steel Strip

The automated gauging of steel-strip thickness demonstrates x-ray absorptiometry at its best. During the thickness measurement, the steel strip may be hot (1500–1750°F), moving (about 2000 ft/min horizontally with vertical vibrations up to several inches in amplitude), and subjected to a spray of cooling water. In 1955, this application was made fully automatic at the U.S. Steel Corporation; that is, the error signal was used to readjust tandem cold reduction mills. Automatic control proved significantly more effective than manual control. Under the drastic operating conditions mentioned above, the guaranteed accuracy of the present guage over its measurement range is within ± 2 mils between 140 and 200 mils; within 1% of thickness between 200 and 400 mils; and within ± 4 mils from 400 mils to the upper limit, 1999 mils. Not bad, considering the conditions!

The gauge proper and its carriage appear in Fig. 13.25. With two ionization chambers as detectors, the gauge continuously compares an unknown (the strip) and a standard (a wedge with its own servomechanism system that permits setting the wedge in such a way that the error signal is zero when the strip is of the same thickness as the intercepted part of the wedge). In automated operation, the error signal is fed into a computer that directs correction of the rolls. For a schematic diagram of the gauge system, see LPWZ, PP. 106–107.

Fig. 13.25.   The General Electric Raymike® 2000 thickness gauge. The device weighs about 7000 lb. The steel strip intercepts a polychromatic x-ray beam as it passes between the jaws of the "C," which are widely separated (84 in.) to keep the gauge from being damaged by the severe operating conditions. A motorized carriage permits positioning of the beam and scanning of the strip. (Courtesy of General Electric Company.)

4.   X-ray Absorptiometry of Artery Walls

Roentgen, who did the first x-ray absorptiometry on the human body, with his wife's hand as the sample, would have been particularly delighted with this and the following application. As always, we deal with source, sample, and detector; but the ancillary instrumentation, especially the computers, raise x-ray absorptiometry to a level of sophistication Roentgen never dreamed of.

Oversimplified into the language of analytical chemistry, "Regression and Progression of Early Femoral Atherosclerosis in Treated Hyperlipoproteinemic Patients"* comes to this: By means of x-ray absorptiometry with polychromatic beams, one may carry out successive radiographic determinations of cholesterol in "hardened arteries" in order to measure changes; remedial treatment may then be based on understanding these changes. The original articles contain detailed information that cannot be given here, but it is important to note that atherosclerosis is less of a one-way street than had been supposed: Among the 25 cases examined, improvement was about as common as further deterioration.

5.   Computerized Axial Tomography

Again, the language used to describe this technique may seem strange to the analytical chemist. We may describe the work as pulsed x-ray absorptiometry with polychromatic beams, done with high-pressure xenon-filled ionization chambers as detectors and human bodies as samples, and with computerized evaluation of the results on an almost unbelievable scale. "Tomo," in Greek, means "section", and this technique involves scanning a section of the human body. It has progressed greatly since 1972, when Houndsfield of EMI first used it to locate tumors by scanning sections of the brain. Today it is possible to scan sections anywhere in the human torso, a task much more difficult than brain scanning, because motion of any sort (e.g., breathing or peristaltic movement of gas in the bowel) can cause blurring and make the results useless. To avoid this, the exposure time has to be drastically reduced, and it has been, to 4.8 sec in the General Electric equipment now to be described.†

---

* Title of an article by Barndt, R., D. H. Blankenhorn, D. W. Crawford, and S. H. Brooks, *Ann. Intern. Med.*, **86**, 139 (1977). For details of the image-processing system and the computer techniques, see Selzer, R. H., D. H. Blankenhorn, D. W. Crawford, S. H. Brooks, and R. Barndt, Jr., "*Computer Analysis of Cardiovascular Imagery*," in *Proc. Caltech/JPL Conference on Image Processing Technology, Nov. 3–5, 1976*, California Institute of Technology, Pasadena, California, 1976.

† Most of the information used here can be found in two General Electric reports: (1) No. 76CRD286, *Five-Second Fan Beam CT Scanner*, by Chen, A. C. M., W. H. Berninger, R. W. Redington, R. Godbarsen, and D. Barrett; and (2) No. 77CRD079, *Fast Scan Computerized Tomography*, by Redington, R. W., A. C. M. Chen, and W. H. Berninger. For copies, write R. Ned Landon, Communications Operation, Corporate Research and Development, General Electric Company, P.O. Box 8, Schenectady, NY 12301, U.S.A. We thank Dr. Redington for his help in preparing this section.

The problem solved with this equipment is a far cry from Roentgen's location of bones in the hand. Imagine the aluminum cell in the simple photometer described above to be filled with oil droplets of different kinds suspended in water, and that they move every 5 sec. Now, use polychromatic x-ray absorptiometry not only to locate the drops in a section of the cell, but to delineate their boundaries and to measure differences in their x-ray absorbances. To make matters worse, the mass-absorption coefficient differs little from one drop to the next, and it is always near that of water. Furthermore, it will be necessary to make absorption measurements on x-ray beams sent into the cell from many directions (remember how a surveyor uses triangulation to fix location). Because *quantitative* results are needed, the intensity–wavelength distribution in the polychromatic beam must be considered. Finally, x-ray exposure must be minimized, and cost kept under control.

With the human body as the sample, the rib cage, backbone, and pelvic bones replace the aluminum cell as regards x-ray absorbance; some oil drops become internal organs, and others might be growths such as tumors. The tradeoffs that had to be made to solve the actual problem cannot be described here, nor can the reasons be given why, for example, pulsed operation of the x-ray tube was preferred to continuous, or why scintillation detectors were rejected in favor of xenon-filled ionization chambers.

A schematic diagram of the "fan beam scanner" that finally evolved apears in Fig. 13.26. In this scanning polychromatic x-ray photometer, the sample (small circle) remains stationary while the source and the 320-detector array are rotated as shown through 360°, each detector providing a different "view" of the sample section being examined. Approximately 10^5 absorbance measurements* are so obtained, from which a television image having 320 × 320 pixels (picture elements) of the section is constructed in 200 sec or less by having the computer perform some 60 million arithmetical operations. The data can also be printed out and stored in conventional fashion.

Resolution is excellent: A tungsten wire 0.15 mm in diameter, surrounded by water, gave a FWHM of 1.7 mm. When contrast is low, as often happens when diagnoses are being made, "noise" (i.e., background) becomes a serious matter. In such cases, low-contrast resolution is evaluated by having a group of observers agree on the diameter of the smallest rod they can see in a matrix of slightly different contrast (absorbance). A 3.5-mm Plexiglas rod was thus seen in ethylene glycol; the contrast difference was 2.5%.

Ultimately, the proof of the pudding is in the eating. Figure 13.27 may fairly be regarded as a pinnacle of achievement in x-ray absorptiometry with

* Each detector takes 288 views, each separated by (360/288)° from its neighbors. The number of measurements consequently is 288 × 320 (views times detectors), or approximately 10^5. The image is reconstructed on a 320 × 320 matrix with 320 × 320, or approximately 10^5, pixels. We thank Dr. R. W. Redington, General Electric Co., for pointing this out to us.

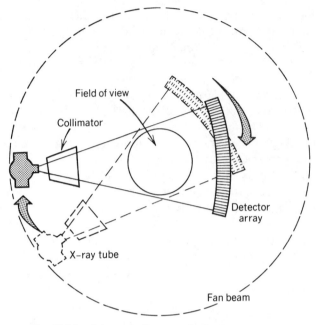

Fig. 13.26.   Schematic diagram of a fan-beam scanner.

polychromatic beams, and a superb demonstration of what the computer can make possible in analytical chemistry.

VII.   X-RAY EMISSION SPECTROGRAPHY

In x-ray emission spectrography, a characteristic x-ray line in the spectrum of each element sought is chosen as the analytical line, is excited somehow in a spectrograph, is identified, and has its intensity measured. This name for the method is operational, unambiguous, and applicable no matter what the means of excitation, which is why we prefer it to *x-ray fluorescence* and its variants. X-ray emission is more important in analytical chemistry than are all other x-ray methods taken together.

The intensity of the analytical line must often be corrected for background. Even the corrected intensity, $(N_T - N_B)$, the difference between total and background counts over the same counting interval Δt, in the simplest case, is usually not proportional to the amount present of the element sought. It may depend on the thickness of the sample, will depend on the amount present of the element sought, and will almost always depend on the matrix (the other elements in the sample). The dependence on thickness is not encountered in ordinary (optical) emission spectrography—certainly not when the sample is destroyed in an arc. Even when the sample is not destroyed, any dependence of intensity on a difference between sample sur-

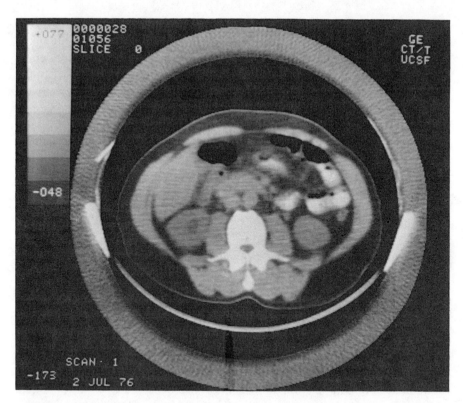

Fig. 13.27. Section of a human torso at the kidney–bowel level as revealed by computerized tomography. Regions of high absorbance (e.g., the spine) appear white. Regions of low absorbance (e.g., bowel gas) appear dark. The two white areas on the right are sections of bowel in which absorbance was increased by a barium compound remaining from a prior fluoroscopic examination. The sharpness of the small black dots (pronounced on the original print) shows that the small gas bubbles they represent did not move during exposure: There was no "motion blurring" during 4.8 sec.

face and interior is entirely different from the *predictable* variation of intensity with sample thickness that characterizes x-ray emission spectrography.

A. VARIATION OF LINE INTENSITY WITH THICKNESS: DETERMINATION OF FILM THICKNESS

The variation of line intensity with thickness is of the utmost importance, because it provides an understanding of the *absorption effects* encountered whenever characteristic lines are emitted from samples of finite thickness, and because it makes possible a unique determination of film thickness.

Consider individual atoms of an element deposited on a thin substrate highly transparent to x-rays—say, atoms of molybdenum on paper. Let a characteristic line (say MoKα) be excited by a polychromatic beam, with the x-ray source and detector both located above the sample. As long as the number of molybdenum atoms is small, they will not noticeably attenuate the incident beam, nor will an x-ray quantum radiated by a molybdenum atom be absorbed by any other. Under these conditions, the intensity of the characteristic line will be proportional, within experimental error, to the number of molybdenum atoms and hence to the thickness of the molybdenum film.

As the film of molybdenum grows thicker, the metal will begin noticeably to filter and to attenuate the incident polychromatic beam, which will become shorter in wavelength and weaker as it penetrates the metal. The intensity of the MoKα line will now increase with thickness at a continuously decreasing rate. If the metal is thick enough, even the shortest-wavelength x-rays will fail at a certain depth to excite Kα quanta at a rate high enough to reinforce the emergent beam measurably, because virtually all such "deep" quanta will be absorbed by the molybdenum on their way to the detector. The depth at which this first occurs is the *critical depth* for the experimental conditions. The critical thickness is the same as this critical depth, and "critical" is used because an increase in thickness (or in depth) beyond the critical level will not measurably increase the intensity of MoKα, which has reached its maximum value for the experimental conditions once the critical value is attained.

Consider the ideal case, in which the x-rays involved are monochromatic, all influences of composition are absent, the simplest x-ray optics obtained, and excitation of a characteristic line in the film by a characteristic line of the substrate does not occur. Suppose now that a beam of intensity I_0 falls upon a metal film d cm thick to excite a characteristic line of intensity I_d. The contribution to I_d of a volume element of constant area and of thickness dx, located at depth x, is

$$dI = k \csc \theta_1 I_0 \exp[-(\mu_1 \csc \theta_1 + \mu_2 \csc \theta_2) px] dx \qquad (23)$$

where k is a proportionality constant that measures the absorption and conversion of incident to characteristic radiant energy, μ_1 and μ_2 are mass-absorption coefficients for the incident wavelength λ_1 and the characteristic wavelength λ_2, ρ is the density, θ_1 and θ_2 are the angles made by the incident and emergent beams with the (plane) sample surface, and x is the depth (distance perpendicular to surface). The integrated equation may be written

$$I_d = k \csc \theta_1 I_0 \int_0^d e^{-a\rho x} dx = \{k \csc \theta_1 I_0[1 - \exp(-a\rho d)]\} ap \qquad (24)$$

where

Table 13.IV
Three Regions in the Determination of Film Thickness

Region	exp $(-a\rho x)$	d or x	dI/dx
Useless	Negligible	Greater than d_c	Approaches zero
Exponential	Significant	Intermediate	Variable
Linear	Near unity	Smaller than x_L	Approaches constancy

$$a = \mu_1 \csc \theta_1 + \mu_2 \csc \theta_2 \qquad (25)$$

At infinite thickness ($d = \infty$),

$$I_\infty = A \csc \theta_1 \frac{I_0}{a\rho} \qquad (26)$$

At the critical thickness, d_c, the ratio

$$\frac{I_d}{I_\infty} = 1 - \exp(-a\rho d) \qquad (27)$$

is indistinguishable from unity. To calculate the critical thickness from equation 27, we arbitrarily choose a value (say 0.99) for this ratio and calculate d (that is, d_c) from the equation.

The preceding oversimplified mathematical treatment really amounts to an evaluation of the absorption effect. The exponential term in equation 23 obviously is a product of two exponential terms, each deriving from Beer's law. One term governs the attenuation of the beam incident upon the volume element in question, and the other governs the attenuation of the characteristic line emerging from this element. The films are so thin that the use of one value each for θ_1 and for θ_2 over the entire film thickness is justified. Finally, one must assume that the intensity measured by the detector remains proportional to the intensity of the source.

On the basis of the mathematical treatment, it is convenient to define three regions, as in Table 13.IV; these are illustrated in Fig. 13.28.

The linear region, as was pointed out, is of particular interest, owing to its simplicity. In equation 23, as x becomes smaller and smaller, the value of the exponential term increases to unity. Once this increase has gone far enough to make the term indistinguishable experimentally from unity, then the term itself may be regarded as constant; whence

$$\Delta I = k \csc \theta_1 I_0 \Delta x \qquad (28)$$

for all values of x below x_L, the value at which the relationship of intensity

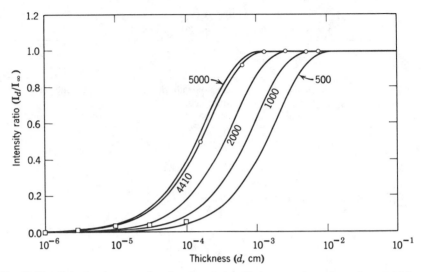

Fig. 13.28.   Calculated curves showing the relationship between intensity ratio and thickness for various values of exponent a. The abscissa scale is logarithmic. The circles represent plated coatings; the squares represent evaporated coatings of chromium on molybdenum. [After Liebhafsky, H. A., and P. D. Zemany, *Anal. Chem.*, **28**, 455 (1956).]

to thickness first becomes linear. This simplicity is a great advantage of x-ray emission spectrography for the determination of traces.

The region of thicknesses above the critical level, which region is useless for the determination of film thickness, is of the utmost importance in the x-ray emission spectrography of bulk samples, for it is the region in which intensities of analytical lines are independent of thickness.

Thicknesses of various kinds of films are being determined successfully in industry on the basis of the considerations discussed above (see LPWZ). A noteworthy example is the measurement of tin-plate thickness by use of the Quantrol, a spectrograph developed by Applied Research Laboratories; it gives excellent results that are concordant with those obtained absorptiometrically.

B.   DETERMINATION OF COMPOSITION: PRELIMINARY DISCUSSION

X-ray emission spectrography, in common with the other x-ray methods, is usually not a complete method of analysis. It cannot identify and determine all the elements present, except in very simple samples. For the more restricted, but highly important, task of identifying and determining with acceptable precision the large number of elements to which it is suited, x-ray emission spectrography is perhaps the best and most powerful method available to the analytical chemist.

The identification of elements—a qualitative procedure—was mentioned earlier and is illustrated in Fig. 13.29 for two kinds of x-ray emission spectrography. This information was obtained in a matter of minutes without changing the bank notes for better or worse.

X-ray emission spectrography is outstanding among analytical methods because of the relative ease with which qualitative information can be transformed into semiquantitative or quantitative data. This transformation may involve determining precise values of peak heights or areas such as those in Fig. 13.29, usually compensating for background and sometimes allowing for absorption and enhancement effects (see below), and interpreting the final intensity values in terms of the amount or proportion of the corresponding element in the sample.

In quantitative work, the data to be interpreted are almost always total counts. Even when the background is negligible or properly corrected for, interpretation may be difficult because of complications arising from sample thickness, composition, heterogeneity, preparation, or handling, or because of complications arising in equipment. Thickness as a factor has already been discussed. For the present, only samples exceeding the critical thickness will be considered; trace determinations will come later.

C. SAMPLE COMPOSITION: ABSORPTION AND ENHANCEMENT EFFECTS

Because samples for x-ray emission spectrography can take many forms (films, evaporated residues, powders, solutions, and massive solids), the question of composition units needs attention. In the limiting case of near-zero thickness, analytical-line intensity is proportional to the number of radiating atoms in a unit volume. This seems to argue for the use of volume units. Actually, this argument is misleading when composition is being determined. The important variable then is the total number of radiating atoms in the path of the beam, and this number is proportional to the weight of the element sought. For the other region of interest, that of "infinite" thickness, the weight fraction is a logical unit also. However, calibration curves often do not show a linear relationship between analytical-line intensity and content of the element sought, no matter what unit is chosen. The weight fraction will be used here because it is a simple, logical unit, and the most generally satisfactory.

Suppose the weight fraction W_E^S of element E in sample S is to be determined by measuring the intensity I_E^S of an analytical line. In the simplest case, we may assume that

$$I_E^S = W_E^S I_E^E \tag{29}$$

where I_E^E is the line intensity for the pure element at infinite thickness. Let us examine qualitatively the effect that the presence in the sample of elements other than E (these other elements are often called the *matrix*) can have on the validity of equation 29.

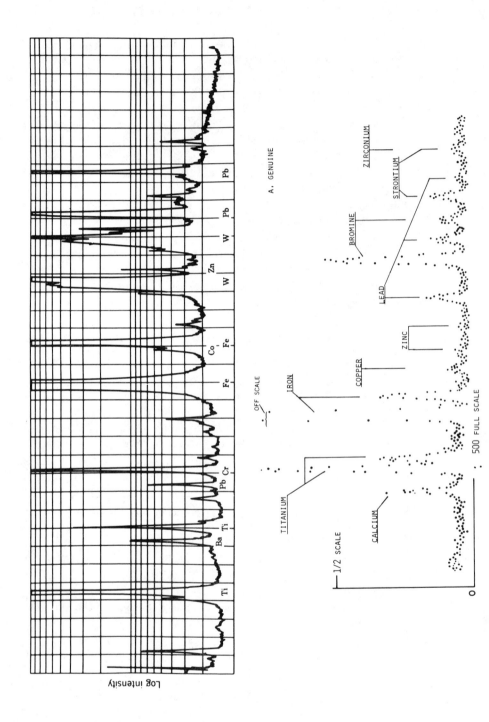

A. GENUINE

Table 13.V
Qualitative Absorption and Enhancement Effects

Case	Sample	Intensity of FeKα	Net effect	Comment
A	Fe	I_{Fe}^{Fe}	None	Reference standard
B	Fe–Al	$> W_{Fe}^{S} I_{Fe}^{Fe}$	Positive absorption effect	μ_{Al} less than ρ_{Fe}
C	Fe–Pb	$< W_{Fe}^{S} I_{Fe}^{Fe}$	Negative absorption effect	μ_{Pb} greater than μ_{Fe}
D	Fe–Co	$W_{Fe}^{S} I_{Fe}^{Fe}$ (approx.)	No pronounced effect[a]	Mass-absorption coefficients comparable; CoKα cannot[a] excite FeKα
E	Fe–Ni	$> W_{Fe}^{S} I_{Fe}^{Fe}$	Predominantly enhancement effect	Mass-absorption coefficients comparable; NiKα excites FeKα

[a] Note that CoKβ, which is of shorter wavelength than CoKα, can excite FeKα. This small enhancement effect was disregarded for the value of simplicity.

The net effect of the presence of these other elements is conveniently assessed by comparing the intensity calculated from equation 29 with the actual intensity of a sample of known composition. The net effect may be to increase the intensity over the calculated (a positive effect) or to decrease it (a negative effect). Individual effects may result from the following causes: (1) presence of an element with an absorption coefficient smaller than that of E (positive absorption effect); (2) the reverse of this situation (negative absorption effect); and (3) presence of an element a characteristic line of which excites the analytical line on absorption of the characteristic line by E (enhancement effect—always positive). The situation may be complicated further by the presence of absorption edges and by the filtering of a polychromatic beam if such is used for excitation.

To illustrate this qualitative discussion, the various effects have been summarized in Table 13.V for the determination of iron, a good choice because the spectra of the nearby transition elements (Fig. 13.30) show why different effects are to be expected. Note that even in pure iron, there is an absorption of FeKα radiation, which will produce an absorption effect depending on thickness, but that this does not appear in Table 13.V because the sample in Case A exceeds the critical thickness.

It is an advantage of x-ray methods that complexities of the kind shown in Table 13.V are qualitatively predictable. Absorption effects can even be calculated from mass-absorption coefficients in the simpler cases; before the

Fig. 13.29. "Signatures" of genuine $20 bills obtained by x-ray emission spectrography as follows: above, via Bragg reflection in conventional spectrography, lines tungsten of (W) from x-ray tube (after LPWZ); below, pulse-height analysis of spectrum excited by radioactive source (5 millicurie $^{109}_{48}$ Cd) as detected by Si(Li), 80 mm in area and 0.5 mm from sample (after Woldseth, R., in *X-ray Energy Spectrometry*, Kevex Corp., Burlingame, California, 1973).

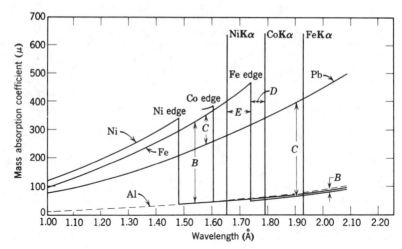

Fig. 13.30.   Spectral data to illustrate absorption and enhancement effects for three transition elements. (To avoid crowding, only part of the cobalt absorption curve is shown.) See Table 13.V. Case *B*: Substitution of Al for Fe decreases the absorption of the incident beam and has little effect on the analytical line; there is a net positive absorption effect. Case *C*: Substitution of Pb for Fe decreases the absorption of the incident beam, but greatly increases the absorption of the analytical line; there is a net negative absorption effect. Case *D*: Note the wavelength relationship indicated in the figure; enhancement is possible. Case *E*: Note the wavelength relationship in the figure; enhancement occurs.

advent of the computer, there was no way to calculate enhancement effects easy enough to be useful.

D.   DEVIATIONS FROM PROPORTIONALITY

Were complications absent from x-ray emission spectrography, one would expect the proportionality between analytical-line intensity and weight fraction of equation 29. That is, one would expect the analytical-line intensity of element E of sample S to be given by the product of the weight fraction of E in S and the intensity of the analytical line for the pure element E. This expectation is realized only in the simplest cases.

Deviations from proportionality have three principal causes:

1.   Absorption and enhancement effects, placed together because they both involve absorption. These have already been discussed and we shall return to them in Section VII.I. They will be called *interelement effects*.

2.   Effects traceable to heterogeneity in the samples, principally surface effects and segregation.

3.   Instability, including drifts and fluctuations, in the spectrograph and in the associated equipment.

Class 3 deviations often increase with the complexity of the electronic circuitry.

The diversity and the effectiveness of the means for coping with deviations from proportionality attest to the great interest in x-ray emission spectrography and the resourcefulness of the analytical chemist. Methods that ensure reliable results in the face of the three classes of deviations are described briefly below. The obvious method of separating from the matrix the element to be determined is omitted.

1. Comparison with a Standard

In this, the comparative method, the ultimate in reliability is attained when unknown and standard are uniform, identical in composition, and identically prepared. Simultaneous observation of standard and unknown, or rapid commutation between the two, is desirable. The method is of great value even when these ideal conditions are not realized, for example, when a working (calibration) curve is used. This method of comparison is satisfactory in many simple analyses.

2. Use of an Internal Standard

An added internal standard must be properly chosen and uniformly distributed. An element in the sample may sometimes serve as a built-in standard, or a scattered line in the background may serve as a reference.

3. Dilution

Dilution with a relatively transparent material is useful in dealing with Class 1 deviations. Water, starch, alumina, and borate glass are representative diluents; the proper choice depends on the sample. The extent of dilution needed to eliminate deviations due to absorption effects often can be estimated from the equations governing absorption. If the diluent contains light elements only, then both absorption and enhancement effects due to heavy elements will be negligible in highly dilute solution. Dilution with heavy elements, such as barium, also can stabilize absorbance, but this procedure always reduces analytical-line intensity and may lead to enhancement effects if the diluent is improperly chosen.

Briquetting, fusion, the use of Lucite molding powder, and solution of the sample all reduce or eliminate Class 2 deviations. These operations usually are accompanied by enough dilution to reduce absorption or enhancement effects.

E. SAMPLE PREPARATION AND HANDLING

The limit on the precision (measured as relative standard deviation in the analytical result) ultimately attainable in x-ray emission spectrography is set by the standard counting error. For reasonable counting rates and counting intervals, this limit lies somewhere below 1 part per thousand in the weight fraction of a major constituent. Only in exceptional cases does the relative standard deviation realized for actual unknowns rise much above 1 part per hundred. General experience indicates that much of this shortfall is often attributable to the sample preparation needed when spectrographic work on the *original* sample cannot be made to give satisfactory results.

Samples are either standards or unknowns. Because the comparative method is the mainstay of x-ray emission spectrography, the technique is at its simplest and best when unknown and standards approach true comparability—the nearer, the better. For example, they should have comparable critical depths and closely similar surfaces and particle-size distributions; if unavoidably heterogeneous, they should be similar in heterogeneity; and they should be presented to the exciting beam in the same way.

The method of excitation can make a great difference; for example, electrons penetrate less deeply than most x-rays, so that factors associated with surfaces carry greater weight when the former are used. Homogeneous glasses and alloys with smooth surfaces make the best samples. Results are best when standard and unknown have virtually identical compositions. How to approach these conditions varies too much from case to case to be discussed here. Important routine determinations are clearly the class of investigation worth taking the most trouble over. With the best equipment now available, sample preparation will limit the reliability of results more often than will equipment errors.

Errors due to sample preparation and handling often can be detected by one or more of the following measures: (1) leaving the sample in the spectrograph and making several independent counts; (2) resetting the goniometer between independent counts; (3) removing and returning the sample to see whether the counting rate changes; (4) scanning different portions, or testing the effect of rotating the sample; (5) counting after altering the surface (e.g., a powdered sample sometimes can be stirred or shaken); and (6) making replicate x-ray determinations on different portions of a sample.

F.   METHODS OF EXCITATION

Analytical lines were first excited by x-rays and electrons. These remain the principal agents used for excitation in x-ray emission spectrography, with electrons regaining the earlier prominence they had lost. Excitation can also be accomplished by any other means supplying an atom with the energy needed to eject an inner-shell electron. Owing mainly to research in nuclear physics, bombardment by protons and by ions of other elements has become available for special purposes. The various emanations from radioactive atoms also serve; these are conveniently subdivided into radioactive *x-ray* sources, such as $^{55}_{26}$Fe, that emit characteristic x-ray lines, and all others, among which γ-ray sources are prominent. Needless to say, *mixed excitation* (e.g., by x-rays plus electrons) often occurs. X-rays and electrons promise to remain the principal methods of excitation for "workhorse" x-ray emission spectrography.

G. THE BACKGROUND PROBLEM

Analytical lines are generally superimposed upon an unwanted background, and sometimes are "swallowed" by it. X-ray emission spectrography cannot deliver analytical results of highest reliability unless the background problem is solved. Then ($N_T - N_B$), the difference between total and background counts for the same counting interval Δt, is the *corrected count*, which normally is used to arrive at the amount present of the element sought. How this is done depends on the situation. The most important background situations are:

1. The background is negligible.
2. No background correction is needed, even though appreciable background exists, as when a satisfactory working (calibration) curve is obtained without such correction.
3. Background correction is needed, but a satisfactory N_B value can be established directly.
4. Background correction is needed, but the background varies with wavelength, owing to spectral interference.
5. The background is so large that statistical fluctuations in N_B preclude satisfactory determination of the element sought.
6. Light elements must be determined.

Except for situation 5, the need for a background correction, although it may add trouble and cost, will seldom vitiate a determination; see LPWZ, pp. 377–386, for details. An unusually complex case is shown in Fig. 13.31.

In summary: background corrections should be made only if they improve the analytical results, and reduction of background to the point where it may be neglected is usually worthwhile.

H. REFLECTORS FOR X-RAY EMISSION SPECTROGRAPHY

Devices other than crystals, for example, gratings, layers of metals, soap films (the last called *Langmuir–Blodgett gratings*), can accomplish the kind of reflection discovered by Bragg, hence the term *reflector*. For a discussion of this important subject, see LPWZ, Chapter 5, "The Selection of X-ray Wavelengths," pp. 198–231.

Over how wide a 2θ range is a Bragg reflector useful? A practical answer to this complex question is: "over the widest range geometry permits." At low 2θ values, a reflector will not intercept all the beam until some angle near 20° is reached. At 2θ values somewhere below 180°, the detector will collide with other components, and reflected intensity often decreases. In Table 13.VI, the bars have been drawn to show geometrical limitations from 8 to 16°, and the absence thereof from 16 to 145°. If a better answer is needed, and more than one crystal is available, the performance of each crystal should be tested on a standard sample.

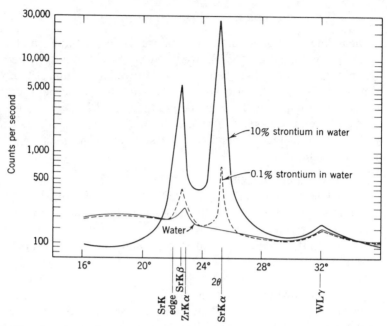

Fig. 13.31.   Complex background in the determination of strontium in aqueous nitrate solutions. Note (1) gradually decreasing background with increasing 2θ as shown by water; (2) unexpected presence of zirconium Kα owing to impurity in glass cell; (3) pronounced decrease in background at low 2θ for 10% solution, caused by high absorbance on short-wavelength side of strontium absorption edge; (4) slight increase in background with increasing strontium on long-wavelength side of edge, caused by increased Compton scattering of SrKα by sample; and (5) peak (almost merged with backgroung) at 30° (2θ) results from scattering of target line by sample. The procedure was to use SrKα as the analytical line, and establish the background correction $F = N_B/N'_B$ on water, taking N'_B near 2θ = 30° on unknown solutions. (H. A. Liebhafsky, H. P. Pfeiffer, E. H. Winslow, and P. D. Zemany, unpublished results, 1958.)

Here, as elsewhere in x-ray emission spectrography, the light elements cause trouble. Bragg's law shows that a Bragg reflector cannot deal with wavelengths exceeding 2d, for sin θ cannot exceed unity. Reflectors with 2d value in the range ~10–100 Å are therefore needed. Furthermore, the analytical lines that lie in this wavelength range will have high mass-absorption coefficients, so the reflector must have a smooth surface if the negative absorption effect is not to be prohibitive. Unfortunately, smooth surfaces are poor Bragg reflectors. The multilayered soap films that we call Langmuir–Blodgett gratings are the best Bragg reflectors for the light elements (more precisely, for wavelengths in the 10–100 Å range).

As x-ray emission spectrograph systems continue to grow more sophisticated, not only are more kinds of flat Bragg reflectors being pressed into service, but curved reflectors (LPWZ, pp. 218–223) continue to gain ground. A flat reflector can accept analytical lines from the *entire area* of a sample that is both broad and long, but (as we have seen) it needs a parallel beam

Table 13.VI
Flat Bragg Reflectors for X-ray Emission Spectrography (An Incomplete List)
A. Spacings and Useful Ranges

Bragg reflector	$2d$ (Å)
PbSt	100
PbMy	80.5
OHM	63.5
KAP	26.63
Gypsum	15.19
ADP	10.65
EDDT	8.808
PET	8.742
Ge	6.532
Si	6.276
NaCl	5.614
LiF (200)	4.028
LiF (220)	2.848
Topaz	2.712

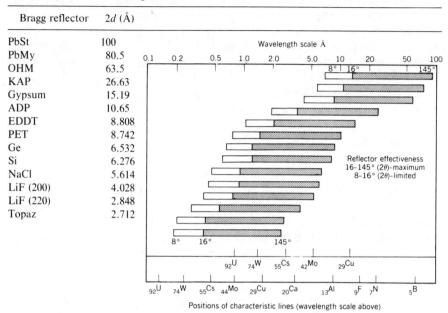

Positions of characteristic lines (wavelength scale above)

B. Further Information (Most reflectors have several names)

Bragg reflector	Name	Remarks
PbSt	Lead stearate	Langmuir–Blodgett grating
PbMy	Lead myristate	All following reflectors are crystals
OHM	Octadecyl hydrogen maleate	Good for determining carbon
KAP	Potassium acid (hydrogen) phosphate	Preferred for determining F or Na
Gypsum	Calcium sulfate dihydrate	May effloresce; store properly
ADP	Ammonium dihydrogen phosphate	Low intensity
EDDT	Ethylenediamine ditartrate	Low intensity
PET	Pentaerythritol	Higher intensity than the preceding two
Ge	Germanium	Both elements: even orders missing
Si	Silicon	in reflection; intensity comparable
NaCl	Sodium chloride	Useful but hygroscopic; store properly
LiF (200)	Lithium fluoride (note Miller indexes)	Best general-purpose crystal
LiF (220)	Lithium fluoride (note Miller indexes)	Good intensity, high resolution
Topaz	Aluminum fluosilicate (hydrated)	Best resolution

and hence cannot escape the huge intensity losses that accompany collimation. Curved Bragg reflectors accept beams that are convergent or divergent; *collimation is not required.* Curved reflectors can give excellent
resolution and form a good image, but they must view x-rays that originate
from a *line* or a *point*, not an *area.* When this requirement makes a slit
necessary, there will be a loss of intensity from an extended sample. When
the analytical lines are generated at a point, as in the x-ray emission spectrograph known as the electron microprobe, curved reflectors are at their
best.

Let us pull together the information already presented by pointing out
that the optical path (air, helium, or vacuum; the last is preferred), Bragg
reflector, detector window (scintillation detectors need more than a simple
window), and detector must all be the best for an element *of given atomic
number* if the analytical-line intensity actually measured is to approach most
closely that at the sample surface. Thus, the Philips PW 1600 x-ray emission
spectrograph system, with an atomic-number range from 9 to 92 (F to U),
has a vacuum path for all solid samples;* curved reflectors (of three kinds)
for the range 9–19 (F to K) inclusive, with a flow proportional detector and
1-μ polypropylene window; a curved LiF(200) reflector for the range 20–27
(Ca to Co) inclusive, with a sealed argon (xenon for Co) proportional detector
and 50-μ beryllium window (150-μ for Co); a flat LiF(200) reflector for the
range 28–39 (Ni to Y) inclusive, with the detector and window just listed
for Co; the reflector just named for the range 40–48 (Zr to Cd) inclusive,
with a scintillation detector; LiF(220) as reflector for $_{49}$In, $_{50}$Sn, and $_{51}$Sb,
with a scintillation detector; and finally, LiF(200) for $_{52}$Te to $_{92}$U inclusive,
with a sealed xenon proportional detector. Not every element from $_9$F to
$_{92}$U is in the Philips list. Up to $_{52}$Fe, the analytical lines are Kα; beyond,
they are Lα, which are more suitable, owing to the high energies and short
wavelengths of the K lines of heavy elements. Philips has achieved a careful
matching of "the time, the place, and the girl" that may be unparalleled in
science and is a tribute to the simplicity of x-rays and the experimental
ingenuity of the human mind.

I. X-RAY EMISSION SPECTROGRAPHY AND THE COMPUTER

No analytical chemist, especially one who remembers what happens, for
example, when concentrated NH_4OH is added to aqueous Cu^{2+}, need feel
embarrassed or intimidated by computer language. But he cannot afford to
ignore the computer itself. Nowhere is this more obvious than in x-ray emission spectrography, where the computer has brought to a climax the changes
originating in physics and in vacuum-tube electronics, and exemplified most
recently by solid-state circuitry and devices.

We have seen the computer in its two principal roles: exercising *operational control* (to keep sheet steel uniform in thickness) and *performing cal-*

---

* To retard the evaporation of liquid samples, a helium optical path is used.

culations (to convert results of absorptiometric measurements in computerized tomography to presentable form). It performs both functions in modern x-ray emission spectrography. We will take the first function for granted, and sketch two ways of carrying out the second. Both ways require standards; hence, both are *comparative methods*, though less obviously so than direct commutation between a standard and an unknown identically prepared and nearly identical in composition. Naturally, information about such standards can be stored for call-up by a computer where needed.

It will be convenient to take the work of Sherman in the 1950s as a point of departure for a jump to the 1970s (see LPWZ for interim developments).

1. Regression Treatments

Regression treatments were described by Sherman in 1953 "as a practical correlation of [analytical-line] intensity and [sample] composition" which could lead to "close linear approximations . . . for limited ranges of composition." The computer has made this assessment far too conservative.

A *multiple linear regression* of Y on the variables $x_1 - x_n$ is represented by

$$Y = a + b_1 x_1 + b_2 x_2 + \cdots + b_n x_n \qquad (30)$$

More complex regression equations containing higher terms (e.g., $x_1 x_2$ or x_1^2) are sometimes needed. In x-ray emission spectrography, we may speak of *calibration* (setting up of a working curve by the use of standards) followed by *determinations* on unknowns (use of the working curve). For calibrations, the Y will be measured intensities of analytical lines; for determinations, the Y have in the past had to be weight fractions of the elements in the unknown. The sets of equations for the two purposes differ in form, and the second set must include the information that all weight fractions considered sum to (approximately) unity.

Obviously, multiple regression is an approach broad enough to include all likely possibilities. The coefficients (the b values) could relate to interelement effects (see below). Uncertainties in standards, measurement errors, and instrumental factors can be taken into account during calibration by increasing the number of measurements and calculating least-square values for the coefficients.

The computer today makes it possible to carry out and to use regression treatments that once were prohibitively elaborate. Clearly, only important routine determinations, such as are needed for the control rather than for the characterization of materials, ordinarily justify the considerable work and expense that comprehensive regression treatments still entail.

2. Calculation of Interelement Absorption and Enhancement Effects

When the number of unknowns of a given kind is not large enough to warrant a regression treatment, the calculation of interelement effects from fundamental parameters (i.e., x-ray data such as mass-absorption coefficients, absorption jump ratios, fluorescent yields, and relative intensities of characteristic lines) is a valuable way of extending the range of unknown compositions over which comparison with a standard can give acceptable results. The comparison still has an empirical basis, but the extension is quasitheoretical, and it is an alternative not open to spectral methods employing radiation longer in wavelength than x-rays.

The calculation of interelement effects is simple when only absorption effects need be considered. Enhancement effects present a more difficult integration problem, because the x-rays responsible for these contributions, instead of being incident on the sample in a narrow, collimated beam, are generated throughout the sample interior. Sherman solved the calibration problem (see above) for alloys of $_{24}Cr$, $_{26}Fe$, and $_{28}Ni$ by calculating the interelement effects from fundamental parameters, but having no computer available, he could not deal with the more difficult determination problem. In his calculations, Sherman assumed an effective wavelength of 1.2 Å (a value near those of the NiLβ) for the excitation of his analytical lines—a crude approximation, because the corresponding absorption edges lie at 2.07 Å (Cr), 1.79 Å (Fe), and 1.49 Å (Ni).

Stephenson,* by use of a computer, solved both calibration and determination problems in an ingenious way that significantly advanced the calculation of interelement effects. He made the key assumption that an individual effective wavelength slightly shorter than that of the absorption edge of each element sought could be used to make the needed intensity calculations—an assumption amply supported by experimental evidence. The computer, which relies on iteration, made it possible for Stephenson to use the same kind of intensity equations for both calibration and determination.

Needless to say, the calculation of interelement effects can at best eliminate only this one cause of deviations from proportionality between analytical-line intensity and weight fraction.

J. A GLANCE AT X-RAY EMISSION SPECTROGRAPHS

For the *characterization* of materials, *sequential* spectrographs, so called because they measure the intensity of only one analytical line at a time, are commonly used. For the *control* of materials, processes, or production, *simultaneous* or *multichannel* spectrograph systems are often used, because they can measure simultaneously the intensities for all analytical lines of interest. We cannot do justice here to the great diversity of high-quality equipment now on the market.

1. Multichannel Systems

During World War II, millions of intensity measurements in the visible and ultraviolet began to be made annually in the aluminum industry on mul-

---

* Stephenson, D. A., *Anal. Chem.*, **43**, 310 (1971).

tichannel spectrographs called quantometers. In these instruments, which were developed by Applied Research Laboratories (ARL), one channel, complete with appropriate Bragg reflector and detector, was allocated to each analytical line, and one channel was reserved for a standard or a monitor. ARL subsequently developed x-ray quantometers along similar lines. The four VXQ models ("V" stands for "vacuum," "X" for "x-ray," and "Q" for "quantometer") made by ARL have been joined recently by the Philips PW 1600 system mentioned earlier.

Such systems need end-window x-ray tubes for the simultaneous service of all (up to 28) channels; targets of tungsten, rhodium, chromium, and platinum are available. They make full use of the computer and (as indicated by the earlier discussion about the Philips PW 1600) incorporate a wealth and diversity of modern equipment.

ARL provides auxiliary equipment for the analysis of slurries (usually of ores) and an automatic powder-briquetting system for on-line process or production control. Detailed applications cannot be described; let us say simply that ARL and competitive analytical systems are used in the mining industry (e.g., to control the processing of ores for the concentration of copper, lead, zinc, molybdenum, and nickel, in the manufacture of glass and refractory silicate materials, in the making of Portland cement—in fact, almost anywhere reasonably precise determinations on powders are needed in sufficient number and in minimum time.

2. Special X-ray Emission Spectrographs

Chapter 14 deals with the electron microprobe, which could logically be placed in the category of spectrographs described here. The meaning of "special" in this context can be inferred from the following descriptions: *Portable spectrographs* make possible rapid determinations in the field. This saves time and trouble, if the results are adequate, and it makes on-the-spot decisions possible. These spectrographs can be useful adjuncts to larger instruments. They are well suited to the classroom.

A portable, battery-operated spectrograph of *total* weight near 15 lb is available from Columbia Scientific Industries. A radioactive source either excites the sample directly, or generates the characteristic line of a "target element" for this purpose. The source, located behind a shutter (diameter, 0.5 in.), is coaxial with the scintillation detector; both are housed in a cylinder ($2\frac{1}{2} \times 8$ in.). The source is so weak (say 10^{-7} times the intensity of an x-ray tube) that no special shielding is needed, and values of 10^3–10^5 cps from pure elements are attainable. Balanced Ross filters provide wavelength selection, and results of the desired sequential readings appear on a simple scaler. The standard counting error is unavoidably large. A special version containing six filters has been used on board ships for determinations of the principal constituents (manganese, nickel, iron, and cobalt) in the manganese

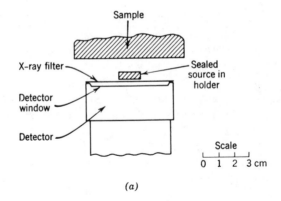

(a)

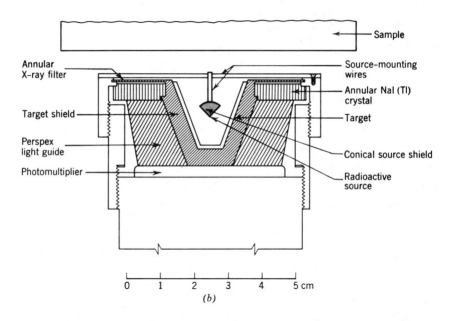

(b)

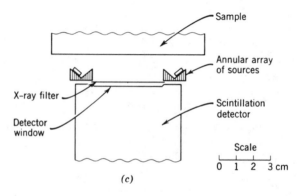

(c)

nodules present in enormous amounts on the floors of the Pacific and Indian Oceans.

The Pitchford Manufacturing Corporation has produced an x-ray emission spectrograph with a collimator, LiF crystal, goniometer, and Geiger detector that can be held on any flat surface or put on rods or tubing by means of self-positioning, adjustable shoes. Alternatively, the unknown and standard may be placed in a two-position sample holder. Various read-outs are possible. The equipment is intended mainly for identification of materials in industrial plants.

The success of the electron microprobe has stimulated a return to the use of *electron excitation* of analytical lines over sample areas of roughly 1 cm². The sample, if not conducting, must be made conducting, as by admixture of graphite. Three versions of these special spectrographs are offered by Telsec Instruments alone. The Telsec Betaprobe B.300 is a fully automated multichannel instrument that gives simultaneous teletype print-outs of all analytical-line intensities; percentage results for up to 12 elements appear 45 sec after sample insertion, the first 30 sec being needed for pump-down. In general, (1) results obtained with these instruments are in accord with what is known about x-rays and electrons; for example, interelement effects are usually absent or much reduced, owing to the low penetrating power of the electrons; (2) light-element determinations are easier to carry out than with x-ray excitation; (3) for heavier elements, x-ray excitation is preferable, because it can use **K** and **L** spectra, whereas the Betaprobes are restricted to **L** and **M**; and (4) the calibration curves are often linear. We have given here only a small sampling of the special spectrographs available from perhaps half a dozen sources. Many use Si(Li) detectors with high resolution, and these can be classified according to their modes of analytical-line excitation: (1) radioactive isotopes; (2) low-power x-ray tubes; and (3) tubes with high bremsstrahlung (see below) for the excitation of characteristic lines from a target.

We have discussed radioisotope sources; Fig. 13.32 illustrates them further. They are low in cost and stable; true, they decay, but at a predictable rate. To minimize radiation hazards, the sources generally used are weak (about 10 mCi, or 10^6 useful photons/sec/cm²), so analytical-line intensity

Fig. 13.32. Three geometrical arrangements of radioactive sources that give analytical-line intensities adequate for many trace determinations. (*a*) "Central source" arrangement of sealed source, sample, and scintillation detector. (From LPWZ, pp. 372–373.) (*b*) Same arrangement (in principle) with characteristic-line generator ("target") added. The spectral purity of the characteristic line generated to excite the analytical line can exceed 90%. A simple target is often a powdered oxide with a minimum amount of binder (epoxy resin) added. Targets are easy to change. (*c*) Annular arrangement for use with a tritium–zirconium source and a scintillation detector. The annular arrangement gives increased intensity. Here the radioactive source is not sealed, but consists of a thin layer of tritiated zirconium on tungsten. With a Si(Li) detector, the annular arrangement *must* be used, because of the small detector aperture. [From Rhodes, J. R., *The Analyst*, **71**, 683 (1966).]

suffers, and sensitivities range from only 10 to 1000 ppm for the elements sought. Collimation ordinarily is out of the question. Finally, not very many satisfactory isotopic sources are available.

Low-power x-ray tubes do not need water cooling, and are therefore suitable for portable spectrographs. They retain to some degree the advantages of Coolidge x-ray tubes: good stability ($\pm 1\%$), high intensity (say, 10^8 photons/sec/cm^2), and low background. The characteristic lines they generate obviously cannot excite the analytical lines for very many elements.

Bremsstrahlung ("braking radiation") results from the slowing of electrons within a tube anode. The resulting continuous spectrum can be used to excite the characteristic lines of a "target" in the manner shown in Fig. 13.32b. Targets can be changed as appropriate to cover satisfactorily a wide range of elements. The relative cost is high, and the equipment cannot be made portable.

The trend is toward "hard-wire" equipment (e.g., multichannel pulse-height analyzers), that is, equipment that cannot be modified by the user. Such equipment can be happily "married" to (interfaced with) computers that perform needed calculations, which can in turn be interfaced with other computers that control processes or production.

3. A Sequential Spectrograph in a Computerized Laboratory

The sequential spectrograph is the most common x-ray spectrograph, and the one best suited to the nonroutine samples formerly analyzed by the wet methods of inorganic chemistry. The capabilities of these spectrographs have been greatly extended since about 1960. Their operation is preferably computer controlled, and computers are also employed to do all the calculations needed (e.g., background correction and calculation of interelement effects) for a print-out of the final analytical results. A comparison of a computerized laboratory with the highly successful Fluo-X-Spec Laboratory reveals what is happening in this field (see LPWZ, pp. 488–489).

In the General Electric Company, the Chemical and Structural Analysis Branch of the Corporate Research and Development Department includes a computerized laboratory* in which a Siemens SRS (Sequenz Roentgen Spektrometer), which is a sequential x-ray emission spectrograph, shares a computer with related instruments.

Complete description of any of the several excellent sequential x-ray emission spectrograph systems available would take too much space. The interested reader is referred to the manufacturers' literature. Some important features of the Siemens instrument are evident from the following description of its computerized operation.

* Ciccarelli, M. F., W. T. Hatfield, and E. Lifshin, *Siemens Rev.*, **41** (1974). H.A.L. thanks Mr. Ciccarelli for a tour of his laboratory, and is grateful to him and to Dr. Lifshin for a critique of this section.

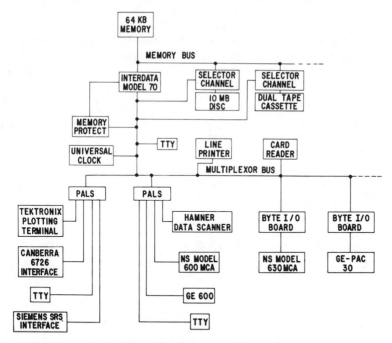

Fig. 13.33. Schematic diagram of the General Electric computerized x-ray laboratory.

Ciccarelli et al. mention three alternatives as regards the choice of a computer for such an operation: (1) a minicomputer dedicated to the spectrograph; (2) time sharing on large, remote computers such as might serve large components or even large organizations; or (3) a small computer that serves the family of instruments in a modern x-ray laboratory. Minicomputers are too often undesirably restrictive. Time sharing is often unsatisfactory, because communication between the analytical instruments and the computer is not adequate over the interval of *real time* needed by the instrument to complete all determinations on the samples inserted in one loading of the instrument. The third alternative avoids the weaknesses of the other two provided that the minicomputer has a large enough *core memory* ("internal" memory with which the computer interacts "instantaneously"), and provided that the minicomputer is adequately supported by peripheral equipment.

These conditions are met by the system diagrammed in Fig. 13.33. Shown in this figure are:

1. An Interdata Model 70 computer.
2. The core memory of the computer, 64 kilobytes.
3. A disc and dual tape cassette, providing bulk storage of information to supplement the core memory. This information is quickly, but not instantaneously, available to the computer. It can be supplied in modular fashion as needed.

4. A Teletype terminal (TTY).
5. Programmable asynchronous line modules (PALs), sophisticated "information switches" that can set priorities for serving interfaced equipment and fix the "communication speed" between the computer and peripheral equipment.
6. A multiplexor bus, which transmits simultaneously the messages from all the PALs connected to it, much as a single telephone line can transmit many conversations at once.
7. A multichannel analyzer (MCA), that is, a multichannel x-ray emission spectrograph.
8. A GE-PAC 30 computer added to increase the computing capability.
9. The SRS interface, which is a logic controller, a box that transmits to the spectrograph the commands (options chosen) by means of which operational control of the spectrograph is exercised. The operator sets on the logic controller the option selected for each of the following components:

Component	Options
Crystal changer	Choice of four Bragg reflectors
2θ wheel	20 angles, from 5° to 147°
Detector	Flow proportional or scintillation
Filter	Filter (e.g., Ti), aperture, or unrestricted incident beam
Collimator	0.15° or 0.4° collimator
Sample rotation	"On" (if desired for nonuniform samples) or "Off"
Sample changer	1–10 samples
Time	1 sec to infinity
Counts	1 to infinity
High voltage	20–40 kV (chromium tube AGCr61)
Current	3–60 MA (chromium tube AGCr61)

A laboratory with its own computer serving all of its instruments offers great advantages. But for these to be realized to the full, the minicomputer must be capable of calculating interelement effects. For this calculation according to Stephenson's program (named CORSET), a minicomputer does not have enough core memory available, since part of this memory must be dedicated to other computer operations. Ciccarelli has resolved this difficulty by shortening and simplifying CORSET to produce QUAN,* which requires only 12 kilobytes of core memory. In the main, he shortened CORSET by deleting less frequently needed operations and by "fitting" the energies of the analytical lines to their absorption edges so that the atomic number

* Ciccarelli, M. F., *Anal. Chem.*, **49**, 345 (1977). Mr. Ciccarelli has already filled over 200 requests for QUAN and is willing to fill more.

of an element is all the computer needs to start the interelement-effect calculations. (In such calculations, energies may of course replace wavelengths.)

QUAN is supported by information in bulk storage. Because QUAN covers all elements that could appear in a sample from $_9$F up, it is not surprising that the program sometimes needs help. For example, it can determine only one element (usually oxygen) by difference; if unsuspected elements are present, and some of these are light elements (e.g., beryllium, boron, carbon, or nitrogen) beyond the reach of the spectrograph, there will be trouble. If the unsuspected elements are heavier, a rapid, semiquantitative 2θ scan will show their presence and reveal what additional information the computer needs for a quantitative result. Furthermore, the computer warns the operator if the energy of the analytical line falls within 10 eV of that of an unsuspected absorption edge from the matrix of the sample; the operator can then take steps to make sure that the computer gets the proper information. Classical wet methods, time consuming though they are, may have to be used as a court of last resort, but the number of such cases will decrease as the x-ray laboratory gains experience and adds to its library of standards, such additions often being samples that were "problem children." The reliability of the x-ray laboratory will thus grow with age and experience as it moves toward the goal of comparing each new unknown with a standard or standards nearly enough identical, both chemically and physically, that the interelement calculation need not cover wide composition ranges, properties such as particle size and homogeneity are closely comparable, and methods of preparation are the same. Homogeneous, smooth-surfaced glasses and alloys make ideal samples for such an approach.

The computerized laboratory just described employs five people and contains equipment that cost about $500,000. In 1976, among other work, it performed 1074 electron-microscope assignments, carried out an average of six determinations on each of 254 samples with the electron microprobe, and did an average of about five determinations on each of 575 samples on the SRS.

VIII. APPLICATIONS OF X-RAY EMISSION SPECTROGRAPHY

Owing to the rapid and still accelerating growth of x-ray emission spectrography, this section—intended to be that most immediately useful to the analytical chemist—will necessarily suffer from incompleteness and be in greatest danger of obsolescence. Consequently, applications have been chosen to illustrate the considerations of the previous section, and references are cited to help the reader to the original literature.

A. QUALITATIVE AND SEMIQUANTITATIVE DETERMINATIONS

Qualitative determinations come so naturally that they need neither definition nor description. Semiquantitative determinations cover much ground,

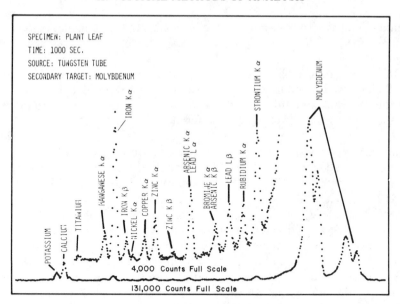

Fig. 13.34.   X-ray spectrum of "orchard leaves" standard reference material no. 1571 (National Bureau of Standards). Spectrum obtained from 100-mg sample excited with tungsten tube–molybdenum secondary target (40 kV/35 mA), x-ray detector with a Si(Li) detector; count time, 1000 sec. (Courtesy Kevex Corp.)

but may be defined operationally as those in which one strives to obtain adequate reliability in a short time with minimum effort. As indicated earlier, energy resolution (dispersion) is particularly well suited to "survey" analyses. An illustration of multichannel analysis is given in Fig. 13.34, showing the x-ray spectrum of the National Bureau of Standards (NBS) "orchard leaves" standard. A similar example is the earlier comparison of x-ray spectra from genuine and counterfeit banknotes (Fig. 13.29). Some qualitative and semiquantitative analyses, in which several characteristic lines and ratios of their intensities are used for identification, are referred to as signature analyses (see, e.g., reference 52).

An interesting semiquantitative approach maximizing the amount of information obtainable rapidly and with a minimum of effort was devised by Anater (3). Compromise standard operating conditions are chosen that depend on the equipment available. Normalization factors (NFs) are next measured on known standards for each element but one within the range of the equipment, the one element (in Anater's case iron) being as a reference. The following equations described the procedure:

$$\frac{\text{NACD}_{\text{Fe}}}{W_{\text{Fe}}} = Q_{\text{Fe}} \qquad \frac{\text{NACD}_{\text{E}}}{W_{\text{E}}} = Q_{\text{E}} \tag{31}$$

$$\frac{Q_{\text{Fe}}}{Q_E} = \text{NF}_{\text{E}} \tag{32}$$

where the W are percentages by weight, and the NACDs (net adjusted chart divisions) measure peak height on the same scale for iron and for element E, with the heights properly corrected for background. Unknowns of all kinds are run as originally received if the sample holder can accommodate them. For any element E in an unknown, W_E can be calculated from these equations if the needed NEs are known and if all elements in the sample give analytical lines within the range of the equipment. Clearly, the unknown need not contain iron, since

$$\frac{Q_E}{Q_{E'}} = \frac{NF_E}{NF_{E'}} \tag{33}$$

When not all elements in the sample can be determined, the assumption

$$W_E + W_{E'} + W_{E''} \cdots + W_{E^{n'}} = 100 \tag{34}$$

is made for the n elements determined, and the results are reported on this basis. They will obviously be too high.

Anater applied this scheme to 27 elements above and including $_{13}Al$. He claims reliability within a factor of 2, with most results precise to $\pm 20\%$ (relative) of the amount present. Representative NFs vary from 80 for Al (least sensitive to the method) to 0.74 for Ca (most sensitive).

B.   DETERMINATIONS ON COMPLEX MATERIALS

Minerals are the prime example of materials studied under this heading. For general reference, the book by Adler (1) should be consulted.

Major constituents can be determined with excellent precision, even in complex cases, provided that standards virtually identical in composition with the unknowns are available for direct comparison. In this approach the intensity of, say, the $K\alpha$ line of element E, to be determined in an unknown U, is compared with the intensity of the corresponding analytical line in a standard St. After background correction of the total counts for unknown and standard, the weight fraction (or weight percentage) of the element present can then be calculated by simple proportion as follows:

$$\frac{W_E^U}{W_E^{St}} = \frac{(\text{corrected counting rate})_E^U}{(\text{corrected counting rate})_E^{St}}$$

The precision that can be achieved is illustrated by the study of granite rocks by Baird and Henke (6). A soft x-ray tube generating CuL lines was used for the excitation of analytical lines in 17 rock samples that had been powdered and compressed into smooth pellets, but not fused. Their results, obtained in comparison with quartz, appear in Table 13.VII. Because these samples were unfused, it was possible to determine oxygen, sodium, and magnesium. The results are even more remarkable when one remembers

Table 13.VII
Precisions of Silicate Analyses by Several Methods (6)

Element	Approx. composition (w/o)	Intralaboratory analytical precision[a]				
		X-ray	Wet chemistry	Emission	Flame	Neutron activation
O	48.0	0.4	0.3	—	—	0.9
Na	2.6	0.02	0.15	0.11	0.07	—
Mg	0.5	0.01	0.15	0.12	—	—
Al	9.3	0.04	0.19	0.53	—	—
Si	30.0	0.10	0.14	1.1	—	—
K	3.7	0.03	0.21	0.15	0.07	—
Ca	2.5	0.01	0.07	0.14	—	—
Ti	0.22	0.003	—	—	—	—
Fe	3.1	0.02	0.21	0.14	—	—
	99.9					

[a] Standard deviation (w/o).

that fusion is normally required for high precision. There is no better evidence of the value of x-ray emission spectrography than this table. It is the only method by which all elements, *oxygen included*, can be directly determined. The standard deviations speak for themselves.

The determination of mixtures of rare earths has always been a thorn in the side of the classical analytical chemist. The chemical similarity of the elements is of course responsible for this, and this, and this similarity makes the isolation of pure compounds difficult. Moseley (42) was consequently motivated to make the first attempt at quantitative rare-earth determinations by x-ray emission spectrography. The x-ray spectra showed that what he had thought was praseodymium contained rare-earth elements in these proportions: lanthanum, 50; cerium, 35; praseodymium, 15. "*Caveat emptor*," he must have thought!

Although x-ray emission spectrography has an overriding advantage here over classical analytical methods, reliable x-ray results presuppose careful work, as Lytle et al. have pointed out (38) (see also LPWZ, pp. 205–206). Kunzendorf and Wollenberg (34) have shown what can be achieved with x-ray emission spectrography by exciting planar rock surfaces with collimated 59.5-keV γ-rays from a $^{241}_{95}$Am source, using a Ge(Li) detector in conjunction with a 1024-channel pulse-height analyzer. By unfolding the complex spectrum obtained from a known powdered mixture of rare-earth oxides, they obtained data by use of which it was possible to reverse the process and calculate a spectrum for a sample that contained lanthanum, cerium, praseodymium, and neodymium. The calculated and measured spectra agreed within 3%. From this result and other work, they concluded that their method could detect rare-earth elements when present near 10 ppm, but that conventional x-ray emission spectrography carried out on powdered samples would give results of higher reliability.

Quite often it is necessary to determine elements with high precision when no external standard is available with which to compare the unknown. Such cases usually are nonroutine, so that obtaining suitable external standards is not worth the effort. The use of internal standards, either added or built in, is then the best way to compensate for deviations from proportionality. In either case, an intensity *ratio* I_E/I_{St} is measured in which I_E and I_{St} refer to an analytical line of element E and the characteristic line of the standard, respectively, lines that differ but little in wavelength. If both lines belong to the same series (e.g., if both are **K** lines), E and the standard will then have to be periodic-table neighbors, as in the classic work of Eddy and Laby (22).

Unfortunately, conditions under which internal standards must be used are seldom ideal, but this situation is partly redeemed because the simplicity of x-ray spectra makes it possible to anticipate most difficulties. It is generally accepted in spectrography that a *ratio* of intensities can be established more certainly than can a *single* intensity. X-ray emission spectrography is no exception, provided that enough counts are taken to make the statistical counting error negligible. If both intensities are measured simultaneously, the internal-standard method is at its most effective, because experience has shown that class 2 and 3 deviations (see Section VII.D) usually are eliminated and class 1 deviations are minimized. Scattered x-rays, if properly selected, can sometimes take the place of lines from internal standards (5).

The usefulness of an added internal standard (sodium bromide) was demonstrated in a method for detecting tungsten (25) in aqueous solution.* An internal standard was needed because sodium tungstate itself generated a pronounced negative absorption effect that was increased by the presence of sodium hydroxide in variable amounts. Bromine as sodium bromide was chosen as the standard Br**K**α (1.041 A) and tungsten Lγ1 (1.099 Å) were selected as the analytical lines best suited for comparison. Accordingly, solutions for the determination of tungsten were prepared to contain 2 mg bromide ion per gram of solution. The bromide ion was added as sodium bromide weighed precisely enough not to reduce appreciably the overall precision of the tungsten determination. The intensity ratio was computed as a ratio of counting intervals. Corrections for background proved unnecessary—a marked advantage. The time for a complete determination, including sample preparation and computation, was 30 min.

* The following points should be considered in connection with the x-ray emission spectrography of solutions: (1) Measures must be taken to ensure that evaporation of solvent does not misalign the beam path during the determination; this has the same effect as a change in geometry. (2) Liquids require cells to contain them and cannot be handled as solids are. (3) Scattering of x-rays by the solvent may increase the background above that observed with solid samples, but the background from liquids is usually very reproducible. (4) Internal standards soluble in liquid samples usually can be found, and uniform distribution of the standard is thus assured. Even when no internal standard is used, homogeneity is a desirable feature of a sample in solution. (5) The critical depth of the solutions may vary considerably, usually being greatest for the pure solvent. (6) The contribution of the solvent to absorption effects must be considered; solvents ordinarily do not produce enhancement effects.

The solutions filled a 3-ml container. To prevent evaporation and to maintain a fixed distance between the x-ray tube window and the sample surface, the beaker section was covered with Mylar film, 0.0025 cm thick; placed in a plastic sample holder; and pressed firmly against the sample drawer. The Mylar film attenuated the x-rays uniformly enough not to affect the precision of the results.

The internal standard proved wholly satisfactory. Linear working curves were obtained, and sodium hydroxide concentrations up to 4 N had no effect on the intensity ratio.

C.   LARGE-SCALE APPLICATIONS

A typical large-scale application is a closed-control-loop system in the manufacture of cement. In normal operation, an x-ray emission system continuously determines four elements (calcium, aluminum, silicon, and iron) in the raw cement mixture as it issues from the ball mill at rates of up to about 2000 tons/day. The computer in the system checks the analytical information every 6–10 min and accumulates it for 1 hr. At the end of the hour, the computer uses this information, properly weighted, to calculate whether the composition of the raw cement mixture corresponds to certain "holding points" for the finished product. Any significant deviations of the calculated results from these holding points are then corrected by appropriate automatic adjustments of the weighers for the raw-material feeds. On-line experience with this x-ray emission control system has been favorable: sampling has been improved and the cost of analyses lowered, manufacturing equipment now lasts longer because process control is better, and more uniform cement is being made at lower cost.

Figure 13.35 shows the more interesting components of the x-ray control system. The reliability of the control procedure is shown in Table 13.VIII.

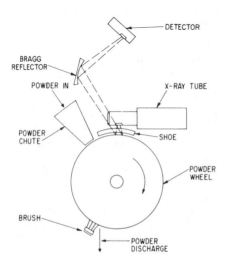

Fig. 13.35.   Sampling in closed-loop system for cement manufacture. The rim of the powder presenter, a rotating wheel, is suitably indented to retain the powder that is the sample. Before the characteristic lines are excited, the powder is compacted by the brake-shoe-like device (center) to make the sample smooth and uniform. The x-ray beams pass through a 1-mil beryllium window in the shoe. (See reference 36.)

Table 13.VIII
Reliability of X-ray Emission Control of Cement Manufacture

Constituent	Percentage by weight	Accuracy[a]	Precision (standard deviation)
CaO	45	±0.5%	±0.15% (1 part in 300)
SiO$_2$	15	±0.5%	±0.15% (1 part in 100)
Al$_2$O$_3$	3	±0.5%	±0.15% (1 part in 20)
Fe$_2$O$_3$	2	±0.1%	±0.025% (1 part in 80)

[a] When errors affecting precision are negligible, the percentage by weight as measured should not differ from the true value by more than the limit shown; for example, if 45% CaO is measured under these conditions, the true value will lie between 44.5 and 45.5% CaO. These are the limits set by experience with actual mixes.

D. TRACE DETERMINATIONS

Trace determinations divide logically into those in which the traces are major constituents (samples small in mass) and those in which they are minor constituents (samples may be large).

Determinations of traces as major constituents are characterized by the absence of class 1 deviations, and hence are governed by considerations applying in the determination of film thickness by x-ray emission spectrography. Examples are a thin film on a suitable substrate, a sample dissolved at low concentration in a solvent transparent to x-rays, or a small sample uniformly dispersed in a similarly transparent medium.

Spot-test techniques (26,47,53) are ideally suited to x-ray emission spectrography, as Fig. 13.36 shows. The slight deviation from proportionality at

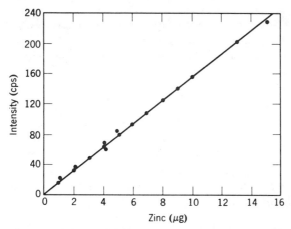

Fig. 13.36.   Counting rate of zinc Kα line for various amounts of zinc on Schleicher & Schuell no. 704E ½-in. disks. Samples were prepared by placing measured amounts of zinc sulfate solution on the paper and drying to remove the water. [After Pfeiffer, H. G., and P. D. Zemany, *Nature*, **174**, 397 (1954).]

the largest amount of zinc may mark the appearance of a detectable negative absorption effect. It is unlikely that the conditions are ideal, because the solution saturates the paper, so the zinc salt residue is neither on the surface as a thin film nor uniformly distributed. That linearity is nevertheless achieved illustrates another advantage of working with small samples: As long as all the sample intercepts the incident beam and absorption effects are negligible, there can be considerable departure from ideal conditions without essential loss of simplicity. With a uniform x-ray beam, for example, distribution of the sample need not be uniform, as it will not be if the sample is the residue from the evaporation of a drop of solution. The method of Fig. 13.36 obviously is at its best when more than one element is present under conditions in which class 1 deviations are small or negligible. Ion-exchange membranes often are a valuable means of isolating and concentrating a trace element from solution. Such membranes can be inserted directly into a spectrograph for trace determinations by x-ray emission (27,54). Other approaches suitable for analysis of species in solution use precipitation, ion-exchange resins processed into pellets (14), or nonspecific chelating agents (e.g., ammonium 1-pyrrolidine dithiocarbamate) for multielement assays (23).

A rapidly growing area is the analysis of particulate matter in air. Detection limits of a few micrograms per square centimeter are readily achieved provided the background (mainly due to scattering) is reduced to a minimum, that is, provided the samples are as thin as possible. Typically the collected air particulate matter weighs less than 1 mg/cm² on a filter (e.g., Nucleopore, Millipore) weighing 1–10 mg/cm². Substantial efforts have been made to overcome limitations due to backing-material background, particulate location, and size effects. Round-robin studies have demonstrated the effectiveness of this technique for characterizing air particulate matter (19). Most of the work in this area has been carried out with energy-dispersive (energy-resolution) systems. Recently, an inexpensive, small spectrometer was designed to measure low-concentration pollutants at on-site locations (11). The instrument uses a 100-W x-ray tube and an air-path spectrometer, and has detection limits below 1 μg for elements above vanadium.

Another field of rapidly growing applications is the analysis of biological specimens. As in studies with air particulate samples, all forms of excitation (electrons, photons, and heavy ions) have been applied in conjunction with wavelength- and energy-dispersive detection systems. Applications cover a broad range; examples are the determination of Cu, Zn, Br, Rb, Sr, and Pb in urine at levels below 1 ppm (with preconcentration on an ion-exchange resin) and the determination of iodine in milk (2,21).

E.   TRACE ANALYSIS IN BULK SAMPLES

The x-ray emission method has been successfully applied for the trace analysis (e.g., of lead in part-per-million concentrations) in bulk samples since the 1930s. Since then, numerous studies, dealing with a great number of specific problems, have been described (12,39). Interesting recent work has concerned the relative advantages of **K** and **L** lines in the determination of barium in geological samples (44) and of **M** and **L** lines for measurement of the actinides (41). The accuracy that can be achieved is illustrated by a study by Brenner et al. (15), who determined Fe, Ti, K, Mn, Cr, Ni, Cu, Zn, Sr, Ba, and Rb in silicate rocks. For elemental concentrations ranging from 5 to 7500 ppm, linear correlation coefficients of 0.988–0.998 were obtained.

F.   DETERMINATION OF LIGHT ELEMENTS

Despite notable progress, determination of elements lighter than fluorine is still beyond the reach of standard x-ray emission analysis systems. Difficulties begin at the x-ray source and continue through the sample, optical path, and Bragg reflector on into the x-ray detection system; alternatively, in energy-dispersive (-resolution) setups, limitations arise from the need for working with detectors with ultrathin windows or no windows at all.

Fundamentally, the problems associated with the detection of light elements are common to *all* elements for analytical lines of long wavelength. But the light elements are unfortunately unique in having *only* such lines to offer. The energy of a characteristic line is the difference in energy between two atomic states. Ordinarily, we take this energy to be (virtually) independent of chemical or physical state. This useful (hence justified) generalization rests on the assumption that x-ray spectra are concerned with *core* (inner-shell) electrons, and not with *valence* electrons. The generalization must fail as atomic number decreases.

Sulfur illustrates the problem. Of the various chemical effects on x-ray absorption, we mention only the wavelength shift of the **K** absorption edge. The wavelength of this edge is shorter the higher the oxidation state of the atom (37); for example, it is 5.0220 Å for sulfide sulfur in Cr_2S_3 and 4.9976 Å for sulfate sulfur in magnesium sulfate. Superimposed on the influence of the state of oxidation is a smaller influence traceable to elements chemically combined with the sulfur—the edge in *zinc* sulfate is found at 5.0156 Å. The shifts in wavelengths are very small. Comparable shifts were observed by Faessler and Goehring (24) in the wavelengths of the S**K** lines (see Fig. 13.37). In this simple case, wavelength shift and line intensity are both of interest. Theory predicts, and experiment establishes, 2:1 as the intensity ratio between Kα1 and Kα2 for sulfur. Where three lines appear in a spectrum in Fig. 13.37, the presumption is that more than one kind of sulfur atom is present, and that an intensity ratio exceeding 2 results from the coincidence of certain lines. The observed intensities in Fig. 13.37 are those to be expected on the basis of this presumption, 2 being the known value of the Kα1/Kα2 intensity ratio for every kind of sulfur atom.

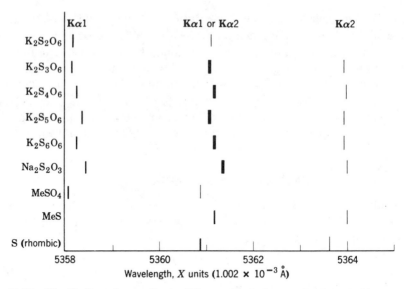

Fig. 13.37. The $K\alpha$ lines from sulfur in different chemical states. In the polythionates, the triplet results from the superposition of the $\alpha1$, $\alpha2$ doublets generated by sulfur of two different oxidation numbers. The maximum wavelength shift is about 0.004 Å. (See reference 24.)

Except in determinations of the lightest elements, these complexities are not important in x-ray emission spectrography as usually done. Analytical lines virtually unaffected by such complexities can generally be found.

Insight into chemical binding, as distinct from analytical chemistry, is a different story. In principle, such insight can be obtained with x-ray absorptiometry (from the positions and fine structure of absorption edges), with photoelectron spectrography (from the binding energies of ejected electrons), and with highly refined x-ray emission spectrography (from the change in wavelengths and intensities of characteristic lines). The last method suffers because it measures energy differences between *two* atomic states, whereas the others can give directly the energy with which an electron is held. The difference ought to be less sensitive to changes in chemical binding than a directly measured value. For a discussion of photoelectron spectrocopy see Chapter 17.

Notable progress has been made in recent years in the excitation and detection of soft x-rays. Henke and coworkers have devised low-energy x-ray tubes (30). Hanser et al. (28) have taken advantage of the high x-ray production cross section for low-atomic-number elements bombarded with charged particles. With the α-particles from a ^{210}Po source and a flow proportional counter with a thin window of polypropylene, they determined the thickness of a thin film containing carbon, nitrogen, oxygen, or fluorine. Windowless Si(Li) detectors are used increasingly for low-energy x-ray measurements. An ingenious approach has been pioneered by Cairns (18). It combines highly efficient and selective excitation with low-energy ion beams as "projectiles" and improved detection with the variable-geometry flow proportional detector mentioned earlier (see Section IV.D.3). Excitation is made selective and efficient by using a projectile with a **K** or **L** electron-binding energy roughly equal to that of the target element. Substantial enhancement of the x-ray production is obtained by proper matching of projectile and target. An interesting study by Cairns deals with the profiling of boron in silicon, including the detection of $BK\alpha$ x-rays with the variable energy flow proportional counter, utilizing the polypropylene window to screen out $CK\alpha$ x-rays (17).

The reader interested in references not be cited here will find the following two compilations useful:

1. Appendix VI, "Bibliography of Element Determinations," in *X-ray Absorption and Emission in Analytical Chemistry*, Wiley, New York, 1960, pp. 328–331 (second printing, 1966).
2. Appendix VI, "Determination of Elements by X-ray Emission Spectrography. A Guide to the Recent Literature (1964–1970)," in LPWZ, pp. 535–549.

IX. RELIABILITY OF X-RAY EMISSION SPECTROGRAPHY

A. PRACTICAL STANDARD COUNTING ERROR

As used here, *reliability* includes both *accuracy* and *precision*. With respect to both, x-ray emission spectrography is, at its best, the most favored among widely used analytical methods.

Accuracy measures the deviation from a "true value": The smaller this deviation for a highly precise result, the more nearly accurate the result. X-ray emission spectrography being a comparative method, no question of its accuracy need arise if the comparison with a satisfactory standard is satisfactorily made.

Under the best conditions, *precision* is no problem, because it is predictable when, as will be the case for such conditions, the total number of counts made follows the *unique* Gaussian distribution to be described below. (This Gaussian distribution must not be confused with the Gaussian distributions of *energies* that were earlier described as being recorded by detectors on which "monochromatic" or "monoenergetic" beams were incident.)

We present here only the experimental approach to obtaining this unique Gaussian distribution; for the theoretical background and applications, see LPWZ, pp. 328–349. With a tungsten sample in the spectrograph, the goniometer was adjusted until a counting rate near 100 cps was obtained. The detector, being a Geiger detector, counted all quanta incident upon it; at 100 cps, coincidence errors were negligible. The time interval required to reach 1024 counts was then measured 393 times in succession. For each of the 393 counting intervals, N, the number of counts that would have occurred

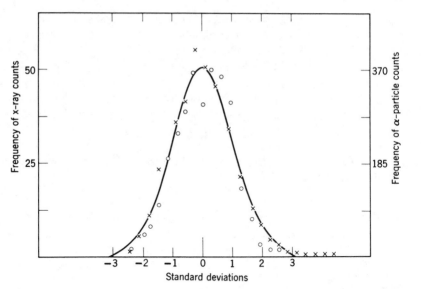

Fig. 13.38.   Experimental proof that x-ray emission spectrography and radioactivity both conform to the unique Gaussian fluctuation curve based on N alone. Crosses represent data of
Rutherford and Geiger; circles represent x-ray emission data; solid line is theoretical Gaussian
curve. [From Liebhafsky, H. A., H. G. Pfeiffer, and P. D. Zemany, *Anal. Chem.*, **27**, 1257
(1955).]

in 10 sec, was calculated by simple proportion. In this way, a body of data
was obtained for which $\Delta t = 10$ sec; $n = 393$ (values of N); and $\overline{N}$ *(the
mean N)* = 1018. From the 393 values of N, the Gaussian curve in Fig.
13.38 was constructed. To it were added data obtained in 1910 by Rutherford
and Geiger using a radioactive source. The two sets of data, obtained in
widely different experiments, conform to the *same* Gaussian distribution.

The standard deviation s is the square root of the variance; graphically,
it is the horizontal distance from the mean (here $\overline{N}$) to the point of inflection
in a well-established distribution curve. The standard deviation for a *random
process*, S_c, equals $\overline{N}^{1/2}$; that is, a truly random process is characterized by
the *unique* Gaussian for which this relationship holds. We shall call s_c the
standard counting error to distinguish it from standard deviations arising
from other sources of error.

For Fig. 13.38

$$s_c = \overline{N}^{1/2} = \sqrt{1018} = 32 \text{ counts} \tag{35}$$

For the individual data,

$$s = \left(\frac{\sum_i (N_i - \overline{N})^2}{393 - 1} \right)^{1/2} = 30 \text{ counts} \tag{36}$$

The agreement is good. The conclusion is that under the conditions studied, x-ray emission from the spectrograph is a random process, as is the emission of α-particles from a radioactive source; *ergo*, the spectrograph was operating at the *minimum* possible s, which is s_c.

Figure 13.38 suggests that x-ray emission spectrography is unique among analytical methods as regards the extent to which precision can be *predicted* and *controlled* in determinations of many elements under conditions not difficult to attain. For example, one can (1) use s_c as a criterion of merit (if $s \gg s_c$, find out why); (2) by the rules for the combination of errors, calculate s_c for more complex cases, such as in corrections for background, and for ratios of counts; and (3) predict precisions at different N values and test the predictions.

B. TRACE DETERMINATIONS

Concern about the environment, poisons, and carcinogens has already increased to an unexpected extent the effort and government funding devoted to trace determinations; eventually, even the average taxpayer may become interested in what trace determinations are, what they cost, and how well they can be trusted.

Trace determinations, both qualitative and quantitative subdivide naturally into two classes: class 1, in which the traces are major constituents; and class 2, in which they are minor. Class 1 samples are obviously minute: Substrates and solvents do not count. Separation of the elements sought can change a class 2 determination into a class 1. In this and similar operations, the techniques of and skills needed for microchemistry should not be neglected, if only to minimize risk of contamination.

Trace determinations and x-ray emission spectrography are a natural combination when the elements sought are high enough in atomic number. Even in 1910, Moseley could write, "The prevalence of lines [in his **K** spectrum of cobalt] due to impurities [traces] suggests that this [x-ray emission spectrography] may prove a powerful method of analysis." In the General Electric Company during the 1930s, trace determinations, some in the form of Feigl's spot tests, became important in connection with the mercury boiler. By 1954, it had been shown that zinc present in microgram amounts in such spots (on filter paper or a similar substrate) could be quantitatively determined via x-ray emission spectrography. By 1955, the technique was extended to include the use of ion-exchange membranes as collectors as well as substrates; 0.001 ppm of cobalt in solution could be determined in this way. As time went on, it became clear that trace determination via x-ray emission spectrography had much in common with the theory of errors and with the determination of film thickness (see LPZW, especially pp. 306–307, 322, 347–351).

Figure 13.39 hints at most of what was learned before 1960. The figure is a calibration curve: The ordinate records $N_T - N_B$ for a series of standard

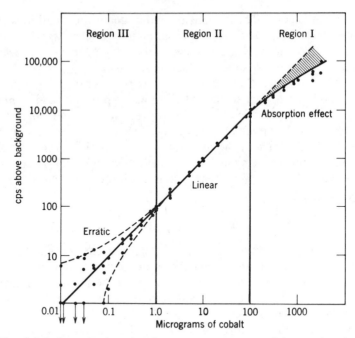

Fig. 13.39. Cobalt determinations that demonstrate occurrence of three regions in trace determinations by x-ray emission spectrography under the simplest conditions. An uneven distribution of cobalt in sample is probably a source of error in region I. Erratic distribution in region III has a statistical basis; the envelope (broken curve) is two standard deviations away from the solid line in this, the region of *qualitative* determinations. For meaning of arrows below region III, see text. (After Liebhafsky, H. A., H. G. Pfeiffer, and P. D. Zemany, *X-Ray Microscopy and X-Ray Microanalysis, Proc. 2nd Int. Symp.*, A. Engström, V. Cosslet, and H. Pattee, Eds., Elsevier, New York, 1960, p. 322.)

cobalt "spots" on filter paper; the abscissa shows against the number of micrograms of cobalt in the spots. The figure divides itself naturally into three regions with indefinite boundaries.

Regions I and II emphatically recall the determination of film thickness by x-ray emission spectrography. Region II is in the *linear region* which merges into the *exponential region* (region I) as the negative absorption effect (shaded area in Fig. 13.39) gains in importance and reduces the usefulness of trace determinations by this method.

In the *erratic region*, III, statistical considerations are needed. Here, determinations grow increasingly imprecise as the amount of cobalt decreases. As $\overline{N}_T - \overline{N}_B$ (*a difference of means*) approaches zero, $N_B > N_T$ becomes possible for individual counts, and was in fact observed for the amounts marked with arrows in Fig. 13.39. In region III, even *qualitative* conclusions can become doubtful, which means that *quantitative* results tend toward complete unreliability. The decisions that must be made have two inherent risks: the *producer risk*, of reporting cobalt absent though it is present (think of an unhappy producer forwarding goods for which he is not paid); and the *consumer risk*, of reporting cobalt present though it is absent (think of an irate consumer paying for goods he does not get).

In all determinations, one must decide at some point whether the sample actually contains the element sought. To eliminate all risk from this decision is to weaken trace determinations by restricting their applicability, for one could not then make such determinations except over concentration ranges where the element sought is *surely* known to be present. We must have recourse to statistics, and we may do this (1) by proving that trace determinations in the limits are random processes in the sense of Fig. 13.38, and (2) by defining statistically a minimum amount guaranteed detectable (MAGD). Both steps have been successfully carried through. For spots containing traces of zinc or strontium, Gaussian distributions have been found, that is, ones for which s was near s_c. For futher details, see LPZW, pp.350-351. Under carefully chosen conditions, a MAGD of 0.003 μg Mn was found for spot samples on which approximately $N_T = 5000$ counts were taken.

In 1985, trace determinations are capable of dealing with amounts far smaller than those studied above. If anything, the MAGD concept deserves greater attention as the mass of the trace decreases, because the uncertainties deriving from the background generally increase as this happens.

X. NEUTRON ABSORPTION AND SCATTERING

The interactions of neutrons with matter are complex, but for many purposes a stream of neutrons may be regarded as made up of individual "bullets" that strike "target" nuclei in any material on which they impinge (31). For each different kind of nucleus in a thin foil of thickness t held perpendicular to a uniform incident beam of neutrons, the *total cross section* (target area) σ is simply related to N (number of nuclei per cubic centimeter of the sample), n (number of neutrons per cubic centimeter in the incoming beam), v (speed of the neutrons in the beam), C (number of "collisions" per square centimeter per second between neutrons and nuclei), and t (in centimeters):

$$\sigma = \frac{C}{nvNt}$$

Formulas for absorption of neutrons by matter may be developed from this equation in a manner similar to that used earlier for absorption by x-rays. Since σ is usually a small number, it is expressed in *barns* (1 barn = 10^{-24} cm^2). One of the advantages of using neutrons is that due to their relatively weak interactions with matter (small σ), they may be used to investigate the absorption characteristics of very large and dense objects.

The cross section σ is actually not a simple quantity. It varies in an often irregular-appearing manner with the atomic number and mass of the nucleus involved in scattering, with the velocity v of the incident neutrons, and, at low velocities, with the physical state and temperature of the sample, as well as with other parameters. The (kinetic) energy of the neutrons, E (in electron volts), or their de Broglie wavelength λ (in angstroms) is usually used to characterize the neutron beam, rather than the velocity v (in centimeters per second) or (in meters per second). This can be done because of the relationship

$$\lambda = \frac{h}{mv} = \frac{h}{\sqrt{2mE}}$$

$$= \frac{0.286}{\sqrt{E}} \text{ Å}$$

where h is Planck's constant and m is the mass of the neutron. Most neutron beams used for analytical purposes are composed of *thermal neutrons*. These have a Maxwellian distribution of velocities, and are characterized by their most probable speed. For reference purposes, the total cross sections for some elements at the particular energy 0.0253 eV, which is equivalent to λ = 1.8 Å, v = 2200 m/sec, are given in Table 13.IX (43). This corresponds to the most probable speed of neutrons from a source at a temperature of 293 K.

Specific advantages of neutron absorption in applications can be envisioned from the numbers in Table 13.IX. For example, relatively large cross sections are possessed by light elements such as hydrogen, lithium, and boron compared with those of some heavy elements such as lead. This makes it possible to analyze for small amounts of strong absorbers, such as organic compounds and water, in the presence of large amounts of weak absorbers such as some of the heavier metals. These absorption characteristics are much different from, and are complementary to, those displayed by x-rays.

In *neutron radiography* (9,10), a divergent beam of neutrons is passed through a sample and attenuated by it. The intensity of the attenuated beam is recorded indirectly, usually by photographic means. In the simplest case, a thin sheet of gadolinium (Gd) is inserted between two sheets of photographic film, and the sandwich is placed perpendicular to the beam of neutrons that has been attenuated by passing through the sample. When Gd is exposed to thermal neutrons, it promptly emits γ-rays, which are converted internally to electrons, which in turn expose the x-ray film. The image on the exposed film then shows the variation in neutron absorption that occurred within the sample. (Weakly exposed film areas correspond to strongly absorbing regions in the sample.) If the sample itself is radioactive, or if it becomes so on exposure to thermal neutrons, the background radiation may be too great for this technique to be used. Instead, a thin sheet of dysprosium (Dy) or indium (In) replaces the photographic sandwich of the previous method. When these materials absorb thermal neutrons, they become radioactive. The foil thus serves as a storage device (memory) that records

Table 13.IX
Total Thermal Neutron Absorption Cross Sections of some
Elements (λ = 1.8 Å)

Element	σ (barns)	Elements	σ (barns)
H	21	Ho	100
He	1	Er	200
Li	71	Tm	130
Be	6	Yb	67
B	760	Lu	85
C	5	Hf	110
N	12	Ta	27
O	4	W	24
F	4	Re	102
Ne	2	Os	26
Na	4	Mn	15
Mg	3	Fe	13
Al	2	Co	44
Si	2	Ni	22
P	5	Cu	12
S	1	Zn	5
Cl	50	Ga	9
Ar	1	Ge	10
K	4	As	11
Ca	[7]	Se	21
Sc	50	Br	13
Ti	10	Kr	32
V	10	Rb	7
Cr	7	Sr	11
In	196	Y	9
Sn	5	Zr	7
Sb	10	Nb	6
Te	10	Mo	8
I	10	Ru	9
Xe	29	Rh	155
Cs	50	Pd	12
Ba	13	Ag	70
La	18	Cd	2,460
Ce	5	Ir	440
Pr	18	Pt	21
Nd	66	Au	108
Pm		Hg	395
Sm	5,800	Tl	13
Cu	4,600	Pb	12
Gd	49,000	Bi	9
Tb	65	Th	200
Dz	1,030	U	16

the neutron-beam intensity by the formation of unstable nuclei, which can release the information at a later time by their decay. After suitable exposure, the irradiated sheet of metal is removed from the neutron beam and placed between two sheets of photographic film. The relatively slow radioactive decay yields an image similar to that which would have been obtained if the prompt response of Gd could have been used. Other recording devices have been proposed and tested, as well as other neutron-energy ranges in the incident beam (10), but photographic methods and thermal neutrons are still used almost exclusively for radiography.

In *neutron gauging* (10) applications, the sample is placed so as to intercept the neutron beam, and a detector is positioned so as to measure the attenuation of the beam directly or to catch the scattered neutrons at some angle from the direct beam. The most useful techniques to date have involved a fast-neutron source (million-electron-volt energy range) and a thermal-neutron detector. In this type of application, advantage is taken of two effects: (1) The cross section for hydrogen is reasonably large (31); and (2) the transfer of energy in a collision is most effective when the mass of the colliding particles is the same. Thus, the overall transformation of high-energy incident neutrons into measured numbers of thermal neutrons is most efficient for hydrogen, and has been used extensively for the determination of H_2O in such diverse materials as soils and rocks (5,8,10,13) and food encased in metal containers (10). The extent of analytical application of gauging has been less than that of radiography, although in some fields gauging measurements are made routinely.

The requirement for an intense, reliable, and inexpensive source of neutrons in a desired energy range has been the major limiting factor on the development of further applications of neutron-absorption methods. The sources available include nuclear reactors and particle accelerators, which are expensive and not portable, and radioactive neutron sources, which are cheaper and may be made portable, but are of lower intensity. The techniques for detection of neutrons (10,49) vary with the energy range, although all methods require the interaction of neutrons with strong absorbers so as to produce some kind of radiation or charged particles, the effects of which are then observed. Neutron methods seem to require both specialized equipment and specialized techniques for their general application. As a result, several commercial laboratories have been formed to perform the necessary services (10).

Specific elemental analysis is not done with neutron-absorption techniques. Because of this, the methods have not been used much by analytical chemists, and the information relevant to applications only rarely finds its way into the usual analytical literature. The best sources for information regarding applications are publications that deal with materials science, nondestructive testing methods, and applications of nuclear energy. Diffraction of neutrons, which is of great importance for other purposes (51), seems to be devoid of routine analytical applications.

XI. CONCLUSION

In the early 1960s, one could safely say "like it or not, the chemistry is going out of analytical chemistry" (53). X-ray emission spectrography can fairly claim to have been the bellwether of this change. Consider this: In 1914, Moseley (42) had already used x-ray emission spectra to show that his "praseodymium" in fact contained $_{57}La$, $_{58}Ce$, $_{59}Pr$ in the proportions 50:35:15. Thirteen years later, Noyes and Bray, in one of the classic works of analytical chemistry (46), were still estimating amounts of $_{59}Pr$ from the formation of brown color in the unavoidable presence of $_{57}La$ and $_{60}Nd$, a method they knew to be defective, and of which they said: "it gives, after the analyst has had some experience with known mixtures . . . , a fair estimate in most combinations . . . [and] is perhaps as good . . . as can be hoped for." In retrospect, what chance had an uncertain brown color against a sharp x-ray line? None whatsoever, provided that instrumentation, mainly electronic, could (as it eventually did) make Moseley's method reliable and easy to use in an analytical laboratory. Now that such laboratories contain solid-state devices and computers, it may be that this retreat of chemistry is nearing an end. How much further can it go?

Chemistry and civilization have both gained enormously by what has happened. But there has also been a loss. Few analytical chemists today remember Noyes, Bray, and the trace elements; Richards, Baxter, and atomic weights; or Hillebrand, Lundell, and silicate rocks. Yet men such as these, who worked hard, brilliantly, and long to achieve what some computers now print out on demand, had strengths the computer cannot assess, and they could find answers to questions that instruments cannot formulate. We cannot afford to lose or ignore all that they had to offer. If we do, the next edition of this *Treatise* will need a different title!

REFERENCES

1. Adler, L., *X-ray Emission Spectrography in Geology*, Elsevier, New York, 1966.
2. Agarawal, M., R. B. Bennett, I. G. Stump, and J. M. D'Auria, *Anal. Chem.*, **47**, 924 (1975).
3. Anater, T. F., *U.S. Atomic Energy Commission Report WAPD-32*, U.S. Atomic Energy Commission, Washington, D.C., 1968.
4. American Society for Testing Materials, *Nuclear Methods for Measuring Soil Density and Moisture*, ASTM Special Technical Publication 293, American Society for Testing Materials, Philadelphia, 1961.
5. Andermann, G., and J. W. Kemp, *Anal. Chem.*, **30**, 1306 (1958).
6. Baird, A. K., and B. L. Henke, *Anal. Chem.*, **37**, 727 (1965).
7. Baker, A. D., and C. R. Brundle, Eds., *Electron Spectroscopy—Theory, Techniques and Applications*, Plenum, New York, 1977.
8. Ballard, L. F., *Instrumentation for Measurement of Moisture*, National Cooperative Highway Research Progress Report 138, Highway Research Board, Washington, D.C., 1973.

9. Berger, H., *Neutron Radiography*, Elsevier, Amsterdam, 1965.

10. Berger, H., Ed., *Practical Application of Neutron Radiography and Gauging*, ASTM Special Technical Publication 586, American Society for Testing Materials, Philadelphia, 1976.

11. Birks, L. S., and J. V. Gilfrich, *NRL Report 7926*, 1975.

12. Birks, L. S., and J. V. Gilfrich, *Anal. Chem.*, **48**, 273R (1976).

13. Black, C. A., Ed., *Agronomy*, Vol. 9: *Methods of Soil Analysis*, Part I, American Association of Agronomy, 1965.

14. Blount, C. W., K. E. Leyden, T. L. Thomas, and S. M. Guill, *Anal. Chem.*, **45**, 1045 (1973).

15. Brenner, I. B., L. Argor, and H. Eldad, *Appl. Spectrosc.*, **29**, 423 (1975).

16. Cairns, J. A., C. L. Desborough, and D. F. Holloway, *Nucl. Instr. Meth.*, **88**, 239 (1970).

17. Cairns, J. A., D. F. Holloway, and R. S. Nelson, in *Advances in X-ray Analysis*, Vol. 14, Plenum, New York, 1971, pp. 173–183.

18. Cairns, J. A., A. D. Marwick, R. S. Nelson, and J. S. Briggs, *Rad. Effects*, **12**, 7 (1970).

19. Camp, D. C., A. L. VanLehn, J. R. Rhodes, and A. H. Pradzynski, *X-ray Spectrom.*, **4**, 123 (1975).

20. Carlson, T. A., *Photoelectron and Auger Spectroscopy*, Plenum, New York, 1976.

21. Crecelins, E. A., *Anal. Chem.*, **47**, 2035 (1975).

22. Eddy, C. E., and T. H. Laby, *Proc. R. Soc.* (*Lond.*), **127A**, 20 (1930).

23. Elder, J. F., S. K. Perry, and P. F. Brady, *Environ. Sci. Technol.*, **9**, 1039 (1975).

24. Faessler, A., and M. Goehring, *Naturwissenschaften*, **39**, 169 (1952).

25. Fagel J. E. Jr., H. A. Liebhafsky, and P. D. Zemany, *Anal. Chem.*, **30**, 1918 (1958).

26. Felten, E. J., I. Fankuchen, and J. Steigman, *Anal. Chem.*, **31**, 1771 (1959).

27. Grubb, W. T., and P. D. Zemany, *Nature*, **176**, 221 (1955).

28. Hanser, F. A., B. Sellers, and C. A. Ziegler, in *Proceedings of ERDA Symposium on X- and γ-Ray Sources and Applications*, CONF-760539, 1976, pp. 238–241.

29. Henke, B. L., in *Advances in X-ray Analysis*, Vol. 12, Plenum, New York, 1969, p. 495.

30. Henke, B. L., in *Advances in X-ray Analysis*, Vol. 5, Plenum, New York, 1962, pp. 285–305.

31. Hughes, D. J., *Neutron Cross Sections*, Pergamon, New York, 1957.

32. Hurley, R. G., and E. W. White, *Anal. Chem.*, **46**, 2234 (1974).

33. Kevex Corp., Burlingame, California.

34. Kunzendorf, H., and H. A. Wollenberg, *Nucl. Instr. Meth.*, **87**, 197 (1970).

35. Liebhafsky, H. A., *Anal. Chem.*, **34**, 23A (1962).

36. Liebhafsky, H. A., D. H. Wilkins, and F. Bernstein, in *XIII Colloquium Spectroscopicum Internationale*, Adam Hilger, 1968, p. 58.

37. Lindstrom, B., *Acta Radiol. Suppl.*, **125**, 34 (1955).

38. Lytle, F. W., J. I. Botsford, and H. A. Heller, *U.S. Bureau of Mines Rept. Invest. No. 5378*, 1957.

39. Macdonald, G. L., *Anal. Chem.*, **50**, 135R (1978).

40. McKay, K. G., *Phys. Rev.*, **76**, 1537 (1949).

41. Miller, A. G., *Anal. Chem.*, **48**, 177 (1976).

42. Moseley, H. G. J., *Philos. Mag.*, **27**, 703 (1914).

43. Mughabghab, S. F., and D. I. Garber, *Neutron Cross Sections, Vol. I: Resonance Parameters*, U.S. Atomic Energy Commission Report BNL-325, 3rd ed., U.S. Atomic Energy Commission, Washington, D.C., 1973.

44. Murad, E., *Spectrochim. Acta, Part B*, **30,** 433 (1975).

45. Musket, R. G., and W. Bauer, *Nuclear Instr. Meth.*, **109,** 449 (1973).

46. Noyes, A. A., and W. C. Bray, *A System of Qualitative Analysis for the Rare Elements*, Macmillan, New York, 1927, p. 219.

47. Pfeiffer, H. G., and P. D. Zemany, *Nature*, **174,** 397 (1954).

48. Policarpo, A. J. P. L., M. A. F. Alves, M. Salete, S. E. P. Leire, and M. C. M. Dos Santos, *Nucl. Instr. Meth.*, **118,** 221 (1974).

49. Price, W. T., *Nuclear Radiation Detection*, McGraw-Hill, New York, 1964.

50. Siegbahn, K., *Pure Appl. Chem.*, **48,** 77 (1976).

51. Willis, B. T. M., Ed., *Chemical Applications of Thermal Neutron Scattering*, Oxford University Press, Oxford, 1973.

52. Woldseth, R., *X-ray Energy Spectrometry*, Kevex Corp., Burlingame, California, 1973.

53. Zemany, P. D., H. G. Pfeiffer, and H. A. Liebhafsky, *Anal. Chem.*, **31,** 1776 (1959).

54. Zemany, P. D., W. W. Welbon, and G. L. Gaines, Jr., *Anal. Chem.*, **30,** 299 (1958).

ELEMENTAL X-RAY MICROANALYSIS OF MATTER BY MEANS OF FOCUSED ELECTRON PROBES

By C. E. FIORI AND C. R. SWYT, *Division of Research Services, Biomedical Engineering and Instrumentation Branch, National Institutes of Health, Bethesda, Maryland*

Contents

I. INTRODUCTION

The instrumentation and methodology to be discussed in this chapter are composites, with contributions from several disciplines, primarily electron optics and x-ray spectrometry; the theory to be discussed draws from these fields. However, the subject matter, electron-beam X-ray microanalysis, properly can be called a discipline in its own right, because of the extensive theoretical and practical methodology developed over the last 35 years for obtaining accurate elemental analyses from a wide variety of specimens.

In electron-beam X-ray microanalysis, one focuses a beam of fast electrons onto a small region of a specimen surface and examines the x-ray signals that emanate due to the electron beam–specimen interaction. The diameter of the electron probe usually is in the range of 10 nm to 1 μ. Depending on the volume of interaction, the total mass analyzed can be anywhere from 10^{-11} to 10^{-16} grams. However, the limits of detection vary from 0.001 to 1%. The technique is thus microanalysis rather than trace analysis. Instruments on which electron-beam microanalysis can be performed are the scanning electron microscope equipped with an energy-dispersive detector, the analytical electron microscope, and the electron-beam X-ray microanalyzer. Present commercial instruments can form a probe less than 10 nm in diameter. Electron-beam microanalysis is thus an important tool, because it provides the investigator with the capability to measure the concentration and distribution of a large number of elements at the nanometer level. The analyst must understand the physical processes involved in the electron beam–specimen interactions if he or she is to choose from the wide range of possibilities the optimum analytical conditions for a particular problem or specimen; the quality of a chemical assay clearly depends on this choice. In this chapter we give a description of the physics germane to the analysis of electron-"thin" (<200 nm) and electron-"opaque" (>10 μ) samples. We do not have the space to cover the considerably more complicated situation of a specimen of intermediate thickness. This is not a "how to"chapter; there are few equations into which one can stick numbers from a specimen, perform some arithmetic, and obtain a chemical answer. Instead, we hope to describe the underlying physical principles in a sufficiently, but not overly, rigorous manner to convey an understanding of the techniques of X-ray microanalysis—the principal purpose of this chapter. However, the subject, by its nature, requires a mathematical treatment. We have attempted to accompany the required equations with sufficient definitions and discussion to make them understandable. The treatment is certainly not exhaustive; we generally discuss only one approach to any given aspect of the overall problem of quantitation. The interested reader should assume that others exist and can be found in the works cited in the reference list. In this chapter we will use the words "specimen" and "standard" in their respective self-explanatory contexts. We will use the word "target" when the subject under discussion applies to both.

II.   THE PRODUCTION OF X-RAYS

X-rays observed in an analytical electron column instrument arise from two types of inelastic or energy-loss interactions between the fast electrons of the beam and target atoms. In one type, a beam electron interacts strongly with a core electron and imparts sufficient energy to remove it from the atom. The ejected core electron can have any energy up to the beam energy less the characteristic shell energy. The beam electron is depreciated in energy by whatever kinetic energy the ejected electron has acquired plus the characteristic energy required to remove the latter electron. A characteristic x-ray is occasionally emitted when the ionized atom relaxes to a lower-energy state by transition of an electron from a given outer shell to the vacancy in the core shell. The x-ray is called "characteristic" because its energy equals the energy difference between the two levels involved in the transition and this difference is characteristic of the element. The energy levels of the shells vary in a discrete fashion with atomic number, so that the difference in energy between shells changes significantly even between consecutive atomic numbers. The transition that fills an inner-shell vacancy can occur from any outer shell and may occur via transitions to intermediate shells. Consequently, characteristic x-rays with different energies are generated from each element. Since the energies of the levels are well known, the photon-energy distribution that results from a statistically meaningful quantity of x-rays from each transition available to an ensemble of atoms of the same atomic number usually suffices to identify unambiguously the atom species. We call such a distribution about a specific x-ray energy a "line." The x-ray lines from all of the transitions possible for an element that terminate in a particular shell, such as the **K**, **L**, or **M** shell, are collectively called the **K**, **L**, or **M** "family." The lines that are generated from all elements by transitions between two particular levels are referred to as a "series." The relationship between the energy of a given x-ray line and atomic number was given in 1913 by Moseley (32):

$$E = a(Z - b)^2 \tag{1}$$

where a and b are constants that are different for each series, Z is the atomic number, and E is the energy of the x-ray line in the series.

Figure 14.1 plots as a function of atomic number several prominent lines that result from the most probable transitions terminating in the **K**, **L**, or **M** shells, respectively. These are the lines most often observed in x-ray spectra. The points are connected for clarity. These "most probable" x-ray lines are also the most intense lines in a given family, because for a large number of ionized atoms of a given atomic number Z, more transitions resulting in the most probable lines occur between shells than transitions between shells producing other lines in the family. The relative intensities of the various lines in a given family can range over several powers of 10. As might be expected, the most intense x-ray lines are the ones most often used as analytical signals. X-ray lines are characterized by a natural width, commonly specified by the full width at half the peak maximum (FWHM). Because usually there is instrumental broadening when an x-ray line is measured, we

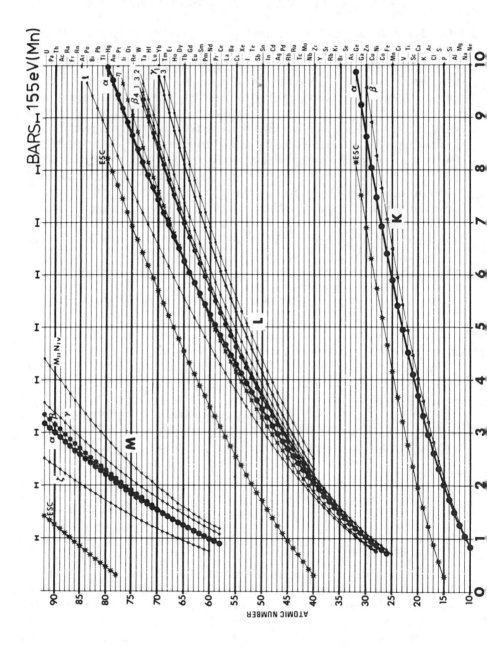

BARS—155eV(Mn)

make a distinction between the measured distribution and the line: We call the measured distribution a "peak." We specify its width also as FWHM. The difference between these distributions is very large when a solid-state energy-dispersive detector (described below) is used. For example, the natural width of the MnKα line* is 1.50 eV. This line typically is broadened to a peak of 155 eV FWHM by a solid-state detector and its associated electronics.

The second type of inelastic interaction we must consider occurs between a fast beam electron and the nucleus of a target atom. A beam electron can decelerate in the Coulomb field of an atom, which consists of the net field due to the nucleus and the core electrons. Depending on the deceleration, a photon is emitted that can have an energy ranging from near zero up to the energy of the beam electron. X-rays that emanate because of this interaction process are referred to in this chapter as *continuum x-rays*. They are also commonly called "background," "white," or "bremsstrahlung" x-rays.

By counting characteristic x-rays we obtain a measure of the number of analyte atoms present in the volume of target excited by the electron beam. By counting continuum x-rays we obtain a measure approximately proportional to the product of the average atomic number and the mass of the excited volume. If the average atomic number of the target is known, the continuum signal gives a measure of target mass.

III. X-RAY DETECTORS

There are two types of x-ray detectors commonly used in microanalysis: the energy-dispersive spectrometer (EDS) and the wavelength-dispersive spectrometer (WDS). Quantitative analysis is possible with both, and each has advantages and disadvantages. Fortunately, the detectors complement each other. The relative cost of each is approximately the same, on the order of 30–50 thousand dollars, including the cost of the ancillary electronics and display system. In this section we will explain how each detector works and the relative merits of each.

A. THE ENERGY-DISPERSIVE SPECTROMETER

The development of the energy-dispersive x-ray detector has made it possible to perform chemical microanalysis in a very straightforward manner with virtually all types of electron column instruments. It is a solid-state

---

* The symbol for an element closed up to a line name denotes the line for that element.

Fig. 14.1. Moseley plot of the energies of the x-ray emission lines observed in the energy range 0.75–10keV (14). The curves marked "ESC" are due to a prominent artifact of the Si(Li) detector.

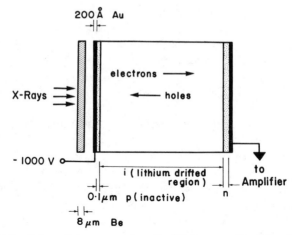

Fig. 14.2.   Schematic representation of a Si(Li) energy-dispersive detector.

device with no moving parts, is relatively inexpensive, and is easy to operate. By far the most common form of the energy-dispersive detector is the so-called lithium-diffused-silicon [or Si(Li)] detector (14,15,20).

A diagrammatic cross section of a Si(Li) detector is shown in Fig. 14.2 that demonstrates the basic operational principles of the device. X-rays originating from the analytical region of the specimen pass through a thin beryllium window into a cooled, reverse-biased p-i-n (p-type, intrinsic, n-type) lithium-drifted-silicon detector. Absorption of each individual x-ray photon leads to the ejection of photoelectrons and Auger electrons that give up most of their energy to the creation of electron–hole pairs. The components of the pairs are swept in opposite directions by the applied bias to form a charge pulse, which is then converted to a voltage pulse by a charge-sensitive amplifier. The magnitudes of the pulses produced by the detector are on the average proportional to the photon energy of each incoming x-ray. For example, if the detector captures one photon having an energy of 5 keV, the total number of electrons swept from the detector is approximately 1300, which represents a charge of $2 \cdot 10^{-16}$ C. This is an extraordinarily small charge. The subsequent circuitry must be capable of amplifying this signal by about 10 powers of 10. Cooling of the detector crystal and first stage of the preamplifier to near liquid nitrogen temperature (~80 K) is essential. Because of various physical considerations beyond the scope of this chapter and the large amplification factor required, the generated x-ray lines which have natural widths of several electron volts are detected as peaks that have widths of the order of 150 eV. Another undesirable side effect of the large amplification factor is that the voltage pulse that results has a duration of many tens of microseconds. The practical result of this is an extremely low throughput. In general, count rates attainable when measuring an analytical x-ray line with this detector are on the order of hundreds of counts per second. Consequently, analytical *precision* of less than 1% is difficult or impossible to obtain with an EDS. Analytical *accuracy* will be discussed in the sections on quantitation.

The signal is further amplified and shaped by a main amplifier, and finally passed to a multichannel analyzer (MCA), where the pulses are sorted according to their voltage. The voltage histogram can be displayed on a cathode ray tube or *X–Y* recorder. The contents of the MCA memory in most recent instruments either reside directly in a computer or can be transmitted to a computer for further processing such as unraveling of spectral overlap and peak identification.

Since the actual transducer, the Si(Li) diode, is small and is contained in a housing that occupies very little space, it is possible to place the detector very close to the specimen. This, together with the almost perfect response of the detector for most of the x-ray energies of interest, results in high quantum efficiency. This is an important property, since it makes possible the use of low-current electron probes. Low-current probes are generally smaller in diameter (<100 nm) and less likely to damage fragile specimens (e.g., biological or polymer specimens) (1,17).

B. THE WAVELENGTH-DISPERSIVE SPECTROMETER

The governing relation of the WDS is Bragg's law of x-ray diffraction in a crystalline solid (20,23),

$$n\lambda = 2d \cdot \sin \theta \tag{2}$$

where n is an integer representing the order of diffraction, λ is the wavelength of the x-ray, and θ is the angle of incidence of the x-rays on the crystal surface.

Figure 14.3 is a schematic representation of a WDS. We present here only the basic concepts; for a detailed description, see Heinrich (23). A fraction of the x-rays emitted from the analytical region of the specimen reach the diffracting crystal; those that have wavelengths that satisfy Bragg's law are diffracted and detected by a gas proportional counter. Those x-rays that do not satisfy Bragg's law either pass through the crystal or are absorbed by the crystal and converted to heat. The proportional counter is a device that converts the energy of each x-ray photon into a charge pulse. This charge pulse is converted to a voltage pulse by a preamplifier, amplified further, and presented to a single-channel analyzer that has two functions. The first is to serve as a voltage discriminator allowing only pulses produced by only one order of diffraction from the analyzing crystal (one integer n in Bragg's law) to be accepted. The second function is to normalize the selected pulses both in amplitude and width so that they are compatible with the electronic components that follow, such as counters and rate meters. Typically, the height of a pulse is 5 V and the width is 1 μsec. This width is in marked

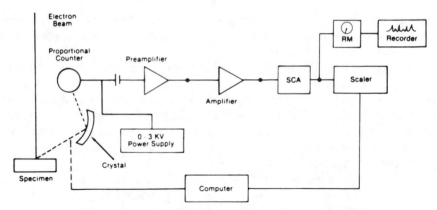

Fig. 14.3.   Schematic representation of a wavelength-dispersive detector.

contrast to the width of the pulses that are obtained from the energy-dispersive detector (which are usually 50 times wider). Consequently, a much higher throughput is possible. Since the analyzing crystal has "preselected" a very narrow energy interval, the improved throughput primarily benefits the analytical signal rather than other parts of the x-ray spectrum. Typical count rates from an analytical x-ray line might be 30,000 counts/sec. Hence, it is often possible to obtain analytical precision on the order of 0.1%.

The spectral resolving power of the WDS is quite high, so the width of a recorded x-ray peak can approach the natural line width, a few electron volts. Consequently, spectral overlap is rarely a problem with this type of spectrometer, in marked contrast to the EDS.

To summarize the analytical advantages and disadvantages of the two detectors: The EDS has high quantum efficiency, but low spectral resolving power and count-rate performance. It is the detector of choice for beam-sensitive samples such as polymers or biological material. Spectral interference, which occurs when the analytical x-ray line of interest is overlapped by an x-ray line from another element in the specimen, is often a problem and mathematical procedures must be used to reduce the effect. Analytical precision below 1% is difficult or impossible to obtain. The WDS, on the other hand, has low quantum efficiency, but high spectral resolving power and count-rate performance. Spectral overlap is rarely a problem and precision below 0.1% is often obtainable. Probe currents two to three powers of 10 higher than are required with the energy-dispersive detector must be used and beam-induced radiation damage can occur in fragile specimens.

X-rays are electromagnetic radiation and have a wavelength λ that is related to their photon energy E by the equation

$$\lambda = \frac{hc}{eE} = \frac{1.2397}{E} \tag{3}$$

where h is Planck's constant, c is the speed of light, e is the charge of the

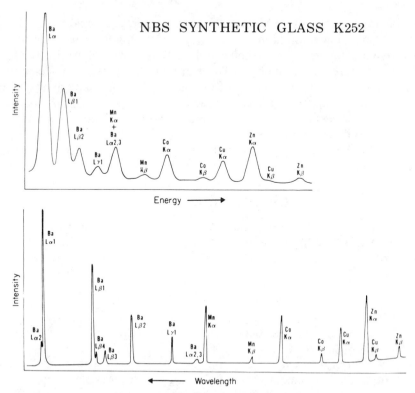

Fig. 14.4.   Partial spectra of a complex glass observed with a WDS (bottom) and an EDS (top). Note higher spectral resolving power of the WDS.

electron, E is the photon energy in kiloelectron volts, and λ is given in nanometers. Since solid-state EDSs convert the energy of an x-ray photon to a voltage pulse that is linearly proportional to the energy, it is convenient to display spectra on a scale calibrated in units of energy. Similarly, since the diffraction of x-rays in a WDS depends on the relationship between the x-ray wavelength and the interatomic distances in the diffracting crystal, it is convenient to use a scale that is linearly proportional to wavelength. The design and construction of a WDS that can generate a spectrum on a scale linearly calibrated in energy is quite complicated. Thus, it is simply for the convenience of linear calibration that we use two reciprocally related properties of a photon. Figure 14.4 is a partial energy scan of a complex glass with the two types of spectrometers.

IV.   THE THIN TARGET

Because relatively few ineleastic collisions suffered by the incident beam electrons in any target result in x-ray production, most inelastic collisions result in the loss of only very small amounts of energy, through mechanisms such as ionization of an atom by the ejection of an outer-shell electron or collective excitation of outer-shell electrons. Also, only a few electrons are multiply scattered in a thin target. Therefore, most scattered beam electrons lose very little energy in total in traversing a thin target, and almost none are absorbed. That this is the case in fact defines a target as thin. We may therefore conveniently assume that the energy of the electrons involved in collisions producing x-rays from a thin target is the beam energy.

Let us first consider in some detail the equation describing the generation of characteristic x-rays.

A.   THE CHARACTERISTIC SIGNAL

We can predict the number of characteristic x-rays generated into 4π steradians from the following relation:

$$I_{\text{ch}} = \left(N_0 \rho \, \frac{C_A}{A_A} \right) Q_A \omega_A F_A N_e \, dz \qquad (4)$$

where N_0 is Avogadro's number; ρ is the density of the target in the analyzed volume; C_A is the weight fraction of the analyte in the volume; A_A is the atomic weight of element A; Q_A is the ionization cross section for the shell of interest and has dimensions of area; ω_A, the fluorescent yield, is the probability that an x-ray will be emitted due to the ionization of a given shell; F_A is the probability of emission of the x-ray line of interest relative to the probability of emission of all the lines that can be emitted because of ionization of the same shell; N_e is the number of electrons that have irradiated the target during the measurement time; and dz is the target thickness in the same units of length as used for Q and ρ.

The quantity in parentheses in equation 4 is the number of atoms of analyte per unit volume of target, and is obtained by the following rationale. By definition the weight fraction of an element A is the mass of A divided by the total mass. The mass of A per cubic centimeter is the weight fraction of A times the density ρ in the analytical volume given in grams per cubic centimeter. We emphasize that density is a measured quantity and refers to the mass of all the atoms per unit of volume. To convert mass to number of atoms we use Avogadro's number, $N_0 = 6.02 \cdot 10^{23}$, the number of atoms in a mole of an element. Therefore, the number of atoms of element A per cubic centimeter is $(C_A \rho)(N_0/A_A)$, where A_A is the gram atomic weight of a mole of element A. The total number of atoms in a volume of 1 cm³ is then the sum over all the elements i in the volume:

$$N = N_0 \rho \sum_i \frac{C_i}{A_i} \qquad (5)$$

The basic functional form of the characteristic cross section is due to Bethe (3,4):

$$Q = 6.51 \cdot 10^{-20} \left(\frac{n_s b_s}{UV_c^2} \right) \ln(C_s U) \qquad (6)$$

where the constant is the product πe^4 (in kiloelectron volts squared times square centimeters; e is the charge of an electron), n_s is the number of electrons that populate the sth shell or subshell of interest (e.g., $n_K = 2$). V_c is the energy required to remove an electron from a given shell or subshell (the critical excitation energy), U is the overvoltage ratio V_0/V_c (where V_0 is the energy of the impinging beam electron), and b_s and C_s are constants for the sth shell or subshell. For the remainder of this chapter we use the symbol V to denote electron energy and the symbol E to denote photon energy. The "area" in square centimeters of Q is essentially the size of the K-, L-, or M-shell "target" that a beam electron must "hit" to produce an ionization of that shell. In general this area is about 100 times smaller than the area of the entire atom. Powell (36,37) reviewed a number of semiempirical cross sections for the beam-energy range commonly found in the electron microprobe (<40 keV). He recommended, for the K shell, values of $b_K = 0.9$ and $C_K = 0.65$ when the energies are expressed in keV. Powell favorably notes a modification of the Bethe formula by Fabre:

$$Q_F = 6.51 \cdot 10^{-20} \frac{n_K \ln(U)}{V_c^2 a(U + b)} \qquad (7)$$

where $a = 1.18$ and $b = 1.32$ are constants for the K shell. The range of the overvoltage ratio he recommended to be $1.5 < U < 25$. The other terms are as above. It was not intended that these cross sections be applied above about 40 keV, and it should cause no great surprise if they fail in the range of beam energies found in the analytical electron microscope.

Several forms have been suggested for the ionization cross section for the energy range 70–200 keV. This is an energy region where the effects of relativity should not be ignored, and a generally useful cross section should accommodate these effects explicitly, or at least provide empirical adjustment. Zaluzec (49,50), for example, has recommended the formulation given by Mott and Massey (33) of the relativistic cross section derived by Bethe and Fermi (5) and Williams (47):

$$Q_z = \left(6.51 \cdot 10^{-20} \frac{n_K a_K}{V_c V_0} \right) [\ln(b_K U) - \ln(1 - \beta^2) - \beta^2] \qquad (8)$$

where all energies are given in kiloelectron volts; $a_K = 0.35$(K shell); $b_K = 0.8U/[(1 - e^{-\Delta})(1 - e^{-\delta})]$, $\Delta = 1250/(V_c U^2)$, $\delta = 1/(2V_c^2)$; and $\beta = v/c = (1 - [1 + (V_0/511)]^{-2})^{1/2}$ is the relativistic correction factor. Normally

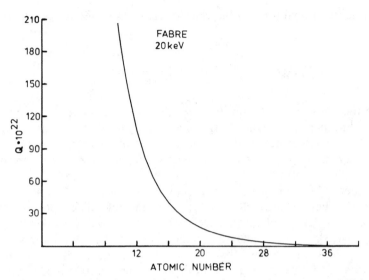

Fig. 14.5.   Plot of the Fabre characteristic cross section for a beam energy of 20 keV as a function of atomic number. Note the rapid decrease in ionization probability with increase in atomic number.

a relativistic correction would be made to the term V_0. However, quantitative data-reduction procedures, as we shall see below, utilize the characteristic cross sections in ratios and this and any error in the value of a_K will cancel.

Figure 14.5 shows the Fabre cross section plotted as a function of atomic number for a 20-keV electron beam (a practical beam energy for the scanning electron microscope and microprobe). In Figure 14.6 is plotted the Bethe–Fermi cross section (with Zaluzec's coefficients) for a 100-keV electron beam (a typical working voltage in the analytical electron microscope) as a function of atomic number. Note the change in the vertical scale between the two plots and the rapid decrease in ionization probability as the atomic number increases.

B.   THE CONTINUUM SIGNAL

The interaction of a large number of beam electrons with a thin foil of a given element produces an emitted continuum spectrum having a distribution approximately proportional to $1/E$, where E is the photon energy. The magnitude of this distribution is proportional to the number of atoms comprised by the foil. Other factors, such as beam current and measurement time, scale the magnitude linearly. The beam energy V_0 and observation angle θ, defined below, affect the overall shape of the distribution in a complex manner. We can predict the number of continuum x-rays generated into a unit steradian from the following relation:

$$I_{co} = \sum_i (N_i Q_i) N_e \qquad (9)$$

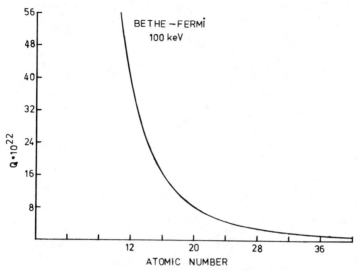

Fig. 14.6.   Plot of the Bethe–Fermi relativistic cross section for a beam energy of 100 keV as a function of atomic number. Note the rapid decrease in ionization probability with increase in atomic number.

where N_i is the number of atoms of element i, Q_i is the continuum cross section for that element and has dimensions of area, and N_e is the number of electrons that have irradiated the N_i target atoms during the measurement time. Q_i is usually assumed to be differential in photon energy and observation angle. Consequently, it is not necessary to include dE and $d\theta$ terms in equation 9. Using the fact that the total number of atoms in 1 cm³ is

$$N_0\rho \sum_i \frac{C_i}{A_i} \tag{10}$$

equation 9 can be rewritten for a thin film as

$$I_{co} = N_0\rho \sum_i \left(\frac{C_iQ_i}{A_i}\right) N_e \, dz \tag{11}$$

where dz is the target thickness in the same units of length as used for Q_i and ρ. We note the following differences between the generation of characteristic lines and that of the continuum radiation. For our purposes, the ejection of a core electron by a fast electron and occasional subsquent emission of a characteristic x-ray photon are independent events. A characteristic photon has an equal probability of being emitted in any direction (isotropy) after the ionization. Relativistic considerations apply only to the probability of ionization. The probability of continuum emission, on the other hand, is intimately related to the probability that a fast beam electron will decelerate

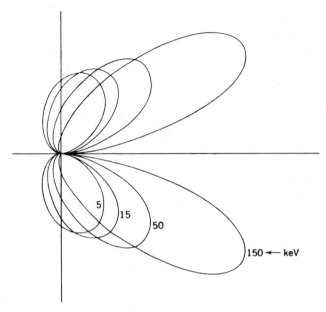

Fig. 14.7. Plot of continuum distribution, demonstrating its anisotropic property. Beam direction is left to right. Continuum anisotropy is a complicated function of not only beam energy, but also photon energy and target atomic number.

in the Coulomb field of an atom. Indeed, the emission is a direct consequence of the deceleration. Furthermore, the probability of continuum emission is directionally dependent, being biased toward the forward direction as defined by the direction of travel of the beam electrons (Fig. 14.7). We define the observation angle to be the angle between the incident-beam direction and the x-ray detector. The "angle" and magnitude of the maximum of each of the lobes shown in the Fig. 14.7 depends on the photon energy being observed, the beam energy, and the average atomic number of the target. Target tilt has no effect (other than to change the apparent thickness of the target). Because of this anisotropy, continuum radiation is usually expressed as emission into a unit steradian at a specified angle, rather than as emission uniformly into 4π steradians, as the characteristic cross section is.

We continue our discussion of the continuum cross section by deriving an expression used in a popular biological correction procedure due to Marshall and Hall (31). We first note two properties of continuum radiation:

1. The maximum energy that can be given up by an incident-beam electron is its kinetic energy, which is numerically equal to the beam voltage V_0. Consequently, there is a highest photon energy E_0 in a continuum spectrum due to beam electrons of energy V_0. This highest energy is the so-called high-energy, or Duane–Hunt, limit.

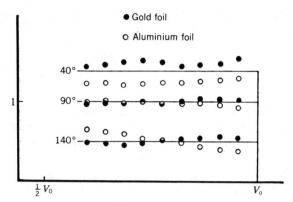

Fig. 14.8.   Plot of emitted continuum energy as function of photon energy for two elements and three observation angles. Note essentially uniform distribution up to beam energy V_0. [From Compton and Allison (9).]

2.   In a thin target both theory and experiment indicate that the amount of emitted energy in an energy interval dE is approximately uniformly distributed from near zero up to the high-energy limit (9) (Fig. 14.8).

From consideration 2, we can define the fraction of the total emitted continuum energy E_t in the interval from E to $E + dE$ as $dE/(E - 0)$. The efficiency of the generation of continuum energy in a thin film is defined to be the total continuum energy (from near zero to E_0) generated by electrons that lose an amount dV of their energy divided by the energy lost. Kirkpatrick and Wiedmann (28) have determined from the theory of Sommerfeld (see Section IV.B.1) that this efficiency of production is $2.8 \cdot 10^{-9} ZV_0$, where V_0 is the beam voltage in kilovolts and Z is the atomic number. We can thus express the total radiated continuum energy as

$$E_t = 2.8 \cdot 10^{-9} ZV_0 \, dV \qquad (12)$$

The fraction of this quantity in the energy interval dE provides us with the number of photons I_{co} of energy E. The number of photons in the vanishingly small interval dE is obtained by dividing the amount of energy in that interval by the photon energy E. Consequently, the number of photons in the energy interval dE is given by the quantity $(E_t \, dE/E_0)/E$ or

$$I_{co} = 2.8 \cdot 10^{-9} \left(\frac{Z}{E}\right) dE \, dV \qquad (13)$$

The energy loss dV of electrons traversing a thin film of a single element of thickness dz is given by the Bethe equation for the slowing down of an electron (2):

$$-dV = \frac{2\pi e^4 N_0 Z \rho}{V_0 A} \ln \left(\frac{1.166 V_0}{J} \right) dz \qquad (14)$$

where J is the mean excitation energy of an atom (20,23), 1.166 is the square root of half the base of the natural logrithm, dz is the target thickness in centimeters if ρ is given in grams per cubic centimeter, and the other terms are as defined earlier. The derivation of the equation is based on the assumption that the energy transferred to the relatively massive nucleus when an electron interacts with an atom is negligible, so that the interaction can be assumed to involve only the atomic electrons. [For a more detailed discussion of the assumptions used by Bethe in deriving the equation and of potential errors from its application see Heinrich (23), pp. 226–232.] We note that although the continuum results from interactions of beam electrons and atomic nuclei and equation 14 applies to energy assumed lost only in interactions between the beam electrons and atomic electrons, the energy loss in equation 14 can be used in equation 13. This is because equation 12 describes only the "efficiency of production"; no connection between the energy loss and the continuum production is implied. Combining these two equations, we obtain the number of continuum photons, I_{co}, with energy E generated in the interval dE by a beam electron with energy V traversing a thin film of thickness dz of one element:

$$I_{co} = \frac{220 Z^2 \rho}{A V_0 E} \ln \left(\frac{1.166 V_0}{J} \right) dz \, dE \qquad (15)$$

The energy terms are in electron volts. Equation 15 can be expressed in units of a cross section, differential in photon energy, by multiplying by the quantity $A/(N_0 \rho \, dz)$, giving

$$Q_{iu} = \frac{3.65 \cdot 10^{-22} Z^2}{V_0 E} \ln \left(\frac{1.166 V_0}{J} \right) dE \qquad (16)$$

The subscripts "iu" on Q distinguish this cross section by its principal characteristics: isotropy, and uniformity in energy distribution from property 2, above. The resulting cross section now has the required dimensions of area in square centimeters per beam electron.

There are several versions of the J factor available in the literature. We present as an example the Sternheimer formulation (2)

$$J = Z(9.76 + 58.82 Z^{-1.19})0.001 \text{ keV} \qquad (17)$$

The use of the J factor in equation 14 permits the application of the equation, originally derived from a quantum-mechanical treatment of the hydrogen atom, to higher-atomic-number elements. Equation 16 is an apparently simple expression of the continuum cross section. The simplicity is, however, a result of the approximations and assumptions used in the derivation. The cross section takes no account of the strong relativistic effects for beam energies above several kiloelectron volts. Furthermore, the equation poorly predicts the continuum distribution as a function of photon energy and, as mentioned above, it takes no account of the strong anisotropy of the continuum. The degree of anisotropy is a complicated function of beam energy, photon energy, and target atomic number. However, despite these considerable shortcomings, the cross section, as used by Marshall and Hall, is remarkably effective in many biological applications—especially when applied in the lower-energy regime of the scanning electron microscope (SEM) and microprobe. Care must be exercised, however, to use the equation only under the limited conditions they carefully specified.

We next consider two continuum cross sections that take more exact account of the physical processes involved in continuum generation. However, it must be noted that these more "exact" cross sections also depend on the use of certain simplifications, since the formulation of the cross section from which these are derived cannot be solved in closed form. There are two fundamentally different types of simplifications of particular relevance. The first is useful for applications in the energy regime of the SEM and electron-beam x-ray microanalyzer (<50 keV) and the second in the energy regime of the analytical electron microscope (≥50 keV). We shall discuss representative cross sections utilizing each (16).

1. The Coulomb Approximation: The Theory of Sommerfeld

Sommerfeld (42) developed a theory of continuum generation that assumes a pure Coulomb field about a point nucleus. Relativity, electron spin, and retardation of potential are neglected, and the incoming electron and the scattered electron are represented by plane waves that occupy all space. Other assumptions, which affect the spectral distribution at both the extreme high- and low-energy limits, are made, but have little effect on the distribution in the range of interest to us (45).

The mathematical form of the Sommerfeld theory is troublesome to evaluate. Numerical estimates for selected atomic numbers and values of photon and electron energies have been made by Kirkpatrick and Wiedmann (28) following the method of Weinstock (46). These estimates are accurate to within 2% of the fully calculated theory. The Sommerfeld theory is valid only for incident-beam energies less than several kiloelectron volts. However, it is possible to apply a relativistic correction factor and extend the range of utility to encompass the range of energies used in the electron probe x-ray microanalyzer and the analytical electron microscope. The neglect of the screening effects of core electrons can be expected to cause some error, especially in heavy atoms.

The algebraic fit, Q_{K-W}, of Kirkpatrick and Wiedmann to the Sommerfeld theory is given by

$$Q_{K-W} = [I_x(1 - \cos^2 \theta) + I_y(1 + \cos^2 \theta)]/E \qquad (18)$$

where

$$I_x = \frac{300Z^2}{V_0} \left[0.252 + a \left(\frac{E}{V_0} - 0.135 \right) - b \left(\frac{E}{V_0} - 0.135 \right)^2 \right] 1.51 \times 10^{-28}$$

$$I_y = \frac{300Z^2}{V_0} \left[-j + \frac{k}{E/V_0 + h} \right] 1.51 \times 10^{-28}$$

$$a = 1.47B - 0.507A - 0.833$$

$$b = 1.70B - 1.09A - 0.627$$

$$A = \exp \left(\frac{-0.223V_0}{300Z^2} \right) - \exp \left(\frac{-57V_0}{300Z^2} \right)$$

$$B = \exp \left(\frac{-0.0828V_0}{300Z^2} \right) - \exp \left(\frac{-84.9V_0}{300Z^2} \right)$$

$$h = \frac{-0.214y_1 + 1.21y_2 - y_3}{1.43y_1 - 2.43y_2 + y_3}$$

$$j = (1 + 2h)y_2 - 2(1 + h)y_3$$

$$k = (1 + h)(y_3 + j)$$

$$y_1 = 0.22 \left[1 - 0.39 \exp \left(\frac{-26.9V_0}{300Z^2} \right) \right]$$

$$y_2 = 0.067 + \frac{0.023}{V_0/300Z^2 + 0.75}$$

$$y_3 = -0.00259 + \frac{0.00776}{V_0/300Z^2 + 0.116}$$

and E and V are expressed in electron volts.

Q_{K-W} can be made differential in energy and angle by multiplying by $d\Omega$ and dE, the differential of solid angle subtended by the detector at the angle θ and the differential of photon energy, respectively.

Motz and Placious (34) have recommended a relativistic correction that permits the use of the Kirkpatrick–Wiedmann cross section in the energy regime of the SEM, the electron-beam X-ray microanalyzer, and to a limited degree, the analytical electron microscope:

$$Q_{K-W} = \left(I_x \left[\frac{1 - \cos^2 \theta}{(1 - \beta \cos \theta)^2} \right] + I_y \left[\frac{1 + \cos^2 \theta}{(1 - \beta \cos \theta)^2} \right] \right) \Big/ E \quad (19)$$

where β, the relativistic correction factor, is as defined for equation 8. (Note that the value of the relativistic correction made to the continuum radiation is zero for a 90° observation angle.) They derived this formulation by comparing the relativistic and nonrelativistic cross sections given by Heitler (reference 25, p. 242, equations 13 and 17). As pointed out by Motz and Placious, Heitler's cross sections are derived with free-particle wavefunctions. Consequently, the corrected cross section can be considered only a rough estimate of the exact cross section that would be obtained with relativistic Coulomb wavefunctions. The Motz–Placious correction has been used by Statham (43) for a lower-beam-energy (20 keV) Monte Carlo study in bulk specimens (photon energies of ~1–10 keV) in which continuum anisotropy is examined. A more involved relativistic correction to the Kirkpatrick–Wiedmann equation that includes approximations for screening and retardation effects is given by Robertson (reference 40, p. 41).

2. The Born Approximation

The Born theory also assumes free-particle wavefunctions. An extensive review and bibliography of continuum cross sections obtained by a number of workers using the Born approximation is given by Koch and Motz (29). As pointed out by these authors, such cross sections are available in a relatively simple analytical form for relativistic energies, with or without screening correction. For the Born approximation to be valid, it is necessary that the initial and final electron kinetic energies in an electron–nuclear interaction be large enough that the following two conditions are satisfied:

$$\frac{2\pi Z}{137\beta_0}, \frac{2\pi Z}{137\beta} \ll 1$$

where

$$\beta_0 \text{ is } \left(\frac{\text{initial velocity}}{c} \right)$$

$$\beta_0 = \frac{v_0}{c}$$

$$\beta = \frac{v}{c}$$

and β is as defined for equation 8. Despite the fact that the beam and photon energies in the analytical electron microscope often do not satisfy these conditions, the theory is useful in predicting the shape of the continuum distribution for applications in the analytical electron microscope.

A representative Born cross section due to Bethe and Heitler (reported in reference 29) and modified by Robertson (40), and differential in photon energy and angle, is given by

$$
Q_{E,\theta} = \frac{dk}{k}\frac{p}{p_0}SZ^2 d\Omega \frac{\beta_0[1 - \exp(-2\pi Z/137\beta_0)]}{\beta\ [1 - \exp(-2\pi Z/137\beta)]}
$$

$$
\times \left\{ \frac{8\sin^2\theta(2E_i^2 + 1)}{p_0^2\Delta^4} - \frac{2(5E_i^2 + 2E_fE_i + 3)}{p_0^2\Delta^2} - \frac{2(p_0^2 - k^2)}{G^2\Delta^2} \right.
$$

$$
+ \frac{4E_f}{p_0^2\Delta} + \frac{L}{pp_0}\left[\frac{4E_i\sin^2\theta(3k - p_0^2E_f)}{p_0^2\Delta^4} + \frac{4E_i^2(E_i^2 + E_f^2)}{p_0^2\Delta^2} \right.
$$

$$
+ \frac{2 - 2(7E_i^2 - 3E_iE_f + E_f^2)}{p_0^2\Delta^2} + \frac{2k(E_i^2 + E_fE_i - 1)}{p_0^2\Delta}\bigg]
$$

$$
- \frac{4\epsilon}{p\Delta} + \frac{\epsilon^G}{pG}\left[\frac{4}{\Delta^2} - \frac{6k}{\Delta} - \frac{2k(p_0^2 - k^2)}{G^2\Delta} \right]\bigg\}
$$

(20)

where

$$\beta_0 = \frac{p_0}{E_i};$$

$$\beta = \frac{p}{E_f}$$

$$k = \frac{E}{511}$$

$$dk = \frac{dE}{511}$$

$$E_i = \frac{V_0}{511}$$

$$E_f = E_i - k$$

$$p_0 = (E_i^2 - 1)^{1/2}$$

$$p = (E_f^2 - 1)^{1/2}$$

$$S = 2.31 \times 10^{-29}\ \mathrm{cm}^2$$

$$L = \ln\left[\frac{(E_fE_i - 1 + pp_0)^2}{k^2 + p_0^2\alpha^{-2}\Delta^{-2}} \right]$$

$$\Delta = E_i - p_0\cos\theta$$

$$\epsilon = \ln \left[\frac{E_f + p}{E_f - p} \right]$$

$$G^2 = p_0^2 + k^2 - 2p_0 k \cos \theta$$

$$\alpha = \frac{108}{Z^{1/3}}$$

$$\epsilon^G = \frac{1}{2} \ln \left[\frac{(G + p)^4}{4(k^2 \Delta^2 + p_0^2 \alpha^{-2})} \right]$$

the subscripts "i" and "f" stand for "initial" and "final," $d\Omega$ is the solid angle subtended by the detector at an angle θ, θ is the observation angle of the detector relative to the electron-beam axis as described above, E is the photon energy (in kiloelectron volts), and V is the energy of the beam electrons (also in kiloelectron volts). Except for the modifications of Robertson, this equation is essentially equation 2BN of reference 29, which is given in units of mc^2. We have included the conversions to beam and photon energies in keV.

The Robertson formulation includes a screening correction due to Gluckstern and Hull (18) and a correction for the Coulomb effect due to Elwert (12). Robertson reports good agreement between his experimental measurements at 100 keV beam energy and the modified Born (as formulated by Bethe and Heitler) equation. It is to be expected that this equation will be more accurate in the energy regime of the analytical electron microscope (70–400 keV) than the Kirkpatrick–Wiedmann equation 18.

We now make a short digression to note the following concerning continuum cross sections: The generalized formulation of the continuum radiation cannot be solved in closed form. As a result, all current cross sections derive from approximations of varying accuracy. However, it is possible to factor out the square of the atomic number from all continuum cross sections. Most forms of the factored cross section will retain additional terms involving Z. In the more complete cross sections the residual terms include a photon-energy dependence. However, in equation 11, which predicts the number of continuum photons from a material for a given electron flux, we note that the square of the atomic number factored from Q_i is divided by the atomic weight. This quantity, Z^2/A, is approximately equal to $Z/2$ throughout the periodic table ($A \approx 2Z$). We stress the word "approximately." As can be seen in Fig. 14.9, the function does not plot as a straight line, and indeed is not even monotonic; that is, the same value of Z^2/A occurs at more than one Z. Consequently, from equation 11, the continuum radiation plotted as a function of atomic number has the undesirable shape displayed in Fig. 14.9. The argument can be put forth that by stating the problem in terms of equation 9 we can obtain a smoothly varying continuum distribution as a function of the square of the atomic number. However, real-world specimens

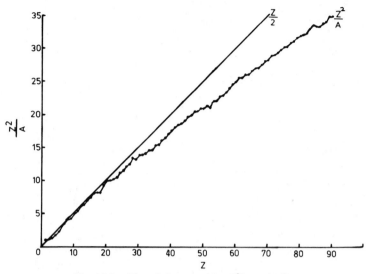

Fig. 14.9.   Plot of the quantities Z^2/A and $Z/2$.

do not have the same number of atoms in any given volume. To accommodate this reality, we use equation 11 for most practical situations.

C.   DERIVATION OF THE HALL PROCEDURE IN TERMS OF X-RAY CROSS SECTIONS

The density of the target within the analytical volume of the electron probe is generally not known unless the material being irradiated is a standard reference material. We also note that the thickness of a target at the site of impact of the electron beam is rarely a known quantity. However, it is obvious that a change in either local density or thickness causes a proportionate change in the number of atoms with which a beam electron is likely to interact in passing through a thin target. Characteristic and continuum generation are equally affected. To treat these changes more easily, density and thickness are frequently combined into a single quantity, "mass thickness," denoted by (ρdz), which has the somewhat confusing dimensions of grams per square centimeter (g/cm^3 · cm = g/cm^2). We must stress that the concept of mass thickness is useful only if the beam energy does not change by more than a very small amount in traversing the target thickness dz. A typical biological target should be less than several thousand angstroms in thickness for the above assumption to be sufficiently accurate. With this caveat, we construct the ratio of equations 4 and 11.

$$\frac{I_{ch}}{I_{co}} = \frac{(N_0 C_A/A_A)Q_A w_A F_A N_e(\rho\ dz)}{N_0 \sum_i \left(\frac{C_i Q_i}{A_i}\right) N_e(\rho dz)} \qquad (21)$$

We note that (ρdz), the mass thickness, Avogadro's number, and the number of electrons incident on the film during the measurement period all cancel in the ratio. To simplify the formula further we take advantage of the fact that for the analysis of a given element, the fluorescent yield and associated relative transition probability are constants that can be gathered into one grand constant k. Thus we have

$$\frac{I_{ch}}{I_{co}} = k \frac{(C_A Q_A / A_A)}{\displaystyle\sum_i \frac{C_i Q_i}{A_i}} \tag{22}$$

where Q_A is the characteristic cross section for a particular element A and Q_i is the continuum cross section for the ith element; the subscript i must span all the elements present in the electron beam–target interaction volume. Equation 22 is the most general formula for the method proposed by Hall.

We now put into equation 22 the cross sections used by Hall, equations 6 and 16, and collect all numerical constants into k to obtain

$$\frac{I_{ch}}{I_{co}} = k \frac{[C_A/(A_A U V_C^2)]\ln(C_S U)}{\displaystyle\sum_i \frac{C_i Z_i^2/A_i}{V_0 E} \ln\left(\frac{1.166 V_0}{J_i}\right)} dE \tag{23}$$

We can simplify this formula further by taking advantage of the fact that the operating voltage V_0 is held constant during the analysis and the energy band dE of continuum measurement, centered at the energy E, is usually not changed during the analysis. Consequently, dE, E and V_0 (and hence U) can be absorbed into k, resulting in the simpler form

$$\frac{I_{ch}}{I_{co}} = k \frac{C_A/A_A}{\displaystyle\sum_i \frac{C_i Z_i^2}{A_i} \ln\left(\frac{1.166 \cdot V_0}{J_i}\right)} \tag{24}$$

where J_i is a function of atomic number such as is given in equation 17. This is the usual representation of the Marshall–Hall (31) equation, which is a more rigorous expression of the continuum correction concept proposed by Hall et al. (22) earlier and often referred to as the Hall method or correction. Usually, the equation is used in the following manner: A measured characteristic-to-continuum ratio from a characterized material is set equal to the right side of equation 24. Since the target being irradiated is a reference standard, one presumably knows the atomic numbers Z_i, atomic weights A_i, and weight fractions C_i. Since all terms except the constant k are known, k can be calculated for the given set of experimental conditions. Next, one holds these conditions constant and measures the characteristic and continuum intensities from the specimen. To calculate a weight fraction C_A of

TABLE 14.I

			Number of Atoms				
	H	C	N	O	P	S	Z^2/A
Water	2.0			1.0			3.67
Fatty acid (oleic)	34.0	18.0		2.0			2.87
Triglyceride	107.0	60.0		12.0			2.98
Glucose	12.0	6.0		6.0			3.40
Deoxyribose	10.0	5.0		4.0			3.33
Nucleic acid cytidine monophosphate	14.0	9.0	3.0	8.0	1.0		3.78
Same without P							3.41
Protein (with S)	112.0	66.7	18.3	25.0		1.0	3.28
Protein (no S)	112.0	66.7	18.3	25.0			3.20
Araldite	8.4	5.8	0.02	1.19		0.08	3.15
Nylon	11.0	6.0	1.0	1.0			3.01
Polycarbonate	14.0	16.0		3.0			3.08

analyte from the specimen one must know the weight fraction of every one of the elements comprised by that part of the specimen irradiated by the electron beam. Crucial to the Hall procedure is the assumption that the quantity $\sum (C_i Z_i^2 / A_i)$ for the biological specimen to be analyzed is dominated by the matrix, so that the unknown contribution of the analyte, C_A to the sum may be neglected. Furthermore, the value of the sum is known or can be estimated from other information about the detail being analyzed. Values of $\sum (C_i Z_i^2 / A_i)$ for a number of typical biological materials are presented in Table 14.I [21].

The Hall procedure works well for "thin" biological specimens in the energy regime of the SEM and electron-beam x-ray microanalyzer and, with care, in the energy regime of the analytical electron microscope. The method does not work when the specimen average atomic number is expected to vary from one to another specimen and to differ from that of the standard. Consequently, most applications in materials science cannot be accommodated by the procedure. In the next section we discuss a method that does not use the continuum radiation and works for a wide variety of arbitrary specimens and standards. This latter procedure works best in the analytical electron microscope.

D.   THE CLIFF–LORIMER PROCEDURE

There are many problems, originating mainly in materials science, for which it is possible to measure a characteristic x-ray signal from each of the elements present in the target. Cliff and Lorimer [8] developed an empirical analytical procedure for this class of problem that applies when the beam voltage is about 100 keV and the x-ray spectrometer is a Si(Li) detector. The method is based on a detector-dependent set of relative elemental sensitivity factors, k_{iR}, for the element i relative to a reference element R. These factors, which must be remeasured periodically, are given by

$$k_{iR} = \left(\frac{C_i}{C_R}\right)\left(\frac{I_R}{I_i}\right) \tag{25}$$

In this equation C_i is the weight fraction of element i in a standard and I_i is the measured characteristic x-ray intensity from element i. These factors should be determined for all the elements for which analyses are to be expected. The response of a detector to x-rays of a particular element is constant over long periods of time (greater than 6 months) and the method exploits this fact. The suite of k_{iR} values is built up by measurements on well-characterized foil standards that satisfy the criterion of "thinness" defined earlier. This suite can be used, for example, to convert the measured intensity ratio of a binary that contains the reference element and one other element into a concentration ratio. By simple rearrangement of equation 25, we obtain

$$\frac{C_A}{C_R} = k_{AR}\frac{I_A}{I_R} \tag{26}$$

Indeed, if either C_A or C_R is known, then the other is known from the defining relation of a binary: $C_A + C_R = 1$.

For the case of specimens containing n elements (including the chosen reference element) we set up the n simultaneous equations

$$\frac{C_i}{C_R} = k_{iR}\frac{I_i}{I_R} \tag{27}$$

and since

$$\sum_{i=1}^{N} C_i = 1 \tag{28}$$

the concentrations C_i can be determined from the energy-dispersive spectrum of a single point in the specimen. Occasionally in a specimen there is one element, such as oxygen, that produces an unmeasurable x-ray line but has a known stoichiometry with the other elements in the specimen. In this case it is possible to modify the forced normalization to include this knowledge.

In many situations the specimen does not contain the chosen reference element. In this case we arbitrarily chose the jth element to be the reference element and use the conversion

$$k_{ij} = \frac{k_{iR}}{k_{jR}} \qquad (29)$$

to obtain the desired relative sensitivity factor.

In practical applications of the technique, complications such as failure of the thin-film criterion occur; recommended solutions to these are discussed by Goldstein (19).

V.   BULK TARGETS

For the purposes of simplifying the explanation of the relevant physics, we begin this section with an admittedly arbitrary definition of a "bulk" target. We require that the target have one planar surface that is oriented at a right angle to the impinging electron beam. All the other surfaces must be sufficiently distant that no incident electron or generated x-ray exits through any of them. The required distance from the point of impact may be as small as several micrometers to as large as several hundred, depending on the nature of the target and operating conditions. The surface onto which the electrons impinge is required to be polished (or otherwise made flat) to a smoothness of under 0.25 μ.

We will not discuss in this chapter the case of bulk specimens and standards that are inclined with respect to the electron beam, because the additional complication obscures the underlying principles we wish to emphasize. We simply caution the reader that in a practical analytical situation with such a configuration, errors in results may occur because of approximations in the quantitation scheme presented. We refer the reader to several publications that discuss inclined targets in great detail (20,23,39).

The field of electron-beam x-ray microanalysis of bulk targets is a mature one and has had an immense impact on several disciplines. The basic concepts of microprobe analysis of bulk specimens were originally presented in 1951 by R. Castaing in his Ph.D. thesis (7). A number of workers have added substantially to Castaing's work, and a large body of literature, including several textbooks, exists for the field. Our approach differs from the standard treatment to some degree, in that we separate the major ideas along the lines of inelastic and elastic scattering of electrons. However, a mature field can be written about in only so many ways, and little is to be gained by trying to be too different. The present authors have leaned heavily on the development of the major concepts presented in an early but superb book by Reed (39) that is now unfortunately out of print.

A. X-RAY GENERATION: GENERAL CONSIDERATIONS

Within the relatively extensive bulk-target volume, beam electrons undergo multiple scattering, both elastic and inelastic. Elastic scattering causes the beam electrons to deviate from their initial direction, but their speed (and therefore their kinetic energy) remains constant. If perfect elastic scattering were the only interaction process, the beam electrons would permeate the entire target and exit with the incident-beam energy from all the surfaces. However, inelastic scattering causes the beam electrons to lose energy, mostly with little change in direction. Indeed, if inelastic scattering were the only interaction between the beam and target, most of the volume of interaction would lie within a truncated cone having a top diameter equal to that of the electron beam and a bottom diameter only somewhat larger. The length of the cone would depend on, among other things, the initial energy of the electrons. Since any given beam electron is scattered many times in a purely random manner by both types of interactions, the shape and extent of the volume of interaction for a large number of electrons are determined by the balance between the elastic and inelastic scattering.

For the beam-energy range of interest to us (5–30 keV), there exist several electron "range" equations that predict the average penetration of electrons into matter. A representative range equation of particular merit is the Kanaya–Okayama expression (27),

$$R = 0.0276A \; \frac{V_0^{1.67}}{Z^{0.889}\rho} \qquad \mu \qquad (30)$$

where V_0 is the beam energy in kiloelectron volts, A is the atomic weight of the target in grams per mole, ρ is in grams per cubic centimeter, and Z is the atomic number of the target.

An "analysis" of a bulk specimen consists of counting x-ray photons characteristic of the analyte element from the specimen and comparing the measured intensity to that obtained from a standard material also containing the analyte. Except in those cases where the composition of the specimen is very close to that of the standard, there is not a simple linear relation between the specimen/standard x-ray intensity and composition ratios. Several corrections must be applied to each of the measured intensities. These are the "stopping power" correction, F_s; a "backscatter" correction, F_b; an x-ray absorption correction, F_A; a secondary characteristic x-ray fluorescence correction, F_f; and the secondary fluorescence correction needed because of the continuum radiation, F_c. They arise from multiple scattering of the electrons and the effect of the target on the intensity of the emitted x-rays. Often, the F_s and F_b corrections are combined into a single correction, F_z, which is called an "atomic number" correction since both are functions of the average atomic number of the target being irradiated. The correction for fluorescence due to the continuum is frequently neglected.

However, in biological applications it is sometimes the most important correction and must not be neglected. A formulation of the F_z, F_A, and F_f corrections such that their composite effect, F, can be obtained by multiplication of the individual corrections is called a "ZAF" correction. The form of the correction is

$$C_A = C_A^* \frac{I_A F}{I_A^* F^*} \tag{31}$$

where I_A/I_A^* is the ratio between the characteristic x-ray signals of element A from the specimen and the standard. The asterisk denotes a quantity that is related to, or obtained from, a standard. We will see that calculation of each of the various terms in F and F^* requires knowledge of the chemical composition of the target. The composition of the standard, of course, we know; the composition of the specimen is the goal of the exercise in the first place. We resolve this seemingly hopeless situation by the procedure of iteration. We estimate a weight fraction for each element present. These values are usually chosen to be the measured intensity ratios of specimen to standard. We then solve equation 31 and obtain new estimates. By repeated application of this procedure we quickly converge on a result. Depending on the iteration algorithm, convergence usually occurs in a few steps.

Let us first consider the physics that leads to the stopping-power correction. From an examination of the equation of the characteristic x-ray cross section, equation 6, we see that the probability of the generation of an x-ray of a given energy is roughly proportional to $\ln(U)/U$, where U is the instantaneous "overvoltage," V/V_C, and V, the energy of the incident electron, depends on the depth in the target. In Fig. 14.10, Q, the cross section for inner-shell ionization (the required first step for the generation of a characteristic x-ray photon), is plotted as a function of electron energy for the element silicon. We note in particular the following characteristics of the curve: The silicon **K** shell cannot be ionized by electrons lower in energy than V_C, the critical excitation energy of the **K** shell (1.84 keV). The probability of ionization is greatest for electron energies of eV_C, where e is the base of the natural logarithms (2.714 . . .). The curve is moderately constant from about 1.5 V_C to over 20 keV. Characteristic cross-section plots for other elements exhibit the same properties and general form. The shape of this curve is important, because understanding it is the first step in acquiring a qualitative appreciation for the distribution of the generation of characteristic x-rays as a function of depth below the surface. We make the assumption that only pure inelastic scattering occurs and ask the question, "What is the average energy of the electrons as a function of depth in the narrow truncated cone hypothesized above?" An approximate answer is obtained by repeated evaluation of the Bethe equation (equation 14). As an

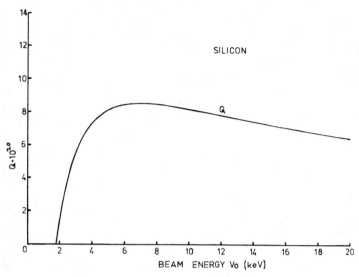

Fig. 14.10.   Plot of the characteristic (Fabre) cross section for the element silicon as a function of beam energy.

example we again use the element silicon. We put in the appropriate numbers and collect the constants to obtain:

$$\frac{-dV}{dz} = \frac{91452}{V} \ln(6.769V) \; \text{keV/cm} \qquad (32)$$

through the first layer of the target. This layer must be sufficiently thin (say 100 nm) that the differential equation 32 can be reasonably approximated by a difference equation. Evaluating equation 32 to find the energy loss for a 20-keV beam electron gives 224 eV. Consequently, the average energy of the beam electrons entering a second 100-nm layer immediately beneath the top layer is 19.7759 keV. Continuing this process and plotting the results (Fig. 14.11) reveals the average electron energy as a function of distance traveled in silicon for a beam of electrons having an initial energy of 20 keV. The curve labeled "ENERGY" in the figure is this distribution, and the left-hand ordinate gives the scale. We emphasize that this curve approximates the average energy as a function of depth. In reality, because of the statistical nature of the scattering process, there is a distribution of energies at any given depth; this effect is called "straggling."

We observe that in the first 2.5 μ of penetration the average beam electron loses only about 5 keV of energy. In the next 2.5 μ, however, it loses virtually all its remaining energy. The curve marked "Q" is the cross section for K-shell ionization as a function of depth below the surface and reveals the distribution of SiK radiation for our idealized situation of pure inelastic scattering. The right-hand ordinate gives the scale for this distribution. We note

H.   OPTICAL METHODS OF ANALYSIS

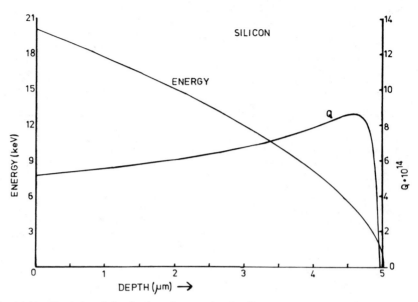

Fig. 14.11.   Dual plot of distribution of energy loss in silicon (ENERGY) for an initial electron energy of 20 keV (left ordinate) and ionization probability Q along this path (right ordinate) as a function of depth.

that as the average beam electron penetrates into the target and loses energy, the likelihood of its causing a K-shell ionization increases until its kinetic energy becomes less than V_C and it thus becomes incapable of removing a K-shell electron from a silicon atom. It is extremely important to keep in mind both here and for the remainder of the chapter the difference between generated and emitted x-rays. Here we refer strictly to the generation process within the target. Indeed, what we have done by neglecting elastic scattering is "straighten out" the electron trajectories. This has permitted us to obtain the energy loss, and hence the distribution of both energy and the probability of ionization along the path more simply, as well as to determine the total distance (range) that an average electron travels. We will consider the effect of elastic scattering only when it is required to understand the difference between generated and emitted x-ray intensities.

At this point we define the concept of stopping power S as

$$S = -\frac{dV}{(\rho dz)} \text{ keV} \cdot \text{cm}^2/\text{g} \tag{33}$$

The stopping power measures the energy lost by an electron in traversing a small mass thickness, that is, in interacting with a given number of atoms. As was shown for the example of silicon, the energy lost depends on the energy of the electron and increases as the energy of the electron decreases.

The formulation of stopping power most often used derives from the Bethe equation (equation 14). To accommodate multielement targets requires extending its applicability to more than one atomic number. We can do this by assuming that the stopping powers of the electrons of various atoms in a multielement target combine additively. Consequently, the multielement stopping-power equation becomes

$$\bar{S} = 7.84 \cdot 10^4 \frac{N_0}{E} \sum_i \frac{C_i Z_i}{A_i} \ln\left(\frac{1.166E}{J_i}\right) \tag{34}$$

where the terms are as explained above. We note that since J increases with increasing atomic number, the logarithmic term decreases with Z. Similarly, as Z increases, the quantity Z/A tends to decrease. The net effect is that the stopping power decreases with an increase in average atomic number. A main reason for defining stopping power in terms of mass thickness is that it is then dependent on the type and number of atoms rather than on the density with which they are packed. Fortunately, it has been possible to develop an entire theory of microprobe quantitation for bulk targets without requiring knowledge of the density at each region of analysis. The use of the variable (ρdz) has permitted this development.

Let us now consider how the stopping-power effect is quantitatively accounted for in the prediction of x-ray intensity. The number of ionizations, dn, of a given shell of an element along some path depends on the number of atoms encountered, or the mass thickness, and the ionization cross section for that shell of the element, as follows:

$$dn = \frac{N_0 Q}{A} (\rho dz) \tag{35}$$

The total number of ionizations caused by a single electron as it decelerates along our hypothetical straight-line path in the pure element is obtained by integrating equation 35. However, Q is a function of energy, which changes as the electron traverses the target. We could parametrically make the connection between energy and mass thickness, but it is more straightforward to change the variable of integration to energy, which gives

$$n = \int_{V_0}^{V_C} \frac{N_0 Q}{A} \left(\frac{d(\rho z)}{dV}\right) dV \tag{36}$$

The integration proceeds from the surface, where the energy is V_0, to a depth at which the electron has lost sufficient energy to be unable to ionize the given shell of the element and $V < V_C$. But since stopping power, S, is defined as $-dV/(\rho dz)$, we write

$$n = -\left(\frac{N_0}{A}\right) \int_{V_0}^{V_C} \frac{Q}{S} dV = \frac{N_0}{A} \int_{V_C}^{V_0} \frac{Q}{S} dV \tag{37}$$

where the integration limits are reversed to eliminate the minus sign. For a multielement target, equation 37 becomes

$$n = \frac{C_A N_0}{A_A} \int_{V_C}^{V_0} \frac{Q_A}{\overline{S}} \, dV \qquad (38)$$

where $\overline{S}$ is defined by equation 34 and Q_A is the ionization cross section for element A in the multielement target. Even though Q_A is not a function of the other elements, the stopping power, and therefore the energy of the electron, clearly is.

It is useful to pause here and note an implication of equation 38. A decrease in stopping power implies an increase in the number of photons generated from element A per beam electron. Since a decrease in stopping power comes about by an increase in the average atomic number, it follows that a given concentration C_A of element A generates more intensity in a higher-atomic-number matrix than in a lower-atomic-number one.

If we form the ratio of the number n of ionizations due to one electron that has decelerated in a specimen containing a weight fraction C_A of analyte A to the number n_0 of ionizations due to one electron that has decelerated in a standard containing the same weight fraction C_A of the analyte, we get

$$\frac{n}{n_0} = \frac{\dfrac{C_A N_0}{A_A} \displaystyle\int_V^{V_0} \frac{Q_A}{\overline{S}} \, dV}{\dfrac{C_A N_0}{A_A} \displaystyle\int_V^{V_0} \frac{Q_A}{\overline{S^*}} \, dV} = \frac{C_A \displaystyle\int_V^{V_0} \frac{Q_A}{\overline{S}} \, dV}{C_A \displaystyle\int_V^{V_0} \frac{Q_A}{\overline{S}} \, dV} \qquad (39)$$

As explained earlier, an asterisk denotes a quantity associated with the standard; the remaining terms are as previously defined. The characteristic ionization cross section for element A, Q_A, is identically the same inside both integrals. The stopping power, however, is not, and this difference must be taken into account. We define the stopping-power corrections as

$$F_s = \frac{1}{\displaystyle\int_{V_C}^{V_0} \frac{Q_A}{\overline{S}} \, dV} \qquad \text{and} \qquad F_s^* = \frac{1}{\displaystyle\int_{V_C}^{V_0} \frac{Q_A}{\overline{S^*}} \, dV} \qquad (40)$$

for element A in the specimen and the standard, respectively.

Before the advent of the fast laboratory digital computer, the above integrations were, for practical purposes, not calculable. Consequently, in many of the computer programs that have been written to perform the conversion of x-ray intensity ratios to elemental concentrations, the approximation $F_s = S$ is made (35). The particular energy at which to evaluate S is usually chosen to be $V = (2V_0 + V_C)/3$. Recently, however, sufficiently fast computers have become commonplace in the laboratory, so the more accurate programs that do these integrations should now be utilized (26).

Our discussion so far has not required any consideration of elastic scattering of the incident electrons, but significant corrections for quantitation in a bulk specimen arise from such scattering. There are two major effects that modify generated x-ray intensities in a bulk specimen. The first is that elastic scattering considerably reduces penetration of beam electrons into a target. The total average distance traveled by a beam electron, the "range," is unchanged, as is the probability of its causing ionization along the path, but the distribution of the characteristic x-rays as a function of depth is dramatically altered from that shown in the example of Fig. 14.11. This will be discussed in Section V.B.

The second effect of elastic scattering, the phenomenon known as "backscattering," is the scattering of electrons out of the target. Many of these electrons have sufficient energy to have caused inner-shell ionizations had they remained in the target. We will see that electron backscattering increases with increasing average atomic number. This "loss" of potentially ionizing electrons as a function of increasing average atomic number very approximately offsets the increase in x-ray generation efficiency in a higher-Z matrix.

Because of the statistical nature of elastic scattering and the complicated nature of backscattering, it is difficult to predict from first principles their effects on the generation of x-rays. Most of the information we have about backscattering comes from experimental measurement. The equations commonly used are the results of exercises in curve fitting and show little evidence of the underlying physical processes. Nevertheless, the algebraic fits have been made to a large body of data from a number of workers, so the empirical nature of the equations should be a cause not for concern, but only caution.

There are two aspects of backscattering that we need to consider: the fraction of incident electrons that is backscattered and the energy distribution for this group of electrons. We examine first the distribution as a function of atomic number of the total fraction of electrons from the beam backscattered at all energies. The traditional symbol for the backscattered fraction is η; its dependence on average atomic number is shown in Fig. 14.12 for two beam energies. Note that only a few electrons incident on a low-atomic-number target are backscattered, but nearly half are backscattered from a high-atomic-number target. The fraction is essentially independent of the energy of the electron beam for energies of 10–40 keV. Outside this range, especially at low beam energies, this is not at all the case. Darlington (10) has shown that η varies strongly with the beam energy between 500 eV and 10 keV. This is one reason among several why low-beam-energy ($<$10 keV) quantitative analysis should be undertaken with great care.

Knowledge of only the fraction of electrons backscattered is not sufficient to correct for the difference in x-ray production efficiency between a spec-

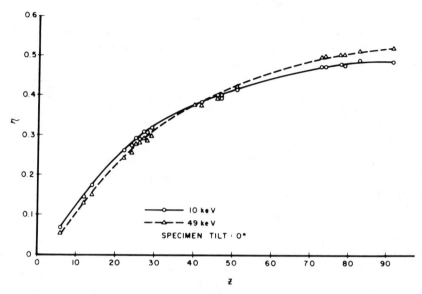

Fig. 14.12.   Variation of the backscatter coefficient as a function of average atomic number for beam energies of 10 and 49 keV. Data of Heinrich (23). [From Goldstein et al. (20).]

imen and standard due to backscatter loss. The electrons that are scattered by an atom near the specimen surface into an angle exceeding 90° can leave the target with essentially the incident-beam energy, but a backscattered electron may have any energy from near zero to the beam energy. If we know the energy of each backscattered electron as it emerged from the target, we could use equation 37 to calculate the number of additional x-rays that would have been produced by the backscattered electron. This value would be the basis of a "backscatter" correction.

Let us proceed along this line of thinking. As noted, both the absolute number and the energy distribution of the backscattered electrons as a function of atomic number are difficult to predict from first principles; we must rely on experimental measurements to obtain them. Bishop (6) made measurements of the number of electrons $d\eta$ backscattered with energies in the small interval of energy dV as a function of energy V in a number of pure elements. The data are usually displayed as a differential energy distribution, $d\eta/dW$, plotted as a function of W, where $W = V/V_0$ and V_0 is, as usual, the beam energy (W is not to be confused with $U = V_0/V_C$, the overvoltage). Figure 14-13 is a plot of some of Bishop's results, which were obtained with an electron-energy spectrometer viewing the target surface at 45°. From the definition of W, the value 1 on the horizontal axis corresponds to a backscattered electron energy equal to the beam energy, 0.5 is half that, and so on. We observe that more than half the electrons backscattered from a gold target (atomic number 79) have energies greater than 0.8 times the incident energy, with the largest number having just less than V_0. In a low-atomic-

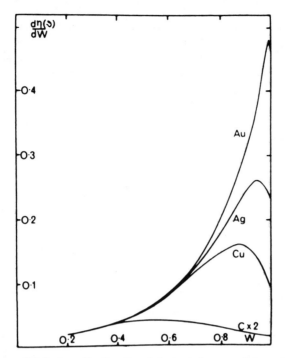

Fig. 14.13. Energy distribution of backscattered electrons for several elements. Measurements were made at a takeoff angle of 45° above the surface. $V_0 = 30$ keV. Data of Bishop (6). [From Goldstein et al. (20).]

number target, however, not only are there far fewer electrons backscattered, but the reduction occurs at the higher energies. This is because the electrons penetrate deeper into the low-atomic-number material, undergoing a greater number of inelastic (energy-loss) events before they are backscattered. For example, fewer than a third of the electrons backscattered from copper exit with greater than 0.8 times the incident energy, with the largest number having about $0.85V_0$ (see Fig. 14.13). For very low-atomic-number elements, such as carbon, relatively few electrons are backscattered at any energy and the differential distribution in energy is quite flat.

Note the difference between Figs. 14.12 and 14.13: Figure 14.12 predicts the fraction of electrons backscattered with all energies from targets of various atomic numbers. Figure 14.13 predicts the fraction of electrons backscattered with a particular energy from a target of a particular atomic number. The total area under each of the curves in this figure contains the same information about the given element as is presented in Fig. 14.12.

The fraction of x-ray intensity lost by backscattering, n_x, is defined as the number of x-rays that the backscattered electrons would produce if they remained in the specimen divided by the number of x-rays that electrons with the incident energy V_0 would produce if there were no backscattering. The latter quantity is obtained by multiplying equation 38 by the number of incident electrons. Similarly, the x-ray intensity that backscattered electrons with energy V would produce if they remained in the specimen is also given by equation 38, if it is multiplied by the number of electrons backscattered with energies in the small energy interval from V to $V + dV$. This number is just the product of the differential energy distribution function at energy V and the number of incident electrons. The total x-ray intensity lost to backscattering is thus the sum of x-ray intensities for all the backscattered-electron energies from V_C to V_0. The backscattered intensity fraction n_x can therefore be written

$$n_x = \frac{\int_{W_0}^{1} \frac{d\eta}{dW} \int_{V_C}^{V} \frac{Q}{S} \, dV \, dW}{\int_{V_C}^{V_0} \frac{Q}{S} \, dV} \tag{41}$$

where $W_0 = V_C/V_0$.

The "backscattered correction" F_b is defined as

$$F_b = \frac{1}{1 - n_x} \tag{42}$$

However, the correction is usually given in terms of the quantity R, a measure of how much the backscattering effect reduces the x-ray intensity, given as $R = 1 - n_x$. Therefore, $F_b = 1/R$. For composite materials it is common practice to use the empirically justified average (38)

$$R = \sum_i C_i R_i \tag{43}$$

where C_i, as above, is the weight fraction of the element i and i spans all the elements in the composite material.

There are several formulas for R in the literature. We present as an example the algebraic fit due to Yakowitz et al. (48) to the R values calculated by Duncumb and Reed (11) from the experimental n_0 data of Bishop (6):

$$R_{ij} = r_1 - r_2 \ln(r_3 Z_j + 25) \tag{44}$$

where

$$r_1 = 8.73 \cdot 10^{-3} U^3 - 0.1669 U^2 + 0.9662 U + 0.4523$$

$$r_2 = 2.703 \cdot 10^{-3} U^3 - 5.182 \cdot 10^{-2} U^2 + 0.302 U - 0.1836$$

$$r_3 = \frac{0.887 U^3 - 3.44 U^2 + 9.33 U - 6.43}{U^3}$$

and U is the overvoltage ratio V_0/V_C. The subscript i denotes the analyte element and the subscript j spans all the elements in the target, including the analyte. Consequently, from equation 43, the backscatter correction for the analyte element i is

$$F_b = \frac{1}{R_i} = \frac{1}{\sum_j C_j R_{ij}} \qquad (45)$$

and

$$F_b = \frac{1}{R_i^*} = \frac{1}{\sum_j C_j^* R_{ij}^*}$$

for specimen and standard, respectively. In general, as the average atomic number of the target increases, the efficiency of the generation of characteristic x-rays increases; however, there is an accompanying loss of generation efficiency because of backscattering. The two effects tend to cancel, but only approximately. A correction must be applied to account for any difference in average atomic number between specimen and standard.

B. X-RAY ABSORPTION

When an x-ray impinges on a thin film of matter, it can pass through the film without a change in energy or direction; be scattered, with or without energy loss; or be annihilated, giving up all its energy by the ejection of a core electron of an atom in the film. For photons in the energy range of interest to us (1–10 keV), only the first and last possibilities are germane to our needs. The annihilation process is called *photoelectric absorption*. It is the cause of the correction to the measured x-ray intensity that we now consider.

Consider a collimated beam of I_0 monochromatic (single-energy) x-rays incident on a single-element thin film. The number of x-rays, I, that pass through a mass thickness (ρz) is given by Beer's law:

$$I = I_0 \exp\left[-\frac{\mu}{\rho}(\rho z) \right] \qquad (46)$$

where μ is the absorption coefficient and ρ is the density.

As the beam energy is increased, the number of x-rays transmitted, I, increases, but at a decreasing rate. An abrupt decrease in I occurs at the critical excitation energy of each shell of the target material; I again increases as the energy of the incident x-rays is increased further. The relatively large decreases in the number transmitted occur at energies equal to the critical

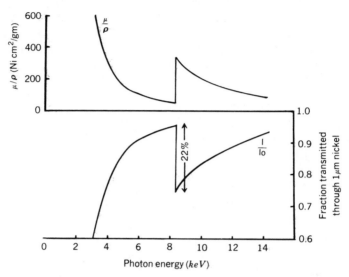

Fig. 14.14.   Curves showing mass-absorption coefficient of nickel for its own **K** radiation (top curve) and fractional transmission through 1-μ nickel foil (lower curve).

excitation energies because the x-rays then have sufficient energy to interact with the electrons of an additional shell through the process of photoelectric absorption. Although the quantity one measures is I, the quantity more often plotted as a function of photon energy is μ/ρ, the mass-absorption coefficient (which has dimensions of area per mass, usually given in square centimeters per gram). The mass-absorption coefficient provides us with a measure of the absorbing power of a target of a given atomic number for photons of given energy. It is independent of the physical state of the material. Figure 14.14 presents plots of μ/ρ and I as functions of photon energy. The upper plot is of the mass-absorption coefficient of nickel in the energy range 3–13 keV, which decreases exponentially with increasing photon energy except at the critical excitation energy, $V_C = 8.3$ keV, of nickel **K** electrons. The lower plot gives, from equation 46, the fractional transmission I/I_0 of x-rays through 1 μ of nickel (density 8.9 g/cm³). We note the abrupt 22% drop in fractional transmission at the **K**-shell binding energy. Beer's law can be applied to composite foils by summing the individual contributions of the constituent elements:

$$\frac{\mu}{\rho} = \sum_i C_i \left(\frac{\mu}{\rho}\right)_i \tag{47}$$

where C_i is the weight fraction of element i.

Beer's law predicts the fraction of incident monochromatic x-rays that emerge from a particular thickness of a material. We can use the equation to determine from I, the number of x-rays measured, the number I_0 of x-rays generated. The number of x-rays generated at any point in the target and the mass thickness that they must traverse depend on the depth beneath the surface. Because of the nonlinearity of equation 46, it is not possible to assign an "average" depth in order to calculate I_0, the total number generated in a particular matrix, from the number detected. Beer's law must be integrated over the distribution in depth of the generation of x-rays of a particular energy if one is to determine the total number of generated x-rays from the number that emerge from the specimen.

Let us review the effects of elastic scattering on the incident electron beam and on the required distribution of x-ray generation over depth. Because of the frequency of large-angle elastic-scattering events, on the average a large number of beam electrons with relatively high energy are directed back toward the surface at the full range of angles. This results in the bombardment of the surface region of the target from below as well as above and thus in an increase in x-ray generation near the surface. This increase occurs, of course, at the expense of the deeper regions of the target that the relatively energetic backscattered electrons would have reached had there been no elastic scattering. Consequently, the distribution over depth of the characteristic x-rays is considerably altered from that shown for the example of Fig. 14.11, where the effects of elastic scattering were neglected.

C. DEPTH DISTRIBUTION OF X-RAYS: THE $\phi(\rho z)$ CURVE

The distribution of the generated characteristic x-rays over depth is given by a function traditionally known as $\phi(\rho z)$. For many of the same reasons as in the case of backscattered electrons, it is not possible to derive the function from first principles. Again, the distribution is determined from experimental measurements. A number of methods are used to make the measurements, but we describe only one, which communicates the essence of most of the techiques. This is the "tracer method" devised by Castaing and used by him and other workers to determine $\phi(\rho z)$ curves. In this method, very thin layers of an element, of known thickness, are vacuum evaporated as free-standing films and placed onto thick substrates of a material of neighboring atomic number. Increasing thicknesses of this substrate material are then deposited onto the thin layers so that they effectively lie at varying depths (ρz) in the substrate material. The intensities of a characteristic line are measured from the embedded thin layers and the corresponding free-standing film. The generated intensity of the line from the thin film at the depth (ρz) in the matrix material is calculated from the measured intensity by applying Beer's law. Plotting the calculated generated characteristic x-ray intensities for the embedded thin layers after these values have been "normalized" by dividing by the measured characteristic intensity from a free-standing film of the same thickness gives the $\phi(\rho z)$ curve. Such a curve for an aluminum matrix with a magnesium tracer is shown in Fig. 14.15. Note that the ordinate is logarithmic and the abscissa is in units of

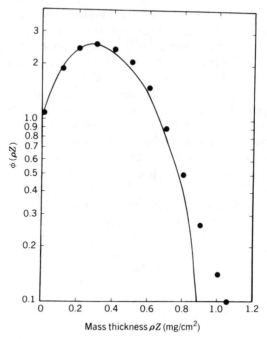

Fig. 14.15. Distribution of the production of MgKα x-rays in aluminum. Solid curve: experimental values of Castaing and Henoc. Dots: Monte Carlo calculations of Newbury and Yakowitz. Note that ordinate is logarithmic. [From Goldstein et al. (20).]

mass thickness. We see that at a "depth" of 0.3 mg/cm², the intensity from a thin film of magnesium in an aluminum matrix is almost 2.5 times that from an equally thick, but free-standing, film. The general shape of the $\phi(\rho z)$ curve is the same for all elements in any matrix. Note that the integral of $\phi(\rho z)$ from $(\rho z) = 0$ to the (ρz) value at which $\phi(\rho z)$ falls to zero is the total normalized generated characteristic intensity from the material.

Despite the fact that we do not have a theoretically derived form for $\phi(\rho z)$, we can formulate the x-ray absorption correction in terms of this function. The next section outlines the procedure.

1. The Absorption Correction

The increment of the total intensity generated in a thin layer of target of mass thickness $d(\rho z)$ at a depth (ρz) relative to that generated in a free-standing film of mass thickness $d(\rho z)$ is

$$dI_0 = \phi(\rho z)d(\rho z) \tag{48}$$

However, as described above, absorption occurs as the x-rays travel toward the detector through the target. We account for the x-ray absorption using Beer's law in the following manner. The x-ray detector is usually located above the target at some angle Ω, called the "takeoff angle." The takeoff angle is specified with respect to the plane of the target surface. X-rays generated in the layer $d(\rho z)$ must traverse a path through the target longer than (ρz) and given by csc $\Omega \cdot \rho z$ to reach the detector. Clearly, from Beer's law, the increment of intensity dI_0 reaching the detector is then

$$dI = \phi(\rho z) \exp \left[- \left(\frac{\mu}{\rho} \right) (\rho z) \csc \Omega \right] d(\rho z) \qquad (49)$$

To simplify notation, we define a quantity χ equal to (μ/ρ) csc Ω. Equation 49 can then be written

$$dI = \phi(\rho z) \exp[-\chi (\rho z)]d(\rho z) \qquad (50)$$

The total normalized intensity of radiation that actually reaches the detector is the sum from all layers of the target from the surface down:

$$I = \int_0^\infty \phi(\rho z) \exp[-\chi (\rho z)]d(\rho z) \qquad (51)$$

This is given as a fraction of the total normalized *generated* intensity I_0 by

$$f(\chi) = \frac{I}{I_0} = \frac{\int_0^\infty \phi(\rho z) \exp[-\chi (\rho z)]d(\rho z)}{\int_0^\infty \phi(\rho z)d(\rho z)} \qquad (52)$$

The fraction $f(\chi)$ is referred to as the *absorption factor*. The range of $f(\chi)$ is 0–1. The absorption factor can be considered the probability that a characteristic x-ray, generated at an angle toward the detector, will escape the target. We strive to maximize $f(\chi)$ in a real analytical situation. More will be said about this shortly. The absorption correction F_a can now be defined to be $F_a = 1/f(\chi)$.

Scattering is a statistical process and the balance between elastic and inelastic scattering of electrons is not amenable to an analytical treatment. Consequently, any attempt to model the depth distribution of the characteristic x-rays requires certain substantial assumptions, omissions, and approximations. A number of authors have engaged in this exercise; a detailed account and derivation of several of the most successful models can be found in Chapter 10 of the recent book by Heinrich (23). All of the models have one thing in common: "fudge factors" to force the model to agree as much as possible with experimental measurements. We mention here one of the models, that of Heinrich and Yakowitz (24), which makes a modest attempt at physical representation but treats the problem more as a mathematical fitting exercise. The accuracy of their model results from the large number of accurate experimental measurements to which the fits were made. Their equation for $f(\chi)$ is

$$f(\chi) = (1 + 1.2 \cdot 10^{-6}\gamma\chi)^{-2} \tag{53}$$

where γ is a parameter that predicts the electron-energy dependence of the absorption factor and is given by

$$\gamma = V_0^{1.65} - V_C^{1.65} \tag{54}$$

where V_0 and V_C are as defined earlier. The model ignores the expected dependence of the absorption factor on the target composition, and indeed the atomic weights and numbers of the target constituents do not appear in the equation. The only parameters are γ, which is the voltage term, and χ, which for a given instrumental configuration involves only the mass-absorption coefficients of the elements of the target.

Given the accuracy of any of the currently available expressions for $f(\chi)$, the analyst should attempt to maintain the condition $f(\chi) > 0.7$ for all elements being measured. If it is not possible to achieve this value, the accuracy of the analysis will deteriorate rapidly as the value of $f(\chi)$ diminishes below 0.7. To maximize $f(\chi)$, the use of low overvoltage ratios should be considered and the takeoff angle should be as high as possible.

In general, the absorption correction is by far the most significant of the corrections that must be made.

D. THE CHARACTERISTIC FLUORESCENCE CORRECTION

In Section V.B it was pointed out that photoelectric absorption of an x-ray consists of an x-ray photon giving up all of its energy to cause the ejection of an inner-shell electron of a target atom. In the process the x-ray is annihilated and the atom is ionized. As noted at the beginning of this chapter, as a first step to returning to the ground state, the ionized atom emits either an Auger electron or a characteristic x-ray. The probability of emission of a particular characteristic x-ray, such as $FeK\alpha$, is the product of the fluorescent yield and the relative transition probability, defined earlier in this chapter.

It should be clear that if we measure, for example, the $K\alpha_1$ line from element A of the target and the K shell of that element significantly absorbs the radiation of an x-ray line of element B, the measured x-ray intensity from element A will be enhanced. This enhancement, *characteristic fluorescence*, can be considerable in materials such as steel, but is almost nonexistent in biological and polymer applications. We call characteristic fluorescence "secondary" to distinguish it from the direct or "primary" excitation process of the electron beam. The target volume in which the fluorescence effect occurs is substantially larger than that in which the primary excitation occurs. Consequently, specimens in which fluorescence is strong and that have steep concentration profiles (relative to the dimensions of the primary excitation volume of the impinging electron beam) produce signals that must be interpreted carefully. Fortunately, this situation is the exception rather than the rule.

Characteristic x-ray fluorescence will always add intensity to a measured peak of an element i, and the form of the characteristic-fluorescence correction F_f manifests this:

$$F_f(i) = \frac{1 + \sum_j \dfrac{I_{ij}^f}{I_i}}{\left(1 + \sum_j \dfrac{I_{ij}^f}{I_i}\right)^*} \tag{55}$$

The quantity I_{ij} is the number of characteristic x-rays of element i produced by the secondary process of fluorescence by characteristic x-rays of element j. The quantity I_i is the number of characteristic x-rays of element i generated directly by the electron beam. We assume a unit of electron-beam current and time, but the exact choice is not relevant, since we will deal exclusively with ratios in this discussion. The total correction factor $F_f(i)$ is the sum of effects from all the characteristic lines from all the elements in the specimen that can produce secondary fluorescence of the measured x-ray line of element i.

The formulation of the ratio I_{ij}/I_i used in most correction schemes today is due to Reed (38). The derivation of Reed's equation is beyond the scope of this chapter; the interested reader is referred to Heinrich (reference 23, pp. 303–328) or Goldstein et. al. (reference 20, pp. 322–325).

E.   THE CONTINUUM-FLUORESCENCE CORRECTION

In a manner similar to the way in which characteristic fluorescence is produced, the continuum radiation ionizes analyte atoms in the target: Continuum photons that have energies greater than the critical excitation energy of the shell responsible for the emission of the measured line can cause the ejection of electrons from that shell. The continuum fluorescence cannot be neglected in a biological application if an x-ray line greater in energy than about 4 keV is used to measure an element that occurs at a concentration of several weight percent or less. An example of this analytical situation is any low-concentration assay using the $K\alpha$ line of any element from vanadium through zinc. (For elements beyond zinc in the periodic table, the analyst can utilize L or M characteristic x-ray lines that have energies less than 4 keV.) In this case, continuum fluorescence produces as least as much of the measured signal as does direct excitation by the beam electrons. In such situations it is important to use standards that are as close as possible in composition to that of the specimen. A particularly useful class of standards for biological microanalysis of bulk specimens are doped lithium borate glasses (13). For a detailed treatment of continuum fluorescence, see Heinrich (reference 23, pp. 328–338).

VI.   CONTINUUM GENERATION IN BULK TARGETS

We have seen how the continuous x-ray spectrum is generated from thin films. The discussion in the next section requires knowledge of the physics of the generation of the continuous spectrum from a bulk target; we here present an admittedly qualitative description of this process. Indeed, the generation of continuum radiation in bulk targets is not completely understood, and is currently a subject of research in several laboratories.

The continuum spectrum generated from a bulk target can be thought of as a linear combination of a succession of thin-target spectra in which the average energy of the impinging electrons that penetrate to the deeper layers is progressively reduced.

We recall from Fig. 14.8 that in a thin film the continuum intensity generated per unit energy interval is distributed approximately uniformly up to the high-energy limit. We also recall that the high-energy limit is equal to the kinetic energy of the impinging electrons. Since the first layer reduces the average energy of the beam electrons, in the second layer the high-energy limit is lower. Following this line of reasoning, we are led to the conclusion that the continuum intensity generated per unit energy interval increases linearly with decreasing photon energy. Indeed, early theoretical work by Kramers (30) demonstrated that the intensity distribution of the continuum as a function of the energy of the photons is

$$I_E \Delta E = k_E \overline{Z}(E_0 - E)\, \Delta E \qquad (56)$$

In this equation $I_E \Delta E$ is the average energy of the continuous radiation produced by one electron in the energy range from E to $E + \Delta E$, E_0 is the incident electron energy in kiloelectron volts, $\overline{Z}$ is the target average atomic number, and k_E is a constant, often called the Kramers constant, that is supposedly independent of $\overline{Z}$, E_0, and E. The "intensity" term I_E in the above discussion and equation refers to the x-ray energy. The number of photons N_E in the energy interval from E to $E + \Delta E$ per incident electron is given by

$$N_E \Delta E = k_E \overline{Z}\, \frac{E_0 - E}{E}\, \Delta E \qquad (57)$$

The difference in shape between the N_E and I_E curves is shown in Fig. 14.16.

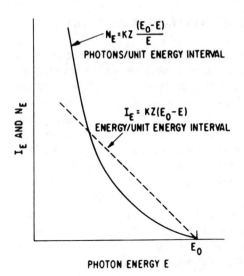

$$N_E = KZ \frac{(E_0 - E)}{E}$$

PHOTONS/UNIT ENERGY INTERVAL

$$I_E = KZ(E_0 - E)$$

ENERGY/UNIT ENERGY INTERVAL

E_0

PHOTON ENERGY E

Fig. 14.16. The theoretical shape of the continuum according to the Kramers equation (equation 57).

The x-rays are generated over a range of depths in the target, and absorption occurs during propagation out of the sample in transit to the x-ray detector, significantly altering the emitted spectrum. Since a larger fraction of the lower-energy x-rays is generated deeper in the excitation volume, and the x-rays of this fraction are heavily absorbed in transit to the surface, it would be expected that the low-energy part of the spectrum is the part most affected. This is indeed the case. The difference between the generated and the emitted spectrum is shown in Fig. 14.17.

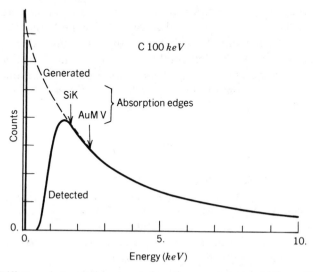

Fig. 14.17. Difference between the generated continuum spectrum and the emitted spectrum. The low-energy roll-off in the emitted spectrum is due to specimen self-absorption of low-energy x-rays.

VII. THE PEAK-TO-LOCAL CONTINUUM METHOD

A recent development that offers considerable promise for the analysis of rough bulk specimens and irregularly shaped particles and fibers that are semiopaque to the penetrating power of the electron probe is the so-called peak-to-local continuum method (41,44). This procedure is based on the observation that the number of characteristic x-rays in a given x-ray line generated per incident-beam electron is roughly proportional to the number of continuum x-rays in a narrow energy band centered at the energy of the characteristic line generated per incident-beam electron. Furthermore, the distribution in depth of the characteristic x-rays is approximately the same as that of the local continuum x-rays. Consequently, the same fraction of local continuum x-rays is absorbed in transit out of the specimen (toward the x-ray spectrometer) as of characteristic x-rays. The use of the ratio of these two signals, then, provides an analytical signal that is essentially independent of the specimen geometry or mess thickness.

VIII. CONCLUSIONS

We have attempted in this chapter to provide an overview of the essential concepts of x-ray microanalysis by means of focused electron probes, with emphasis on the physics rather than on practical methods or applications. We hope that the presentation of the material has given the reader an appreciation of the principles that must be taken into account to do successful electron-beam x-ray microanalysis. The interested reader is encouraged to go to the large body of literature on this subject for detailed treatments of specific subjects that we have touched on and for methods of applying the technique to particular samples and of quantifying results. To those workers who wish to attempt to do elemental analysis in the analytical electron microscope, we recommend caution. This is not a push-button operation; the worker must make a commitment to understand the subtleties of electron scattering and x-ray generation in this machine.

REFERENCES

1. Barnard, T., and L. Seveus, *J. Microsc.*, **112**, 281 (1977).
2. Berger, M. J., and S. M. Seltzer, *National Research Council Publication 113*, National Academy of Sciences, Washington, D.C., 1964, p. 205.
3. Bethe, H., *Ann. Phys. (Leipzig)*, **5**, 325 (1930).
4. Bethe, H. A., in *Handbook of Physics*, Vol. 24, Springer, Berlin, 1933, p. 273.
5. Bethe, H., and E. Fermi, *Z. Physik*, **77**, 296 (1932).
6. Bishop, H. E., in R. Castaing, P. Deschamps, and J. Philibert, Eds., *Proc. 4th Int. Cong. on X-ray Optics and Microanalysis*, Hermann, Paris, 1966, p. 153.
7. Castaing, R., Thesis, Univ. Paris, Paris, 1951.

8. Cliff, G., and G. W. Lorimer, in *Proc. 5th European Congress on Electron Microscopy*, Institute of Physics, Bristol, 1972, p. 140.

9. Compton, A., and S. K. Allison, *X-rays in Theory and Experiment*, 2nd ed., Van Nostrand, New York, 1935.

10. Darlington, E. H., *J. Phys. D, Appl. Phys.*, **8**, 85 (1975).

11. Duncumb, P., and S. Reed, in Heinrich, K. F. J., Ed., *Quantitative Electron Probe Microanalysis*, NBS Special Publication 298, "The Calculation of Stopping Power and Backscatter Effects in Electron Probe Microanalysis." National Bureau of Standards, Washington, D.C., 1968, p. 133.

12. Elwert, G., *Ann. Physik*, **34**, 178 (1939).

13. Fiori, C. E., and D. L. Blackburn, *J. Microsc.*, **127**, (Pt. 2), 223 (Aug 1982).

14. Fiori, C. E., and D. E. Newbury, "Artifacts Observed in Energy-Dispersive X-ray Spectrometry in the Scanning Electron Microscope," in Om Johari, Ed., *SEM/1978*, Vol. 1, SEM Inc., AMF O'Hare, Illinois, 1978, pp. 401–422.

15. Fiori, C. E., and D. E. Newbury, "Artifacts in Energy Dispersive X-ray Spectrometry in the Scanning Electron Microscope (II)," in *SEM/1980*, Vol. 2, SEM Inc., AMF O'Hare, Illinois, 1980, pp. 250–258.

16. Fiori, C. E., C. R. Swyt, and J. R. Ellis, "The Theoretical Characteristic to Continuum Ratio in the Analytical Electron Microscope," in K. F. J. Heinrich, Ed., *Microbeam Analysis—1982*, San Francisco Press, San Francisco, 1982, pp. 57–71.

17. Glasser, R. M., "Radiation Damage with Biological Specimens and Organic Materials," in J. J. Hren, J. I. Goldstein, and D. C. Joy, Eds., *Introduction to Analytical Electron Microscopy*, Plenum, New York, 1979, pp 423–436.

18. Gluckstern, R. L., and M. H. Hull, *Phys. Rev.*, **90**, 1030 (1953).

19. Goldstein, J. I., "Principles of Thin Film X-ray Analysis," in J. J. Hren, J. I. Goldstein, and D. C. Joy, Eds., *Introduction to Analytical Electron Microscopy*, Plenum, New York, 1979, pp. 83–120.

20. Goldstein, J. I., D. E. Newbury, P. Echlin, D. C. Joy, C. E. Fiori, and E. Lifshin, *Scanning Electron Microscopy and X-ray Microanalysis*, Plenum, New York, 1981.

21. Hall, T. A. "Problems of the Continuum-Normalization Method for the Quantitative Analysis of Sections of Soft Tissue," in C. Lechene and R. Warner, Eds., *Microbeam Analysis in Biology*, Academic, New York, 1979.

22. Hall, T. A., A. F. Hale, and V. R. Switsur, "Some Applications of Microprobe Analysis in Biology and Medicine," in T. D. McKinley, K. F J. Heinrich, and D. Wittry, Eds., *The Electron Microprobe*, Wiley, New York, 1966, p. 805.

23. Heinrich, K. F. J., *Electron Beam X-ray Microanalysis*, Van Nostrand Reinhold, New York, 1981.

24. Heinrich, K. F. J., and Yakowitz, H., *Anal. Chem.*, **47**, 2408 (1975).

25. Heitler, W., *The Quantum Theory of Radiation*, 3rd ed., Oxford Univ. Press, New York, 1954.

26. Henoc, J., K. F. J. Heinrich, and R. L. Myklebust, *NBS Technical Note 769*, "A Rigorous Correction Procedure for Quantitative Electron Probe Microanalysis [COR2]," National Bureau of Standards, Washington, D.C., 1973.

27. Kanaya, O., and S. Okayama, *J. Phys. D, Appl. Phys.*, **5**, 43 (1972).

28. Kirkpatrick, P., and L. Wiedmann, *Phys. Rev.*, **67**(11 and 12), (1945).

29. Koch, H. W., and J. W. Motz, *Rev. Mod. Phys.*, **31**(4), 920 (1959).

30. Kramers, H. A., *Philos. Mag.*, **46**, 836 (1923).

31. Marshall, D. J., and T. Hall, "A Method for Microanalysis of Thin Films," in R. Castaing, P. Deschamps, and J. Philibert, Eds., *X-ray Optics and Microanalysis*, Hermann, Paris, 1966, p. 374.

32. Moseley, H. G. J., *Philos. Mag.,* **26,** 1024 (1913); **27,** 703 (1914).

33. Mott, N. F., and H. S. W. Massey, *The Theory of Atomic Collisions,* 3rd ed., Oxford Univ. Press, New York, 1965.

34. Motz, J. W., and R. C. Placious, *Phys. Rev.,* **109**(2), 235 (1958).

35. Poole, D. M., and P. M. Thomas, *J. Inst. Metals,* **90,** 228 (1961–62).

36. Powell, C. J., *Rev. Mod. Phys.,* **48,** 33 (1976).

37. Powell, C. J., in K. F. J. Heinrich, D. E. Newbury, and H. Yakowitz, Eds., *Use of Monte Carlo Calculations in Electron Probe Microanalysis and Scanning Electron Microscopy,* "Evaluation of Formulas for Inner-shell Ionization Cross Sections," NBS Special Publication 460, U.S. Govt. Printing Office, Washington, D.C., 1976, pp. 97–104.

38. Reed, S. B. J., *Br. J. Appl. Phys.,* **16,** 913 (1965).

39. Reed, S., *Electron Microprobe Analysis,* Cambridge Univ. Press, Cambridge, 1975.

40. Robertson, B. W., Ph.D Thesis, Univ. of Glasgow, Glasgow, U.K., 1979.

41. Small, J. A., K. F. J. Heinrich, C. E. Fiori, R. L. Myklebust, D. E. Newbury, and M. F. Dilmore, "The Production and Characterization of Glass Fibers and Spheres for Microanalysis," in *SEM/1978,* Vol. 1, SEM Inc., AMF O'Hare, Illinois, 1978, p. 445.

42. Sommerfeld, A., *Wellenmechanik,* Frederick Ungar, New York, 1950, Chap. 7; *Ann. Physik,* **11**(5), 257 (1931).

43. Statham, P. J., *X-ray Spectr.,* **5,** 154 (1976).

44. Statham, P. J., and J. B. Pawley, "A New Method for Particle X-ray Microanalysis Based on Peak Background Measurements," in *SEM/1978,* Vol. 1, SEM Inc., AMF O'Hare, Illinois, 1978, p. 469.

45. Stephenson, S. T., in *Encyclopedia of Physics,* Vol. XXX, Springer-Verlag, Berlin, 1967, pp. 337–369.

46. Weinstock, R., *Phys. Rev.,* **61,** 585 (1942); **65,** 1 (1944).

47. Williams, E. J., *Proc. R. Soc.,* **139,** 163 (1933).

48. Yakowitz, H., R. Myklebust, and K. F. J. Heinrich, *NBS Technical Note 796,* "FRAME: On-line Correction Procedure for Quantitative Electron Probe Microanalysis," National Bureau of Standards, Washington, D.C., 1973.

49. Zaluzec, N. J., "An Analytical Electron Microscope Study of the Omega Phase Transformation in a Zirconium–Niobium Alloy," Ph.D. Thesis, Univ. of Illinois, Urbana–Champaign, 1978.

50. Zaluzec, N. J., "Quantitative X-ray Microanalysis: Instrumental Considerations and Applications to Materials Science," in J. J. Hren, J. I. Goldstein, and D. C. Joy, Eds., *Introduction to Analytical Electron Microscopy,* Plenum, New York, 1979, p. 121.

Chapter 15

CHEMICAL MICROSCOPY

By Walter C. McCrone, *McCrone Research Institute, Chicago, Illinois, and* Lucy B. McCrone, *McCrone Associates, Chicago, Illinois*

Revised by John G. Delly, *McCrone Associates, Chicago, Illinois*

Contents

I. INTRODUCTION

There are many fields of application for the microscope—bacteriology, cytology, pathology, metallography, and so forth—and microscopists in these fields are called, respectively, bacteriologists, cytologists, patholo-gists, metallographers, and so forth. There is, however, a field of interest for the microscopist not covered above that is, by far, the broadest appli-cation of all—the field of chemical microscopy. Microscopists versed in this broad field are called, of course, chemical microscopists.

To be a chemical microscopist, one must be broadly trained. One must be, first of all, a good chemist, and especially a good physical chemist. One must be a crystallographer, since most of the things one looks at, whether animal, vegetable, or mineral, are crystalline. One must have a deep-seated curiosity about the normally unseen. The dyed-in-the-wool microscopist will continually collect fibers, powders, scales, seeds, bugs, suspensions, emul-sions, chemicals, and so on for a closer look with the microscope.

At the same time a microscopist who knows only the microscope and how things look under one will wither on the vine in the industrial research laboratory. The industrial microscopist must understand the industrial re-search *modus operandi*. One's clients and fellow workers know very little about one's tools and techniques and generally leave one very much alone. One must seek out the problem, study it from a "research director's" point of view, visualize one's role, and sell it. One must know what and where to sample to obtain significant results. One must know how to translate results into answers meaningful to the "client." One must spend hours of "unnecessary" time taking pictures, writing reports, and explaining results. One may spend less than 5% of one's time on any given problem actually looking through the microscope. If one cannot visualize a colleague's prob-lem and one's part in it, no one else will. One will find that one has to do far more than chemical microscopy to complete a job. If the industrial mi-croscopist can do this, he or she will soon need an assistant; if not, the use of the microscope in the company will not flourish, nor will the microscopist.

There is almost no problem in chemistry, especially industrial chemistry, that cannot, at least to a small extent, be helped by a good microscopist; often the problem can be completely solved by microscopy.

The problem of preparing a useful short chapter on chemical microscopy involves a number of personal choices as to the importance of the various possible topics to be covered. In choosing the material to be discussed, the authors have completely eliminated such important subjects as phase mi-croscopy, electron microscopy, microchemistry, metallography, prepara-tion techniques, x-ray microscopy, and many others. This was done in the hope that those subjects the authors believe to be fundamental would be treated more completely.

A knowledge of microscope optics is, of course, essential and is carefully covered here. The analytical use of the microscope is based on the recog-nition of the morphological characteristics of finely divided materials, usu-ally crystalline in nature; hence a thorough background in crystal mor-phology and optics is particularly essential. The authors confess a prejudice

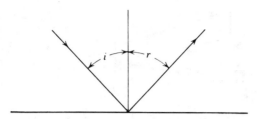

Fig. 15.1.   Specular reflection.

against instrumental refinements in microscopy when they are used as a substitute for a sound basic training in the use of the polarizing microscope.

II. OPTICS

An understanding of elementary optics is essential to the proper use of the microscope. The microscopist will find that unusual problems in illumination and photomicrography can be handled much more effectively once the underlying ideas of physical optics are understood.

A. REFLECTION

A good place to begin is with reflection at a surface or interface. Specular (or regular) reflection results when a beam of light leaves a surface at the same angle at which it reached the surface. This type of reflection occurs with highly polished smooth surfaces. It is stated more precisely as the law of reflection: The angle of incidence i is equal to the angle of reflection r (Fig. 15.1). Diffuse (or scattered) reflection results when a beam of light strikes a rough or irregular surface and different portions of the incident light are reflected from the surface at different angles. The light reflected from a piece of white paper or a ground glass is an example of diffuse reflection.

Strictly speaking, of course, all reflected light, even diffuse, has to obey Snell's law. Diffuse reflected light is made up of many rays, each specularly reflected from a tiny element of surface, and appears diffuse when the reflecting elements are very small. The terms "diffuse" and "specular," referring to reflection, describe not so much a difference in the nature of the reflection as a diference in the type of surface. A polished surface gives specular reflection; a rough surface gives diffuse reflection. It is also important to note and remember that specularly reflected light tends to be strongly polarized in the plane of the reflecting surface. This is due to the fact that those rays whose vibration directions lie closer to the plane of the reflection surface are more strongly reflected. This effect is strongest when the tangent of the angle of incidence is equal to the refractive index of the reflecting surface. The particular angle of incidence at which this occurs is called the Brewster angle.

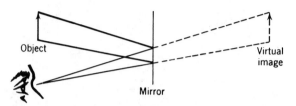

Fig. 15.2.   Image formation by a plane mirror.

B.   IMAGE FORMATION BY REFLECTION

In considering reflection by mirrors, we find (Fig. 15.2) that a plane mirror forms a virtual image, reversed right to left, but of the same size as the object, behind the mirror. The word "virtual" means that the image appears to be in a given plane, but that a ground-glass screen or a photographic film placed in that plane would show no image. The converse of a virtual image is a real image.

Spherical mirrors are either convex or concave, with the surface of the mirror representing a portion of the surface of a sphere. The center of curvature is the center of the sphere and the focus lies halfway between the center of curvature and the mirror surface.

The construction of an image by a concave mirror (Fig. 15.3) follows from two premises:

1.   A ray of light from an object that is parallel to the axis of the mirror must pass through the focus after reflection.

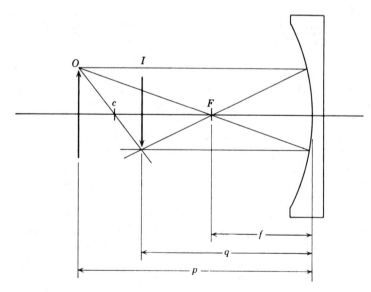

Fig. 15.3.   Image formation by a concave mirror. O, object, I, image; F, focal point; f, focal length.

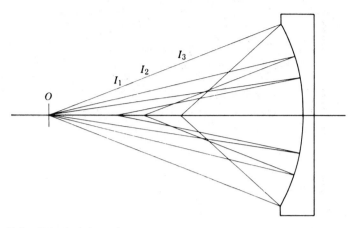

Fig. 15.4.   Spherical aberration by a spherical mirror. O, object; I_1, I_2, I_3, images.

2. A ray of light from an object that passes through the center of cur-
    vature must return along the same path.

A corollary of the first premise is:

3. A ray of light from an object that passes through the focus is reflected
    parallel to the axis of the mirror.

The image from an object can be located by using the familiar lens formula
for concave mirrors of small aperture (diameter):

$$\frac{1}{p} + \frac{1}{q} = \frac{1}{f}$$

where p is the distance from the object to the mirror, q is the distance from
the image to the mirror, and f is focal length.

C.   SPHERICAL ABERRATION

No spherical surface can be perfect in its image-forming ability. The most
serious of the imperfections, spherical aberration, occurs in spherical mirrors
of large aperture (Fig. 15.4). The rays of light making up an image point
from the outer zone of a spherical mirror do not pass through the same point
as the more central rays. This type of aberration is reduced by blocking the
outer-zone rays from the image area with a diaphragm, or by using aspheric
surfaces.

D.   REFRACTION OF LIGHT

Turning now to transparent rather than opaque reflecting materials, we
find that the most important characteristic is refraction. Refraction refers

TABLE 15.I

Refractive Indexes of Common Materials
Measured with Sodium D Light

Material	Refractive index
Vacuum	1.0000000
Air[a]	1.0002918
CO_2[a]	1.0004498
Water	1.3330
Crown glass	1.48–1.61
Rock salt	1.5443
Diamond	2.417
Lead sulfide	3.912

[a] At 18°C and 1 atm.

to the change of direction and/or velocity of light as it passes from one medium to another. The ratio of the velocity in air (or, more correctly, in a vacuum) to the velocity in the medium is called the refractive index (n). Some typical values of refractive indexes measured with monochromatic light (sodium D line) are listed in Table 15.I.

The refractive index generally increases with atomic number. Iodides, for example, have higher indexes than chlorides. A high density for a compound also results in a higher refractive index. Conjugated aromatic hydrocarbons have higher densities and higher refractive indexes than do straight-chain aliphatic hydrocarbons. Substitution in either type of molecule by bromine, iodine, or even nitro groups can cause large increases in refractive index.

1.   Temperature Coefficient of the Refractive Index

The refractive indexes of all substances vary with temperature. The degree of variation depends on the composition and the state of aggregation, that is, on whether the substance is solid or liquid. The temperature coefficient of the refractive index, $-dn/dT$, for liquids is usually more than 100 times as large as for solid substances. Most solids show coefficients of the order of 0.000001–0.00001, whereas most liquids show coefficients of 0.0003–0.0009. In general, the higher the refractive index, the greater the coefficient.

Organic liquids show widely varying coefficients depending on their compositions. Straight-chain paraffin hydrocarbons have coefficients near the minimum, 0.0003, whereas coefficients for aromatic molecules are at least 0.0004, and much higher if they are conjugated or if the molecule has nitro, bromine or iodine substituents.

As an example, consider an immersion liquid with a refractive index of 1.6600 for the sodium D line at 25° C and a temperature coefficient of 0.00042. The refractive index of the liquid will decrease by 0.00042 for every 1°C increase in temperature. In general, temperatures higher than 25°C will result in lower refractive indexes for this liquid, and temperatures lower than 25°C

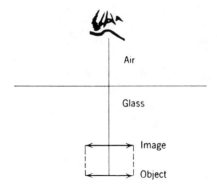

Air

Glass

Image

Object   Fig. 15.5.   Refraction of light at an interface.

will result in higher refractive indexes. At 20°C the above liquid will show a refractive index for sodium light of 1.6600 + (5 × 0.00042), or 1.6621.

2.   Refraction at Normal Incidence

Light striking a surface at normal (right-angle) incidence changes in velocity only; there is no change in direction. Refraction causes an object immersed in a medium of higher refractive index than that of air to appear closer to the surface than it actually is (Fig. 15.5). This effect may be used to determine the refractive index of a liquid with the microscope. A flat vial with a scratch on the bottom (inside) is placed on the stage. The microscope is focused on the scratch and the fine-adjustment-micrometer reading is noted. A small amount of the unknown liquid is added, the scratched is again brought into focus, and the new micrometer reading is taken. Finally, the microscope is refocused until the liquid surface is in sharp focus, and the micrometer reading is taken again. The refractive index may then be calculated from the simplified equation

$$\text{refractive index} = \frac{\text{actual depth}}{\text{apparent depth}}$$

Notice that this equation also tells us that if we try to determine the depth or thickness of a given transparent crystal by focusing on the lower and upper surfaces of that crystal, taking the difference in fine-adjustment drum reading, and multiplying by the micrometers-per-division value of the drum, we will *not* have the correct thickness. We will have the apparent thickness only; not until we multiply the apparent thickness by the refractive index of the specimen will we obtain the actual thickness.

3.   Refraction at Other Than Normal Incidence

Light striking a surface at any angle other than normal incidence changes in direction as well as velocity (Fig. 15.6). A light ray traveling from a region of lesser to greater index is bent *toward* a normal to the interface; a ray traveling from a region of greater to lesser index is bent *away* from the

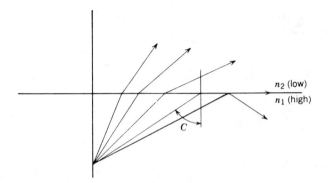

Fig. 15.6.   Reflection at the critical angle (*C*).

normal. In 1621, Snell discovered that the sine of the angle of incidence *i* divided by the sine of the angle of refraction *r* equals the refractive index of the refracting material (assuming one medium to be air):

$$n_r = \frac{\sin i}{\sin r}$$

This is now known as *Snell's law*. Recognizing that the change in angle depends on the refractive indexes of the two substances forming the interface, we can write the expression in a more general form:

$$\frac{n_r}{n_i} = \frac{\sin i}{\sin r}$$

where n_i and n_r are the refractive indexes of the incident and refracting media, respectively.

When the above situation is reversed and a ray of light from a medium of high refractive index passes through the interface to a medium of lower index, the ray is refracted until a critical angle is reached beyond which all of the light is reflected from the interface (Fig. 15.6). This critical angle, *C*, has the following relationship to the refractive indexes of the two media:

$$\sin C = \frac{n_2}{n_1}$$

where $n_2 < n_1$. When the second medium is air, the formula becomes

$$\sin C = \frac{1}{n_1}$$

TABLE 15.II
Dispersion of Refractive Indexes of Several Common Materials

	Refractive index		
Material	F line (blue) 486.1 nm	D line (yellow) 589.3 nm	C line (red) 656.3 nm
Carbon disulfide	1.6523	1.6276	1.6182
Crown glass	1.5240	1.5172	1.5145
Flint glass	1.6391	1.6270	1.6221
Water	1.3372	1.3330	1.3312

E. DISPERSION

Dispersion is another very important property of transparent materials. It is the variation of a property such as refractive index with the color (or wavelength) of light. When white light passes through a glass prism, the light rays are refracted by different amounts and separated into the colors of the spectrum. This spreading of light into its component colors is due to dispersion, which, in turn, is due to the fact that the refractive indexes of transparent substances, both liquids and solids, are lower for long wavelengths than for short wavelengths.

Because of dispersion, the determination of the refractive index of a substance requires designation of the particular wavelength used. The light from the sodium lamp has a strong, closely spaced doublet with an average wavelength of 589.3 nm, called the D line, which is commonly used as a reference wavelength. Table 15.II illustrates the change of refractive index with wavelength for a few common substances.

The dispersion of a material can be defined quantitatively as

$$\text{Dispersion} = v = \frac{n_D - 1}{n_F - n_C}$$

F. LENSES

It is now time to apply the concepts discussed so far to the study of lenses. There are two classes of lenses, converging and diverging (called also convex and concave, respectively). The focal point of a converging lens is defined as the point at which a bundle of light rays parallel to the axis of the lens appears to converge after passing through the lens. The focal length of the lens is the distance from the lens to the principal focus (Fig. 15.7).

G. IMAGE FORMATION BY REFRACTION

Image formation by lenses (Fig. 15.8) follows rules analogous to those already given above for mirrors, namely:

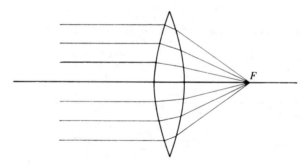

Fig. 15.7. Convergence of light at focal point (*F*) for a thin lens.

1.   Light traveling parallel to the axis of the lens will be refracted so as to pass through the focus of the lens.
2.   Light traveling through the geometrical center of the lens will be unrefracted.

The position of the image can be constructed by remembering that a light ray passing through the focus *F* will be parallel to the axis of the lens on the opposite side of the lens, and that a ray passing through the geometrical center of the lens will be unrefracted.

The magnification *M* of an image of an object produced by a lens is given by the relationship

$$M = \frac{\text{image size}}{\text{object size}} = \frac{\text{image distance}}{\text{object distance}} = \frac{q}{p}$$

where *q* is the distance from image to lens, and *p* is the distance from object to lens.

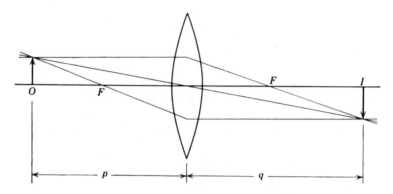

Fig. 15.8. Image formation by a convex lens. *O*, object; *I*, image; *F*, focal point.

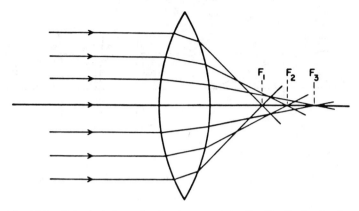

Fig. 15.9. Spherical aberration by a convex lens. F_1, F_2, F_3, focal points.

H.   ABERRATIONS OF LENSES

Lenses have several types of aberration that, unless corrected, cause loss of detail in the image. Spherical aberration (Fig. 15.9) is especially apparent in lenses having spherical surfaces. Light paths near the center of the lens focus at different points from the one at which light paths near the periphery focus. This problem can be reduced by diaphragming the outer zones of the lens or by grinding special aspherical surfaces in the lens system.

Chromatic aberration (Fig. 15.10) is caused by variation of the refractive index with wavelength (dispersion). Thus, a lens receiving white light from an object will form a blue image closest to the lens, and a red one farthest away. Achromatic lenses, employed to minimize this effect, are combinations of two or more lens elements made of materials having different dispersive powers, for example, crown and flint glasses (Table 15.II).

Fig. 15.11 shows two general methods for the minimization of chromatic aberration. In Fig. 15.11a the two lenses must be of the same material and be separated by one-half the sum of their focal lengths. In Fig. 15.11b the central, diverging lens must have a dispersion higher than that of the two identical outer lenses. Such a system could use, for example, one flint and two crown lenses. The use of monochromatic light, an obvious way to elim-

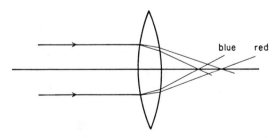

Fig. 15.10.   Chromatic aberration.

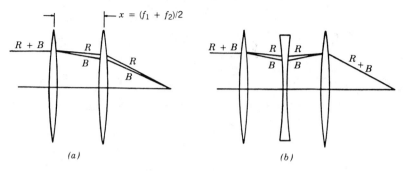

Fig. 15.11.   Achromatizing lens system. R, red light; B, blue light.

inate chromatic aberration, often leads to a loss of conformation and detail when viewing colored objects.

Other important lens aberrations are astigmatism, coma, and field curvature. Field curvature is a natural result of using lenses with curved surfaces. The image plane produced by such lenses will be curved (Fig. 15.12). This kind of image occurs in microscopy unless plano (flat-field) objectives are used. The image across the entire field of view cannot be in good focus at any one focus setting. In Fig. 15.12, either the center of the image is in focus and the edges blurred at F_2 or the edges are in focus and the center is blurred at F_1, where F_1 and F_2 represent two different focus settings. This kind of aberration is particularly noticeable and troublesome in photomicrography, although most manufacturers now produce excellent microscope objectives and lens systems corrected to give very flat fields.

I.   INTERFERENCE PHENOMENA

Interference and diffraction are phenomena that are due to the wave characteristics of light. The superposition of two light rays arriving simultaneously at a given point will give rise to interference effects, which will cause the intensity at that point to vary from dark to bright depending on the phase differences between the two light rays.

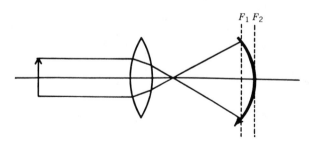

Fig. 15.12.   Field curvature. F_1, F_2, two different focus settings.

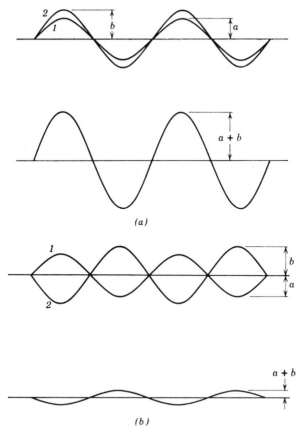

Fig. 15.13. (a) Constructive interference of light rays (1 and 2) of the same frequency but different amplitudes (a and b). The rays are in phase in the upper diagram. In the lower diagram, rays 1 and 2 interfere to give a single wave of the same frequency with an amplitude equal to the sum of the amplitudes of the two former waves. (b) Rays 1 and 2 are now 180° out of phase and interfere destructively. The resultant wave, in the bottom diagram, is of the same frequency but reduced amplitude (a is negative and is subtracted from b).

The first requirement for interference is that the light must come from a single source. The light may be split into any number of paths, but must originate from the same point (or coherent source). Two light waves from a coherent source arriving at a point in phase agreement will reinforce each other (Fig. 15.13a). Two light waves from a coherent source arriving at a point in opposite phase will cancel each other (Fig. 15.13b).

The reflection of a monochromatic light beam by a thin film results in two beams, one reflected from the top surface and one from the bottom surface. The distance traveled by the latter beam in excess of the first is twice the thickness of the film, and its equivalent air path is $2\,nt$, where n is the refractive index and t is the thickness of the film.

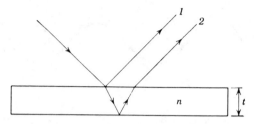

Fig. 15.14.   Interference in a thin film. The two reflected lines, 1 and 2, actually overlie each other, since the film is thin compared with the width of the incident beam. The lines are drawn as though separate in this and the following figure for ease of visualization. t, thickness.

The second beam, however, on reflection at the bottom surface undergoes a half-wavelength shift, and now the total retardation of the second beam with respect to the first is given as

$$\text{retardation} = 2\,nt + \frac{\lambda}{2}$$

where λ is the wavelength of the light beam. When the retardation is exactly an odd number of half wavelengths, destructive interference will take place, resulting in darkness. When it is zero or an even number of half wavelengths, constructive interference occurs, resulting in brightness (Fig. 15.14).

A simple interferometer can be made by partially silvering a microscope slide and coverslip (12) (Fig. 15.15). A preparation between the two partially silvered surfaces will show interference fringes when viewed with mono-chromatic light, either transmitted or from a vertical illuminator. The fringes will be close together in a wedge-shaped preparation and will reflect re-

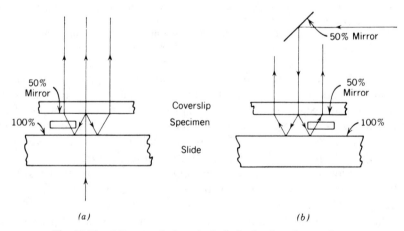

Fig. 15.15.   Microscopical method of viewing interference images.

fractive-index differences due to temperature variations, concentration differences, different solid phases, and so on. This method has been used to measure quantitatively the concentration of solute around a growing crystal (12). Each dark band represents an equivalent air thickness of an odd number of half wavelengths. Conversely, each bright band is the result of an even number of half wavelengths.

The presence of a transparent object with a refractive index different from that of the medium in the microscope field causes, if interference illumination is used, (1) a change of light intensity of the object, if the background is uniformly illuminated (parallel coverslip); or (2) a shift of the interference bands within the object, if the background consists of bands (tilted coverslip).

The relationship of the refractive indexes of the surrounding medium and the object is

$$n_s = n_m\left(1 + \frac{\theta\lambda}{360t}\right)$$

where n_s is the refractive index of the specimen, n_m is the refractive index of the surrounding medium, θ is the phase shift of the two beams in degrees; λ is the wavelength of the light; and t is the thickness of the specimen.

III.   THE SIMPLE MAGNIFYING GLASS

The compound microscope is an extension in principle of the simple magnifying glass; hence it is essential to understand fully the properties of this simple lens system.

A.   IMAGE FORMATION

The apparent size of an object is determined by the angle that is formed at the eye by the extreme rays of the object. By bringing the object closer to the eye, that angle (called the visual angle) is increased, which also increases the apparent size. However, the limit of accommodation of the eye is eventually reached, at which distance the eye can no longer focus. This limiting distance is about 25 cm. It is at this distance that the magnification of an object observed by the unaided eye is said to be 1. The eye can, of course, be focused at shorter distances, but not usually in a relaxed condition.

A positive, or converging lens, can be used to permit placing of the object closer than 25 cm to the eye (Fig. 15.16). By this means the visual angle of the object is increased (along with its apparent size), while the image of the object appears to be 25 cm from the eye, where it is best accommodated.

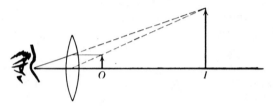

Fig. 15.16.   Virtual-image formation by a convex lens. *O*, object; *I*, image.

B.   MAGNIFICATION

The magnification *M* of a simple magnifying glass is given by

$$M = \frac{25}{f} + 1$$

where *f* is the focal length of the lens in centimeters.

Theoretically, the magnification can be increased with shorter-focal-length lenses. However, such lenses require placing the eye very close to the lens surface and have much image distortion and other optical aberrations. The practical limit for a simple magnifying glass is about $20\times$.

IV.   THE COMPOUND MICROSCOPE

To obtain higher magnifications, a compound microscope is required. In the compound microscope, two lens systems are used to form an enlarged image of an object (Fig. 15.17). This is accomplished in two steps: The first step of magnification is performed by a lens called the objective and the second step, by a lens known as the eyepiece (or ocular).

A.   THE OBJECTIVE

The objective is that lens (or lens system) that is closest to the object. Its function is to reproduce an enlarged image of the object in the bodytube of the microscope. Objectives are available in various focal lengths to give different magnifications (Table 15.III). The magnification is calculated from the focal length by dividing the latter into the tube length, usually 160 mm. The numerical aperture is a measure of the ability of the object to resolve detail. This is more fully discussed in the next section. The working distance is the free space between the objective and the coverslip and varies slightly for objectives of the same focal length, depending on the degree of correction and the manufacturer.

There are three basic classifications of objectives: achromats, fluorites, and apochromats, listed in order of increasing complexity. Achromats are good for routine work, whereas fluorites and apochromats offer additional

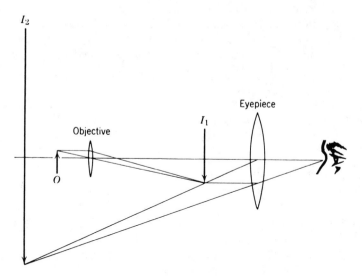

Fig. 15.17. Image formation in a compound microscope. O, object; I_1, I_2, images.

optical corrections to compensate for spherical, chromatic, and other aberrations.

Another system of objectives uses reflecting surfaces in the form of spherical concave and convex mirrors. Reflection optics, because they have no refracting elements, do not suffer from chromatic (color) aberrations, as ordinary refraction-type objectives do. Based entirely on reflection, reflecting objectives are extremely useful in the infrared and ultraviolet regions of the spectrum. They also have much longer working distances than refracting objectives.

The objective support may be of one of two kinds: an objective clutch changer or a rotating nosepiece. The objective clutch changer permits the mounting of only one objective at a time on the microscope. It has a centering arrangement, so that each objective need be centered only once with respect

TABLE 15.III
Characteristics of the Usual Microscope Objectives

Focal length (mm)	Magnification	Numerical aperture	Working distance (mm)
56	3×	0.08	40.0
32	5×	0.09	25.4
16	10×	0.25	6.8
8	20×	0.50	1.3
4	40×	0.66	0.73
1.8[a]	90×	1.25	0.12

[a] Oil immersion.

to the stage rotation. The changing of objectives with this system is some-what awkward. The rotating nosepiece allows three or four objectives to be mounted on the microscope at one time (there are some nosepieces that accept five and even six objectives). In this system, the objectives are usually noncenterable and the stage is centerable. Several manufacturers provide centerable objective mounts so that each objective on the nosepiece need be centered only once with respect to the fixed rotating stage.

B.   THE OCULAR

The eyepiece, or ocular, performs the second step in the magnification process. The eyepiece functions as a simple magnifier viewing the image formed by the objective.

There are three classes of eyepieces in common use: Huyghenian, compensating, and flat field. The Huyghenian (or Huyghens) eyepiece is designed to be used with achromats, whereas the compensating type is used with fluorite and apochromatic objectives. Flat-field eyepieces, as the name implies, are employed in photomicrography or projection and can be used with most objectives, although today it is more common to use so-called flat-field or plano objectives in which flatness is achieved in the objective itself. It is best to follow the recommendations of the manufacturer as to the proper combination of objective and eyepiece.

The usual magnifications available in oculars run from about $5 \times$ up to 25 or $30 \times$. The $5 \times$ magnification is generally too low to be of any real value, and the 25 and $30 \times$ oculars give slightly poorer imagery than do the medium-power oculars and have a very low eyepoint (exit pupil). The most useful eyepieces lie in the $10-20 \times$ magnification range.

C.   MAGNIFICATION

The magnification engraved on an objective is the initial magnification of the specimen at the intermediate image plane before subsequent ocular magnification. The total magnification of the microscope is obtained by multiplying the objective magnification by the ocular magnification. Thus, a $10 \times$ objective and $10 \times$ ocular yield a $100 \times$ microscope magnification. In some of the larger microscopes, where the mechanical tube length is much longer than the standard 160 or 170 mm, lens systems are introduced to optically shorten the long tube. Additional magnification is one of the necessary consequences of using these lens systems. This additional magnifying component, called the *tube factor*, must be taken into account when computing total microscope magnification. The tube factor for these long-bodytube microscopes is commonly 1.25. Thus, a $20 \times$ objective and $10 \times$ ocular, when used on a microscope with a tube factor of 1.25, produce a final image magnification of $20 \times 10 \times 1.25$, or $250 \times$. The magnifications engraved on objectives are usually only nominal. A "$100 \times$" objective, for example, may actually have a magnifying power of $98.8-100.6 \times$; a "$40 \times$" objective may

TABLE 15.IV

Microscope Magnifications ($\times$) Calculated for Various Objective–Eyepiece Combinations

Objective		Eyepiece					
Focal length (mm)	Magnification	5$\times$	10$\times$	15$\times$	20$\times$	25$\times$	MUM (1000 NA)
56	3	15	30	45	60	<u>75</u>	80
32	5	25	50	75	<u>100</u>	<u>125</u>	100
16	10	50	100	150	<u>200</u>	<u>250</u>	250
8	20	100	200	300	400	<u>500</u>	500
4	40	200	400	600	<u>800</u>	1000	660
1.8	90	450	900	<u>1350</u>	1800	2250	1250

MUM, maximum useful magnification; NA, numerical aperture.

actually be 39.2–40.7$\times$. In cases where *exact* magnification is required, the final image size should be determined by viewing the image of a stage-micrometer scale on the microscope with a positive (Ramsden) ocular. If the actual dimensions of the ocular graticule are known, the magnification of the stage scale by the objective can be measured directly.

A convenient working rule to assist in the proper choice of eyepieces states that the maximum useful magnification (MUM) for the microscope is 1000 times the numerical aperture (NA) of the objective. The MUM is related to resolving power in that magnification in excess of MUM gives little or no additional resolving power and results in what is termed "empty" magnification. Table 15.IV shows the results of such combinations and a comparison with the 1000 $\times$ NA rule. The underlined values show the magnification nearest to the MUM and the eyepiece required with each objective to achieve the MUM. From this table it is apparent that only the higher-power eyepieces can give full use of the resolving power of the objectives. A 10$\times$, or even a 15$\times$, eyepiece gives insufficient magnification for the eye to see detail actually resolved by the objective.

D.   FOCUSING THE MICROSCOPE

The coarse adjustment is used to position the bodytube (in some newer microscopes, the stage) roughly to bring the image into focus. The fine adjustment is used after the coarse adjustment to bring the image into perfect focus and to maintain the focus as the slide is moved across the stage.

The student of the microscope should first learn to focus in the following fashion, to prevent damage to a specimen or objective.

1.   Raise the bodytube and place the specimen on the stage.
2.   Never focus the bodytube down (or the stage up) while observing.
3.   Lower the bodytube (or raise the stage) with the coarse adjustment while carefully observing the space between the objective and slide

Fig. 15.18.   Two-lens Abbe condenser. (Drawing courtesy of Wild Heerbrugg.)

and permitting the two to come close together without touching. The lower-power objectives, of course, need not come as close to the object as the higher-power objectives.

4.   By looking through the microscope and turning the fine adjustment in such a way as to move the objective away from the specimen, bring the image into sharp focus.

The fine adjustment is usually calibrated in 1- or 2-μm steps to indicate the vertical movement of the bodytube. This feature is useful in making depth measurements, but should not be relied on for accurate measurements.

E.   THE SUBSTAGE CONDENSER

The substage comprises, as the name suggests, the components beneath the stage. These components are the condenser, the substage (aperture) iris diaphragm, the polarizer, and the mirror.

The condenser provides a converging cone of light that illuminates the specimen. The ordinary condenser is a two-lens Abbe type (Fig. 15.18), though there is also a three-lens Abbe condenser. Probably 90% of all bright-field condensers are of the two-lens type.

1.   Numerical Aperture

The condenser NA should be equal to or greater than the highest objective NA—usually about 1.25–1.32 for the 100× objective. The effective NA, and therefore the resolving power of the system, is, at best, no better than one-half the sum of the objective and condenser apertures. If the condenser NA is less than 1.00 (some, for example, are 0.95), the condenser is not intended for immersion. It may be an excellent dry condenser, but the resolving power of an NA 1.32 objective can never be realized with this condenser. Its effective NA is only 0.95 if the substage diaphragm is wide open (which is seldom the case). If the condenser is marked with an NA greater than 1.0, say 1.3, and if the top lens of the condenser is not oiled to the bottom of the slide, the effective NA will be that of the air space, 1.00.

In summary, to get full resolving power from objectives of NA greater than 1.0, one must use a condenser with an NA matching or exceeding the objective. If the objective has an NA greater than 1.00, one must oil the condenser to the bottom of the slide. Some condensers are provided with interchangeable top lenses to change the NA. Other lenses may be screwed or unscrewed in the condenser forming several different combinations to provide different NAs. Still others have a lever-operated, flip-out top lens, so that there are at least two maximum NAs.

2. Corrections

An ordinary two-lens condenser will form an image of the field diaphragm in the field of view that will be somewhat fuzzy and surrounded with color fringes. This is due to chromatic aberration in such condensers and to the fact that the rays are not all focused in the same plane. Condensers that focus light in one plane are termed *aplanatic*. There may be three, four, or even five lenses in such a condenser. In *achromatic–aplanatic* condensers, chromatic aberrations are also corrected. Such condensers may have five, six, seven, or more elements and are essential for the highest form of critical microscopy (Fig. 15.19). A two- or three-lens condenser is perfectly adequate, however, for a student microscope, for one used only occasionally, or for noncritical routine procedures.

If there is a choice of condenser mount, one should definitely consider the *centering* mount, because this is the only type that permits perfect alignment of the condenser on the microscope's optical axis.

Special-purpose condensers are usually required for dark-field, interference-transmission, interference-contrast, phase-contrast, and long-working-distance applications and strainfree condensers for use with polarized light. Quartz condensers and reflection condensers are made for use with ultraviolet light.

A mirror is provided as part of the substage when a built-in light source is not included. The mirror is about 50 mm in diameter and usually has two sides, one planar and the other concave. The mirrors are second-surface mirrors that will form at least three images of the field diaphragm. The planar side of the mirror is always used when a condenser is in position. The concave side of the mirror is used for very-low-NA objectives when a condenser is *not* used at all. First-surface mirrors are available; these may be polished

Fig. 15.19. Achromatic–aplanatic condenser. (Drawing courtesy of Wild Heerbrugg.)

stainless steel or evaporated aluminum protected with a thin silicon mon-oxide layer.

Polarizing microscopes have a graduated, rotating polarizer in the sub-stage position. If polarizers are purchased for qualitative polarized light work, they should be selected for their color. Those nearest to absolute gray and not green or brown serve best where color photomicrography is im-portant.

F. THE BODYTUBE

The microscope bodytube supports the objective at the bottom (over the object) and the ocular at the top. Some metallographs and other inverted microscopes have the positions of objective and ocular reversed. The tube length is maintained at 160 mm except in Leitz instruments, which have a 170-mm tube length (recently made Leitz microscopes have a 160-mm tube length also), and metallographs, which generally have a 215-mm tube length.

The bodytube may simply be a cylinder that supports the viewing head on the top end and the nosepiece on the lower end, or it may be a complex assembly of lenses, adjustment knobs, and slots. In polarizing microscopes, the bodytube contains the Bertrand lens, which may be centerable and/or focusable, and which may be inserted or removed from the optical axis by means of a knob at the side of the bodytube. The bodytube also contains the analyzer, which can be taken out of the light path and which may be graduated and rotating. Compensators are also introduced into the bodytube. Finally, a lens just below the compensator renders the light parallel to prevent tube length changes when the compensator is introduced. A second lens restores the light beam to its original path. All of these additions make the polarizing microscope more expensive than other microscopes, but the additional information they provide more than compensates for the increased cost.

Magnification changers may be incorporated in the bodytube. These may have zoom-type lenses or lenses providing stepped increases in magnification. Such systems include the Bausch and Lomb Dynazoom, Leitz Variotube, and Zeiss Optovar.

G. STANDS

It is important to emphasize the difference between biological or medical microscopes on the one hand and chemical, polarizing, or petrographic microscopes on the other. We must also emphasize that the latter should be used for the study of small particles.

The biological microscope is basically a stand on which a condenser, objective, and ocular, all of varying degrees of sophistication, are mounted. The polarizing microscope has, in addition, two polarizing elements, a rotating stage and a Bertrand lens. All of these extras are useful in studying crystalline substances and especially in characterizing and identifying small particles. The Bertrand lens is also very useful for observing the back focal plane of the objective in order to determine or improve the quality of illumination.

Any microscopist planning to identify small particles must use the polarizing microscope. Instruments are available for any sum from $300 to over $20,000. A good microscopist can undoubtedly do excellent work with the lowest-priced instruments. In fact, the beginning microscopist will have less difficulty doing good work with the lower-priced instruments. To do a good job with the more sophisticated polarizing microscopes requires care, study, and a basic understanding of microscope optics and illumination. The more sophisticated instruments can, in trained hands, give better and faster results.

Figs. 15.20 and 15.21 show two different versions of a polarizing microscope. The Olympus shown in Fig. 15.20 is used by the McCrone Research Institute for many of its intensive courses in microscopy. The Zeiss Ultraphot III shown in Fig. 15.21 is clearly a superior microscope, but is not meant for the use of most beginners.

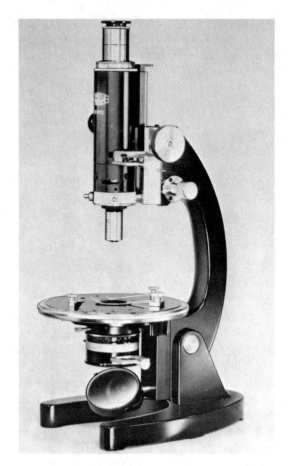

Fig. 15.20.   A simple, but wholly adequate, polarizing microscope (courtesy Olympus, New York).

H.   THE MICROSCOPE STAGE

The stage of the microscope supports the specimen between the condenser and objective. The manufacturer may offer a mechanical stage as an attachment to provide a means of moving the slide methodically during observation. The polarizing microscope is fitted with a circular rotating stage to which a mechanical stage may be added. The rotating stage, which is used for object orientation to observe optical effects, will have centering screws if the objectives are not centerable, and vice versa. It is undesirable to have both objectives and stage centerable, because this does not provide a fixed reference axis for the microscope.

I.   THE POLARIZING ELEMENTS

A polarizer is fitted to the condenser of all polarizing microscopes. In routine instruments the polarizer may be fixed with its vibration direction

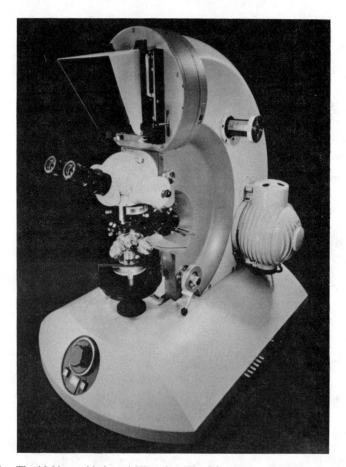

Fig. 15.21. The highly sophisticated Ultraphot III of Carl Zeiss (Courtesy Carl Zeiss, New York).

oriented north–south or east–west, although in research microscopes the polarizer can be rotated. Polarizing filters such as those made by Polaroid are used as the polarizer in modern instruments, replacing the older calcite prisms. The polarizing filter is preferred because it offers (1) lower cost, (2) no maintenance problem, and (3) use of the full condenser aperture.

An analyzer of the same construction as the polarizer is fitted in the bodytube of the microscope on a slider so that it may be easily removed from the optical path. It is oriented with its plane of vibration in a direction perpendicular to the corresponding direction of the polarizer.

J. THE BERTRAND LENS

The Bertrand lens is usually found only on polarizing microscopes, although some manufacturers include it on phase microscopes. It is located in the bodytube above the analyzer on a slider (or pivot) to permit its quick

Fig. 15.22.   Transmitted light stereo-binocular microscope, reflected-light capability is available (courtesy Olympus, New York).

removal from the optical path. The Bertrand lens is used to observe the back focal plane of the objective. It is convenient for quick checking of the type and quality of illumination and for observing interference figures of crystals.

K.   THE COMPENSATOR SLOT

The compensator slot receives compensators (quarter wave, first-order red, and quartz wedge) for the observation of the optical properties of crystalline materials. It is usually placed at the lower end of the bodytube just above the objective mount, and is oriented 45° from the vibration directions of the polarizer and analyzer.

V.   THE STEREOSCOPIC MICROSCOPE

The stereoscopic microscope (Fig. 15.22), also called the binocular, widefield, dissecting, or Greenough binocular microscope, is in reality a combination of two separate compound microscopes. The two microscopes, mounted in one body, have their optical axes inclined from the vertical by about 7°, and from each other by double this angle. When an object is placed on the stage of a stereoscopic microscope, each optical system views the object from a slightly different angle, thus presenting a stereoscopic pair of images to the eyes. The eyes fuse these two images into a single three-dimensional image.

The objectives are supplied in pairs, either as separate units to be mounted on the microscope or, in the new instruments, built into a rotating drum.

TABLE 15.V
Approximate Sizes of Several Common
Particulates

Material	Size (μm)
Ragweed pollen	25
Fog droplets	20
Power-plant flyash	2–5
Tobacco smoke	0.2 (200 nm)
Foundry fumes	0.1–1 (100–1000 nm)

Many instruments have a zoom-lens system that gives a continuous change in magnification over the full range. Objectives for the stereomicroscope run from about 0.4 to 12×, well below the magnification range of the objectives available for single-objective microscopes.

The eyepieces supplied with the stereoscopic microscope run from 10 to 25× and have a wider field than their counterparts in the single-objective microscope.

Because of mechanical limitations, the stereomicroscope is limited to about 200× magnification. It is most useful at relatively low powers for observing shape and surface texture; the study of greater detail should be reserved for the ordinary microscope. The stereomicroscope is also helpful in manipulating small samples, separating ingredients of mixtures, preparing specimens for detailed study at higher magnifications, and performing various mechanical operations (e.g., micromanipulation) under microscopical observation.

VI. RESOLVING POWER AND MICROMETRY

Linear distances and areas can be measured with the microscope. This permits the determination of particle size and the quantitative analysis of physical mixtures. The usual unit of length for microscopical measurements is the micrometer (μm), 1/1000th of a millimeter or about 1/25,000th of an inch. Measuring particles in electron microscopy requires an even smaller unit, the nanometer, 1/1000th of a micrometer. Table 15.V shows the approximate average sizes of a few common airborne materials.

The practical lower limit of accurate particle-size measurement with the light microscope is about 0.5 μm. The measurement of particles smaller than this with the light microscope leads to errors that, even under the best of circumstances, increase to about $\pm 100\%$ (usually $+$) at 0.1–0.2 μm.

A. RESOLVING POWER

1. Diffraction

A knowledge of diffraction will help us understand much better how the microscope actually works. A microscopic object will be resolved only if at least one of the diffracted rays enters the objective aperture along with the direct ray. We can see at once (see Fig. 15.24) why oblique rays from the condenser are necessary for resolution of the finest detail.

We should return now to diffraction itself. In geometrical optics, it is assumed that light travels in straight lines, but this is not always true. A beam passing through a slit toward a screen creates a bright band wider than the slit, with alternate bright and dark bands appearing on either side of the central bright band. The band intensity decreases as a function of the distance from the center. This phenomenon, diffraction, limits image reproduction. For example, the image of a pinpoint of light produced by a lens is not a pinpoint, but rather a somewhat larger patch of light surrounded by dark and bright rings. The diameter D of this diffraction disc (to the first dark ring) is

$$D = \frac{1.22 \, \lambda}{\sin \theta}$$

where λ is the wavelength of the light and θ is one-half the lens angular aperture. For a small diffraction disc to be maintained at a given wavelength, the lens aperture should be as large as possible. It should be noted that a shorter wavelength produces a smaller-diameter diffraction disc.

2.   Resolution

If two small objects are to be distinguished in an image, their diffraction discs must not overlap by more than one-half their diameters. The ability to distinguish such image points is called *resolving power* and is expressed as one-half the diameter of the diffraction disc. The theoretical limit of resolution of two discrete points a distance d is (83).

$$d = \frac{0.61\lambda}{NA}$$

where λ is the wavelength of the light and NA is the numerical aperture of the objective. Using this equation, we can graph resolution (in micrometers) as a function of numerical aperture (Fig. 15.23).

Substituting a wavelength of 450 nm and a numerical aperture of 1.3, the practical limits for visible-light-wavelength and oil-immersion objectives, into this equation, we find that two points about 200 nm (or 0.2 μm) apart can be seen as two separate points. Further increases in resolving power can be achieved for the light microscope only with shorter-wavelength light. Ultraviolet light (wavelength near 200 nm) lowers the limit to about 0.1 μm, the lower limit for the light microscope. The resolution of detail actually observed in a microscopical image also depends on the illuminating condi-

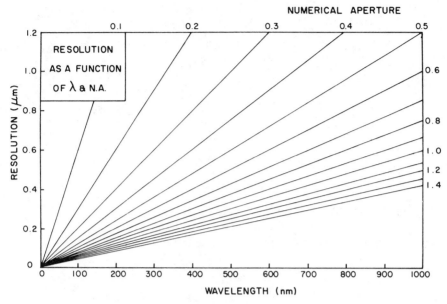

Fig. 15.23. Resolution as a function of numerical aperture and wavelength.

tions, specimen contrast, condition of the optics, and quality of the observer's eye.

Under the very best illumination and contrast conditions, some unaided human eyes can resolve two points only about 60 μm apart. Under less-optimal conditions, 120 μm might better be used, and under average conditions, 300 μm may be the lower limit. Considering, then, the lower limit of resolution of the oil-immersion objective with white light, 0.2 μm, one can calculate the total magnification necessary under these three sets of conditions by dividing 0.2 μm into 60, 120, and 300 μm. This gives 300, 600, and 1500×. This signifies that the ocular should furnish enough magnification to bring the total magnification up to the level required under the conditions at hand. Table 15.VI shows the ocular magnification required by each of the common objectives under these three sets of conditions.

The above reasoning has led to the generally helpful rule of thumb that useful magnification does not exceed 1000 times the numerical aperture of the objective. Although somewhat higher magnification may be used in specific cases, usually no additional detail will be seen.

It is curious, considering the figures in the table, that most, if not all, microscope manufacturers routinely furnish a 10× ocular as the highest power. A 10× ocular is useful, but critical work requires a 15–25× ocular; 5–10× oculars are best used for scanning, although it is nowadays possible to purchase wide-field oculars of high power, which are very suitable for scanning.

TABLE 15.VI

Total and Eyepiece Magnification Necessary to Resolve Detail Shown by Objective

Objective			Total (and eyepiece) magnification required		
NA	d^a	Magnification (×)	200 NA[b]	400 NA[c]	1000 NA[d]
0.08	3.8	2.5	16 (7)	32 (14)	80 (35)
0.10	3.0	5	20 (4)	40 (8)	100 (20)
0.25	1.22	10	50 (5)	100 (10)	250 (25)
0.50	0.61	20	100 (5)	200 (10)	500 (25)
0.66	0.47	43	132 (3)	264 (6)	660 (15)
0.85	0.36	45	170 (4)	340 (8)	850 (20)
1.30	0.20	90	260 (3)	520 (6)	1300 (15)

[a] Resolving power.

[b] Excellent preparation, excellent illumination, excellent eye.

[c] Good preparation, good illumination, good eye.

[d] Average preparation, average illumination, average eye.

3. Abbe's Theory of Resolution

One of the most cogent theories of resolution is that of Ernst Abbe (1840–1905). He suggested that microscopic objects act as diffraction gratings (Fig. 15.24) and that the angle of diffraction therefore increases with fineness of detail. He proposed that a given microscope objective would resolve a particular detail if at least two of the three transmitted rays (one direct ray and two diffracted rays) entered the objective. In Fig. 15.24 the detail shown would be resolved in a and c but not in b. This theory, borne out by simple experiment, shows how to improve resolution. Since shorter wavelengths will give a smaller diffraction angle, they offer a better chance of resolving fine detail. Also, since only two of the transmitted rays are needed, oblique light and a high-numerical-aperture condenser will aid in resolving fine detail.

4. Improving Resolving Power

The following list summarizes the practical approaches to obtaining higher resolution with the light microscope:

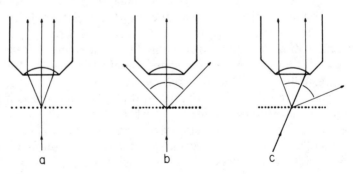

Fig. 15.24. Schematic representation of Abbe's theory of resolution.

1. The specimen should be illuminated with either critical or Köhler illumination (see Section VII).
2. The condenser should be well corrected and have a numerical aperture as high as that of the objective to be used.
3. An apochromatic oil-immersion objective should be used, with a compensating eyepiece of at least 15× magnification.
4. The manufacturer's recommended immersion oil should be placed between the condenser and slide and between the coverslip and objective, and the preparation itself should not be dry.
5. The illumination should be reasonably monochromatic and as short in wavelength as possible. An interference filter transmitting a wavelength of about 480 nm is a very suitable answer to this problem. Ideally, of course, ultraviolet light should be used.

The practical effect of many of these factors is critically discussed by Loveland in a paper (57) on the optics of the object space.

B. MICROMETRY

One of the principal uses of high resolving power is in the precise measurement of particle size. There are, however, a variety of useful approximate procedures as well. The measurement of particle size can be accomplished by the following methods:

1. If the microscope magnification (product of the magnifications of objective and ocular) is known, particle size can be estimated. For example, with a 10× ocular and a 16-mm (or 10×) objective, the total magnification is 100×. A particle that appears to the eye to be 10 mm in diameter in the microscope has an approximate actual size of 10 mm divided by 100, that is, 0.10 mm or 100 μm. This is in no sense an accurate method, but it does permit quick estimation of particle size; the error in this estimate is usually 10–25%. It can be made more accurate, with a monocular microscope, by holding a millimeter rule 10 in. before the left eye while looking through the microscope with the right eye. The images of rule and particle can then be superimposed and the particle-image diameter can be measured. Dividing this size by the magnification of the microscope gives the particle diameter.

2. Another approximate method is also based on the use of known data. If we know approximately the diameter of the microscope field, we can estimate the percentage of that diameter occupied by the object to be measured and calculate the object's approximate size. The field size depends on both the objective and the ocular, although the latter is a lesser influence. Field size should be determined with a stage-micrometer scale for each objective and ocular. If this is done, estimates of sizes can be accurate to 5–10%.

3. The movement of a graduated mechanical stage can also be used for rough measurement of diameters of large particles. Stages are usually grad-

uated (with a vernier) to read to 0.1 mm, or 100 μm. In practice, the leading edge of the particle is brought to one of the cross lines in the ocular and a reading is taken of the stage position. Then the particle is moved across the field by moving the mechanical stage in the appropriate direction until the particle's trailing edge just touches the cross line. A second reading is taken, and the difference in the two readings is the distance moved or the size of the particle. This method is especially useful when the particle is larger than the field, or when the optics give a distorted image near the edge of the field.

4.   The above methods can be extended to projection or photography. The image of the particles can be projected on a screen with a suitable light source or may be photographed. The final linear magnification M on the projection surface (or film plane) is given approximately by

$$M = \frac{D}{25} \times \text{OM} \times \text{EM}$$

where OM is the objective magnification, EM is the ocular magnification, and D is the projection distance from the screen or film in centimeters. The image detail can then be measured in centimeters and the actual size computed by dividing by M. This method is usually accurate to within 2–5%, depending on the size range of the detail measured.

5.   Stated magnifications and/or focal lengths of microscope optics are nominal and vary a bit from objective to objective or ocular to ocular. For accuracy, a stage micrometer is used to calibrate each ocular–objective combination. The stage micrometer is a glass microscope slide that has accurately engraved in the center a scale, usually 2 mm long, divided into 200 parts; each part represents 0.01 mm (10 μm). When this scale is observed, projected, or photographed, the exact image magnification can be determined. For example, if five divisions of the stage micrometer measures 6 mm when projected, the actual magnification is

$$\frac{6}{5\,(0.01)} = 120\times$$

This magnification figure can be used to improve the accuracy of method 4.

6.   The simplest procedure is based on the use of a micrometer ocular. Since the ocular magnifies a real image from the objective, one can place a transparent scale in the same plane as the image from the objective and thus superimpose the scale on the image. This is done by first placing an ocular micrometer-scale graticule in the ocular. The ocular micrometer scale is arbitrary and must be calibrated with each objective used. Measurement methods using ocular scales are among the most accurate. The accuracy of the measurement depends on recognition of the edge of the particle when

observing an image. This image may be slightly out of focus, and certainly involves errors introduced by the kind and quality of the illumination as well as by the optical system itself. Even the refractive index of the mounting medium relative to the index of the particle being measured has an effect on the apparent size. In all cases, the error will be minimized if the magnification of the optical system is sufficient to image the particle over at least 10 ocular scale divisions. At worst, under these conditions, the edge of the particle can be measured accurately to within ±0.25 divisions. Hence the overall error in measurement, considering two sides of the particle, could be ±0.5 divisions, or 10%, for a 10-division particle. This is quite conservative, and one might expect to do better than this. One should not expect, however, to do better than ±2% with the filar ocular scale. The error depends, of course, on the particle size, and increases rapidly as the size decreases to the 1-μm range.

7. The filar micrometer ocular provides one of the simplest and most direct ways of measuring the diameter of a single particle. This device is a scale in the ocular superimposed on the image of the object particle in the field of view. If this scale has been calibrated by means of a stage scale, so that the number of micrometers per ocular-scale division is known, particles can be measured directly by comparison with the ocular scale. Accurate measurement of the particle is facilitated in this case if the ocular has a movable line that can be set on opposite edges of the particle by means of a rotating drum that is itself divided into fractions of an ocular-scale division. The difference between the drum readings after carefully setting the movable line on opposite edges of the particle can then be expressed in fractions of a scale division. The actual dimension in micrometers is then calculated from the calibration data for that magnification (i.e., the number of micrometers per ocular-scale division).

8. The filar ocular is very simple to use and more accurate than measurement against an ocular scale without the filar movement. It is probably not quite as accurate as the image-splitting or -shearing ocular. The latter is a very clever arrangement involving a binocular system in which one observes two superimposed images of the particle, one of which is displaced by known micrometer distances from the other image. If the images are displaced to the point that the two images appear to be just in contact, with neither overlap nor a gap between them, the displacement of the image measures precisely the diameter of the particle. Although setting the image displacement depends somewhat on illumination and optical imagery, the setting is more accurate than with the filar ocular. One can expect with such a device to measure particle diameters down to 20 μm to within ±1%, perhaps ±2% down to 10 μm, ±5% down to 5 μm, ±10% down to 2 μm, and no better than ±25% below 2 μm. The error goes up quickly to around ±100% below 1 μm.

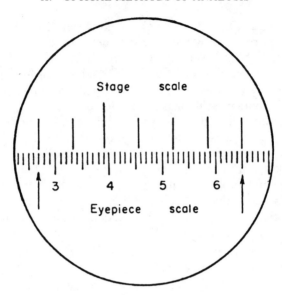

Fig. 15.25. Comparison of stage scale with ocular scale.

C. CALIBRATION OF THE OCULAR SCALE

Each stage scale has divisions 100 μm (0.1 mm) apart; one or more of these are usually subdivided into 10-μm (0.01-mm) divisions. These form the standard against which the arbitrary divisions of the ocular scale are calibrated. One must calibrate each objective separately by noting the correspondence between the stage scale and the ocular scale. First focus the ocular scale with the eye lens of the measuring ocular; then, starting with the lowest-power objective, focus on the stage scale and make the two scales parallel (Fig. 15.25). It should be possible to find the number of ocular divisions exactly equal to some whole number of divisions of the stage scale, expressed in micrometers.

The calibration consists of calculating the number of micrometers per ocular-scale division. To make the comparison as accurate as possible, a large part of each scale must be used. Assume that, with a 16-mm objective, six large stage-scale divisions (ssd) equals 38 ocular-scale divisions (osd). Hence

$$38 \text{ osd} = 600 \text{ μm}$$

$$1 \text{ osd} = \frac{600 \text{ μm}}{38}$$

Thus when that ocular and scale are used with that 16-mm objective at that microscope tube length, each division of the ocular scale equals 15.8 μm, and the scale can be used to measure accurately any object on the microscope

TABLE 15.VII

Example of Calibration Data for an Ocular Scale

Objective	Tube length	ssd/osd	μm/osd
32 mm (4×)	160 mm	18/44	1800/44 = 50.9
16 mm (10×)	160 mm	6/38	600/38 = 15.8
4 mm (45×)	160 mm	1/30	100/30 = 3.33

stage. A particle, for example, observed with the 16-mm objective and measuring 8.5 divisions on the ocular scale is 8.5 × 15.8 or 134 μm in diameter.

Each objective on a microscope must be calibrated in this manner. Table 15.VII illustrates a convenient way to record the necessary data and to calculate micrometers per osd.

Linear distances on other styles of ocular graticules are also calibrated against the stage scale, by similar procedures.

D. DETERMINATION OF PARTICLE-SIZE DISTRIBUTION

The physical bulk properties of a particulate material usually depend on the particle-size distribution, that is, on the relative numbers of each size of particle making up the material. The bulk behavior can be studied and predicted from the average sizes, which are calculated from the distribution data and from the size-frequency data themselves.

The measurement of size distribution differs from the measurement of individual particles in several important aspects. First, it is essential that the sample be representative and that those particles counted also be representative. It is particularly important to represent adequately the larger sizes, since most of the weight of the sample could be in a relatively small number of larger particles. It is essential that sufficient particles be counted for good sampling in all size ranges. This is one of the major limitations of the microscopical procedures for particle-size distribution measurement. One must count far too many particles in the lower or middle size range if one is to have a sufficient sample in the larger sizes.

Second, the measurement of any one particle diameter does not have to be as accurate as any one measurement of a single particle in Section VI.B, above. This is because any error due to rapid setting of the particle boundary will be offset by corresponding errors in the opposite direction with subsequent particles. This makes it possible to measure size distribution of particles in a given sample much more rapidly than would appear possible from the single-particle-measurement discussion. Whereas several minutes might be spent making sure of the diameter of a single-particle sample, one might measure as many as 500 particles/hr in making a size-distribution analysis.

The number of particles to be measured in a sample to obtain accurate and representative data depends on the variety of sizes and the diversity of

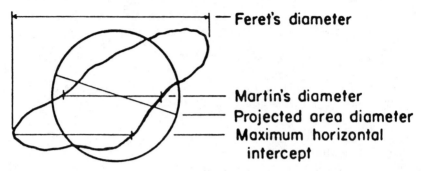

Fig. 15.26.   Various statistical diameters.

shapes. If the shape is highly irregular or elongated and there is a wide diversity of sizes, it may be necessary to count several thousand particles. On the other hand, if the samples are regular in shape and not too diverse in size, only 100 or so particles may be sufficient. The determination is tedious, however, and should be undertaken only as a calibration exercise for other automatic procedures or if only a few such analyses are contemplated.

1.   Statistical Diameters of Particles

The measurement of particle size varies in complexity depending on particle shape. The size of a sphere is defined by its diameter. The size of a cube may be expressed by the length of an edge or a diagonal, and the surface area, volume, and weight (if the density is known) of a cube or sphere can be calculated directly from such a dimension. Horvath (38) has described shape factors. However, if the particles are irregular, weight and other properties cannot be calculated directly, and the particle "size" must include information about the shape of the particle. The expression of this shape thus takes a more complicated form.

An irregularly shaped particle has a number of different dimensions that might be measured as "diameters." Figure 15.26 shows three statistical diameters commonly used in determining particle-size distribution, and a circle equal in area to the particle.

Feret's diameter is the distance between imaginary parallel lines tangent to the particle profile and perpendicular to the ocular scale. It is usually measured with a filar ocular.

The *maximum horizontal intercept* is the longest diameter of the particle from edge to edge, parallel to the ocular reference line. This statistical diameter is the one obtained with image-splitting or -shearing oculars. It always falls between Feret's and Martin's diameters or is equal to one of them.

Martin's diameter is one of the diameters most often used for particle sizing. It is measured as the dimension (parallel to the ocular scale) that

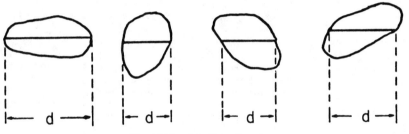

Fig. 15.27. Martin's diameter.

divides a randomly oriented particle into two equal projected areas (Fig. 15.27).

Projected-area diameters are found by comparing the projected area of the particle with the areas of reference circles on an ocular graticule. Two adjacent circles, one larger and one smaller in area that the particle, define a size class into which the particle fits. The statistical diameter of the size class (and particle) is the arithmetic mean of the diameters of the two circles.

2. Use of Martin's Diameter

Martin's diameter is the simplest means of measuring and expressing the diameters of irregular particles, and is sufficiently accurate when averaged for a large number of particles. The more particles that are counted, the more accurate will be the average particle size. Platelike and needlelike particles should have a correction factor applied to account for the third dimension, since all such particles are restricted in orientation on the microscope slide. When particle size is reported, the general shape of the particles as well as the method used to determine the "diameter" should be noted.

The determination of particle-size distribution is carried out routinely by moving a preparation of particles past an ocular scale in such a way that their Martin's diameters can be tallied. All particles whose centers fall within two fixed divisions on the scale are tallied. Movement of the preparation is usually accomplished by means of a mechanical stage, but may be carried out by rotation of an off-center rotating stage. A sample tabulation appears in Table 15.VIII. The ocular and objective are chosen so that at least six, but not more than 12, size classes are required, and sufficient particles are counted to give a smooth curve. The actual number tallied, which may vary from 200 to 2000, depends on the regularity of the particle shape and the range of sizes. The size tallied for each particle is that number of ocular-scale divisions most closely approximating Martin's diameter for that particle.

TABLE 15.VIII

Particle-Size Tally for a Sample of Starch Grains

Size class (osd)	Number of particles	Total
1	卌 卌 卌 1	16
2	卌 卌 卌 卌 卌 卌 卌 卌 卌 卌 卌 卌 卌 卌 卌 卌 卌 卌 卌 111	98
3	卌 卌 卌 卌 卌 卌 卌 卌 卌 卌 卌 卌 卌 卌 卌 卌 卌 卌 卌 卌 卌 卌	110
4	卌 卌 卌 卌 卌 卌 卌 卌 卌 卌 卌 卌 卌 卌 卌 卌 卌 卌 卌 卌 卌 11	107
5	卌 卌 卌 卌 卌 卌 卌 卌 卌 卌 卌 卌 卌 卌 1	71
6	卌 卌 卌 卌 卌 卌 卌 卌 卌	45
7	卌 卌 卌 卌 1	21
8	11	2
		$\overline{470}$

3. Calculation of Size Averages

The size data may be treated in a variety of ways; one simple, straightforward treatment is shown in Table 15.IX. For a more complete discussion of the treatment of particle-size data, see Chamot and Mason (16), page 436.

Cumulative percentages by number, surface, and weight (or volume) may also be plotted from the data in Table 15.IX. For example, the percentage of the total weight or volume of the sample that is finer (or coarser) than any given diameter is calculated for each diameter as

$$\left[\sum_{d=0}^{3} nd^4 \Big/ \sum_{d=0}^{8} nd^4 \right] \times 100 = \% \text{ finer than } d = 3$$

The calculated percentages for the cumulative weight or volume curve are plotted against d in micrometers.

Finally, the specific surface S_m, in square meters per gram, may be calculated if the density D is known; the surface average, $\bar{d}_3$, is used. If $D = 1.1$,

$$S_m = \frac{6}{d_3 D} = \frac{6}{13.8\,(1.1)} = 0.395 \text{ m}^2/\text{g}$$

4. Use of Projected-Area Diameter (British Standard Method)

The British Standard method (10) for determining particle-size distribution uses the projected-area diameter, illustrated in Fig. 15.26. Particles are sized

TABLE 15.IX
Calculations for Particle-Size Averages

d (Average diameter in osd)	n	nd	nd^2	nd^3	nd^4
1	16	16	16	16	16
2	98	196	392	784	1568
3	110	330	990	2970	8910
4	107	428	1712	6848	27392
5	71	355	1775	8875	44375
6	45	270	1620	9720	58320
7	21	147	1029	7203	50421
8	2	16	128	1024	8192
	$\overline{470}$	$\overline{1758}$	$\overline{7662}$	$\overline{37440}$	$\overline{199194}$

by comparing their projected areas with the areas of reference circles on a standard graticule in the ocular. Each particle is assigned to a size class defined by two adjacent circles that represent the size limits of that class. Thus the distribution of sizes is obtained in terms of the diameters of circles having the same projected areas as the particles. The British Standard graticule is shown in Fig. 15.28.

The circle areas double progressively; hence the diameters progress by $\sqrt{2}$, so that the size classes can form a continuation of the standard series

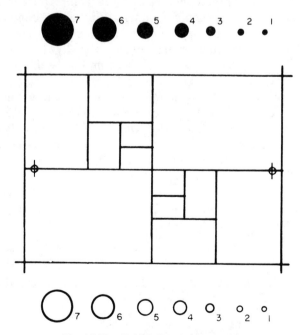

Fig. 15.28. British Standard graticule.

of sieves for particle sizing. The microscope method covers particles between 0.8 and 150 μm, and it usually is necessary to change the objective magnification during a size determination. The rectangular grids of the graticule are used to interrelate the magnifications used and to define the size and number of fields of particles to be counted and sized. The reference circles in the graticule are matched to recommended size classes by adjusting the microscope's tube length for each magnification used—if the tube length cannot be adjusted, other size classes must be calculated.

The size distributions with respect to number and to weight (or volume, if the sample is not homogeneous) are determined separately. The number of particles that must be sized to get statistically valid results can be calculated after preliminary scans have been carried out. Final results are calculated as cumulative percentages by number and weight.

For elongated (acicular) particles the Standard recommends use of a graticule on which rectangles replace the reference circles. Other graticules employing circles have been designed; the best known are the original "globe and circle" by Patterson and Cawood (69), the Porton graticule (21), and several graticules designed by Fairs (25). Cadle (13) discusses their use.

E.   COUNTING ANALYSES

Particulate mixtures can often be quantitatively analyzed by counting the total number of particles of each component in a representative sample. The calculations are, however, complicated by three factors: the average particle size, shape, and density of each component. If all of the components were equivalent in particle size, shape, and density, then the weight percentage would be identical to the number percentage. Usually, however, it is necessary to determine correction factors to account for the differences in size, shape and density.

When properly applied, this method can be accurate to within ± 1% and, in special cases, even better. It is often applied to the analysis of fiber mixtures, and is then usually called a *dot count*, because the tally of fibers is kept as the preparation is moved past a point or dot in the ocular.

A variety of methods can be used to simplify recognition of the different components. These include chemical stains or dyes and enhancement of optical differences such as refractive indexes, dispersion, or color. Often, however, one relies on the differences in morphology, for example, in counting the percentage of rayon fibers in a sample of "silk."

Example

A dot count of a mixture of fiber glass and nylon shows the following result:

nylon   262
fiber glass   168

The percentage, by number, of nylon is therefore

$$\% \text{ nylon} = \frac{262}{262 + 168} \times 100$$

$$= 60.9\%$$

However, although both fibers are smooth cylinders, they have different densities and, usually, different diameters. To correct for diameter, one must measure the average diameter of each type of fiber and calculate the volume of a unit length of each.

	Average diameter (μm)	Volume of 1-μm slice (μm^3)
nylon	18.5	268
fiber glass	13.2	117

The percentage by volume is then

$$\% \text{ nylon} = \frac{262 \times 268}{262 \times 268 + 168 \times 117} \times 100$$

$$= 78.1\%$$

We must still take into account the density of each to calculate the weight percentage. If the densities are 1.6 for nylon and 2.2 for glass, then the percentage by weight is

$$\% \text{ nylon} = \frac{262 \times 268 \times 1.6}{262 \times 268 \times 1.6 + 168 \times 117 \times 2.2} \times 100$$

$$= 72\%$$

Example

A dot count of quartz and gypsum shows the following results:

$$\begin{array}{ll} \text{quartz} & 283 \\ \text{gypsum} & 467 \end{array}$$

To calculate the percentage by weight, we must take into account the average particle size, the shape, and the density of each material. The average particle size with respect to weight, $\bar{d}_4$, must be measured for each and the shape factor must be determined. Since gypsum is more platelike than quartz, each particle of gypsum is thinner. The shape factor can be roughly calculated by measuring the actual thickness of a number of particles. We might find, for example, that gypsum particles average 80% of the volume of the average quartz particle; this is our shape factor. The final equation for the weight percentage is

$$\% \text{ quartz} = \frac{283 \times \pi \bar{d_4^3}/6 \times D_q}{283 \times \pi \bar{d_4^3}/6 \times D_q + 467 \times \pi \bar{d}^*_43/6 \times 0.80 \times D_g}$$

where D_q and D_g are the densities of quartz and gypsum, respectively; 0.80 is the shape factor, and $\bar{d_4}$ and $\bar{d_4^*}$ are the average particle sizes with respect to weight for quartz and gypsum, respectively.

F. AREAL ANALYSIS

The relative areas of two or more substances as presented by a polished surface or thin section are directly related to the percentages of each by volume. If the various densities are known, the percentages by weight can also be calculated. This technique is usually applied to metals, but has been applied to the analysis of cast high-explosive mixtures such as amatol (trinitrotoluene plus ammonium nitrate), of the percentage of filler in plastics, of pore size in filters, and of other systems. The relative amounts of particles can also be measured accurately if the particulate mixture is first cast in plastic, and then cut and polished to give a plane section. It must be possible to differentiate between the different phases in the thin section by the differences in optical properties, or in the polished surface, when necessary, by use of an appropriate etching agent.

The relative areas may be measured directly in the microscope if a cross-ruled ocular reticle is used. The areas may also be measured on a photomicrograph or a drawing made with a camera lucida (drawing camera) and a planimeter.

Many types of reticles and graticules have been designed for various methods of areal, counting, and particle-size analysis. Delly (21) reviews this subject succinctly.

G. MICROSTEREOLOGY

Microstereology is a technique for deducing the internal structure of a three-dimensional body from the microscopical study of two-dimensional sections through it (77). By applying statistical methods, line-intersection or point counts can be used to determine complicated structures quantitatively. The technique is very tedious, but advances in microstereology are now being spurred by the application of automated counting and measuring methods.

H. AUTOMATED SIZE ANALYSIS

Several authors (39,49) have made an effort to simplify the job of recording data from microscopical particle-size determination. These procedures are based usually on setting movable crosslines in the ocular to the particle diameter and then pressing a button to record that setting. These procedures seem to have little advantage over a tally counter for each size range.

A slightly different system, promoted by Zeiss, is applied to photomicrographs or electron micrographs (24). An iris diaphragm is oriented over each particle in turn and set to an equivalent area. Pressing a button records the diaphragm setting, and hence the equivalent diameter. This system works quite rapidly, and many hundreds of particles per hour can be measured by a patient operator.

Integrating stages are mechanical stages with several identical spindles, any of which can be used to traverse a sample. The distance moved using any one spindle is shown by a scale on that spindle. Such stages are used for linear analysis of mixtures: By traversing each component with its own spindle, the accumulated distances for each are measured separately and recorded automatically on the spindle.

Finally, there are completely automatic systems, such as the Texture Analyzer (Leitz), πMC (Millipore), Quantimet 720 (IMANCO, Metals Research), Micro-Videomat (Zeiss), Telecounter (Schaefer), Digiscan (Kontron), and QMS (Bausch & Lomb), which employ highly sophisticated electronics to take all of the drudgery out of the microscopical procedure (28,42,59). The microscope is still used, but the image itself is analyzed by electronic means. Particles can be counted and measured at very high rates. These instruments can be programmed to count particles, measure projected areas, or measure any of the standard diameters such as Martin's, Feret's, or the projected-area diameter. They eliminate the problem of representative sample size, because thousands of particles can be counted and sized in seconds.

Automatic image analyzers still have the problem of accurate imagery of the particle before measurement, and are especially prone to error at high magnification, when some particles are usually out of focus. The microscopist normally focuses up and down for best focus on each particle, whereas the automatic image analyzer measures all particles in the field of view with one focus setting. Nevertheless, one can, with these precautions in mind, obtain very accurate particle-size distribution data. These analyzers also, of course, quickly calculate any of the averages or percentages needed, thus encouraging mathematical analysis of the data, which would be too tedious without automatic computation.

I. ANGULAR MEASUREMENT

The rotating stage of the polarizing microscope is graduated in degrees and can be read to some fraction of a degree (usually a tenth) with a vernier. To measure an angle in a specimen, the stage is rotated to bring each side of the angle parallel to the crossline in turn. Depending on how the specimen was rotated and which crosslines were used, 90° may need to be added to or subtracted from the difference in the readings to compute the angle. Several sets of readings should be taken and averaged.

If rotating the specimen is undesirable or impossible, goniometer oculars are available in which the crosslines are rotated through angles read on a graduated scale at the periphery of the ocular.

VII.   PHOTOMICROGRAPHY

Photomicrography, as distinct from microphotography, is the art of taking pictures through the microscope. A microphotograph is a small photograph; a photomicrograph is a photograph of a small object. Photomicrography is a very valuable tool in recording the results of microscopical study. It enables the microscopist to:

1. Describe a microscopic field objectively without resorting to written descriptions.
2. Record a particular field for future reference.
3. Make particle-size counts and areal analyses easily and without tying up a microscope.
4. Enhance or exaggerate the visual microscopic field to bring out or emphasize certain details not readily apparent visually.
5. Record images in ultraviolet and infrared microscopy that are otherwise invisible to the unaided eye.

There are two general approaches to photomicrography: One requires only a plate or film holder supported above the eyepiece of the microscope with a light-tight bellows; the other utilizes any ordinary camera with its own lens system supported with a light-tight adaptor above the eyepiece. It is best, in the latter case, to use a reflex camera, so that the image can be carefully focused on the ground glass. Photomicrography of this type can be regarded simply as replacing the eye with the camera-lens system. The camera should be focused at infinity, just as the eye is for visual observation, and it should be positioned close to and over the eyepiece.

The requirements for photomicrography, however, are more rigorous than for visual work. The eye can normally compensate for varying light intensities, curvature of field, and depth of field. The photographic plate, however, lies in one plane; hence the greatest care must be used to focus sharply on the subject plane of interest and to select optics that give minimum amounts of field curvature and chromatic aberration.

With black and white film, color filters may be used to enhance the contrast of some portions of the specimen while minimizing chromatic aberrations of the lenses. In color work, however, filters cannot usually be used for this purpose and better optics may be required.

Photomicrographic cameras that fit directly into (or onto) the microscope are available for film sizes from half-frame 35 mm to 4 × 5 in. Others are

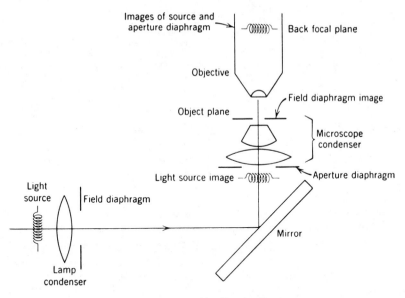

Fig. 15.29. Köhler illumination.

made that accommodate larger film sizes and have their own stands and uprights upon which the microscope is mounted. The former are preferred for ease of handling and lower cost, but the latter types are preferred for greater flexibility and versatility and for lack of vibration. The Polaroid camera has many applications in microscopy and can be used on the microscope directly, but because of its weight, should be used only when the microscope has a vertically moving stage for focusing rather than a focusing bodytube.

A. KÖHLER ILLUMINATION

Great care should be exercised in aligning the illuminating system. Köhler illumination (Fig. 15.29) is preferred over other types of illumination, since it gives a uniformly bright field and good control of the size of the field illuminated. Köhler illumination is distinguished from Nelsonian illumination by the differences shown in Table 15.X.   The basic requirements for Köhler illumination are as follows:

1. One set of conjugate foci of the field diaphragm, including (*a*) the front focal plane of the ocular with any reticles located in that plane; (*b*) the field of view; and (*c*) the retina, or film plane.
2. A second set of conjugate foci of the lamp filament, including (*a*) the eyepoint or Ramsden disc, (*b*) the objective back focal plane, and (*c*) the aperture diaphragm.

Only by meeting these basic requirements will the best possible illumination be achieved. With a built-in illuminator, the following steps are necessary to set up Köhler illumination:

TABLE 15.X
Comparison of Nelsonian and Köhler Illumination

	Köhler	Nelsonian
Lamp filament	Any type	Ribbon filament
Condensing lens on illuminator	Required	Required
Field diaphragm on illuminator	Required	Not required
Ground glass at illuminator	None	None
Image of light source	At condensor front aperture	In plane of specimen
Image of field diaphragm	In plane of specimen	Not required

1. Focus the microscope on any preparation.
2. Close the field diaphragm (in or on the base) and focus sharply on the iris leaves by racking the substage condenser up or down.
3. Center the field diaphragm using centering screws on the substage condenser; open the iris to the edge of the field of view.
4. Insert the Bertrand lens or a phase telescope (or remove the ocular) to observe the objective back focal plane; focus the filament image by sliding the lamp housing or bulb along its axis.
5. Center the image of the filament with the bulb centering screws.
6. Remove the Bertrand lens (or replace the ocular) and adjust the sub-stage-aperture diaphragm for optimum contrast and resolving power.

With an external illuminator:

1. Center the lamp filament on the axis of the lamp condenser (use bulb centering screws).
2. Focus the lamp filament on the center of the microscope mirror.
3. Tilt the mirror to throw some light into the microscope.
4. Focus on any preparation.
5. Close the field diaphragm (on the lamp); tilt the mirror if necessary to keep its image in the field of view.
6. Focus the image of the field diaphragm (move substage condenser up or down).
7. Center the field diaphragm (tilt the mirror).
8. Perform step 4 of the instructions for a built-in illuminator.
9. Center the lamp-filament image by tilting or swiveling the lamp.
10. Remove the Bertrand lens (or replace the ocular); recheck step 7.
11. Adjust the substage aperture diaphragm for optimum contrast and resolving power.

If any of these steps are impossible, you will be unable to obtain the best image your condenser, objective, and ocular are capable of delivering. If you can, remove any diffusers; if you cannot, try to achieve the most intense but uniform illumination of the object back focal plane.

B. DETERMINATION OF CORRECT EXPOSURE

Correct exposure determination can be accomplished by trial and error, by relating new conditions to previously used successful conditions, and by the use of photometers.

Most photomicrographs are properly exposed when a white area within the frame is fully, but not over, exposed. A spot meter reading that white area would be appropriate; however, occasionally there may be no white area (with slightly uncrossed polars, for example). In other situations, such as dark-field illumination, the white "spot" is a point or line not measurable with the spot meter. Full bright-field methods are relatively easily handled by either a spot meter or a full-field meter. One must remember that the area of the background obstructed by detail in the field will lower the full-field meter reading. The proper exposure, however, is unchanged by the number of, say, coal particles in a field of view. There are two approaches to the problem of obtaining a proper meter reading: (1) The preparation can be removed so that the meter sees an empty white field of view; or (2) the meter reading on the field ready to be photographed can be increased based on the estimated percentage of the field obstructed. The first solution is the more reliable, especially when the field is obstructed with colored particles as well as with dark areas and edges.

The problem of exposure determination when the background is black, gray, or colored rather than white is somewhat more difficult. Imagine, for example, a single birefringent fiber between crossed polars compared with the same field with 20 birefringent fibers. The meter readings will differ by a factor of about 20, yet the proper exposure time for both pictures is identical. The situation is similar for dark-field, central stop dispersion staining, or any other non-bright-field illumination. In these situations the solution is to take the meter reading on an equivalent bright field. With crossed polars, for example, the meter reading can be taken after removing the analyzer; with parallel polars, with a compensator inserted or with a special preparation showing a full field of high-order white polarization colors.

No matter which method one chooses, the same method must be used for all subsequent photomicrographs of the same type. An exposure series is taken based on a meter reading on a bright field, and the photomicrographs are taken with the corresponding non-bright-field arrangement. The best exposure is then judged and the time is multiplied by the corresponding bright-field meter reading to give an exposure factor K. Subsequent exposure times (by that non-bright-field method) are then calculated by dividing that constant K by the meter reading taken on the field of view with the corresponding bright-field illumination. Measurement of the meter reading on the corresponding bright field brings us back to the original problem of obstruction of the field of view by object detail; hence one must remove the preparation or estimate the degree to which the meter reading is low and make that correction before dividing it into K to get the proper exposure time.

Top lighting is also easily handled by this general approach of metering a corresponding bright field. An easy solution is to set up the best lighting for the subject, then measure the meter reading on a readily available white surface (e.g., a 3 × 5 in. file card, a notebook page, or typewriter bond). An exposure series with that lighting will yield the exposure factor to be used subsequently.

This constant is, of course, dependent on the film speed, and indeed is inversely proportional to the ASA rating:

$$\text{new } K = \frac{\text{old } K \text{ (old ASA)}}{\text{new ASA}}$$

To use this system of exposure control one simply decides on the most expedient choice of corresponding bright field to be used for each method of illumination.

Some conditions for exposure control in photomicrography include the following:

Exposure	Meter reading	K^a	f-stop
one or no polar	same	60	4
crossed polars	one polar	200	7
slightly uncrossed polars	one polar	200	7
dispersion staining			
largest annular stop[b]	no stops	60	4
largest central stop[b]	no stops	250	8
Hoffman modulation contrast	same	60	4
top light	on bond paper	500	11
dark field	frosted slide	300	9.5
conoscopic (40× objective, 6× ocular)	orthoscopic, one polar	500	11

[a] Determined using Eastman Kodak Ektachrome, ASA 160; a different film with a different ASA would require other K values (see text).
[b] The dispersion-staining objective has several different-sized stops. Using different stops would require other K values.

There is no chance of over- or underexposing one's pictures using this system if one follows the prescribed procedure. It is important to remember that any series of K values is unique to the microscope and camera on which they were determined. The above table lists the conditions for successful exposures using an Olympus POS polarizing microscope and an Olympus PM-6 35-mm photomicrographic camera with a tungsten 6-V illuminator fitted with a clear daylight glass filter. It goes without saying that the illumination was strictly Köhler. The exposure meter was a Science and Mechanics Model 102. Although it is a simple matter to divide mentally the appropriate K value by the meter reading, a graph can be made by use of

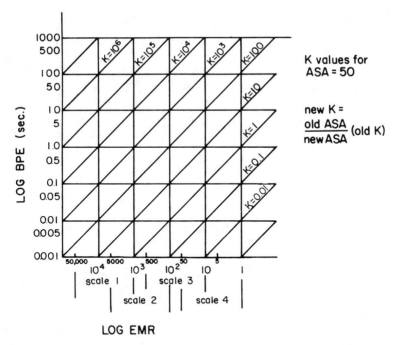

Fig. 15.30.   Best possible exposure time (BPE) as a function of exposure-meter reading (EMR) using a Science and Mechanics meter.

which the exposure-meter reading and the proper K yield the exposure time in seconds (for a film speed of 160 ASA in this example) (see Fig. 15.30).

A second method of calculating the exposure time from the meter reading preferred by some microscopists who have the Science and Mechanics meter is the use of f-stops instead of the K value. The psychology is unchanged in terms of the preceding tabulation of conditions; however, the meter reading and the determined best exposure time are used with the circular slide rule on the meter to determine an f-stop value. Subsequently, a meter reading with the appropriate f-stop will give the best exposure time. The f-stop corresponding to the K values are also listed in the tabulation. The K-value method may be preferred because one can make the simple calculation needed mentally; for example, if EMR is the exposure-meter reading, $K/\text{EMR} = 200/2000 = 1/10$ sec is the best exposure. The K/EMR ratio can be changed to yield exact available exposure time by using neutral density filters or by very small changes in aperture diaphragm or lamp voltage. In practice, given a calculated exposure time of $\frac{1}{10}$ sec, a pessimist might elect to bracket this exposure with $\frac{1}{5}$ and 1/25-sec exposures. It is, however, possible to control this procedure with sufficient care that a single calculated exposure will always give the best possible exposure.

There are other systems used for photomicrography, but the general idea illustrated here should be applicable in other situations. For example, many

TABLE 15.XI
Totally Absorbing Particles

Anthracite	Carbon black
Magnetite, Fe_3O_4	Asphalt
Metals and alloys	Coke
Wolframite, (Fe, Mn) WO_4	Oil soot
Franklinite, (Fe, Mn, Zn) $(FeO_2)_2$	Graphite
Spent catalyst	Stibnite, Sb_2S_3
Copper oxide, CuO	Potassium permanganate, $KMnO_4$
Lead suboxide, Pb_2O	Pyrolusite, MnO_2
Molybdenum tetraiodide, MoI_4	Mercurous oxide, Hg_2O
Nickel iodide, NiI_2	Niobium nitride, NbN
Platinum oxides, PtO and PtO_2	Niobium oxides, NbO and NbO_2
Platinum hydroxide, $Pt(OH)_2$	Osmium oxides, OsO and OsO_2
Platinum iodide, PtI_2	Strontium carbide, SrC_2

35-mm cameras can be adapted for photomicrography simply by clamping the camera onto the microscope bodytube. Most 35-mm-camera manufacturers furnish such an adapter as an accessory. Any such camera with a through-the-lens meter can be used to control the exposure time. The Pentax MX or KX camera, for example, has a built-in meter reading exposure time directly. If photomicrographs are to be taken with crossed polars, the exposure time is read with no analyzer and the crossed-polar exposures are taken with 6 times the indicated exposure time (3 stops) with one polar. Similarly, with dispersion staining, the exposure time is read with the annular stop and the actual exposure time with the central stop is also 6 times (3 stops) longer. Photomicrographs taken with a "590" gelatin filter require one stop (2 times) more exposure than the indicated time read with white light.

C.   INCIDENT-LIGHT ILLUMINATION

The particle analyst has a real need for reflected-light examination (top lighting). Without it, one cannot completely describe an opaque particle. In fact, without top light one can say only that an opaque particle was *observed*—not very satisfactory if the particle must be identified.

Many large, settled dust particles can be examined with a stereobinocular microscope equipped for top lighting. Direct examination of particles on membrane filters or impactor slides is easy with such an instrument. When particles are smaller, however, better resolving power and higher magnification are needed. Also, the problem becomes one of observing any given particle preparation with transmitted polarized light, perhaps simultaneously with incident light, or at least in rapid sequence, with the same microscope.

Strangely enough, many transparent substances appear opaque when viewed in transmitted light with a polarizing microscope. Opacity can be due to total internal reflection (Table 15.XI) or to a combination of total

TABLE 15.XII

High- and Low-Refractive-Index Particles That Often Appear Opaque When <1 μm in Diameter

High refractive index	Low refractive index
Copper oxide, Cu_2O	Sodium fluoride, NaF
Thallium iodide, TlI	Potassium silicofluoride, K_2SiF_6
Zinc sulfide, ZnS	Potassium fluoride, KF
Chromium oxide, Cr_2O_3	Ammonium silicofluoride, $(NH_4)_2SiF_6$
Cadmium sulfide, CdS	Lithium fluoride, LiF
$CaFe_2O_4$	Rubidium fluoride, RbF
Mercuric sulfide, HgS	Sodium silicofluoride, Na_2SiF_6
$AgAsS_2$	Potassium fluoborate, KBF_4
Arsenic tri-iodide, AsI_3	Magnesium silicofluoride, $MgSiF_6 \cdot 6H_2O$
Lead oxide, PbO	Potassium fluoride, $KF \cdot 2H_2O$
Silicon carbide, SiC	Cesium fluoborate, $CsBF_4$
Aluminum carbide, Al_4C_3	Manganese silicofluoride, $MnSiF_6 \cdot 6H_2O$
$Pb_3Cr_2O_9$	Cobalt silicofluoride, $CoSiF_6 \cdot 6H_2O$
Mercuric iodide, HgI_2	Iron silicofluoride, $FeSiF_6 \cdot 6H_2O$
Antimony iodide, SbI_3	Zinc silicofluoride, $ZnSiF_6 \cdot 6H_2O$
Mercuric sulfide, HgS	$CoF_2 \cdot 5HF \cdot 6H_2O$
Copper oxide, CuO	Sodium cyanate, NaCNO
Iron oxide, Fe_2O_3	NH_4HF_2
Selenium	Nickel silicofluoride, $NiSiF_6 \cdot 6H_2O$
Diamond	$NiF_2 \cdot 5HF \cdot 6H_2O$
Titanium dioxide, TiO_2	Sodium sulfate, $Na_2SO_4 \cdot 10H_2O$

reflection and refraction (Table 15.XII). Highly refractive solids, such as rutile, diamond, silicon carbide, realgar, and lead oxide (PbO), reflect and refract so much light in Aroclor and the lower-index media that they may seem opaque, especially when very small. The extent to which such substances appear opaque depends on the refractive-index difference between the solid particle and its surroundings. Even very low-index solids (Table 15.XII) may appear opaque in Aroclor. The particles will reflect light to the same degree as high-index particles.

The particles listed in Table 15.XII are often thought to be opaque when they are very small, especially when many such particles are held together in a matrix. A dried white-paint particle or a white sidewall tire particle is a good example (Figs. 15.31 and 15.32). Such particles are dense dispersions of colorless, transparent, but highly refractive particles of titanium dioxide, zinc oxide, basic lead carbonate, or a similar substance in a transparent, colorless, organic binder. These particles, unless very thin, are "opaque"— no light is transmitted through the particle into the microscope objective. Even air bubbles, or perhaps especially air bubbles, can appear opaque by this mechanism. Ethyl cellulose particles with and without a top light illustrate this. These polymer particles are very porous, that is, filled with air bubbles. The refractive-index difference between air and ethyl cellulose is great, and even though the index difference between ethyl cellulose and

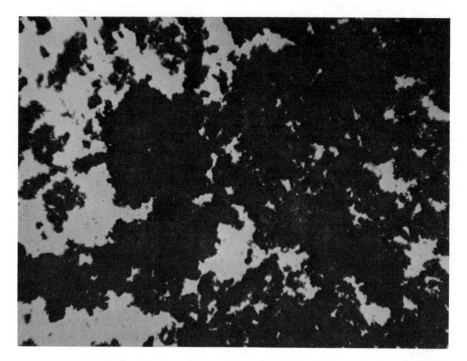

Fig. 15.31.   Particles from buffing white sidewall automobile tires, transmitted light only.

Aroclor is not too great, the particles seem opaque (Figs. 15.33 and 15.34). In the smallest particles of polymer, individual "opaque" air bubbles can be observed. Glassy flyash particles often contain enough small air bubbles to render them "opaque." Some substances are, of course, truly opaque.

Two particles that appear opaque are shown in Figs. 15.31–15.34, first with transmitted light only, then with a top light added. The need for top light is thus graphically illustrated.

Obviously, the particle analyst must use top lighting to fully characterize and identify any particle. What are the various approaches to this problem and which should be used?

There are two general procedures for illuminating the top surfaces of particles. One makes use of built-in or attachable light sources to obtain either incident bright-field illumination or incident dark-field illumination (Fig. 15.35). This may be called *vertical illumination*, although strictly, only the incident bright field uses vertical light rays. The second general procedure for top lighting makes use of auxillary illuminators; this is generally called *unilateral illumination*.

D.   UNILATERAL ILLUMINATION

An auxiliary light source may be added to any microscope, since one or more external illuminators may be placed at the side with the light beams

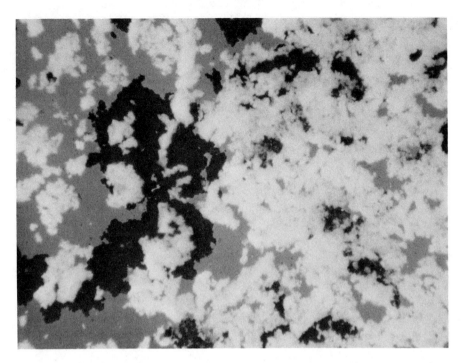

Fig. 15.32. Same as Fig. 15.31, but with top light as well as transmitted light.

directed downward from above the microscope stage, past the objective, onto the top of the particles. Such lamps must deliver a concentrated spot of light and they must be used at as high an angle with respect to the horizontal as possible. The light, in any case, will undergo some reflection from the upper surface of the coverslip. Less light is lost if the angle of incidence is high. There is, of course, no such problem if the microscope slide preparation is uncovered. Particles on a membrane filter or spread loosely on a slide can be examined best by top lighting. A very dilute collodion "paint" (ordinary 3% collodion solution diluted 100-fold with ethyl or amyl acetate) of the particles may be prepared, spread on a slide and allowed to dry. The particles are thus attached to the slide, but no visible collodion film remains. This permanent dry preparation may be stored, or it may be examined at once. The collodion does not interfere with examination of particles mounted in media of various refractive indexes. Any desired liquid is added, a coverslip is placed over the preparation, and the sample is examined in transmitted light. The liquid can be washed away with benzene and the resulting dry preparation is ready for a second liquid, for dry examination with top lighting, or for storage.

The objective may obscure the light path if the illuminator is set too high or if objectives with short working distances are used. Special objectives designed to permit passage of high-angle incident light, however, are now

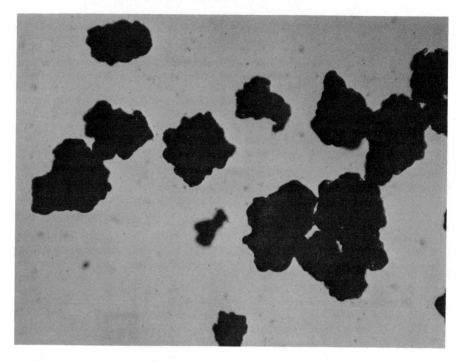

Fig. 15.33.   Ethyl cellulose, transmitted light only.

on the market. Usually, examination with top light is limited to 16-mm 10X objectives or those of lower power. This means a limiting total microscope magnification of about 200X. Anyone who wishes to use top lighting for particles too small to be observed properly at 200X has a choice: (1) Ignore the fact that the high-power objective obscures the direct light, and use it anyway; (2) use special long-working-distance objectives; or (3) use vertical illuminators incorporated into the microscope. Each of these requires some further explanation, especially the first. It turns out that the light will enter the narrow slit between the objective, say a 4-mm 40X, and the coverslip, where it will be reflected back and forth. Enough light bounces around and finally finds its way to the particles that the upper surfaces can be studied. It helps to use two auxiliary illuminators, and it may be necessary to turn off the transmitted light beam to see the reflected light.

A variation of this technique that does not require an auxiliary illuminator involves intense, high-numerical-aperture, substage illumination with crossed polars. The objective reflects enough light to illuminate the tops of opaque particles. If the field itself is bright, the intensity of illumination on the particles is too low to register on the retina, but if the field is darkened by crossing the polars, the color and shape of opaque particles can be seen with the 4-mm 40X objective.

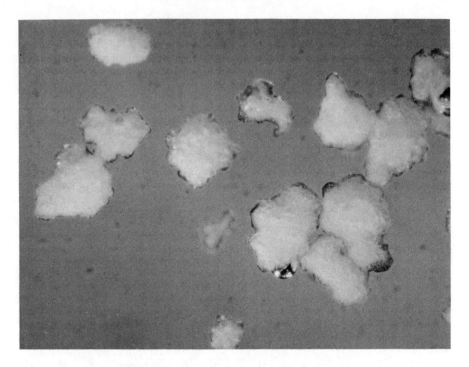

Fig. 15.34. Ethyl cellulose, top light as well as transmitted light.

Why, then, use any auxiliary or special vertical illuminators? The answer is twofold: Top lighting is less intense with the multiple-reflection method, and this method is less convenient. Properly arranged top lighting of sufficient intensity permits simultaneous observations with transmitted polarized and top lighting and thus saves time.

Vertical   Illuminators

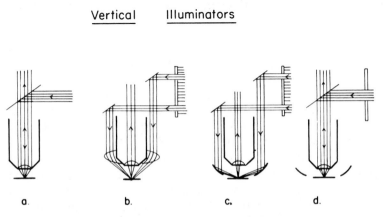

a.   b.   c.   d.

Fig. 15.35. Vertical illuminators.

Long-working-distance objectives. Several companies manufacture long-working-distance objectives. These are usually either 8 mm 20X or 4 mm 40–45X. In every case, the working distance is equal to or greater than the working distance of a standard 16-mm 10X objective. It is therefore possible to obtain excellent top lighting using external light sources with these objectives.

Auxiliary illuminators. Many manufacturers produce suitable spotlight auxiliary illuminators. The essential requirement is a focusable spot image of the filament about 8–10 cm from the lamp. The spot size should not be more than 5–10 mm across. An interesting possibility is the use of a light pipe, usually constructed from polymethylmethacrylate (Lucite, Plexiglas, and Perspex are trade names). This material, in a rod of any desired diameter, can be used to focus the beam from an illuminator. When warmed to about 100°C, the rods can be bent and drawn down to small diameter. The lamp can then be any distance away and located anywhere relative to the microscope. The light pipe gathers the light, transmits it to the microscope, and "focuses" it on the specimen.

Vertical illuminators. Many manufacturers furnish attachments for the microscope that fit between the objective and bodytube so that light can enter the bodytube from one side and be reflected and focused through or around the objective. In some, such as most metallurgical objectives, a planar glass plate reflects the light straight down through the objective for incident bright-field illumination (Fig. 15.35a). Others, such as the Leitz Ultropak (Fig. 15.35b), reflect the light around the outside of the objective and then refract it onto the specimen for incident dark-field illumination. The light is refracted by a lens with a hole cut in the center, through which the objective protrudes. This condensing lens is focusable. Still other vertical illuminators (e.g., the "Epi" type) allow either bright- or dark-field illumination. These have some advantages over the other types in that either incident dark-field or incident bright-field illumination is possible with a single objective. Only the reflecting mirror above the objective is changed (Figs. 15.35c, d).

All objectives intended for use with vertical illuminators are corrected for use *without* a coverslip, as they are intended primarily for uncovered polished metal specimens. The incident dark-field objectives may be used with success with covered specimens, just as with unilateral illuminations. Incident bright-field objectives, however, cannot be used with covered specimens, as the light coming from the objective will be reflected from the top of the coverslip, filling the field with an intense glare.

One sets up illumination with incident bright-field objectives as for Köhler's method with transmitted light. The vertical illuminator will have two diaphragms, one a field diaphragm and the other an aperture diaphragm. When setting up the illumination it is best to use a planar, specular surface as a specimen. The field diaphragm is closed somewhat and the specimen

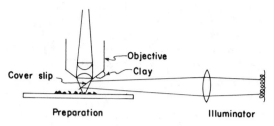

Fig. 15.36. A makeshift illuminator for low-power magnification.

is brought into focus. It will be seen that the field diaphragm is is in sharp focus—the objective lens is also acting as the light-condenser lens. The field diaphragm is centered and then opened until it is just outside the field of view. The lamp filament is focused on the aperture stop (a ground glass will aid this step if the iris is not conveniently located). Using the Bertrand lens, or removing the ocular, the aperture stop is centered in relation to the objective aperture. Some vertical illuminators have provision for tilting the planar glass reflector. The specimen is now observed and the aperture stop set according to the requirements of the specimen (resolution and contrast).

For use in quantitative polarized light work, a right-angle prism is used instead of the planar glass reflector. This gives more light and does not depolarize the beam, but because it occludes part of the objective aperture, it may reduce the resolution. With this reflector, the aperture stop is closed and centered on the prism.

Objectives for use with vertical illuminators, in addition to being corrected for use without a coverslip, are corrected for a longer tube length because of the interposition of the reflecting elements. This distance may be 180–210 mm (185 mm for the Ultropak).

A crude but effective vertical-illumination system can be set up quickly by setting about half of an 18-mm coverslip in a bit of modeling clay and fixing it at a 45° angle to the edge of the objective (Fig. 15.36). The coverslip may also be fixed with clay to the slide rather than to the objective. A very small segment of a coverslip will fit beneath 10X objectives.

Another interesting top light is a 1-in.-diameter ring of "neon" tubing. Aristo Grid Lamp Products, Inc. (65 Harbor Road, Port Washington, L.I., NY 11050), can furnish one of these with a transformer, or you can have a signmaker make one for you (request the signmaker to give you a green glow rather than the red neon glow). It can be arranged around the objective so that light strikes the particles from all sides and is reflected back into the objective. Such an illuminator gives a bright image but without shadows. This "flattens" the particles and may not allow good shape interpretation. Along the same lines, some microscopists (1) use half a ping-pong ball lying on the stage over the preparation, with an opening in the top through which the objective peers. One or more auxiliary spot illuminators can be focused on the outer surface of the ping-pong ball to illuminate the specimen. All shadowing is, of course, lost and the particles are flattened; much of the surface texture is also lost.

The choice of system depends, to some extent at least, on one's budget. Entirely satisfactory results can be obtained with either built-in vertical illumination or auxiliary illumination. Reflected-light intensity is less with auxiliary illumination, but this causes inconvenience only in photomicrography. Since the proper exposure for top light is usually longer than for transmitted light, a double exposure is usually required.

With incident illumination there should be no problem in differentiating paint spray from oil soot, or metal shavings from insect parts. To fully identify small particles, it is necessary to use all obtainable information. Often the use of top lighting makes unnecessary the use of more sophisticated tools, such as electron microscopy, x-ray diffraction, or electron microprobe analysis.

VIII.   CRYSTAL MORPHOLOGY

Crystals are characterized by the arrangement of their constituent atoms in a repeating three-dimensional array quite unlike the random arrangement of atoms in a liquid or other noncrystalline material (Fig. 15.37). The orderly three-dimensional arrangement of atoms can vary, but only according to definite rules of crystallography. Many metals, all of the alkali halides, and many other compounds have a very simple arrangement of atoms, called *cubic*, and all such materials are said to be in the *cubic system*. There are six such systems in all, differing as to the arrangement of atoms (Fig. 15.38). To show the three-dimensional arrangement, a top, front, and side view are shown for each system in Fig. 15.38.

A.   THE SIX CRYSTAL SYSTEMS

The arrangement of atoms in the different crystal systems can be described as follows:

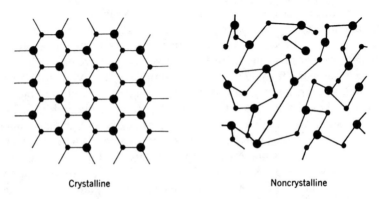

Crystalline   Noncrystalline

Fig. 15.37.   Crystalline and noncrystalline materials.

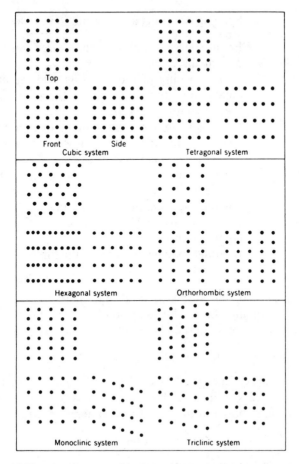

Fig. 15.38. Arrangement of lattice points in the six crystal systems.

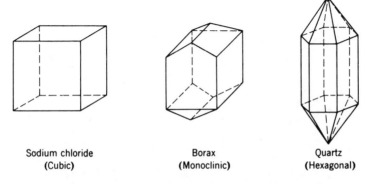

Sodium chloride	Borax	Quartz
(Cubic)	(Monoclinic)	(Hexagonal)

Fig. 15.39. Some typical crystal habits.

System	Spacing of atoms and orientation of axes
Cubic	Equal spacing of atoms along three mutually perpendicular directions
Tetragonal	Equal spacing of atoms along two mutually perpendicular directions; normal to the plane thus formed, the spacing is different
Hexagonal	Equal spacing of atoms along three directions 120° apart in the same plane; normal to the plane thus formed, the spacing is different
Orthorhombic	Unequal spacing of atoms along three mutually perpendicular directions
Monoclinic	Unequal spacing of atoms along two mutually oblique directions; normal to the plane thus formed, the spacing is different
Triclinic	Unequal spacing of atoms along three mutually oblique directions

The arrangement of the atoms in solid materials is obviously very important, since it explains, first, why some solids are crystalline, and second, why different crystalline materials have different shapes (Fig. 15.39).

B.   THE CUBIC SYSTEM

The external faces of a crystal are directly related to the internal arrangement of atoms, the crystal lattice. The possible positions of crystal faces on a crystal lattice are shown in two and three dimensions in Fig. 15.40.

1.   The Cube Form

In the cubic system there are three common crystal forms: the cube, the octahedron, and the rhombic dodecahedron. The cube (Fig. 15.41) results

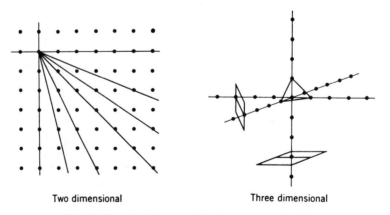

Two dimensional   Three dimensional

Fig. 15.40.   Possible crystal faces on a crystal lattice.

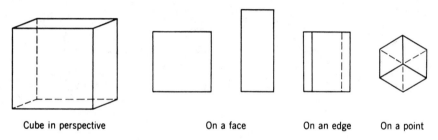

Cube in perspective   On a face   On an edge   On a point

Fig. 15.41. Cube seen in various positions.

when all of the faces on the crystal lie parallel to two of the three axes and cut only one axis.

2. The Octahedron

The octahedron (Fig. 15.42) has eight faces, all of which intersect all three crystal axes.

3. The Rhombic Dodecahedron

The rhombic dodecahedron (Fig. 15.43) has 12 faces, all of which intersect two of the three crystallographic axes.

Figures 15.41–15.43 illustrate another behavior of crystals—the growth of distortions. In each of these three figures, the sketch on the left shows the ideal crystal, on which all of the faces are equally developed. The sketches of crystals lying on a face in each figure are distortions in which at least one face has developed more than the others. At this point, it is very important to notice that all of these crystals, both ideal and distorted, have identical crystal angles. In any distortion the crystal faces move only parallel to themselves, thus changing the general crystal shape, but not the crystal angles. This is an expression of the law of constancy of interfacial angles.

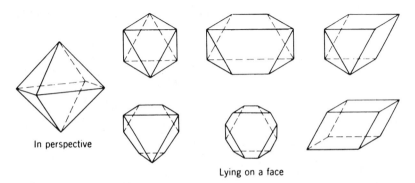

In perspective

Lying on a face

Fig. 15.42. Octahedron seen in various positions.

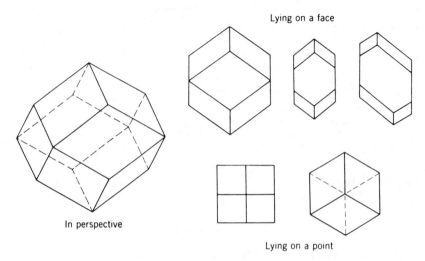

Fig. 15.43.   Rhombic dodecahedron seen in various positions.

C.   CRYSTAL HABIT

The general shape of a crystal is usually referred to as the *crystal habit*. Figure 15.44 shows some of the more common habits. The same descriptive terms are used for crystal habits in any of the crystal systems. Words like *cubes, hexagons, prisms, rhombs* and so on are best not used unless the crystal system is known.

D.   CRYSTAL CLEAVAGE

Another important feature of many crystalline substances is cleavage. Some minerals, such as limestone (calcite), have very excellent cleavage properties, so the powdered mineral breaks up into quite well-formed small crystals called rhombohedra (Fig. 15.45). Few other minerals show as perfect cleavage as calcite, and many show no regular cleavage. In fact, quartz, a crystalline material, and glass, a noncrystalline material, show almost iden-

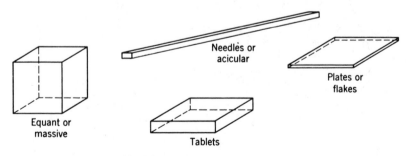

Fig. 15.44.   Common crystal habits.

Ideal

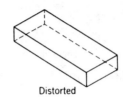

Distorted

Fig. 15.45. Rhombohedra.

tical cleavages, usually called *conchoidal*. These two materials, when finely ground, cannot be differentiated microscopically except by optical tests.

Crystalline materials are usually described morphologically in terms of crystal habit, form, and angles, and, when possible, the crystal system. The crystal angles used most are the so-called *profile angles*, which are the angles shown by the crystal when observed lying on a crystal face. A cube, for example, shows 90° profile angles; an octahedron, 60° or 120° angles; a rhombic dodecahedron, 70°32' and its supplement, 109°28'; and a calcite rhombohedron, 101°54' and its supplement, 78°6'.

E. LAWS OF CRYSTALLOGRAPHY

Crystal forms and habits are governed by two laws of crystallography. The first is the law of *constancy of interfacial angles*, which states that in all crystals of the same substance, the angles between corresponding faces are constant. The second is the law of *rational indexes*, which states that the ratios of the intercepts of different crystal faces on the different crystallographic axes are small integers.

F. CRYSTAL SYMMETRY

One of the easiest ways to differentiate between crystals by system is through the use of external crystal symmetry. The elements of external symmetry include the plane (P), the axis (A), and the center (C). The axis of symmetry may be twofold (A_2), threefold (A_3), fourfold (A_4), or sixfold (A_6). Each of the crystal systems has different symmetry requirements (Table 15.XIII). Some of these symmetry elements are illustrated in Fig. 15.46.

G. THE ASSIGNMENT OF CRYSTALLOGRAPHIC AXES

The assignment of crystallographic axes is based on the positions of the external symmetry elements and on the internal lattice spacings as follows:

Cubic   Three mutually perpendicular identical directions are chosen as the three a axes such that the required $4A_3$ axes form body diagonals of a cube defined by the three a axes.

Tetragonal   The required A_4 axis is chosen as c and two mutually perpendicular directions in the plane perpendicular to c are chosen as the two a axes.

TABLE 15.XIII
Symmetry Elements in the Crystal Systems

| | External symmetry | |
System	Maximum	Minimum[a]
Cubic	$3A_4, 4A_3, 6A_2, 9P, C$	$4A_3$
Tetragonal	$A_4, 4A_2, 5P. C$	A_4
Hexagonal	$A_6, 6A_2, 7P, C$	A_6
Rhombohedral	$A_3, 3A_2, 3P, C$	A_3
Orthorhombic	$3A_2, 3P, C$	$2A_2$
Monoclinic	A_2, P, C	A_2
Triclinic	C	None

[a] Sphenoids and other hemisymmetric forms may show less symmetry.

Twofold symmetry (A_2); two planes (tetragonal, orthorhombic)

Threefold symmetry (A_3); three planes (cubic, trigonal, hexagonal)

Fourfold symmetry (A_4); four planes (cubic, tetragonal)

Sixfold symmetry (A_6); six planes (hexagonal)

Single plane (P); (monoclinic)

Center (C); (triclinic)

Fig. 15.46.   Elements of symmetry.

Hexagonal or rhombohedral	The A_3 or A_6 axis is chosen as c and three axes 120° apart in the plane perpendicular to c are chosen as the three a axes.
Orthorhombic	The intersection of two planes of symmetry becomes one axis; the other two axes lie in the two planes and perpendicular to the intersection. If the crystal has three planes of symmetry, then the three intersections become the three axes. The three axes are chosen as a, b, and c, and $c < a < b$ as determined by diffraction or, if necessary, by interfacial angles.
Monoclinic	The unique axis perpendicular to the plane of symmetry is b. Axes a and c are chosen from directions in the plane of symmetry such that $c < a$ and such that simple forms result.
Triclinic	Three prominent directions are chosen as a, b, and c such that $c < a < b$ and such that simple forms result.

H. CRYSTAL FORMS

Similar external faces of a crystal are classified as *forms*; for example, the six faces of a cube are similar and constitute the cube form. In general, a single face having no equivalent is called a *pedion*; two parallel faces having no equivalents are called a *pinakoid*. A form made up of two nonparallel faces is called a *dome* if symmetrical about a plane of symmetry, and a *sphenoid* is not symmetrical about a plane of symmetry (except in the tetragonal systems, where all such faces are called sphenoids). *Prisms* are forms made up of three, four, six, eight, or 12 similar faces and are called *trigonal* with three, *tetragonal* or *rhombic* with four, *hexagonal* with six, *ditetragonal* with eight, and *dihexagonal* with 12. A group of three, four, six, eight, or 12 similar faces intersecting in a point is called a *pyramid*; the modifying adjectives are the same as for the corresponding prisms.

I. MILLER INDEXES

The notation usually used for naming crystal faces consists of the so-called Miller indexes, which have the form $\{hkl\}$. This notation is based on the above-described assignment of axes and an expression of the intercepts of the face on the three (or four, in hexagonal systems) axes. The axes are taken in the order a_1, a_2, a_3 (cubic); a_1, a_2, c (tetragonal); a_1, a_2, a_3, c (hexagonal); or a, b, c (orthorhombic, monoclinic, and triclinic). Crystals are always oriented so that a_1 or a lies front (positive) to back (negative); a_2 or b lies right (positive) to left (negative), and a_3 or c lies top (positive) to bottom (negative) (Fig. 15.47).

If we take the simplest face that intersects all of the axes (Fig. 15.48), we can express the intercepts as OA:OB:OC (these may be obtained by calculation from appropriate interfacial angles or, preferably, by x-ray diffrac-

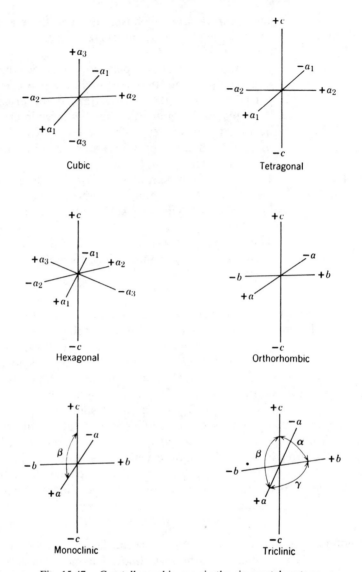

Fig. 15.47.   Crystallographic axes in the six crystal systems.

tion determination of the actual lattice spacings). In any case, the Miller index *hkl* of another face on the same crystal having intercepts OH, OK, and OL (Fig. 15.48) is

$$hkl = \frac{OA}{OH} : \frac{OB}{OK} : \frac{OC}{OL} = \tfrac{1}{4}:\tfrac{1}{3}:\tfrac{1}{2} = 3:4:6 \text{ or simply } 346$$

where *hkl* are expressed as the simplest whole numbers. OA, OB, and OC

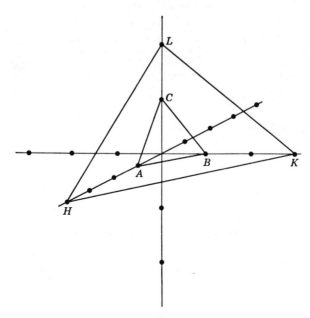

Fig. 15.48. Miller indexes: *ABC* face = 111; *HKL* face = 346.

are usually referred to as a_0, b_0, and c_0, the lattice parameters. When h, k, or l is 0, that face is parallel to the corresponding axis. Pedions and pinakoids are usually 100, 010, or 001; domes and prisms are usually $hk0$, $h0l$, or $0kl$, and bipyramids and sphenoids are usually hkl. A single symbol, hkl, is a face symbol; {hkl} is a form symbol and includes all of the faces in that form; and [hkl] is a zone symbol that includes all faces having parallel intersection edges.

There are a number of forms in each crystal system. The most important to the chemical microscopist are tabulated in Table 15.XIV. In each case the form symbol {hkl} is the generalized Miller index of that face in the form that most closely fits the description—"in front, on top, to the right" (in that order) when the crystal is "set up" properly. When properly set up, the crystal is visualized with the c axis vertical (a_3 vertical in the cubic system), the a axis front to back (a_1 in the cubic and tetragonal systems), and the b axis right to left (a_2 in the cubic and tetragonal systems). The positive ends of these axes are, respectively, top, front, and right. Crystals in the hexagonal system are set up with a_1 and $-a_3$ symmetrically arranged pointing to the front, a_1 slightly to the left, and $-a_3$ slightly to the right; a_2 then lies right to left, with $-a_2$ to the left.

J. TWINNING

Crystals are often twinned. In most cases, twinning results in formation of re-entrant angles, as in the examples shown in Fig. 15.49. Generally, but

Table 15.XIV
Forms and Faces in the Six Crystal Systems

System	Form symbol	Miller indexes of all faces
Cubic		
Holosymmetric forms		
Cube	{100}	100, $\overline{1}$00, 010, 0$\overline{1}$0, 001, 00$\overline{1}$
Rhombic dodecahedron	{110}	110, $\overline{1}$10, 1$\overline{1}$0, 1$\overline{1}$0, 101, $\overline{1}$0$\overline{1}$, $\overline{1}$01,
		10$\overline{1}$, 011, 0$\overline{1}\overline{1}$, 0$\overline{1}$1, 01$\overline{1}$
Octahedron	{111}	111, $\overline{1}\overline{1}\overline{1}$, $\overline{1}$11, 1$\overline{1}\overline{1}$, 1$\overline{1}$1, $\overline{1}$1$\overline{1}$, 11$\overline{1}$,
		$\overline{1}\overline{1}$1
Trapezohedron	{*hkk*}	211, 121, 112, $\overline{2}\overline{1}\overline{1}$, $\overline{1}\overline{2}\overline{1}$, $\overline{1}\overline{1}\overline{2}$, $\overline{2}$11,
		$\overline{1}$21, $\overline{1}$12, 2$\overline{1}\overline{1}$, 1$\overline{2}\overline{1}$, 1$\overline{1}\overline{2}$, 2$\overline{1}$1, 1$\overline{2}$1,
		1$\overline{1}$2, $\overline{2}$1$\overline{1}$, $\overline{1}$2$\overline{1}$, $\overline{1}$1$\overline{2}$, 21$\overline{1}$, 12$\overline{1}$, 11$\overline{2}$,
		$\overline{2}\overline{1}$1, $\overline{1}\overline{2}$1, $\overline{1}\overline{1}$2

Hemisymmetric forms (one half of faces of corresponding holosymmetric form are missing)

System	Form symbol	Miller indexes of all faces
Pyritohedron		
positive	{210}	210, 2$\overline{1}$0, 102, 10$\overline{2}$, 021, 02$\overline{1}$, $\overline{2}\overline{1}$0,
		$\overline{2}$10, $\overline{1}$0$\overline{2}$, $\overline{1}$02, 0$\overline{2}\overline{1}$, 0$\overline{2}$1
negative	{120}	120, $\overline{1}$20, 201, 20$\overline{1}$, 012, 0$\overline{1}$2, $\overline{1}\overline{2}$0,
		1$\overline{2}$0, $\overline{2}$0$\overline{1}$, $\overline{2}$01, 0$\overline{1}\overline{2}$, 01$\overline{2}$
Tetrahedron		
positive	{111}	111, $\overline{1}\overline{1}$1, 1$\overline{1}\overline{1}$, $\overline{1}$1$\overline{1}$
negative	{1$\overline{1}$1}	$\overline{1}\overline{1}\overline{1}$, 11$\overline{1}$, $\overline{1}$11, 1$\overline{1}$1

NOTE: Additional forms possible in the cubic system are seldom, if ever, observed microscopically; the pyritohedra and trapezohedra are very rarely observed.

System	Form symbol	Miller indexes of all faces
Tetragonal		
Holosymmetric forms		
Prism		
first order	{110}	110, $\overline{1}\overline{1}$0, $\overline{1}$10, 1$\overline{1}$0
second order	{100}	100, $\overline{1}$00, 010, 0$\overline{1}$0
Dipyramid		
first order	{111}	111, $\overline{1}\overline{1}\overline{1}$, $\overline{1}$11, 1$\overline{1}\overline{1}$, 1$\overline{1}$1, $\overline{1}$1$\overline{1}$, 11$\overline{1}$,
		$\overline{1}\overline{1}$1
second order	{101}	101, $\overline{1}$0$\overline{1}$, $\overline{1}$01, 10$\overline{1}$, 011, 0$\overline{1}\overline{1}$, 0$\overline{1}$1,
		01$\overline{1}$
Pinakoid	{001}	001, 00$\overline{1}$
Hemisymmetric forms		
Disphenoids	{*hk*1}	111, $\overline{1}\overline{1}$1, 1$\overline{1}\overline{1}$, $\overline{1}$1$\overline{1}$
	{*h$\overline{k}$*1}	1$\overline{1}$1, 11$\overline{1}$, $\overline{1}$11, $\overline{1}\overline{1}\overline{1}$
	{*h*01}	101, $\overline{1}$01, 01$\overline{1}$, 0$\overline{1}\overline{1}$
	{0*k*1}	011, 0$\overline{1}$1, $\overline{1}$0$\overline{1}$, 10$\overline{1}$

NOTE: More complex forms, e.g., scalenohedra and the *hk*1 and *h*01 dipyramids and disphenoids, are extremely rare in microscopy.

System	Form symbol	Miller indexes of all faces
Hexagonal		
Holosymmetric forms		
Prism		
first order	{10$\overline{1}$0}	10$\overline{1}$0, $\overline{1}$010, 0$\overline{1}$10, 01$\overline{1}$0, $\overline{1}$100, 1$\overline{1}$00
second order	{11$\overline{2}$0}	11$\overline{2}$0, $\overline{1}\overline{1}$20, 2$\overline{1}\overline{1}$0, $\overline{2}$110, $\overline{1}$2$\overline{1}$0, 1$\overline{2}$10

Table 15.XIV (*continued*)

System	Form symbol	Miller indexes of all faces
Dipyramid		
first order	$\{h0\bar{h}1\}$	$h0\bar{h}1,\ \bar{h}0h1,\ 0\bar{h}h1,\ 0h\bar{h}1,\ \bar{h}h01,\ h\bar{h}01,$ etc.
second order	$\{hh\cdot\overline{2h}\cdot1\}$	$hh\cdot\overline{2h}\cdot1,\ \overline{hh}\cdot2h\cdot1,\ 2h\cdot\overline{hh}1,\ \overline{2h}\cdot hh1,$ etc.
Pinakoid	$\{0001\}$	$0001,\ 000\bar{1}$
Hemisymmetric forms		
Pyramid		
first order, positive	$\{h0\bar{h}1\}$	$h0\bar{h}1,\ \bar{h}0h1,\ 0\bar{h}h1,\ 0h\bar{h}1,\ \bar{h}h01,\ h\bar{h}01$
first order, negative	$\{h0\bar{h}\bar{1}\}$	$h0\bar{h}\bar{1},\ \bar{h}0h\bar{1},\ 0\bar{h}h\bar{1},\ 0h\bar{h}\bar{1},\ \bar{h}h0\bar{1},\ h\bar{h}0\bar{1}$
second order, positive	$\{hh\cdot\overline{2h}\cdot1\}$	$hh\cdot\overline{2h}\cdot1,\ \overline{hh}\cdot2h\cdot1,\ 2h\cdot\overline{hh}1,\ 2h\cdot\overline{hh}\cdot\overline{hh}1,$
second order, negative		$hh\cdot\overline{2h}\cdot\bar{1},\ \overline{hh}\cdot2h\cdot\bar{1},\ 2h\cdot\overline{hh}\bar{1},\ \overline{2h}\cdot hh\bar{1},$ etc.
Dipyramid		
first order	$\{h0\bar{h}1\}$	$h0\bar{h}1,\ h\bar{h}01,\ 0\bar{h}h1,\ h0\bar{h}1,\ \bar{h}h0\bar{1},\ 0\bar{h}h\bar{1}$
second order	$\{hh\cdot\overline{2h}\cdot1\}$	$hh\cdot\overline{2h}\cdot1,\ \overline{2h}\cdot hh1,\ h\cdot\overline{2h}\cdot h1,\ hh\cdot\overline{2h}\cdot1,$ $\overline{2h}\cdot hh\bar{1},\ h\cdot2h\cdot h\bar{1}$

Rhombohedral subsystem (hexagonal)
Hemisymmetric forms

System	Form symbol	Miller indexes of all faces
Rhombohedron		
positive	$\{h0\bar{h}1\}$	$h0\bar{h}1,\ h\bar{h}01,\ 0\bar{h}h1,\ hh\cdot\overline{2h}\cdot1,\ \overline{2h}\cdot hh\bar{1},$ $h\cdot\overline{2h}\cdot h\bar{1}$
negative	$\{0h\bar{h}1\}$	$0h\bar{h}1,\ \bar{h}0h1,\ 0h\bar{h}1,\ \overline{hh}\cdot2h\cdot\bar{1},\ 2h\cdot\overline{hh}\bar{1},$ $h\cdot\overline{2h}\cdot h\bar{1}$
Pedion	$\{0001\}$	0001
	$\{000\bar{1}\}$	$000\bar{1}$

Ogdosymmetric forms (three-fourths of faces of corresponding holosymmetric form are missing)

System	Form symbol	Miller indexes of all faces
Pyramid		
first order, positive	$\{h0\bar{h}1\}$	$h0\bar{1},\ \bar{h}h01,\ 0\bar{h}h1$
first order, negative	$\{h0\bar{h}\bar{1}\}$	$h0\bar{h}\bar{1},\ \bar{h}h0\bar{1},\ 0\bar{h}h\bar{1}$
second order, positive	$\{hh\cdot\overline{2h}\cdot1\}$	$hh\cdot\overline{2h}\cdot1,\ \overline{2h}\cdot hh1,\ h\cdot\overline{2h}\cdot h1$
second order, negative	$\{hh\cdot\overline{2h}\cdot\bar{1}\}$	$hh\cdot\overline{2h}\cdot\bar{1},\ \overline{2h}\cdot hh\bar{1},\ h\cdot\overline{2h}\cdot h\bar{1}$

NOTE: Again, the more complex forms possible in the hexagonal system are rarely observed microscopically.

Orthorhombic
Holosymmetric forms

System	Form symbol	Miller indexes of all faces
Prism	$\{hk0\}$	$hk0,\ \bar{h}k0,\ h\bar{k}0,\ \bar{h}\bar{k}0$
	$\{h01\}$	$h01,\ \bar{h}0\bar{1},\ \bar{h}01,\ h0\bar{1}$
	$\{0k1\}$	$0k1,\ 0\bar{k}\bar{1},\ 0\bar{k}1,\ 0k\bar{1}$
Dipyramid	$\{hk1\}$	$hk1,\ \bar{h}\bar{k}\bar{1},\ \bar{h}k1,\ h\bar{k}1,\ \bar{h}k\bar{1},\ h\bar{k}\bar{1},\ hk\bar{1},$ $\bar{h}\bar{k}1$
Pinakoid	$\{100\}$	$100,\ \bar{1}00$
	$\{010\}$	$010,\ 0\bar{1}0$
	$\{001\}$	$001,\ 00\bar{1}$
Hemisymmetric forms		
Pyramid		
positive	$\{hk1\}$	$hk1,\ \bar{h}\bar{k}1,\ \bar{h}k1,\ h\bar{k}1$
negative	$\{hk\bar{1}\}$	$hk\bar{1},\ \bar{h}\bar{k}\bar{1},\ \bar{h}k\bar{1},\ h\bar{k}\bar{1}$

Table 15.XIV (continued)

System	Form symbol	Miller indexes of all faces
Disphenoid		
right	$\{hk1\}$	$hk1$, $\bar{h}\bar{k}1$, $\bar{h}k\bar{1}$, $h\bar{k}\bar{1}$
left	$\{\bar{h}k1\}$	$\bar{h}k1$, $\bar{h}\bar{k}\bar{1}$, $hk\bar{1}$, $h\bar{k}1$
Dome	$\{0k1\}$	$0k1$, $0\bar{k}1$
	$\{0k\bar{1}\}$	$0k\bar{1}$, $0\bar{k}\bar{1}$
	$\{h01\}$	$h01$, $\bar{h}01$
	$\{h0\bar{1}\}$	$h0\bar{1}$, $\bar{h}0\bar{1}$
Pedion	$\{001\}$	001
	$\{00\bar{1}\}$	001
Monoclinic		
Holosymmetric forms		
Prism	$\{hk0\}$	$hk0$, $\bar{h}\bar{k}0$, $\bar{h}k0$, $h\bar{k}0$
	$\{0k1\}$	$0k1$, $0\bar{k}\bar{1}$, $0\bar{k}1$, $0k\bar{1}$
	$\{hk1\}$	$hk1$, $\bar{h}\bar{k}\bar{1}$, $\bar{h}k\bar{1}$, $h\bar{k}1$
	$\{\bar{h}k1\}$	$\bar{h}k1$, $h\bar{k}\bar{1}$, $hk\bar{1}$, $\bar{h}\bar{k}1$
Pinakoid	$\{001\}$	001, $00\bar{1}$
	$\{010\}$	010, $0\bar{1}0$
	$\{100\}$	100, $\bar{1}00$
	$\{h01\}$	$h01$, $\bar{h}0\bar{1}$
	$\{\bar{h}01\}$	$\bar{h}01$, $h0\bar{1}$
Hemisymmetric forms		
Sphenoid		
first order, right	$\{0k1\}$	$0k1$, $0k\bar{1}$
first order, left	$\{0\bar{k}1\}$	$0\bar{k}1$, $0\bar{k}\bar{1}$
third order, right	$\{hk0\}$	$hk0$, $\bar{h}k0$
third order, left	$\{h\bar{k}0\}$	$h\bar{k}0$, $\bar{h}\bar{k}0$
fifth order, right negative	$\{hk1\}$	$hk1$, $\bar{h}k\bar{1}$
fifth order, right positive	$\{\bar{h}k1\}$	$\bar{h}k1$, $hk\bar{1}$
fifth order, left negative	$\{h\bar{k}1\}$	$h\bar{k}1$, $\bar{h}\bar{k}\bar{1}$
fifth order, left positive	$\{\bar{h}\bar{k}1\}$	$\bar{h}\bar{k}1$, $h\bar{k}\bar{1}$
Dome		
first order, upper positive	$\{0k1\}$	$0k1$, $0\bar{k}1$
first order, lower negative	$\{0k\bar{1}\}$	$0k\bar{1}$, $0\bar{k}\bar{1}$
third order, front	$\{hk0\}$	$hk0$, $h\bar{k}0$
third order, back	$\{\bar{h}k0\}$	$\bar{h}k0$, $\bar{h}\bar{k}0$
fourth order, upper negative	$\{hk1\}$	$hk1$, $h\bar{k}1$
fourth order, upper positive	$\{\bar{h}k1\}$	$\bar{h}k1$, $\bar{h}\bar{k}1$
fourth order, lower negative	$\{\bar{h}k\bar{1}\}$	$\bar{h}k\bar{1}$, $\bar{h}\bar{k}\bar{1}$
fourth order, lower positive	$\{hk\bar{1}\}$	$hk\bar{1}$, $h\bar{k}\bar{1}$

Pedion 010, 100, 001, $0\bar{1}0$, $\bar{1}00$, $00\bar{1}$, $h01$, $\bar{h}0\bar{1}$, $h01$, $h0\bar{1}$

Triclinic

Holosymmetric forms

Pinakoid $\{hk1\}$, $\{\bar{h}k1\}$, $\{h\bar{k}1\}$, $\{hk\bar{1}\}$, $\{h01\}$, $\{\bar{h}01\}$ $\{0k1\}$, $\{0\bar{k}1\}$, $\{hk0\}$, $\{h\bar{k}0\}$, $\{100\}$, $\{010\}$, $\{001\}$

Hemisymmetric forms

Pedion (each of the individual pinakoid faces may appear alone as a pedion)

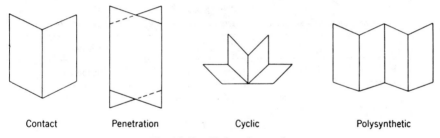

| Contact | Penetration | Cyclic | Polysynthetic |

Fig. 15.49. Twinned crystals.

not always, the different portions of the twinned crystal can be differentiated by optical tests. It is important to reserve the use of the words *twins* and *twinning* to those situations in which the two or more interpenetrating crystals bear a definite, unique, and characteristic angular relationship. The common composition plane, or twin plane, will always coincide with a possible crystallographic face, nearly always a face with simple Miller indexes, and usually a pinakoid face. Randomly intergrown crystals are not twins.

K. POLYMORPHISM

Polymorphism is a phenomenon observable with most, and probably all, elements and compounds. A compound showing polymorphism can crystallize with different internal lattices, thereby giving correspondingly different external crystal morphologies and internal physical properties. The crystal systems of the two (or more) modifications of the compound are usually, but not always, different. A few of the most common examples of polymorphism are carbon, which crystallizes as diamond (cubic) and graphite (hexagonal); calcium carbonate, which crystallizes as calcite (rhombohedral), aragonite (orthorhombic), and vaterite (hexagonal); the red (tetragonal) and yellow (orthorhombic) forms of mercuric iodide; and the monoclinic and orthorhombic forms of sulfur. Further information on this property of crystals is included in Section X (Fusion Methods).

L. ISOMORPHISM

Isomorphism is, in a sense, the opposite of polymorphism; instead of different crystalline modifications of the same compound, isomorphism involves different compounds with closely similar lattices as well as closely similar forms and habits. It is usually observed when the molecules of several different substances are nearly the same size and shape. The groups in Table 15.XV would be expected to be isomorphous from known atomic radii, and are actually observed to be so.

In each of these series the crystals of each pure compound appear to be identical; however, on careful measurement the corresponding interfacial angles are found to be slightly different. It is also a property of isomorphous

TABLE 15.XV
Examples of Iosomorphous Series of Closely Related Compounds

Series 1	Series 2	Series 3	Series 4
$KClO_4$	$NH_4H_2PO_4$	$La_2(C_2O_4)_3 \cdot 10H_2O$	p-Dichlorobenzene
$KMnO_4$	KH_2PO_4	$Ce_2(C_2O_4)_3 \cdot 10H_2O$	p-Dibromobenzene
NH_4ClO_4	$NH_4H_2AsO_4$	$Pr_2(C_2O_4)_3 \cdot 10H_2O$	
NH_4BF_4	KH_2AsO_4	etc.	

compounds that they form mixed crystals (usually in all proportions) with intermediate crystal angles and optical properties.

IX.   CRYSTAL OPTICS

The spacing of atoms along the crystallographic axes determines not only the crystal system, shape, and angles, but also the optical properties. Each of the crystals in the cubic system, for example, has identical spacing of the atoms along its three mutually perpendicular crystallographic axes; hence it should not be surprising that the optical properties of all cubic crystals are identical in all directions within the crystal.

A.   COLOR

Solids may show absorption colors or reflection colors. Substances that absorb transmitted light very strongly or are opaque may show a characteristic color or luster in reflected light. Some substances show one color by transmission and another by reflection; fuchsin, for example, is red by transmission, green by reflection.

Some substances are pleochroic; that is, they show different colors in different directions. This phenomenon is characteristic of colored anisotropic substances. The colors shown change with orientation and can be seen only with polarized light. Colored fibers, when they are pleochroic, show different light absorptions parallel to their lengths than crosswise. Tourmaline is pleochroic, as are azurite and malachite. Sphalerite, garnet, and other cubic crystals are not pleochroic, though they are colored, because they are isotropic. The strongest absorption almost invariably occurs parallel to the vibration direction showing the highest refractive index. The difference in absorption is also proportional to the birefringence of the substance.

B.   THE REFRACTIVE INDEX

The most important optical property to the microscopist is the refractive index, n. The refractive index of any substance can be visualized as the degree of slowing of light as it passes through that substance. Light travels

TABLE 15.XVI
Refractive Indexes of Some Compounds Arranged in Order
of Increasing Molecular Weight

Compound	Molecular weight	Refractive index
Sodium fluoride	42.00	1.336
Sodium chloride	58.45	1.5442
Sodium bromide	102.91	1.6412
Sodium iodide	149.92	1.7745

most rapidly in a vacuum, but almost as rapidly in air. In water it is slowed
by about 25% ($n = 1.33$), and in general,

$$n_s = \frac{c_0}{c_s}$$

where c_0 is the velocity of light in a vacuum and c_s is the velocity of light
in a substance whose refractive index is n_s.

Thus glass, with a refractive index of slightly more than 1.50, must slow
light by about 33%. Higher-atomic-weight elements are more effective in
slowing light, and therefore compounds containing high-atomic-weight elements usually have higher refractive indexes. Table 15.XVI illustrates this.

C. MEASUREMENT OF THE REFRACTIVE INDEX

The refractive index of any solid may be determined microscopically by
immersing fine particles of the solid in liquids of known refractive index until
in one such liquid the particles become invisible. When this occurs the refractive index of the invisible particles must be identical with that of the
liquid medium in which they are immersed. Similarly, the index of the solid
may be determined by varying the index of the mounting liquid until it
matches that of the solid, and then measuring the index of the liquid. Variation of the index may be achieved by mixing liquids or by varying the
temperature and/or the wavelength of the light.

1. Liquids

There are a number of methods for measuring refractive indexes of liquids,
and they vary considerably in complexity of instrumentation, precision, and
size of sample required. A detailed review of these methods is given by
Weissberger (78). Several procedures are adaptable to the microscope, and
these methods are summarized below.

a. IMMERSION METHODS

The simplest and most obvious method of determining the refraction index
of a liquid is through use of the immersion method mentioned above. This

method depends on the use of solids of known refractive index; when the liquid causes complete disappearance of a particular glass powder or crystalline solid, the two must have identical refractive indexes. Sets of glass powders having carefully determined refractive indexes are commercially available; these cover the index range from 1.33 to 1.68 in approximately 0.01-unit increments. Many inorganic salts and minerals, such as those in Table 15.XVI, also can be used.

b. CELL OF KNOWN THICKNESS

This method, as described by Johannsen (reference 44, p. 238), really goes back to 1767 to the Duc de Chaulnes, who used it to measure the average refractive index of parallel mineral plates. For liquids (or melted solids), a cell 2 mm deep and 1 mm or less in diameter is used with a cover glass. Volumes of 0.002 ml or less are sufficient, but accuracy of better than ±0.01 is seldom achieved.

c. MEASUREMENT OF VERTICAL DISPLACEMENT

If the vertical displacement of the image by equal thicknesses of liquids of known refractive indexes is measured and plotted on a curve [see Chamot and Mason (16), pp. 329, 330], the refractive index of an unknown liquid may be read from the curve if the displacement is measured under the same conditions of thickness, temperature, and wavelength. Depth of the cell need not be known, but may be increased to 3 or 4 mm to increase accuracy. Precision is ±0.005 and may be improved with practice.

d. NICHOLS STAGE REFRACTOMETER

Several refractometers designed for use on the stage of a microscope consist of a cell containing a small glass prism mounted on a slide. The slide is engraved with a reference mark. When the cell is filled with liquid the image of the reference line is displaced by a distance related to the refractive index of the liquid and of the glass prism.

The Nichols refractometer (Fig. 15.50) is the most sensitive apparatus of this type, having an accuracy of ±0.001 with white light and of ±0.0004 with monochromatic light and suitable temperature control. The device consists of a metal plate provided with two cylindrical cells. Each cell contains two small glass prisms arranged as shown in Fig. 15.50b; a reference mark is engraved on the top surface of the bottom of the cell. The prisms in one cell have a refractive index of 1.52 and those in the other an index of 1.72. The former is used for liquids of $n < 1.40$ and $n = 1.65–1.85$; the latter for liquids of $n = 1.40–1.65$ and $n > 1.85$. The prisms are closely surrounded by their mounts so that the volumes of the cells are minimized.

When a cell is filled with liquid the image of the reference line is refracted by each of the two liquid–glass prisms, and two displaced images are seen in the microscope eyepiece (Fig. 15.50c). Because of the arrangement of the prisms the displacements are toward opposite sides of the cell and the dis-

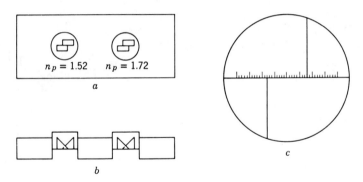

$n_p = 1.52$ $n_p = 1.72$

a

b

c

Fig. 15.50. Nichols refractometer.

tance between the lines depends on the relationship between the refractive index of the liquid and that of the glass prisms. The relative distance between the lines is measured with a micrometer eyepiece.

Each cell is first calibrated with liquids of known index. A curve for each is plotted showing refractive index versus eyepiece micrometer divisions between the lines. Indexes for unknown liquids can then be read from the curves. The micro model of the Nichols refractometer requires only 0.006–0.008 ml of sample.

e. JELLEY MICROREFRACTOMETER (41)

This instrument is well suited to the needs of the microscopy laboratory. It can be made easily, except for the microprism, which may be purchased from any of several optical supply houses or from E. Leitz, Inc. (Cat. no. 640 004).

The instrument is illustrated in Fig. 15.51. The illuminated slit in the scale, which is graduated in refractive-index units, is viewed through the aperture. The flat glass plate is held over the aperture with clips. A 45° glass microprism is positioned so that it partly covers the aperture, and is held in place by capillary attraction when a micro drop of liquid is placed between the prism face and the glass plate. If less than 0.01 ml of liquid must be used, the prism may be held with a cement. The liquid forms a compound liquid–glass prism and, as shown in Fig. 15.52, light from the slit is deviated by the glass–liquid interface. The virtual refracted image of the slit is seen superimposed upon the undeviated image of the scale. The slit is located on the scale at the position corresponding to n_D for the glass of the prism, so that the slit itself may be seen through the prism. The positions of the scale divisions may be located by calibration with liquids of known index or by calculation using the refractive indexes of known liquids, the refractive index and angle of the glass prism, the trigonometric relationships apparent in Fig. 15.52, and a correction factor for the thickness of the glass plate. The calculation method is more accurate.

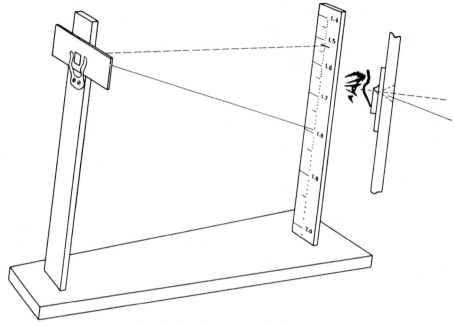

Fig. 15.51.   Jelley microrefractometer.

For volatile liquids a hollow prism can be used, and heating devices for temperature control have been devised. The Jelley instrument can determine the index of 0.0001 ml of liquid with an accuracy of ± 0.0001 and has a range of 1.330–2.000.

f. REFRACTOMETERS MEASURING THE CRITICAL ANGLE

Of the various instruments measuring refractive index by means of the angle of refraction, the Abbe and Pulfrich refractometers are the most familiar. Determinations are made by measuring the angle of emergence of a

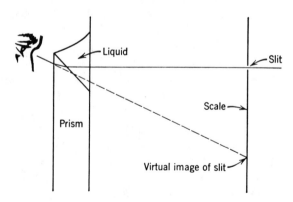

Fig. 15.52.   Light path in the Jelley microrefractometer.

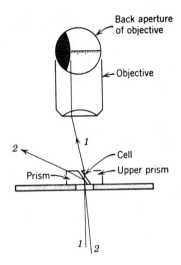

Fig. 15.53.   Wright stage microrefractometer.

beam of light that passes from air through the liquid and then through a glass prism and out into the air again. The critical angle of refraction (see Section II) is used to make observation of the emergent beam easier and more accurate.

Instruments of the Abbe type operate on the same optical principle as do the Pulfrich devices. The liquid sample is held between two prisms, one for refraction and the other primarily for holding and illuminating the sample. The scale reads directly in n_D, and white light is normally used for illumination. The sample size required is about 0.05 ml, but may be decreased considerably by filling a portion of the space between the prisms with a small ($\frac{1}{4} \times \frac{1}{4}$ in.) square of lens tissue soaked with the sample.

Instruments are available with ranges of n = 1.30–1.70 and 1.45–1.84; their accuracy of measurement is ±0.0001. "Precision" Abbe instruments are also available that afford accuracies of ±0.00002 to ±0.00006. The precision Abbe refractometer made by Bausch and Lomb can easily be used to measure refractive indexes for a variety of wavelengths, and the ordinary instrument (equipped with Amici prisms) is arranged to give the refractive index for sodium light, plus a measure of dispersion. Instruments are available without Amici prisms that require monochromatic light for illumination. The refractive indexes of solids also can be measured, and certain models of the Abbe instrument provide for determining indexes of opaque solids by use of the critical angle of reflection (which is equal to the critical angle of refraction).

The Wright stage microrefractometer consists of a pair of miniature prisms mounted on a microscope slide as indicated in Fig. 15.53 so that a small gap lies between them. The angle of the prisms is 60°. The surface of the lower one is ground, whereas that of the upper one is polished. A drop of liquid is placed in the gap and a low-power objective is focused on the top of the cell. With the use of monochromatic light the back aperture of the objective is observed with the Bertrand lens in place. The field is divided into dark and light portions separated by a sharp dividing line; the position of this critical boundary depends on the index of the liquid and is measured with a filar eyepiece micrometer. The apparatus is calibrated with liquids of known refractive index.

g. SPECTROMETER METHOD

If a hollow prism is filled with a liquid, the angle through which a beam of light is deviated on passing through the prism can be measured with a spectrometer. The angle of deviation is a function of the refractive index of the liquid and the angle of the prims. With specially constructed prisms [see Hartshorne and Stuart (37), p. 263], liquid samples of 0.05–0.1 ml are sufficient. The advantages of using the spectrometer are that the measurable range of indexes is unlimited and an accuracy of ± 0.000001 is possible. The chief disadvantage is that more care is necessary than is required for instruments of the Abbe and Pulfrich types. The indexes of solids capable of being shaped into prisms also can be measured.

h. TEMPERATURE AND WAVELENGTH CONTROL

Any of the more complex instruments will usually have built-in provisions for temperature control. For simple equipment the microscopist may have to design devices for this purpose. The average liquid has a temperature coefficient of refractive index of about 0.0005 units/°C. The temperature must be controlled, therefore, to ± 1°C if one is to measure the refractive index to ± 0.0005.

Control of the wavelength of illumination may be achieved with monochromators, with special light sources, or with filters. For the Jelley instrument and for microscope-stage refractometers, a simple and satisfactory way of obtaining monochromatic illumination is to use interference filters with a source of strong white light. These filters are available with half widths of about 11 nm for any desired wavelength. The interference filter with a peak at about 589 nm is a good substitute for a sodium lamp. Monochromators are now available based on the use of interference wedges.

An excellent table on page 735 of reference 78 lists refractometers recommended for assorted conditions, together with the usual accuracy of each instrument under those conditions.

2. Solids

In microscopical immersion methods for determining the refractive indexes of solids, the medium whose index matches that of a given solid particle must be found by trial and error. There are, however, two tests that speed this process considerably: the Becke line test and the oblique-illumination test. Both tests enable the microscopist to tell whether the index of a given medium is greater or less than that of a particle immersed therein.

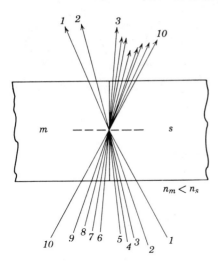

Fig. 15.54.   Origin of Becke line.

a.   BECKE-LINE TEST

The Becke line is a bright halo near the boundary of a transparent particle that moves with respect to that boundary as the microscope is focused up and down. The halo will always move to the higher-refractive-index medium as the position of focus is raised. The halo crosses the boundary to the lower-refractive-index medium on downward focusing. The particle must be illuminated with axial light which may be obtained by closing the iris diaphragm of the condenser until only a small aperture remains.

Becke's explanation of this test is based on the fact that light is reflected at an interface when it approaches from the high-refractive-index side. This is illustrated schematically in Fig. 15.54, where it can be seen that although an equal number of axial (very slightly converging) rays from each side of the vertical interface enter m and s, a larger number emerge from s, the higher-index substance, because rays 3 and 4 are reflected back into s at the interface. The bright halo of the Becke line is thus caused by the concentration of emerging rays, and this concentration will appear to move toward the center of s as the focus is raised. This physical picture of the phenomenon becomes less clear as a basis for the Becke line's apparent movement across the interface into m on lowering of the focus—the usual explanation depends on backward projection of the refracted and reflected rays. Further treatment of this aspect of the subject is found in reference 43, pages 272–275.

Most particles and crystals are somewhat lens-shaped, being thinner at the edges and thicker in the center. Figure 15.55 illustrates the formation and appearance of the Becke line for such particles.

It is useful to remember the obvious fact that the more nearly the refractive indexes of the particle and the medium are matched, the less contrast will be shown by the particle boundary. Hence, the Beck-line test tells

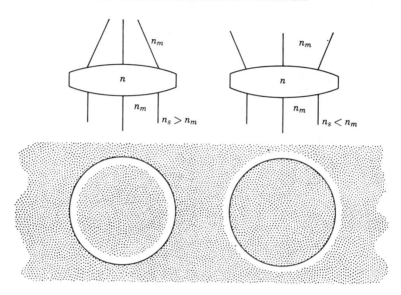

Fig. 15.55.   Becke lines from lens-shaped particles. n_m, refractive index of medium; n_s, refractive index of sample.

whether the particle or the medium has the higher index, and the degree of contrast tells how much higher.

The accuracy of the Becke test is ordinarily about ± 0.001, but it can be brought to ± 0.0004 by using monochromatic light and temperature control. Phase-contrast illumination permits improvement of the accuracy to ± 0.0001 or better for isotropic materials.

b. Oblique-Illumination Test

Occasionally, two bright lines will be seen in the Becke test that move in opposite directions as the objective is raised, making it difficult to decide which is the "true" Becke line. Crystal fragments having irregularly sloping edges may give rise to this effect. The sensitivity of the Becke test is thus reduced if the edges of the particles are sharply inclined. In such cases the oblique-illumination test may be preferred.

If particles immersed in a medium of different refractive index are illuminated with unilateral oblique transmitted light, they will appear asymmetrically shaded. The side of the particle on which the shading is seen depends on whether the index of the particle is lower or higher than that of the medium.

Unilateral oblique illumination is obtained by covering one side of the *fully opened* iris diaphragm of the substage condenser with a finger or piece of cardboard. The condenser should be racked up close to the stage to focus on the preparation (if the condenser is lowered so as to focus much below the level of the preparation, the shadowing effect is reversed). When the

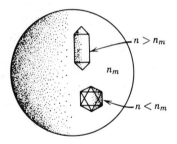

Fig. 15.56. Oblique-illumination test for refractive-index determination.

condenser and shade are properly positioned, about half the field appears dark. A particle of higher index than that of the mounting liquid will appear shaded along its inner edge on the same side as the darkened side of the field. A particle of lower index will be shaded on the side opposite to the dark side of the field (Fig. 15.56).

If the indexes of particle and liquid are the same, no shadow will appear. Shading of the particle is due to convergence or divergence of the oblique light by the particle, which behaves as a rough lens. One side of the particle tends to transmit light in such directions that it can enter the microscope objective, whereas the light from the other side of the particle is lost. The side transmitting to the objective will, of course, appear bright. Figure 15.57 shows the paths of the light rays, greatly exaggerated. Since the microscope gives an inverted image, the effects shown are reversed as observed.

Most liquids used in refractive-index work have greater dispersive power than do solids. When, as in Fig. 15.57c, solid and mounting liquid have the same index for a wavelength of light in the middle of the visible spectrum (e.g., sodium light), the solid will have the higher index for blue light. In

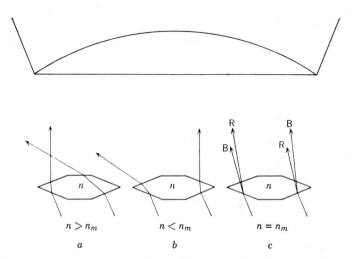

Fig. 15.57. Light paths in the oblique-illumination test. B, blue light; R, red light.

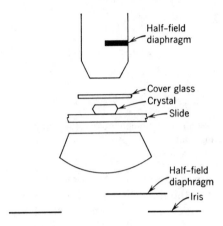

Fig. 15.58.   Saylor's double-diaphragm method.

white light the particle will tend to converge red light and diverge blue, as in Fig. 15.57c, and in oblique light opposite edges of the particle will be tinted with these colors. In axial light the particle will show a reddish fringe on the inside and a bluish fringe on the outside. Occurrence of such color fringes at the edges of the image of the particle indicates that the solid and the liquid have the same index for light in the middle of the spectrum. If more accurate results are necessary, monochromatic light must be used. This phenomenon is known as the Christiansen effect and is the basis for dispersion staining methods.

Sensitivity of the oblique-illumination test may be improved by Saylor's double-diaphragm method (72). An additional diaphragm is inserted at the back focal plane of the objective, on the same side as the diaphragm already placed below the condenser (Fig. 15.58). These two aperture diaphragms can be adjusted to give what approaches unilateral dark-field illumination and thus extend the accuracy of the oblique-illumination test to ± 0.0001.

c. Index-Variation Immersion Methods

These methods are applied to the basic Becke or oblique-illumination tests to reduce the number of liquids needed and for other special purposes.

(1) Mixing of Two Liquids

Instead of mounting particles of a substance successively in single liquids of known refractive index, the particles may be mounted in a liquid of somewhat higher index and small increments of a lower-index liquid may be added until a match in index is observed (by the Becke-line or oblique-illumination test). The index of the mixed liquid is then determined with a microrefractometer.

The procedure has the advantage that after one or two initial tests to determine the range of indexes involved, only two liquids need to be used.

TABLE 15.XVII

TABLE 15.XVII
Dispersion of Refractive Index for Several Common Compounds

Compound	Refractive Index	Wavelength (nm)		
		486.1[a]	589.3[b]	656.3[a]
Sodium chloride	n	1.5534	1.5443	1.5407
Mercuric iodide	ω	—	2.748	2.600
Magnesium silicofluoride	ϵ	1.3634	1.3602	1.3587
Water	ω	1.3133	1.3090	1.3071
Periclase	n	1.7475	1.7376	1.7335
Zincite	ϵ	2.081	2.020	2.000
Quartz	ω	1.5497	1.5422	1.5419
Potassium bicarbonate	α	1.383	1.380	1.379
Potassium perchlorate	γ	1.4812	1.4769	1.4750
Barium sulfate	β	1.641	1.636	1.634

[a] Lines in the mercury arc.
[b] Sodium doublet; average wavelength 589.3 nm.

The procedure uses a minimum of sample and permits back-titration if the end point is overreached.

(2) Variation of Wavelength of Light

As stated earlier, variation of the index of a substance with the wavelength of light is called dispersion of the refractive index. All refractive indexes of solids vary with wavelength, and this variation can be used to assist in the identification of unknown solids (Table 15.XVII).

Variation of the wavelength may be achieved by using a succession of filters, spectrally limited lamps (sodium, hydrogen, mercury, etc.), or a monochromator. By varying the wavelength, a greater range of refractive-index values is obtained from a single immersion liquid.

Normally, of course, a sodium lamp, or at least a yellow filter, is used for the measurement of the refractive index; hence, the indexes for sodium light are of most interest. Consider what is observed, however, when a solid is mounted in a medium having a different dispersion. Even though the refractive indexes may be identical at one wavelength, they will be different at all other wavelengths. In particular, consider quartz particles mounted in nitrobenzene, which has a refractive index with red light identical with the index of quartz with red light. With blue light, the two indexes (of quartz and nitrobenzene) are different. The result observed with dark-field illumination is a blue outline around each quartz crystal, because only the light that is scattered at the crystal–liquid interface is seen. This procedure is based on a very old observation by Christiansen, who found he could make up emulsions of colorless components and, by proper choice of the components, vary the refractive index and dispersion to obtain colored gels. These "chromatic emulsions," as he called them, also changed color as the

TABLE 15.XVIII
Color Obtained in Dispersion of Staining Various Media with
Several Common Compounds

Compound	Medium	Color
Corundum	Methylene iodide	Yellow–orange
Quartz	Fenchyl oil	Yellow–orange
Orthoclase	Fenchyl oil	Blue
Quartz	Nitrobenzene	Blue
Orthoclase	Nitrobenzene	White
Kaolin	Nitrobenzene	Yellow–orange
Hornblende asbestos	α-Bromonaphthalene	Yellow
Hornblende asbestos	Nitrobenzene	Blue–green

temperature was changed, since this also changed the wavelength at which the two components had identical indexes. This same phenomenon, without the temperature change, is called "dispersion staining" by Crossmon (18), who has applied it microscopically.

The method is most effective for those compounds that are isotropic or that do not vary greatly in refractive index or dispersion with direction in the crystal. This includes many minerals (see Table 15.XVIII).

It can, however, be used for other minerals, for example, calcite, which has two widely different refractive indexes (Table 15.XIX). In this case, the

TABLE 15.XIX
Refractive Indexes of a Few Common Tetragonal and
Hexagonal Crystals

Compound	Index	
	Omega (ω)	Epsilon (ϵ)
Quartz, SiO_2	1.544	1.553
Beryl, $3BeO \cdot Al_2O_3 \cdot 6SiO_2$	1.581	1.575
Saltpeter, $NaNO_3$	1.587	1.336
Apatite, $CaF_2 \cdot 3Ca_3P_2O_3$	1.634	1.632
Calcite, $CaCO_3$	1.658	1.486
Dolomite, $CaCO_3 \cdot MgCO_3$	1.682	1.503
Willemite, Zn_2SiO_4	1.694	1.723
Magnesite, $MgCO_3$	1.700	1.509
Corundum, Al_2O_3	1.768	1.760
Siderite, $FeCO_3$	1.875	1.633
Zircon, $ZrSiO_4$	1.924	1.968
Zincite, ZnO	2.008	2.029
Anatase, TiO_2	2.554	2.493
Rutile, TiO_2	2.616	2.903
Silicon carbide, SiC	2.654	2.697
Cinnabar, HgS	2.854	3.201

characteristic color (yellow in α-bromonaphthalene) will be shown only when the crystal is oriented so that the omega index is being observed.

(3) Variation of Temperature

Refractive indexes of both liquids and solids decrease as the temperature increases. The change per degree Celsius, $-dn/dt$, is much greater for liquids than for solids. Hence, the index change during heating or cooling of such solids can be neglected while the change in index of the immersion liquid is varied thermally to provide a larger index range from a single liquid. The method is not generally so well suited for use with organic compounds, since for these the value of $-dn/dt$ may be relatively large.

The term $-dn/dt$ is called the temperature coefficient of the refractive index. It remains essentially constant for each liquid over the normal range of temperatures and, once determined, can be used to calculate the refractive index of the liquid at any other temperature.

Measurements of the temperature coefficient of the refractive index can be made conveniently on a microscope hot stage. Solid particles of known refractive index are mounted in a liquid having an index sufficiently higher to make them visible at room temperature, and then are heated until an index match is obtained. Repeating this determination with a solid of slightly lower refractive index (about 0.02–0.03) will give a second, higher temperature at which a match is obtained. From the two temperatures and the refractive indexes of the two solids the coefficient can be determined.

Because the value of $-dn/dt$ for mounting liquids is appreciable, the room temperature during an ordinary refractive-index determination (not by a variation method) should always be noted and, if necessary, used to correct the observed refractive index.

(4) Double-Variation Method

An even greater range of index values can be obtained from a single liquid by combining the techniques of wavelength and temperature variation. This procedure is known as Emmons's double-variation method (23).

d. LIQUIDS FOR MEASURING REFRACTIVE INDEX

Lists of standard immersion liquids for refractive-index determinations are found in Chamot and Mason (16), page 331, and other texts (37,43). A list is also given in the Eastman Organic Chemicals Catalog. Directions for preparing sets of mixed liquids are given by Needham (68), pages 201–203.

Sets of calibrated mixed liquids can be purchased from R. P. Cargille Laboratories, Inc. These liquids are available with indexes from 1.30 to 2.11 and are packaged in $\frac{1}{4}$-oz. bottles marked with the temperature coefficient and dispersion for each liquid. There are also high-index sets with indexes from 2.12 to 2.31.

A difficulty common in index determination of organic compounds is that such solids may dissolve in the Cargille liquids. If the solubility of the solid

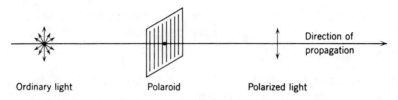

Fig. 15.59. Polarization of light.

is low, the determination usually can be made by working quickly. Otherwise a liquid having less solvent power for the compound must be found. Solutions of potassium iodide or potassium mercuric iodide in water will cover the refractive index range up to 1.73, and the same salts dissolved in glycerine are useful up to an index of 1.79. If a more viscous medium is desired, as when crystals may need to be rolled into better positions, these salts can be dissolved in Aquaresin, a glyceroborate sold by Glyco Products; such mixtures will give indexes up to about 1.77.

e. Isotropic and Anisotropic Materials

Only glass, some plastics, and compounds in the cubic system show single refractive indexes; they are isotropic. Most transparent particles have several different refractive indexes depending on the vibration direction of light through the particle; they are anisotropic. Anisotropic materials show either two or three "principal" refractive indexes in particular directions, but any refractive index between these principal values may be observed in intermediate directions.

f. Polarized Light

Polarized light must be used to study anisotropic crystals. Ordinary light can be polarized by means of a Polaroid filter, most chemical microscopes are equipped with such filters. Polarized light differs from ordinary light in that it has a single vibration direction, perpendicular to the direction of propagation. The polarization of ordinary light by a Polaroid filter is shown in Fig. 15.59. If the filter were rotated 90°, the plane of vibration of the polarized light would be rotated correspondingly.

g. Uniaxial Crystals

Crystals of the tetragonal and hexagonal systems have one unique crystallographic direction and either two (tetragonal) or three (hexagonal) directions that are alike and perpendicular to the unique direction (Fig. 15.47). The unique direction corresponds to the c axis and the others to the a axes. The arrangement of structural units (atoms, ions, etc.) around the unique axis is completely symmetrical; hence light travels along this axis (propagation direction) with the same velocity for vibration directions in any of the directions perpendicular to the unique axis.

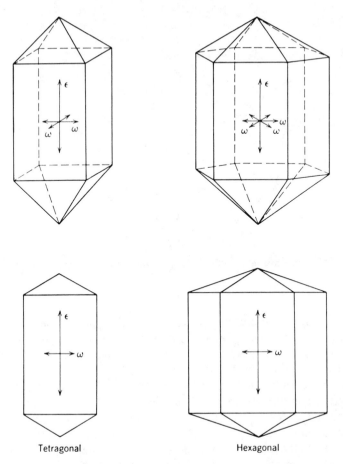

Tetragonal   Hexagonal

Fig. 15.60.   Refractive indexes of tetragonal and hexagonal crystals.

If, now, a crystal such as one of those shown in Fig. 15.60 is observed with polarized light, the refractive index observed will be an epsilon, omega, or intermediate index, depending on the orientation of the crystal relative to the vibration direction of the polarized light. The refractive index actually observed can therefore be controlled by rotating the crystal to the proper position relative to the vibration direction of the polarizer.

If a given crystal is mounted in a liquid medium having a refractive index equal to the omega index (ω) of the crystal, then on rotation of the stage, the crystal will disappear two times (180° apart) during each complete rotation. At the two positions 90° from those positions in which the omega index is observed, the crystal boundaries will appear most strongly (greatest contrast), since here the epsilon refractive index (ϵ) is being observed (Fig. 15.61).

All of the crystals shown in Table 15.XIX will appear and disappear as the stage is rotated if they are mounted in a medium having either one of

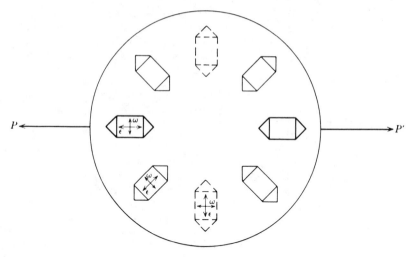

Fig. 15.61.   Variation in contrast of uniaxial crystal in liquid of refractive index matching the omega index of the crystal.

their two principal refractive indexes and if they are properly oriented. In contrast to this behavior, each of the isotropic compounds listed in Table 15.XX has a single refractive index and disappears in all positions when mounted in a liquid having that refractive index.

If the two crystals in Fig. 15.60 are observed in other orientations, different refractive indexes are observed (Fig. 15.62). It is important to notice, however, that every view of this crystal shows the omega refractive index, even though most of the views show only a component of the epsilon. The view parallel to the length, in fact, shows only the omega index in all positions of rotation of the stage. If a sample is suspected of containing crystals of calcite, a small portion should be mounted in a liquid having the omega refractive index for calcite (1.658). All crystals of calcite, no matter how they are tipped, will disappear completely every 180° during rotation of the stage when mounted in such a liquid. There is one such refractive index for all tetragonal and hexagonal crystals; this is not true, however, for crystals in all crystal systems.

The epsilon refractive index can be observed only when the axis of the microscope is exactly perpendicular to the optic axis of a uniaxial crystal, that is, when the crystal lies on a face parallel to c. As illustrated in Fig. 15.61, both ϵ and ω may be determined for a crystal in this position.

Uniaxial crystals are described as optically positive or negative, depending upon whether ϵ or ω is the greater: If $\epsilon > \omega$, the crystal is positive $(+)$; if $\omega > \epsilon$, the crystal is negative $(-)$.

If the position of the optic axis is known (e.g., from morphology), the optic sign, or sign of double refraction, can be determined by the use of compensators, as discussed above. If the position of the optic axis is not

TABLE 15.XX
Refractive Indexes of a Few Isotropic Crystals

Crystal	Refractive Index	Crystal	Refractive Index
NaF	1.326	$Sr(NO_3)_2$	1.586
K_2SiF_6	1.339	$NaBrO_3$	1.616
KF	1.352	NH_4Cl	1.640
$(NH_4)_2SiF_6$	1.370	NaBr	1.641
KCN	1.410	CsCl	1.642
Na alum	1.439	RbI	1.647
K alum	1.456	KI	1.667
NH_4 alum	1.459	CsBr	1.698
KCr alum	1.481	NH_4I	1.703
NH_4Fe alum	1.485	NH_4Br	1.485
KCl	1.490	MgO	1.736
RbCl	1.494	As_2O_3	1.755
$NaC_2H_3O_2 \cdot UO_2(C_2H_3O_2)_2$	1.504	NaI	1.775
$NaClO_3$	1.515	$Pb(NO_3)_2$	1.782
NaCl	1.544	CsI	1.788
RbBr	1.553	AgCl	2.07
KBr	1.559	TlCl	2.25
$Cr_2(SO_4)_3 \cdot 18H_2O$	1.564	AgBr	2.25
$Ba(NO_3)_2$	1.571	TlBr	2.42

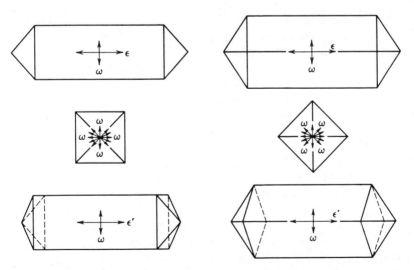

Fig. 15.62.   Refractive indexes shown by a tetragonal crystal as it is tipped through 90°.

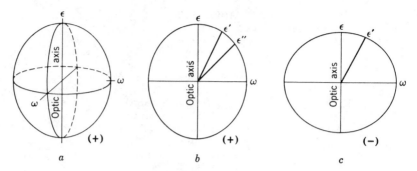

Fig. 15.63.   Uniaxial indicatrix.

obvious from morphology, the sign must be determined from the interference figure.

Figure 15.62 illustrates the effect on the observation of refractive indexes of tilting a uniaxial crystal. No matter what position the crystal assumes, ω will always be visible because the ω vector component has the same velocity in all vibration directions through the crystal. The velocity of the ϵ component, however, varies for different directions through the crystal and is at a maximum (for negative crystals) or minimum (for positive crystals) along the optic axis. If a crystal lies on a face parallel to the optic axis, the observed value of ϵ does not change as the crystal is rotated about the optic axis. But if the optic axis is tipped through 90° until it stands vertical to the microscope stage, ϵ will be observed to go through a series of values (ϵ') that approach ω until, in the vertical position, only ω is observed in both (or all) directions.

These relationships can be conveniently represented by a three-dimensional construction called the indicatrix. Radii proportional in length to refractive-index values may be used to represent indexes observed in all directions of light propagation through a crystal. Such radii then describe solid geometrical forms whose surfaces define refractive-index values observed for any orientation of a crystal. The radius for ω is constant and describes a circle. The radii for ϵ vary from ϵ through ϵ' to ω. When the ellipse and the circle are placed at right angles to each other and rotated about ϵ, the resulting form is an ellipsoid of revolution whose maximum (for optically negative crystals) or minimum (for optically positive crystals) dimension is the same as the diameter of the circle resulting from ω. The uniaxial indicatrix thus produced is shown in Fig. 15.63 a; b and c are two-dimensional sections of the indicatrixes for positive and negative crystals and show the relationships of ϵ, ϵ', and ω for each.

h. BIAXIAL CRYSTALS

There is a third group of crystals, also anisotropic, that have three principal refractive indexes instead of one (cubic, isotropic) or two (tetragonal or hexagonal, uniaxial). These crystals are all in the orthorhombic, mono-

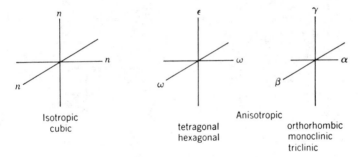

Fig. 15.64. Refractive indexes of cubic, uniaxial, and biaxial crystals.

clinic, and triclinic systems and are termed *biaxial*. These three systems differ optically as well as morphologically from crystals in the other systems, reflecting the fact that biaxial crystals have unequal spacing of atoms along the three different crystallographic axes (Fig. 15.64). A representation of the optical differences is also shown in Fig. 15.64.

Most of the compounds of interest to the chemist fall in this third group. The three principal refractive indexes (Table 15.XXI) are designated α, β, and γ. The lowest refractive index is always α, γ is the highest, and β is the intermediate refractive index.

TABLE 15.XXI

Refractive Indexes of a Few Common Orthorhombic, Monoclinic and Triclinic Crystals
(Listed in Order of Increasing β)

Compound	α	β	γ
Borax, $Na_2B_4O_7\cdot10H_2O$	1.447	1.469	1.472
Gypsum, $CaSO_4\cdot2H_2O$	1.521	1.523	1.530
Orthoclase, $K_2O\cdot Al_2O_3\cdot6SiO_2$	1.518	1.524	1.526
Albite, $Na_2O\cdot Al_2O_3\cdot6SiO_2$	1.525	1.529	1.536
Biotite, $(K, H)_2(Mg, Fe)_2(Al, Fe)_2 (SiO_4)_3$	1.541	1.574	1.574
Anhydrite, $CaSO_4$	1.569	1.575	1.613
Anorthite, $CaO\cdot Al_2O_3\cdot2SiO_2$	1.576	1.583	1.589
Pyrophyllite, $Al_2O_3\cdot4SiO_2\cdot H_2O$	1.552	1.588	1.600
Talc, $3MgO\cdot4SiO_2\cdot H_2O$	1.539	1.589	1.589
Muscovite, $K_2O\cdot3Al_2O_3\cdot6SiO_2\cdot2H_2O$	1.561	1.590	1.594
Topaz, $(Al, F)_2SiO_4$	1.619	1.620	1.627
Wollastonite, $CaSiO_3$	1.616	1.629	1.631
Barite, $BaSO_4$	1.637	1.638	1.649
Diopside, $CaMg(SiO_3)_2$	1.664	1.671	1.694
Olivine, $(Mg, Fe)_2SiO_4$	1.662	1.680	1.699
Aragonite, $CaCO_3$	1.530	1.681	1.685
Malachite, $CuCO_3\cdot Cu(OH)_2$	1.655	1.875	1.909
Sulfur, S	1.950	2.038	2.241
Cerussite, $PbCO_3$	1.804	2.076	2.078
Realgar, As_2S_2 (Li light)	2.46	2.59	2.61
Stibnite, Sb_2S_3	3.194	4.046	4.303

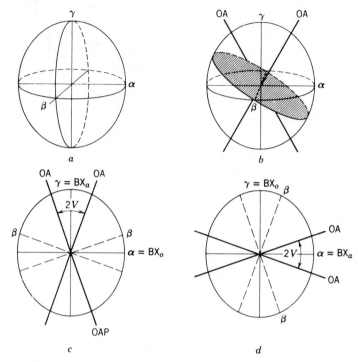

Fig. 15.65. Biaxial indicatrix. BX_a, acute bisectrix; BX_o, obtuse bisectrix; OA, optic axes; OAP, optic axial plane; $2V$, optic axial angle.

The indicatrix for such crystals is a triaxial ellipsoid, rather than an ellipsoid of revolution as with uniaxial crystals. There are, however, two sections through the triaxial ellipsoid that are circular, and light traveling perpendicular to these sections splits into two components of equal velocity corresponding to the β refractive index. The two directions of travel are the two optic axes of biaxial crystals (Fig. 15.65b). The three axes of the ellipsoid are α, β, and γ (Fig. 15.65a).

The two optic axes form an acute angle, designated the optic axial angle ($2V$), in the plane of α and γ. The optic axial angle is a constant for any particular substance and can be measured and used to characterize that substance. Figures 15.65c and d are sections of the biaxial indicatrix showing $2V$. The angle $2V$ is always acute. In Fig. 15.65c it is bisected by γ; in Fig. 15.65d, by α. The bisector of the optic axial angle is called the acute bisectrix. In Fig. 15.65c it is γ; in Fig. 15.65d it is α. It is also apparent in Fig. 15.65c that the angle supplementary to $2V$ is obtuse and is bisected by α; α is therefore called the obtuse bisectrix in 15.65c; in Fig. 15.65d, the obtuse bisectrix is γ. The plane containing the optic axes as well as α and γ is called the optic axial plane.

Biaxial crystals have two isotropic directions, the optic axes. Primed values of all three refractive indexes may be observed. If the crystal is oriented

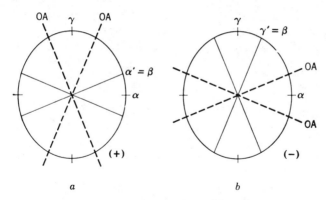

Fig. 15.66.   Sections of biaxial indicatrix in α–γ plane. OA, optic axes.

on the microscope stage so that one is looking directly parallel to β (i.e., looking down β), the principal values of α and γ are observed. Since the principal value of β is intermediate between those of α and γ, β can also be represented on the section of the biaxial indicatrix showing the α–γ plane. In Fig. 15.66a, β is closer in numerical value to α than to γ and is equal to α'. In Fig. 15.66b, the numerical value of β is closer to γ and therefore equal to γ'. The dividing line between α' and γ' is the numerical midpoint between the values of α and γ. This is a special situation involving the α–γ plane only; in general, the primed value will be named for the nearest numerical index: α, β, or γ. It may also be seen in Fig. 15.66 that the lines representing the principal value of β are perpendicular to the optic axes.

The refractive indexes actually observed for the compounds in Table 15.XXI depend on the orientation of the crystal; each different view will present a different maximum and minimum refractive index. If the orientation of the crystal is known, the two refractive indexes for that view are known (Fig. 15.67). Since, by convention, all crystals having three principal refractive indices have α as their lowest index and γ as their highest, no crystal can show, in any orientation, a refractive index lower than α or higher than γ. This is the basis for a procedure used by mineralogists for the identification of unknown minerals. Assume that the unknown could be a compound in any of the six crystal systems. The mineralogist starts by mounting a few crystals of the sample in any medium with a refractive index in the neighborhood of 1.60. By checking the Becke line and the degree of contrast in all positions of rotation of several crystals, one can judge what refractive-index medium to choose as the second mountant. One then proceeds by successive approximation to find the two refractive-index media that represent the two extreme indexes for that compound.

If *all* of the crystals disappear in a single medium in all positions of rotation of the stage, the compound is isotropic and therefore cubic (if not a glass). If *all* of the crystals disappear in a given medium in a given position during rotation of the stage, but are visible in other positions, they are tetragonal

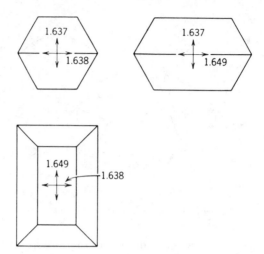

Fig. 15.67.   Refractive indexes shown by various views of BaSO$_4$.

or hexagonal and have two indexes—ω and ϵ. Finally, if only a very few crystals in the preparation show the highest or lowest refractive index on rotation of the stage, the compound is in the group of crystal systems that show three principal refractive indexes, and the two measured are α, the lowest index observed, and γ the highest index observed.

i.   USE OF THE ANALYZER

Other optical tests can be used to assist in the identification of an unknown, and some of these are based on the use of crossed polars. Thus far, we have used only a single polar to get plane-polarized light in order to be able to pick out and measure a particular refractive index for an anisotropic crystal. If we now insert a second polar, called an *analyzer*, into the field of view between the objective and the eyepiece, we shall see that the field becomes black if there is no specimen on the stage (the vibration directions for light in the two polars must be perpendicular to each other). If we place a preparation of a mixture of crystalline materials on the stage between the two polars, we shall see that some of the crystals appear colored, some appear white, and some are invisible against the black background. The last of these are isotropic; therefore, if crystalline, they must be in the cubic system. The crystals that appear white or colored are anisotropic and must have at least two principal refractive indexes.

j.   EXTINCTION

If the orientation of the crystals that appear white or colored between crossed polars is changed by rotating the stage, all of these crystals will be observed to disappear (become black) four times during complete rotation of the stage. These positions, 90° apart, are called extinction positions and

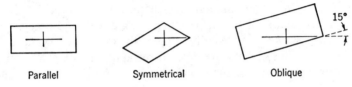

Parallel   Symmetrical   Oblique

Fig. 15.68.   Extinction in anisotropic crystals.

tell us the positions of the vibration directions of each crystal. These directions are parallel to the vibration directions of the two polars when the crystal
is extinct and are represented on drawings of the crystal by two arrows
crossing at right angles (see, e.g., Figs. 15.60–15.62 and 15.67).

Uniaxial and biaxial crystals may exhibit any of the three types of extinction: parallel, symmetrical, or oblique. These are observed with the cross
hairs of the eyepiece aligned parallel to the vibration directions of the polarizer and analyzer (so that they represent these directions) and the crystal
centered in the field at the intersection of the cross hairs. The stage is rotated
until the crystal appears dark, the analyzer is removed, and the position of
the cross hairs with respect to the long side or prominent angles of the crystal
is noted. As illustrated in Fig. 15.68, extinction is *parallel* if a cross hair
parallels the long direction of the crystal; *symmetrical* if it bisects a prominent angle; and *oblique* if the cross hair is oblique to the long direction of
the crystal. The angle between the nearer cross hair and a prominent direction of the crystal is called the *extinction angle*, and it never exceeds 45°.

This test is very useful when it is necessary to locate a particular refractive
index of a crystal. Take calcite as an example (Fig. 15.69). Crystals of calcite
will all show extinction between crossed polars when the bisectors of the
rhomb angles are parallel to the vibration directions of the polars. The vibration directions in the calcite crystal are therefore parallel to the bisectors
of the rhomb angles. Either of the crystals shown in Fig. 15.69 will show
extinction when the two arrows are parallel to the cross hairs in the eyepiece,
and will show color in all intermediate positions. To observe the omega
index, the crystal is rotated to any extinction position and that polar is removed whose vibration direction bisects the obtuse rhomb angle. The mi-

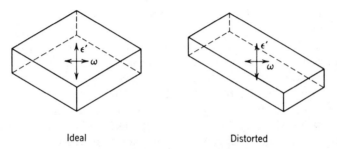

Ideal   Distorted

Fig. 15.69.   Extinction in rhombs of calcite.

croscope field is now illuminated with light whose vibration direction is parallel to the bisector of the acute rhomb angle (i.e., the omega vibration direction). In practice, it is usually simpler to rotate the crystal to the extinction position that orients the desired refractive-index direction parallel to the vibration direction of the polarizer and then to remove the analyzer.

k. BIREFRINGENCE

Differences in atomic, ionic, or molecular arrangement in the principal crystallographic directions of anisotropic crystals give rise to differences in optical properties such as refractive index in these same directions. An anisotropic crystal may be pictured as interacting with light to differing extents along which the arrangements of atoms are different. This interaction results in a slowing of light (or an increase in refractive index) as it travels through the crystal.

Polarized light is limited to a single vibration direction for any setting of the polarizer (see Fig. 15.59). Therefore, a crystal can be oriented so that the directions of its crystallographic axes are known with respect to the vibration direction of the polarized light, and a particular crystallographic direction can be set parallel to the vibration direction of the polarized light. Observation of refractive indexes by means of such alignment has been described above.

When polarized light enters an optically anisotropic crystal, the light is resolved into components vibrating in planes that are perpendicular to each other. The splitting of plane-polarized light into two vector components is called double refraction. The components follow two principal vibration directions having different refractive indexes, and therefore travel at different rates through the crystal and emerge with one retarded with respect to the other by a definite amount that depends on the difference in the two refractive indexes, $n_2 - n_1$, and the thickness. The actual distance of one behind the other is called the *retardation*.

If the crystal is oriented so that one of its principal refractive indexes is parallel to the vibration direction of the polarizer, the second vector component becomes zero. All of the light emerging from the crystal has the same vibration direction as the polarizer and all is cut off (absorbed) by the analyzer, whose vibration direction is perpendicular to that of the polarizer. Crystal and field, therefore, both appear dark, and the crystal is said to be at extinction.

If the crystal is not so oriented, the vector components emerging from the crystal will be recombined in the vibration plane of the analyzer. Since one component is retarded with respect to the other, the image will appear colored because of interference between the two components. Some wavelengths of light will be destroyed by interference and their colors will be subtracted from white light. The remaining light will consist of other colors that form the image of the crystal. These are called *interference* (or *polarization*) colors, and substances exhibiting them are termed *birefringent* (or doubly refractive). The actual colors depend on the amount of retardation, which, in turn, depends on the thickness and the difference in refractive indexes (birefringence). If the crystal is of nonuniform thickness, several colors may be observed.

The polarization colors are governed by the quantitative expression for birefringence:

$$\text{birefringence} = n_1 - n_2 = \frac{\text{retardation}}{1000 \times \text{thickness}}$$

where n_1 and n_2 are the two refractive indexes for any particular view of the crystal. Because retardation is measured in nanometers and thickness in micrometers, a factor of 1000 is placed in the denominator.

If a wedge of regularly increasing thickness of an anisotropic material, such as quartz, is turned to the 45° position and observed between crossed polarizers a definite sequence of interference colors is seen. The sequence is known as Newton's series and is divided into "orders" by the red bands that occur periodically as the thickness increases. The first-order colors are gray, white, yellow, orange, and red, in the order in which they occur as thickness increases. The higher orders include blues and greens in place of gray and white, and the colors become paler until they approach "high-order white" at about the 10th order. The color series is shown in the Michel–Levy chart reproduced at the end of Chamot and Mason's book (16). The chart includes scales relating thickness, birefringence, and interference color that permit determination of any one of these variables when the other two are known.

With some experience the microscopist becomes adept at estimating birefringence from the interference color and can use it for quick checks of the difference between the refractive indexes of a substance and for distinguishing between substances in a mixed preparation of similar-sized particles. Birefringence can be measured as a numerical constant by measuring the thickness and retardation of a given crystal.

1. COMPENSATORS

When two anisotropic substances are superimposed, addition or subtraction of retardation may occur (Fig. 15.70). If both substances are placed in a position of brightness (45° away from the vibration directions of polarizer and analyzer) and their slower components are vibrating in the same direction, the retardation of one substance is added to that of the other. The polarization color resulting from addition is of a higher order than that of either substance alone and is equivalent to the numerical sum of the retardations indicated on the Michel–Levy chart. The slower component corresponds to the higher refractive index and should be indicated by the shorter arrow. If the slow components are at right angles to each other, subtraction of retardations occurs. The resultant interference color is of a lower order

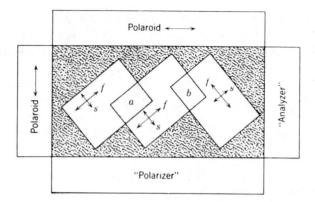

Fig. 15.70.   Addition and subtraction of retardation. *s*, slow component; *f*, fast component; *a*, addition of retardations, resulting in interference color of higher order than that of either overlapping plate alone; *b*, subtraction of retardations, resulting in interference color of lower order than that of at least one of the overlapping plates.

than that of one of the substances and corresponds to a net retardation numerically equivalent to the difference in the retardations of the two substances.

The fast and slow vibration directions of a crystal, that is, the lower and higher refractive indexes, respectively, thus may be located by using a compensator. Compensators are of several types, but basically they represent one of the anisotropic substances in Fig. 15.70; a crystal on the microscope stage represents the other substance. The slow vibration direction of the compensator is usually marked on the holder and fixed by inserting the compensator in a slot in the body tube at an angle 45° from the vibration directions of the polars. By rotation of the stage, the crystal can be aligned so that its slow vibration direction is successively oriented parallel to and perpendicular to the slow direction of the compensator, and the interference colors produced by addition or subtraction of retardations can be observed.

The "first-order red" plate compensator is made of a layer of selenite or quartz of the proper thickness to produce a retardation equivalent to first-order red. It is especially useful with very weakly birefringent materials (i.e., those that are gray or white when observed by themselves). For these, addition of retardation is indicated by the appearance of second-order blue; subtraction, by first-order yellow.

The quartz wedge also is used as a compensator. The variable retardation of the wedge extends to several orders of interference colors. The retardation of the wedge that exactly compensates that of a crystal can be found by pushing in the wedge until it reaches a position at which the interference colors of the crystal, after going through successively lower orders, appear dark gray or black. Compensation of retardation is possible only when the slow components of the crystal and the wedge are perpendicular to each other. If the crystal shows successively higher colors as the retardation of the wedge is increased, the slow directions of crystal and compensator are parallel and the stage must be turned 90° before compensation can be achieved.

Compensators of the Berek type consist of a small plate of quartz or calcite that can be tilted by means of a micrometer screw so that progressively thicker sections of the plate are placed in the light path. The effect is the same as that of the quartz wedge and the micrometer permits precise measurement of the retardation. The Berek compensator is a standard tool in fiber microscopy. Special purpose compensators include the Babinet, Sénarmont, Ehringhaus, and Brace–Köhler types.

m. INTERFERENCE FIGURES

The observation of an object microscopically in the usual manner is termed *orthoscopic observation*. There is another way of using the microscope that involves the study of the objective back focal plane by removing the eyepiece, by using the Bertrand lens, or by examining with a magnifier the image at the eyepoint above the eyepiece. Such observations are called *conoscopic* to differentiate them from orthoscopic observations. This term quite truthfully connotes that the observation is associated with a cone of light. It has already been pointed out that the observation of the back focal plane of the objective under ordinary conditions shows the angular aperture of illumination being used. Each point in the field of the back focal plane is associated with a particular direction of propagation of the illuminating radiation. The point in the center of the field corresponds to light traveling parallel to the axis of the microscope. All points at the edge of the field represent light traveling through the specimen at the maximum angle (highest angular or numerical aperture) relative to the axis of the microscope. Intermediate angles, of course, are represented by points in the back aperture of the objective lying between the center and the edge of the field.

Considering now the optical indicatrix of a uniaxial crystal, assume that the crystal is oriented with its c axis (ϵ vibration direction or optic axis) parallel to the axis of the microscope. In this orientation of the crystal observed orthoscopically between crossed polars, the birefringence is zero; that is, the crystal is isotropic. If the crystal is now tipped a few degrees in any direction, the birefringence is no longer zero but is still very small and the polarization color is low-order gray. Further tilting of the crystal will result in an increase in birefringence and higher-order interference colors. Note that this tilting of the crystal is equivalent to maintaining the position of the crystal and changing the direction of light passing through the crystal.

By conoscopic observation of a crystal placed in the center of the microscope field, a pattern of interference colors corresponding to the full cone of directions by which the crystal is illuminated will be observed in the back focal plane of the objective, each direction showing its own interference color. Superimposed on this pattern of interference colors will be the pattern

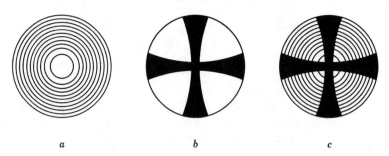

a   b   c

Fig. 15.71. Components of uniaxial optic axis interference figure. (a) Interference color pattern. (b) Extinction pattern. (c) Combined pattern.

of extinction positions. The combination of these two patterns is the *interference figure*.

Figure 15.71 shows the interference-color and extinction patterns characteristic of a uniaxial crystal observed parallel to the c axis, as well as the corresponding interference figure. Of course, only the final, combined figure is observed.

The interference-color pattern varies with the thickness of the crystal and with the difference in the two principal refractive indexes ($\epsilon - \omega$); an increase in either increases the order of the interference colors at any given point in the figure; more rings will be visible in the field. On the other hand, the extinction pattern is unchanged by changes in thickness (except that the arms of the dark cross become narrower with increased thickness and/or birefringence).

All uniaxial crystals (tetragonal and hexagonal) show interference figures of this type when viewed parallel to the c axis. Figure 15.72 shows how the interference figure varies with orientation of the crystal for one particular hexagonal compound (δ-HMX).

Biaxial crystals show a different type of figure, but one related to the uniaxial figure (Fig. 15.71). Figure 15.73a is the typical uniaxial figure; Figs. 15.73b and c are typical biaxial interference figures showing the positions of the extinction cross. In Fig. 15.73b the extinction brushes are opened up, corresponding to a crystal that is optically nearly uniaxial (meaning that two of the three principal refractive indexes, α, β, and γ, are very close together). In Fig. 15.73c the brushes are opened up more, corresponding to more nearly equal values of $\gamma - \beta$ and $\beta - \alpha$, to a larger optic axial angle.

Interference figures are very useful in the characterization and identification of compounds, although their effective use is not for dilettantes. They are useful in determining the orientation of the principal refractive-index directions in a crystal, in determining the optic sign, in studying dispersion, in determination of crystal system and, of course, in characterizing and identifying compounds.

Additional detail is beyond the scope of this brief account. The study of Hartshorne and Stuart (37) is suggested as a source of cogent discussions

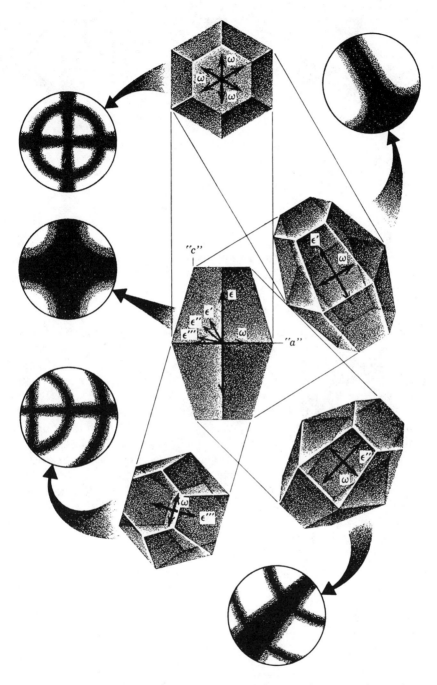

Fig. 15.72. Interference figures seen with various orientations of a hexagonal crystal.

| *a* | *b* | *c* |

Fig. 15.73.   Centered uniaxial (*a*) and biaxial (*b*, *c*) interference figures. Only the extinction brushes are shown; the interference-color fringes have been omitted because they vary with thickness.

on this subject. Table 15.XXII is a convenient summary of the important morphological and optical characteristics of crystals in each of the six systems.

X.   FUSION METHODS

Fusion methods is the name given to all microscopical thermal analyses, in which the behavior of substances is studied while their temperature is altered, usually through the use of hot stages and cold stages. Fusion methods were developed in the United States (62) somewhat later than the *Mikromethoden* of the Koflers (46), although apparently independently. Since these two schools of thought developed along different lines with different results, it is interesting to make a systematic comparison. Both groups were familiar almost from the start with the publications of Otto Lehmann who, in his volume titled *Die Kristallanalyse*, dated 1891 (54), described many of these techniques, including the mixed fusion (*Kontaktmethode*) for qualitative determination of the phase diagram (as a basis for the determination of identity or lack of identity of two compounds).

A fundamental difference between the two groups resulted from the dependence of the Kofler group on the highly precise Kofler hot stage. Most of the early work in this field in the United States had as its objective the development of simple, rapid techniques applicable with modest technical background and rudimentary equipment. Most observations were made, therefore, on preparations that had been melted over a microflame as they cooled on the bare microscope stage. The Kofler group, in general, made their observations during heating of the preparation; the United States group, during cooling.

A.   APPLICATIONS

Although Lehmann's efforts in the field of fusion methods were almost completely ignored for nearly 50 years, they have now been extended to

TABLE 15.XXII
Summary of Optical and Morphological Properties of Solids

	Isotropic		Anisotropic					
			Uniaxial		Biaxial			
	Glasses (supercooled liquids)	Cubic crystals	Tetragonal	Hexagonal	Orthorhombic	Monoclinic	Triclinic	
	Random arrangement of atoms; no faces, no angles	Regular arrangement of lattice points — Three mutually perpendicular directions having identical spacings	Unique c axis normal to plane of three mutually perpendicular a axes having identical spacing	Unique c axis normal to plane of three a axes 120° apart and with identical spacing	Regular arrangement of lattice points — Three mutually perpendicular axes, a, b, c, with different spacing and $c < a < b$	One axis, b, perpendicular to plane of 2 mutually oblique axes with all spacings different and $c < a$	Three mutually oblique axes with different spacing $c < a < b$	
	Single refractive index for all directions		Two refractive indexes: ϵ, parallel to c; ω, for all directions perpendicular to c		Three refractive indices: $\alpha < \beta < \gamma$			
					α, β, γ parallel to a, b, c, but not necessarily respectively	α, β, or γ parallel to b; other two in plane of a and c	α, β, and γ mutually perpendicular, but oblique to a, b, and c	
			Extinction: parallel or symmetrical		Parallel or symmetrical (except on hkl faces)	Parallel or symmetrical when observed in a–c plane; otherwise oblique	Oblique	
Symmetry:								
Minimum: none	$4A_3$		A_4^a	Rhombohedral: A_3; hexagonal: A_6	$2A_2^b$	A_2 or P	None	
Maximum: none	$3A_4$, $6A_2$, $4A_3$, $9P$, C		A_4, $4A_2$, $5P$, C	Rhombohedral: A_3, $3A_2$, $3P$, C; hexagonal: A_6, $6A_2$, $7P$, C	$3A_2$, $3P$, C	A_2, P, C	C	

[a] May be "composite" A_4.
[b] May be "composite" A_2.

cover many applications in organic and physical chemistry. A partial list of these applications follows:

1.0　Determination of purity
2.0　Analysis of mixtures
3.0　Characterization and identification of fusible compounds and mixtures
4.0　Determination of composition diagrams
　　　4.1　Two-component systems
　　　4.2　Three-component systems
5.0　Investigations of polymorphism
6.0　Measurement of physical properties
　　　6.1　Molecular weight
　　　6.2　Rates of crystal growth
　　　6.3　Crystal morphology
　　　6.4　Crystal optics
7.0　Study of boundary migration
8.0　Study of kinetics of crystal growth
9.0　Correlation of physical behavior with crystal properties (e.g., thermal stability of decomposable compounds with lowering of melting point, change in refractive index of the melt, etc.)

B.　SCOPE

Before considering further differences in the two schools of fusion methods, the following list of properties determinable by fusion methods, showing their wide scope, should be considered:

1.0　During heating
　　　1.1　Sublimation
　　　　　1.11　Degree of sublimation
　　　　　1.12　Nature of sublimate
　　　　　　　　1.121　Liquid globules
　　　　　　　　1.122　Liquid plus crystals
　　　　　　　　1.123　Badly formed crystals
　　　　　　　　1.124　Well-formed crystals of stable form
　　　　　　　　　　　　1.1241　Profile angles
　　　　　　　　　　　　1.1242　Form, habit, system
　　　　　　　　　　　　　　　　1.12421　Isotropic or anisotropic
　　　　　　　　　　　　　　　　1.12422　Uniaxial or biaxial
　　　　　　　　　　　　1.1243　Extinction
　　　　　　　　　　　　1.1244　Sign of elongation
　　　　　　　　　　　　1.1245　Principal refractive index of indexes
　　　　　　　　　　　　1.1246　Conoscopic observations
　　　　　　　　　　　　　　　　1.12461　Optic axial plane
　　　　　　　　　　　　　　　　1.12462　Optic axial angle

```
                        1.12463   Dispersion
                        1.12464   Sign of double refraction
                 1.1247   Pleochroism
            1.125   Well-formed crystals of unstable form
            1.1251–1.1257   (same as 1.1241–1.1247)
      1.2   Polymorphic transformation
            1.21   Temperature at first discernible transformation
            1.22   Monotropic transformation
            1.23   Enantiotropic transformation
                 1.231   Transition temperature
      1.3   Loss of water (or solvent) of crystallization
            1.31   Temperature at first discernible loss
      1.4   Decomposition
            1.41   Temperature at first discernible decomposition
            1.42   Color of decomposition products
2.0   During melting
      2.1   Melting point
      2.2   Refractive index of the melt
      2.3   Temperature coefficient of refractive index of the melt
      2.4   Dispersion of refractive index of the melt
3.0   During cooling
      3.1   Supercooling of melt
            3.11   Slight supercooling
            3.12   Readily supercools
                 3.121   Mobility of supercooled melt at room temperature
      3.2   Rate of growth of crystals as a function of temperature
      3.3   Form of crystal front
            3.31   Badly formed crystals
                 3.311   (sams as 1.1242–1.1247)
            3.32   Well-formed crystals
                 3.321   (same as 1.1241–1.1247)
      3.4   Polymorphic transformation
            3.41   (Same as 1.21–1.231)
      3.5   Gas bubbles
      3.6   Shrinkage cracks
      3.7   Mechanical twinning
      3.8   Anomalous polarization colors
      3.9   Characterization of unstable polymorphs
            3.91   (same as 1.1241–1.1247, 3.1, 3.2, 3.3, and 3.5–3.8)
4.0   Meltback (leaving some crystalline material unmelted as seed)
      4.1   (same as 1.0, 2.0, and 3.0, omitting 1.1, 2.2, 2.3, and 3.1)
5.0   After cooling to room temperature
      5.1   Supercooled liquid
            5.11   Refractive index
            5.12   Temperature coefficient of refractive index
```

C.   CHARACTERIZATION AND IDENTIFICATION METHODS

On cursory examination, the above outline seems to include an abundance of quantitative physical properties that would serve to uniquely characterize any possible compound. Unfortunately, however, too many compounds show only a few or none of these necessary properties. some, such as hexachloroethane, sublime completely before melting; others, such as sucrose, decompose completely before or during melting, or crystallize to give badly formed, very small crystals without crystallographic character. Another difficulty is that whereas one group of compounds may grow well from the melt to give a number of quantitative characteristics, a second group may give an equal number of numerical characteristics, but of a different type (e.g., refractive index of the melt, temperature coefficient and dispersion of the refractive index as compared with profile angles, extinction angles, and optic axial angles). Obviously the properties to be chosen for classification purposes must be measurable for all compounds to be included.

The requirements for the ideal physical property with which to characterize all compounds studied might be listed as follows. Each property should (1) be measurable for all compounds under study, (2) be easily and quickly determined without extraordinary background and training being required

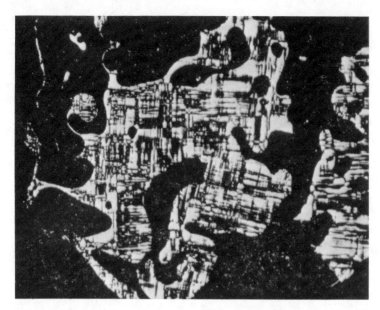

Fig. 15.74.   Fusion preparation of distearyl ethylenediamine.

of the investigator, (3) be measured with high accuracy and precision and be expressed numerically, (4) vary widely from compound to compound, and (5) be tabulated readily so that an unknown can be identified quickly.

Consideration of the properties determined by fusion methods as listed above shows that few, if any, properties satisfy all of these criteria. For example, anomalous polarization colors, dispersion of refractive index, and pleochroism defy ready tabulation; profile angles do not vary greatly from compound to compound; the presence of gas bubbles, shrinkage cracks, and the like cannot be expressed numerically; conoscopic observations cannot be made with rudimentary background and equipment; and none of them can be determined for all fusible compounds. Even with these limitations, individual compounds can be quickly recognized (see, e.g., Fig. 15.74).

The Koflers (46) have been especially ingenious in the manner in which they have surmounted these difficulties. A few listings from their identification tables illustrate their approach to this problem (Table 15.XXIII).

In these tables, which include more than 1000 compounds, the primary tabulating characteristic is the melting point, or the best possible substitute for the melting point for those compounds that decompose or sublime before melting. Then, since most compounds are stable somewhat below the melting point, the Koflers have introduced the eutectic melting point with two standard compounds. The eutectic melting point is as easy to measure as the melting point of the pure compound and nearly as valuable diagnostically.

This gives three numerical constants characteristic of each compound; yet the Koflers, to make identification even more certain, have developed

TABLE 15.XXIII

Excerpt from the Koflers' Tables for the Identification of Organic Compounds (46)

Melting point (°C)	Compound	Eutectic temperature with		Refractive index (n) of melt	Temperature (°C)	Special characteristics
		Phenacetin	Benzanilide			
151–153	α-Benzoinoxime	113	133	1.5609	147–149	Sublimes at 120°C to give kernels, rods, and droplets; solidifies to a glass; on rewarming gives a spherulitic mosaic
153	Diphenylthiourea	116	133	1.6128 1.6010	155 167	Sublimes at 130°C to give needles and plates
150–158	β-Nitroso-α-naphthol	111	129	Decomposes		Brownish-green; sublimes at 130°C to give needles, rods, and later droplets; melts sluggishly with decomposition
151–158	Boldin	116	130	Decomposes		Brown; melts to give a brown melt that solidifies to a glass; with 50% phenacetin, n = 1.5700 at 113–115°C; n = 1.5609 at 124–126°C

a very clever technique for measuring the refractive index of the melt. It involves the use of a set of glass-powder standards covering the range from 1.34 to 1.67 in increments of about 0.01. The refractive-index measurement is made by determining the temperature at which the glass-powder standard has the same index as the melt. This temperature is, of course, unique, although the compound must be very pure before the determination is made. The Koflers have taken care of this requirement by developing the absorption method of purifying compounds, which involves soaking up the eutectic melt in filter paper as the temperature rises. The unmelted residue just before complete melting is the pure compound.

Occasionally, when the compound decomposes it is possible at least to bracket the refractive index between two of the glass-powder standards (e.g., this can be done with α-benzoinoxime) or, in some cases, to determine the refractive index in the usual manner not on the pure compound, but on an accurately weighed mixture of that compound and a standard (e.g., Boldin with phenacetin).

The characterization and identification of fusible compounds and their mixtures have developed quite differently in the United States from their development in Europe. Most of the value of this means of identification has come in the study of relatively small groups of compounds, for example, substituted aminoquinolines (33), sterols (31), high explosives (65), and hexachlorobenzenes (2). In these cases the analyst has worked with each of the compounds almost daily, so he has become familiar with each compound by sight. Identification is usually made on the basis of some outstanding morphological or optical characteristic, such as anomalous polarization colors, unique shrinkage cracks or gas bubbles, odd crystal habit, or transformation mechanism. Obviously such characteristics cannot be recorded in tabular fashion, and can be described verbally only with difficulty. The analyst must remember each characteristic, perhaps with the help of photomicrographs or tabulated suggestions as to the proper key property of each substance. In spite of these limitations, there seems to be no difficulty in applying this method to the analysis of groups of as many as compounds. In one research program on high explosives (65), the analyst had no difficulty remembering the key properties of a group of nearly 70 high explosives and mixtures thereof. As a result, any one member of that group could be recognized unequivocally in 1–2 min, including time for sample preparation. Actual observation time through the microscope averaged probably under 10 sec for conclusive identification; often a glance sufficed.

This system has the serious limitation that only those analysts trained in the recognition of each compound can make the analysis, although a second analyst may learn the distinctive characteristics very quickly. Also, the analyst must maintain his ability by constant examination of compounds in the system; otherwise he might forget the key properties. In spite of these limitations, the method has obvious application to research and analysis.

The problem has also been considered of the general analysis of a number of compounds by fusion properties assembled into tables analogous to the Kofler tables. Rather than melting points and refractive indexes of the melts, however, crystallographic properties are used as identifying characteristics. The use of crystallographic properties has the advantage that present tabulations of data (20,81,82) can be used directly, but unlike the Kofler system, it has the serious disadvantage that the analyst must have an extensive background in crystallography. It is possible, however, to determine both optical and morphological properties by fusion methods, which offers another advantage: that the unknown need not be purified before analysis. The general methods involves successive steps, as outlined in Section X.B, in the classification of materials by properties determinable by fusion methods. The method is suitable for compounds that decompose if recrystallized on the microscope slide from thymol, nitrobenzene, benzyl alcohol, Aroclor, or a similar high-boiling liquid having high surface tension. This technique is, by the way, an excellent means of obtaining otherwise unobtainable unstable high-temperature polymorphs; it is for this reason that in the identification scheme the original crystals of the unknown are not usually melted completely, so that they remain as seeds during cooling.

Laskowski (52,53) characterized polynuclear aromatic compounds by tabulating the melting points of all solid phases formed during the mixed fusion with 2,4,7-trinitrofluorenone. Since each system includes at least one eutectic, an addition compound, a second eutectic or a transition temperature (peritectic reaction), and the compound itself, there are at least four easily determined melting points to use for identification purposes. This scheme approaches the Kofler procedure based on eutectic melting points with two standard compounds and the refractive index of the melt for each compound. However, the 2,4,7-trinitrofluorenone method is restricted to compounds forming addition compounds with that reagent.

D. POLYMORPHISM

Finally, the use of fusion methods for the study of polymorphism (see Fig. 15.75) and of two- and three-component phase diagrams (see Fig. 15.76) has been developed to a high degree (46,62). Most organic compounds possess several polymorph, and the phase diagrams for organic systems are fully as complex as those obtained for metals. The discovery of quasieutectic syncrystallization, in which the supercooled melt appears to crystallize as a eutectic over a wide range of binary composition, was made as a result of microscopical study of crystallization from the melt. The stabilization of lattices that are unstable for a pure compound by forming suitable solid solutions is a curious and important procedure, and one that is difficult, if not impossible, by classical macro methods.

Fig. 15.75. Polymorphic transformation in cholesteryl acetate.

E. DETERMINATION OF PURITY

The determination of purity is usually carried out by semiquantitative observation of the crystallization process on cooling of the melt. The amount of eutectic melt remaining after crystallization is an indication of the amount

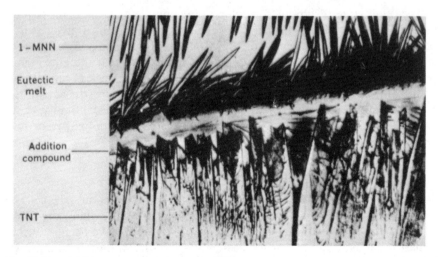

Fig. 15.76. Addition compound formed in fusion preparation of trinitrotoluene (TNT) and mononitronaphthalene.

of impurity. The absence of eutectic melt is possible only with pure compounds, when the impurity shows solid–solid solubility, or in the trivial case when the two melts are completely immiscible. Observation of the melt itself before crystallization will reveal the last of these situations, as well as the presence of higher-melting, undissolved solid components. Melting-point depression in known two-component systems, change of optical properties in isomorphous mixtures, refractive-index differentiation of impurity from the major component, change of crystal habit, and changes in the rate of crystal growth have all been used to indicate purity. Some of these techniques can be made quantitative, and then become useful as methods for the analysis of mixtures.

F.   ANALYSIS OF MIXTURES

The analysis of mixtures has been carried out by measurement of the optic axial angle or other optical property in isomorphous system (67), by measurement of the rate of crystal growth of the principal component of the mixture (11), and by determination of the melting-point depression. The first and third of these methods are suitable only for binary mixtures. The method based on rate of crystal growth is suitable for polycomponent mixtures if the impurities, no matter how many or even whether they are known, are always present in the same ratio. A given impurity may increase or decrease the rate of growth of the major component; hence the method has to be carefully standardized in terms of the nature of the impurity compounds.

G.   DETERMINATION OF COMPOSITION DIAGRAMS

The qualitative technique for determining composition diagrams is often used as a means of proving lack of identity or identity of a given pair of compounds (62). In this test two compounds are fused and allowed to crystallize in contact across a narrow zone of mixing, so that the entire composition diagram is shown in the composition gradient. Lack of any discontinuity (including discontinuity in growth rate) during growth across this gradient indicates identity of the two components. A discontinuity in the rate of growth indicates a difference in purity (one discontinuity) or nonidentity (two discontinuities).

H.   MECHANISM OF CRYSTAL GROWTH

The use of rate of crystal growth in determining the amount of p,p'-DDT in technical DDT has created an interest in learning more about the mechanism of crystal growth by fusion methods. The results have been published in part (32,67). Additional work has shown that the rate of crystal growth of a given compound in a binary mixture is directly related to the viscosity of the mixture. It is hoped that further work on the kinetics of crystal growth may result in the calculation of heats of fusion from crystal growth-rate data.

Boundary migration (61) is a phenomenon occurring with a few organic compounds that show anisotropy of elasticity, permitting them, in a sense, to be under unidirectional stress. As a result, crystals of such compounds show unidirectional growth into and through each other. This is strictly an orientation effect, such that one particular face grows into a different face of another identical crystal in contact with it. This is a most interesting effect that can be observed and studied only by fusion methods.

I. MISCELLANEOUS APPLICATIONS OF FUSION METHODS

The use of fusion methods to measure other physical properties of fusible materials also has been studied. For example, the decomposition of organic compounds on heating can be followed quantitatively and the kinetics of decomposition determined by fusion techniques. Decomposition will result in a lowering of the equilibrium melting point and in a change of the refractive index of the melt. Either property can be used, depending on the temperature at which the study is to be made and the melting point of the compound under study. The density of cast high explosives is related to the manner in which they are crystallized, since gases dissolved in the melt and shrinkage cracks contribute to lowering of the density. Microscopical fusion methods are, therefore, a logical means of studying this problem.

Finally, fusion methods have been extended to high temperatures (4,15,34) and to the study of inorganic systems (4,34,79). This is done by the direct application of fusion methods at high temperatures by means of special hot stages and furnaces, and with auxiliary lens systems to prevent damage to the microscope optics. It is also done indirectly by use of low-melting organic reagents, such as 8-hydroxyquinoline, to give characteristics precipitates with the various inorganic ions (79).

XI. MORPHOLOGICAL ANALYSIS

Speaking generally, the use of the microscope in analysis is limited to heterogeneous samples in which the desired component, even when present in trace amounts, maintains its identity as a separate phase. In fact, its use is based almost entirely on the recognition of particular portions of a physical mixture. The component present in trace amounts must be separated or concentrated mechanically, or it must be distinguishable to the eye of the microscopist.

Solutions may be analyzed morphologically if the desired component can be crystallized from the melt, solution, or vapor, or separated out by precipitation.

Various physical concentration procedures may be used to increase the sensitivity or ease of morphological analysis. These range from the Pasteur technique of mechanically picking out the desired component to the use of a magnet. Other methods of separation include sublimation, chromatogra-

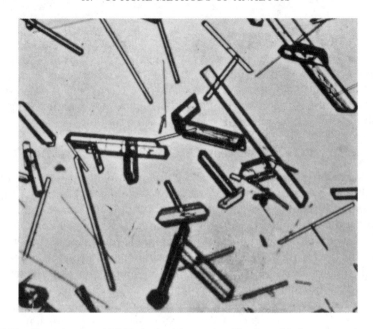

Fig. 15.77.   Crystals of p,p'-DDT obtained by recrystallization of 1–2 mg from thymol on a microscope slide.

phy, ion exchange, flotation, centrifugation, and differential solubility. Figuratively speaking, the desired component may be "isolated" by immersion of the sample in a medium of the proper refractive index, by use of polarized light, by crossed polars, or by appropriate staining methods.

A.   ADVANTAGES AND DISADVANTAGES

The major advantages of the microscope in morphological analysis are (1) small sample size, (2) high sensitivity, (3) that results are obtained in terms of compounds present rather than ions, (4), that additional facts are obtained about the sample (e.g., heat treatment, polymorphic form, particle size and shape, nature of dispersion), and (5) speed of analysis.

The major disadvantage is the general inapplicability of the microscope to the study of trace constituents in homogeneous systems.

Many materials can be almost instantaneously identified microscopically by observation of their morphological characteristics. All of the natural or synthetic fibers can be quickly identified, as can bacterial cells, pollen grains, metals and alloys, many crystalline materials, plant and animal tissues, and many other materials (Figs. 15.77–15.80).

When necessary, the electron microscope can be used to go beyond the limit of resolving power of the light microscope. The latter is limited, in terms of recognition of a small particle, to a minimum size of about 1 μ,

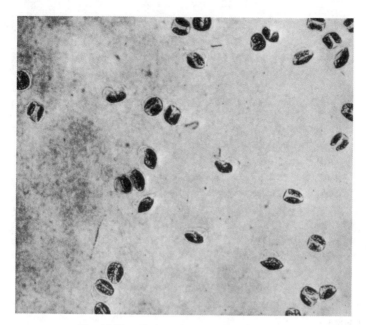

Fig. 15.78.   Grains of pollen from a pine tree.

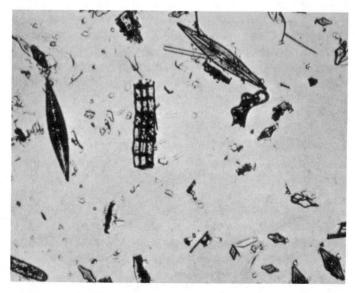

Fig. 15.79.   A sample of diatomaceous earth.

Fig. 15.80.   A polished and etched section of steel showing lamellar plates of cementite. Fe_3C, in the eutectic with ferrite.

whereas the corresponding limit for the electron microscope is about 100 times smaller.

The virtue of the microscope as a tool in trace analysis lies in the ability it gives the microscopist to recognize and identify fine particles having weights under 10^{-6} grams when such materials constitute a very minor portion of the entire sample.

B. BASIS FOR MORPHOLOGICAL ANALYSIS

The study of morphology is the study of external form or shape. The human eye and brain are constantly using the morphologies of objects, both animate and inanimate, in everyday life to identify friends and relatives, to choose the proper key to unlock the front door, to find a car in the parking lot, to choose the best piece of filled chocolate from the candy box, and so on. The method is not infallible, as everyone knows who has tried in vain to unlock a car or house almost identical with one's own or who has greeted someone presumed to be an old friend, only to be rebuffed by a cold look of nonrecognition. Still, with training to develop the powers of observation and with a good memory, a person can recognize and name with certainty a surprisingly large number of people, cars, airplanes, trees, flowers, animals, and so on.

In the same way, a microscopist is able to identify a surprisingly large number of metals and alloys, bacteria, plant and animal fibers, pure crystalline compounds, animal cells and tissues, diatoms, starchh grains, pollen grains, and other items. The masses of these individual particles or fibers range from about 10^{-6} down to perhaps as low as 10^{-18} grams. Each single particle, embodying all of the characteristic morphological features of the material, is readily identified. Furthermore, with a knowledge of its density and measurement of its physical dimensions, the weight usually can be calculated with great accuracy. The accuracy of this weight figure is greater the more nearly the physical shape resembles a regular geometrical form and the more accurately the density is known.

The microscopical identification of textile and paper fibers is a good example of the use of morphology in analysis. The scales on wool fibers; nodes on flax; serrated cells in esparto; pitted baglike cells in nonconiferous wood-pulp; smooth, highly birefringent cylinders of nylon; and isotropy in glass-wool fibers enable the microscopist to identify these fibers almost at a glance. In a similar way, metals and alloys can be quickly recognized by the metallographer, bacteria by the bacteriologist, human tissue sections by the pathologist, and so on.

C. REFINEMENTS OF THE METHOD

Occasionally the analyst (metallographer, pathologist, etc.) will make the job of identification easier by using the electron microscope, since many metal structures, diatoms, and, especially, bacteria, are too find to be easily resolved with the use of the light microscope. The analyst also may use staining as an aid in identifying tissues, fibers, or bacteria, or he may employ one or more optical methods to identify a given component or to check a mixture for the presence of a given component.

The polarizing microscope, the phase-contrast microscope, and the interference miscroscopes may be used to improve the visibility of certain preparations and to bring out details that might otherwise be nearly invisible. The polarizing microscope is especially valuable for differentiation between isotropic and anisotropic components, and even for the detection of one component of very different birefringence in a mixture of birefringent substances. The phase-contrast microscope and the interference microscope both increase contrast, although the latter is more useful for quantitative determination of minute masses.

Proper choice of a mounting medium also can be of great help in picking out minor components of a mixture. In the analysis of diamond powder, for example, because diamond has a very high refractive index ($n_D = 2.417$), it is possible to check very quickly for contamination by mounting the sample in a sulfur–selenium mixture having a refractive index near 2.41 (about 60% selenium and 40% sulfur is a suitable medium). All the usual contaminants (adulterants), such as quartz, corundum, glass, silicon carbide, and even rutile ($\epsilon = 2.903$, $\omega = 2.616$) will appear with strong contrast. Mounting in a medium having an index near 1.65 will quickly reveal quartz, glass, and corundum by their low contrast compared with diamond; however, silicon carbide and rutile also would have high contrast in this medium. As an added check in this case, the preparation can be examined quickly with crossed polars, since diamond is isotropic and most mineral contaminants are anisotropic.

Another specialized technique for rendering small percentages of a minor component readily visible is dispersion staining. This technique is based on the use of a mounting medium having a dispersion of the refractive index very different from that of the component being analyzed, but with an identical refractive index at one wavelength. All light of this one wavelength is then transmitted by the particle, whereas light of other wavelengths is partly scattered. The net result is a particle with a colored border. Very small percentages of particular compounds can be identified in this way, although, of course, a specific mounting medium must be prepared for each component; for example, quartz mounted in nitrobenzene gives a blue "dispersion staining" of all quartz particles.

Obviously, even with all these optical aids, the use of morphology alone as a means of identification has limitations. The procedure functions best when one has an atlas of photomicrographs for each system of components. To be able to identify a given particle, the microscopist must either have seen one before, or, at least, a good photomicrograph of it. The aim is to associate such microscopical characteristics as shape, surface, transparency, color, birefringence, and refractive index with the name of that substance. A photomicrographic atlas is an ideal solution to the problem because a microscopist can remember the morphological characteristics of only so many different substances. If the miscroscopist can remember a large number, he or she is still the only one who can make the analyses. An atlas supplements the microscopist's memory and helps other microscopists learn to identify particles. A number of photomicrographic atlases have been published on specific classes of materials. These include wood (29), papermaking fibers (3,14), fur (9), hair (55), textile fibers (58,60), pollens (40,48), cement clinkers and slags (47), boiler effluent (36), and pharmacological substances (75,76).

A more general work, *The Particle Atlas* (66), includes representative substances from these and other categories. Now in a greatly enlarged second edition, it contains more than 1400 full-color photomicrographs and a detailed description of 1000 substances. Also included are careful descriptions of the techniques necessary for collecting, manipulating, and characterizing particles, as well as analytical tables for the identification of specific particles after a simple preliminary classification. With this atlas, an interested person willing to invest the necessary time could become expert in identifying common particles found anywhere in the world.

In certain cases, supplementary numerical properties, such as interfacial angles and refractive indexes, can be measured. When this is possible the analytical method becomes more generally useful. Tables of crystallographic data are often based entirely, or almost entirely, on morphological properties. The five volumes of Groth (35), for example, include the interfacial angles for nearly every one of the 5000 compounds listed; Winchell's compilations of crystal data for both organic (81) and inorganic (82) compounds and, especially, the *Barker Index* (71) are basically compilations of morphological data for crystals. Even the more recent book by Donnay and Nowacki (22), although it is based on x-ray determination of unit cell parameters, is useful in this connection, since interfacial angles can be readily calculated from the unit cell dimensions.

The axial ratios are always expressed as $a:b:c$ for the biaxial systems (orthorhombic, monoclinic, and triclinic) and $a:c$ for the uniaxial systems (tetragonal and hexagonal). Ammonium dihydrogen phosphate, a tetragonal crystal, has an axial ratio $a:c$ of $1:1.002$. From this ratio we know that the distance between atoms along the c axis is 1.002 times the distance between atoms along the a axis. The actual spacings, as measured by x-ray diffraction, are $a = 7.51$ Å and $c = 7.53$ Å. Either the axial ratio or the cell dimensions can be used to calculate the interfacial angle by means of simple trigonometry. The tangent of one-half the angle about c is $1/1.002$ (or $7.51/7.53$), or 0.998. The observed interfacial angle is then $2(44.94°) = 89.88°$. This angle will be constant for all crystals of reasonably pure ammonium dihydrogen phosphate, wherever, or however, crystallized.

The law of constancy of interfacial angles is a basic law of crystallography. Crystals of a given compound may, and usually do, vary widely in external shape (habit), but each and every crystal of a given compound will show the same angle between corresponding faces and edges. Other crystals, even though tetragonal and, at first inspection, identical, will be found to show significantly different axial ratios and interfacial angles. Potassium dihydrogen phosphate, isomorphous with ammonium dihydrogen phosphate, has an interfacial angle of 93.60°, corresponding to $a:c = 1:0.939$; another isomorph, ammonium dihydrogen arsenate, has a corresponding angle of 90.24°, with a ratio $a:c$ of $1:1.004$.

X-ray diffraction data, either single-crystal or powder diffraction, can be obtained on very small amounts of material. Although the sufficient amount depends, to some extent, on the scattering power of the elements present in the sample, at least one sample of powder (actually a product of lattice decomposition on the surface of a single crystal), weighing about 10^{-10} grams, gave sufficient powder diffraction lines for certain identification. With mixtures, of course, larger amounts, ranging up to something like 10^{-8} grams, are required. Single-crystal data can be obtained on any crystal large enough to orient microscopically on the x-ray goniometer, that is, about 10^{-8} grams.

D. SIZE AND WEIGHT RELATIONSHIPS FOR SMALL PARTICLES

It is often a surprise to calculate the actual weight of a single particle or crystal on which all of the necessary crystallographic properties can be de-

TABLE 15.XXIV
Weights in Grams of Cubes of Various Sizes and Densities

Cube edge	Density			
	1	2	4	8
10^{-2} cm (100 μm)	10^{-6}	2×10^{-6}	4×10^{-6}	8×10^{-6}
10^{-3} cm (10 μm)	10^{-9}	2×10^{-9}	4×10^{-9}	8×10^{-9}
10^{-4} cm (1 μm)	10^{-12}	2×10^{-12}	4×10^{-12}	8×10^{-12}
10^{-5} cm (0.1 μm)	10^{-15}	2×10^{-15}	4×10^{-15}	8×10^{-15}
0.01μm (100 nm)	10^{-18}	2×10^{-18}	4×10^{-18}	8×10^{-18}
1/nm	10^{-21}	2×10^{-21}	4×10^{-21}	8×10^{-21}

termined (Table 15.XXIV). A single particle measuring as little as 5 μ in diameter can be manipulated at will with practice, without a micromanipulator. Its refractive indexes may be measured if intermediate washings are done to eliminate traces of the previous index medium; it can be moved from slide to slide; it can be photographed, x-rayed, measured, and, after all this, saved for later study. Its weight, if it were a sphere having a density of 1, would be 6.5×10^{-11} grams.

In many cases, single particles a few micrometers in maximum diameter can be identified directly by morphology or by simple microtests. In one such case, a single crystal (10^{-10} grams) of ignited oxide was dissolved in water on a slide and immediately reprecipitated to give many minute calcite crystals 1–2 μ in size and between 10^{-11} and 10^{-12} grams in weight. These were identified by obvious morphology (rhombohedra) and refractive index; the original particle was determined to be CaO.

E.   CONCENTRATION PROCEDURES

Often the morphology of a given component may not be unique, or the component may be too thoroughly dispersed throughout the sample to be visible as discrete particles. In these cases, concentration by crystallization (usually precipitation) may be useful. The procedures to be described presently, it should be noted, apply to systems wherein the desired component is homogeneously dispersed throughout the sample, as, for example, ions in a solution. The procedures are not, in themselves, highly sensitive, since the sample must be relatively concentrated; however, the total weight of the sample may be extremely small, usually less than a microgram.

Chamot and Mason (17) have described a number of precipitation reactions covering all the common and most of the uncommon cations and anions. These procedures involve addition of standard and specific reagents to a small aqueous test drop, with resulting precipitation of characteristically shaped crystalline precipitates. The basis for identification in each case is the specificity of the precipitation reaction and comparison of the crystals thus obtained with the written description and photomicrograph. Similar precipitation tests have been described for organic compounds (6).

Corresponding procedures have been applied to some organic systems, usually using reagents that form addition compounds. The test sample and the reagent usually can be melted immediately adjacent to and touching each other on a slide and under a cover glass (mixed fusion). Picric acid forms deeply colored molecular addition compounds with most of the polynuclear aromatic hydrocarbons; hence very small samples containing a few percent of either such a hydrocarbon or picric acid can be characterized as such by melting in contact with picric acid or a polynuclear aromatic, respectively, as a reagent. 2,4,7-Trinitrofluorenone has also been used by Laskowski (52,53) as a test reagent for many aromatic derivatives.

The common requirement of these precipitation reactions is careful standardization of concentrations, times, and temperatures, since otherwise the precipitate might not form at all or, at best, the morphology might be changed and thereby become unrecognizable. Other methods of crystallization not requiring a specific reagent also can be used: Recrystallization from a solvent or from the melt or vapor (always under controlled conditions) gives characteristic shapes easily recognized and remembered. The latter methods (recrystallization from the melt and vapor states) have been described under the general title *Fusion Methods in Chemical Microscopy* by McCrone (62). These methods have found wide application, because they are rapid and require little technical background and small samples (see Section X).

During World War II, fusion methods were successfully applied to the identification of high explosives (65). After careful study of each of the many possible systems, an analyst familiar with the appearance of the crystals growing from the melt was able to identify any one of 60-odd high-explosive compounds or mixtures in a few seconds using samples weighing a fraction of a milligram.

XII. MICROCHEMICAL TESTS

A. INTRODUCTION

In the day of such sensitive microanalytical instruments as the electron microprobe, the scanning electron microscope (SEM) with energy-dispersive detector, and the ion microprobe, one might assume that microchemical tests under the light microscope would have been superseded. On the contrary, the need for routine analysis of subnanogram samples (5) has sparked interest in any and all means of ultramicroanalysis. Microscopical microchemical tests have several assets:

1. They detect beryllium and lithium, most anions, and many organic compounds, for which purposes the electron probe and SEM are insensitive or incapable.

2. They are inexpensive, sensitive, reliable, and quick.
3. In a few fields, like drug detection, they are widely used and the literature has grown fairly continuously.

In Chamot's day it made sense to use an analytical scheme based entirely on microchemical tests, and he devised a fairly complete scheme for inorganic ions (17). Today instrumental methods are faster, more quantitative, and sometimes more sensitive than they were then (63); they should be used when a complete unknown must be analyzed. Microchemical tests are best used to check for specific ions or molecules.

The most generally useful microchemical tests are based on morphology, not of the original particles, for of precipitates formed by adding a reagent to an aqueous solution of the sample.

B.   INORGANIC QUALITATIVE ANALYSIS

1.   Chamot's System

In the Chamot system, basically, each chemical element or ion can be precipitated from a drop of solution by addition of one or more of a group of specific reagents, and each gives a characteristic crystalline precipitate. Most of the tests require only a minute or two, and not only is identification reliable, but with practice, a fairly accurate estimate of the percentage of the element can be made.

The test is usually performed by mixing a small drop of the reagent with a small test drop on a microscope slide. In general, the reagent is dissolved in a small drop of water and made to run quietly into contact with the test drop containing the unknown. Occasionally it is best to add the reagent crystals directly to the test drop or to draw a reagent drop across the dried residue from the test drop. When the mixture has stood for a few seconds or minutes, characteristic crystals of the resulting compound can be observed. The specific reagent and conditions for each precipitation can be found in Volume II of the *Handbook of Chemical Microscopy* (17). Some of the most useful of these tests are summarized in Table 15.XXV. The details of each test should be studied in Chamot and Mason's book (17) before any actual testing.

Later important contributors to inorganic microscopical qualitative analysis include Schaeffer (73) and Keune (44). Feigl's spot tests (26,27) are also useful in special cases, although these tests are not microscopical.

2.   Scanning for Individual Ions

Special procedures have been developed for scanning a sample microchemically for particular ions. These tests are based on the use of gelatin-coated microscope slides or similar substrates. First proposed by Crozier and Seely (19), this technique requires the gelatin or other substrate to be impregnated with a reagent specific for the test ion. Particles of the unknown

TABLE 15.XXV
Microscopical Tests for Inorganic Ions

Ion	Reagents
Sodium	Uranyl acetate, zinc uranyl acetate
Potassium	Chloroplatinic acid, perchloric acid
Ammonium	Chloroplatinic acid, iodic acid
Beryllium	Chloroplatinic acid, potassium oxalate
Calcium	Ammonium carbonate, sulfuric acid
Strontium	Ammonium carbonate, iodic acid
Barium	Ammonium carbonate, potassium ferrocyanide
Magnesium	Ammonium hydroxide and phosphate
Zinc	Oxalic acid, potassium mercuric thiocyanate
Copper, lead	Zinc, potassium mercuric thiocyanate
Mercury	Potassium bichromate, zinc
Aluminum	Ammonium bifluoride, cesium sulfate
Tin	Zinc, cesium chloride
Arsenic	Ammonium molybdenate, cesium sulfte, potassium iodide
Chromium	Silver nitrate, lead acetate
Uranium	Thallous nitrate, ammonium carbonate
Fluoride	Sodium fluosilicate
Iron	Potassium ferrocyanide, potassium thiocyanate
Carbonate	Silver nitrate, calcium acetate
Nitrate	Silver nitrate, nitron[a] sulfate
Cyanide	Silver nitrate, ammonium sulfide, ferric chloride
Chloride	Silver nitrate, thallous nitrate

[a] "Nitron" is the common name for diphenylenedianilohydrotriazole.

are made to impinge on the gelatin surface, where, if they are soluble at the pH of the gelatin, they dissolve and may react chemically with the reagent in the gelatin.

Assume, for example, that the gelatin is impregnated with silver nitrate and the particles impinging on the gelatin surface consist at least partly of sodium chloride. On standing for a short time, the sodium chloride particles or, for that matter, the particles of any soluble chloride present dissolve and begin to react with the silver nitrate. The result is a halo of crystalline silver chloride at the side of each chloride-containing particle. These halos have diameters, when fully developed, of 8–10 times the size of the original particle; hence they are easily observed microscopically.

This test has been developed for a variety of cations and anions and could be applied to almost any ion. Different substrates, such as Millipore filters, have been used by Lodge and others. Lodge and Fanzoi (56) have given the following directions for applying this test to the detection of soluble calcium compounds: Purified Eastman Kodak pigskin gelatin (2.5 grams), in 8 ml glycerol and 5 ml 30% aqueous ammonium ferrocyanide, is heated in an oven at 85°C until the gelatin dissolves. Microscope slides are then coated as described by Pidgeon (70). After several days, particles containing soluble calcium react with the ferrocyanide in the gelatin to form white microcrystalline halos consisting of large single crystals.

This procedure has the advantage that it more directly associates the test ion with the particles, since only those particles containing the test ion in a soluble form will produce halos. It also, when calibrated, indicates the size of the reacting particle, and gives visible halos with particles of sizes at or even below the resolving power of the light microscope. It is, therefore, particularly useful for the characterization of suspended particles (1 μ), in contrast to the larger settled dust particles, which can, in most cases, be better handled by morphological analysis.

The most serious disadvantage of the gelatin procedure is the required solubility of the particle containing the test ion. If the test ion as combined in the parent particle is less soluble than the precipitate formed by that ion and the reagent, no halo will form. It is, of course, possible to increase the rate of reaction by storing the slides in a moist atmosphere. The solubility may also be enhanced by exposing the slide, with the particles on it, to moist hydrochloric acid or ammonium hydroxide vapors. All in all, the procedure is quite useful, particularly when the particles to be tested are smaller than 1 μ or cannot be identified by morphology.

C.   ORGANIC QUALITATIVE ANALYSIS

Much of the emphasis in the development of microcrystal tests for organic compounds has been on methods for drugs. Fulton's monumental effort (30) is a compendium of such tests. Less complete, but easier to use, is the English translation of the book of Bhreens and Kley (6), which is oriented toward organic compounds more generally.

A different approach to organic microchemistry was advanced by Laskowski, who found reagents that form molecular addition compounds with various classes of organic compounds (51–53). The individual compound is identified by physical data such as the melting points of addition compounds and eutectics.

As noted earlier, Fiegel's spot tests (27) may be useful in special cases.

D.   MINIATURIZATION

Most of the microchemical methods considered thus far require milligrams or, at best, micrograms of sample. The impregnated gelatin slide technique can work for individual 1-μ particles, however. Other means of reducing the quantity of sample needed are discussed below.

The electron microscope may be used to observe the characteristic morphology of a reaction product (74,80). Although optical properties such as color birefringence and refractive index cannot be determined, electron diffraction can be applied to identify reaction products.

Instead of depending on increased microscope resolution to observe smaller amounts of product, one can reduce the size of the system in which the reaction occurs. Microchemical tests have been successfully scaled down to include picogram particles. Several approaches are possible, but all involve a means of restricting either the drop size or the area of precipitation. By doing this electrolytically with very fine-tipped cathodes, Brenneis (8) succeeded in detecting 50 pg copper or silver. Benedetti-Pichler and Rachelle (7) localized the precipitate and controlled the size of the droplets by placing two very small droplets of reagent and test solution close together in a drop of oil. The two drops coalesce to give a visible precipitate.

By treating the microscope slide with silicone oil, one can keep reagent drops from spreading, so that aqueous or organic liquids can be confined to very tiny droplets (50). Precipitation from aqueous solutions on the smallest scale requires control of humidity. This has been achieved by using small cells that can be flushed with moist or dry air at will. Picogram quantities of lithium and beryllium have been thus detected, and the method is general (64).

More specifically, a test precipitate containing a small weight percentage of the test ion in as small a test drop as possible will give the highest sensitivity. Once the test substance and reagent are chosen, there remains the problem of microminiaturizing the test drop. The ideal way to accomplish this is also the simplest. It requires a special recess (well) slide with an optional V-shaped air gap open to the atmosphere. Sliding a coverslip along the V allows more or less air to enter, thus controlling the rate of evaporation of water from the test drop hanging on the underside of the coverslip.

To make a test, a 1–2 mm circle is marked with a carbide stylus near the center of a 1-in. coverslip. The cover slip is then inverted and lightly coated with silicone oil, and the solid test substance is placed at the center of the circle (but on the opposite site). A corresponding quantity of reagent in solid form is then placed immediately adjacent to the test substance, touching it if possible, but definitely within 10 μ of it. The cover slip is next placed, sample side down, over the well depression in the slide. The slide is placed on an electrically coated heating slide connected to a variable voltage control. On slight heating of the slide, the pickup of moisture by the test substance is accelerated and controlled.

The microscope is focused on the sample, using an appropriate objective: $10\times$ for large (nanogram) samples and high magnification, non-immersion for smaller (picogram) samples. With the sample centered in the field of view and in good focus, a drop of water is added to the V slot so that it runs into the space between the coverslip and the top surface of the well slide. By sliding the slide, one can make the cover-slip cover the V and the well, thereby more or less sealing the cell. It is not difficult to grip the edges of a 25-mm coverslip with finger and thumb across the slide and hold it in position, maintaining the sample in the field of view as the slide is moved beneath it with the other hand.

The sample is observed as it picks up moisture to form a droplet of solution in which both reagent and test substance dissolve and react. The drop can be enlarged to any desired size, growth can be stopped, or evaporation can be begun by controlling the position of the coverslip over the V opening and the temperature. As mentioned, the precipitate can be partly dissolved as more water condenses in the drop, then grown to larger size by slow evaporation.

One advantage of this ultramicro technique is that most of the Chamot tests can be adapted to the same procedure.

Another advantage of the ultramicro test is that it converts a destructive to, in effect, a nondestructive test. Often, the total sample is a single nano- to microgram particle. In the usual Chamot test the entire sample would be dissolved for testing. With the ultramicro procedure, however, such a sample, if crushed between two clean slides, is sufficient for 100 or more tests. Even after several Chamot tests, sufficient sample should remain for x-ray diffraction, optical crystallographic, or other techniques.

E.   QUANTITATION

Benedetti-Pichler and Rachelle (7) felt that the sensitivity limit for microchemical tests with the light microscope is in the femtogram range, but at this level the precipitated crystals would have to be very fine needles or very thin plates to be visible. For most quantitative analysis, more material is necessary. Benedetti-Pichler provides a very complete treatment of ultramicrochemical analysis in his book *Identification of Materials via Physical Properties, Chemical Tests and Microscopy* (Academic, New York, 1964).

Klimeš and Janák (45) performed photometric analysis of picogram samples using silica-gel particles as cells in which the color reactions take place. Well-established color reactions, such as the dimethylglyoxime reaction for nickel, have been used at the picogram level.

XIII.   CONCLUSION

The last section in this chapter, Section XI, Morphological Analysis, is really a description of chemical microscopy as it is practiced in the industrial laboratory. The principal application of chemical microscopy, which is analytical, depends on the microscopist's ability to recognize by sight a large variety of different substances under the microscope, to identify by refined optical or microchemical tests those substances not recognized instantly, to recognize the effects of various physical treatments (e.g., milling, calendaring, heating, washing, quenching, grinding, digestion, and recrystallization) on a substance, and to relate the performance of a substance to its physical appearance under the microscope.

ACKNOWLEDGMENTS

The portions of this chapter dealing with the microscope and its optics are based on notes originally prepared by Marvin A. Salzenstein. Many of the figures were drawn by Warner Hudson, who did much valuable checking of data and collating of materials.

GENERAL REFERENCES

The following may be useful as general references in addition to the works cited in the detailed reference list (16,37,43,46,62,68,82).

A. BOOKS

Allen, T., *Particle Size Measurement*, Chapman and Hall, London, 1968.

Bambauer, H. V., F. Taborszky, and H. D. Trochim, *W. E. Tröger Optical Determination of Rock-Forming Minerals, Part 1: Determinative Tables*, E. Schweizerbartsche, Verlags-buchhandlung, Stuttgart, 1979.

Benedetti-Pichler, A. A., *Identification of Materials via Physical Properties, Chemical Tests, and Microscopy*, Academic, New York, 1964.

Bloss, F. D., *An Introduction to the Methods of Optical Crystallography*, Holt, Rinehart & Winston, New York, 1961.

Bloss, F. D., *The Spindle Stage: Principles and Practice*, Cambridge University Press, New York, Cambridge, London, 1981.

Bunn, C. W., *Chemical Crystallography*, 2nd ed., Oxford, Clarendon 1961.

Burrells, W., *Industrial Microscopy in Practice*, Morgan and Morgan, Hastings-on-Hudson, New York, 1964, and Fountain, London, 1961; 2nd ed., (as *Microscope Technique—A Comprehensive Handbook for General and Applied Microscopy*), Fountain, London, 1977.

Cosslett, V. E., *Modern Microscopy*, G. Bell and Sons, London, 1966.

Delly, J. G., *Photography Through the Microscope*, Eastman Kodak Company Publication P-2, Eastman Kodak Co., Rochester, New York, 1980.

El-Hinnawi, E. E., *Methods in Chemical and Mineral Microscopy*, Elsevier, Amsterdam, 1966.

Fleischer, M., R. E. Wilcox, and J. J. Matzko, *Microscopic Determination of the Nonopaque Minerals*, U.S. Geological Survey Bulletin 1627, U.S. Department of the Interior, Washington, D.C., 1984 (revision of E. S. Larsen, and H. Berman, U.S. Geological Survey Bulletin 848, 1934).

Hallimond, A. F., *The Polarizing Microscope*, 3rd ed., Vickers Instruments, York, England, 1970.

Hartley, W. G., *Hartley's Microscopy*, Senecio, Oxford, 1979.

Hartshorne, N. H., and A. Stuart, *Practical Optical Crystallography*, 2nd ed., American Elsevier, New York, 1969.

Heinrich, E. W., *Microscopic Identification of Minerals*, McGraw-Hill, New York, 1965.

Loveland, R. P., *Photomicrography: A Comprehensive Treatise*, 2 Vols., Wiley, New York, 1970; Reprinted by Robert E. Krieger Publishing Co., Inc., Melbourne, Florida, 1981.

Martin, L. C., *The Theory of the Microscope*, Blackie, London, 1966.

Mason, C. W. *Handbook of Chemical Microscopy*, Vol. 1, 4th ed., Wiley, New York, 1983.

McCrone, W. C., *The Asbestos Particle Atlas*, Ann Arbor Science Publishers, Ann Arbor, Michigan, 1980.

McCrone, W. C., L. B. McCrone, and J. G. Delly, *Polarized Light Microscopy*, Ann Arbor Science Publishers, Ann Arbor, Michigan, 1978.

McLaughlin, R. B., *Special Methods in Light Microscopy*, Microscope Publications, Chicago, 1977.

Michel, K., *Die Wissenschaftliche und Angewandte Photographie*, Springer, Vienna, 1967.

Muir, I. D., *The 4-Axis Universal Stage*, Microscope Publications, Chicago, 1981.

Needham, G. H., *The Practical Use of the Microscope, Including Photomicrography*, Thomas, Springfield, Illinois, 1958; 2nd ed., 1977.

Rochow, T. G., *Light-Microscopical Resinography*, Microscope Publications, Chicago, 1983.

Rogers, A. F., *Introduction to the Study of Minerals*, 3rd ed., McGraw-Hill, New York, 1937.

Shelly, D., *Manual of Optical Mineralogy*, Elsevier, Amsterdam, and Oxford, New York, 1975.

Shillaber, C. P., *Photomicrography*, Wiley, New York, 1944.

Stoiber, R. E., and S. A. Morse, *Microscopic Identification of Crystals*, Ronald, New York, 1972.

Vaughan, J. G., *Food Microscopy*, Academic, New York, 1979.

Vickers, A. E. J., *Modern Methods of Microscopy*, Butterworths, London, 1956.

Wahlstrom, E. E., *Optical Crystallography*, 5th ed., Wiley, New York, 1979.

White, G. W., *Introduction to Microscopy*, Butterworths, London, 1966.

B. ARTICLES

Altukhov, A. M., and T. I. Yatskova, "Determination of Refractive Index Using a Diffraction Grating," *Opt. Spektrosk.*, **54**, 1102 (1983) (in Russian).

Delly, J. G., "Microscopy's Color Key," *Indust. Res.*, **44** (October 1973).

Dunne, D. P., and N. F. Kennon, "Crystallographic Measurements Using Optical Microscopy," *Met. Forum*, **5**, 24 (1982).

Feklichev, V. G., "Procedure for the Optical Study of Minerals," *Diagn. Svoistva Miner.*, pp. 131–75, 240–1 (1981) (in Russian).

Fong, W., "Rapid Microscopic Identification of Synthetic Fibers in a Single Liquid Mount," *J. Forensic Sci.*, **27**, 257 (1982).

Graves, W. J., "A Mineralogical Soil Classification Technique for the Forensic Scientist," *J. Forensic Sci.*, **24**, 323 (1979).

Hamer, P. S., "Pigment Analysis in the Forensic Examination of Paints III. A Guide to Motor Vehicle Paint Examination by Transmitted Light Microscopy," *J. Forensic Sci. Soc.*, **22**, 187 (1982).

Heuse, O., and F. P. Adolf, "Nondestructive Identification of Textile Fibers by Interference Microscopy," *J. Forensic Sci. Soc.*, **22**, 103 (1982).

Holik, A. S., "Refractive Index Measurement by NDIC," *Microscope,* **31**, 223 (1983).

Kubic, T. A., J. E. King, and I. S. DuBey, "Forensic Analysis of Colorless Textile Fibers by Fluorescence Microscopy," *Microscope,* **31**, 213 (1983).

Liva, M., "New Candidates for Mounting Media," *Microscope,* **31**, 231 (1983).

McCrone, W. C., "Soil Comparison and Identification of Constituents," *Microscope*, **30**, 17 (1982).

McCrone, W. C., "Particle Characterization by PLM," Parts 1–3, *Microscope*, **30**, 185, 315 (1982); **31**, 187 (1983).

McCrone, W. C., "Microanalytical Tools and Techniques for the Characterization, Comparison and Identification of Particulate (Trace) Evidence," *Microscope*, **30**, 105 (1982).

McCrone, W. C., and W. Hudson, "The Analytical Use of Density Gradient Separations," *J. Forensic Sci.*, **14**, 370 (1969).

McCrone, W. C., and S. A. Skirius, "Dispersion Staining in the IR and UV," *Microscope*, **27**, 75 (1979).

Nicastro, A. J., "Polarized Light Microscopy in the Study of Liquid Crystals," *Am. Lab.*, **14**, 12 (1982).

Saylor, C. P., "A Study of Errors in the Measurement of Microscopic Spheres," *Appl. Optics*, **4**, 477 (1965).

Schmidling, D., "High Numerical Aperture Dispersion Staining Techniques," *Microscope*, **29**, 121 (1981).

Schram, R. R., "Light-Scanning Photomacrography," *Microscope*, **29**, 13 (1981).

Virta, R. L., K. B. Shedd, and W. J. Campbell, "Identification and Quantification of Asbestos in Construction Materials Using Polarized Light Microscopy: The Need for Standards," *NBS Spec. Publ.*, **619**, 34 (1982).

Wills, W. F., Jr., "A Search for Microchemical Reagents for Iron, Copper and Cobalt," *Microscope*, **29**, 87 (1981).

CITED REFERENCES

1. Albertson, C. E., *Microscope*, **14**(7), 253 (1964).
2. Arceneaux, C. J., *Anal. Chem.*, **23**, 906 (1951).
3. Armitage, F. D., *An Atlas of the Commoner Paper Making Fibres, An Introduction to Paper Microscopy*, Guildhall, London, 1950.
4. Bauman, H. N., Jr., *Bull. Am. Ceram. Soc.*, **277**, 267 (1948).
5. Bayard, M. A., *Microscope*, **15**, 26 (1967).
6. Behrens, T. H., and P. Kley, *Organische Mikrochemische Analyse*, L. Voss, Leipzig, 1922; English translation by R. Stevens, Microscope Publications Ltd., Chicago, 1969.
7. Benedetti-Pichler, A. A., and J. R. Rachelle, *Ind. Eng. Chem., Anal. Ed.*, **12**, 233 (1940).
8. Brenneis, H. J., *Mikrochemie*, **9**, 385 (1931).
9. Brevoort, H. L., *Fur Fibers as Shown by the Microscope*, Royal Microscopical Society, London, 1886.
10. *British Standard 3406, Part 4. Optical Microscope Method for the Determination of Particle Size of Powders*, British Standards Institution, London, 1964.
11. Bryant, W. M. D., *J. Am. Chem. Soc.*, **60**, 1394 (1938).
12. Bunn, C. W., *Discussions Faraday Soc.*, **5**, 132 (1949).
13. Cadle, R. D., *Particle Size: Theory and Industrial Applications*, Reinhold, New York, 1965.
14. Carpenter, C. H., L. Leney, H. A., Core, W. A., Côté, Jr., and A. C. Day, *Papermaking Fibers*, Technical Publication 74, State University College of Forestry, Syracuse, New York, 1963.
15. Cech, R. E., *Rev. Sci. Instr.*, **21**, 747 (1950).
16. Chamot, E. M., and C. W. Mason, *Handbook of Chemical Microscopy*, 3rd ed., Vol. I, Wiley, New York, 1958.
17. *Ibid.*, 2nd ed., Vol. II, 1940.
18. Crossmon, G. C., *Anal. Chem.*, **20**, 976 (1948).
19. Crozier, W. D., and B. K. Seely, in *Proceedings of First National Air Pollution Symposium*, Stanford Press, Stanford, California, 1950, p. 45.
20. "Crystallographic Data," *Anal. Chem.*, March 1948–March 1961; *Crystal Front*, March 1961, *et seq.*
21. Delly, J. G., *Particle Analyst*, No. 89, Ann Arbor Science Publishers, Ann Arbor, Michigan, 1968.

490 H. OPTICAL METHODS OF ANALYSIS

22. Donnay, J. D. H., and W. Nowacki, *Crystal Data*, Geological Society of America, New York, 1954.

23. Emmons, R. C., *Am. Mineralogist*, **11**, 115 (1926); **13**, 504 (1928); **14**, 414, 441, 482 (1929).

24. Endter, F., and H. Gebauer, *Optik*, **13**, 97 (1956).

25. Fairs, G. L., *Chemistry and Industry*, **62**, 374 (1943); *J. R. Microsc. Soc.*, **71**, 209 (1951).

26. Fiegl, F., *Spot Tests in Inorganic Analysis*, 5th ed., American Elsevier, New York, 1958.

27. Fiegl, F., *Spot Tests in Organic Analysis*, 7th ed., Elsevier, New York, 1954.

28. Fisher, C., *Microscope*, **19**, 1 (1971).

29. Forest Products Research Laboratory, *An Atlas of End-Grain Photomicrographs for the Identification of Hardwoods*, Bulletin 26, Her Majesty's Stationery Office, London, 1953.

30. Fulton, C. C., *Modern Microcrystal Tests for Drugs*, Wiley-Interscience, New York, 1969.

31. Gilpin, V., *Anal. Chem.*, **23**, 365 (1951).

32. Gilpin, V., W. C. McCrone, A. Smedal, and H. Grant, *J. Am. Chem. Soc.*, **70**, 208 (1948).

33. Goetz-Luthy, N., *J. Chem. Ed.*, **26**, 159 (1949).

34. Grabar, D. G., and W. C. McCrone, *J. Chem. Ed.*, **27**, 649 (1950).

35. Groth, P., *Chemische Kristallographie*, Engelmann, Leipzig, 1906–1919.

36. Hamilton, E. M., and W. D. Jarvis, *Identification of Atmospheric Dust by Use of the Microscope*, Central Electricity Generating Board, London, 1963.

37. Hartshorne, N. H., and A. Stuart, *Crystals and the Polarising Microscope*, 4th ed., Arnold, London, 1970.

38. Horvath, H., *Staub Reinhaltung der Luft*, **34**, 197 (1974) (in English).

39. Humphries, D. W., *Microscope*, **15**, 267 (1966).

40. Hyde, H. A., and K. F. Adams, *An Atlas of Airborne Pollen Grains*, St. Martin's, New York, and Macmillan, London, 1958.

41. Jelley, E. E., *J. R. Microsc. Soc.*, **54**, 234 (1934).

42. Jesse, A., *Microscope*, **19**, 21 (1971).

43. Johannsen, A., *Manual of Petrographic Methods*, 2nd ed., McGraw-Hill, New York, 1918.

44. Keune, H., *Bilderatlas zur Qualitativen Anorganischem Mikroanalyse*, VEB Deutscher Verlag für Grundstoffindustrie, Leipzig, 1967.

45. Klimeš, L., and J. Janák, *Microchem. J.*, **13**, 534 (1968).

46. Kofler, L., and A. Kofler, *Mikromethoden zur Kennzeichnung Organischer Stoffe und Stoffgemische*, Wagner, Innsbruck, 1948.

47. Konovalov, P. F., B. V. Volkonskii, and A. P. Khashkovaskii, *Atlas of Microstructures of Cement Clinkers, Refractory Materials and Slags*, Gos. Izd. Lit. po Stroit, Arkhitekt, i Materialam, Leningrad, 1962.

48. Kremp, G. O. W., *Morphologic Encyclopedia of Palynology*, University of Arizona, Tucson, 1965.

49. Lark, P. D., *Microscope*, **15**, 1 (1965).

50. Laskowski, D. E., *Anal. Chem.*, **37**, 174 (1965).

51. Laskowski, D. E., and O. W. Adams, *Anal. Chem.*, **31**, 148 (1959).

52. Laskowski, D. E., D. G. Grabar, and W. C. McCrone, *Anal. Chem.*, **25**, 1400 (1953).

53. Laskowski, D. E., and W. C. McCrone, *Anal. Chem.*, **26**, 1497 (1954); **30**, 542 (1958).

54. Lehmann, O., *Die Kristallanalyse*, Engelmann, Leipzig, 1891.

55. Lochte, T., *Atlas der menschlichen und tierischen Haare*, Schöps, Leipzig, 1938.

56. Lodge, J. P., Jr., and H. M. Fanzoi, *Anal. Chem.*, **26**, 1839 (1954).

57. Loveland, R. P., *J. R. Microscop. Soc.*, **79**, 59 (1960).

58. Luniak, B., *Identification of Textile Fibers*, Pitman, London, 1953.

59. Manalen, D. A., and J. S. Glass, *Am. Lab.*, Vol. 2, p. 45 (October 1970).

60. Mauersberger, H. R., Ed., *Matthews' Textile Fibers*, 6th ed., Wiley, New York, 1954.

61. McCrone, W. C., *Discussions Faraday Soc.*, **5**, 158 (1949).

62. McCrone, W. C., *Fusion Methods in Chemical Microscopy*, Interscience, New York, 1957.

63. McCrone, W. C., *Particle Analyst*, No. 49, Ann Arbor Science Publishers, Ann Arbor, Michigan, 1968.

64. McCrone, W. C., *Microscope*, **19**, 235 (1971).

65. McCrone, W. C., J. H. Andreen, and S. M. Tsang, *Office Sci. Res. and Develop., Rep. No. 3014*, NTIS, Washington, DC August 1, 1944.

66. McCrone, W. C., and J. G. Delly, *The Particle Atlas*, 2nd ed., Vols. I–IV, Ann Arbor Science Publishers, Ann Arbor, Michigan, 1974; *ibid.*, Vols. V–VI, 1978.

67. McCrone, W. C., A. Smedal, and V. Gilpin, *Anal. Chem.*, **18**, 578 (1946).

68. Needham, G. H., *The Practical Use of the Microscope*, Thomas, Springfield, Illinois, 1959.

69. Patterson, H. S., and W. Cawood, *Trans. Faraday Soc.*, **32**, 1084 (1936).

70. Pidgeon, F. D., *Anal. Chem.*, **26**, 1832 (1954).

71. Porter, M. W., and R. C. Spiller, Eds., *Barker Index*, Heffer, Cambridge, 1951–1956.

72. Saylor, C. P., *J. Res. Natl. Bur. Std.*, **15**, 277 (1935).

73. Schaeffer, H. F., *Microscopy for Chemists*, Van Nostrand, New York, 1953 pp. 114–157, reprinted by Dover, New York, 1966.

74. Tufts, B. J., and J. P. Lodge, Jr., *Anal. Chem.*, **30**, 300 (1958).

75. Wallis, T. E., *Analytical Microscopy*, 3rd ed., Little, Brown, Boston, 1965.

76. Wallis, T. E., *Textbook of Pharmacognosy*, 5th ed., J. & A. Churchill Ltd., London, 1967.

77. Watts, J. T., *Microscope*, **18**, 35 (1970).

78. Weissberger, A., Ed., *Techniques of Organic Chemistry, Vol. I: Physical Methods of Organic Chemistry*, Interscience, New York, 1945.

79. West, P. W., and L. Granatelli, *Anal. Chem.*, **24**, 870 (1952).

80. Wiesenberger, E., *Mikrochim. Acta*, 506 (1957).

81. Winchell, A. N., *Optical Properties of Organic Compounds*, 2nd ed., Academic, New York, 1954.

82. Winchell, A. N., and H. Winchell, *The Microscopic Characters of Artificial Minerals*, 3rd ed., Academic, New York, 1964.

83. Zieler, H. W., *Microscope*, **17**, 249 (1969).

Part 1
Section H

Chapter 16

ELECTRON MICROSCOPY

By George G. Cocks, *Los Alamos National Laboratory, Los Alamos, New Mexico**

Contents

* Professor Emeritus of Chemical Microscopy, Cornell University, Ithaca, New York.

I. INTRODUCTION

Microscopes are image-forming instruments. Most microscopes operate in one of two distinct modes of image formation: (1) the "whole-beam" or "fixed-beam" mode, in which all parts of the image are formed simultaneously; and (2) the probe or scanning mode, in which the image is built up point by point in some sequence. The beams or probes may consist of various types of electromagnetic radiation (e.g., light, x-rays, or radio waves), of particles (e.g., electrons, ions, or neutrons), or of sound waves. This chapter is concerned with electron-beam and electron-probe instruments.

Electron microscopes possess unique characteristics that are of considerable benefit in carrying out chemical analyses. It is the purpose of this chapter to point out the analytical uses to which the electron microscope has been or could be put, in the hope of stimulating both its increased use in anaysis and the further exploration and development of analytical techniques.

The theory and practice of electron microscopy will be discussed in order to acquaint those who are not familiar with the instrument with it and to help them use it effectively. Because it is not possible to include all of the information necessary to accomplish this goal, pertinent references to the extensive literature on electron microscopy will be given. The skill necessary to operate an electron microscope that is in good condition and properly aligned is not difficult to acquire. With careful reading of the instruction manual that comes with the microscope, and with a minimum of instruction, it should be possible for the user to focus the microscope, adjust the magnification, and obtain good electron micrographs at low or intermediate magnification. The other tasks necessary for routine use of the electron microscope are the alignment of the optical system and simple maintenance, including replacement of filaments, cleaning, adjustment of apertures, and so on. Here again, reference to the literature and some instruction will suffice.*

The quality of the results obtained in microscopy depends strongly on the ability of the microscopist to prepare the specimen properly. For transmission electron microscopy, this is considerably more difficult to acquire than is the ability to operate the instrument. To become adept in specimen preparation may require many months, depending on the particular techniques to be mastered. Although there are general classes of techniques, such as replication, thin sectioning, dispersion of particulate matter, and shadowing, any technique selected for a given problem usually must be modified and adapted to the particular specimen being examined. Unless the microscopist confines his studies to a very narrow class of specimens, specimen preparation is not likely to become a routine matter. For scanning electron microscopy, specimen preparation is much simpler, although still very important.

---

* Serious problems in maintenance or adjustment of the microscope are usually not corrected by the microscopist unless he happens to be trained in electron optics, electronics, and vacuum technology. In general, the manufacturers of microscopes make available the services of qualified technicians who will maintain the microscope in good operating condition.

The literature of electron microscopy contains many descriptions of specimen-preparation techniques. However, it is usually difficult for the novice to carry out these techniques simply by following the published directions, which are not always given in sufficient detail. In general, it is advisable to learn specimen-preparation techniques from an experienced microscopist. This can be done by visiting working laboratories and by attending courses in electron microscopy. Short intensive courses in electron microscopy are offered by a number of universities and colleges, research laboratories, and companies that manufacture microscopes. Attendance at technical meetings of societies for electron microscopy, or at meetings where the results of electron-microscopical studies are reported, provides valuable opportunities to learn techniques both by listening to the presentations and by discussions with experienced microscopists.

Effective use of the electron microscope requires more than manipulative skill. The most important part of microscopy is the interpretation of results. The ability to interpret is acquired only after considerable experience, and is based on knowledge of the field in which the investigation is being made.

There is a tendency for those not familiar with microscopical investigation to feel that when they have obtained a satisfactory image by some technique, they have exhausted the possibilities of microscopy. They also tend to feel that anyone experienced in interpreting visual images, and this includes most people, is capable of interpreting microscope images. But interpretation of microscopical images requires, in addition to basic knowledge of the field of investigation, knowledge of the effects of specimen preparation, of the interaction of the beam with the specimen, and of the characteristics of the microscope itself.

It was stated above that the skills necessary to take micrographs at low and intermediate magnifications are not difficult to acquire. However, if the electron microscope is to be used effectively at its maximum resolution, the microscopist must have developed interpretive ability, specimen-preparation technique, and operative skills to the highest degree.

The electron microscope is usually used at relatively high magnifications compared with the magnifications used with the light microscope. The specimens being examined may have structures that are entirely unfamiliar. In addition, the total area of a specimen that can be examined is relatively small and sampling problems are acute. In this situation, experience has shown that much time can be saved and results can be more accurately interpreted if light microscopy is used to provide a basis for the examination at high magnification. It is usually profitable to examine and select specimens and to follow the various stages of specimen preparation by using various light microscopes. Many of the specimen manipulations can be carried out under a stereomicroscope, and the replicas, sections, or other electron-microscopical specimens can be evaluated with metallurgical or transmission microscopes.

A. HISTORY OF THE DEVELOPMENT OF ELECTRON MICROSCOPY

The first electron microscopes were built in 1931 and 1932. These instruments were based on the earlier development of the science of electron optics, which dated from the discovery of cathode rays by Plucker in 1859 (62). Wiechert, in 1899 (62), was the first to use a solenoid to concentrate cathode rays, and in 1903 Wehnelt (62) used electrostatic means for the same purpose. Geometrical electron optics was originated by Busch in 1926 (56) when he treated a magnetic coil as a lens and derived lens equations for it. The electromagnetic lens was further developed by Gabor (105), who constructed an iron-clad lens in 1926–1927. Electrostatic lenses were studied by Knoll around 1929, and Knoll and Ruska also contributed to the development of the electromagnetic lens (260).

Physical electron optics was founded by de Broglie in 1924 (82) when he first discussed the wave nature of the electron. The wave property associated with a moving electron was established experimentally by Davisson and Germer in 1927 (80) when they succeeded in observing the diffraction of electron beams.

These studies in electron optics culminated in the building of an emission electron microscope by Knoll and Ruska (173). This instrument, which produced an image of the electron source, utilized electromagnetic lenses. Another emission microscope using electrostatic lenses was built by Brüche and Johannson (50). A third emission microscope was built by Marton (195). All three of these instruments were finished in 1931–1932. The first transmission electron microscope (TEM) was also built by Knoll and Ruska in 1931 (172). These early instruments were improved rapidly, and the first commercial TEM, designed by Von Borries and Ruska, was produced by Siemens and Halska in 1939. Under the direction of Professor E. F. Burton at the University of Toronto, C. E. Hall built the first electrostatic emission microscope. These instruments are described in Hall's M.A. dissertation (University of Toronto 1936). After leaving Toronto, Hall went to the Eastman Kodak Company in Rochester, New York, and built the first electron microscope in the United States. In 1939, again under Professor Burton's direction, Prebus and Hillier developed the first TEM built in North America (55,101,242). In 1941 Zworykin, Hillier, and Vance at RCA developed the first commercial TEM in North America.

In 1935 Knoll (170) proposed a scanning-type electron microscope, and in 1938 von Ardenne (308) reported on the first scanning transmission electron microscope (STEM) built. In 1942 Zworykin, Hillier, and Snyder (336) reported on the first "reflection" scanning electron microscope (SEM) suitable for examining the surfaces of thick specimens by means of secondary electrons emitted from the specimen. McMullan, a graduate student working with Oatley at Cambridge University, built a SEM in 1948 (198). This instrument was developed further by Smith (283). Later Smith designed and had built another instrument, for the Pulp and Paper Research Institute of

Canada (284). A major advance in the development of the SEM occurred when Everhart and Thornley (97) modified the electron detector so as to greatly improve its signal-to-noise ratio. Pease and Nixon (233,234) built a SEM that was the prototype for the first commercially available instrument produced by the Cambridge Scientific Instrument Company. Since that time, the SEM has developed rapidly. The SEM is a particularly valuable analytical tool because it can be readily combined with the electron probe microanalyzer, making it possible to do qualitative and quantitative elemental analyses.

After its early introduction (308), the STEM was more or less neglected until about 1966, when Crew and his coworkers built such an instrument (77,78). This instrument made use of a field emission electron gun described by Crew et al. in 1969 (79) and by Butler (57). Since that time scanning transmission electron microscopy has developed steadily, and at present a number of manufacturers offer STEMs. The application of this instrument can be expected to develop rapidly, especially in view of its analytical capabilities. More information on the early history of electron optics and electron microscopy can be found in an article by Rudenberg (259); in Von Borries's *Die Ubermikroskopie* (312); in the *Encyclopedia of Microscopy* (62); in an article by Calbick (58); in Heidenreich's *Fundamentals of Transmission Electron Microscopy* (137); in *The Early History of the Electron Microscope* by Marton (196); and in articles by Mulvey (219), Freundlich (102), Gabor (109), and Franklin et al. (101). More recent historical developments can best be found in the books and articles mentioned in Section I.D. on the General Literature of Electron Microscopy.

B.   ADVANTAGES OF ELECTRON MICROSCOPY IN CHEMICAL ANALYSIS

All microscopes are designed to form an enlarged image of a specimen and to present this image in such a way that it can be perceived visually. The eye can perceive the shape of an object, its color, and the intensity of the light transmitted by or reflected from it. Therefore, microscopes make it possible to observe these three qualities for objects too small to be visible to the unaided eye. However, microscopes differ as to the completeness with which these qualities can be observed. The electron microscope normally cannot reveal the color of an object, although a few highly specialized instruments have been constructed to produce the analog of color at electron wavelengths.

The primary advantage of the electron microscope in chemical analysis is that it makes possible the analysis of very small objects. The size of objects visible in the electron microscope ranges, at the present time, from a few angstroms (Å) to about 25 μ for a TEM or STEM. For a SEM the range is from about 50 Å to several centimeters. The electron microscope not only permits the analysis of very tiny objects, but also reveals the spatial or structural relationships between the components analyzed. This latter property is the most important aspect of the use of electron microscopy in chemical analysis.

Chemical analyses with the electron microscope may be either qualitative or quantitative. Qualitative methods may be of many kinds, as will be discussed below. Quantitative methods often depend on counting and measuring the sizes of recognizable constituents in mixtures. Mixtures of various phases, which cannot be analyzed by chemical means in terms of the compounds present, can be so analyzed by microscopical techniques. In addition to phase analysis, modern microscopes are capable of performing elemental analyses in many sophisticated ways.

C. EXAMPLES OF CHEMICAL ANALYSIS USING ELECTRON MICROSCOPY

A few examples may serve to illustrate the value of electron microscopy in chemical analysis. Aerosols collected in air-pollution studies have been analyzed by chemical microscopical means (303). Particles of halides of the order of 50 Å in diameter and particles of sulfates 1000 Å in diameter were identified by precipitation of compounds recognizable in the electron microscope. In addition to the qualitative identification of these ions, it was shown that the size of a region of precipitate was directly related to the amount of ion in the particle, and on the basis of this relationship, the size distributions of the halide and sulfate in the original aerosol were deduced.

In light microscopy there is a well-developed science of histochemistry, which is concerned with the localization of organic and inorganic compounds in living cells and tissues. Some progress has been made toward adapting histochemical techniques to electron microscopy. For example, it has been demonstrated (15) that methylmercuric chloride reacts specifically with protein-bound sulfhydryl groups to produce compounds that can be seen in the electron microscope. This staining reaction has been used to identify these functional groups in the squamous epithelial cells involved in the formation of keratin.

D. GENERAL LITERATURE OF ELECTRON MICROSCOPY

The literature of electron microscopy has become very large, so large that one is almost forced to restrict one's reading to several specialized fields. However, there are a number of books, journals, and articles of interest to nearly all microscopists. One of the early books was *Electron Optics and the Electron Microscope*, by Zworykin et al. (337). Hall's book *Introduction to Electron Microscopy* (130) was published in 1953 and revised in 1966. It is still an excellent introductory text. *Practical Electron Microscopy for Biologists* (200) by Meek is an excellent introductory text. Although the title indicates that it is for biologists, it is good for anyone who needs an introduction to electron microscopy. Haine and Cosslett's book *The Electron Microscope: The Present State of the Art* (125) is concerned primarily with the theory of electron microscopy and the construction of microscopes. It does contain a section on specimen-preparation techniques and applications.

Several excellent books and sets of books deal with electron-microscopical techniques and methods of specimen preparation. One of the early books edited by Kay (165) is now out of print, but it is a valuable book and may be available in libraries or private collections. Reimer has written a second edition of his *Elektronenmikroskopische Untersuchungs- und Präparationsmethoden* (254), and Glauert has edited a five-volume work entitled *Practical Methods in Electron Microscopy* (117). Hyatt has written a book on *Basic Electron Microscopy Techniques* (154) and edited a series entitled *Principles and Techniques of Electron Microscopy* (153).

There have been a number of international conferences on electron microscopy. The first of these was held in Delft in 1949 (244). This meeting was followed by one in Paris in 1950 (69). In 1954, a third meeting was held under the auspices of the Joint Commission on Electron Microscopy of the International Council of Scientific Unions (73). Some confusion may arise as to the numbering of these conferences. The published proceedings of the Delft meeting were not numbered. The proceedings of the Paris meeting were published under the title *Comptes Rendus du Premier Congres International de Microscopie Electronique*. However, the London meeting was labeled the *Third International Conference on Electron Microscopy*, recognizing the Delft meeting as the first. In 1958 the Fourth International Conference was held in Berlin (18), and the Fifth International Conference was held in Philadelphia in 1961 (45). The Sixth International Congress on Electron Microscopy was held in Kyoto in 1966 (264), the seventh at Grenoble in 1970 (100), the eighth at Canberra in 1974 (266), and the ninth at Toronto in 1978 (289).

Starting with the Fourth International Congress in Berlin, all of these congresses have been sponsored jointly by the International Federation of Societies for Electron Microscopy (IFSEM) and by the microscopy society of the host country. IFSEM is the direct descendent of the Joint Commission on Electron Microscopy mentioned above. The proceedings of these meetings provide an invaluable reference for electron microscopists.

The proceedings of several other European conferences have also been published (25,116,145,281,299,300,305). These publications are very similar to the proceedings of the international conferences described above and are similarly useful. All of these meetings, starting with the meeting in Stockholm in 1956, were also cosponsored by IFSEM. The Electron Microscopy Society of America (EMSA) has held an annual meeting since its establishment. In the early years abstracts of the meetings were published in the *Journal of Applied Physics*. Since 1967 the proceedings of the meetings have been published as separate bound volumes (7,16). Siegel and Beaman have compiled selected papers from an EMSA meeting dealing with electron microscopy and microbeam analysis (275). Hren has edited the proceedings of a workshop on analytical electron microscopy (146). The Electron Microscopy and Analysis Group (EMAG) of the Institute of Physics in Great Britain has been meeting for many years. Beginning in 1971, the proceedings of these meetings have been published (203,220,225,306).

The publications mentioned above are concerned with microscopy in general. Although both TEM and SEM are covered, they lean more heavily toward TEM.* Textbooks on SEM have been written by Oatley (226), Hearle et al. (135), Holt et al. (144), and Goldstein and Yakowitz (120). Since 1968 Johari has been organizing annual meetings on SEM. The proceedings of these meetings (159,160) afford a unique means of keeping abreast of this rapidly advancing field.

Several bibliographies on electron microscopy are available. Cosslett's *Bibliography of Electron Microscopy* (70) and a bibliography of the same name compiled by Marton and coworkers (191–194,253) cover the literature from the beginning of electron microscopy through 1949. The New York Society of Electron Microscopists has published a bibliography in the form of punched cards covering the period 1950–1961 (223). This bibliography was also published in book form. The *Bulletin Signaletique* (52) abstracts many articles of interest to electron microscopists; abstracts of articles on electron microscopy appear in Section 6. *Analytical Chemistry* has published a series of reviews of electron microscopy, starting in 1949. These reviews appear in a special issue of the journal, which appears in April of even-numbered years. Other journals of general interest to microscopists include the *Journal of Microscopy* (Blackwell, Oxford), the *Journal de Microscopie* (CNRS, Paris), the *Journal of Electron-microscopy* (Japanese Society for Electron Microscopy), *Mikroskopie* (Vienna), and *Zeitschrift für die Wissenschaft Mikroskopie* (Stuttgart).

II. THEORY OF ELECTRON MICROSCOPY

A. PHYSICAL BASIS OF ELECTRON MICROSCOPY

It has been pointed out above that there are two modes of image formation commonly used in electron microscopes, the whole-beam mode and the scanning mode. This classification relates to the type of beam used to form the image. Microscopes can be further classified as emission microscopes, transmission microscopes, or reflection microscopes. This classification refers to the manner in which the image-forming beam approaches and leaves the specimen.

In an emission microscope the image-forming beam is generated within the specimen. For example, in a thermionic emission microscope the specimen is heated until it gives off the electrons that are used to form the image.

___

* In this chapter, we shall use the abbreviation ''TEM'' for both ''transmission electron microscope'' and ''transmission electron microscopy,'' and likewise for ''SEM'' and ''STEM.''

In a transmission microscope, electrons are supplied by an illuminating system. These electrons are transmitted through the specimen and the emergent beam is used to form the image.

In a strict sense, the image-forming electrons in a reflection microscope are supplied by an illuminating system and are reflected from the specimen.

Whole-beam instruments can be emission, transmission, or reflection types, and the same is true for scanning instruments. There are some instruments that are more complex; for example, they may simultaneously be both transmission and reflection types.

There is still another classification scheme that is used in microscopy. This scheme relates to various methods of manipulating the illuminating and image-forming beams. Commonly, bright-field microscopy is used, but dark-field, phase-contrast, interference, and other types of microscopy can be employed.

Because microscopes did not develop historically according to this or any other classification scheme, the nomenclature of microscopy is often confusing and inadequate. "Scanning microscopy" is a good descriptive term, but "whole-beam microscopy" and "fixed-beam microscopy" are somewhat awkward. "Transmission microscopy" is a descriptive term in most cases. "Reflection microscopy" often is ambiguous. In light microscopy, this type of illumination is usually called "vertical illumination" or "incident illumination." "Incident illumination" is a fairly descriptive term, but it does not distinguish between reflection microscopy and transmission microscopy. "Vertical illumination" is widely used, but inappropriate, because the light rays approaching the specimen form a cone and hence approach from a number of directions. Even the axis of the cone may not be vertical, because such an illumination will work in any orientation. Finally, "vertical" does not distinguish between reflected and transmitted or emitted image-forming rays.

"Reflection microscopy" is a widely used term. It is a good technical term and distinguishes between reflected illumination and transmitted or emitted image-forming rays. However, the distinction between truly reflected image-forming rays and image-forming rays that are emitted when the illuminating beam strikes the specimen should be kept clearly in mind. Examples of emission caused by the illuminating beam are fluorescence (cathodoluminescence and x-ray emission) and the emission of secondary or Auger electrons.

1. Interaction of Electrons with the Specimen

When light interacts with a specimen, the light can be partially or completely transmitted, partially or complete absorbed, or scattered by reflection and refraction. Absorption may result in fluorescence, phosphorescence, heating of the specimen, the ejection of photoelectrons, or combinations of these. Reflection and refraction may, in addition to scattering the rays, result in phase shifts among the various scattered rays. Light microscopes have been designed to utilize all of these specimen–light interactions to produce contrast in microscopical images. These interactions are also used for chemical analysis, for example, by microspectrophotometry, by fluorescence analysis, and by determination of optical properties using polarizing microscopy. These methods are, of course, in addition to analysis based on morphology.

In a closely analogous manner, the interactions of electrons with the specimen can be used for chemical analysis. Again, as in light microscopy, morphological analysis is of paramount importance. Indeed, with electron-microscopical resolution approaching atomic dimensions, morphological anaysis may become identical with chemical analysis.

a. TRANSMISSION WITHOUT SCATTERING

Figure 16.1 illustrates schematically the ways in which electrons interact with a specimen. If the specimen is thin enough for TEM (i.e., thinner than about 0.1 μ), there is a fairly high probability that the electrons will be transmitted through the specimen without interacting with it. Such electrons are referred to as unscattered transmitted electrons. An atom is mostly empty space, and it is not surprising that electrons can easily penetrate hundreds of layers of atoms without colliding with either an atomic nucleus or the electrons orbiting around it.

b. ELASTIC SCATTERING (137,330)

(1) Incoherent Scattering

If the electron passes near a nucleus, it may be deflected from its original path. As a result of a single encounter with a nucleus, an electron may be deflected through an angle of up to 180°; however, deflections through smaller angles are more probable. Even though the electron may be scattered through a relatively large angle, the amount of energy transferred to the nucleus, and thus lost by the electron, is very small (less than 0.1 eV), too small to be measured readily, because the nucleus is so much more massive than the electron. Electrons that are scattered through relatively large angles without an appreciable loss of energy are called elastically scattered electrons.

(2) Coherent Scattering (Electron Diffraction) (130,137,204,289)

In the preceding discussion, electrons and nuclei were treated as charged particles. However, because moving electrons also have a wave nature, they are diffracted by the specimen. The diffraction of a coherent beam of electrons by various types of specimens can be diagrammed by using Huygens's construction. According to Huygens's principle, the electron waves striking a nucleus are scattered and a new wave train is formed around each scattering point. The wave front of the new wave train will be spherical, with the

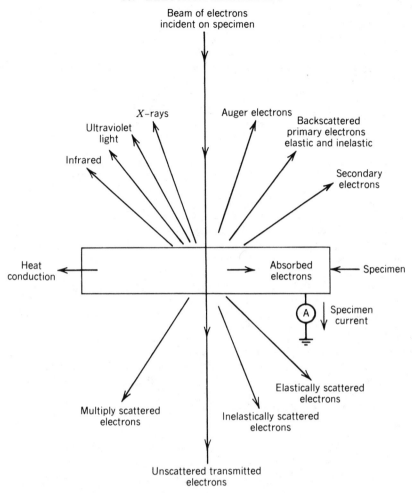

Fig. 16.1. Interactions of electrons with a specimen.

scattering point (in this case a nucleus) at the center of the sphere. Consider three somewhat idealized specimens: In Fig. 16.2a the specimen is an atomically flat amorphous material and the electron beam strikes this flat surface at some arbitrary angle θ. As each wave front passes a nucleus, a new spherical wave front is generated. Figure 16.2a is a cross section of the specimen, so the scattered spherical wave fronts appear as circles. Where these circles coincide, the waves are in phase. Thus an envelope (consisting, e.g., of the wavelets scattered by the various atomic nuclei as the incident wave front passes) forms a new wave front, the wave front of the reflected ray, in this drawing. The only other set of envelopes (wave front) that can be seen in this drawing is the transmitted wave front. Each of the two new rays consists of wave fronts generated when a given incident wave front interacts with the several nuclei. Note that the angle of incidence equals the

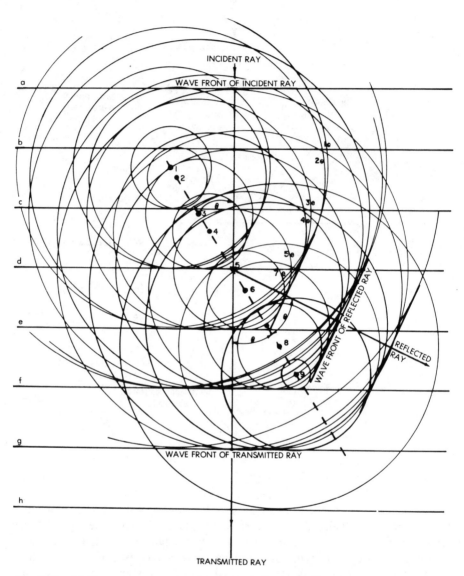

Fig. 16.2. (*a*) Huygens construction for atoms randomly distributed on a two-dimensional plane. The figure shows the plane in cross section.

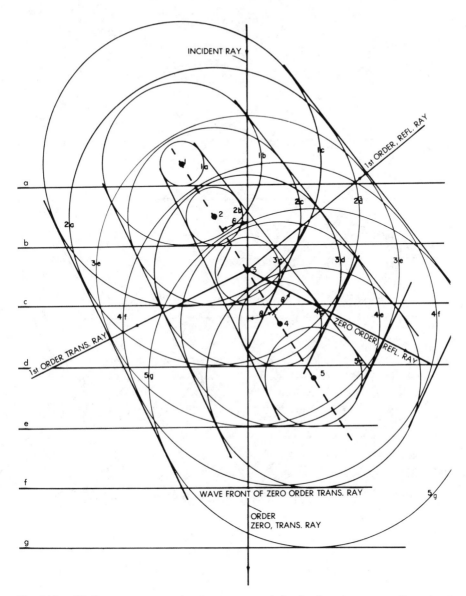

Fig. 16.2. (b) Huygens construction for atoms regularly distributed on a two-dimensional lattice plane.

angle of reflection. The transmitted beam is not deviated, because in the construction of the diagrams the refractive index is assumed to be 1.0.

If the specimen is crystalline, the atoms have long-range order and in a cross section cut perpendicular to the surface of the specimen the atoms will be arranged regularly. If, for simplicity, it is assumed that the atoms are equally spaced and that the specimen is very thin (one atomic layer),

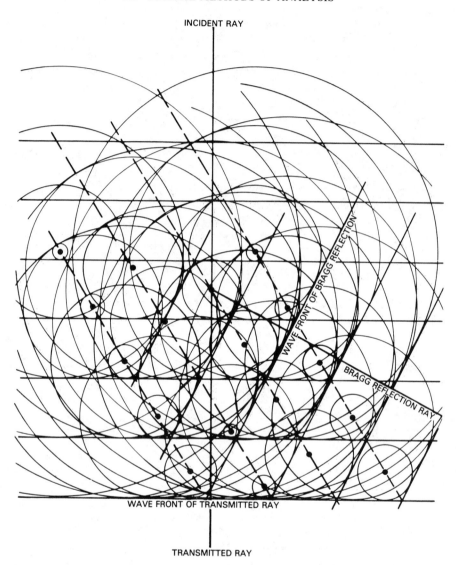

Fig. 16.2. (c) Huygens construction for a three-dimensional crystal lattice. Again the figure shows the lattice in cross section.

the Huygens construction would be as shown in Fig. 16.2b. Here we see that the incident ray is split into an ordinary reflected ray (labeled "ZERO ORDER REFL.") and the ordinary transmitted ray (labeled "ZERO ORDER TRANS."), just as in Fig. 16.2a. However, because of the regularity of the atomic array, two new diffracted rays appear, one labeled "1st ORDER REFL." and a transmitted ray labeled "1st ORDER TRANS."

In Figure 16.2b there are just two transmitted rays and two reflected rays. The number of rays depends on the atomic spacing and the wavelength of

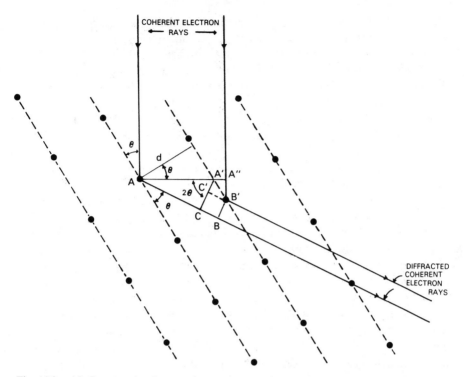

Fig. 16.2. (d) Construction for the derivation of Bragg's law, derived from the relationships between the electron waves and the crystal lattice shown in c.

the electron beam. If the atomic spacing remains constant and the wavelength is made shorter and shorter, more reflected and transmitted rays appear when the ratio of wavelength to atomic spacing reached certain values. The rays are arranged in pairs; for example, two first-order diffracted rays appear symmetrically arranged about the zero-order reflected ray. The same is true of the transmitted rays. Many of these diffracted orders will appear as the ratio of wavelength to atomic spacing becomes progressively smaller. As the ratio becomes larger, the number of rays diminishes, and when a value of 1 is reached, only the zero-order reflected and transmitted rays remain.

The angles at which the reflected rays appear depend also on the angle at which the incident ray strikes the atomic plane. The angles of the transmitted diffracted rays are also dependent on the angle of incidence and on the refractive index of the material. For the electron beams ordinarily used in microscopes, the refractive indexes of materials are very nearly 1, and the transmitted ray is very nearly parallel to the incident ray.

Most crystalline specimens are not as thin as illustrated in Fig. 16.2b, and the diffraction phenomena are different. Figure 16.2c is the Huygens construction for a thick crystal. Here, as in Fig. 16.2a, there are only a reflected and a transmitted ray, and as the specimen is made progressively thicker, the intensity of the transmitted ray becomes smaller.

There is another and more important difference between planar arrays and three-dimensional arrays of atoms with respect to their interactions with electron waves. The Huygens constructions in Figs. 16.2a and b demonstrate that reflection occurs because the wavelets scattered from a planar array of atoms by a single incident wave front are in phase in a certain direction. When the specimen is three-dimensional, the reflected wave fronts from the various parallel atomic planes also must be in phase. This places a further restriction on the rays diffracted by a crystal, namely that the rays reflected from each of a set of atomic planes must be in phase. A diffracted ray will emerge from a crystal only if a set of lattice planes has a specific orientation with respect to the incoming beam. This orientation depends on the wavelength of the waves (λ), the distance between atomic planes (d), and the angle of incidence θ of the incident beam relative to these planes. The interrelationship among these parameters is given by Bragg's law, $n\lambda = 2d \sin \theta$, where n is a whole number.

Drawing Huygens wavelets is a graphical method of determining where diffracted rays go when they strike a properly oriented crystal. The problem can also be solved geometrically, and the result is Bragg's law. Figure 16.2d shows the same lattice as Fig. 16.2c. Two rays are shown entering the lattice and being scattered by nuclei in two adjacent atomic planes. We know that if these rays are to be reflected coherently the path difference between the two rays must be one wavelength (λ) or some multiple $n\lambda$ of the wavelength. We also know that the angle of incidence is equal to the angle of reflection (θ). AA″ is the wave front of the incoming ray and is perpendicular to the rays. If we construct BB′ perpendicular to the reflected rays, this represents the wave front for the reflected ray. If we construct CC′ parallel to BB′ and extend it to point A′, then $AB - A''B' = n\lambda$. But $A''B' = C'B' = CB$; therefore, $AC = n\lambda$, $\cos \theta = d/AA'$, and $\sin 2\theta = AC/AA'$. Because $AA' = d/\cos \theta$ and $AC = AA' \sin 2\theta$, $AC = d \sin 2\theta/\cos \theta$. But $\sin 2\theta = 2 \sin \theta \cos \theta$ and $AC = 2d \sin \theta \cos \theta/\cos \theta = 2d \sin \theta$. Since $AC = n\lambda$, $n\lambda = 2d \sin \theta$, which is Bragg's law.

Electron diffraction can be used for phase identification and a wide variety of crystallographic investigations. Understanding diffraction helps one to understand contrast in electron images and the resolving power of microscopes.

c. Inelastic Scattering

(1) Nuclear Collisions (Bremsstrahlung)

Inelastically scattered electrons are electrons that have lost all or part of their energy as a result of an encounter with an atom of the specimen (277,330).

Electrons that undergo inelastic collisions with the nucleus of an atom may lose all or only a part of their energy. The energy lost by the electron may be converted into one or more x-ray photons. These x-ray photons may vary in energy from some maximum, dependent on the voltage used to accelerate the electron, down through a range of lower energies. The maximum-energy photon results from a collision that completely stops the electron and converts its entire energy into an x-ray photon. A lower-energy x-ray photon results from a collision in which only part of the electron's kinetic energy is converted to an x-ray photon. The overall result is the production of the so-called white radiation on the continuous x-ray spectrum. This type of radiation is also called *bremsstrahlung*, German for "braking radiation," because it is radiation caused by braking or deceleration of the electrons.

(2) Electronic Collisions

Electrons that undergo inelastic collisions with the orbital electrons of an atom may also lose part of their energy. The collision usually results in the ejection of the orbital electron. The resulting vacant orbital position is quickly filled by electrons from orbits further from the nucleus. If a vacancy in the K shell is filled by an electron from the L shell, an x-ray photon is emitted. This emission is called K_α, and its energy is the difference in energy between the two orbital positions, $E_{K\alpha} = E_K - E_L$. If the vacancy in the K shell is filled by an electron from the M shell, the emission is called K_β and its energy is $E_{K\beta} = E_K - E_M$. In a similar manner, electrons ejected from the L shell are replaced by electrons from other shells and a series of emissions called L_α, L_β, and so on are produced. The same is true of the M shell.

As a result of these processes, a sample that is bombarded with electrons will give off an x-ray spectrum consisting of a continuum plus lines representing the K, L, and M transitions. Because the energies involved in the K, L, and M transitions vary with the atomic species, each element gives off a characteristic spectrum that can be used for qualitative and quantitative analysis of the elements in the sample.

(3) Cathodoluminescence (144)

For a given series of lines, for example the $K_{\alpha 1}$ series, the energy of the photon increases regularly with the atomic number of the element. However, for a given element, the L series is less energetic than the K series, the M series is still less energetic than the L series, and so on. The wavelength λ associated with a photon varies inversely with the energy according to the equation $\lambda = hc/eE$, where h is Planck's constant, c is the velocity of light, e is the charge on the electron, and E is the energy of the photon in electron volts. Substituting into this equation the values of h, c, and e yields the equation $\lambda = 12.398/E$. When electrons collide with the outermost orbital electrons of some of the elements in the middle of the periodic table, the photons resulting from refilling of the vacancies may have energies corresponding to ultraviolet, visible, or infrared radiation. The emission of ultra-violet, visible, or infrared radiation on bombardment with electrons (cathode rays) is called cathodoluminescence.

The mechanism of cathodoluminescence is not fundamentally different from that of the production of x-rays. However, the outermost electrons are often the valence electrons, and thus are involved in chemical bonds. There-fore, cathodoluminescence gives information about the chemistry of the ele-ments in the phases forming the specimen.

In addition to allowing identification of some elements by their charac-teristic color, cathodoluminescence can be used to observe crystal defects and impurities, to distinguish among polymorphs of crystals (e.g., between anatase and rutile), and to study semiconductors. Biological materials treated with certain reagents may exhibit cathodoluminescence.

(4) Auger-Electron Production

When an electron strikes and removes an electron from an inner shell, the vacancy is usually filled by an electron from an outer shell. The energy that results can produce an x-ray photon, as discussed above. However, this energy may also be dissipated by the ejection of an electron from an outer shell of the atom. For example, the incident electron may knock an electron from the K shell. This vacancy may be filled by an electron from the L shell. The energy available from this transition may result in the ejec-tion of an electron from the M shell. Thus, this is a three-electron process. The available energy is $E_K - E_L$. The energy of the ejected electron (E_a) is therefore $E_K - E_L - E_M$, where E_M is the energy required just to eject the electron from the M shell of the atom.

If this process occurs near enough to the surface of the specimen, the ejected electron, called an Auger electron (307), may escape from the spec-imen without appreciable energy loss. Auger electrons range in energy from 10 to 1000 eV, and the corresponding escape depths range from 1 to 10 Å. Auger electrons generated too deep in a specimen may lose their charac-teristic energy partially or completely, but those that have not lost appre-ciable energy are characteristic of the elements that produced them. The Auger-electron spectra consist of broad peaks, which are usually superim-posed on the secondary and scattered electron distribution. This makes them difficult to observe; the usual procedure is to take the derivative $dn(E)/dE$ of the electron distribution curve, a procedure that greatly increases the signal-to-noise ratio.

Auger electrons can be produced by all of the elements except hydrogen and helium, which have only a K shell. However, the yield of Auger electrons is high for elements with atomic numbers from 3 to 15. Auger analysis may be especially valuable for these light elements, because it is difficult to use x-ray emission for elements in this range. Auger analysis is also very useful for the analysis of surfaces, because the depth from which Auger electrons can escape is 10 Å or less.

d. Energy-Loss Electrons

The primary electrons that have interacted with the specimen have suffered energy losses. If the specimens are thin enough that the primary electrons each suffer only one collision, then analytical information can be obtained by studying the energy losses resulting from these encounters (67,162).

Energy losses suffered by primary electrons can be divided into four categories:

1. *Energy losses resulting from elastic collisions.* Although these losses are very small (less than 0.1 eV), and cannot be measured with existing equipment, they do cause the atom involved to vibrate in phonon modes. The increased vibration of the atoms increases the heat energy of the specimens.

2. *Energy losses resulting from inelastic collisions with core or inner-shell electrons.* These losses are characteristic of the elements. Their values lie between 50 and 2000 eV. The energy-loss spectrum in this range can be used for analysis of the elements in the specimen.

3. *Energy losses resulting from inelastic encounters with the valence electrons.* In a solid the valence electrons are involved in the bonds that hold the atoms together, and can be considered to belong to the specimen as a whole. The losses suffered by the primary electron in encounters with these electrons are called plasmon losses. Plasmon losses most often are broad peaks in the energy-loss spectrum with energies between 1 and 50 eV. However, for some elements the plasmon losses occur as narrow characteristic peaks that can be used for elemental analysis.

4. *General background intensity in an energy-loss spectrum.* This causes some difficulty by reducing the signal-to-noise ratio. The origin of this background is not yet known.

e. Radiation Damage

Whenever energy is deposited in a specimen by a collision, the specimen may be damaged. Inelastic collisions are therefore the prime cause of radiation damage. All of the three categories of inelastic events, namely collisions with the nucleus, collisions with the inner or core electrons, and collisions with the valence electrons, can result in radiation damage.

Collisions of an electron with a nucleus can result in the atom being knocked out of its position in the structure. The magnitude of this effect is relatively small, because of the rather large energies required to displace atoms. Lighter atoms are more easily displaced, and at ordinary accelerating voltages (up to 100 kV), hydrogen, carbon, and similar low-atomic-number atoms may be lost from a structure. At higher accelerating voltages, this effect becomes more important. It is interesting to note that most atomic displacements are caused by the impact of heavier ions such as 0^{2-}, which come from water in the microscope column.

The interaction of incident electrons with the orbital electrons, particularly the valence electrons, results in the breaking or rearrangement of the chemical bonds. Polymerization can also take place. In general these effects are irreversible.

Regardless of the mechanism of the energy loss suffered by a primary electron, the specimen is heated. The resulting temperature rise depends on the rate at which heat is lost to the surroundings, which is controlled by the heat capacity, the thermal conductivity, and radiation. On a poorly conducting support, temperature increases of 3000°C have been observed, although for metallic specimens temperature rises of 60°C or less can be expected. High temperatures can result in local melting and other phase changes including sublimation or boiling. Annealing and recrystallization also can occur.

Radiation damage is not well understood at present. It is the principal obstacle in the effort to image atoms. Image formation depends on the specimen interacting with the electron beam, and this interaction seems almost certain to cause radiation damage. It is possible that methods of minimizing damage may be devised. It also may be possible to understand radiation damage and to interpret structures in spite of such damage (288).

f. SPECIMEN CONTAMINATION

Gases and vapors in the vacuum system, such as those arising from the greases and oils used, also suffer radiation damage. The vapors in the specimen chamber tend to polymerize on the specimens, because that is where the electron beam is concentrated. This can and does result in the buildup of a varnishlike coating on the specimens and nearby surfaces, especially apertures. This contamination layer interferes with the imaging of the specimen.

Contamination can be minimized by heating the critical surfaces to about 200°C or by surrounding the specimen with cold fingers or cold traps. It can also be reduced by keeping the system clean and by designing the system in such a way as to eliminate the sources of contamination.

g. MULTIPLE SCATTERING

(1) Kikuchi Patterns

Up to this point in the discussion, the encounters between an electron and an atom or molecule have been single events. In these encounters the electrons generally lose energy, but the amount of energy lost varies. In elastic encounters, very little energy is lost. In inelastic events, considerable but varying amounts are lost. In some collisions with an atomic nucleus, all of the electron's energy may be lost.

Fig. 16.3.   An electron-diffraction pattern observed in the transmission mode. This pattern shows a possible distribution of the electrons about the centrally located unscattered transmitted beam. Many of the electrons, especially the inelastically scattered ones, cluster close to the transmitted beam, whereas the elastically scattered electrons form rings and spots at relatively large angles with respect to the unscattered transmitted beam. (Pattern taken by DoSuk Lee.)

Electrons that have lost only part of their energy may escape from the surface of the specimen, but more often they travel further within the spec- imen and undergo multiple encounters with the atoms of the specimen.

Specimens prepared for transmission microscopy should be thin enough that an electron will undergo only one encounter as it travels through the specimen. The electrons that emerge from such a specimen are traveling in all directions, although most of them are concentrated near the primary beam of unscattered electrons. This distribution of electrons can be calculated (79,130), but a qualitative idea can be gotten by looking at electron-diffrac- tion patterns (see Fig. 16.3). If the specimen is crystalline and thick enough that the average electron is scattered more than once, it is possible that the second or higher-number collisions will result in a coherent, Bragg-type reflection. The pattern appearing in the final image under these conditions is called a Kikuchi pattern, after its discoverer (168). Figure 16.4 shows a Kikuchi pattern. It consists of a general background with dark and light lines. Such patterns give information on crystallographic orientation and crystal perfection (297).

Similar patterns called Kossel patterns are formed by x-ray diffraction phenomena. Kossel patterns can be obtained from appropriately designed

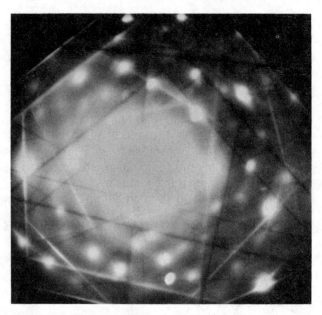

Fig. 16.4.   A Kikuchi pattern. Kikuchi patterns are observed in the transmission mode and show the distribution of multiply scattered electrons that results when the specimen is a relatively thick single crystal. The Kikuchi pattern consists of dark and light lines. This Kikuchi pattern is superimposed on the spot pattern that arises from singly scattered electrons. (Pattern taken by DoSuk Lee.)

electron microscopes. The information derived from Kossel patterns is similar and in some ways complementary to that obtained from Kikuchi and channeling patterns (see below) (144,327).

(2) Electron-Channeling Patterns

For thick specimens, patterns that look like Kikuchi patterns are formed by backscattered and secondary electrons. These channeling patterns are also called Coates patterns or psuedo-Kikuchi patterns. Channeling patterns are obtained from SEMs of the common "reflection" type. As the beam scans a specimen, the angle of incidence of the beam relative to a crystal in the specimen changes. At some angles the electron beam can penetrate deeply into the crystal, and the scattered electrons do not escape from the surface, but rather are multiply scattered within the crystal. At other angles the beam cannot penetrate deeply within the crystal, and the singly scattered electrons, and those scattered only a few times, may escape from the surface of the specimen. This is shown in a schematic way in Fig. 16.5. The result is the appearance of bright and dark bands in the image that are analogous to Kikuchi patterns (Fig. 16.6) and can be used in the same way (100,120,144).

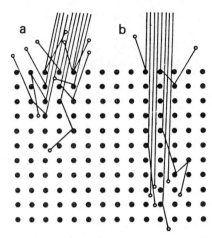

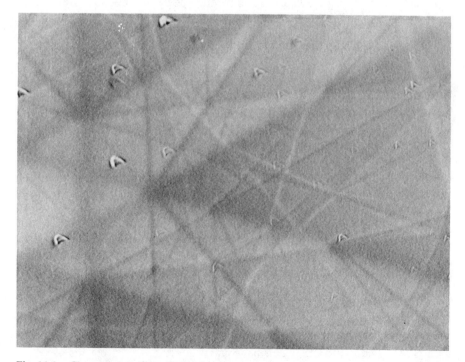

Fig. 16.5. Schematic diagram of the channeling effect. This figure shows how the number of backscattered electrons can vary depending on the angle at which the incident electron beam strikes the surface of a crystal. (From reference 120.)

Fig. 16.6. Channeling pattern of a single crystal of silicon. Because the amount of scattering depends on the angle of incidence of the beam on the specimen (see Fig. 16.5), such patterns are taken by "rocking" the beam through an angle while the image is formed. In this case, the rocking angle was ±3.3°.

h.   ABSORPTION OF ELECTRONS

As the specimen becomes thicker, the incident electrons that penetrate it may undergo many collisions with the atoms of the specimen. These collisions may be of any of the types discussed above. Some incident electrons are backscattered out of the specimen after one or more collisions. Some secondary electrons also escape from the surface of the specimen, but others suffer multiple collisions within the specimen and finally come to rest.

(1) Heat and Charging

Each collision involves some energy loss, and, as discussed in Section II.A.1.e, each energy loss heats the specimen. In addition, if the specimen happens to be electrically insulated, it will become negatively charged, and the charge may result in parts of the specimen moving or jumping to some part of the microscope. Charging also deflects the incident beam of electrons and thus interferes with image formation. The specimen should therefore be electrically grounded and thermally connected with a heat sink.

(2) Absorption- (Specimen-) Current Generation

The number of electrons absorbed by the specimen depends on the composition of the specimen and on the crystallographic orientation of the materials in the specimen. Therefore, when the specimen is electrically insulated from the microscope but connected directly to the proper circuit of a SEM, the absorbed current can be used to form an image of the specimen.

The absorbed current, also called the specimen current, is equal to the beam current of the incident electron beam minus the various emission currents. The emission currents are the electron currents resulting from secondary-electron, backscattered-electron, and Auger-electron emission from the specimen. Auger emission is low compared with the secondary and backscattered emissions, so effectively the specimen-current image is complementary to the combined secondary and backscattered images. Thus the specimen-current images generally contain information on topography (such as is associated with secondary-electron images) and on atomic number (such as is associated with backscattered-electron images) (144).

By biasing the specimen, secondary emission can be suppressed, resulting in a specimen-current image that is complementary to the backscattered image and thus contains primarily atomic-number information.

i.   CHARGE-COLLECTION EFFECTS

When the specimen is a semiconductor, the charging and the resultant current flows induced by the incident beam can be used to form an image in a SEM. The effects that can be imaged depend on the circuit of which the specimen is a part. In the absence of any external biasing, voltage currents will flow if the specimen exhibits electron-voltaic effects of either the barrier or bulk electron type (144,166).

Images formed as a result of charge-collection effects yield information about the functioning of solid-state electronic devices. However, charge-collection effects and hole–electron pair-generation effects can be useful for examining most nonmetallic specimens.

j. EFFECTS OF SPECIMEN MAGNETISM

If the specimen has magnetic properties, the transmitted or "reflected" electron beams may be affected. Thus the electron image will give some information regarding these properties (120,137,289). The magnetic properties of a specimen may be examined by using either transmission or scanning microscopy. The techniques are well developed, and are collectively called Lorentz microscopy.

k. EFFECTS OF THICKNESS OF THE SPECIMEN

As a beam of electrons penetrates a sample, the various interactions of the electrons discussed above occur. If the sample is thick enough, the beam electrons will be scattered many times and as they go deeper into the specimen the beam will tend to spread. Duncumb and Shields (86,120) have shown this schematically (see Fig. 16.7). The pattern varies with the accelerating voltage of the beam and with the atomic number of the specimen. Murata et al. (221) have calculated the trajectories of electrons penetrating various specimens.

As the electrons undergo various types of reactions with the atoms of the specimen, Auger electrons, backscattered electrons, secondary electrons, x-rays, and other types of electromagnetic radiation are produced. Goldstein (119,120) has summarized these effects as shown in Fig. 16.8 for a pure element of low to medium atomic number. In this figure "R" is the maximum range of electrons, "R(x)" is the range of x-rays, and "X_d" is the calculated depth of complete diffusion. The spatial resolutions for x-rays and backscattered electrons are shown. The spatial resolutions for secondary and Auger electrons are approximately the same as the probe diameter. Figure 16.1 is another type of summary of the interactions of electrons with the specimen.

Although Fig. 16.8 was developed for explaining phenomena occurring in SEM, it helps in understanding transmission and scanning transmission microscopy as well.

(1) Specimens for Transmission Electron Microscopy

As pointed out above, transmission specimens are ideally thin enough that on the average only one collision occurs as a beam electron traverses the specimen. Thicknesses of transmission specimens ranges from a few to several hundred angstroms. In general only electrons scattered within a narrow angle about the unscattered beam, whether elastically or inelastically, contribute to image formation. However, Auger electrons and secondary electrons, along with x-rays and other electromagnetic radiation, are gen-

ELECTRON BEAM-SPECIMEN INTERACTION

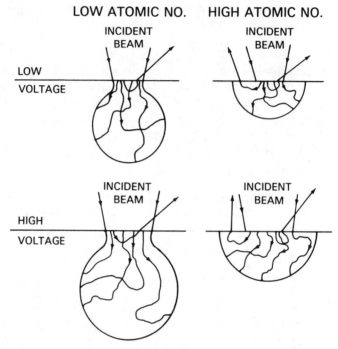

Fig. 16.7.   Interactions of the electron beam with a thick specimen. These diagrams show variations caused by variations in atomic number and beam voltage. (From reference 86.)

erated by the collisions and may escape the specimen from either of its surfaces. These radiations may be used for chemical analyses.

Resolution does not depend on the diameter of the incident beam. The factors that affect spatial resolution in transmission microscopes will be discussed later.

Coherently elastically scattered electrons are also transmitted through thin specimens and form diffraction patterns that can be used for crystallographic studies and phase identification.

If the specimen becomes thicker, the electrons will be scattered more than once, and Kikuchi and Kossel patterns may be formed. Such patterns are useful for crystallographic investigations.

(2)   Specimens for Scanning Transmission Electron Microscopy

Specimens for STEM are comparable in thickness to or somewhat thicker than TEM specimens. Therefore the same interactions between the beam and the specimen occur with STEM as with TEM.

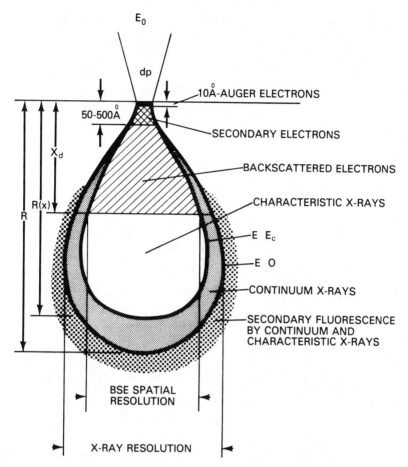

Fig. 16.8. Interaction of the electron beam with the specimen. This diagram shows how the electron beam produces different effects as it penetrates different regions of the specimen. BSE, backscattered electrons. (From reference 120.)

Although spatial resolution does depend on probe-beam size, this in turn depends on the same factors that control the resolution of a TEM. They will be discussed later.

(3) Specimens for Scanning Electron Microscopy

Specimens for SEM are usually thick enough that all electrons are either scattered back out of the surface being struck by the beam or stopped, absorbed, within the specimen. Images may be formed by using the various "reflected radiations" that result, including secondary, backscattered, or Auger electrons. Images may be formed also by using x-rays, cathodoluminescence, or specimen currents or charging. Many types of chemical and physical analysis can be done using the various reflected radiations.

Spatial resolution depends on the probe size and on the spreading of the electrons in the specimen. Figure 16.8 shows spatial resolutions for the various radiations.

2. Transmission Electron Microscopy

a. IMAGING SYSTEM

The transmission electron microscope (TEM) and its characteristics can be understood by comparison with the more familiar light microscope. Figure 16.9 is a schematic diagram of the essential image-forming elements of a compound microscope. Suppose that electrons or photons emanate from all points on the object and travel in all directions. To clarify the diagram, only a few of the rays emanating from two points on the object are shown. Three of the rays from each of these object points are shown passing through the objective lens and through the rest of the microscope. The rays selected are those that pass through the center of the objective and through the outer edges of the usable portion of the objective. It must be borne in mind that in an ideal system all the rays that pass through the objective lens from these and other points on the object contribute to the formation of the final image. The cones of rays that originate from object points and pass through the objective are limited by the aperture diaphragm or stop. The aperture diaphragm may be either the edge of a lens or a diaphragm (the objective aperture) especially designed to limit the useful area of the objective. The limiting aperture of a microscope may be in any of several positions, but it is usually placed in one of the two positions labeled "objective aperture stop" and "back focal plane" in Fig. 16.9. The vertex angle of the cone of rays emanating from the object point on the optic axis and passing through the objective aperture is defined as the angular aperture of the objective lens. (See also Fig. 16.19).

An ideal objective lens focuses all the rays emanating from a point on the object and passing through the lens to a corresponding point in the intermediate image. Thus, points A' and B' in the intermediate image correspond to points A and B in the object. This intermediate image is inverted and magnified. The rays that form this real image continue onward and pass through the ocular or projector lens. This lens projects an image of the intermediate image on a suitable receptor. Thus, the intermediate image serves as an object for the projector lens, which forms the final image; this is again inverted and enlarged as compared with the intermediate image. The angular aperture of the image-forming rays passing through the projector lens is determined by the diameter of the aperture of the objective lens and the distance of the intermediate image from the objective.

The diagram in Fig. 16.9 shows the simplest type of compound microscope without an illuminating system. Practical light microscopes may have a field lens above the intermediate image. Also, the individual lenses, particularly the objectives, comprise a number of elements and air spaces; however, the

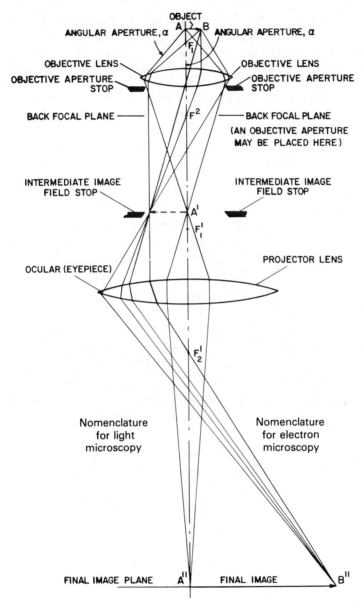

OBJECT

ANGULAR APERTURE, α   A   B   ANGULAR APERTURE, α

F'

OBJECTIVE LENS   OBJECTIVE LENS
OBJECTIVE APERTURE   OBJECTIVE APERTURE
STOP   STOP

BACK FOCAL PLANE —— F^2 —— BACK FOCAL PLANE
                                  (AN OBJECTIVE APERTURE
                                  MAY BE PLACED HERE)

INTERMEDIATE IMAGE   INTERMEDIATE IMAGE
FIELD STOP   FIELD STOP

A'
F'_1

OCULAR (EYEPIECE)   PROJECTOR LENS

F'_2

Nomenclature   Nomenclature
for light   for electron
microscopy   microscopy

FINAL IMAGE PLANE   A''   FINAL IMAGE   B''

Fig. 16.9.   Schematic diagram showing the arrangement of image-forming elements of light and electron microscopes and the paths of a few selected rays through these elements.

523

complex lenses act as one thick lens. Electron microscopes almost always
include an intermediate lens, and may have two or more elements in the
objective and/or projector lenses. The intermediate lens may be used in the
same manner as a field lens, but it is also commonly used as a third stage
of magnification. The additional lenses can be used for special purposes,
such as electron diffraction, dark-field microscopy, and x-ray microscopy.
These special arrangements will be discussed later.

One important difference between light and electron microscopes is that
a vacuum must be maintained inside the latter to reduce the scattering of
the electron beam by the gases in the column. This fact strongly affects the
types of specimens that can be examined. For example, chemical reactions
occurring in solutions cannot be carried out easily and simultaneously be
observed in the electron microscope. The examination of living cells also is
very difficult. Several attempts have been made to overcome the limitations
imposed on specimens by the necessity of operating in a vacuum (229,298).
High accelerating voltages make it possible to examine thicker specimens,
and thus make it more convenient to examine organisms or reactions in
solutions.

b. Illuminating Systems

In Fig. 16.9 the electrons or photons are shown emanating from the object.
In some types of microscopy (emission microscopy) this is actually the case.
However, in general, the specimen is illuminated by electrons or photons
generated in an illuminating system. Two arrangements of illuminating systems for electron microscopes are shown in Fig. 16.10. In this type of microscope the specimen is illuminated by a beam of electrons that penetrates

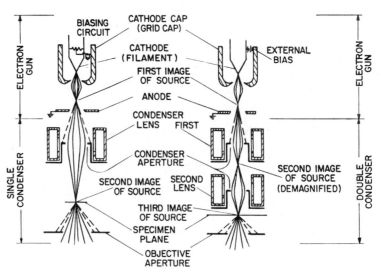

Fig. 16.10.   Illuminating systems for TEMs.

the specimen and is modified by the structures within the specimen. This modified beam of electrons then continues through the imaging system as shown in Fig. 16.9. Just as in light microscopy, it is possible to adjust the illuminating system so that either bright-field or dark-field conditions exist. The electron analogs of phase-contrast and interference microscopy also can be obtained.

In the illuminating systems used for transmission microscopy (Fig. 16.10) the electrons emitted by the filament are accelerated toward and through the anode aperture by the potential gradient that exists between the anode and cathode. The cathode is usually kept at some negative potential of the order of 50–100 kV. (Both higher and lower voltages are used for special purposes.) A cathode cap or grid (also called the Wehnelt cylinder) surrounds the filament. This cap is usually biased negatively with respect to the filament. The bias is commonly obtained as the voltage across a resistor in series with the electron current, as shown in the left-hand drawing of Fig. 16.10, but in some instruments the bias voltage is supplied from a separate source, as shown in the right-hand drawing. The bias voltage on the cathode cap causes it to act as a lens that focuses the electron beam. The cross section of this focused beam at its minimum diameter (the so-called crossover) serves as the effective electron source for the rest of the condenser lens system. The crossover does not coincide exactly with the focused image.

The condenser lens consists of an ironclad electromagnet, symmetrical about its optic axis, with the gap in the iron located in the bore of the lens. Electron lenses will be described more fully in a later section. The magnetic field in the condenser lens deflects the electrons in the beam, and if the strength of the magnet is properly adjusted, an image of the source will be formed at the specimen position. Under these conditions the intensity and brightness at the specimen position are at a maximum. This is because all the electrons that pass through the condenser aperture are concentrated in the image of the source at the specimen plane. The angular aperture of the illuminating beam also is at a maximum. The condenser aperture should generally be chosen so that the angular aperture of the illuminating beam is equal to or slightly less than the angular aperture of the objective lens. Under these conditions the resolving power of the microscope is at a maximum and image contrast in a focused image is at a minimum.

As the condenser lens is defocused by increasing or reducing the current in the lens coil, the image of the source is formed above or below the specimen plane, and the intensity, brightness, and angular aperture of the illuminating beam are reduced. As a result the resolving power is reduced, but the image contrast is improved.

The drop in intensity at the specimen plane when the condenser lens is defocused results from the increase in the diameter of the electron beam. Thus, although the same number of electrons pass through the condenser aperture and through the specimen plane, they are spread over a larger area and intensity is decreased. If the electrons are absorbed in dense parts of the specimen and specimen support, a heating problem may arise because the paths by which the specimen dissipates heat are fixed by the mechanical characteristics of the microscope. Thus, it is desirable to illuminate only that area of the specimen that is to appear in the final image formed by the microscope. In light microscopy this is accomplished by placing a field stop or diaphragm in a suitable location in the illuminating system. This field diaphragm is variable in size, and with it the area of the specimen illuminated can be controlled. In earlier electron microscopes, no means other than focusing and defocusing was provided for controlling the area of the illuminating beam at the specimen plane. However, in more recent instruments this control is accomplished by use of a double condenser or a smaller electron source.

A schematic diagram of a double-condenser system is shown in the right-hand drawing of Fig. 24.10. It controls the area of the specimen illuminated by demagnifying the image of the source. The demagnification is accomplished in two stages, rather than one, because such a system provides adeqeuate working space near the specimen and decreases the overall length of the illuminating system. An additional advantage is that the size of the focused image can be varied. By use of this system the maximum brightness can be maintained at the specimen while the total beam current striking the specimen is minimized. Thus, some of the specimen-heating problems can be overcome and only those portions of the specimen being observed need be exposed to the electron beam.

c. IMAGE-TRANSFORMING AND -RECORDING SYSTEMS

Since the eye is sensitive to light, the final image of a light microscope can be formed directly on the retina of the eye. The image formed in an electron microscope, however, must be transformed into a light image before it can be observed. Ordinarily, TEMs are equipped with a fluorescent screen on which the final image is formed. The phosphor with which this screen is coated absorbs electrons and emits visible fluorescent light. Thus the final image can be observed visually, enabling the operator to examine the specimen and to carry out such necessary operations as focusing the image for observation and photography, and alignment of the microscope.

(1) Photography

Photographic emulsions are sensitive to electrons, so the electron image can be recorded directly on film. In general, photographic films react to electrons much as they do to light; however, there are some differences. A photographic film or plate consists of a photographic emulsion spread in a thin layer on the surface of a sheet of polymer or glass. The emulsion is a dispersion of crystals of silver halides in gelatin. For a typical emulsion suitable for electron microscopy, the silver halide crystals may have an average diameter of 2 μ.

The sensitivity of a photographic emulsion is generally described by a *characteristic* curve. This type of curve is also called an ''H and D'' curve,

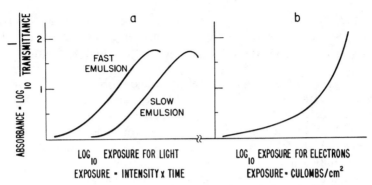

Fig. 16.11. Characteristic curves contrasting the reactions of photographic films to (*a*) light and (*b*) electrons.

because it was introduced by Hurter and Driffield. To produce a characteristic curve one plots absorbance (or perhaps density or optical density) against the log of the exposure. Figure 16.11 shows several such curves. Characteristic curves plotted in this way correlate quite well with the response of the human eye.

The sensitivity of photographic emulsions varies widely with the wavelength of the light to which they are exposed. Silver halide emulsions without sensitizing dyes are sensitive only to blue light. Photographic emulsions are sensitive to electrons of all wavelengths commonly used in electron microscopy. These variations in sensitivity are related to the energies of the photons or electrons. Photons of red light are not energetic enough to expose an emulsion such as that described above. Ultraviolet photons are sufficiently energetic to expose such an emulsion. When a silver halide crystal in the emulsion is struck by a few (approximately 10–100) photons of blue light, it is rendered developable; that is, the silver halide crystal can be completely reduced to metallic silver by exposure to the reducing agents in the photographic developer solution. All electrons commonly used in microscopy are sufficiently energetic that a single electron can render a silver halide crystal developable. Thus the quantum efficiency of photographic emulsions for electrons is nearly 100%, making photographic film a nearly perfect image-recording medium for electron microscopy.

Factors others than photon or electron energy affect photographic sensitivity. Variations in crystal size strongly affect the sensitivity of an emulsion to light. If a single crystal must collect several photons to become developable, then for a given intensity of light a large crystal will collect enough photons sooner than will a small one. Thus sensitivity to light increases as the cube of the crystal diameter, and emulsions that are fast with respect to light contain silver halide crystals larger than those of slow emulsions. For electrons this is not true, because one electron will sensitize one crystal regardless of size over a wide range of sizes. However, the crystals can be made so small (less than about 0.1 μ) that an electron may pass through the crystal without transferring any energy to it (see Section II.B.4). For example, some spectroscopic emulsions have very small crystals and are slower with respect to electrons than are the usual photographic emulsions.

The thickness of the emulsion layer and the distribution of crystals within the layer can also affect the film speed for electrons. Emulsion layers can be made thin enough that electrons, especially high-voltage electrons, pass through without losing their energy to the silver halide crystals. Electrons in the range 50–100 kV penetrate 10 μ or more into the emulsion. The effect of the distribution and concentration of crystals is obvious, as the electron must strike a crystal on its way through the emulsion layer.

One measure of photographic image contrast is the slope of the characteristic curve, that is, the ratio of the log of the exposure to the resultant density. The characteristic curve for light has a long relatively straight portion, which is the region ordinarily used for photography. For electrons the contrast changes continually. Thus contrast in electron micrographs depends on exposure, with longer exposure giving higher contrast. For photography with light, films can be made that have various contrasts when fully developed; however, for exposure to electrons the contrast does not vary much with different films or with different electron energies. Contrast can be varied for both light and electrons by controlling development. Films developed for less than the recommended optimum time will be lower in contrast, whereas films that are overdeveloped will be higher in contrast.

An important characteristic of a photographic film is its resolving power (see also Section II.A.1.a.). Resolving power is limited by the size of the silver halide crystals, by migrations of the silver grains that may occur during development, and by scattering of electrons parallel to the film's surface. Films that are fast for light have large silver halide crystals and poor resolution, so films chosen for electron microscopy should have crystals that are as small as they can be while still being able to absorb electrons. Negatives taken using such films can usually be enlarged 7–10 times without losing resolution. Usually this is adequate even for very high-resolution microscopy (274).

Photographic emulsions can be coated on glass plates or on sheets of polymer. Glass plates are advantageous because they are dimensionally stable and because they are easily outgassed in the vacuum. For many years glass plates were the more popular; however, they are more expensive than film, harder to store, and subject to breakage. In the past emulsions were usually coated on cellulose acetate, although this material was hard to outgas and was not dimensionally stable. However, it was less expensive than glass plates, easily stored, and not subject to breakage. Recently emulsions for electron microscopy have been coated on a polyester backing such as Estar®, a material that is more dimensionally stable than cellulose acetate and is fairly easily outgassed. Such film is satisfactory for most electron microscopy, and it is becoming more widely used.

To summarize, most photographic films can be used for electron microscopy. There usually is no point in choosing fast emulsions, because for electrons the speed is almost the same for all emulsions. Light-slow blue-sensitive emulsions are the most suitable, because they are fine grained and can be handled with relatively bright safelights. Contrast can be controlled by exposure and development.

More detailed discussions of the photographic process as it is applied in electron microscopy can be found elsewhere (48,130,274,329).

(2) Viewing Screens

The fluorescent viewing screen mentioned above is coated with a phosphor that is somewhat coarse grained. Thus its resolving power is rather poor compared with that of a photographic image. Also, it does not give contrast as high as can be obtained in a micrograph. Its primary use is in surveying the specimens, choosing appropriate areas for micrography, and focusing the image. In spite of these deficiencies it is sometimes an advantage to produce a micrograph by photographing the fluorescent screen. For example, electron micrographic prints can be produced in a very short time by photgraphing the screen. Motion pictures can also be made by photographing the fluorescent screen, but some microscope manufacturers offer a better appraoch: cine attachments that go inside the photochambers of their microscopes.

Electron microscopes may also be equipped with image intensifiers and/ or television display systems. In the attachment of an image intensifier the ordinary fluorescent screen is replaced by one viewed in transmission. This screen is then optically coupled to an image intensifier by a lens system having a large angular aperture or by a fiber-optical system. The coupling system usually brings the image out of the vacuum system so that the image intensifier can operate at atmospheric pressure. Such a system may increase the brightness of the image by a factor of 10^3. This means that the beam current through the specimen can be reduced by the same factor. Thus a specimen that is very sensitive to beam damage can be observed and photographed. The image intensifier can also be used for microscopy at high magnifications, which usually results in a very low image brightness on the ordinary fluorescent screen.

Television display systems may be coupled with the microscope in several different ways. A television camera may be used to pick up the image from the fluorescent screen. This is the simplest system, but it has low sensitivity and poor resolution because of the characteristics of the fluorescent screen. A better way is to use the TV camera to observe the image of an image intensifier. This system is much more sensitive, and its resolution can approach the ultimate resolution of the microscope. Another possibility, which may be still better, is to place the elements of the image intensifier and the TV photocathode directly in the vacuum system of the microscope. If this is done the pressure must be maintained at 10^{-8} torr or less.

There are several advantages to using a TV display system. Image brightness can be amplified electronically. Contrast can be increased, decreased, or reversed electronically. By use of a long-persistence screen on the viewing cathode ray tube, the number of electrons used to form the image can be increased, and thus random noise in the image can be minimized. Finally, images can be stored on video tape for subsequent viewing or processing by computers.

3.   Scanning Transmission Electron Microscopy

a.   OPTICAL SYSTEM

A scanning transmission electron microscope (STEM) forms an image in a fashion quite different from that of ordinary light microscopes or TEMs. A diagram showing the optical system of a STEM is given in Figure 16.12.

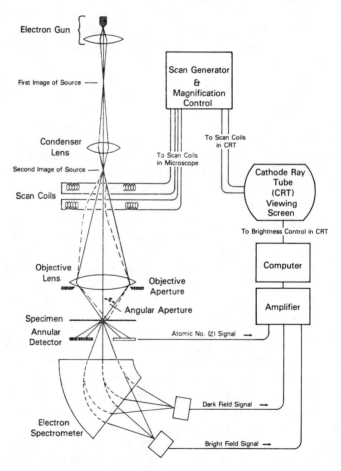

Fig. 16.12.   Schematic diagram of a STEM.

For simplicity only the outermost rays that arise from a point on the optic axis and finally reach the object are shown. However, a more complex diagram like that shown in Fig. 16.9 could have been drawn.

The optical system from the electron source to the specimen is designed to produce a very small probe beam striking the specimen. The source itself is very small and the first image of the source is approximately the same size. The condenser produces a demagnified second image of the source, and the objective further demagnifies this image and focuses it on the specimen. In recent instruments the size of the final image of the source may be approximately 0.3 nm (3 Å).

This very small probe beam can be scanned across the specimen by means of scan coils. The dashed lines in Fig. 16.12 show a deflected probe beam. There is, of course, another set of scan coils arranged at right angles to those shown to allow the probe beam to be scanned across the specimen in a rectangular raster.

The scan coils are driven by a scan generator, which also produces a synchronous raster on the cathode ray tube (CRT). The raster on the specimen is quite small compared with the raster on the CRT. The ratio of the linear size of the CRT raster to the raster on the specimen is the magnification of the microscope. The scan generator can be adjusted to vary this ratio, thus changing the magnification of the microscope.

The specimens for this type of microscope must be thin. In general they are comparable to specimens used for TEM. The electron probe beam interacts with the specimen as discussed above. The elastically and inelastically scattered electrons emerging from the specimen are collected, and the resulting current is amplified and used to control the brightness of the beam in the CRT. As a result an image of the specimen is formed on the viewing screen of the CRT.

The detectors that collect the electrons coming through the specimen may be of various types and configurations. Electron detectors will be discussed in Section II.A.4.b. In the simplest case a single detector can be used to collect all of the transmitted electrons. However, more complex collector systems can increase the amount of information obtainable. Figure 16.12 shows three separate collectors. An annular detector collects electrons that are scattered at relatively large angles. These are generally elastically scattered electrons, and the heavier the atomic nuclei that cause the scattering, the greater the number of electrons scattered at high enough angles to be collected by the annular detector. The strength of the signal from the annular detector is therefore related to atomic number.

The electrons that pass through the annular detector enter an electron spectrometer. The elastically scattered electrons have lost very little energy and are focused in a relatively small area. When these electrons are used to produce an image on the viewing screen, the image is like that produced by a TEM operating in the bright-field mode. This type of image contains primarily morphological information.

Inelastically scattered electrons are focused by the electron spectrometer in various locations depending on the amount of energy they have lost. An image formed by using these electrons is somewhat analogous to a dark-field image. Such an image contains information on both morphology and chemical composition.

One of the major advantages of the STEM is its ability to produce various types of images. The computer in the system allows these various images to be combined in various ways to yield still more useful information. A STEM may also be used to provide energy-dispersive x-ray analyses and energy-loss spectra (85).

b. ILLUMINATING SYSTEM

The resolution of the STEM depends on the diameter of the probe beam that is focused on the specimen. This in turn is dependent on the source size and on the optical system that focuses the source onto the specimen. The limitations imposed by the optical system will be discussed in Section II.B.3. The electron source (or gun) may be of the types shown in Figs. 16.10 and 16.16. However, these sources are rather large, and high-resolution STEMs are equipped with field emission guns similar to the gun shown in Fig. 16.13 (57,78,79).

A field emission gun consists of a sharply pointed cathode, usually made of tungsten. The radius of curvature of the point of such a cathode ranges from a few tens to a hundred nanometers. This sharp point and the anode create a very high field gradient in the vicinity of the cathode, and as a result electrons are extracted from the cathode without heating of the cathode being required. Such a gun must be operated at low pressures, about 10^{-10} torr or lower. In addition to being very small, such a source has a very high brightness.

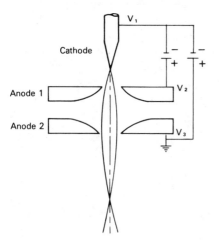

Fig. 16.13. Field emission electron gun. (Adapted from references 57, 78, and 79.)

The remainder of the illuminating system consists of one or two electromagnetic lenses, which are arranged to demagnify the first image of the source, as shown schematically in Fig. 16.12.

c. IMAGE-TRANSFORMING AND -RECORDING SYSTEMS

A STEM produces an image of the specimen on a CRT (a television monitor). This image can be viewed directly, or it can be recorded photographically or on a video recorder.

4. Scanning Electron Microscopy

a. OPTICAL SYSTEM

The scanning electron microscope (SEM) is similar in many respects to the STEM. However, thick specimens are used and the image-forming signals are collected from above the surface irradiated by the probe beam. Figure 16.14 is a schematic diagram of the optical system of a SEM.

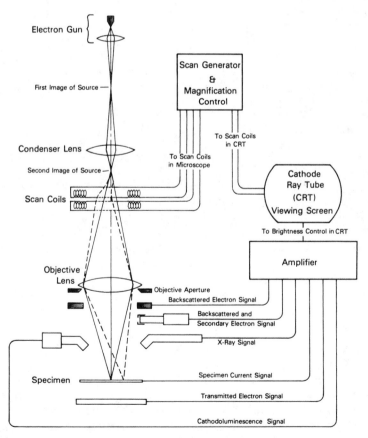

Fig. 16.14. Schematic diagram of a SEM.

Electrons are produced by an electron gun, which forms an image of the electron source. The condenser lens forms a demagnified second image of the electron source, and the objective again demagnifies and focuses an image of this second image of the source on the specimen. As in the TEM, the diameter of this probe beam at the specimen is determined by the diameter of the electron source and by the optical system between the source and the specimen.

Again as in the STEM, there are scan coils and a scan generator. When both sets of scan coils are operated, the probe beam describes a rectangular raster on the surface of the specimen. When only one set of scan coils is used, the beam describes a line across the specimen, and when the scan coils are turned off, the probe beam is focused on a spot on the specimen.

The probe beam interacts with the specimen as described in Section II.A.1. The various emissions from the specimen are collected by one or more of the detectors shown in Fig. 16.14. The signals are fed through amplifers and used to modulate the intensity of the electron beam in the CRT, thus producing an image on the viewing screen whenever the scan generator is operating to produce a raster.

As indicated in Fig. 16.14, a variety of signals can be used to produce an image. The information carried by these signals varies widely. The uses of the various signals are discussed in Section II.A.1.

b. SIGNAL DETECTORS

(1) Electron Detectors

Nearly all SEMS are equipped with an electron-detector system developed by Everhart and Thornley (97,120). This system consists of a scintillator that emits light when struck by an electron. The scintillator is connected with a photomultiplier tube by means of a light pipe. The scintillator is coated with a thin layer of metal (usually aluminum), and this metal forms an electrode, which is maintained at a potential of about +23 kV to attract and accelerate low-energy electrons so that they enter the scintillator and cause a light pulse. This enters the photomultiplier and produces a signal that is in turn amplified by the photomultiplier. This signal is amplified further and used to modulate the electron beam in the CRT. The scintillator is surrounded by a Faraday cage, which can be maintained by a potential of −50 to +250 V. When positively charged, this Faraday cage, called the "collector," serves to collect the secondary electrons emitted by the specimen. It also serves to protect the probe beam from the +12-kV potential on the scintillator.

The Everhart–Thornley detector can be used to detect both secondary and backscattered electrons. Only those backscattered electrons traveling directly toward the scintillator are detected. When the collector is charged at +250 V, it attracts secondary electrons—not only those traveling toward the scintillator, but also those traveling in other directions. As the collector potential is lowered, fewer and fewer secondary electrons are attracted, and as the potential is reversed and reaches -50 V, nearly all the secondary electrons are prevented from reaching the scintillator.

Another type of electron detector is the solid-state detector. Such detectors consist of flat semiconductor chips, and can be arranged about the specimen in various ways to increase the efficiency of detection of the emitted electrons and to investigate directional emission. Solid-state detectors are energy sensitive—the higher the energy of the incident electron, the greater its production of electron–hole pairs and the greater the signal.

Backscattered electrons are more energetic than secondary electrons, and they are easily and efficiently detected by solid-state detectors. Secondary electrons have little effect unless they are accelerated by a suitable potential, as in the Everhart–Thornley detector.

Currents generated by solid-state detectors must be amplified more than those generated by the Everhart–Thornley detector, which contains a photomultiplier.

(2) Specimen-Current Detectors

Absorption of electrons generates a specimen current, which may be amplified and used as an image-forming signal. Specimen currents can be collected if the specimen is conductive and is insulated from the microscope. Collection of this signal is quite efficient, but the currents are small and must be greatly amplified to be useful. Both the specimen-current signal and the signal produced by a solid-state detector are somewhat noisy, because they must be highly amplified, and very high-gain amplifiers are difficult to make.

(3) Cathodoluminescence Detectors

The detector for cathodoluminescence usually consists of a photomultiplier with a light pipe attached to its window. Such an arrangement is not very efficient for collecting the luminescence from the specimen, because the light-pipe window is relatively small. The efficiency can be increased by placing an elliptical mirror over the specimen in such a position that the specimen is at one focus of the ellipse and the light-pipe window is at the other focus.

(4) X-ray Detectors

When the specimen is struck by electrons, x-rays are generated (see Section II.A.1.c). The energies of the x-ray photons, and therefore their wavelengths, are characteristic of the elements in the specimen. The principal use of x-ray emissions in SEM is for elemental analysis, and therefore x-ray-detector systems used on SEMs are spectrometers.

Two types of spectrometers are in common use: wavelength-dispersive spectrometers and energy-dispersive spectrometers (120,144).

Wavelength-dispersive spectrometers are in many ways similar to grating spectrometers used for spectroscopy in the ultraviolet, visible, and infrared regions of the electromagnetic spectrum. The grating consists of a crystal having interplanar spacings suitable for diffracting the particular x-ray wavelengths to be investigated [see Section II.A.1.b.(2)]. Three to five different crystals are required to cover the range of wavelengths to be expected in elemental analysis. The spectrometer is arranged so that these crystal gratings can be interchanged.

Because no suitable lenses for forming x-rays are available, the grating crystals are bent so that focusing action occurs. The source of the x-rays, the bent crystal, and the slit for the x-ray detector must all lie on a circle, called the Rowland circle, if the system is to work at its highest efficiency. The radius of curvature of the bent crystal is twice that of the Rowland circle. Because the x-ray wavelengths that are generated when the electron beam strikes the specimen vary with the elements present, the spectrometer must be made so that the bent crystal and the detector slit move along the Rowland circle. Also, because the angle of incidence is equal to the angle of diffraction, the slit must move around the circle twice as fast as the bent crystal, and at the same time, the bent crystal must always be tangent to the Rowland circle. The mechanism for accomplishing all of this is quite complex, but these spectrometers, called fully focusing spectrometers, have the best spectral resolution and x-ray-collection efficiency.

Sometimes the spectrometer mechanism is simplified and the conditions specified above are not completely met. Such a semifocusing spectrometer does not have as high a spectral resolution and efficiency as does the fully focusing type; however, the simpler mechanism may allow greater accuracy and reproducibility in locating the spectral lines.

The detector most frequently used with a bent-crystal spectrometer is the gas-flow proportional counter. Such detectors are modified Geiger-type counters. To detect soft (low-energy) x-rays, a thin window is required. Because it is difficult to seal such a thin window permanently, the ionizing gas flows continuously through the counter. Such a counter not only counts each x-ray photon entering it, it also produces a signal that is proportional to the energy of the photon.

The second type of x-ray detector is the energy-dispersive detector. This type of detector produces signals (pulses) proportional to the energy of the x-ray photons that strike it. If the pulse-height distribution of x-ray photons striking the counter can be analyzed, relative proportions of the various photon energies can be determined. From the relative proportions of the various photon energies, the relative proportions of the elements giving off the photons can be determined.

Gas-flow proportional counters such as are used in wavelength-dispersive systems can measure electron energies; however, their spectral resolution is inadequate for use in energy-dispersive systems. Solid-state detectors having adequate resolution are available. Commonly these detectors are plates of germanium or silicon that have been treated with lithium. These detectors must be kept at liquid nitrogen temperatures to prevent migration of the lithium and to reduce thermal noise. (A short time at room temperature will destroy this type of detector). Solid-state detectors are therefore immersed in Dewar flasks having thin beryllium windows.

The signals emerging from the solid-state counter are fed into a multi-channel pulse-height analyzer, where they are separated into a few hundred channels according to pulse height and accumulated. The instrument can then recall and display the pulse-height distribution on request.

Energy-dispersive x-ray detectors can be placed close to the specimen, making their collection efficiency high. They also admit, count, and classify x-ray photons of a wide range of energies (wavelengths) simultaneously. The wavelength-dispersive spectrometer must scan the Rowland circle to detect x-rays of various wavelengths; this requires considerable time. Also, because the bent crystal is some distance from the specimen, the collection efficiency may be lower. However, at wavelengths longer than 0.07 nm, the spectral resolution of the wavelength-dispersive spectrometer is superior to that of the energy-dispersive type, and the background intensity is less, allowing more accurate quantitative analysis. Without special equipment, neither type of detector can analyze well for elements that have atomic numbers less than that of sodium.

c. ILLUMINATING SYSTEMS

SEMs commonly have thermionic emission electron guns similar to those used in TEMs. Some microscopes can be equipped with field emission guns such as those used in STEMs. Many SEMs may be equipped with an electron gun that is partly a thermionic emission gun and partly a field emission gun. A schematic diagram of such a gun developed by Broers and others (4,5,48) is shown in Fig. 16.15.

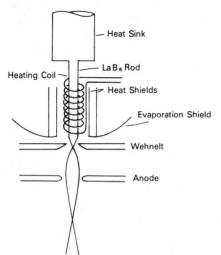

Fig. 16.15. Electron gun with LaB$_6$ cathode. (Adapted from reference 48.)

In Broers's gun the electron source is a sharply pointed LaB_6 rod. The rod is heated to between 1600 and 1680°C. In the gun in the diagram, heating is accomplished by means of a heater coil, but the rod can be heated by an electron beam or laser radiation. The electrons are drawn from the LaB_6 tip by a combination of thermionic and field emission. Such a gun requires a pressure lower than 10^{-6} torr. This is a higher vacuum than is commonly available in a SEM, so an auxilliary vacuum pump is generally required.

d. IMAGE-TRANSFORMING AND -RECORDING SYSTEMS

In SEM, the signal for each point in the image is produced sequentially as the beam scans the specimen. Thus it is possible to manipulate the signal from each point in the image by electronic means. As was mentioned above, most signals must be amplified before they are fed into the CRT to produce a visual image. If the amplifier is linear in its response, a "normal" image is formed. However, the amplifier can be selected to have various nonlinear responses.

The use of a nonlinear amplifier is one example of signal manipulation. Some of the common types of signal manipulation are:

1.   Nonlinear (also called gamma) amplification.
2.   Black-level suppression.
3.   Signal differentiation.
4. y-Modulation.

Nonlinear amplification is used to produce more contrast in darker parts of the image as compared with the brighter parts. Black-level suppression subtracts the dc component of the image signal from the overall signal, which consists of both ac and dc components. Because the ac components contain most of the information, subtracting the dc component gives more information. In signal differentiation, a derivative of the original signal is fed into the CRT. This has the effect of enhancing edges of objects in the specimen. In y-modulation each line in the raster on the CRT is a plot of signal intensity versus distance along the line. The signal intensity is plotted in the y-direction; hence the name "y modulation."

This is a very brief description of signal manipulation. Many more possibilities exist, including combining signals in various ways. In addition to signal manipulation, it is possible to process the image itself by using either analog or digital computers. With computer processing it is possible to obtain information such as grain size and amounts of the phases present (180). More limited image-storing circuits have been developed that allow the image to be stored and manipulated (283).

The most common method of storing a SEM image is to photograph the CRT. In general, self-developing direct-positive photographic processes such as that developed by Polaroid are used.

B. BASIC LAWS

1. Electron Sources

Most electron microscopes in use today derive the electrons employed in their imaging systems from thermionic emission (see Figs. 16.10 and 16.16). Usually the source consists of a filament of tungsten wire, bent to form a "V," which is heated by passing an electric current through it. The apex of the "V" is centered behind the opening in the cathode cap, and the electrons used in image formation are those emitted from the outside surface of the apex. This is only a small part of the total number of electrons emitted by the entire filament. Because the use of a source whose image on the specimen is larger than the field of view causes specimen damage and difficulties with heating, the area of the effective source is kept as small as possible. The most important characteristic of a source is its brightness or current density per unit solid angle. The brightness is made as high as possible consistent with reasonable filament life.

The electron emission I_s, in amperes per square centimeter, is a function of the absolute temperature T of the filament. It can be calculated using the equation

$$I_s = AT^2 \exp\left(-\frac{b}{T}\right)$$

where A and b are constants. The value of A for pure metals, which may be used as filaments, is 60.2. However, the constant A has relatively little effect compared with b, which appears in the exponent and is proportional to the work function of the material. The value of b for tungsten is 52,400.

The brightness β_s of an electron gun is theoretically invariant and is limited by thermodynamic considerations. The theoretical value is computed according to the equation

$$\beta_s = \frac{eI_s\phi}{\pi kT}$$

where I_s is the current density emitted by the source as computed above, ϕ is the accelerating voltage, k is Boltzmann's constant (8.6×10^{-5} eV/K), and e is the charge of the electron.

The current density I_i at an image plane produced by any train of lenses is related to the current density I_s at the cathode by Abbe's sine condition (130)

$$\frac{I_i}{I_s} = M^2 = \frac{n_i^2 \sin^2 \alpha_i}{n_s^2 \sin^2 \alpha_s}$$

Table 16.I

Electron gun type	Brightness (A/cm²/steradian)	Vacuum needed (torr)	Source size (μ)
Thermionic	6×10^4 @ 25 kV	$\sim 10^{-5}$	25–100
LaB$_6$	1×10^7 @ 75 kV	$< 10^{-6}$	10
Field emission	2×10^9 @ 23 kV	$\sim 10^{-10}$	10^{-3}

where M is the magnification of the lens train, α_i is the angular aperture of the beam at the image, α_s is the angular aperture of the beam leaving the source (see Figures 16.16 and 16.19), and n_i^2/n_s^2 is the ratio of the energy in the image space to the energy at the source, or $e\phi/kT$. Thus

$$I_i = I_s \frac{e\phi \sin^2 \alpha_i}{kT \sin^2 \alpha_s}$$

For electron microscopes, α_i is small and the solid angle of the cone of electrons impinging on the image plane can be written as $\pi \sin^2 \alpha_i$. The brightness β_i at the image plane is thus

$$\beta_i = \frac{I_i}{\pi \sin^2 \alpha_i} = \frac{I_s e\phi \sin^2 \alpha_i}{kT\pi \sin^2 \alpha_s \sin^2 \alpha_i} = \frac{I_s e\phi}{\pi kT \sin^2 \alpha_s}$$

For a complete discussion of brightness, its importance, and the factors that affect it, see Haine and Cosslett (125).

Thermionic emission electron guns are the most common type used in TEMs and SEMs. Field emission guns (see Fig. 16.13) are used in SEMs and may also be used in the other types of microscopes. Guns using a combination of field and thermionic emission are often used in SEMs and may be used in TEMs. Table 16.I gives a reasonable set of values for various gun parameters.

Louis de Broglie (82) postulated that charged particles in motion behave as if they were wavelike. In considering the theoretical aspects of electron microscopy, it is convenient to treat electrons as electromagnetic waves. The wavelength λ of a stream of particles is obtained as follows:

$$\lambda = \frac{h}{mv}$$

where h is Planck's constant, m is the mass of the particles, and v is their velocity. When an electron is accelerated by an electrostatic potential, its velocity can be calculated by setting the kinetic energy of the moving electron equal to the potential energy of an electron placed in the electrostatic field; thus,

$$\tfrac{1}{2}mv^2 = e\phi$$

where e is the charge of the electron.

$$v = \frac{2e\phi}{m}$$

and

$$\lambda = \frac{h}{2em\phi}$$

Therefore, λ (in Å) $= 150/\phi$. If the accelerating potential is 50 kV, $\lambda = 0.055$ Å.

2. Electron Lenses

To form an image using light, electrons, or any form of radiation, it must be possible to bend the rays in question at will. In light optics, images are formed by using lenses of transparent materials. Light is refracted at the interface between two transparent materials in accordance with the equation

$$n_1 \sin i = n_2 \sin r$$

where n_1 and n_2 are the refractive indexes of the first and second transparent media, i is the angle of incidence, and r is the angle of refraction measured from the normal to the interfacial surface.

Electrons can be bent in a similar fashion by electrostatic or magnetic fields. Electrostatic lenses are easier to understand and will be discussed first. An electron in an electrostatic field is acted on by a force and is accelerated in the direction of positive potential. Figure 16.16 is a diagram of a self-biased gun. A self-biased gun is an electrostatic lens that produces a highly intense, slightly divergent beam of electrons suitable for the illuminating system of the microscope. In this figure a potential of $-80,000$ V is assumed to be supplied to the filament. The anode is at ground potential. The electrostatic field resulting is distributed as shown by the equipotential surfaces. The electrons driven from the filament as a result of thermionic emission are attracted toward the anode. The electrons will cross an equipotential surface in a direction perpendicular to the surface unless they already possess a velocity in another direction. The thermionic electrons considered here do possess various low velocities and directions as they leave the surface of the filament, but to a first approximation these may be neglected.

Ray paths through a lens can be plotted in a semiquantitative manner by considering the equipotential surfaces as the boundaries between regions of different indexes. The more positive the potential relative to the ground, the

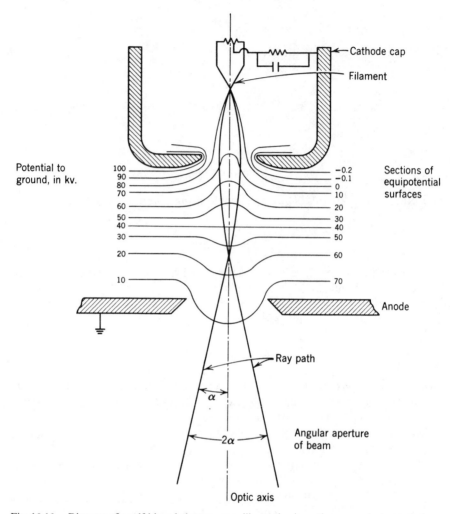

Fig. 16.16.   Diagram of a self-biased electron gun, illustrating its action as an electrostatic lens.

higher the refractive index. The effective refractive indexes can be calculated as follows:

$$\frac{\sin i}{\sin r} = \frac{n_2}{n_1} = \frac{\phi_2}{\phi_1}$$

In the particular situation diagrammed in Fig. 16.16, the equipotential surfaces near the filament form a highly convergent lens, causing the electrons to cross over and focus within the lens field. This crossover of the beam serves as the effective source for the rest of the illuminating system.

After passing through the focus, the electrons continue onward, the beam diverging as it leaves the gun.

It is interesting to consider how the ray paths change as the spacing of the cathode cap and its bias relative to the filament are changed. As the cathode cap is moved nearer to the filament, the equipotential surfaces become less convex, its strength as a converging lens diminishes, and the focal point and crossover move away from the filament. Decreasing the bias has a similar effect. At low bias the crossover point is far from the filament. In fact, with the older unbiased guns, the crossover was at specimen level or below, and the location of the crossover was controlled by adjusting the filament height. At high bias, near the cutoff point, the equipotential surfaces are highly curved; under these conditions, the crossover point is close to the filament. In this case, lens aberrations are so high that many electrons are deflected out of the useful aperture of the microscope and the brightness level falls off. At some intermediate bias these aberrations are reduced, and the gun then gives the theoretical brightness.

The effects of varying the temperature of the filament also are of interest in connection with the resulting effects on the operation of the self-biased gun. A rise in the temperature of the filament causes greater emission of electrons, which results in a higher beam current and a higher potential drop across the bias resistor, and thus a rise in the effective bias on the cathode cap. When the temperature is raised still further, the bias becomes larger and tends to diminish the beam current. As the beam current diminishes, so does the bias; therefore, the system is self limiting. As a result, when the filament current is raised, the brightness of the image rises to a maximum, called saturation, and then levels off.

The electrostatic lens in the electron gun is highly specialized. Both the object (the filament) and its image are immersed in the lens. It also takes electrons from a high negative potential and delivers them into a space at ground potential. Under this latter condition, an electrostatic lens can be a negative lens, producing a diverging beam without first causing the electrons to pass through a focus. Normally, however, an electron lens takes a beam of electrons traveling through space at ground potential and delivers them into a similar space at the original velocity. Under these conditions an electron lens is always a positive lens.

Figure 16.17 shows a schematic cross section of an electrostatic electron lens. It consists of three perforated disks. The two outer disks are usually connected together and are at ground potential, whereas the central disk may be connected to the cathode supply of the microscope. (If the polarity of the disks is reversed, the lens still works in much the same way.) The equipotential surfaces are indicated in the diagram. The electrons are traveling through the microscope at the velocity given them by the electron gun. Three rays that have passed through the tip of the object arrow O are traced along paths through the lens. The three rays chosen are (1) rays traveling parallel to the optical axis, (2) rays passing through the geometrical center of the lens, and (3) rays passing through the front focal point F_1. The bending of the electron path at the equipotential surfaces can be approximated as described above. As the electron progresses through the lens, it is slowed as it passes to lower and lower potential. After it passes the center of the lens, its speed increases and it leaves the lens at the same velocity with which it entered. In the meantime each ray is deflected from its original path in the manner shown in the diagram. Thus, all rays diverging from the point of O converge to form its image at I.

If the ray paths outside the lens are extended into the lens, the principal planes P_1 and P_2 of the lens can be located. The lens formula for a thick lens expresses the relationships between these cardinal points just as it does in light optics:

$$\frac{F_1P_1}{OP_1} + \frac{F_2P_2}{IP_2} = 1$$

$$\frac{OF_1}{F_1P_1} = \frac{F_2P_2}{F_2I}$$

and

$$M = \frac{F_1P_1}{OF_1}\frac{F_2I}{F_2P_2}$$

where M is the magnification, F_1P_1 is the distance between points F_1 and P_1, and likewise for OP_1, F_2P_2, and so on. For symmetrical lenses such as the one described here, $F_1P_1 = F_2P_2$; that is, the front and rear focal lengths are equal and the lens formula can be expressed in the more familiar form

$$\frac{1}{OP_1} + \frac{1}{P_2I} = \frac{1}{F_1P_1} = \frac{1}{F_2P_2}$$

Most electron microscopes use electromagnetic rather than electrostatic lenses. Both types of lenses have been investigated extensively from both the theoretical and the experimental standpoint, and electromagnetic lenses have been found to be superior for use in microscopes.

The ray paths through an electromagnetic lens are much more complex than those through electrostatic or light optical lenses. Electrons passing through an electromagnetic lens such as the one shown diagrammatically in Fig. 16.18 are accelerated not only toward the optical axis of the lens, but also in a tangential direction; thus their actual path as viewed along the axis is a complex spiral. The analysis of the exact paths is very complex and will not be discussed. However, it may help to follow two rays in a qualitative manner. The surfaces of equal magnetic field strength are not shown in the diagram, because they have no simple analogy to refractive indexes, as they have in the case of the electrostatic field. Instead, a few magnetic lines of

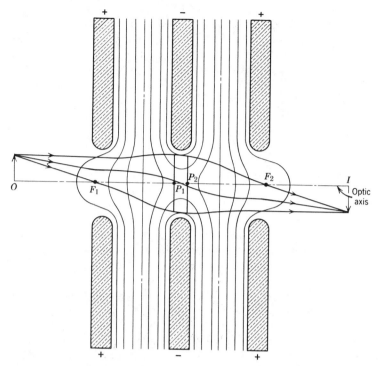

Fig. 16.17. Diagram of an electrostatic electron lens, showing ray paths of the electrons originating at the tip of the object arrow O and forming an image at I. P_1 and P_2 are the intersections of the principal planes with the optic axis, and F_1 and F_2 are the focal points.

force are shown and the ray tracing is considered from the standpoint of a particle moving in a magnetic field. Since the magnetic field is symmetrical about the optical axis, the magnetic field vector at any point can be considered as the resultant of a radial vector toward the optical axis and an axial vector parallel to the optical axis.

Consider an electron traveling parallel to the optical axis through the tip of the object arrow (Fig. 16.18). When it enters the magnetic field, it is traveling parallel to the axial component of the field and is therefore unaffected by it. The electron does move across the radial component of the field and, as a result, is subjected to a force perpendicular to the plane of the section. This force curves its path upward out of the plane of the section, that is, in a tangential direction. As soon as the electron moves in a tangential direction, it interacts with the axial field, and as a result is forced toward the optical axis of the system. The addition of the tangential and axial force vectors to the velocity vector of the electron as it enters the field causes the electron to move in a spiral path. The particular electron path being traced here intersects the optical axis at the focal point F_2 and continues onward outside the lens field through the point of the arrow image.

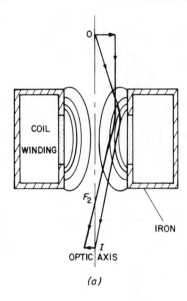

(a)

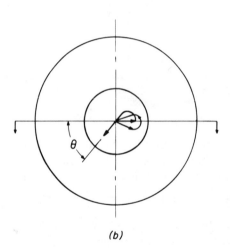

(b)

Fig. 16.18. An electromagnetic electron lens. The imaging action is like that for a light lens or an electrostatic lens, but is more complicated, because the electrons spiral through the lens. Part b is a projection of Part a as viewed along the optic axis.

The second electron path traced in Fig. 16.18 is that passing through the tail of the object arrow O at some angle with the optical axis. When this electron enters the magnetic field, it is acted on by the same array of force vectors described above for the electron traveling parallel to the optical axis. The resultant path is shown passing through the image point at I. Other electrons passing through the object and modified by it are affected by the same forces and spiral through the lens to form the image. The lens formulas used for light and electrostatic lenses are applicable to the magnetic lens.

Images formed by light or electrostatic electron lenses are inverted with respect to the object. Because of the spiral path of the electrons in an electromagnetic lens, the image in this case is rotated about the optical axis as well as being inverted. This is shown most clearly in Fig. 24.18b. The amount of rotation depends on the conditions, particularly on the current flowing in the lens coil and the velocity of the entering electrons.

The relationship between the optical parameters of an electromagnetic lens and its physical parameters has been the subject of extensive investigation. The results of this work are discussed by Haine and coworkers (125,126). The focal length f of a lens is given by the expression

$$f = \frac{25V_r(S + D)}{(NI)^2}$$

where V_r is the relativistically corrected electron energy, S is the pole-piece gap, D is the diameter of the lens bore, and NI is the number of effective ampere turns in the lens coil. This formula is based on the assumption that the lens acts as a thin lens.

In practical electron microscopes the electron beam is quite narrow and occupies only the center of the electron lens. In this region the equipotential and equal-field surfaces are essentially segments of spheres, so that electron lenses are spherical lenses. As a result, they have spherical aberrations; that is, the focal length for the central zone of the lens is not at the same point as the focal length for the outer zones of the lens. The useful aperture of a lens is limited by this spherical aberration. In glass lenses for light optics, this type of aberration can be corrected by combining lenses. There is at present no practical way of correcting electron lenses for spherical aberration, and it is this factor that limits the resolving power of the electron microscope. The spherical aberration constant, C_s relates the radial error dr of a ray in image space to the angle a between the ray and the optic axis in object space and to the magnification M. The magnitude of C_s is calculated from the equation

$$dr = MC_s a^3$$

The spherical aberration of a weak lens is given by the formula

$$\frac{C_s}{f} = 5 \left(\frac{f}{S + D}\right)^2$$

Another important lens error is chromatic aberration. The chromatic aberration constant C_c is defined as the variation df in focal length with small changes in the velocity (wavelength) of the electron or with small changes in lens current:

$$df = C_c \left(\frac{d\phi}{\phi} - \frac{2dI}{I}\right)$$

where ϕ is the accelerating potential and I is the lens current. To prevent chromatic aberration from limiting the resolving power of the microscope, the value of df must be kept smaller than the depth of focus of the microscope. Practically, this means that the accelerating voltage and lens current power supplies must be stable to about 2 ppm.

Astigmatism is a third aberration of direct concern to the microscopist. It results when the equipotential or equal-field-strength surfaces become toroidal rather than spherical, so that the focal length of electrons for one azimuth of the lens is different than that for electrons in the azimuth at 90° to the first. This type of aberration usually results from inhomogeneities in lens parts. Most modern microscopes have provisions for correcting astigmatism, so it need not limit the resolving power of the microscope.

A number of other aberrations affect electron lenses. For a complete discussion of aberrations, see Hall's book *Introduction to Electron Microscopy* (130).

3.   Resolving Power of the Microscope

As has already been pointed out, the microscope is primarily designed to extend human vision to very small objects. Therefore, the most important quality of a microscope is its ability to make visible fine detail in the specimen. This ability to see fine detail is called the resolving power of the microscope.

Resolving power can be measured in a number of ways, for example, in terms of the ability to detect small particles or the ability clearly to observe the shape of polygonal crystals. Probably the most commonly used criterion is the distance d by which two points must be separated in order to be seen as two points in the microscopical image. The resolving power of an aberrationfree lens is related to the characteristics of the objective lens and the wavelength λ of the illumination by the equation

$$d = \frac{0.6\lambda}{\text{NA}}$$

where NA, the numerical aperture, is $n \sin \alpha$, n being the refractive index of object space relative to image space and 2α being the angular aperture of the lens. The angular aperture of a lens is the angle subtended at the specimen by the aperture of the objective lens. Figure 16.19 shows a schematic arrangement of the condenser and objective lenses and their respective apertures. It is assumed in the formula for resolving power given above that the angular apertures of the condenser and objective are equal.

In an electron microscope the potentials on both sides of the condenser and objective lenses are the same, and therefore $n = 1$. The wavelength of 50-kV electrons has been shown to be 0.055 Å. The limit of resolution, or resolving power, of an electron microscope thus depends on the angular

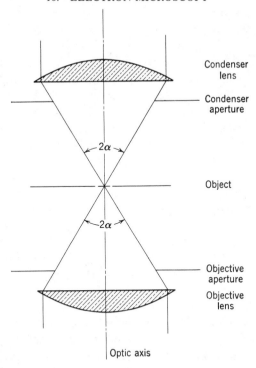

Fig. 16.19. Schematic diagram of condenser–objective relationships in a microscope. 2α is the angular aperture of the beam.

aperture of the objective lens. Lenses used for light microscopy have been highly developed over the years and the angular aperture of a high-quality dry lens approaches π radians (180°). However, electron lenses have not yet been developed to the point at which high angular apertures can be used. At present, the spherical aberration of electron lenses limits their usable angular aperture. The optimum aperture is given by the equation (125)

$$\alpha = 1.4 \left(\frac{\lambda}{C_s} \right)^{\frac{1}{4}}$$

Placing this value of α into the expression for resolution, we obtain

$$d = \frac{0.6\lambda}{\alpha} = \frac{0.6\lambda}{1.4(\lambda/C_s)^{\frac{1}{4}}}$$
$$= 0.43 C_s^{1/4} \lambda^{3/4}$$

For 50-kV electrons and optimum lens design, d is approximately 2.7 Å. The angular apertures α used in most microscopes vary from 0.002 to 0.005 radians, and the actual resolving power of the microscope is usually somewhat poorer than that calculated on a theoretical basis.

The formulas and calculations for the resolving power of a microscope apply to TEMs, both scanning and fixed beam, and to light microscopes. For fixed-beam TEMs a necessary condition is that the angular aperture of the condenser be equal to that of the objective. In SEMs the so-called objective lens is really a condenser lens, and there are no imaging lenses. For SEMs the resolving power of the instrument is closely related to the diameter of the probe beam at the specimen plane—the spot size.

A number of factors affect the spot size: the size of the electron source, the demagnification of the lens system, the diffraction effects, and the effects of spherical and chromatic aberration. The spot size d is calculated by Joy (144,163) as

$$d^2 = d_0^2 + (\tfrac{1}{2}C_s\alpha^3)^2 + \left(C_c\alpha\,\frac{\delta E_0}{E_0}\right)^2 + (\alpha\delta z)^2 + \left(\frac{1.22\lambda}{\alpha_i}\right)^2$$

where d_0 is the Gaussian spot size and depends on the demagnification of the lens system, C_s is the spherical aberration coefficient, α is the angular aperture of the beam emerging from the gun (see Fig. 16.16) α_i is the angular aperture of the beam converging on the specimen, C_c is the chromatic aberration coefficient, E_0 is the mean energy of the beam, δE_0 is the energy spread of the beam passing through the objective lens, and δz is the spacing of the two stigmatic foci.

Thus $C_s\alpha^3$ is the diameter of the spot due to spherical aberration, $C_c\alpha\delta E_0/E_0$ is the diameter of the spot due to chromatic aberration, $\alpha\delta z$ is the diameter of the spot due to astigmatism, and $1.22\lambda/\alpha_i$ is the diameter of the spot due to diffraction effects.

In a well-set-up microscope the contributions of astigmatism and chromatic aberration are small compared with those of the other factors, and the equation can be simplified to

$$d^2 = d_0^2 + (\tfrac{1}{2}C_s\alpha^3)^2 + \left(\frac{1.22\lambda}{\alpha_i}\right)^2$$

The current I_b in the probe spot is given by

$$I_b = 2.5\beta\alpha^2\left[d^2 - (\tfrac{1}{2}C_s\alpha^3)^2 - \frac{1.22\lambda^2}{\alpha_i}\right]$$

The resolution improves with small spot size, but making the spot size too small may reduce the current below the minimum useful value.

C. EXPERIMENTAL FACTORS THAT LIMIT SENSITIVITY

1. Resolution

The theoretical resolving power of TEMs and SEMs at the present stage of development is of the order of 2 Å for accelerating potentials of 50–100 kV. The calculated resolving power indicates the capabilities of the instrument, but the resolution obtained in a micrograph depends also on the properties of the specimen, the condition of the microscope, and the ability of the operator.

Isolated particles of materials of high density compared with their surroundings have been detected in micrographs even though they were considerably smaller than the resolving power based on the separation of particle images. On the other hand, the *shape* of an individual particle cannot be distinguished unless the particle is larger than the resolving power. The relationship between particle shape and resolving power has been discussed by Von Borries and Kausche (314), who have stated that a particle of n sides ($n \geq 6$) must be $4n/\pi$ times larger than d before it can clearly be recognized as a polygon.

The effect of these limits of resolution on the application of the electron microscope to chemical analysis is currently negligible. Other factors, having to do with image contrast and specimen-preparation techniques, limit such methods at present. If, for example, a specific reaction yielding a precipitate is carried out, the limits of detection of the precipitate particles would theoretically be of the order of 10 Å or less, depending on the density. However, the problems of preparing and locating the sample are such that the resolving power of the microscope is no limitation. Tufts and Lodge (303) have identified particles 50 Å in diameter using this type of technique.

The theoretical resolving power of the SEM is closely related to the probe-beam spot size, as discussed in Section II.B.3. However, because the electron beam penetrates the specimen and because the specimen is usually thick enough that no transmission of electrons occurs, the resolution is larger than the spot size. Figures 16.7 and 16.8 show the spreading of the beam as it penetrates the specimen. Because of this spreading, the image-forming radiations arise from a volume having dimensions larger than the diameter of the probe-beam spot.

Therefore the resolution of the SEM depends on probe-beam spot size, the accelerating voltage, the atomic number of the specimen material, the type of radiation used to form the image, and in some cases the thickness of the specimen. Determining the resolution of a SEM is somewhat difficult, but for most microscopes and samples the resolution is 100 Å or higher. With favorable specimens, some instruments may have resolutions of less than 100 Å.

2. Image Contrast

The contrast of a photographic image can be expressed as the ratio of the optical density of the lightest part of the image to the optical density of the darkest part of the image. The contrast C of a given object in an image can be expressed as

$$C = \frac{D_i - D_b}{D_b} \times 100$$

where D_i is the optical density of the image of the object and D_b is the optical density of the image of the background immediately surrounding the image of the object. Unless an object has a contrast of at least 2–5%, it cannot be detected by the human eye.

Contrast arises from the interaction of the image-forming radiation with the specimen. The optical system of the microscope must be designed to use the interaction of the specimen with the beam to produce contrast in the image. To obtain good contrast, the microscopist must be able to use the optical system properly; in addition, he must be able to prepare the specimen so as to take maximum advantage of the specimen–beam interactions.

In light microscopy, image contrast depends on differential absorption of light by the specimen, differential phase shifts caused by variations in the optical thickness of the specimen, differential scattering of the light rays caused by reflection and refraction, and birefringence of the specimen. All light microscopes make use of absorption, reflection, and refraction effects to provide contrast in the image. More specialized instruments exploit the other phenomena to provide image contrast. Polarizing microscopes exploit the birefringence of materials, phase-contrast and interference microscopes exploit the phase shifts, and dark-field microscopes exploit reflection and refraction effects.

In an analogous manner, the electron microscopes exploit the various specimen–electron beam interactions to provide image contrast. Transmission electron microscopes form images by using elastically scattered electrons. Image contrast arises because the more highly scattered electrons are prevented from contributing to the image. (If all the electrons scattered by the specimen could be used to form the image, there would be no contrast in the image.) The electrons scattered beyond a chosen angle can be prevented from arriving at the image by means of an objective aperture.

As described in Section II.B.3, the optimum angular aperture α is a function of the spherical aberration of the objective and the wavelength of the electrons. The angular aperture of a microscope is determined by the size and position of the objective aperture (see Fig. 16.19). Thus contrast can be improved at the cost of loss of resolution by choosing an objective aperture that is smaller than the optimum. If the objective aperture is removed entirely, an image will still be seen, because spherical aberration will prevent highly scattered electrons from arriving at the proper point in the image.

Inelastically scattered electrons will generally not arrive at the proper image point, that is, the point conjugate to the object point from which the electron was scattered, because they will have undergone a change in wavelength and the chromatic aberration of the objective lens will have caused them to deviate from the conjugate image point. Electrons that arrive in the image at locations other than their conjugate image point merely cause an increase in background and a concomitant loss of contrast. Specimens that are too thick lack contrast, primarily because they produce large numbers of inelastically scattered electrons and the chromatic aberration of the lens causes these electrons to deviate from their conjugate image points.

In light microscopy, image contrast depends on several factors: differential absorption in the object, differential phase shifts in the light waves due to refractive-index differences, interference phenomena, and polarization phenomena. The electron intensity at any point in the image is proportional to the number of electrons transmitted through or scattered by the corresponding object point, which subsequently pass through the usable aperture of the objective. The number of electrons so scattered, I, can be calculated as

$$I = I_0 \exp(-N\delta x)$$

where I_0 is the number of electrons in the incident illuminating beam, N is the number of scattering atoms per unit volume of a specimen whose thickness is x, and δ is their scattering cross section. The scattering cross section can be broken into two terms:

$$\delta = \delta_e + \delta_i$$

where δ_e refers to elastic and δ_i to inelastic cross sections. These cross sections, or comparable ones defined somewhat differently, have been calculated by a number of investigators (72,125,130,182,326,331).

Inelastic-scattering angles are usually quite small, less than 10^{-2} radians, whereas elastic-scattering angles are mostly greater than 10^{-2} radians. Except for the light elements, elastic scattering exceeds inelastic scattering. The effect on image contrast of changing the size of the objective aperture is primarily the result of changes in the number of inelastically scattered electrons arriving at the image. Thus, when specimens are thick and of low atomic number, the use of an objective aperture is indicated.

Under the conditions prevailing in electron microscopy, the contrast in the image is primarily dependent on the density and thickness of the specimen. Since there is a 20-fold variation in density among the elements, it may be possible to use image-contrast and thickness measurements to identify or at least to classify the elements in a specimen. The problem of identifying biologically important compounds has been discussed by Zeitler and Bahr (331,333).

For thin specimens with fine detail, for which there is considerable unaffected background illumination, phase effects (125,333) cause contrast in the images of specimens. However, with thick specimens, the transmitted beam loses coherence and phase effects are negligible.

Inelastically scattered electrons in the beam that passes through a specimen may be classified into two groups: those that have undergone characteristic energy losses, and those that have a continuous distribution of energy losses (see Section II.A.1.c.). The characteristic energy losses are the result of the transferral of sufficient energy to electrons in the inner orbits of the specimen atoms to move them to higher energy levels. These characteristic energy losses therefore correspond to the energies of the x-ray spectra of the specimen atoms. In general, the characteristic energy losses are of the order of 10–15 eV.

The noncharacteristic energy losses suffered by transmitted electrons are much larger, ranging to several hundred electron volts. The cross section of the noncharacteristic energy losses is relatively low.

The inelastically scattered electrons have wavelengths that are different from those of the primary beam. Therefore, because of the chromatic aberration of the objective lens, these electrons tend to increase the size of the Airy diffraction disk and thus lower the resolution of the microscope image. The noncharacteristically scattered electrons may contribute markedly to loss of image contrast when thick specimens are being examined.

The STEM makes use of the elastically and inelastically scattered electrons to produce contrast in the image. The annular detector (see Fig. 16.12) collects electrons scattered at relatively large angles. These are generally elastically scattered. The contrast derived from this detector is related to atomic number (see Section II.A.3.a). The bright-field signal produces image contrast related to the morphology of the specimen. The dark-field signal produces contrast related to both the morphology and the chemistry of the specimen. A STEM can also be set up to produce images in which contrast is derived from x-ray emission (again see Section II.A.3.a).

Image contrast in the SEM may be produced by signals produced by backscattered electrons, secondary electrons, x-rays, specimen currents, specimen charging, Auger electrons, transmitted electrons, and cathodoluminescence. These types of contrast are related to both the morphology and the chemistry of the specimen (see Fig. 16.14 and Section II.A.4).

D. FUNDAMENTAL TYPES OF INFORMATION OBTAINABLE

The information derived from electron microscopy can be classified into four fundamental types: (1) physical structure (morphology), (2) the scattering and absorption of electrons by materials, (3) diffraction effects (elastic scattering), and (4) characteristic emissions. Each of these types of information can be used for qualitative and quantitative analyses.

1. Physical Structure (Morphology)

Electron microscopes can be used to observe the morphology of a specimen. Therefore, methods of analysis that depend on recognition of morphological structures are common. There are two ways in which chemical analysis can be carried out by using the physical structure of the specimen as an analytical criterion of the presence or absence of an element or compound.

In one method, a specific reaction that produces a characteristic solid phase can be used. The presence or absence of this solid phase among the reaction products is indicative of the presence or absence of the element or compound for which the reagent used is specific. This method is essentially an extension of the spot-testing techniques developed by Feigl and colleagues. The principal advantage of the electron microscope for this type of analysis lies in its ability to reveal very tiny amounts of a solid phase. Thus, it can be used to identify tiny particles or the constituents of dilute solutions. For example, as mentioned earlier, Tufts and Lodge (303) have detected the chloride ion in sodium chloride particles as small as 50 Å on a side. The spot-test technique also can be used in a semiquantitative manner. Because the size of the spot or ring formed in the reaction is indicative of the amount of the element present in the original particles, the spot-size distribution as determined in the microscope can be used to determine the number of particles and the distribution of the element in question in the particles of the sample. This technique was described by Tufts (302), who originally used the method for light microscopy.

A second procedure by which morphology can be used in the identification of phases in the electron microscope is through direct recognition of shape or crystal habit. Shape recognition can be used in two ways. If the system being examined is well known, it may be possible to recognize the various constituents present in a specimen and to obtain an estimate of their relative concentrations. This method is particularly applicable to multiphase solids such as metal alloys, mineral aggregates, ceramics, and similar materials. Polished and etched sections of such materials can be analyzed quantitatively by areal, linear, or point-count methods if the phases present can be recognized. The method can be used also for the examination of mixtures of powders, slurries, and the like if suitable specimen-preparation techniques can be developed for the specific application.

Internal standardization, a technique used in light microscopy, also can be used for the electron-microscopical analysis of powders and slurries. Williams and Backus (322) have used the internal-standardization technique to determine the molecular weight of the bushy stunt virus.

Direct recognition of shape or crystal morphology as a means of identification has been reviewed by Suito and Uyeda (291). In many cases, the practical use of this method has depended on the establishment of a correlation between morphology and electron diffraction.

The recognition of crystal morphology can be used in still another way for qualitative chemical analysis. If a reagent reacts with the members of a group of ions to form characteristic crystalline precipitates, then the microscopical observation of these characteristic crystals can be used to identify the ions of the group present in an unknown material. This method is an extension of the methods of chemical microscopy developed by Behrens and Kley (22), Emich (92), Chamot and Mason (60), and others. Similar techniques have been modified for use in electron microscopy by Gulbransen et al. (124), Fischer and Simonsen (99), and Wiesenberger (319).

Although the methods of analysis that depend on morphology are the most commonly used in electron-microscopical analysis, they are still in the early stages of development. At present, most of these uses depend on establishing familiarity with the particular systems being analyzed. This is done by comparing morphological examinations with other, independent methods of analysis such as diffraction. Much more work can and should be done to establish systematic methods of qualitative analysis for inorganic ions and organic functional groups. For these purposes, extension of the methods of chemical microscopy and spot testing seems most promising.

2.   Scattering of Electrons by Atoms

The scattering phenomena described previously can be used in chemical analysis in several ways. As was pointed out, the apparent opacity of a specimen depends largely on its physical density. Thus, there is a possibility that the elements might be identifiable by use of TEM or STEM if their thicknesses in the specimen could be measured reliably. In biological work, the image density has been related to the mass thickness of specimens (53,277,332).

Electron staining is used in chemical analysis, particularly in histochemistry, to identify the constituents of a specimen. The technique used is to treat the specimen with a reagent specific for the constituent sought. The scattering of electrons is increased wherever the reagent reacts with the specimen. This technique is particularly effective if very dense elements are included in the reagent.

Elastically scattered electrons also carry information about the atomic number of the specimen. The higher the atomic number, the higher the elastic scattering. Scanning TEMs produce a Z signal related to atomic number. Scanning electron microscopes can produce a backscattered electron image, which is also related to atomic number. At present, atomic-number-related images are used in a qualitative fashion; for example, a backscattered electron image of a polished specimen can show contrast among the various constituents and the relative atomic numbers of the constituents can be judged from it. There is the possibility that this type of analysis can be made more quantitative.

By attaching an electron spectrometer to the optical system of a TEM or a STEM, it is possible to produce a spectrum of the inelastic (energy-loss) electrons. Some of these electrons have energies characteristic of the elements by which they were scattered (see Section II.A.1.d).

An electron optical system can be arranged so that an image can be formed with the electrons emerging from the electron spectrometer. Images of the specimen can then be formed wth selected energy-loss electrons. In this way maps of the distribution of elements in the specimen can be produced.

3.   Secondary Emissions

a.   SECONDARY ELECTRONS

The information obtained from secondary-electron images in the SEM is morphological.

b.   X-RAYS

Electron microscopes can be equipped with energy-dispersive and/or wavelength-dispersive spectrometers to analyze the x-rays given off when an electron beam strikes a specimen (see Section II.A.1.c). These x-rays are characteristic of the elements producing them, so elemental analyses can be performed by means of microscopes equipped with x-ray spectrometers. Both qualitative and quantitative analyses can be made. Analysis of elements having atomic numbers below 11 (sodium) are very difficult and usually are not possible.

c.   AUGER ELECTRONS

Auger electrons are characteristic of the element that produces them (see Section II.A.1.c). Electron microscopes to be used for Auger analysis must be equipped with special detectors and must be capable of operating at pressures of 10^{-7} torr instead of the more usual 10^{-5} torr. Auger electrons are produced by all elements except hydrogen and helium. The Auger-electron yield is especially high for elements of atomic numbers 3–15. Auger-electron emission occurs to depths of 10 Å or less, so the technique is especially useful for surface analysis.

d.   CATHODOLUMINESCENCE

By equipping an electron microscope with a photomultiplier detector, secondary emission in the ultraviolet, visible, or infrared region can be used to form electron images (see Fig. 16.14 and Section II.A.1.c). Cathodoluminescence gives information about the chemistry of the specimen. It can also be used to observe crystal defects and impurities, to distinguish among polymorphs, and to study semiconductors and biological materials.

4.   Diffraction Effects

Diffraction of electrons by crystalline specimens is a special case of elastic scattering. It is particularly important in the analysis of crystalline samples by electron microscopy. Electron diffraction is also carried out on instruments other than electron microscopes, but electron microscopes have the advantage of making it possible to observe the fine details of the specimen and to select small areas of the specimen for the preparation of diffraction patterns. This latter procedure is called selected-area electron diffraction.

The use of selected-area electron diffraction (see Chapter 17 and Section II.A.1.b) makes it possible to correlate the morphology of a specimen with its chemical composition. It is also possible to get some idea of the distribution of a crystalline compound in a transmission specimen. This is accomplished by forming the microscopical image using the diffracted electrons only, a specialized dark-field illumination in which the primary beam is prevented from contributing to the image. Thus, crystals of a particular material that are properly oriented are seen as bright against a dark background. This is the so-called selected diffraction microscopy technique (62,127,250,251,272,294). It can be used for the identification of the constituents of, for example, thin sections (both biological and nonbiological), thin films, and dispersions of particles.

Kikuchi patterns obtained with TEMs and electron-channeling patterns obtained with SEMs give information on crystallographic orientation and crystal perfection (see Section II.A.1.g). Kossel patterns, which give the same kind of information as do Kikuchi and channeling patterns, can also be obtained with electron microscopes.

5. Emission Microscopy

In emission microscopy the specimen is induced to give off electrons, which are subsequently used to form an image of the specimen. The TEM can be used, although inefficiently, to image the filament of the electron gun. More efficient arrangements for examining the electron emissions from specimens usually make the specimen a part of an electrostatic immersion lens. Such an arrangement accelerates the electrons and focuses them to form an intermediate image, which is then further magnified by another electron lens.

The fundamental analytical information to be derived from this type of microscope depends on the means used to cause emission from the specimen. If the specimen is heated, the emission depends on the temperature and the work function of the constituents of the specimen. At a suitably chosen temperature it is possible to use the differences in work function to obtain contrast differences among the microconstituents, and thus make elementary analyses in some cases (268–271).

Specimens also can be made to emit electrons by bombarding them with other electrons, with ions of various sorts, and with other forms of radiation. The analytical uses of these methods have not yet been explored fully.

The β-rays from radioactive specimens also can be used to form an image of the specimen. This technique has several possible analytical uses. It may be possible to identify the radioactive species in an unknown from its electron image, but of more direct interest is the microscopy of specimens treated with tracers. It is also possible to prepare autoradiographs, which can be viewed directly in the electron microscope for location of the radioactive species (32,111,280).

E. RELATED TECHNIQUES

1. X-ray Microscopy

Enlarged images of specimens can be formed using x-rays. Several such techniques exist, including contact microradiography (231), point projection x-ray microscopy (224,236), true x-ray microscopy using lenses (177), and x-ray diffraction microscopy (63). With the exception of the x-ray diffraction microscopy, all of these are transmission techniques. In their analytical applications they depend on absorption or diffraction phenomena. The absorption of x-rays by a specimen is well understood; therefore, the applications of these techniques to the identification of constituents in a specimen are much more advanced. These techniques are discussed in Chapters 13 and 14 of this treatise, as well as in several other publications (62,74,75,95).

2. Field Electron and Field Ion Microscopy

Although it is in principle a simple type of electron microscope, the point emission microscope has the highest resolution of any present-day electron microscope. There are two modifications of this microscope: the field electron microscope and the field ion microscope. It was invented by E. W. Muller in 1936 and subsequently developed by him to its present state (209–211). A similar apparatus was also described in 1936 by Johnson and Shockley (161), although their instrument did not give a complete image.

In this device, a sharply pointed cathode is placed opposite a fluorescent screen in a highly evacuated vessel. The screen is lightly coated with a conductor and is made the anode of a high-voltage supply. If the voltage gradient is of the order of 40 MV/cm at the cathode, electrons are emitted and travel radially from the cathode to the screen. This is accomplished with a reasonable applied voltage ($\sim$20 kV) by making the cathode radius very small ($\sim$1 μ), with the result that the field gradient near the tip is very steep and an image of the cathode tip can be observed on the screen at a high magnification. The magnification is approximately equal to the ratio of the cathode–screen distance to the radius of the tip. A typical magnification is 500,000$\times$. The theoretical resolving power of the microscope is 20 Å, and the actual resolving power approaches this value.

The image formed by a field electron microscope with a clean cathode tip shows the crystallographic faces of the metal in the tip. The intensity of electrons at an image point is a function of the field strength and the work function of the cathode material at the corresponding object point. Molecules of material adhering to the surface of the cathode also appear in the image, provided they are large enough to be resolved. Smaller molecules and atoms can be detected because of their effect on the work function of the cathode tip. This makes the field electron microscope ideally suited to studies of adsorption, surface migration, and surface reactions.

The field ion microscope (212–217) is quite similar in construction to the field electron microscope. The polarity of the electric field is reversed, making the sharply pointed tip the anode. The potential drop required is 450 MV/cm. Whereas the field electron microscope requires a very good vacuum system, the field ion microscope is operated with gas pressures of ~0.001 torr. The best gas is helium, although other gases can be used. The tip is maintained at very low temperatures, with liquid hydrogen or liquid helium used as the coolant.

The field ion microscope has a theoretical resolving power of 1.5 Å, and resolutions of 2.3 Å have been obtained visually. This is the first microscope to make the individual atoms visible. Each atom in the many facets of a hemispherical anode tip is observed, and vacancies, dislocations, and other crystal imperfections can be examined. Also, the individual atoms of impurities or adsorbed materials can be seen. The surface layers of the anode can be evaporated at will by increasing the field strength, and loss of individual atoms can be observed. At present, field ion microscopy is applicable only to the more refractory metals.

3. Ion, Neutron, and Other Particle Microscopes

Several microscopes operating on the same principle as the electron microscope but using other charged particles have been built. Ions can be focused in the same way and with the same types of lenses as those used in electron microscopy. They have the potential advantage that their de Broglie wavelengths are much shorter than those of electrons and therefore the resolving power should be higher. A number of ion microscopes using Li^+ have been built (29,73,76,110). These microscopes have not attained the resolution of the electron microscope. Protons also have been used for microscopy (61,185,186), and they should have advantages over other ions because of their small size.

Ion microscopes would, in general, be used in chemical analyses in much the same way that electron microscopes are used.

4. Reconstruction Wave-Front Microscopy (Holographic Microscopy)

Gabor (106–108) has proposed the construction of a microscope operating on a principle considerably different from that of the ordinary electron microscope. Holographic microscopy has also been called diffraction microscopy.

The limitation on the resolving power of the electron microscope is imposed by the aberrations of the objective lens and the stability of the electronic system. The information necessary for forming an image of a specimen is contained in the beam before it enters the objective lens. This information is in the form of amplitude and phase relations. Gabor suggested that the limitations set by the aberrations of electron lenses could be circumvented by recording these phases and amplitude relationships before the beam entered the objective lens. Then, by subsequently viewing this pattern using light waves and highly corrected light optics, an image of the original object could be reconstructed.

To do this, Gabor proposed illuminating a specimen with a coherent source of electrons and recording the pattern (or hologram, as it was called) on photographic film. This pattern could then be viewed with a telescope and the reconstructed image seen.

Experimental work has shown that this system will work, but there are limitations such that much developmental work will be required before the resolving power of this instrument surpasses that of the conventional electron microscope (125). The applications of this instrument to chemical analyses are obvious. It could be used in the same manner as an ordinary electron microscope, but if it could resolve the atomic structure of the specimen many analyses could be carried out directly.

III. PRACTICE OF ELECTRON MICROSCOPY

A. TYPES OF MICROSCOPES

Figure 16.20 shows a recent-model TEM. The column in the center is the electron optical system and consists of the electron source at the top, the magnetic lenses and specimen chamber in the center, and a viewing screen at the bottom. Beneath the viewing screen is a photographic camera chamber. The controls for operating the microscope are arranged on the panels on either side of the column. The lens-current supplies and the high-voltage supply are located in the cabinets below the table. The vacuum pumps are separate. Attached to the column is an energy-dispersive x-ray analyzer. Wavelength-dispersive x-ray analyzers can be attached if desired. This microscope is pumped with a turbomolecular pump and an ion pump, which gives the advantage of keeping pump-oil vapors from contaminating the microscope or the specimen.

Figure 16.21 is a sectional view of an electron microscope with a typical arrangement of lenses. Of particular interest are the means provided for alignment of the optical components. The gun can be tilted and translated. All of the individual lenses also can be translated, and the illuminating system can be tilted as a unit.

A variety of TEMs are available commercially. In the early days of electron microscopy, a number of laboratories built microscopes. However, except for laboratories interested primarily in electron optics and microscope

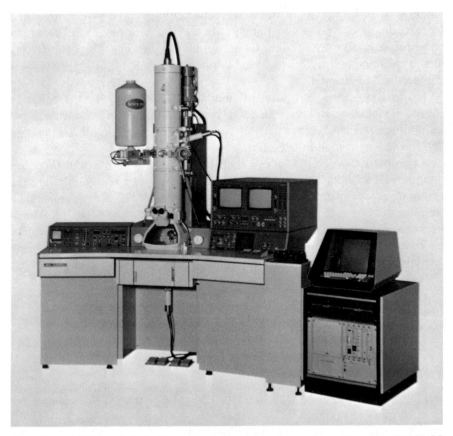

Fig. 16.20.   A TEM equipped with an energy-dispersive x-ray analyzer. (Courtesy of JEOLCO USA Inc.)

design, this is no longer practical unless some specialized instrument is de-sired.

Because there is a variety of instruments available and because they have numerous special features, it is not practical to describe each of them fully.

It is difficult to specify the features necessary or desirable for a micro-scope to be used for chemical analysis. Most of the techniques applied in chemical analysis by electron microscopy are far from routine, and as a result, it is better to have as versatile an instrument as possible. One feature, namely selected-area diffraction, is a necessity for analytical work. High accelerating voltages are very useful for penetrating thick specimens. The range of resolving powers available is rather small, and in general all are adequate for analytical work. Special tilting stages, heating stages, and cool-ing stages may prove useful.

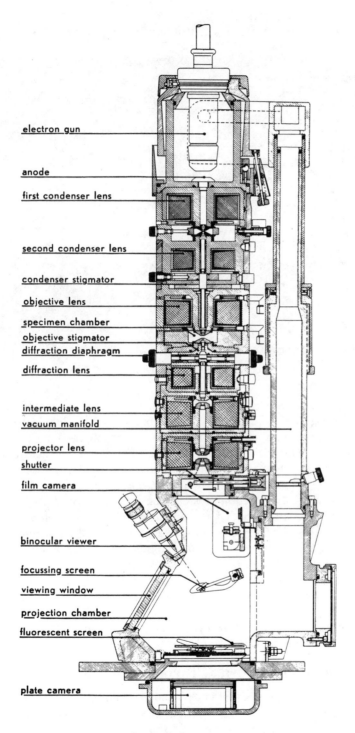

electron gun

anode

first condenser lens

second condenser lens

condenser stigmator

objective lens

specimen chamber

objective stigmator

diffraction diaphragm

diffraction lens

intermediate lens

vacuum manifold

projector lens

shutter

film camera

binocular viewer

focussing screen

viewing window

projection chamber

fluorescent screen

plate camera

Fig. 16.21. Cross section of an electron microscope column. (Courtesy of Philips Electronic Instruments.)

Fig. 16.22. A SEM equipped with an energy-dispersive x-ray analyzer.

Figure 16.22 shows a SEM. The microscope column is mounted on a separate cabinet. The electron source is at the top of the column, the lenses are in the center section, and one of the specimen chambers is at the bottom. In this particular instrument another specimen chamber is located part way up the column. Attached to the column is an x-ray analyzer. Some of the electrical system is located in the table on which the column is mounted. For this microscope the vacuum system consists of an oil diffusion pump backed by a mechanical pump. The diffusion pump is in the table supporting the column, but the mechanical pump is separate from the column.

The console, which is separate from the column table, contains most of the electrical and electronic circuits. The controls are on the panel above the table. There are three CRTs on this instrument; two are for viewing the image and the third is for photography. The camera on the right of the panel covers this third CRT.

STEMs have external arrangements that are similar to those used for SEM, although they may be more elaborate.

B. ELECTRON-MICROSCOPICAL LABORATORIES AND ACCESSORIES

The electron microscope should be placed in a room that can be darkened. Water for cooling purposes and electrical power must be available. Vibrations can interfere with the best operation of the microscope, so if it is not possible to choose a vibrationfree area, special provisions should be made to isolate the instrument as completely as possible from the source of the vibrations. It is also important that the area be free of strong and variable magnetic and electric fields.

If only one person is to use the microscope, it may be practical to have the specimen-preparation area in the same room as the microscope. However, it is generally much better to have a separate specimen-preparation laboratory. Air conditioning, specifically humidity control, is desirable, although not absolutely necessary. Many replication procedures are affected adversely by high atmospheric humidity.

Darkroom facilities for handling and developing the photographic materials are also a necessary part of an electron-microscopical laboratory. Again it is possible, but not desirable, to use space in the darkenable microscope room for this purpose. It is also convenient to have facilities for making prints and enlargements from the negative micrographs.

A vacuum evaporator for "shadowing" specimens and preparing substrate films is a necessary part of every electron-microscopical laboratory. Equipment for thinning specimens for TEM depends on the type of specimens to be analyzed by the laboratory. A microtome is necessary for preparing animal- and plant-tissue thin sections. It can also be used for other soft materials, and with a diamond knife it can cut thin sections of hard plastics, metals, and ceramics. Electropolishing and etching equipment is used for preparing thin sections of metals, and ion mills can be used for minerals, ceramics, metals, and many other materials.

It is generally true that light and electron microscopy should be combined. In any case, some light microscopes, such as stereomicroscopes, are necessary for the preparation and preliminary examination of specimens for electron microscopy.

C. PREPARATION OF SAMPLES

The results of electron-microscopical investigations and the accurate interpretation of these results depend on proper preparation of the specimen. Although there are several basic specimen-preparation techniques, it is generally true that each specimen presents its own unique problems.

Samples may be arbitrarily classified into two groups: those that are placed directly into the microscope for viewing, and those that for some reason, such as opacity, cannot be used in the microscope. The so-called replication procedures must be used for the second class of specimen.

The first group can be further subdivided into specimens that are particulate or fibrous and those that are in the form of or can be prepared as thin films. For particulate materials some form of support must be provided. This support must be mechanically strong enough to hold the specimen and it must be attachable to the stage mechanism. Small disks of copper or nickel mesh (usually 200 meshes/in.) are most commonly used to provide mechanical support and means of attachment to the stage. Some particulate materials, such as carbon black, smokes, and fine filaments of materials, can be placed directly on these specimen grids. The portion of the specimen that protrudes over the screen openings is then observable in the electron microscope. Most finely particulate materials cannot be handled in this fashion. To support these materials a thin membrane is stretched over the openings in the specimen grid and the specimen is placed on this support film.

1. Transmission Electron Microscopy

a. PREPARATION OF SUPPORT FILMS

Support films for TEM must be strong enough to hold the specimen in the electron beam, but they should be as thin as possible to avoid undue scattering and absorption. Cellulose nitrate, cellulose acetate, polyvinyl formal, silica, silicon monoxide, beryllium, and aluminum are among the materials that were commonly used in the past. Although these materials are still used, they have largely been replaced by evaporated carbon films.

Plastic support films may be made in two general ways: They can be cast on the surface of clean water, or they can be cast on a smooth solid surface such as glass or mica and subsequently floated off. When a film is to be cast on water, clean water is placed in a suitable dish (a low dish 4 or 5 in. across and 2 in. or so deep is suitable), the dish is made level, and then the dish is filled to overflowing. Any floating surface contaminants can then be swept off with a clean strip of paper or other scraper, using techniques devised for surface-tension and monomolecular-film work. A drop of a solution of the plastic (about 0.5–3% plastic) in a suitable solvent is placed at the center of the dish and allowed to spread over the surface of the water. It will help to clean the surface of the water if the first film cast is skimmed away and discarded. A suitably thin film will have gray interference colors when viewed by reflected light. For observing interference colors, the light source should be broad and uniform, and it is helpful to have the dish sitting on a black surface. The film can be scooped from the surface of the water directly onto specimen grids held with tweezers, but it is simpler and more efficient to place a number of specimen grids face downward on the gray area of the film and to pick up all of them at one time. To pick up a group of specimen grids in this way, hold a piece of paper several inches square with your fingers behind it and place it gently on top of the specimen grids. Then press the paper, grids, and film downward through the water with a sweeping motion and lift them free of the dish. If this is successfully done, the grids will be sandwiched between the film and the paper. The sandwich can then be allowed to dry and the specimen grids with their support films can be lifted from the paper with tweezers as they are needed. Glass slides or pieces of wire screening (about 80 mesh) may be substituted for the paper.

An alternative method involves placing a number of specimen grids on a wire mesh tray beneath the surface of the water before the film is cast. The tray can then carefully be lifted up so that the film covers the grids and is lifted free of the water surface, or the water can be drained or siphoned from the dish, allowing the film to settle on the grids.

If specimen-support films are to be cast on glass, the glass must be clean and free of contamination. This can be accomplished by scrubbing the slide with a solution of detergent or a slurry of a mild abrasive such as MgO. Rubber gloves are usually worn during these operations to prevent contamination of the slide with oils from the skin. After washing, the slide is carefully rinsed with distilled water and dried with a blast of warm air. New microscope slides can sometimes be cleaned adequately by polishing with lintless paper such as a lens tissue. The criteria for cleanliness are as follows: (1) The solution of plastic flows over the slide and wets it uniformly; (2) the dried plastic support film can be floated free of the slide; and (3) no structure is visible in the support film when it is examined in the electron microscope.

After the slide is clean and dry, it is coated with a solution of the plastic to be used and the solution is allowed to dry. The solution may be applied in a number of ways, such as dipping the slide and holding it in a vertical position until it drains and dries or simply flowing the solution over one surface with a pipet, draining, and drying. The thickness of the resulting support film depends on the concentration of the solution of plastic, the way in which the slide is drained, and the speed of drying. It is usually helpful to touch the corner of the draining slide to an absorptive paper tissue to aid in draining. Drying may be speeded through the use of an infrared heat lamp. A concentration of approximately 1% by weight of the plastic in a suitable solvent has been found to be appropriate, although the optimum concentration depends on the particular plastic and solvent and on the conditions of preparation. The conditions of preparation can be determined by trial and error, and standardized so that good support films can be produced consistently.

After the film has dried, the portion of the film on the front surface of the slide is separated from any film on the edges or back of the slide by scraping the edge with a razor blade. The slide is then carefully lowered into a dish of water so that the slide surface makes an angle of 45° or less with the surface of the water. If this is done properly, the plastic film will float away from the slide and remain on the surface of the water. Sometimes it is necessary to tease the edges of the film away from the slide with a needle to initiate a smooth separation. When the film is floating on the water surface it can be picked up on specimen screens as described above. It is also possible to score the film into squares of 2–3 mm on a side while the film is still on the glass slide. The squares are then floated off and scooped up separately on specimen screens.

The materials most commonly used for support films in the past were cellulose nitrate and polyvinyl formal (Formvar). Cellulose nitrate appears to work somewhat better for casting on water, whereas Formvar seems better for casting on a glass surface. These plastics are available in various grades dependent on the degree of polymerization and, in the case of cellulose nitrate, on the degree of nitration. The plastics used for support films should be clean and free from foreign matter. Cleanliness varies with the source of supply. Therefore, it is best to check the suitability of a given material by preparing blank support films and inspecting them in the microscope.

The solvents used to prepare solutions of the plastics should also be pure and dry. The presence of water in the solution causes holes to form in the support films. The solvents used for cellulose nitrate are amyl or butyl acetate, although a mixture of diethyl ether and ethyl alcohol also can be used. Formvar is usually dissolved in ethylene dichloride or dioxane. It is desirable to protect the solutions from light, which may cause decomposition.

At times it is desirable to produce support films that are full of small holes. The specimens can then span the holes and thus may be examined without interference by the support film. Such films can be prepared by introducing water into the film before it dries. If the films are cast on glass in a humid atmosphere, they will generally pick up enough moisture to cause holes to form. Breathing on the film or directing steam onto the film before it dries is a more certain way of producing holes.

Other types of support films have been used. These include other plastics prepared in the same general way; evaporated films of SiO, SiO_2, beryllium, and Be–Al alloys; and films of Al_2O_3 prepared by anodizing aluminum. However, the most popular support film in use is the evaporated carbon film. Carbon films are very strong, resistant to nearly all chemicals, and transparent to electrons. They are electrical conductors, so they are not subject to distortion and disruption due to accumulated electrostatic charges. They are also black, and therefore much more easily seen and handled than are the transparent support films. The use of carbon films was originally proposed by Bradley (35,37).

In the preparation of carbon support films, the carbon is evaporated onto a structureless substrate from which it can be removed by stripping techniques such as those described above. Figure 16.23 is a drawing of an apparatus for evaporating carbon in a vacuum evaporator. In this drawing a small piece of $\frac{1}{16}$ in. carbon rod is being evaporated. Alternatively, one or both of rods A and B can be pointed. The use of the $\frac{1}{16}$ in. rod makes it easier to control the amount of carbon evaporated by varying the length of the rod. Low voltages (5–10 V) and high currents (20–50 A) are needed to carry out this type of evaporation. It is desirable to degas the carbon by heating it below the point of evaporation for a time.

A number of different substrates have been used for the evaporation of carbon support films. Among these are microscope slides coated with thin layers of plastic and freshly cleaved mica. Carbon can be stripped directly from a microscope slide if the surface of the slide is in suitable condition. A new microscope slide can be polished with lintless paper and used directly. Slides also can be cleaned with detergent before polishing with paper, but slides that have been cleaned with acid, strong bases, or other solvents that

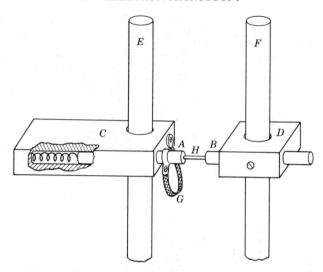

Fig. 16.23. Apparatus for the evaporation of carbon or carbon–platinum mixtures. A and B are ¼-in.-diameter carbon rods. A slides freely in block C and is forced outward by a light spring. B is prevented from sliding by a set screw. Electrical connections are made through the support rods E and F, and through the flexible connection G. H is a short piece of ¹⁄₁₆-in.-diameter spectroscopic carbon inserted in a snug hole in A. H is evaporated by passing a current through the assembly.

may attack the glass often do not release the carbon or produce a carbon film having structure.

A method of preparing substrate films of crystalline graphite or mica has been described by Fernandez-Moran (98). A crystal of graphite or mica suitably mounted is coated with a lacquer. The lacquer is allowed to dry, stripped from the surface of the crystal, and dissolved in a container of suitable solvent. Thin single-crystal flakes of graphite or mica that adhered to the lacquer are left floating in the solvent. These flakes can then be scooped from the solvent onto a specimen grid.

b. PREPARATION OF PARTICULATE MATERIALS

Specimens can be deposited on support films for microscopical examination in any of several ways, depending on the state of the sample to be examined. Dry powders can simply be placed on the support film with a small spatula or camel's hair brush; after the excess powder is shaken or blown off, the specimen can be examined. Such a method may result in mechanical separation of the components of the sample. An aerosol of the dry powder can be made with an atomizer and the particles can then be allowed to settle on the support. Aerosol particles can also be deposited on the support film by gravity, or by electrostatic or thermal precipitators (85).

If the dry particles of a powder have a tendency to agglomerate, the simple methods of preparation described above are not satisfactory. Specimens of such powders can be prepared by milling or grinding them in a liquid. The

Fig. 16.24.   Arrangement for washing specimens or dissolving plastics from composite films.

higher the viscosity of the liquid, the stronger the dispersing action. For example, the powder may be placed in a drop of a solution of plastic (such as is used to prepare support films) on a microscope slide. The mixture is then rubbed with a glass rod, another microscope slide, or a spatula. The shearing forces set up when the liquid is rubbed tear apart agglomerates and disperse the particles throughout the liquid. As the solvent evaporates, the plastic solution becomes more and more viscous, thus increasing the shear forces. After dispersion is complete, enough solvent is added to permit spreading the slurry over the glass slide. The solvent is then allowed to evaporate and the plastic film containing the particles can be floated off the slide and picked up on specimen screens for viewing in the microscope.

This method of preparation leaves the particles embedded in the plastic film, and in general this results in loss of contrast and resolution. The contrast cannot be improved by the method of shadow casting because the particles are embedded. To overcome this difficulty, various procedures have been devised; for example, the plastic film containing the particles can be coated with a different plastic dissolved in a solvent that does not affect the first plastic. The composite film can then be scored into small squares, floated off the microscope slide, and picked up on specimen grids. The plastic in which the particles are embedded can then be dissolved away, leaving the particles dispersed on the second plastic film.

The dissolution of the plastic in which the particles are embedded must be done carefully to avoid washing the particles away or causing them to reagglomerate. Washing operations often can be carried out by carefully allowing solvent to flow from a small pipet over the specimen screen. A more convenient method is illustrated in Fig. 16.24. The specimen grids are placed on small pieces of lens tissue, and these in turn are placed in a Petri dish on a support made of stainless steel screen. The solvent is poured into the dish until it wets the paper but does not flow over it. The cover of the Petri dish can then be replaced and the dissolution allowed to proceed. Dissolution will occur even when the plastic to be dissolved is above the support film. Rinsing in fresh solvent can be carried out by pipetting off the used solvent and replacing it with fresh; however, it is much easier to pick up the paper with its specimen grids and transfer it to another dish. The grids adhere very well to the wet paper.

The washing technique described above can also be used to carry out reactions with the specimen on the support film. The solvent is replaced by the desired reagent. When carrying out such a reaction, care must be taken that the specimen grid and the other supporting materials do not interfere with the reaction.

Particulate materials are often received in the form of a suspension in a liquid, or it may be convenient to suspend a supplied dry powder in a liquid. If the liquid is volatile, specimens can be prepared by placing a droplet of the suspension on a support film and allowing the liquid to evaporate. If such samples have a slight tendency to agglomerate, they can often be spread in a thin layer on a microscope slide, using a glass rod or another microscope slide. Often, when this is done, the layer is so thin that the liquid evaporates before agglomeration can occur. The process can be speeded by heating the slide. Such a preparation can then be removed from the slide by coating it with a dilute plastic solution such as is used for support films. This incorporates the particles in the film and subsequent treatment is then carried out as described above. If the microscope slide is coated with plastic before the dispersion is made, the film with the particles on its surface can be floated off, picked up on grids, and shadowed for examination in the microscope.

One of the important quantitative analytical uses of the microscope is for particle-size analysis. The most difficult and critical step in a particle-size analysis is the preparation of a specimen that is representative of the material being examined. The dispersion of the particles on the support film must be such that agglomeration is minimized and the individual particles can be counted and measured. Incorporating the sample into the film is likely to produce the most representative specimens, although such an approach is undesirable for other reasons, as indicated previously. Samples in well-stirred suspensions are usually uniformly distributed throughout the liquid, so a droplet of the suspension dried on a support film should provide a representative sample. However, it is necessary to examine the contents of the entire droplet in the microscope to be sure that no segregation occurred when the liquid evaporated. If relatively large droplets are used, the dried contents will cover more area of the support film than can be examined in the microscope. Thus, it is desirable to use small droplets for counting purposes. If the entire contents of the dried droplets are in the usable area of the specimens, there still remains some uncertainty regarding the areas obscured by the grid wires. A method of spraying a suspension with a high-pressure spray gun has been used to produce droplets fine enough that their entire contents are visible in one field at high magnification (13).

Counting and measuring all of the particles in each of a number of fine droplets will generally give good quantitative particle-size distributions. The total number of droplets that must be counted depends on the size and size range of the particles and their concentration in the droplet. If the volume of the original droplet can be determined, the number of particles and the volume of the particles per unit volume of the original suspension can be determined. Then, if the density of the particulate material is known, the weight percentage of the particulate material in the liquid can be determined. It also may be possible to determine the mass per unit volume of the suspension independently by drying a known volume and weighing the residue. From these data the density of the particulate matter can be determined. If the particles counted are single molecules, their molecular weight can be determined.

The volume of the original spray droplet can be determined through the use of an internal standard. This is the common technique (60), and it involves the addition of a known number of recognizable particles to the original suspension so that the number of these recognizable particles per unit volume of the suspension can be calculated. Then, all that is necessary to determine the volume of the original droplet is to count the number of added particles. If the internal-standard technique is used for determination of the molecular weight of visible macromolecules, it is necessary only to count the numbers of the molecules and the internal-standard particles and to determine the mass per unit volume of the original suspension by drying and weighing (253).

There are many other techniques for preparing specimens of particulate materials, which cannot be described here. However, they are generally combinations of or variations on the techniques that have been described. For specific information, see the recommended references in Section I.D. and references 85, 165, and 267.

c. PREPARATION OF FIBROUS MATERIALS

The preparation of specimens of fibrous materials is similar in many ways to the preparation of particulate materials. The fibrous materials referred to here are those of electron-microscopical dimensions, such as collagen, cellulosic fibrils, and asbestos fibers. Textile fibers are too large to be treated as particulate material and must be sectioned, replicated, or prepared in some other manner. Some textile fibers and biological materials can be dispersed into a finely fibrous or particulate state by mechanical fragmentation. Waring blenders and ultrasound are used for this purpose.

d. SHADOW-CASTING TECHNIQUES

One of the primary goals of specimen-preparation techniques is the enhancement of contrast in the image of the specimen. Often the various fea-

Fig. 16.25. Vacuum evaporation unit. The bell jar and safety cover have been lifted to show the arrangements for evaporation. The specimen holder is one designed for microscope slides. It holds 12 slides in a circle about the filament. The shadow angle can be adjusted for each slide.

tures of a specimen are indistinguishable because they affect the electron beam in almost the same way and to the same extent. For example, virus particles lying on a support film are very difficult to distinguish because the particles and the support film differ very little in their ability to scatter electrons and because the virus particles may well be thinner than the support film. If such low-contrast specimens have characteristic surface contours, these contours can be brought out in the image by a shadow-casting procedure (323–325). Shadow casting is used almost universally, and a vacuum evaporator for carrying it out is indispensable in the electron-microscopy laboratory.

There are many different designs of vacuum evaporators, but they all consist of a vacuum chamber equipped with supports for the attachment of filaments, carbon rods, specimens, and so forth. Of course, they include vacuum pumps with the necessary piping and valves, gauges for measuring the pressure in the vacuum chamber, and a power supply for heating filaments. Figure 16.25 is an overall view of a vacuum evaporator; Fig. 16.26 shows an arrangement of specimen screens and a filament for shadow casting. A schematic diagram illustrating the principle of the shadow-casting

Fig. 16.26.   View of an arrangement for shadow casting with the bell jar removed. The posts serve as mechanical supports as well as electrical leads. The filament is a shallow ''V'' with the shadow metal, in the form of wire, wrapped around its apex. The specimen holder is designed to hold six specimen screens, and can be adjusted to the desired shadow angle.

arrangement is shown in Fig. 16.27. An example of the enhancement of contrast resulting from shadow casting is shown in Fig. 16.28.

The metal atoms radiating from the molten shadow metal on the filament travel in straight lines from the filament to the specimen. The thickness of shadow metal deposited depends on the amount of metal being evaporated and on the angle of incidence of the metal on the specimen plane. Shadow metal is absent in the area behind the particle. If the shadow angle is known,

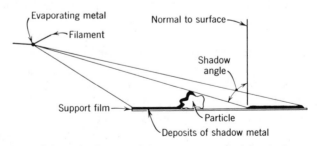

Fig. 16.27.   Schematic diagram of the arrangement used for shadow casting.

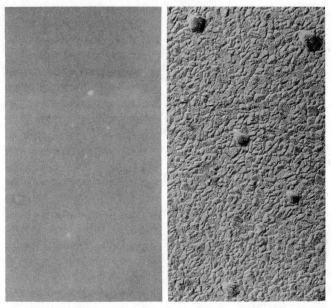

Fig. 16.28.   Replica of the surface of an aluminum film without shadows (left) and with shadows (right).

the height of the top of the particle above the surrounding support film can be calculated from the length of the shadow. If the surface of the surrounding support film is not perfectly level, the shadow lengths are not truly indicative of the height of the object. Local shadow angles can be measured if spherical particles can be added to the specimen before shadowing. Some of the quantitative aspects of the use of shadow casting to measure particle size are discussed by Backus and Williams (12) and in references 85 and 130. Shadowing not only enhances contrast, but also gives a three-dimensional appearance to the micrograph of the specimen, which is of considerable help in the interpretation of the micrograph.

A number of materials have been used for shadowing. Among these are chromium, nickel, germanium, manganese, palladium, platinum, gold, uranium, and U_3O_2. Alloys such as platinum–palladium and mixtures such as platinum–carbon have also been used. The most effective materials are those with high scattering powers for electrons. Platinum and uranium are most effective from this standpoint. Another important characteristic of a shadow material is its tendency to grow small crystallites (to granulate), rather than to remain as a uniform structureless layer. The platinum–carbon mixture has been shown to have the least granularity (40), and therefore the highest resolution is obtained in specimens shadowed with this material. Another factor to be considered in the selection of a shadowing material is ease of evaporation. Platinum is somewhat difficult to evaporate, because its melting point is high and it alloys readily with tungsten. However, small amounts of it can be evaporated successfully from a tungsten filament. Uranium also alloys slightly with tungsten. For further information on shadow casting, see references 59, 228, 252, and 321, as well as sections in the recommended books (42,85,130) and the reports of conferences (18,45,69,73,100,145,244, 266,281,289).

e. PREPARATION OF THIN FILMS AND SECTIONS

Many electron-microscopical specimens occur as, or can be prepared in the form of, thin films, which can be examined by transmission microscopy. Corrosion films or thin films resulting from oxidation or other chemical re- actions are often found on metals or other materials. In the early stages of their formation they are usually thin enough for examination by transmission microscopy. Many fields of scientific work involve the use of thin evaporated films of metals and other material, and much useful information can be ob- tained by examining these films microscopically. Films are often condensed from the vapor phase during distillation, pyrolytic, or vapor-phase reactions. Biological tissues, too, often occur in the form of thin films, and in some cases, tissues that ordinarily occur in bulky form can be induced to grow as thin tissues by means of tissue-culture techniques. A few crystalline ma- terials form layer structures that can be cleaved into thin sheets or films; examples are mica and graphite. The advantages of examination of speci- mens in the form of electron-transparent films can be exploited in the study of bulky specimens through the preparation of thin sections. Metals, bio- logical tissues, ceramic materials, and polymers such as textiles can be cut, chemically thinned, or ion milled to form thin electron-transparent layers.

Corrosion films and films resulting from the reaction of a solid material with its environment must be removed from the bulk of the solid material before examination in the microscope. This can sometimes be accomplished by mechanical stripping. A thick plastic lacquer can be applied to the film, and after drying, the lacquer can be peeled away, carrying the film with it. If the film is too adherent for this technique, the substrate is usually dissolved away, leaving the film free for mounting and examination. This method re- quires a solvent for the substrate that does not affect the film being studied. Although the search for such a solvent often requires considerable effort, a suitable solvent usually can be found.

Stripped reaction films may be strong enough to support themselves on a specimen screen in the microscope, but it is often necessary to provide additional support. It may be possible to pick them up on regular specimen screens covered with support films, but it is often more convenient to apply the support film directly to the sample. The support film with the specimen clinging to it can then be stripped mechanically or chemically for exami- nation. Mechanical stripping of such a combination can be accomplished by backing with a lacquer or pressure-sensitive tape that is not soluble in or whose solvent does not affect the support film itself. After stripping, the backing material can be removed by dissolving it away, using the apparatus and techniques described in Section III.C.1.a (see Figure 16.24). Chemical stripping by dissolution of the substrate can sometimes be carried out by applying the solvent directly on top of the support film. The film is often porous enough to permit the penetration of the solvent. However, it is usually convenient, and sometimes necessary, to score the film plus support into small squares with a needle. This allows the solvent to attack the substrate at the needle marks and to undermine the film that is being removed. Other techniques for stripping fragile reaction films are usually combinations or elaborations of those already described. For example, the film may be stripped with a thick lacquer, the support film applied to the film–substrate interface, the backing lacquer dissolved away, and the specimen shadowed for examination.

Evaporated films and films deposited in other ways are treated much the same as reaction films; however, their handling may be made easier by the possibility of forming them directly on a support film. In fact, many studies of evaporated films have been carried out by evaporation of the film in the microscope, thus making it possible to study the process of film deposition. It is also possible to observe certain solid–gas reactions, such as oxidation, in the microscope. At present, such studies are limited to the few materials that oxidize under the conditions existing in the microscope. However, a number of specimen holders have been developed that permit control of the atmosphere surrounding the specimen (287), and in the future the observation of chemical reactions undoubtedly will be extended to many systems.

In recent years the preparation of thin foils from bulk specimens by chemical dissolution has been used extensively. A specimen of a bulk material is essentially machined away by chemical or electrochemical means until a reasonable area of it becomes thin enough to transmit electrons. It is then mounted in a specimen holder and examined in the microscope. This technique is applicable to many materials, although at present it is used most extensively for metallurgical samples. The techniques used for metals are described in detail by Thomas (296) and by Kelly and Nutting (67). The methods used for metal can in many cases be adapted for thinning other materials.

Ion milling is another technique used for thinning materials (17,117,238). It can be used for thinning many materials, but it is especially useful for metals and ceramics or minerals. Chemical polishing can be used for thinning these materials, and electropolishing is especially appropriate for metals. However, chemical or electropolishing may require searching for the appropriate solvents and conditions, whereas ion milling usually works for all materials.

Bulk samples of biological tissues are usually cut into thin slices with a microtome. Microtomes especially designed for electron microscopy are used for this purpose (177,279,280). The specimen-fixation and -embedding methods also have been adapted or specifically worked out for electron microscopy. Generally the specimens are fixed and subsequently dehydrated in a series of alcohol-in-water solutions, ending with pure alcohol. The specimen is then impregnated with a resin such as butyl methacrylate, epoxy resins (e.g., Araldite) (118), or polyethylene (265). This embedded specimen is then trimmed to a small size (approximately 0.5 mm square) and mounted in the microtome for cutting. Specimens are usually collected in a trough of liquid attached to the knife. They can then be picked up on specimen grids and examined in the microscope. The section also may be picked up on a support film and the embedding resin dissolved away before examination. The details of the preparation and sectioning of tissues are quite complex and must be carried out carefully to produce satisfactory specimens. Instructions on sectioning can be found in a number of reference books (117,130,177,232,280), as well as in the recent literature of the biological sciences.

Materials other than biological tissues often can be cut into thin sections with a microtome. Plastics, textiles, paper, and metals are among the materials that have been sectioned successfully. In general, sectioning methods are more appropriate for resilient (usually noncrystalline) materials than they are for materials such as metals that can be plastically deformed. Brittle materials are difficult to section, although they may be cleaved or chipped into thin sections with some success. The parameters involved in cutting metals have been investigated by Phillips (235). It is important to mount the specimen to be sectioned in an embedding material having a hardness comparable to that of the specimen. The area of the section must also be kept quite small. Knives for microtomy are generally of steel, glass, or diamond. The diamond knives are generally the most satisfactory, if care is taken to avoid chipping or otherwise mistreating them. Glass knives (117,179) are easily made and can be discarded when they become dull. Steel knives require careful sharpening and become dull rather rapidly.

f. STAINING OF SAMPLES

Specimens are stained primarily to enhance contrast within the specimen. In general, the contrast is enhanced because some parts of the specimen are chemically different from the other parts and thus absorb and react with the stain more readily. It is obvious that this phenomenon can be used for analytical purposes; that is, the components of the sample can be identified by treating the sample with appropriate staining reagents. For example, ferritin can be conjugated with antibody globulins; these can then be used to locate the sites of antigen–antibody interactions in biological tissues with a high degree of specificity. The iron in the ferritin provides the high degree of electron scattering needed (257). Staining has been most extensively employed in the examination of biological specimens (19,113,117,276); however, it is applicable to many other systems, particularly those containing organic compounds, such as textiles, plastics, and paper.

For staining of biological materials and similar specimens, the stain must have a high electron scattering power, a quality associated with the heavy elements. However, there is another analytical method that can be considered staining that does not require reagents of high scattering power. The surface of a bulk material, for example, glass or a metallographically polished specimen, can be treated with a reagent that reacts with the constituents of the specimen to form an insoluble precipitate. After the reaction is complete, the specimen is rinsed and dried, and a surface replica is prepared. The precipitate formed may be stripped off with the replica or it may cling to the specimen, but in either case examination of the replica will reveal the distribution of the reactive component in the surface of the specimen (201).

A third method of staining is so-called negative staining. This method is used for morphological studies. The stain, which is a dense material such as phosphotungstic acid, does not penetrate the fine structure being examined, but rather surrounds it and fills any pores that may be present. Thus, the components of the specimen are outlined by a heavy, highly electron-scattering material, and are therefore outlined with high contrast. Because of its high resolution, this method of staining makes it possible to examine very fine organic structures such as viruses and bacteriophages.

g. Replication

The surfaces of bulk samples are usually examined by means of replicas. Such surface or topological studies may be directed toward understanding a naturally occurring surface or they may be used to investigate the internal structures of a specimen that has been sectioned and etched. Replicas suitable for electron-microscopical examination of surfaces are prepared by depositing a material in intimate contact with the surface, removing this deposited material from the surface, and examining it in a TEM. Satisfactory replicas must be thin, for good electron transmission, and strong, to withstand the electron beam; they must also have high resolving power. They must be capable of reproducing the surface contours of the specimen with a high degree of accuracy.

The preparation of good replicas is an art as well as a science, and considerable experience is required before good replicas can be prepared consistently. As a result many variations of replica methods have been worked out for specific applications. Only the basic classes of replicas and their preparation can be discussed here.

The simplest type of replica is the thin negative plastic replica. In the preparation of this type of replica, a dilute lacquer is poured over the surface of the specimen and the solvent is allowed to evaporate, leaving a thin film of the plastic on the surface of the specimen. If the lacquer wets the surface of the specimen an intimate contact is established and the interface of the resulting dried plastic with the specimen will reproduce the surface of the specimen with a degree of perfection that depends on the type of plastic used. The back or free surface of this type of replica tends to be quite smooth and flat, because of the effect of surface tension operating during the evaporation of the solvent. Thus, a bump or elevation on the surface of the specimen results in a thinning of the replica and a pit or depression causes a thickening of the replica. If the surface contours of the specimen are not too rough, this type of replica gives fairly good results, especially if it is shadowed to enhance the contrast. Interpretation of the replica is fairly easy because pits and bumps can be easily recognized by differences in opacity. Micrographs of such replicas tend to have poor resolution, not because of inaccurate reproduction of the specimen surface, but because the plastic is usually too thick.

A second type of replica is exemplified by the evaporated carbon replica. To prepare this type of replica, one evaporates carbon onto the surface, as described in Section III.C.1.a; the surface is thus covered with a uniformly thick layer of carbon. When stripped from the specimen, the interfacial surface of this type of replica is a negative reproduction of the surface of the specimen. However, the back of the replica, rather than being flat, is a somewhat less faithful positive reproduction of the specimen surface. These replicas have been called "double" or "oxide type" replicas. The latter name arose because one of the early techniques for producing this type of replica was through the anodic oxidation of an aluminum surface. It is somewhat more difficult to interpret the structures revealed by this type of replica, but its resolution is higher because it is thin in all areas.

Often positive replicas of either the evaporated or plastic type are prepared by replicating the surface of a thick plastic negative replica. The thick plastic replica is produced when a more concentrated lacquer is used to prepare the negative replica. Such a replication process is often advantageous, because a thick plastic negative can be removed from the original specimen surface easily and without altering the surface of the specimen, and because it makes it possible to produce any number of replicas of the same surface. Again, the positive replica is freed for examination by dissolving away the thick negative replica. In general, these second-generation replicas do not appear to have lost resolution; at least, the loss of resolution due to the two-stage process seems to be minor compared with the inherent resolving power of the replication process.

There are several ways of depositing replica materials on the surface of a specimen. Plastic replicas are often prepared by coating with lacquer as described above, but they also may be prepared by softening the surface of a sheet of plastic and pressing the sheet against the specimen surface. Softening may be accomplished by treatment with a solvent or by heating. Plastic replicas can also be prepared by polymerizing a monomer while it is in contact with the specimen surface or by allowing a molten plastic to solidify in contact with the specimen. Thick negative replicas also have been prepared by casting molten materials other than plastic and by pressing soft, ductile materials such as aluminum against the surface. Temperatures and pressures must be chosen that do not alter the specimen itself.

The second ("double") type of replica can be deposited on the specimen by vacuum evaporation, by thermal or chemical decomposition of a gaseous phase, and by oxidation of metals. In depositing vacuum-evaporated materials on a specimen, care must be taken to produce a uniform coating rather than a shadow-type coating. This can be done through the use of multiple evaporation sources or through rotation of the specimen. However, with certain materials, notably carbon, the atoms tend to migrate around the irregularities in the specimen or bounce from the surface of the vacuum chamber; as a result, a uniform coating tends to form even if a single source is used and the specimen is stationary. This effect is enhanced if the pressure in the vacuum chamber is allowed to rise above that usually used for shadowing (5×10^{-5} torr or lower).

True oxide replicas are prepared by oxidation of the metal surface to be examined. Only metals that can form thin, continuous, structureless oxide films can be replicated in this way. Aluminum is the metal most widely used for the preparation of oxide replicas, although other metals such as stainless steel can be replicated in this way. The oxide replica is usually formed by an anodization process. The use of oxide replicas is not confined to self-replication, that is, replication of the oxide-forming metal. Useful replicas can be formed by evaporating aluminum onto the surface to be replicated, stripping off the aluminum film, oxidizing it, removing the excess aluminum, and mounting the resulting free oxide film on a specimen grid for viewing in the microscope.

Replicas are removed from the surface being replicated in several ways. The simplest, in principle, is mechanical stripping. Thick negative replicas can usually be peeled from the surface of the specimen by lifting an edge or corner, grasping it with tweezers, and stripping it away. Thin replicas are usually backed with another material to give them sufficient strength to withstand the stripping action. The most common backing material is pressure-sensitive tape. The procedure is as follows: Specimen screens are placed on the replica over the areas to be examined. These screens are usually covered with a small disk of paper. The replica is then moistened with the preparer's breath, the tape is pressed on over the specimen grids, and the composite is stripped away. The grids can then be removed by picking them up with a pair of tweezers. The area of the replica above the grid will usually tear away from the surrounding replica and remain on the grid. (If the replica does not adhere to the grid, the grid can be coated with a thin layer of adhesive before it is placed on the back of the replica.) Replicas removed in this way are usually relative thick. Thinner negative replicas can be stripped by backing them with another lacquer, allowing it to dry, then stripping the composite, cutting it into squares, mounting the squares on grids, and dissolving away the backing material with a suitable solvent. Again the two plastic materials must be chosen with due regard to the criterion of mutual insolubility in the respective solvents.

Stripping methods are not usually destructive of the surface of the specimen. However, many replicas must be removed by dissolution of the specimen itself. Oxide replicas are good examples of this type. After the metal surface has been anodized, it can be scribed into squares of appropriate size and the entire surface immersed in a reagent that dissolves the metal but not the oxide. In the case of aluminum, $HgCl_2$ solution can be used. The small squares of oxide are then fished from the solution with specimen grids for viewing in the microscope. Vacuum- or vapor-deposited replicas can be removed from thin substrates in the same way.

Thin replicas often can be floated from the specimen surface by immersing the specimen slowly and at an angle into water. If the leading edge of the replica can be teased up, the water will penetrate beneath the replica and it will float off, remaining on the surface of the water, from which it can be picked up. The water may or may not cause deterioration of the specimen surface.

A type of replica that is of special interest to the analyst is the extraction replica. When a replica is removed from a multiphase system by dissolution of the specimen, some of the phases present may not be attacked by the solvent. In this case the unaffected phases may cling to the replica. An example of such a replica is shown in Fig. 16.29. Selected-area electron-diffraction patterns taken of such a region can be used to identify the phases clinging to the replica and to determine their orientations. The replica itself gives some information about the matrix phase and the distribution of the insoluble phases in the matrix. When preparing extraction replicas it is generally possible to place the solvent directly on the replica material covering the specimen. Apparently the types of replicas thus far examined are porous to the solvent. The solvents chosen must not attack the replicas if the extraction-replica technique is to be successful.

The method of mounting the replica on the specimen grid depends on the method used for preparing and stripping the replica from the sample. When stripping a thin plastic negative with pressure-sensitive tape is done as described above, the replica is in place on the specimen grid and it is necessary only to place the grid in the microscope for viewing. When a thin positive replica is made from a thick negative or when a thin negative replica is stripped by backing it with a thick plastic, a composite replica results. These composite replicas are usually cut into small pieces, which are placed on specimen grids. The thick plastic replica or backing is then dissolved away with a suitable solvent. The method of dissolving away the plastic backing is described in Section III.C.1.a (Fig. 16.24). Care must be taken to place the composite film on the specimen grid so that the replica surface is properly oriented, that is, so that the electron beam penetrates the replica from the back. The location of the thick plastic with respect to the solvent does not affect the dissolution of the plastic appreciably. Other methods of dissolving the thick plastic involve refluxing with the solvent or gently washing with solvent dispensed from a pipet or buret.

Replicas removed from the surface of the specimen by dissolution of the specimen are found floating in or on the surface of the solvent. Those that are floating can be picked up by several methods, some of which are de-

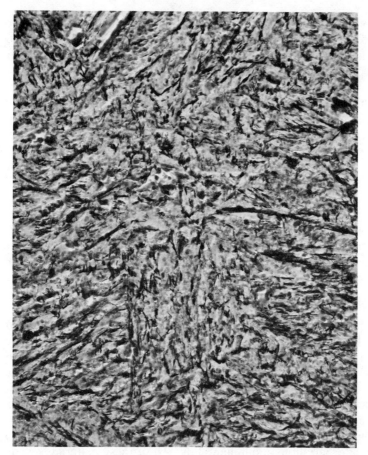

Fig. 16.29. An extraction-type replica of a martensitic steel. The carbide particles are embedded in and clinging to the replica.

scribed in Section III.C.1.a for picking up support films. Another method involves picking up the replica on a lens of the solvent in a small wire ring or a loop. This loop can then be lowered over a specimen grid mounted on a post so that the replica is positioned on the grid. The grid is picked up and dried by touching it to an absorbent paper; it is then ready for examination in the microscope. When picking up a replica floating in a liquid, it is often very difficult to see the replica, particularly if it is made of a transparent material. Oblique lighting and a dark background increase the visibility of the replica. Carbon replicas are more easily seen, because they are fairly dark in the thickness used for replicas. (When one is working with carbon replicas, the background should be white.)

Frequently it is desirable to prepare and mount replicas of the specimen so that a chosen area is available for viewing in the microscope. This involves choosing a field, usually by light microscopy, replicating it, and mounting the replica so that the chosen area is centered on the grid and is not obscured by grid wires. Many techniques for doing this have been described in the literature (31,36,51,132,152,222). Some of these methods have been summarized by Bradley (43).

Microscope images of replicas are usually low in contrast. Shadowing of the replicas is amost invariably used to improve contrast. The shadowing materials may be deposited directly on the replica after it has been mounted on the grid. However, it is frequently convenient to shadow the specimen itself or the thick negative replica before the final replica is deposited. Shadowing the specimen directly usually is preferable, as it does not introduce any unnecessary steps that could result in loss of resolution. However, it may be difficult to separate the shadow from the sample.

The resolution of the replica is often limited by the granularity of the shadow metal. At present, replicas formed by the simultaneous evaporation of carbon and platinum to form a shadowed replica have demonstrated the highest resolving power (39). The methods of shadow casting and replication have been reviewed by Bradley (41).

h. Materials for Support Films and Replicas

Support films or replicas that are examined in the microscope must be thin so that they do not scatter electrons excessively, but they must be strong enough to support themselves over the open areas of specimen grids. They also must be relatively free of structure. Materials used for support films should be highly transparent to the electron beam, and replica materials must be able to conform to the surface being replicated. In general, the same materials are used for support films and final replicas, the most common of these being cellulose nitrate, polyvinyl formal (Formvar), carbon, SiO, SiO_2, Al_2O_3, and beryllium–aluminum alloy. All of these materials except Al_2O_3 and the plastics are deposited by vacuum evaporation. The evaporated materials need not be especially pure as long as the impurities present do not cause undue mechanical trouble, such as sputtering or decrepitation of the material during evaporation. Impurities may, of course, cause troubles not recognized at present, such as excessive crystallization, structure in the film, or loss of strength of the film.

The plastics deposited from solution are more directly sensitive to impurities. Some materials may be deliberately added as plasticizers for the plastics, but materials that are insoluble or that separate from the plastic solution during drying must be absent if good replicas or support films are to be obtained. Therefore, only the purest possible plastics should be used and the solvents should be freshly distilled and thoroughly dried. Formvar may be dissolved in ethylene dichloride or dioxane. Cellulose nitrate is usually dissolved in amyl or ethyl acetate.

Materials for thick negative replicas or for backing thin replicas for stripping must have some additional characteristics. The solvent and the plastic itself must not react with the final replica. This criterion is not difficult to meet when the secondary replica is made of carbon, SiO_2, SiO, or a similar evaporated material. However, when the final replica is a plastic, only a few materials are available for the thick negative or backing film. Cellulose nitrate in amyl acetate works quite well with Formvar. Watersoluble plastics such as polyvinyl alcohol can be used with either Formvar or cellulose nitrate.

Materials for thick negative replicas must conform to the surface to be replicated; that is, they must flow into cracks and corners of the sample and they must have a minimum of molecular surface structure on the replica interface. They must also be easily removable from the sample. They should be flexible and have good "memory" so that re-entrant angles and openings are replicated faithfully even if the replica was deformed in the act of being stripped from the sample. Thick negative replicas or backing films should not swell excessively when they are being dissolved or the final replica may be cracked and torn.

Thick negatives or backing films may be formed by applying a thick lacquer and allowing it to dry. Sheets or tapes also may be used by moistening one side with a solvent, pressing the sheet onto the surface, and allowing it to dry. Sheet materials are particularly suitable for the replication of large areas for the purpose of storing information on surface topography for future study. For example, a large sheet replica may be made of a corrosion test specimen before corrosion starts. Then after corrosion has started, the structures of the surface where corrosion was initiated may be examined on the replica to determine the cause of the initial attack. Cellulose acetate or cellulose nitrate moistened with suitable solvents can be used for large-area replicas.

At times it may be desirable to produce a rigid replica of a surface. This can be done in a number of ways. If a thick slab of plastic, instead of a flexible sheet, is moistened, pressed against the sample surface, and allowed to dry, a rigid replica will be formed. Such replicas also can be formed by melting the replica material and casting it against the sample surface with or without pressure. Such replicas are usually difficult to separate from the sample by mechanical parting. Portions of the plastic that enter re-entrant openings or flow beneath overhanging portions of the sample surface may be broken off when the replica separates. Separation of such a replica often requires the use of a parting layer or dissolution of the sample.

2. Scanning Transmission Electron Microscopy

Specimens for the STEM must be quite similar to those intended for transmission microscopy. Therefore most of the methods of specimen preparation described above are applicable.

Specimens more than 0.1 μ thick generally do not allow many electrons to pass through them without losing energy. The electrons that have suffered energy losses do not form good, clear images in a TEM because of the chromatic aberration of the objective lens. A STEM does not have an objective lens after the specimen; therefore there is no chromatic aberration and thicker samples (up to several micrometers) can be used.

3.   Scanning Electron Microscopy

The SEM forms images of the surfaces of samples. Thus there is no need to make the very thin specimens required by transmission microscopes. This makes specimen preparation very easy. The surfaces of quite large specimens (up to 100 mm in diameter and thickness) can be examined in many microscopes. Samples that are smaller, such as thin sections on regular microscope slides, can be accommodated easily in the specimen chamber. Still smaller specimens—powders, small pieces, and the like—can be mounted on specimen stubs, that is, small metal holders.

Surfaces to be examined can be prepared in many ways: by simple fracturing, by polishing of a cut surface, by grinding the material to a powder and mounting the powder, and so on.

In most cases specimens should be conductive, so that the electrons sticking to their surfaces will be conducted away, preventing charging. Specimen charging in general interferes with image formation, because the electron probe beam is repelled. With metals this is usually not a problem unless a dielectric coating covers the metal. Dielectric specimens can be treated in three ways to prevent or ameliorate the effects of charging. First, the specimen can be coated with a conducting layer, usually a metal or carbon. Coating can be done by vacuum evaporation or by sputtering. The entire surface should be coated if possible. Carbon tends to migrate over the surface as it is evaporated, so it has some advantages in the coating of chunks, particles, or specimens with re-entrant angles. Rotating the specimen in as many azimuths as possible during vapor deposition also is helpful. If only some surfaces can be coated, it is usually possible to clamp the specimen so that an electrical contact is made or to paint a conductor onto the specimen using colloidal silver paint or graphite. Specimens are often cemented to conductive surfaces with these paints.

The second way to minimize charging effects is to use a lower accelerating voltage. The third way is to minimize exposure to the electron beam by using a long-persistence screen in the CRT and permitting only a single frame to impinge on the screen; thus the beam spot passes only once over any point on the sample.

D.   OPERATION OF MICROSCOPES

There are many different microscopes in operation today, and new models appear on the market regularly. It is not possible to describe in detail the techniques of operating all these instruments. Therefore, this section will be general in nature. Information on the operation of specific instruments can be obtained from the manufacturer of the instrument or through training courses and general textbooks.

1. Transmission Electron Microscopes

a. ALIGNMENT OF THE MICROSCOPE

There are two reasons why an electron microscope must be properly aligned to give the best image. As in light optical systems, image aberrations are at a minimum on or near the optic axis of the system. Thus, even with perfect stability of lens current and accelerating potential, the best image is obtained when the optic axes of all lenses are coincident and pass through the electron source and the center of the viewing screen or photographic film. Quite often, off-axis aberrations are less important than the blurring of the image that results from fluctuations in lens currents and accelerating potential. The motion of a focused image point that results from electrical instabilities increases rapidly with the distance of the image point from the optic axis. In well-designed microscopes of current manufacture, all other aberrations are made less than the spherical aberration, the factor that limits the resolving power of present-day microscopes. The electrical stability can also be held to such a value that it does not limit resolving power. Therefore, to obtain images with resolutions approaching the resolving power of the instrument, it is necessary that the microscope be properly aligned.

The conditions for proper alignment of a three-stage magnetic TEM with a single condenser are given by Haine and Cosslett (125) as follows:

1. The diverging beam from the electron gun must be centered on the condenser-lens aperture.
2. The focused illumination spot must be centered on the objective-lens axis.
3. The illumination must be parallel to the objective-lens axis.
4. The axis of the second projector lens must intersect the viewing screen near its center.
5. The axis of the second projector lens must be aligned with the optical center of the second projector lens so that the center of rotation of the image, for varying objective current, is in the center of the screen.
6. The first projector (or intermediate lens) must be aligned symmetrically with the objective-lens axis.
7. The objective aperture must be aligned centrally around the objective axis.

Strictly, conditions 5 and 6 should specify that the axes of the first and second projector lenses should be coincident with the objective axis, but practically the conditions as specified are sufficient to give the best performance.

The criteria used to adjust the microscope to satisfy these conditions are the rotation and sway, or translation of image points, when the current in the magnetic lens is changed. If an object point lies on the optic axis of a lens, the image point also will lie on the axis, regardless of the strength of the lens. Thus, the location of the optic axis can be determined on the basis of the spiraling of the off-axis image points about the optic axis as the lens current is varied. Also, if the beam of electrons that forms the image is not symmetrical about the optic axis and if the object and the viewing planes are fixed, the image will sway as it goes in and out of focus. When a lens is properly aligned, the image expands and contracts with a spiral motion about its center as the current in the lens is varied.

Usually one begins alignment by placing all of the lenses and the electron source in their mechanical centers; the instrument is set for low magnification. The condenser-lens current is set for overfocus or underfocus, so that the illuminated area at the specimen is large. A specimen is put into position and its image is focused on the viewing screen. The specimen can then be removed so that it will not interfere with the observation of the centering criteria. The condenser lens is then focused, while the screen is kept illuminated by translating the electron source, if necessary. When the image of the source is in focus, it should be centered in the field. The image of the source should appear symmetrical if the beam from the gun is symmetrical about the lens axis. In many instruments the electron source can be tilted, or the filament and grid cap can be translated with respect to the anode so that the beam is symmetrical and the image of the filament also appears to be symmetrical. When properly aligned, the image of the source should expand and contract symmetrically about its center as the condenser-lens current is raised or lowered. In instruments having centerable condenser apertures, the aperture must be centered before this criterion can be met. The operation described above satisfies alignment condition 1.

Alignment conditions 2 and 3 are somewhat interdependent and can be satisfied at the same time. A specimen is placed in the microscope and focused. Some recognizable feature of the image is centered in the field of view by moving the stage. The objective lens is then defocused, and unless the illuminating beam is parallel to the objective-lens axis, the noted feature will move away from the center of the image. If one then restores this feature to the center by tilting the illuminating system, the illuminating beam should be parallel to the objective-lens axis. It may then be necessary to recenter the image of the filament on the objective-lens axis by translating the illuminating system.

Haine and Cosslett (125) describe another method of satisfying conditions 2 and 3. An image feature is again selected and carefully focused. If the current in the objective is then reversed, the focus will not change but image rotation will occur. The axis of rotation, that is, the optic axis of the objective, intersects the image plane at some point on the perpendicular bisector of the line connecting the positions of the selected feature before and after lens-current reversal. This perpendicular bisector is visualized and the image feature is moved to some point on it. The current is reversed again and the image feature will again move unless it was accidentally placed on the center of rotation. If the image feature moves, the perpendicular bisector of the two positions is again imagined and the intersection of the first and second perpendicular bisectors is located. This is the center of rotation and the intersection of the optic axis of the objective with the image plane. This point is then centered in the image field. Now, if a feature in the center of the image sways when the objective current is changed, the illuminating beam is not parallel to the optic axis of the objective. The illumination is then made parallel by tilting the illuminating system until the swaying no longer appears.

Alignment condition 4 is usually set by the manufacturer of the instrument, and no means for adjusting it is provided.

Condition 5 is met by translating the objective lens and illuminating system as a unit until the image remains centered while the projector-lens current is varied. In some instruments there is no provision for varying projector-lens current. Instead, the magnification is changed by varying the current in the first projector (intermediate) lens.

Condition 6 is met by either translating the first projector (intermediate) lens or by tilting the upper part of the column about the center of the second projector lens until no image motion is observed when the current in the first projector is varied.

Usually it is necessary or desirable to repeat the alignment process at least twice to attain better alignment. With the above procedures, the instrument is aligned with respect to its magnetic properties. It is convenient, in practice, to align in this way because the image remains centered when the instrument is focused or the magnifcation is changed. However, because of stray magnetic fields, image sway may be present if the accelerating voltage is varied. For the best resolution, it is usually desirable to align the instrument with respect to voltage. This is done in a similar manner, except that the criterion for alignment is that the image expands and contracts symmetrically when the accelerating voltage is varied.

The above alignment steps are carried out without an objective aperture. If an objective aperture is to be used, it can be inserted and centered about the beam using the following method. Adjust the intermediate lens to the diffraction position. A specimen should be in position and it should be thick enough to give appreciable electron scatter. The illuminating beam is defocused to give parallel illumination. Under these conditions it is possible to observe a bright spot with an image of the objective aperture around it. The objective aperture can then be centered about the bright spot.

The alignment of a double-condenser system can be carried out in much the same manner as the alignment of the first projector lens described above. When properly aligned, the illumination remains centered on the objective-lens axis whether the first condenser is energized or not. In some instruments there is no provision for the separate adjustment of the first condenser lens. In most instruments equipped with a double condenser there is also an astigmatism corrector. This stigmator can be adjusted by simply making the focused image of the source circular; however, it is more convenient to remove the condenser-lens aperture and observe the caustic image of the filament. When properly adjusted, the caustic takes the form of a three-pointed star.

More complete information on the principles and practice of alignment can be found in references 48, 117, 125, 130, and 165.

b. Checking of Microscope Stability, Cleanliness, Astigmatism, and Resolution

The resolving power of a properly aligned TEM is limited by one of two general classes of factors. The highest-quality microscopes are at present limited by the spherical aberration of the objective lens. Most instruments are apparently limited by the stability of the electrical and electronic components even when all of these components are operating at their rated performances. This class of factors is inherent in the design of the microscope. The other class of factors, the members of which are not inherent in the design of the microscope but may limit resolving power, includes mechanical vibrations, drift of the specimen caused by mechanical or thermal motions, stray magnetic or electrical fields, the accumulation of static charges on surfaces near the electron beam, and defective electronic or electrical components.

It is desirable to have some method of testing the microscope to establish its resolving power when it is in peak operating condition and to detect and distinguish among the defects that may arise during operation. One obvious method of testing the microscope is to prepare a micrograph of some reliably reproducible test specimen. Since resolution is generally defined in terms of the separation of object points, it is desirable to have a test object with structures separated by distances comparable to the expected resolving power. These structures must have adequate contrast and ideally should have identical shapes. To fulfill these requirements would require structures of atomic dimensions; thus far, no completely satisfactory test specimen has been found. One of the most successful test objects consists of a dispersion of heavy-metal particles on a thin carbon support film (Fig. 16.30). These specimens are usually prepared by evaporation of metal onto the carbon support, followed by heating in the electron beam to recrystallize the metal into small granules. The resolution is obtained by measuring the minimum separation of particles that can be observed in the micrograph and dividing this value by the magnification of the micrograph. Usually several micrographs of the same area are taken to make certain that the particular particles measured are not accidental clumps of grains in the photographic film. This test measures the resolution of the microscope for the particular sample and operating conditions. The resolving power of the microscope therefore must

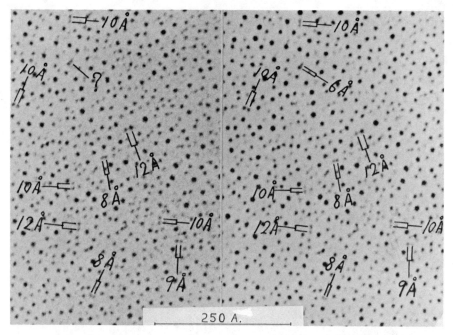

Fig. 16.30. Electron micrograph of heavy-metal particles on a carbon support film. This type of specimen can be used to test resolution. This particular micrograph shows a resolution of 6 A. (Courtesy of JEOLCO USA, Inc.)

be at least as good as the resolution shown in the micrograph; it may be better.

Another test specimen is the lattice of a single crystal of a suitable material. Copper phthalocyanine is commonly used. This crystal essentially consists of planes of heavy copper atoms separated by lighter atoms, namely hydrogen, carbon, and nitrogen. Figure 16.31 shows a copper phthalocyanine crystal with the (001) spacing resolved. The separation of the planes of copper atoms is 12.6 Å. The (100) spacing of potassium chloroplatinate (6.99 Å) also has been resolved, as has the (220) spacing of gold, 1.44 Å. The resolution of such crystal lattices is particularly interesting because it can be shown that in accordance with Abbe's theory of resolution, the angular aperture of the objective must be large enough that at least two diffracted beams contribute to image formation.

Both of the resolution tests described above are dependent on the properties of the specimen as well as on the resolving power of the microscope. A method of testing the resolving power of the microscope independent of the specimen has been developed by Haine and Mulvey (69,73,125). A description of the method can also be found in Kay's book (165). The specimen used is a support film having tiny circular holes. An overfocused image of a hole will exhibit Fresnel fringes around its perimeter, as shown in Fig.

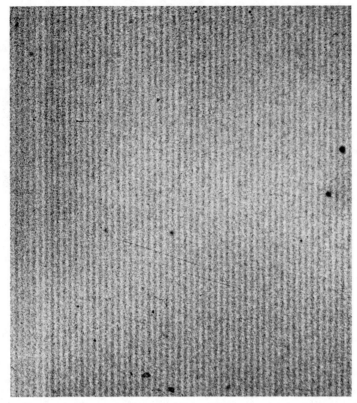

Fig. 16.31.   Copper phthalocyanine crystal with the (001) spacing resolved. The spacing is 12.6
Å.

16.32. The width of the fringe (measured from the center of the bright ring
to the center of the dark ring) is a direct measure of resolution. The resolution
d is calculated according to the formula

$$d = 1.2ym\delta$$

where ym is the mean fringe width and δ is the asymmetry of the fringe
width (125). Since it is difficult to detect asymmetry of less than 10% of the
mean fringe width, the test becomes less sensitive as the mean fringe width
increases, and finally the asymmetry is completely masked. Thus the fringe
test is most sensitive near focus, where the fringe width is small and the
asymmetry is most clearly shown.

The asymmetry of the Fresnel fringe results from the astigmatism of the
lens. Therefore, the asymmetry of the fringes can be used to measure astigmatism and to serve as a criterion for correcting it. However, the fringes
are affected by instrumental defects other than astigmatism. For example,
mechanical vibrations and specimen drift cause blurring of the fringes in the

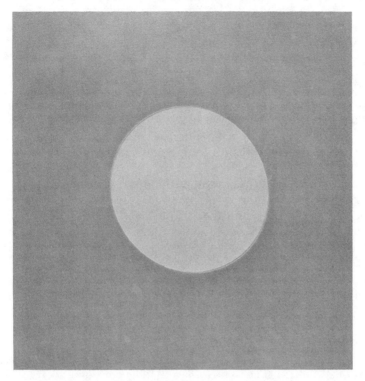

Fig. 16.32.   Micrograph of a hole in a support film. The microscope is slightly overfocused and a Fresnel fringe is visible at the periphery of the hole. Astigmatism is indicated by the asymmetry of the fringe.

direction of motion. Fluctuations in lens current cause rotation of the image, and variations of accelerating potential cause expansion and contraction of the image about the voltage center. Both of these defects can be detected by their effect on the fringes. Too large an electron source will also blur the fringes because of the loss of coherence of the electron beam.

The Fresnel-fringe test can be used to detect any of the factors that limit the resolving power of the microscope, providing the fringe width is made comparable to the desired resolving power. No attempt to correct astigmatism by using the criterion of fringe asymmetry is practical unless the condition of the microscope is adequate for the formation of sharp fringes.

c. COMPENSATION OF THE OBJECTIVE

A stigmator for the correction of objective-lens astigmatism is one of the adjustments provided on most present-day microscopes. The stigmator is an apparatus that makes it possible to alter the magnetic field of the lens. With it the orientation and degree of alteration of the lens field can be varied at will. The stigmator may be an electrostatic or electromagnetic lens, or it may consist simply of pieces of ferromagnetic materials that can be positioned as desired. The stigmator compensates for asymmetry of the field in the lens; thus, it can be used to eliminate or minimize any astigmatism that happens to exist in the objective.

For microscopes with externally adjustable stigmators, compensation of the objective is relatively simple. An image of a circular or spherical object is located and focused in the microscope. Carbon black particles have been used, but a better test object is a hole in a support film. The image is observed through a viewing microscope or telescope and the focus is adjusted so that Fresnel fringes appear. If the fringes are not of uniform width and sharpness around the circumference of the test object, the strength and orientation of the stigmator are adjusted. During the final stages of compensation the fringe is photographed, because it is difficult to observe visually. In older microscopes, compensation controls consist of screws in the objective pole piece spacer. These are inside the vacuum system of the microscope. With this arrangement compensation is a tedious process, since the adjustments can be made only by removing the pole piece. Usually it is made still more difficult by the low magnification of the microscope, necessitating the use of photomicrographs to check each change in the compensating screws.

d. ACCELERATING POTENTIAL

Commercially available electron microscopes commonly have accelerating potentials ranging from 25 to 120 kV. In selecting the voltage suitable for a given specimen, several factors need to be considered. Over this range of voltage there is very little change in the resolving power of the microscope. The principal advantage of high accelerating potentials is the resulting increase in the penetrating power of the electrons. Thus, the higher accelerating voltages permit the use of thicker specimens, and since it is often difficult to produce thin specimens of a given sample this is of considerable advantage. At the same time, specimen contrast is decreased. The inherent contrast in many specimens, particularly organic materials, is low. If staining, shadowing, or other methods of enhancing the contrast of the specimen are applicable, it may be possible to use high voltages and still retain sufficient contrast. For any given specimen the selection of the accelerating potential will be a compromise between penetrating power and image contrast.

High-voltage microscopes having acceleration potentials of 100 kV to several megavolts are now available, but these are special instruments.

e. SELECTION OF LENS APERTURES

The angular aperture of an ideal objective lens limits the resolving power of the microscope. Practically, the maximum useful angular aperture for modern objective lenses is a compromise between angular aperture and spherical aberration, as discussed in Sections II.B.2 and II.B.3. In the absence of a physical stop or aperture, electrons that are scattered beyond this maximum by the specimen or are directed beyond it by the illuminating system are not focused at their conjugate image point. Instead, those that arrive in the image area contribute to a general background and thus reduce image contrast. The physical objective aperture, a metal disk with a hole of the appropriate size, is mounted in one of the positions shown in Fig. 16.9. The upper position is actually between the pole pieces of the objective lens and is at the center of the magnetic field of the lens. The lower position is in the back focal plane of the objective lens. The exact location of the back focal plane varies with the design of the lens and with the current flowing through it. However, for a given design operated under normal conditions, the change in position with lens current is not great and a satisfactory fixed location along the optic axis can be chosen. In most commercial microscopes, the location of the objective aperture along the optic axis is not under the control of the operator. The operator must, however, center the aperture about the optic axis, and mechanical means for carrying out this task are provided in all microscopes.

Objective apertures placed in the center of the magnetic field limit the area of the specimen visible at low magnification, whereas a back-focal-plane location does not impose such a limit. Also, an aperture in the field center is subject to a more rapid contamination from bombardment by the electron beam. The approximate size of the hole in the aperture can be calculated by multiplying the focal length of the lens by the angular aperture desired. The optimum angular aperture for a typical high-resolution lens is of the order of 10^{-2} radians. Thus, for a focal length of 4 mm, the size of the aperture would be 40 μ. For specimens of low contrast, apertures smaller than the optimum may be used, but the resolving power of the lens is reduced accordingly.

Condenser apertures are usually placed in the magnetic field of the condenser lens. For double condensers, they are located in the second condenser lens. The size of the aperture opening can be approximated by using the specimen-to-aperture distance in the formula for calculating objective-aperture size. In general, the angular aperture of the condenser should not exceed the optimum aperture of the objective lens. The angular aperture of the illuminating beam can be reduced by defocusing the condenser lens.

Contamination of the objective aperture, either from deposition of carbon during electron bombardment or from improper handling, usually results in serious distortions of the electron beam. The contaminants are usually dielectrics, and when bombarded with the electron beam, they accumulate a charge. The resulting electrostatic field then deflects the image-forming electrons as they pass through the aperture. Therefore, clean apertures are necessary for good image formation.

In the past, platinum apertures were in general use. However, this material is difficult to clean and has been replaced to some extent by molybdenum, tantalum, or copper apertures. Platinum apertures may be cleaned by several methods, including immersion in fused potassium hydroxide, immersion in hot sulfuric acid–chromic acid mixtures, or immersion in fused potassium sulfate followed by boiling concentrated sulfuric acid. Molybdenum apertures can be cleaned by heating them to 1200°C on a molybdenum heater in a vacuum. Copper apertures can be discarded when they become contaminated. New copper apertures can be made by perforating a sheet of the metal with a sharp needle and subsequently etching the hole to the proper size using 10% nitric acid.

Most microscopes have apertures in addition to the condenser and objective apertures. These apertures are placed in strategic locations in the column to prevent scattered electrons from reaching the image plane. Unless they are highly contaminated, they do not affect the resolving power of the microscope. Their prime purpose is to maintain image contrast.

f. CALIBRATION OF MAGNIFICATION

The magnification used for quantitative electron microscopy depends on the specimen detail that must be resolved. Ordinarily, in quantitative work, electron micrographs are taken; it is advantageous to take the micrographs at as low a magnification as possible and to enlarge the negatives optically. This procedure maximizes the area that can be seen in the micrograph and thus makes it somewhat easier to obtain representative sampling of the specimen. Therefore, care must be taken to ensure that the detail to be resolved is larger than the resolution limit of the photographic emulsion.

The magnification of the electron microscope usually is measured by taking micrographs of a replica of a ruled diffraction grating. From measurements taken at the different accelerating voltages and lens-current settings, calibration curves can be constructed. A grating having 15,000 lines/in. is convenient for this purpose. At low magnification a replica of such a grating is quite satisfactory as a length standard. At higher magnification the lines are so widely separated in the image that they cannot be used. At these magnifications fine structures between the grating lines can be used, the distance between these structures being determined by comparison with the grating spacing as observed at lower magnification.

Other standards of length can be used. Any suitable specimen can be measured in the light microscope by comparing it with a stage micrometer. This specimen can then be transferred to the electron microscope and used as a standard. Another length standard, much used in the past, was a particular batch of polystyrene latex that was found to be very uniform in particle size (11). The average diameter of these particles was determined many times (112) and found to be consistently 2590 ± 25 Å.

Crystal lattices are excellent standards for the calibration of the microscope at high magnifications. Figure 16.31 shows the (001) lattice spacing of copper phthalocyanine. This spacing is accurately known from x-ray and electron diffraction determinations to be 12.6 Å. Thus this and other crystals can be used for length calibration in the microscope.

Calibration charts can be incorrect for various reasons. Hysteresis in the lenses can cause appreciable error unless care is taken to standardize the magnetic history of the lens. This can be done by turning the lens current off and on several times or by going through a definite procedure in varying the lens current so that the focal point is always approached from the same direction. Errors in calibration also can occur as a result of changes in the electrical circuits, and therefore calibration should be checked frequently. Errors in the original calibration can occur if the replica of the grating becomes distorted during preparation or viewing. This effect can be checked by using the replica as a grating in a spectroscope and calculating the grating spacing from the resulting data. Even if the average is determined spectroscopically, it must be remembered that the spacing of the individual lines can vary. Finally, there can be errors in the measurements of the micrographs used in calibration. In general, calibration charts are not more accurate than $\pm 15\%$. In quantitative work a calibration should be made at the time the micrograph is taken. With care, magnifications accurate to better than $\pm 2\%$ can be obtained.

g. PHOTOGRAPHIC TECHNIQUES

A brief discussion of the theory of photography in electron microscopy is given in Section II.A.2.c. In the practical operation of the microscope, several additional points need to be considered. Since the range of sensitivity of emulsions to electrons is rather small, the choice of emulsions to be used depends more on resolving power and on the ease of handling, developing, and storing the negatives.

Plates are less subject to shrinkage than is film and are therefore superior to film for quantitative work involving measurements. Also, they are usually more easily degassed by pumping in the vacuum system. They are, however, fragile and bulky to store. Whether plates or films are used, it is worthwhile to provide a means for drying them in a vacuum desiccator before they are put into the microscope.

Exposures can be determined in various ways. It is quite common to determine exposure time by trial and error or to estimate it from previous experience. There are a number of exposure meters described in the literature, and some microscopes are equipped with exposure meters by the manufacturer. When large numbers of exposures are to be made and developed in a batch, it is particularly necessary for all exposures to be correct.

Because it is usually advisable to take electron micrographs at low initial magnification and subsequently to enlarge them optically, it is necessary that care be taken to use the proper developers and to use good darkroom techniques.

h. MODES OF OPERATION OF THE MICROSCOPE

(1) Bright-Field Transmission Microscopy

Bright-field transmission microscopy is the most common mode of operation of the electron microscope. In most of the discussion thus far, this is the method of operation that has been tacitly assumed. There are, however, a number of other methods of operation. These are best illustrated by ray diagrams, such as are shown in Fig. 16.33. In this figure there are two ray diagrams for bright-field transmission microscopy. These are designated "Low magnification" and "High magnification." High-magnification electron microscopy may be either two-stage or three-stage magnification. (Some instruments have still another stage of magnification.) In a given instrument these modes of operation differ only in the excitation of the objective, intermediate, and projector lenses. In bright-field microscopy, the field of view is uniformly illuminated with electrons if no specimen is present. The introduction of the specimen modifies this situation by scattering electrons out of the image. Thus, the image is dark against a bright field.

(2) Dark-Field Transmission Microscopy

Dark-field transmission microscopy is obtained when the electron beam is tilted or modified by stops so that only electrons scattered by the specimen contribute to the formation of the image. There are four methods of attaining this effect (130). In the first method shown in Fig. 16.33, an annular aperture is placed in the condenser, producing a hollow cone of illuminating electrons. An objective aperture of the proper size will then prevent any undeviated electrons from contributing to the image. Under these conditions only scattered electrons pass through the objective aperture to form an image. To attain this type of dark-field illumination is difficult. The second method involves tilting the illuminating system about the specimen so that the unscattered electrons are again stopped by the objective aperture (230). Some microscopes provide the proper tilting mechanism for using this type of illuminating system. The deflected beam method is effectively the same as the tilted beam method but instead of mechanically tilting the column the beam is deflected electromagnetically. The easiest method of attaining dark-field illumination is through the displacement of the objective aperture. However, this method does not produce a good image, because the imaging electrons are those that pass through the edges of the useful lens field and are adversely affected by the aberrations of the lens. A modification (294,295) of this last method makes use of an annular aperture that can be moved along the optic axis of the microscope in the space between the objective and its back focal plane. This permits the selection of various hollow cones of electrons having different angular apertures, and it permits the formation of dark-field images of objects scattering at the selected aperture angles.

Theoretically, the resolving power of the microscope in dark-field operation should be better than that for the bright field. The annular condenser aperture and the tilted illuminating systems should be superior to the displaced objective aperture, since they form the image with the paraxial electrons. However, even with these systems, inelastically scattered electrons are adversely affected by the chromatic aberrations of the lenses.

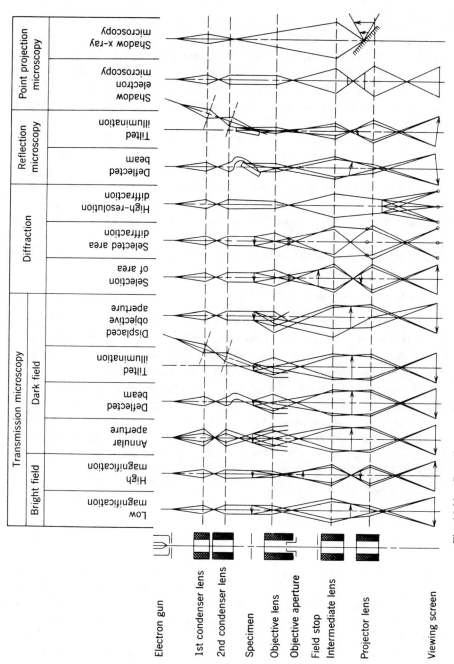

Fig. 16.33. Ray diagrams for various modes of operation of the electron microscope.

Transmission microscopy								Diffraction			Reflection microscopy		Point projection microscopy	

Bright field — Low magnification, High magnification

Dark field — Annular aperture, Deflected beam, Tilted illumination, Displaced objective aperture

Diffraction — Selection of area, Selected area diffraction, High-resolution diffraction

Reflection microscopy — Deflected beam, Tilted illumination

Point projection microscopy — Shadow electron microscopy, Shadow x-ray microscopy

Electron gun

1st condenser lens

2nd condenser lens

Specimen

Objective lens

Objective aperture

Field stop

Intermediate lens

Projector lens

Viewing screen

599

Dark-field images can be formed using either elastically or inelastically scattered electrons. With amorphous materials the inelastically scattered electrons contribute appreciably to the dark-field image. Elastically scattered electrons from crystalline materials can be selected for dark-field imaging by arranging the apertures and/or tilting the illuminating system so that selected diffraction spots or rings form the image. Images formed in this manner appear as a dark field with all of the crystals contributing to the selected diffraction spot or ring, which is bright against the general dark background (62,130,230,294).

Dark-field microscopy using diffracted electrons at selected Bragg angles (also called selected diffraction) is potentially a powerful analytical technique. The distribution of crystalline materials in a matrix or a mixture of particles can be determined. However, it should be recognized that only those crystals that are oriented to give Bragg reflection that contributes to the image will be shown. Interpretation of the results of this type of examination must be made with care.

(3) Electron Diffraction

Electron diffraction used in conjunction with electron microscopy is a powerful method of analytical microscopy. The microscope can be set up for diffraction analysis in a number of ways, some of which are shown in Fig. 16.33. Diffraction was discussed in Section II.A.1.b in this chapter and is the subject of Chapter 17. However, the importance of selected-area diffraction to chemical analysis cannot be overemphasized. With this technique, it is possible to analyze very small crystallites or groups of crystallites that have been selected visually. This is done by means of an adjustable field-limiting aperture. The portion of the specimen that is not of interest is masked out with the aperture and the diffraction pattern of the area of interest is then examined and analyzed.

(4) Reflection Microscopy

Reflection microscopy is used to examine specimens too thick to transmit electrons. Two methods of reflection microscopy are commonly used. In one of these the illuminating system is tilted and the specimen is arranged so that electrons strike the surface of interest at some selected angle (Fig. 16.33). The image is formed by electrons that are reflected through the aperture of the objective lens. In the other arrangement for reflection microscopy, the illuminating system remains aligned with the rest of the microscope, but the illuminating beam is deflected first away from the optic axis and then back against the specimen by means of magnetic fields (Fig. 16.33). This latter system has the advantage of simplicity with regard to changing from transmission alignment to reflection alignment.

Reflection microscopy is seldom used since the advent of the SEM.

(5) Shadow Microscopy

Shadow microscopy (26) is a technique by means of which a shadow or silhouette of the specimen is produced on the viewing screen or film of the microscope. To accomplish this an electron beam of relatively high angular aperture is focused to a small spot. The specimen is placed in the diverging beam beyond but close to the focal spot. A shadow image is then formed at a magnification determined by the relative distances between the focal spot, the object, and the image. The resolution depends on the magnification and on the size of the focal spot. It is, of course, possible to magnify the primary image with electron lenses, as shown in Fig. 16.33.

(6) X-ray Projection Microscopy   This is another type of shadow microscopy. An electron beam is focused to a tiny spot by means of an electron optical system. A suitable metal target is placed at the focal point. This arrangement produces a highly intense point source of x-rays. The specimen, placed between this point source of x-rays and an x-ray detecting system, casts a shadow image on the detection system. Again the magnification depends on the relative distances between source, object, and receptor, and the resolution depends on magnification and spot size.

The shadow electron microscope is at present not a very useful analytical tool because of the complex relationships between electron scattering and the elements making up the specimen. The x-ray projection microscope, however, is a powerful analytical tool, for several reasons: (1) The interaction of x-rays with matter is much better understood, and absorption methods can be used for identification of elements; (2) it is possible to make use of the fluorescence excited in the specimen by the x-ray beam to analyze for the elements present in the specimen; and (3) special methods of microdiffraction can be used to identify crystalline constituents in areas selected by means of the x-ray image.

X-ray projection microscopy is mentioned here because it is possible to modify the electron microscope so that it is an x-ray projection microscope. Some commercial microscopes have made provision for this mode of operation, and there are kits available that can be used to modify some microscopes not so provided. However, most x-ray projection microscopy is carried out with apparatus especially designed for the purpose. Therefore, the method will not be discussed here. Reference can be made to Chapters 13 and 14 for x-ray methods. There are also many references (62,74,75,96,309) to this method in the scientific literature.

i. Microscope Accessories

There are many microscope accessories that are necessary for or useful in the analytical use of the electron microscope. These include hot and cold stages, goniometer or universal stages, high-vacuum stages, and reaction stages for gaseous reactions. These and other accessories have been discussed and reviewed by Agar and Horne (2) and by Reisner (255,256).

2.   Scanning Transmission Electron Microscopes

The author has had no experience with the operation of a STEM. These instruments are relatively new, and anyone interested in this type of microscopy should consult with the company producing the microscope and with the published literature.

3.   Scanning Electron Microscopy

a. ALIGNMENT OF THE MICROSCOPE

The principles of SEM alignment are much the same as those described in Section III.D.1.a. Because the electron optical system of the SEM operates as a condenser only, alignment is somewhat easier. If the microscope is equipped with a light microscope, a fluorescent specimen can be put into the microscope and alignment can be accomplished by observing the image and manipulating lens currents and mechanical movements as described in III.D.1.a. If a light microscope is not attached, alignment can be accomplished by maximizing specimen current. On many microscopes the lenses are fixed and only the electron gun needs alignment. As with TEMs of recent manufacture, alignments and tilts can be done electromagnetically rather than mechanically. This method uses electromagnets to deflect or displace the electron beam and thus allow alignment of the microscope.

b. CHECKING OF MICROSCOPE STABILITY, CLEANLINESS, AND ASTIGMATISM

Checking the SEM involves examining an image of some well-known object both visually on the CRT and photographically. The holey-film type of test object is not suitable for SEM. There are no test objects that are entirely satisfactory. Electrodeposited gold screens or grids make fairly satisfactory objects for low magnifications. For high magnifications, metals such as gold can be deposited by vacuum evaporation on a hot carbon substrate. This results in a granular deposit.

With a properly aligned microscope the field of view should be evenly illuminated and the part of the object in the center of the field should remain there during focusing and during changes in spot size. If astigmatism is present, the object will elongate in different directions as one focuses. Astigmatism is corrected by adjusting to the best focus, then turning the astigmatism-correction controls until the best possible image is formed, and then refocusing and repeating the correction until no further improvement can be detected.

Vibration can be detected by selecting a straight edge (e.g., the edge of a crystal), arranging the edge perpendicular to the raster lines, and observing it either visually or photographically. Vibration causes the straight edge to become rough.

Straight edges can also be used to detect contamination. If a small raster is used and if contamination builds up, the deposit can be observed as a bump on the straight edge. Contamination can also be seen on flat surfaces.

A microscope that is dirty produces a poor image, so detection of dirt depends on recognizing the poor quality of the image of a test object. Micrographs taken of a test object when the instrument is aligned and clean can be kept for comparison with later micrographs of the same object. When one is taking micrographs for purposes of checking instrument operation, areas of the specimen contaminated during previous observation should be avoided.

c. RESOLUTION

Resolution is checked by photographing a suitable test object. Granulated metal films are fairly satisfactory. In principle the test is carried out as described in Section III.D.1.b, but the results are usually much less satisfactory than for transmission microscopy. Better test specimens are needed.

d. ACCELERATING POTENTIAL

The selection of accelerating potential depends on the specimen, on the mode of operation, and on the purposes of the microscopist. High-atomic-number materials generally produce better images at higher potentials. However, if charging occurs, one way of overcoming it is to use a lower accelerating potential. If analyses by x-ray spectroscopy are being performed, the accelerating potential must be high enough to excite the desired atomic transitions. These are only a few of the considerations to be made in selecting the proper accelerating potential.

e. CALIBRATION

The calibration of the magnification of the microscope is in general as described in Section III.D.1.f. However, the standard used must be suitable for the SEM. Ruled gratings or replicas thereof are satisfactory. A standard is available consisting of a cross section of a series of electrodeposited metal layers.

f. PHOTOGRAPHIC TECHNIQUES

Most SEMs are equipped with a camera that uses Polaroid or similar quick-developing processes. Negative films can of course be used. The setting of suitable brightness and contrast is at first a trial and error process. After suitable exposures are obtained, the meter readings can be recorded and future micrographs can be taken using the meters.

g. MODES OF OPERATION

The modes of operation are described in Section II.A.4 along with the type of information obtainable from each mode.

E.   INTERPRETATION OF DATA

Electron-microscopical data accumulated for analytical purposes must be interpreted in the light of all the factors involved in carrying out the experiment. The electron microscope is an instrument for extending human vision, and people generally feel competent to interpret images even though they may not be familiar with the particular object being imaged. It must be kept in mind that the imaging system of the electron microscope and the necessary preparation of the electron-microscopical specimen for viewing have a very important bearing on the interpretation of the image. To interpret a microscopical image the microscopist must be familiar with the properties of the microscope and with the effects of the specimen-preparation technique.

1.   Sampling

The amount of a sample that can be examined in a reasonable time is quite small, and a major problem in analytical microscopy is the preparation of representative samples. Samples for analytical microscopy should be prepared with all the precautions generally exercised in good analytical work, but there are some additional precautions that should be taken. Perhaps the most important is a thorough examination of the sample, and the specimen prepared from it, in the light microscope. All stages in the preparation of the specimen should be followed by using light microscopes.

2.   Effects of Electron Irradiation

The interaction of visible light with matter has been investigated for many years and is fairly well understood. However, information on the interaction of electrons with matter is not as extensive. There are a number of effects that are well known and commonly observed in the electron microscope. If a specimen is thin, so that it does not absorb too many electrons, its temperature does not rise appreciably above ambient. Thick specimens absorb electrons and are heated, sometimes to very high temperatures. The effects of heating are usually observed as melting, boiling, sublimation, or curling in the beam. Dislocations form or move about in the specimen, and crystallites may coalesce or grow, anneal, or recrystallize. Even though the experiments may not involve work with metal or inorganic materials, grain growth may manifest itself as a granulation of shadow metal. Some heating effects are difficult to observe because the entire specimen may be heated before it is brought into the field of view. It is usually best to increase the intensity of the beam slowly when beginning observation of a new specimen, so that heating effects can be seen. There may, of course, be heating effects of beam intensities below the threshold of visibility. Use of the double condenser confines heating effects to small areas and reduces the heating of the specimen. Heating may also cause mechanical trouble, such as stage or aperture movements.

In addition to heating the specimen, the absorption of electrons may result in the accumulation of an electrostatic charge. Charged particles may be repelled by the beam and suddenly jump off the specimen support or vibrate so that the image is poor. Charging generally can be minimized by shadowing with metals or carbon and checking that good electrical contact exists between the specimen and the specimen holder.

Another effect of electron irradiation commonly observed is the decomposition of oil vapors from the vacuum system. This decomposition results in the relatively rapid growth of a layer of carbon where the beam strikes the specimen or parts of the microscope. This layer can seriously interfere with interpretation of micrographs taken at high magnification. Deposition of such a layer can be eliminated or minimized in several ways. The specimen can be heated to about 200°C, at which temperature the deposition of the layer is prevented. The vacuum in the vicinity of the specimen can be partially freed of vapors by the use of cold fingers. Cold fingers are metal surfaces, cooled usually with liquid nitrogen, which are near and often surrounding the specimen. Contaminating vapors condense on these surfaces rather than on the specimen. These effectively condense the vapor and improve the vacuum locally. Another method of reducing deposition is the introduction of gas (e.g., oxygen) at low pressures into the specimen region.

There are also differences between the optics of electrons and light that must be considered in the interpretation of electron images. The very small angular apertures of electron optical lenses result in a relatively large depth of field. The dimensions of the object parallel to the optic axis are not shown as they are in light microscopes and normal vision. This dimension generally must be brought out by shadowing or by stereoscopy. Diffraction phenomena also play a somewhat different role in electron-image formation. For example, images of crystalline materials may contain Bragg reflections, which move during focusing and may interfere seriously with the formation of a usable image. It is important to realize that the contrast of crystalline specimens may depend on Bragg reflections resulting from the orientation of the individual crystallites. In these cases tilting the specimen may change the appearance of the specimen drastically. Images of dislocations are also strongly dependent on the crystallographic orientation of the crystal in which they occur. Here again, tilting of the specimen may alter the image completely.

Another characteristic of electron microscopes is the necessity of placing the specimen in a vacuum. The effect of this may be negligible or very important, depending on the material to be examined. Drying of biological tissues has a profound effect on the tissue, and consequently on the methods of preparing the specimen and interpreting its image.

3. Effects of Specimen Preparation

Because of the difficulties involved in the preparation of specimens for electron microscopy, the specimens often contain defects that seriously in-

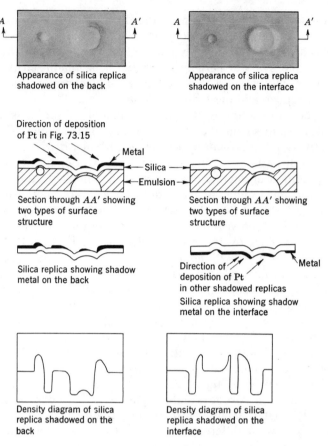

Fig. 16.34.   Relationship between a structure, its replica, and the appearance of the replica in the electron microscope.

terfere with the interpretation of micrographs prepared from them. Even when specimen preparation proceeds perfectly, there are qualities inherent in the resulting specimen that must be allowed for to obtain a valid interpretation of the microscope image. For example, samples of animal or plant tissues must be fixed and dehydrated before they can be sectioned for observation. The reactions involved in these operations inevitably change the structures in the tissue. Until a method is devised for viewing living tissue without harming it, fixation and the resulting structural changes must be recognized in image interpretation.

It is not possible to discuss even a limited number of preparation techniques with regard to their influence on image interpretation. Therefore, as an example, some of the factors involved in replication of a specimen will be examined. Figure 16.34 shows the effects of two methods of shadowing on the appearance of a replica of a two-phase system. The appearance of the final image is shown at the top. The essential steps in replication are shown in the center, and a plot of photographic density of the image is shown at the bottom. In general it is not desirable to shadow the back of a replica, as shown in the first column, because of the possibility that the back is not a faithful replica. However, neglecting this possibility, it makes interpretation simpler if the shadow is deposited on a positive replica surface, that is, one that has the same contours as the original specimen. Visual interpretation can be made still easier if a reverse print of the micrograph is prepared. In such a print the shadows are dark and the micrograph closely approximates an image of a similar structure lighted from the side and photographed with visible light. Understanding of the appearance of a microscopical image can be attained only when the effects of all of the stages of specimen preparation on the image are understood.

Defects caused by the use of improper preparation techniques or through accidents are common in electron-microscopical specimens. They must be recognized or eliminated before the interpretation of analytical data can be carried out reliably. With particulate materials, losses of particles may occur. Often portions of the sample are lost preferentially because of their size or composition, making the specimen nonrepresentative. Particles are also likely to form agglomerates, or agglomerates may be dispersed during specimen preparation. The medium in which the particles may be suspended may contain soluble components that coat the particles and obscure their structure.

Sections of materials prepared by cutting or thinning are subject to their own peculiar defects. Microtome sections may be compressed or torn, or contain knife marks. Improper fixation is a common cause of defective tissue sections. Sections that are thinned chemically or electrolytically may have reaction products left on their surfaces, or inclusions may be dissolved out preferentially. Such sections are very thin and may easily become distorted or broken through handling. Replica defects are many and varied. Plastic replicas or replicas prepared from plastic intermediates may contain strain marks or may exhibit the remains of tiny bubbles caused by impurities in the plastic. These specimen defects are some of the most prevalent types encountered in electron microscopy.

In spite of the large number of possible complicating factors, the interpretation of electron-microscopical data is fairly well understood. Confidence in interpretation can be built up through experience gained by carrying out investigations of the following types: comparison of electron-microscopical data with light-microscopical data; study of results of studying well-known structures, such as the surface of a glass microscope slide; comparison of micrographs of the same specimen prepared by a variety of techniques; comparison of results obtained by different microscopists in the same or in different laboratories; and comparison of data obtained through electron microscopy with data obtained by any other analytical method that can be brought to bear on the problem.

IV. ANALYTICAL USES OF ELECTRON MICROSCOPY

Because the microscope is used primarily to extend human vision to very small objects, the first uses of the electron microscope were in morphological studies. Among the materials examined were biological specimens, metal films, metal smokes and oxides, and soils. Many of these studies had more or less remote applications to chemical analysis. This brief review will include only those considered to have direct significance. Undoubtedly many references of value to the chemical analyst have been omitted. Also because the subject has become so extensive and complex this review has not been extended beyond that published in the previous edition of this treatise.

A. ANALYSIS BASED ON MORPHOLOGY

It is obvious that if the constituents of a mixture have characteristic shapes and if these shapes are known to the analyst, the approximate composition of the mixture can be determined by counting methods. Such methods have been used in light microscopy for many years (22,60,92). Much of the early work in electron microscopy as related to chemical analysis consisted in examining the morphology of chemical compounds, particularly those occurring in collodial systems. Among the first such materials to be studied extensively were soils and clay minerals. The hexagonal platy habit of kaolin was described by Eitel et al. (90) in 1938. They also observed the changes in crystal shape caused by heating the specimen. The morphology of a number of clays was described by Von Ardenne, Endell, and Hoffman (311) in 1940. In the same year, Jakob and Loofman (158) investigated several soil minerals and suggested that the electron microscope could possibly be used to detect clay minerals. In 1941, Humbert and Shaw (149) and Bujor (51) used electron microscopy to distinguish between California montmorillinite and Wyoming bentonite on the basis of crystal shape. There is no difference between these clays with respect to chemical analysis or to crystal structure as determined by x-ray methods, and yet their colloidal properties are different. In the same year, Endell (94) made a study of the fine structures of clays, and pointed out the significance of the electron microscope for identifying the clays (93). Humbert (148) described new techniques for preparing clays for examination and published electron micrographs of a number of clays. In 1943, Alexander et al. (3) proposed the name endellite for the compound $Al_2Si_5O_5 \cdot (OH)_4 \cdot 2H_2O$ and characterized it by light and electron microscopy as well as by diffraction methods. Bujor (51) identified montmorillonite, kaolinite, and silica gel in certain fractions of Tomesti clay. In 1947, Moore (207) discussed the applications of electron microscopy to geology, with particular reference to the examination and identification of clays. The American Petroleum Institute sponsored a number of research projects designed to characterize completely a carefully selected suite of clay samples. A series of reports was issued on this work. Report No. 6 of this series was concerned with the electron-microscopical examination of these clays, and the report includes electron micrographs of each of them. The complete set of reports was published in 1951 (6) as a single volume. In 1952, Bramao and his coworkers (44) published an article on the characterization of clays. In this work the same set of samples was examined by several methods, including differential thermal analyses, x-ray diffraction, and electron microscopy. From these examinations they established criteria for the characterization of kaolinite, halloysite, and related minerals found in soils. Suito and Uyeda also studied the clay minerals by electron microscopy and diffraction and published their results in 1956 (291). At the Sixth National Conference on Clays and Clay Minerals, Bates and Comer (21) discussed the classification of the serpentine group of minerals according to their platy or fibrous habit. The proceedings of this conference were published in 1959. Also in 1959, Bates (20) authored a paper, entitled "Morphology and Crystal Chemistry of 1:1 Layer Lattice Silicates," in which 64 silicate minerals of layered structure were described. Comer (68) has written a general article on the electron microscope in the study of minerals and ceramics that gives an excellent picture of the present status of clay microscopy.

The history of the electron-microscopical analysis of clays has been described fairly completely because it illustrates a type of analysis that can be done only with the electron microscope. Clays consist of very fine crystalline particles too small to be resolved by light microscopy. Some clays are chemically identical although they have different physical properties. The electron microscope can be used to distinguish among these clays and to explain the variation in physical properties.

Cements have also been the subject of electron-microscopical analysis for many years. Probably the first work in this field was carried out by Radczewski and his coworkers (246–249) and was published in 1939. This and most of the analytical electron microscopy in the field of cements was directed toward studying hydration reactions. The constituents of cement were examined separately so that the hydrates formed could be characterized by their morphology. This knowledge of crystal morphology was then used to study the setting of the cement. Eitel, in the years 1941–1943, published a number of articles on various aspects of cement and silicate research (87–89). Von Ardenne, Endell, and Hoffman discussed cements as well as clays in their 1940 paper (311). Sliepcevich, Gildart, and Katz (282) studied the crystals that form during the hydration of Portland cement. Swerdlow, McMurdie, and Heckman (292) investigated the hydration of tricalcium silicate and reported their results in the Third International Conference on Electron Microscopy, held in London in 1954; at the more recent Fourth Congress (Berlin, 1958) Groth and Schimmel (121) discussed the influence of fluosilicates on the hydration of cement. A number of other investigators have published articles of interest to the cement industry (8,23,30,65,123).

During the years since the first electron microscope was put into operation, large numbers of materials have been studied from the standpoint of the morphology of thin crystalline components. Much of the information resulting from these studies can be used for identification of the crystalline components of unknown systems. Turkevich and Hillier (304) and Prebus (241) have reviewed much of the early work on colloidal systems. The number of published accounts of morphological crystallographic investigations has increased considerably since that time, so a comprehensive review is not possible in this chapter. References to systems of interest can in most cases be found in the recommended general literature, described in Section I.D.

The usefulness of crystal morphology in chemical analysis is greatly increased when systematized. The use of interfacial angles in crystal analysis is well known, and extensive compilations of such data have been published (122,240). The application of the goniometric methods of identifying crystals in microscopical analyses is complicated by the fact that it is difficult to observe anything but silhouette angles in the microscope, whereas the diagnosis depends on interfacial angles. Donnay and O'Brien (83) have shown that silhouette angles observed in the light microscope can be correlated with the true interfacial angles and axial elements of a crystal. These methods were extended to electron microscopy by Kirkpatrick and Davis (169). Using the methods of crystal drawing and spherical projections, including stereographic and cyclographic projections, the necessary calculations can be made from angles measured on electron micrographs of a crystal. A universal stage for an electron microscope has been built (165). The use of such stages should greatly facilitate electron-microscopical crystallographic studies. The number of independent crystallographic angles that may be measured on a given crystal is a function of the crystal system to which it belongs. This method cannot distinguish among compounds that crystallize in the cubic system, since all interfacial angles in the cubic system are identical regardless of chemical composition. For all other systems the number of independent crystallographic angles that may be measured is equal to the number of angles required to calculate the axial elements. The tetragonal and hexagonal systems require only one angle; the orthorhombic, two; the monoclinic, three; and the triclinic, five. Thus, the electron microscope is capable of providing crystallographic data that can be used to identify many compounds for which well-formed crystals can be obtained.

The morphological analyses discussed thus far are limited to chemical compounds existing as well-formed crystals. The electron microscope also has been demonstrated to be useful in more systematic qualitative chemical analyses, such as those described by Chamot and Mason (60). In these methods the unknown material is reacted with reagents that form characteristic crystalline precipitates if certain ions are present in the unknown. Thus, by an appropriate choice of reagents, a systematic analysis can be carried out for many elements and ions. Gulbransen, Phelps, and Langer (124) first investigated the applicability of these classical light-microscopical methods to electron microscopy in 1945. They devised techniques for carrying out the reactions and preparing the resultant precipitates for viewing in the electron microscope. Using this technique, they examined a number of precipitates and compared the crystals observed in the electron microscope with crystals from the corresponding light-microscopical test and with descriptions of macroscopic crystals of the same composition. The observed habits were in good agreement with the micro and macro observations. They also made an estimate of the detection limits of one of the tests and used electron diffraction to identify some of the crystals formed. This development of chemical electron microscopy was carried further by Fischer and Simonsen (99), who investigated metallorganic precipitates used in inorganic analysis. The nickel, palladium, and bismuth derivatives of dimethyglyoxime were described.

In 1955, Wiesenberger (317) published the first of a series of papers dealing with electrolytic methods of precipitating crystals suitable for identification of certain elements. The electrolytic method was originally suggested by Emich (92) and was developed by Brenneis (47) for use in light microscopy. Wiesenberger adapted this microelectrolysis to electron microscopy and showed the usefulness of the method by applying it to the identification of copper, lead, and silver. Wiesenberger has continued to publish in this field (318–320), and has also investigated precipitation effected by diffusion of reagents and unknowns through a membrane (318,319). Precipitation reactions also have been used for the chemical electron-microscopical analysis of small particles of aerosols. Lodge and Tufts (184,301–303) have led in the development of techniques applicable to this field of investigation. The aerosol particles are collected on membrane filters of Formvar films and are placed in contact with a drop of reagent. The mass of the resulting precipitate is proportional to the amount of the element originally present in the particle, and particle-size distributions of the original particles can be obtained by counting and sizing the precipitate clumps. Koenig (174) has worked out a similar method for detecting silver ions. This method uses photographic developers to reduce the ion to metallic silver. This method is capable of detecting 3×10^{-17} grams of AgI.

The methods of analysis discussed thus far have been essentially qualitative. Since the electron microscope is capable of producing highly magnified images, it is natural that it should be used for quantitative measurements of size and shape. The quantitative microscopical methods developed for light microscopy (60) have in general been transferred successfully to electron microscopy. These methods include particle-size counting, counting the constituents of heterogeneous mixtures, areal analysis, linear analysis, and point-count analysis. The history of electron microscopy is full of applications of these methods and there is no point in attempting to cite them all. Instead, some of the methods particularly applicable to electron microscopy will be discussed.

The dimensions of a microscopical object lying in the plane perpendicular to the optic axis of the microscope are easily measured if the magnification is known. Dimensions parallel to the optic axis, however, are more difficult to measure; several means of making such measurements have been employed. Thickness might be measured in terms of the degree of absorption of the electrons in the specimen, but specimen–beam interactions are not yet well known and this method can be applied only in special cases. This subject will be discussed in the next section.

Marton and Schiff (197) in 1941 developed stereoscopy as a method of measuring specimen thickness. Various methods of making stereomicrographs were used by other investigators until, in 1944, Heidenreich and Matheson (138) described methods of producing and interpreting stereomicrographs that are essentially those used today.

Another method of measuring the height of a microscopic object is shadow casting. This process, which has been described more completely above, was first used by Muller (218) in 1942. However, it was Williams and Wyckoff (323–325) who developed the method to a practical stage of usefulness. Many other workers have improved the shadow-casting process; for example, Preuss (243) investigated the mode of propagation of the metal vapors, Hibi and Yada (139) made use of nozzles to direct the metal vapors, and Bradley (39) combined shadow casting with replication. Bradley's method is probably the best method of shadow casting in use today.

From the beginning of electron microscopy it has been the hope of microscopists that they would ultimately be able to see molecules and atoms. In 1940, Von Ardenne (310) reported seeing metal particles as small as 10–15 Å and organic molecules having diameters of 40 Å. He estimated the molecular weights of these particles and found the values to be in fair agreement with values measured by other methods. In the same year, Husemann and Ruska (150,151) carried out the first systematic investigation of the determination of molecular weights by electron microscopy. In 1942, Cohen (66) and Stanley (286) reported on their work on the molecular size and shape of the nucleic acid of tobacco mosaic virus. In 1944, Kropa, Burton, and Barnes (175) investigated the molecular weights of various polystyrenes. In 1945, Boyer and Heidenreich (34), in a report on an extensive study of the electron-microscopical determination of molecular weight, attempted to explain the difficulties of the method and the discrepancies between values obtained with the electron microscope and those obtained by other means. One of the principal problems in particle-size work is the preparation of specimens that are representative of the material being examined. Many methods of sample prepration have been used (140,157,267), and the choice of a method for a particular material depends on that material's properties and condition. One of the surest ways of obtaining representative samples is through the use of the spraying methods developed by Backus and Williams (13). In this method, a suspension of the particles is sprayed onto the support film and all of the particles in a single droplet are counted. Internal standards can be used in this method, and if the particles are individual molecules, good molecular weight values can be obtained. Another method of measuring molecular weights of larger molecules can be used when crystals of these molecules can be grown. Hall (130) describes a number of such cases. In this method, electron micrographs are prepared that show the arrangement of molecules in the crystal. From the lattice spacings and the density of the dry crystals, excellent molecular weight values can be calculated. Finally, estimates of molecular weight have been made on polymeric materials using resinographic methods developed by Rochow (258). These methods are analogous to the method using crystals, but because the regularity of most polymers is low, the method is quite difficult and usually not precise.

B. METHODS OF ANALYSIS BASED ON THE ABSORPTION AND SCATTERING OF ELECTRONS

As mentioned earlier, one possible method for measuring the thickness of a specimen in the electron microscope involves the measurement of the attenuation of the electron beam on passage through the specimen. This kind of measurement can be made only if the composition of the specimen and its characteristic attenuation of beam intensity at the image plane are known. On the other hand, if the thickness and characteristic attenuation were known, the composition of the specimen might be determinable. The staining of specimens to enhance contrast and to identify the chemical constituents of a specimen also depends on selective attenuation of the electron-beam intensity.

In hopes of understanding and making use of attenuation phenomena, a number of investigators have examined the electron optics of the microscope and the electron–specimen interactions. The first to study these phenomena were Marton and Schiff (197), who published their work in 1941. In 1946 (27) and 1947 (28), Boersch discussed image contrast in terms of elastic and inelastic scattering and the effects of crystal lattices and phase-contrast phenomena. The theory of image formation was carried still further by Von Borries (313) and Hall (128,130). In 1952 Glaser (114) considered image formation from the standpoint of wave optics. Haine has also considered the problem on a wave-optical basis, and by using a simplified theory he evaluated the scattering from single atom. Haine's work is summarized in his book (cowritten with V. E. Cosslett) on the present state of electron microscopy (125).

The theory of electron–specimen interactions has been applied to the quantitative measurement of mass and thickness in the electron microscope. The theory was discussed from this standpoint by Kruger-Thiemer (176) in 1955 and by Zeitler and Bahr (331–333) in 1957, 1959, and 1960. Experimental measurements of mass have been reported by Hall (129) in 1955, Bahr (14) in 1957, and Burge and Silvester in 1960 (53,54). Thickness measurements of thin films have been made by Bachmann and Siegel (10), De (81), Halliday and Quinn (133), and Reisner (256). There is some possibility that differences in scattering power of the elements may provide a means for identifying elements by measurements of scattering power. This possibility was discussed from the standpoint of elastic-scattering theory by Cosslett in 1958 and 1959 (71,72).

For many years biologists have been staining specimens to enhance their contrast for light microscopy. Histochemical techniques for the identification of substances in biological tissues have been highly developed. Many of these techniques have been used by light microscopists for nonbiological problems. Electron staining, that is, the use of stains to enhance contrast in electron-microscopic specimens, has been used extensively in electron microscopy (24,129,131,143,208,239,316). A symposium on the subject was held in 1959 (293). Most of the work has been done with nonspecific stains, but a number of attempts have been made to develop specific stains. Although this has proven difficult, there is a strong need for such stains and the further development of electron histochemistry is highly desirable. The application of staining techniques to nonbiological microscopy is, of course, also desirable.

The attenuation of the electron beam by the specimen as considered thus far has been an overall attenuation manifested by a change in the intensity of the final image. However, electrons that strike the atoms of a specimen material may transfer some of their energy to the atom that was struck. Thus, if the electrons in a beam passing through a sample all originally have the same momentum (i.e., are monochromatic), the emerging electrons will have various momenta. Rutheman (261–263) studied this phenomenon beginning in 1941, and found that energy transfers of 1–100 eV occurred when a beam of 5.3 keV was passed through a thin film of aluminum. Moreover, as expected from quantum considerations, the energy losses were discrete, corresponding to differences between discrete energy levels in the specimen. These discrete energy losses correspond to the energies of the x-rays that result when electrons strike the metal. In 1943, Hillier (141) described a method of identifying the elements in an electron-microscopical specimen by analyzing the velocities of the electrons transmitted by the specimen. This method was further developed by Hillier and Baker (142) and by Ellis (91). Marton, in 1944 (188), proposed another, similar apparatus for carrying out this kind of analysis. In recent years this type of analysis has been more or less superseded by the electron probe microanalyzer, which identifies the elements in the specimen by analyzing the characteristic x-rays given off when the probe beam strikes the specimen (74,75,95).

C.   CHARACTERISTIC EMISSIONS: THERMIONIC, PHOTOELECTRIC, AND NUCLEAR

The earliest electron microscopes were of the thermionic emission type (189,259). In these instruments the specimen itself was heated until it emitted electrons, which were then accelerated in a potential field and focused to form an image of the source on an appropriate receptor (130,187,199,253). The number of electrons emitted from a heated specimen is a function of the characteristics of the material, the temperature, and the conditions existing in the instrument. The specimen will begin to emit electrons only after a certain minimum temperature has been reached. This minimum temperature depends on the work function of the material of the specimen, if other conditions are held constant. Thus, it is possible to identify materials by measuring their work functions. Combining this method with electron microscopy makes it possible to locate some of the constituents of a specimen by observing the image formed in a thermal emission microscope. Scott and Packer (268–271) were able to use this method to locate magnesium and calcium in biological tissues. To accomplish this, they incinerated muscle and other tissues very carefully on a metal substrate and subsequently used the substrate, with the ash in place, as the source in an emission microscope.

Electrons can be ejected from the surface of a specimen by several means other than electron bombardment. In 1953 Mollenstedt and Duker (205) described the use of ion bombardment to produce secondary electrons for microscopy. In 1954 Mollenstedt and Keller (206) reported further progress in this work. In 1933 Brüche (49) described an electron microscope that formed an image by using photoelectrons ejected from the specimen when it was exposed to light. At about the same time Pohl (237) described an instrument operating on the same principles. Many of the recent experiments in the use of secondary electrons for microscopy have been reported at the meetings of the International Federation of Societies for Electron Microscopy (45,100,264,266,289), and in the European Regional Conferences (25,116,145,281,299,300,305), and in the publications of the Electron Microscopy Analysis Group (195,220,225,306).

When the nuclei of radioactive materials disintegrate, β-particles (electrons) are often a product. These β-particles can be made to form an image by using techniques similar to those used in thermionic emission microscopes. Marton and Abelson (190) first reported on this type of electron microscope in 1947. The resolution of the microscope proved to be quite low because of variations in the speed of the emitted electrons. The chromatic aberrations of electron lenses are such that high-resolution images cannot be formed with a beam composed of electrons having various velocities. This type of electron microscope could be very useful in chemical analysis if these difficulties could be overcome, because the distribution of radioactive species would be shown directly in the image.

Another approach to the problem of locating radioactive species in a sample has been taken (32,111,134,183,227,245). Autoradiograms are prepared by either placing the specimen in contact with a nuclear emulsion or applying the nuclear emulsion directly to the specimen, allowing the combination to stand in the dark until the exposure is adequate, developing the emulsion using regular development methods, and finally viewing the autoradiogram in the electron microscope. When the emulsion is applied directly to the specimen, the composite is placed in the microscope and the tracks in the emulsion can be traced to their points of origin. Thus, the location of the radioactive species can be determined quite accurately.

REFERENCES

1. Agar, A. W., "The Operation of the Electron Microscope," in D. Kay, Ed., *Techniques for Electron Microscopy*, Thomas, Springfield, Illinois, 1961, p. 1.

2. Agar, A. W., and R. W. Horne, "Ancillary Apparatus and Special Attachments for the Electron Microscope", in D. Kay, Ed., *Techniques for Electron Microscopy*, Thomas, Springfield, Illinois, 1961, p. 32.

3. Alexander, L. T., G. T. Faust, S. B. Hendricks, H. Insley, and H. F. McMurdie, *Am. Mineralogist*, **28**, 1 (1943).

4. Ahmed, H., "A High Brightness Boride Emitter Electron Gun" in Glauert, A. M., Ed., *Proceedings of the Vth European Congress on Electron Microscopy*, Institute of Physics, London, 1972, p. 10.

5. Ahmed, H., and W. C. Nixon, "Boride Guns for High Signal Level SEM," in, Johari, O., Corvin, I, *Proceedings of the Vth Annual SEM Symposium,* Illinois Institute of Technology Research Institute, Chicago, 1973, p. 217.

6. American Petroleum Institute, *Reference Clay Minerals*, Columbia Univ. Press, New York, 1951.

7. Arceneaux, C. J., Ed., *Proceedings of the Annual Meetings of the Electron Microscopy Society of America*, (25th Annual Meeting in 1967 to 32nd Annual Meeting in 1974), Claitors Publishing Div., Baton Rouge, Louisiana.

8. Arkosi, K., "Electron Microscopical Research on Natural Hydraulic Binders," in reference 18, Vol. I, p. 730.

9. Aumonier, F. J., and R. Ross, Eds., *J. R. Microscop. Soc.*, **79**, Series III (1960).

10. Bachmann, L., and B. M. Siegel, "Determination of the Thickness of Thin Films by Inelastic Scattering of Electrons," in reference 145, p. 157.

11. Backus, R. C., and R. C. Williams, *J. Appl. Phys.*, **19**, 1186 (1948).

12. Backus, R. C., and R. C. Williams, *J. Appl. Phys.*, **20**, 224 (1949).

13. Backus, R. C., and R. C. Williams, *J. Appl. Phys.*, **21**, 11 (1950).

14. Bahr, G. F., *Acta Radiol. Suppl.*, **147** (1957).

15. Bahr, G. F., and G. Moberger, *Exp. Cell Res.*, **6**, 506 (1954).

16. Bailey, G. W., Ed., *Proceedings of the Annual Meetings of the Electron Microscopy Society of America* (33rd annual meeting in 1975 to 37th annual meeting in 1979), Claitors Pub. Div., Baton Rouge, Louisiana.

17. Barber, D., *J. Mater. Sci.*, **4**, 18 (1970).

18. Bargmann, W., Möllenstedt, G., Niehrs, H., Peters, D., Ruska, E., Wolpers, C., Eds., *Vierter Internationaler Kongress für Elektronen-Mikroskopie, Berlin, 1958*, Vol. I, Springer-Verlag, Berlin, 1960.

19. Barnett, R. J., "The Combination of Histochemistry and Cytochemistry with Electron Microscopy for the Demonstration of the Sites of Succinic Dehydrogenase Activity," in reference 18, Vol. II, p. 91.

20. Bates, T., *Am. Mineralogist*, **44**, 78 (1959).

21. Bates, T., and J. J. Comer, "Further Observations on the Morphology of Chrysotile and Halloysite," in *Proceedings of the Sixth National Conference on Clays and Clay Minerals*, Pergamon, New York, 1959, pp. 237–249.

22. Behrens, H., and F. Kley, *Mikrochemische Analyse*, L. Voss, Leipzig and Hamburg, 1921.

23. Bernal, J. D., in *Proceedings of the Third International Symposium on Chemistry of Cements, London, 1952*, Cement and Concrete Association, London; Reinhold, New York, 1954, Paper 9.

24. Bielig, H. J., G. A. Kausche, and H. Haardick, *Z. Naturforsch.*, **4b**, 30 (1949).

25. Bocciarelli, D. S., Ed. *Electron Microscopy 1968, 4th European Regional Conf. Sept. 1–7, 1968, Rome*, Vol. 1 (Physical Science), Vol. 2 (Biological Science), Tipografia Poliglotta Vaticana, Rome, 1968.

26. Boersch, H., *Naturwissenschaften*, **27**, 418 (1939).

27. Boersch, H., *Monatsschrift*, **76**, 86, 163 (1946).

28. Boersch, H., *Monatschrift*, **78**, 163 (1947).

29. Boersch, H., *Experientia*, **4**, 1 (1948).

30. Bogue, R. H., *Chemistry of Portland Cement*, 2nd ed., Reinhold, New York, 1955, Chapter 5, p. 26.

31. Booker, G. R., *Br. J. Appl. Phys.*, **5**, 349 (1954).

32. Botty, M. C., *Anal. Chem.*, **32**, 92R (1960).

33. Bowen, D. K., and C. R. Hall, *Microscopy of Materials—Modern Imaging Methods Using Electron X-ray and Ion Beams*, Wiley, New York, 1975.

34. Boyer, R. F., and R. D. Heidenreich, *J. Appl. Phys.*, **16**, 621 (1945).

35. Bradley, D. E., *Br. J. Appl. Phys.*, **5**, 65, 98 (1954).

36. Bradley, D. E., *Br. J. Appl. Phys.*, **6**, 430 (1955).

37. Bradley, D. E., *J. Appl. Phys.*, **27**, 1399 (1956).

38. Bradley, D. E., *Mikroskopie*, **12**, 257 (1957).

39. Bradley, D. E., *Br. J. Appl. Phys.*, **10**, 198 (1959).

40. Bradley, D. E., "A New Approach to the Problem of High Resolution Shadow Casing: The Simultaneous Evaporation of Platinum and Carbon," in reference 18, Vol. 1, p. 428.

41. Bradley, D. E., *J. R. Microscop. Soc.*, **79**, 101 (1960).

42. Bradley, D. E., "Replica and Shadowing Techniques," in reference 62, p. 229.

43. Bradley, D. E., "Techniques for the Optical Selection of Specimens," in D. Kay, Ed., *Techniques for Electron Microscopy*, Thomas, Springfield, Illinois, 1961, p. 138.

44. Bramao, L., J. G. Cady, S. B. Hendricks, and M. Swerdlow, *Soil. Sci.*, **73**, 273 (1952).

45. Breese, S. S., Ed., *Proceedings of the Fifth International Congress for Electron Microscopy, Philadelphia, 1962*, Academic, New York, 1962.

46. Breiger, E. M., and A. M. Clanert, in reference 73, p. 330.

47. Brenneis, H. J., *Mikrochemie*, **95**, 385 (1931).

48. Broers, A. N., *J. Sci. Instr.*, **2**, 273 (1969).

49. Brüche, E., *Z. Physik*, **86**, 448 (1933).

50. Brüche, E., and H. Johannson, *Naturwissenschaften*, **20**, 353 (1932).

51. Bujor, D. J., *Neues Jahrb. Mineral. Abhandl.*, **55**, 35 (1943).

52. *Bulletin Signaletique*, Centre Nationale de la Recherche Scientifique, Paris.

53. Burge, R. E., and N. R. Silvester, *J. Biophys. Biochem. Cytol.*, **8**, 1 (1960).

54. Burge, R. E., and N. R. Silvester, "Mass Measurement in the Electron Microscope," in reference 145, p. 161.

55. Burton, E. F., J. Hillier, and A. Prebus, *Phys. Rev.*, **56**, 1171 (1939).

56. Busch, H., *Ann. Physik.*, **81**, 974 (1926); *Arch. Elektro Tech.*, **18**, 583 (1927).

57. Butler, J. W., "Digital Computer Techniques in Electron Microscopy," in reference 264, p. 191.

58. Calbick, C. J., *J. Appl. Phys.*, **15**, 685 (1944).

59. Caldwell, W. C., *J. Appl. Phys.*, **12**, 779 (1941).

60. Chamot, E. M., and C. W. Mason, *Handbook of Chemical Microscopy*, 3rd ed., Vol. I, 1958; 2nd ed., Vol. II, 1940, Wiley, New York.

61. Chanson, P., and C. Magnan, *Compt. Rend.*, **233**, 1436 (1951).

62. Clark, G. L. Ed., *The Encyclopedia of Microscopy*, Reinhold, New York, 1961.

63. Clark, G. L., "Diffraction Microscopy," in reference 62, p. 569.

64. Clark, G. L., "Reflection Microscopy," in reference 62, p. 672.

65. Clark, G. L., P. M. Bernays, and J. P. Tordella, "Progress in Lime Investigation with X-rays and the Electron Microscope," in *National Lime Association, Proceedings Twenty-Fourth Convention*, National Lime Association, Washington, D.C., 1942.

66. Cohen, S. S., "The Molecular Size and Shape of the Nucleic Acid of Tobacco Mosaic Virus," *Fed. Proc.*, **1**, No. 1 (1942).

67. Colliex, C., and P. Trebbia, "Electron Energy Loss Spectroscopy in the Electron Microscope. Present State of Affairs," in J. M. Sturgess, Ed., *Electron Microscopy, 1978*, Vol. III, Microscopical Society of Canada, Toronto, 1978, p. 268.

68. Comer, J. J., in *ASTM Special Technical Publication 257*, American Society for Testing Materials, Phioladelphia, 1959, p. 94.

69. *Comptes Rendus du Premier Congres International de Microscopie Electronique Paris, 1950*, Soc. Francaise de Microscopie Theorique et Appliquee, Paris, 1953.

70. Cosslett, V. E., Ed., *Bibliography of Electron Microscopy*, Edward Arnold, London, 1950.

71. Cosslett, V. E., *J. R. Microscop. Soc.*, **78**, 18 (1958).

72. Cosslett, V. E., *J. R. Microscop. Soc.*, **78**, 1 (1959).

73. Cosslett, V. E., et al., Eds. *Proceedings of the Third International Conference on Electron Microscopy, London, 1954*, Royal Microscopical Society, London, 1956.

74. Cosslett, V. E., A. Engstrom, and H. H. Pattee, Eds., *Proceedings of the Cambridge Symposium on X-ray Microscopy and Microradiography*, Academic, New York, 1957.

75. Cosslett, V. E., and W. C. Nixon, *X-ray Microscopy*, Cambridge Univ. Press, London, 1960.

76. Couchet, G., M. Gauzit, and A. Septeir, *Compt. Rend.*, **233**, 1087 (1951).

77. Crew, A. V., *Science*, **154**, 729 (1966).

78. Crew, A. V., D. N. Eggenberger, J. Wall, and L. M. Welter, *Rev. Sci. Instr.*, **39**, (4), 576 (1968).

79. Crew, A. V., J. Wall, and L. Welter, *J. Appl. Phys.*, **39**, 5861 (1969).

80. Davisson, C. J., and L. H. Germer, *Nature*, **119**, 558 (1927).

81. De, M. L., *Nature*, **192**, 547 (1961).

82. de Broglie, L., *Philos. Mag.*, **47**, 446 (1924); *Ann. Phys. (Paris)*, **3**, 22 (1925).

83. Donnay, J. D. H., and W. A. O'Brien, *Ind. Eng. Chem., Anal. Ed.*, **17**, 593 (1945).

84. Drummond, D. G., Ed., *J. R. Microscop. Soc.*, **70**, (III), 1 (1950).

85. Drummond, I. W., *Am. Lab.* April, 83 (1976).

86. Duncumb, P., and P. K. Shields, *Br. J. Appl. Phys.*, **14**, 617 (1963).

87. Eitel, W., *The Electron Microscope as an Instrument for Quantitative Measurements in Silicate Research*, W. de Gruyter, Berlin, 1941, pp. 48–66.

88. Eitel, W., *Agnew. Chem.*, **54**, 185 (1941).

89. Eitel, W., et al. *Abhandl. Preuss. Akad., Wiss. Math-Naturw. Kl.*, **5**, 5, 13, 21, 37 (1943).

90. Eitel, W. H., H. O. Muller, and O. E. Radezeiwski, *Ber. Dtsh. Keram. Ges.*, **20**, 165 (1938).

91. Ellis, S. G., *J. Appl. Phys.*, **19**, 1191 (1948).

92. Emich, F., *Lehrbuch der Mikrochemie*, J. C. Bergman, Munich, 1926.

93. Endell, J., *Tonind. Ztg.*, **65**, 69 (1941).

94. Endell, J., *Keram. Rundschau*, **49**, 23 (1941).

95. Engstrom, A., V. E. Cosslett, and H. H. Pattee, *X-ray Microscopy and Microanalysis*, Van Nostrand, New York, 1960.

96. Engstrom, A., H. H. Pattee, and V. E. Cosslett, Eds., *Proceedings of the Stockholm Symposium on X-ray Microscopy and X-ray Analysis*, Elsevier, Amsterdam, 1960.

97. Everhart, T. E., and R. F. M. Thornley, *J. Sci. Instr.*, **37**, 246 (1960).

98. Fernandez-Moran, H., in *Electron Microscope Society of America Meeting*, Milwaukee, Wisconsin, 1960, in J. Appl. Phys., **31**, 1840 (1960).

99. Fischer, R. B., and S. H. Simonsen, *Anal. Chem.*, **20**, 1107 (1948).

100. Favard, P., Ed., *Microscopie Electronique 1970: Proceedings of the Seventh International Congress on Electron Microscoy, Grenoble, 1970*, Société Française Microscopie Electronique, Paris (1970).

101. Franklin, U. M., G. C. Weatherly, and G. T. Simon, "A History of the First North American Electron Microscope," in reference 289, Vol. III, p. 4.

102. Freundlich, M. M., *Science*, **142**, 185 (1967).

103. Fryer, J. R., *The Chemical Applications of Transmission Electron Microscopy*, Academic, New York, 1979.

104. Fuller, H. W., and M. E. Hale, *J. Appl. Phys.*, **31**, 238 (1960).

105. Gabor, D., *Forschungsh. Stud. Gres. Hochtspann.*, Anlag. No. 1 (1927).

106. Gabor, D., *Nature*, **161**, 777 (1948).

107. Gabor, D., *Proc. R. Soc. (London), Ser. A*, **197**, 454 (1949).

108. Gabor, D., *Proc. Phys. Soc. (London), Ser. B*, **64** (6), 449 (1951).

109. Gabor, D., "The History of the Electron Microscope from Ideas to Achievements," in reference 266, Vol. 1, p. 6.

110. Gauzit, M., *Compt. Rend.*, **233**, 1586 (1951).

111. George, L. A. II. *Science*, **133**, 1423 (1961).

112. Gerould, C. H., *J. Appl. Phys.*, **21**, 183 (1950).

113. Gersch, I., "Selective and Cytochemical Staining of Frozen-Dried Preparation: Study With the Electron Microscope," in reference 18, Vol. II, p. 167.

114. Glaser, B. W., *Grundlagen der Elektronoptik*, Springer-Verlag, Vienna, 1952.

115. Glauert, A. M., "The Fixation, Embedding and Staining of Biological Specimens," in D. Kay, Ed., *Techniques for Electron Microscopy*, Thomas, Springfield, Illinois, 1961, p. 167.

116. Glauert, A. M., Ed., *Proceedings of the Fifth European Congress for Electron Microscopy*, Institute of Physics, Manchester, 1972.

117. Glauert, A. M., Ed., *Practical Methods in Electron Microscopy*, Vols. 1–5, North Holland, Amsterdam; American Elsevier, New York, 1972–1977.

118. Glauert, A. M., and R. H. Glauert, J. Biophys. Biochem. Cytol., **4**, 191 (1958).

119. Goldstein, J. I., "Metallography—A Practical Tool for Correlating Structure and Properties of Materials," in *ASTM Special Technical Publication 557*, American Society for Testing Materials, Philadlephia, 1974, p. 86.

120. Goldstein, J. I., and H. Yakowitz, *Practical Scanning Electron Microscopy*, Plenum, New York, 1975.

121.   Groth, H., and G. Schimmel, "On the Hydration of Cement and the Influence of Fluosilicates," in reference 18, Vol. I, p. 728.

122.   Groth, P., *Chemische Krystallographie*, W. Engelmann, Leipzig, 1906–1919.

123.   Grudemo, A., in *Proceedings of the Swedish Cement and Concrete Research Institute*, Vol. 26, Stockholm, 1955, pp. 1–103.

124.   Gulbransen, E. A., R. T. Phelps, and A. Langer, *Ind. Eng. Chem., Anal. Ed.*, **17**, 646 (1945).

125.   Haine, M. E., and V. E. Cosslett, *The Electron Microscope: The Present State of the Art*, Interscience, New York, 1961.

126.   Haine, M. E., and T. Mulvey, *J. Sci. Instr.*, **31**, 326 (1954).

127.   Hall, C. E., *J. Appl. Phys.*, **19**, 198 (1948).

128.   Hall, C. E., *J. Appl. Phys.*, **22**, 755 (1951).

129.   Hall, C. E., *J. Biophys. Biochem. Cytol.*, **1**, 1 (1955).

130.   Hall, C. E., *Introduction to Electron Microscopy*, 2nd ed., McGraw-Hill, New York, 1966.

131.   Hall, C. E., M. A. Jakus, and F. O. Schmitt, *J. Appl. Phys.*, **16**, 459 (1945).

132.   Halldal, P., J. Markali, and T. Naess, *Mikroskopie*, **9**, 197 (1954).

133.   Halliday, J. S., and T. Quinn, *Br. J. Appl. Phys.*, **11**, 486 (1960).

134.   Harford, C., and A. Hamlin, *Nature*, **189**, 505 (1961).

135.   Hearle, J. W. S., J. T. Sparrow, and P. M. Cross, *The Use of the Scanning Electron Microscope*, Pergamon, and Oxford, 1972.

136.   Heidi, H. G., *Naturwissenschaften*, **14**, 313 (1960).

137.   Heidenreich, R. D., *Fundamentals of Transmission Electron Microscopy*, Interscience, New York, 1964.

138.   Heidenreich, R. D., and L. A. Matheson, *J. Appl. Phys.*, **15**, 423 (1944).

139.   Hibi, T., and K. Yada, "Recent Improvements in Metallic Shadow Casting," in reference 73, pp. 460–462.

140.   Hillier, J., "The Electron Microscope in the Determination of Particle Size Characteristics," in *Symposium on New Methods for Particle Size Determination in the Subsieve Range*, Am. Soc. Testing Materials, Philadelphia, 1941, p. 90.

141.   Hillier, J., *Phys. Rev.*, **64**, 318 (1943).

142.   Hillier, J., and R. F. Baker, *J. Appl. Phys.*, **15**, 663 (1944).

143.   Hillier, J., S. Mudd, and A. G. Smith, *J. Bacteriol.*, **57**, 319 (1949).

144.   Holt, D. B., M. D. Muir, P. R. Grant, and T. M. Boswarva, *Quantitative Scanning Electron Microcopy*, Academic, New York, 1974.

145.   Houwink, A. L., and B. J. Spit, Eds., *European Regional Conference on Electron Microscopy, Delft, 1960*, Vol. I. De Nederlandse Vereniging voor Electronenmicroscopie, Delft, 1961.

146.   Hren, J. J., Ed., *Analytical Electron Microscopy, 1979 (Aug. 13–14) Workshop, San Antonio, Texas, Proceedings*, Plenum, New York, 1979.

147.   Hsiao, C. C., *Nature*, **181**, 1527 (1958).

148.   Humbert, R. P., *Am. Ceram. Soc. Bull.*, **21**, 260 (1942).

149.   Humbert, R. P., and B. Shaw, *Soil Sci.* **52**, 481 (1941).

150.   Husemann, E., and H. Ruska, *J. Prakt. Chem.*, **156**, 1 (1940).

151.   Husemann, E., and H. Ruska, *Naturwissenschaften*, **28**, 534 (1940).

152.   Hyam, E. D., and J. Nutting, *Br. J. Appl. Phys.*, **3**, 173 (1952).

153.   Hyatt, M. A., Ed., *Principles and Techniques of Electron Microscopy*, Vols. 1–5, Van Nostrand-Reinhold, New York, 1970–1975.

154. Hyatt, M. A., *Basic Electron Microscopy Techniques*, Van Nostrand Reinhold, New York, 1972.

155. Induni, G., *Helv. Phys. Acta*, **20**, 463 (1947).

156. Isaacson, M., "Specimen Damage in the Electron Microscope," in M. A. Hyatt, Ed., *Principles of Electron Microscopy*, Vol. 7, Van Nostrand Reinhold, New York, 1977, p. 1.

157. Jacobson, A. E., and W. F. Sullivan, *Ind. Eng. Chem.*, **18**, 360 (1946).

158. Jakob, A., II, and H. Loofmann, *Bodenk Pflanzenernahr*, **66/67**, 666 (1940).

159. Johari, O., Ed., *Scanning Electron Microscopy, Proceedings of the Annual Meetings*, Illinois Institute of Technology Research Institute, Chicago, 1968–1977.

160. Johari, O., Ed., *Scanning Electron Microscopy, Proceedings of the Annual Meetings*, Scanning Electron Microscopy Inc. (SEM, Inc.) AMF O'Hare, Illinois, 1978 and 1979.

161. Johnson, R. P., and W. Shockley, *Phys. Rev.*, **49**, 436 (1936).

162. Jouffrey, B., et al., "On Chemical Analysis of Thin Films by Energy Loss Spectroscopy," in J. M. Sturgess, Ed., *Electron Microscopy 1978*, Vol. III, Microscopical Society of Canada, Toronto, 1978, p. 292.

163. Joy, D. C., in Johari, O., Corvin, I., Ed., *Proceedings of the VIth Annual SEM Symposium*, Illinois Institute of Technology Research Institute, Chicago, 1973, pp. 743–750.

164. Kausche, G. A., and H. Ruska, *Naturwissenschaften*, **28**, 303 (1940).

165. Kay, D., Ed., *Techniques for Electron Microscopy*, Thomas, Springfield, Illinois, 1961, p. 352.

166. Kazan, B., and M. Knoll, *Electron Image Storage*, Academic, New York, 1968, pp. 114, 147, 237, 360, 663.

167. Kelly, P. M., and J. Nutting, *J. Inst. Metals*, **81**, 385 (1959).

168. Kikuchi, S., *Proc. Imp. Acad., Japan*, **4**, 271 (1928).

169. Kirkpatrick, A. F., and E. G. Davis, *Anal. Chem.*, **20**, 965 (1948).

170. Knoll, M., *Z. Techn. Phys.*, **16**, 467 (1935).

171. Knoll, M., and G. Lubszinsky, *Z. Physik*, **34**, 671 (1933).

172. Knoll, M., and E. Ruska, *Ann. Phys.*, **12**, 607, 641 (1932); *Z. Techn. Phys.*, **18**, 389 (1931).

173. Knoll, M., and E. Ruska, *Z. Physik.*, **78**, 318 (1932).

174. Koenig, L. R., *Anal. Chem.*, **31**, 1732 (1959).

175. Kropa, E. L., C. J. Burton, and R. B. Barnes, in *Electron Microscope Society of America Meeting*, New York, January 11, 1944, J. Appl. Phys., **15** (1944).

176. Kruger-Thiemer, E., *Z. Wiss. Mikroskop.*, **62**, 44 (1955).

177. Lacy, D., Ed., *The Microtomist's Vade Mecum*, Churchill, London, 1960.

178. Langmore, J. P., "Electron Microscopy of Atoms," in M. A. Hayatt, Ed., *Principles of Electron Microscopy*, Vol. 9, Van Nostrand Reinhold, New York, 1977.

179. Latta, H., and J. F. Hartman, *Proc. Soc. Exp. Biol. Med.*, **74**, 436 (1950).

180. Lebiedzik, J., et al., in Johari, O., Ed., *Scanning Electron Microscopy/1974*, Illinois Institute of Technology Research Institute, Chicago, 1974, p. 121.

181. Lehmpfuhl, G., "Convergent Beam Electron Diffraction," in J. M. Sturgess, Ed., *Electron Microscopy 1978*, Vol. III, Microscopical Soc. Canada, Toronto, 1978, p. 304.

182. Lenz, F., *Z. Naturforsch.*, **9a**, 185 (1954).

183. Liquier-Milward, J., *Nature*, **177**, 619 (1956).

184. Lodge, J. P., Jr., *Chemical Methods of Identification of Individual Particulates*, U. S. Dept. of Health, Education, and Welfare, Washington, D. C., 1957.

185. Magan, C., *Nucleonics*, **4**, 52 (1949).

186. Magan, C., and P. Chanson, "Note sur le Contraste des Images en Microscopie Protonique du a la Diffusion des Protons et au Processus de Perte et Capture de Charge dans les Objets de Faible Poids Atomique," in reference 73, p. 294.

187. Mahl, H., Z. Techn. Physik, **23**, 117 (1942).

188. Marton, C., Phys. Rev., **66**, 159 (1944).

189. Marton, C., "History of Electron Optics," in reference 62, p. 155.

190. Marton, C., and P. H. Abelson, Science, **106**, 69 (1947).

191. Marton, C., and S. Sass, J. Appl. Phys., **14**, (October), 522 (1943).

192. Marton, C., and S. Sass, J. Appl. Phys., **15**, 575 (1944).

193. Marton, C., and S. Sass, J. Appl. Phys., **16**, 373 (1945).

194. Marton, C., S. Sass, M. Swerdlow, A. Van Bronkhorst, and H. Meryman, Natl. Bur. Std. Circ., **502**, (1950).

195. Marton, L., Ann. Bull. Soc. R. Med. Nat. Bruxelles, **92**, 106 (1934).

196. Marton, L., The Early History of the Electron Microscope, San Francisco Press, San Francisco, 1968.

197. Marton, L., and L. I. Schiff, J. Appl. Phys., **12**, 759 (1941).

198. McMullan, D., Ph.d. Dissertation, Cambridge University, Cambridge, 1952; Proc. I.E.E., **B100**, 245 (1953).

199. Mecklenburg, W., Z. Physik, **120**, 21 (1942).

200. Meek, G. A., Practical Electron Microscopy for Biologists, 2nd ed., Wiley, New York, 1976.

201. Melton, C. W., and C. M. Schwartz, J. Appl. Phys., **32**, (8), 1636 (1961).

202. Mikroskopie (Zentralblatt für Mikroskopische Forschung und Methodik), Verlag George Fromme and Company, Vienna.

203. D. L. Misell, Ed., Misell, D. L., "Developments in Electron Microscopy and Analysis, 1977," in D. Misell, Ed., Proc. EMAG, Glasgow, 1977, Conf. Series No. 36, Institute of Physics, London and Bristol, 1977.

204. Misell, D. L., J. Phys. D: Appl. Phys., **10**, 1085 (1977).

205. Mollenstedt, G., and H. Duker, Optik, **10**, 192 (1953).

206. Mollenstedt, G., and M. Keller, "Direkte Übermikroskopische Sichtbarmachung von Oberflächen mittels Ionenausgelöster Elektronen," in reference 73.

207. Moore, C. A., Proc. Oklahoma Acad. Sci., **27**, 86 (1947).

208. Mudd, S., and T. F. Anderson, J. Exp. Med., **76**, 103 (1942).

209. Muller, E. W., Z. Physik, **37**, 838 (1936).

210. Muller, E. W., Z. Physik, **106**, 132 (1937).

211. Muller, E. W., Z. Physik, **106**, 541 (1937).

212. Muller, E. W., "Point Projection Microscopes," in National Bureau of Standards, Symposium on Electron Physics, Washington, D.C., November 5–7, 1951.

213. Muller, E. W., Sci. Am., **196**, 113 (1952).

214. Muller, E. W., Phys. Rev., **102**, 618 (1956).

215. Muller, E. W., J. Appl. Phys., **27**, 474 (1956).

216. Muller, E. W., "Field Emission Microscopy," in W. G. Berl, Ed., Physical Methods in Chemical Analysis, Academic, Vol. 3. New York, 1956, pp. 135–181.

217. Muller, E. W., and K. Bahadur, Phys. Rev., **102**, 624 (1956).

218. Muller, H. O., Kolloid-Z., **99**, 6 (1942).

219. Mulvey, T., Proc. R. Microsc. Soc., **2**, 201 (1967).

220. Mulvey, T., Ed., "Electron Microscopy and Analysis, 1979," in *Proc. EMAG, Univ. of Sussex, Brighton, 1979*, Conf. Series No. 52, Institute of Physics, Bristol, and London, 1980.

221. Murata, K., Matsukawa, T., Shimizu, R. *Jpn. J. Appl. Phys.*, **10**, 678 (1971).

222. Nankivell, J. F., *Br. J. Appl. Phys.*, **4**, 141 (1953).

223. *The International Bibliography of Electron Microscopy*, The New York Society of Electron Microscopists, New York, 1950–1961.

224. Nixon, W. C., "Point Projection X-Ray Microscopy," in reference 62, p. 647.

225. Nixon, W. C., Ed., *Proceedings of the 25th Anniversary Meeting of EMAG (Electron Microscopy and Analysis Group)* Institute of Physics, Cambridge and London, 1971.

226. Oatley, C. W., *Scanning Electron Microscopy*, Cambridge Univ. Press, London, New York, 1972.

227. O'Brien, R. T., and L. A. George II, *Nature*, **183**, 1461 (1959).

228. Olsen, L. O., C. S. Smith, and E. C. Crittenden, *J. Appl. Phys.*, **16**, 425 (1945).

229. Parsons, D. F., V. R. Matricardi, J. Subjek, I. Udyess, and G. Wray, *Biochim. Biophys. Acta*, **290**, 110 (1972).

230. Pashley, D. W., and A. E. B. Presland, *J. Inst. Metals*, **87**, 419 (1958).

231. Pattee, H. H., "Contact Microradiography," in reference 62, p. 561.

232. Pease, D. C., *Histological Techniques for Electron Microscopy*, Academic, New York, 1960, p. 274.

233. Pease, R. F. W., Ph.D. Dissertation, Cambridge University, Cambridge, 1963.

234. Pease, R. F. W., and W. C. Nixon, *J. Sci. Instr.*, **42**, 81 (1965).

235. Phillips, R., *Br. J. Appl. Phys.*, **12**, 554 (1961).

236. Poen, O. S., "Projection Microscopy," in reference 62, p. 661.

237. Pohl, J., *Z. Techn. Physik.*, **15**, 579 (1934).

238. Pohl, M. W., and G. Burchard, *Scanning*, **3**, 251 (1980).

239. Porter, K. R., and F. Kallman, *Exp. Cell Res.*, **4**, 127 (1953).

240. Porter, M. W., and R. C. Spiller, Eds., *The Barker Index of Crystals*, W. Heffer & Sons, Cambridge, 1951.

241. Prebus, A. F., "The Electron Microscope," in J. Alexander, Ed., *Colloid Chemistry*, Vol. 5, Reinhold, New York, 1944, pp. 152–235.

242. Prebus, A., and J. Hillier, *Can. J. Res.*, **A17**, 49 (1939).

243. Preuss, L. E., *J. Appl. Phys.*, **24**, 1401 (1958).

244. *Proceedings of the Conference on Electron Microscopy, Delft, 1949*, Nijhoff, The Hague, 1950.

245. Przyblski, R. J., *Exp. Cell Res.*, **24**, 181 (1961).

246. Radczewski, O. E., H. O. Muller, and W. Eitel, *Naturwissenschaften*, **27**, 837 (1939).

247. Radczewski, O. E., H. O. Muller, and W. Eitel, *Zement*, **28**, 693 (1939).

248. Radczewski, O. E., H. O. Muller, and W. Eitel, *Zement*, **29**, 1 (1939).

249. Radczewski, O. E., H. O. Muller, and W. Eitel, *Veroffentl. Kaiser Wilhelm Inst. Silikatforsch.*, **10**, 139 (1949).

250. Rang, O., and F. Schleich, *Z. Physik*, **136**, 547 (1954).

251. Rang, O., and H. Schluge, *Optik*, **9**, (10), 463 (1952).

252. Rathburn, M. E., M. J. Eastwood, and O. M. Arnold, *J. Appl. Phys.*, **17**, 759 (1946).

253. Recknagel, A., *Z. Physik*, **117**, 689 (1941).

254. Reimer, L., *Elektronenmikroskopische Untersuchungs- und Präparationsmethoden*, 2nd ed., Springer-Verlag, Berlin, 1967.

255. Reisner, J. H., *J. Appl. Phys.*, **31**, 1626 (1960).
256. Reisner, J. H., *Sci. Instr. News, RCA*, **6**, 1 (1961).
257. Rifkind, R. A., C. Morgan, and H. M. Rose, *J. Appl. Phys.*, **31**, 1838 (1960).
258. Rochow, T. G., "Resinography," in reference 62, p. 525.
259. Rudenberg, R., *J. Appl. Phys.*, **14**, 434 (1943).
260. Ruska, E., and M. Knoll, *Z. Techn. Phys.*, **12**, 389, 488 (1931).
261. Rutheman, G., *Naturwissenschaften*, **29**, 648 (1941).
262. Rutheman, G., *Naturwissenschaften*, **30**, 145 (1942).
263. Rutheman, G., *Ann. Physik*, **2**, (6), 113 (1948).
264. Ryozi, U., Ed., *Proceedings of the Sixth International Congress on Electron Microscopy, Kyoto, 1966*, Vols. VI and VII, Maruzen, Tokyo, 1966.
265. Ryter, A., and E. Kellenberger, "Inclusion on Polyester," in reference 18, Vol. II, p. 52.
266. Sanders, J. V., and D. J. Goodchild, Ed., *Proceedings of the Eighth International Congress on Electron Microscopy, Canberra, Australia*, Vols. 1–3, Australian Academy of Science, Canberra, 1974.
267. Schuster, M. C., and E. F. Fullam, *Ind. Eng. Chem., Anal. Ed.*, **18**, 653 (1946).
268. Scott, G. H., and D. M. Packer, *Science*, **89**, 227 (1938).
269. Scott, G. H., and D. M. Packer, *Anat. Rec.*, **74**, 17 (1939).
270. Scott, G. H., and D. M. Packer, *Anat. Rec.* **74**, 31 (1939).
271. Scott, G. H., and D. M. Packer, *Proc. Soc. Exp. Biol. Med.*, **40**, 301 (1939).
272. Scott, R. G., *J. Appl. Phys.*, **28**, 1089 (1957).
273. Shaw, B. T., and R. P. Humbert, *Soil Sci. Soc. Am., Proc.*, **6**, 147 (1941).
274. Siegel, B. M., *Modern Developments in Electron Microscopy*, Academic, New York, 1964.
275. Siegel, B. M., and D. R. Beaman, Eds. *Physical Aspects of Electron Microscopy and Microbeam Analysis*, Wiley, New York, 1975.
276. Sikorski, J., and W. S. Simpson, *Nature*, **182**, 1235 (1958).
277. Silcox, J., "Inelastic Electron Matter Interactions," in J. M. Sturgess, Ed., *Electron Microscopy 1978*, Vol. III, Microscopical Society of Caanada, Toronto, 1978, p. 259.
278. Silvester, N. R., and R. E. Burge, *Nature*, **188**, 641 (1960).
279. Sitte, H., "Physikalische Probleme Bei der Herstellung von Dunnschnitten," in reference 18, Vol. II, p. 63.
280. Sjostrand, F. S., in *Physical Techniques in Biological Research*, Vol. III, Academic, New York, 1956, Chapter 6.
281. Sjostrand, F. S., and J. Rhodin, Eds., *Electron Microscopy: Proceedings of the Stockholm Conference, September, 1956*, Academic, New York, 1957.
282. Sliepcevich, C. M., L. Gildart, and D. M. Katz, *Ind. Eng. Chem.*, **35**, 1178 (1943).
283. Smith, K. C. A., Ph.D. Dissertation, Cambridge Universtiy, Cambridge, England, 1956.
284. Smith, K. C. A., *Pulp and Paper Magazine, Canada*, **60**, T366 (1959).
285. Smith, K. C. A., "Scanning," in reference 62, p. 251.
286. Stanley, W. M., *J. Biol. Chem.*, **144**, 589 (1942).
287. Stojanowa, G., "Ein Kammer fur die Untersuchung von Objekten mit Gasumgebung," in reference 18, Vol. I, p. 82.
288. Sturgess, J. M., "Radiation Damage," in J. M. Sturgess, Ed., *Electron Microscopy 1978*, Vol. III, Microscopical Society of Canada, Toronto, 1978, p. 27.
289. Sturgess, J. M., V. I. Kalnins, F. P. Ottensmeyer, G. T. Simon, Eds., *Proceedings of the Ninth International Congress on Electron Microscopy, Toronto, Aug. 1–9, 1978*, Microscopical Society of Canada Toronto, 1978.

290. Suito, E., *Studies of Micro-Crystals by the Electron Micro Diffraction Method*, Institute for Chemical Research, Kyoto University, Takatsuki, Japan, 1957.

291. Suito, E., and N. Uyeda, *Proc. Jpn. Acad.*, **32**, 117 (1956).

292. Swerdlow, M., H. F. McMurdie, and F. A. Heckman, "Hydration of Tricalcium Silicate," in reference 73, pp. 500–503.

293. "Symposium on Electron Staining," *J. R. Microscop. Soc., Series III*, **78**, Parts I and II (1959).

294. Talbot, J. H., "Identification of Minerals Present in Mine Dusts by Electron Diffraction and Electron Microscopy," in reference 281, p. 353.

295. Talbot, J. H., "Selected Diffraction," in reference 62, p. 251.

296. Thomas, G., *Transmission Electron Microscopy of Metals*, Wiley, New York, 1962.

297. Thomas, G., "Kikuchi Electron Diffraction and Applications," in S. Amelinckx et al., Eds., *Modern Diffraction and Imaging Techniques in Materials Science*, North Holland, Amsterdam, 1970, p. 159.

298. Tighe, N. J., "Experimental Techniques," in H. R. Wenk, Ed., *Electron Microscopy in Mineralogy*, Springer-Verlag, New York, 1976, pp. 144–171.

299. Titlbach, M., Ed., *Electron Microscopy 1964, Third European Congress for Electron Microscopy, Prague, 1964*, Vol. A (Non-Biology), Vol. B, (Biology), Czechoslovak Academy of Sciences, Prague, 1965.

300. Trillat, J. J., Ed., *Congress de Microscopie Electronique, Toulouse, 1955*, Centre Nationale de la Recherche Scientifique, Tolouse, 1956.

301. Tufts, B. J., *Anal. Chem.*, **31**, 238 (1959).

302. Tufts, B. J., *Anal. Chem.*, **31**, 242 (1959).

303. Tufts, B. J., and J. P. Lodge, *Anal. Chem.*, **30**, 3003 (1958).

304. Turkevich, J., and J. Hillier, *Anal. Chem.*, **21**, 475 (1949).

305. Vandermeerssche, G., Ed., *Rapport Europees Congress Togepaste Electronenmicroscopie, Ghent.*, 1954, Rijksuniversiteit, Ghent, Belgium, 1954.

306. Venables, J. A., Ed., "Developments in Electron Microscopy and Analysis," in *Proc. EMAG, Univ. of Bristol, 1975*, Academic, New York, 1976.

307. Venables, J. A., and A. P. Janssen, "Developments in Scanning Auger Microscopy," in J. M. Sturgess, Ed., *Electron Microscopy 1978*, Vol. III, Microscopical Society of Canada, Toronto, 1978, p. 280.

308. Von Ardenne, M., *Z. Techn. Phys.*, **109**, 407 (1938); **109**, 553 (1938).

309. Von Ardenne, M., *Naturwissenschaften*, **27**, 485 (1939).

310. Von Ardenne, M., *Z. Physik. Chem. (A)*, **187**, 1 (1940).

311. Von Ardenne, M., K. Endell, and E. Hoffman, *Ber. Dtsch. Keram. Ges.*, **21**, 209 (1940).

312. Von Borries, B., *Die Ubermikroskopie*, Editio Canto, Aulendorf, Wurttemberg, 1949.

313. Von Borries, B., *Z. Naturforsch.*, **4a**, 51 (1949).

314. Von Borries, B., and G. A. Kausche, *Kolloid-Z.*, **90**, 132 (1940).

315. Von Heimendahl, M., *Electron Microscopy of Materials, An Introduction*, (Translation by Ursula E. Wolff, of *Einfuhrung in die Elektronenmikroskopie*, Academic, New York, 1980.

316. Watson, M. L., *J. Biophys. Biochem. Cytol.*, **3**, 1017 (1957).

317. Wiesenberger, E., *Z. Wiss. Mikroskop.*, **62**, 163 (1955).

318. Wiesenberger, E., *Mikrochim. Acta Wien*, **H. 3–4**, 527 (1957).

319. Wiesenberger, E., "The Use of Microchemical Detection Methods in Electron Microscopy," in reference 13, Vol. I, p. 769.

320. Wiesenberger, E., "Electron Microscopical Tests for Identification of Electrolytically Precipitated Cu With Picric Acid," in reference 145, pp. 320–325.

321. Williams, R. C., and R. C. Backus, *J. Appl. Phys.*, **20**, 98 (1949).

322. Williams, R. C., and R. C. Backus, *J. Amer. Chem. Soc.*, **71**, 4052 (1949).

323. Williams, R. C., and R. W. G. Wyckoff, *J. Appl. Phys.*, **15**, 712 (1944).

324. Williams, R. C., and R. W. G. Wyckoff, *Science*, **101**, 594 (1945).

325. Williams, R. C., and R. W. G. Wyckoff, *Proc. Soc. Exp. Biol. Med.*, **58**, 265 (1945).

326. Wywich, H., and F. Lenz, *Z. Naturforsch.*, **13a**, 515 (1958).

327. Yakowitz, H., "A Practical Examination of Kossel X-ray Diffraction Technique," in C. A. Anderson, Ed., *Microprobe Analysis*, Wiley, New York, 1973, p. 383.

328. Yew, N. C., and D. E. Pease, "Signal Storage and Enhancement Techniques for SEM," in Johari, O., Ed., *Scanning Electron Microscopy, 1974*, Illinois Institute of Technology Research Institute, Chicago, 1974, p. 191.

329. Zeitler, E., *Adv. Electr. Electr. Phys.*, **25**, 277 (1968).

330. Zeitler, E., "Electron Matter Interaction," in J. M. Sturgess, Ed., *Electron Microscopy 1978*, Vol. III. Microscopical Society of Canada, 1978, pp. 29–39.

331. Zeitler, E., and G. F. Bahr, *Exp. Cell Res.*, **12**, 44 (1957).

332. Zeitler, E., and G. F. Bahr, *J. Appl. Phys.*, **30**, 940 (1959).

333. Zeitler, E., and G. F. Bahr, *Sci. Instr. News, RCA*, **5**, (1), 5 (1960).

334. *Z. Wiss. Mikroskopie*, S. Heirzel Verlag, Stuttgart.

335. Zworykin, V. K., *J. Franklin Inst.*, **15**, 535 (1933).

336. Zworykin, V. K., J. Hillier, and R. L. Snyder, *ASTM Bull.*, **117**, 15 (1942).

337. Zworykin, V. K., G. A. Morton, E. G. Ramberg, and J. Hillier, *Electron Optics and the Electron Microscope*, Wiley, New York, 1954.

**Part I
Section H**

Chapter 17

ELECTRON SPECTROSCOPY

By Giorgio Margaritondo *University of Wisconsin, Madison, Wisconsin* and John E. Rowe, *Bell Laboratories, Murray Hill, New Jersey*

Contents

I.   INTRODUCTION

The term "electron spectroscopy" identifies a class of experimental techniques for the study of atoms, molecules, solids, and solid surfaces. These techniques have in common the use of electrons as probes of the system under investigation (48). They also have a common goal: the study of the electronic states of the system. The information about electronic states can in turn provide by-product information about the atomic structure and the chemical composition. Each electron spectroscopy technique can be either an "electron emission" technique or an "electron scattering" technique. In the first kind of techniques, electrons are emitted by the system on excitation by photons, ions, or other electrons. In the second kind of techniques, a primary beam of electrons is scattered by the system. In both kinds of techniques, the electrons emerging from the system are analyzed to characterize their state as much as possible. In a one-electron picture, the final state of the process, $| f \rangle$, is that of a free electron—that is, in quantum-mechanical terms, it is identified by its vector momentum. The final state is related by a scattering or emission process to an initial state $| i \rangle$. In a scattering process $| i \rangle$ is also a free-electron state, whereas in an emission process it is a bound state.

The above general description fits a number of different experimental techniques. Many of these techniques went through a recent period of accelerated development (89). The causes for this acceleration were substantial improvements in the instrumentation and in the theoretical background required to relate the $| i \rangle \rightarrow | f \rangle$ process to the electronic structures of atoms, molecules, and solids. Several techniques that were at an initial development stage a few years ago are now reliably used an increasingly large number of nonspecialized users. Among these, the leading ones are photoemission spectroscopy and Auger-electron spectroscopy, to which our review will dedicate the most emphasis. We emphasize that although these two techniques have made great progress in recent years, neither one has completed its development. New results actually emerge in the forefront of these fields every year. We shall make in this review a reasonable effort to include recent developments as well as established experimental approaches.

Our review begins in Section II with a general classification of the electron-spectroscopy techniques. Section III is dedicated to instrumentation problems. We discuss in particular primary beam sources, electron analyzers, ultrahigh-vacuum hardware, and equipment for sample handling, preparation, and characterization. Sections IV and V are dedicated to photoemission techniques using uv or x-ray photons is exciting primary particles. Section VI discusses the different aspects of Auger-electron spectroscopy. Some considerations about general and specific limits of the electron-spectroscopy techniques are discussed in Section VII. In the final Section VIII we identify the future trends and foreseeable developments in this rapidly evolving field. A detailed theoretical discussion of different electron-spectroscopy techniques can be found in the Appendix.

In all types of electron-spectroscopy techniques, the $| i \rangle \rightarrow | f \rangle$ transition is excited by the interaction between the electromagnetic field of the incoming particle—electron, photon, or ion—and the electrons in the system. This interaction depends in turn on the quantum-mechanical state of the electrons. The identification of this state is the primary aim of electron spectroscopy. The "electronic" state of the system can be defined rigorously for atoms and approximately, that is, within the limits of the adiabatic approximation, for molecules and solids. In the latter case the wavefunction describing the state of the system, $\Phi(\mathbf{r}_i, \mathbf{R}_j)$ (where $\mathbf{r}_i$ are the electronic coordinates and $\mathbf{R}_j$ the nuclear coordinates), is separated into an electronic component and a nuclear component:

$$\Phi(\mathbf{r}_i, \mathbf{R}_j) = \psi(\mathbf{r}_i)\chi(\mathbf{R}_j).$$

The wavefunction $\psi(\mathbf{r}_i)$ identifies the "electronic state" of the system. Notice that $\psi(\mathbf{r}_i)$ is a function of *all* the electron coordinates at once; and that the electrons interact with each other. The description of the electronic state is further simplified by neglecting the electron–electron effects in a one-electron approximation. This approximation is justified by the sharpness of the Fermi–Dirac distribution at all practical temperatures. The many-body effects neglected in the one-electron approximation, however, play an important role in most electron-spectroscopy techniques. Therefore, a complete treatment of electron spectroscopy requires one to go beyond the one-electron limits, as discussed in detail in the Appendix.

In the case of solids the quantum-mechanical description of the electronic states is complicated even in the one-electron approximation, due to the extremely large number of nuclei. For crystalline solids, however, the periodic arrangement of the nuclei simplifies the description because of the *Bloch theorem*. The periodic arrangement of the nuclei is defined in real space by the lattice vectors $\mathbf{L} = n_1\mathbf{a}_1 + n_2\mathbf{a}_2 + n_3\mathbf{a}_3$ ($\mathbf{a}_i$ is the primitive lattice vector; n_i are integers) and in k-space by the G-vectors defined by $\mathbf{G} \times \mathbf{L} = 2n\pi$ (n is an integer). According to the Block theorem, the one-electron wavefunctions in a crystalline solid are combinations of Bloch waves:

$$\psi_B(\mathbf{r}) = u_{\mathbf{k}}(\mathbf{r})e^{i\mathbf{k}\cdot\mathbf{r}} \tag{1}$$

where $u_{\mathbf{k}}(\mathbf{r})$ is a periodic function with the same periodicity as the nuclear array in the crystal. Therefore $u_{\mathbf{k}}(\mathbf{r})$ can be expanded in a Fourier series:

$$u_{\mathbf{k}}(\mathbf{r}) = \sum_{\mathbf{G}} c_{\mathbf{G}} e^{i\mathbf{G}\cdot\mathbf{r}} \tag{2}$$

The Bloch theorem is rigorously valid only for perfect periodicity, that is, for an infinite, perfect crystal. In practice it is valid as long as surface effects and other periodicity breakdowns can be neglected. The surface effects cannot be neglected, however, in solid-state electron-spectroscopy investigations. A general property of all electron-spectroscopy techniques is that they probe only a ultrathin region at the surface of the sample, where surface effects are fundamental (8,68). The physical cause of the surface sensitivity of electron spectroscopy is schematically illustrated by Fig. 17.1, which shows a representative plot of the mean free path for low-energy electrons

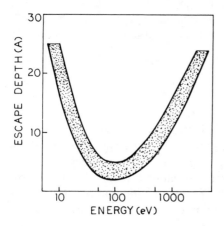

Fig. 17.1.   Kinetic-energy dependence of the escape depth for electrons in solids (68). The shaded area contains all the experimental points given by different kinds of materials. Notice the minimum occurring at 50–150 eV.

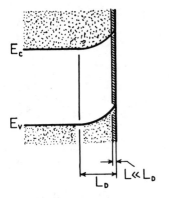

Fig. 17.2. Electron-spectroscopy techniques probe a thin region at the surface of solids. In a semiconductor the thickness of this active region, L, is much smaller than the distance over which the band bending occurs, L_D (93). Therefore even bulklike electronic states in the thin active region will *not* have the same energy as their counterparts in the bulk.

in a solid (68). We see that electrons of 10–200 eV cannot travel in a solid for more than 2–20 Å without losing energy. This gives an idea of the thickness of the region probed by electron spectroscopy. This thickness is 1–4 times the magnitude of the vectors a_i. Even in a first approximation, therefore, no long-range periodicity can be probed in a region so thin—at least not in the direction perpendicular to the surface.

The surface sensitivity of electron spectroscopy has fundamental consequences for the kind of electronic states probed in a solid (93). This is illustrated schematically in Fig. 17.2 for a semiconductor surface. The shaded area corresponds to bulklike valence-band and conduction-band states with edges E_v and E_c, respectively. Notice the *band bending* caused by the presence of the surface charge and extended over a region much thicker than that probed by electron spectroscopy. As a consequence the position in energy of the bulklike states probed by electron spectroscopy is different from that of the bulk. Furthermore, nonbulk states are also present in the thin region probed. In most cases they are caused by impurity atoms, but even on ultraclean surfaces the periodicity breakdown produces nonbulk, "surface" states. The above surface sensitivity is a great advantage in electron spectroscopy: A study of surface properties (e.g., surface chemistry) becomes possible without the overwhelming background bulk signal unavoidable in other experiments.

Each electronic state of the system under investigation and its corresponding wavefunction are related to specific values of several physical quantities, the *quantum numbers*. Rigorously speaking, only the total energy is a "good" quantum number for all stationary states of all systems. The angular momentum is defined rigorously only for the states of the hydrogen atom. Nevertheless, an angular-momentum labeling of the states $(s,p,d,f, \ldots)$ is often used for atoms other than hydrogen and to identify atomic components of molecular and solid-state wavefunctions. The leading role of the energy parameter explains why many electron-spectroscopy investigations are limited to measurements of the electronic-state energy.

Within those limits the electronic structure of a system is characterized by its *density of states* (DOS), $\rho(E)$. The function $\rho(E)$ is defined by the equation

$$\rho(E)dE = dn(E) \qquad (3)$$

where $dn(E)$ is the number of electronic states of the system with energy between E and $E + dE$. The determination of the DOS is the first and often only target of many electron-spectroscopy experiments.

We emphasize that a knowledge of the electronic states not limited to the energy involves properties that are not strictly "electronic." For example, the charge-density distribution in crystalline solids, $\mid \psi \mid^2$, has, according to equations 1 and 2, the same periodicity as the array of nuclei. Therefore the knowledge of $\mid \psi \mid^2$ yields information about the nuclear positions as well. In general information about the electronic states always has built in information about the atomic positions in the system (85) A great deal of effort has been dedicated in recent years to obtaining indirect atomic-structure information from the results of different kinds of electron-spectroscopy techniques (85,89). Some of these efforts will be described in Section V. Particularly important are the techniques for identifying chemisorption geometries, one of the leading topics of research in solid-state chemistry and physics.

II. GENERAL CHARACTERISTICS OF ELECTRON SPECTROSCOPY

In this section we shall outline the measurement techniques of different kinds of electron spectroscopies and the information they deliver (48). A general overview of the electron-spectroscopy techniques is shown in Table 17.I. Details about the theoretical backgrounds behind several of these techniques are discussed in the Appendix.

A. PHOTOEMISSION SPECTROSCOPY

Photoemission spectroscopy (8,89) is based on the photoemission process, which is schematically illustrated in Fig. 17.3. A photon is absorbed by the system, and it transfers energy to one of the electrons. The resulting energy of the excited electrons may be sufficient to overcome the barrier linking the electron to the system. This requires that the energy of the excited electron be higher than a threshold value known as vacuum level. If this is the case, the electron can leave the system and be detected. Depending on the energy of the photons, different kinds of electronic states can be probed. When ultraviolet photons are used, the excited electrons come primarily from the "shallow" levels in atoms and from valence states in molecules and solids. This technique is called ultraviolet photoemission spectroscopy (UPS). When x-ray photons (energy $h\nu > 30$ eV) are used, the electrons can be excited from deeper-lying core levels. Even in molecules and solids these

TABLE 17.I
The Different Kinds of Electron-Spectoscopy Techniques

Acronym	Technique	Input particles	Experimental method
	Emission techniques		
UPS, XPS (or ESCA)	uv, x-ray photoemission spectroscopy	Photons	Electron-direction and energy analysis
AES	Auger-electron spectroscopy	Electrons (or photons)	Electron energy analysis
FES	Field-emission spectroscopy	(Applied external field)	Electron-energy analysis
INS	Ion-neutralization spectroscopy	Ions	Electron-energy analysis
LEED	Low-energy electron diffraction	Electrons	Electron-direction analysis
	Scattering techniques		
ELS	Electron energy loss	Electrons	Electron-energy analysis
APS	Appearance-potential spectroscopy	Electrons	Detection of secondary electrons (or of x-ray emission)

levels have nearly atomic character, since they are not directly involved in the formation of chemical bonds. The presence of the corresponding core levels is an excellent fingerprint for the presence of a given chemical species; hence this technique can be a powerful method of chemical analysis. Furthermore, the small indirect effects on the core levels of the formation of chemical bonds are reasonably easy to detect and provide information about the bonding process itself. The name of this technique is x-ray photoemission spectroscopy (XPS); it is also known as electron spectroscopy for chemical analysis (ESCA).

In its most conventional version, photoemission spectroscopy is carried out in the following way: Photons of fixed energy $h\nu$ are sent into the sample. These photons excite electrons and some of these electrons become photoelectrons. The photoelectrons are collected and analyzed by a suitable

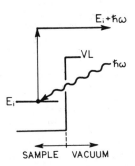

Fig. 17.3. Simple scheme of photoemission process. Electrons of energy E_i are excited by photons. They can leave the solid if their final energy is above the vacuum level (VL) and if the excitation occurred at a depth from the sample surface less than the escape depth.

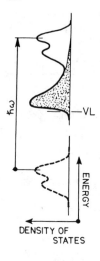

Fig. 17.4. A simplistic view of photoemission spectra. The density of states (DOS) gives the distribution in energy of the electrons in the system. This distribution is shifted to higher energies by the photon-absorption process. Thus the photoemission spectrum should correspond simply to the shifted distribution superimposed to the smooth secondary electron distribution (shaded area), which gives the peak near the vacuum level (VL). A better interpretation of photoemission spectra requires a more sophisticated theoretical background, as discussed in the text—but it still exhibits some of the features of this simplistic view.

analyzer. This instrument gives the number of collected electrons as a function of their kinetic energy, the so-called energy-distribution curves (EDCs). A simplistic interpretation of the technique relates the EDCs obtained in this way to the DOS of the system, $\rho(E)$. As shown in Fig. 17.4, the assumption is that the distribution in energy of the electrons inside the system is simply shifted in energy by $h\nu$ on absorption of photons of that energy. The distribution of photoelectrons outside the solid gives, in this simple picture, the DOS shifted by $h\nu$. This interpretation is an oversimplification for at least two fundamental reasons. The distribution is energy of the electrons is not only shifted, but also distorted, during the excitation process. This point is discussed in detail in the Appendix. Furthermore, the excited electrons in the solid have a good chance of losing energy before leaving the system due to one of the processes discussed later in this section in connection with energy-loss spectroscopy. They may thus become unable to leave the solid, and even if they do leave the solid, they may have less energy than they absorbed from the photon. As a result the spectrum will be a superposition of a *primary* electron distribution—due to unscattered or elastically scattered electrons—on a *secondary* electron distribution due to electrons that lost part of their excitation energy. This superposition is shown in Fig. 17.4 for a typical EDC. Secondary electrons are sometimes a problem, since they introduce a spurious although smooth background. But they also are the key factor in several new photoemission techniques, as we shall see in Section IV.

The above mode of running photoemission experiments was originally dictated by the instrumentation limits. In particular, the use of a fixed photon energy was required by the conventional UV and x-ray sources, which emit strong intensities concentrated in *lines* at fixed photon energies. These limitations have been removed in the past 10 years by the widespread use of synchrotron radiation as a spectrally continuous and tunable photon source

(89). New photoemission techniques have been made possible by these advances in instrumentation, as we shall discuss in Section IV. The new techniques study not only the density of occupied states in the system, but a variety of other parameters, such as the density of *unoccupied* states, the symmetry of occupied and unoccupied states, and the structural properties of the system (89).

B. AUGER-ELECTRON SPECTROSCOPY

Auger-electron spectroscopy (AES) is probably the most advanced electron spectroscopy as far as routine analytic use is concerned (48). It is based on the Auger process, which is illustrated in Fig. 17.5. The Auger process itself is preceded by an excitation process, which leaves a hole in a core level of the system. This preliminary excitation process can be stimulated by absorption of energy either from a primary electron beam or from a primary photon beam; in most analytic Auger experiments, the former is preferred. After the excitation process, the core hole can be filled by one of several different kinds of processes, of which Auger recombination is the most probable. In the Auger recombination the core hole is filled by an electron occupying a higher energy level, that is, either a shallower core level or a valence state. This electron must lose energy in the process, and the energy is transferred to another electron, also either a shallower core electron or a valence electron. This last electron may receive enough energy to leave the system, thus becoming an Auger electron. Auger electrons can be detected by an analyzer that measures their energy. We shall see that the energy of Auger electrons is an excellent fingerprint for the presence of a given chemical species, and AES therefore is a powerful analytic-chemistry probe.

The whole process—excitation plus Auger recombination—is usually labeled using the conventional spectroscopic notations for the initial states of the three electrons involved. For example, if the excitation involves a hole in a $1s$ state and this hole is filled by an electron in the $2s$ state that transfers its energy to another electron in the $2p$ state, then the entire process is labeled KL_1L_2. In this notation the valence states are denoted by the letter V. It is clear that the final energy of the Auger electron is linked to the energies of

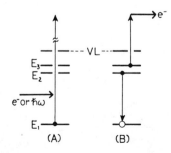

Fig. 17.5. Scheme of the Auger process. In a preliminary step (A) a hole is created in the system through excitation by a photon or an electron, which transfers all its energy to a core electron of energy E_1. The core hole is then filled by the Auger process (B) which involves two other electronic states of the system.

the three initial states. Therefore the emission of Auger electrons at that energy reveals the presence of certain energy levels in the system. Since those levels are characteristic of a given element, it also reveals the presence of that element in the system; hence the analytical capability of AES. The exact link between the kinetic energy of the emitted Auger electrons and the core-level energies is complicated by factors such as electronic relaxation on formation of a core hole and final-state multiplet interactions. Nevertheless, the uncorrected binding energies of the three electrons are the most important factor in determining the Auger-electron kinetic energy. This is what makes it possible to identify elements easily, without complicated calculations.

Neglecting all factors except the uncorrected binding energies, and calling these energies E_1, E_2, and E_3^*, for the first excited electron, the electron that fills the core hole, and the Auger electron, respectively, the kinetic energy of the latter can be written (48)

$$E_k = E_3^* + E_2 - E_1 \qquad (4)$$

This equation is derived by estimating the energy of the two-hole final states in two subsequent steps, one creating a hole in state 2 and the other creating a hole in state 3. The energies in equation 4 are measured from the zero of the kinetic-energy scale, that is, from the vacuum level. Notice the asterisk in the binding-energy term E_3^*; it means that this is *not* the unperturbed binding energy, but rather the binding energy in the presence of another core hole. In processes involving valence electrons, the energies E_2 and E_3^* are not as well defined as for core levels. In fact, either one or both of these energies can be anywhere in the energy range occupied by the valence states, within certain limits that we shall discuss in Section VII.

Figure 17.6 shows an example of Auger spectrum (123). Notice that the spectral features are emphasized by plotting the first derivative of the energy distribution. The presence of their corresponding features immediately identifies germanium, carbon, nitrogen, and chlorine as chemical components

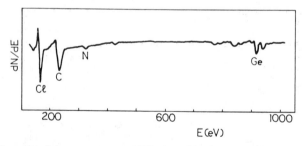

Fig. 17.6.   Example of Auger spectrum (123) plotted in the first-derivative mode to enhance its structure. The system was a surface of germanium contaminated by Cl and by hydrocarbons. The spectrum exhibits the characteristic peaks of Cl, C, N, and Ge.

of this system—which is indeed a contaminated germanium surface. The chemical composition detected by AES is that of the surface, due to the high surface sensitivity of AES, (see Section I). One can see that *qualitative* chemical analysis of a surface is very simple with AES. Quantitative analysis is also possible, but it must rely on a number of approximations, as we shall see in Section VI. One of the most interesting applications of quantitative Auger analysis is the depth-profiling technique, in which surface atoms are progressively removed by ion bombardment while the composition is continuosly probed by AES (12). Other uses of AES will be discussed in Section VI.

C. ENERGY-LOSS SPECTROSCOPY

A typical electron energy-loss experiment is schematically illustrated in Fig. 17.7. The primary electrons are *inelastically* scattered by the system, which in most cases is a solid surface. Part of their kinetic energy is transferred to the surface through excitation of quasi-one-electron transitions, of collective electronic oscillations, or of surface vibrations. An example of an energy-loss process is illustrated by the energy-level diagram in Fig. 17.7.

From a formal point of view the energy-loss processes are described (59) in terms of the complex dielectric function $\epsilon(\omega,\mathbf{k}) = \epsilon_1(\omega,\mathbf{k}) + \epsilon_2(\omega,\mathbf{k})$ of the system. This establishes a link between energy-loss spectroscopy and optical spectroscopy, which is also interpreted in terms of the dielectric function. The number of electrons scattered after having lost an energy E is proportional to the "loss function" $L(E)$ (59). The dielectric-function approach to energy loss shows that

$$L(E) = - \mathrm{Im}\left(\frac{1}{\epsilon}\right) \tag{5}$$

where "Im" indicates the imaginary part of the complex function $1/\epsilon$. The presence of the surface and therefore the limits to space averaging make it

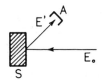

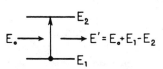

Fig 17.7. Top: Simple scheme of an electron energy-loss experiment. The primary-beam electrons of energy E_0 are inelastically scattered by the system (S), thus losing an energy $E = E_0 - E'$, and they are detected by the analyzer A. Bottom: One possible energy-loss mechanism is by one-electron transitions of the system from a level E_1 to a level E_2. The energy lost by the primary electron that excites the transition is $E_2 - E_1$.

difficult to define the function ϵ appearing in equation 5—indeed, the concept of dielectric function requires space averaging. One ordinarily—but arbitrarily—assumes that the surface and the bulk effects are linearly combined in the loss function, so that the total loss function is a superposition of surface and bulk losses, both given by an equation like equation 5. In the case of the bulk losses, the function ϵ that must be used is the bulk dielectric function of the material. Equation 5 can be written

$$L(E) = \frac{\epsilon_2}{\epsilon^2} \tag{6}$$

where $\epsilon^2 = \epsilon_1^2 + \epsilon_2^2$. This form immediately gives some insight in the physical processes responsible for the energy loss. Maxima in the bulk $L(E)$ function arise either from zeros of ϵ^2 or from maxima in ϵ_2, with the two effects sometimes interfering with each other (59). The condition $\epsilon(E_p) = 0$ defines the energies E_p at which collective oscillations of the electrons in the system are excited (117). The frequency of these oscillations is the *plasma frequency*, and the corresponding quantum of energy $E_p = h\nu_p$ is called the *plasmon*. It should be emphasized that collective electron oscillations are *longitudinal* excitations, and therefore cannot be excited directly by photons, which are transverse excitations. In contrast, electrons in energy-loss experiments do excite plasmons. Besides this difference, there are also similarities between the "optical" transitions excited by photons and the energy-loss transitions excited by electrons. For example, the optical absorption coefficient is proportional to ϵ_2 and therefore maxima in ϵ_2 correspond to maxima in the absorption coefficient. And according to equation 6 they correspond to maxima in the loss function as well. This similarity must be taken *cum grano salis*, however. In a photon-excited process the photon does not carry an appreciable momentum with respect to the electrons that are excited, whereas in an energy-loss process the primary electron does. Thus the momentum-conservation rules have different consequences for the two kinds of processes and lead to differences in the corresponding ϵ_2's. Nevertheless, the physical processes causing maxima in these ϵ_2's are fundamentally the same. These physical processes can be of two different types: one-electron (or quasi-one-electron) transitions, and vibrational transitions. In the first type, one electron in the system is excited from an allowed state to another allowed state, absorbing energy from the primary-beam electron. In the second, the lost energy is used by the system to increase the vibrational energy level of one of its normal modes.

The surface-originating energy losses can be described by an appropriate surface dielectric function (59). Although a rigorous definition of this function is quite difficult, its qualitative features can be understood from a simplified discussion. Once again, the maxima in the loss function arise either from excitation of plasmons or from quasi-one-electron or vibrational transitions. Processes of this last kind are similar to their bulk counterparts, but they involve surface electronic and vibrational states. The zeros in the die-lectric function will again define the excitation frequencies and energies for plasmons. In an elementary boundary-electromagnetic approach (59,121) the surface dielectric function can be written

$$\epsilon_s = \frac{\epsilon + 1}{2} \tag{7}$$

where ϵ is again the bulk dielectric function. For energies not too far from the plasmon resonance, ϵ has the form $1 - v_p^2/v^2$ (which indeed becomes zero for $v = v_p$) and therefore

$$\epsilon_s = \frac{v_p^2}{2v^2} \tag{8}$$

The zero of this function occurs for

$$v = v_s = 2^{1/2} v_p \tag{9}$$

This frequency corresponds to a collective excitation of the electrons lo-calized to the surface—the so-called surface plasmon. In real cases the factor linking v_p and v_s always deviates from $2^{1/2}$. For silicon, for example, $v_p = 17.1$ eV and $v_s = 11.5$ eV (61).

A practical energy-loss spectrum is of course a superposition of all the above processes, with possible interference among them. Figure 17.8 shows the spectrum of an ultraclean silicon surface (111) taken with a primary electron energy of 100 eV. Bulk plasmon losses, surface plasmon losses, bulk one-electron transitions, and surface one-electron transitions are all present in the spectrum, and they are identified in the caption. Notice that the spectral structure is emphasized in Fig. 17.8 by plotting the second de-rivative of the experimental energy-loss spectrum rather than the spectrum itself. The energy resolution of this experiment is not sufficient to see vi-brational energy losses. Vibrational losses can be observed in the high-res-

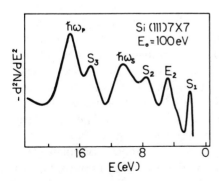

Fig. 17.8. A classic example of an electron en-ergy-loss spectrum (121). The spectrum was taken on a Si(111) surface with 7 × 7 recon-struction using a primary beam of energy $E_0 = 100$ eV. The spectrum is plotted in a negative-second-derivative form to enhance the spectral features (notice that peaks in the undifferen-tiated energy distribution correspond to peaks in $- d^2N/dE^2$. Peaks w_p and w_s are due to the creation of collective excitations, the bulk and surface plasmons. Peak E_2 is a bulk Si one-elec-tron transition. Peaks S_1, S_2, and S_3 are one-electron transitions involving the surface states of the clean Si(111) 7 × 7 surface.

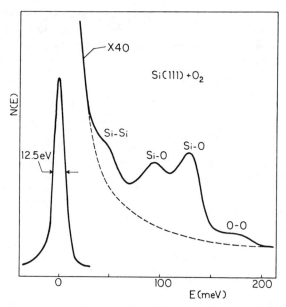

Fig 17.9. High-resolution, low-energy-loss spectrum of oxygen-covered Si(111) (60,126). The spectrum was taken after a moderate exposure to O_2 and before the formation of SiO_2. The peaks in this high-resolution spectrum correspond to the excitations of surface vibrational modes; their labels identify the different modes. Notice in particular the O—O stretching mode, which is evidence for nondissociative chemisorption of the O_2 molecule.

olution energy-loss spectrum of Fig. 17.9, taken on an oxygen-covered silicon surface (60,126). The different vibrational transitions are identified in the caption.

D. APPEARANCE-POTENTIAL SPECTROSCOPY

Appearance-potential spectroscopy (APS) is closely related to energy-loss spectroscopy, but differs with respect to the detection method (58,104). Instead of measuring the inelastically scattered electrons, one measures the secondary products (electrons or photons) of the de-excitation process following the energy-loss-induced excitation of the system. The excitation processes investigated in this way are typically one-electron transitions having a core level for an initial state. One of these transitions is schematically shown in Fig. 17.10. The excitation process leaves a core hole, which is then filled by a subsequent de-excitation process. For example, the core hole can be filled by an electron previously occupying a higher-in-energy core level on emission of a photon (see Fig. 17.10). The typical photon energy is in the soft-x-ray range, and the corresponding acronym for this technique is "SXAPS" or "soft-x-ray APS." The term "appearance" originates from the typical mode of operation, in which the energy of the primary beam is

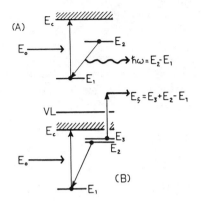

Fig. 17.10. Scheme of the processes involved in APS. In both cases a primary beam of particles of energy E_0 excites a transition from a core level E_1 to states above the threshold E_c. The occurrence of the transition is revealed by the products of the process that fills the core hole. This can be either a transition with emission of a x-ray photon (A) or an Auger process involving two other energy levels, E_2 and E_3, and giving an Auger electron of energy E_f (B).

scanned until a core-level excitation-energy threshold is reached and photons are collected ("appear") at the detector.

An alternate de-excitation mechanism is also shown in Fig. 17.10. In this case the energy lost by the electron that fills the core hole is transferred to another electron through an Auger process. The second electron may have enough energy to leave the system and be detected. This technique is called Auger-electron APS (AEAPS). A third, related technique is called disappearance-potential spectroscopy (DAPS). It consists of detecting the onset of one-electron transitions from a core hole by monitoring the corresponding *decrease* in the number of *elastically* scattered electrons.

E.   OTHER SURFACE TECHNIQUES

1.   Ion-Neutralization Spectroscopy

Ion neutralization spectroscopy (INS) (46,53) provides information similar to that given by uv photoemission using a process conceptually similar to an Auger process. As in AES, Auger electrons are created on neutralization of a core hole. The Auger neutralization involves valence electrons. The core hole is created not by a preliminary excitation process, but by the arrival of an external ion on the surface of the sample. The corresponding process is illustrated in Fig. 17.11. The kinetic energy of the Auger electron is given by an equation similar to equation 4 for the Auger case:

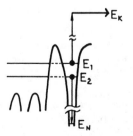

Fig. 17.11. Scheme of an ion-neutralization process at a solid surface. The electrons in the levels E_2 and E_1 can tunnel through the potential barrier and neutralize the ion by an Auger process. The entire process is revealed by the emission of an Auger electron of energy E_k.

$$E_k = E_1 + E_2 - E_n \tag{10}$$

Here E_n is the neutralization energy of the ion at the surface and E_1 and E_2 are the initial-state energies for the two electrons. Notice that the E_n does *not* coincide with the neutralization energy in free space, since this energy is substantially lowered at the surface by the image-charge force. The correction is of the order of 2 eV.

As shown in Fig. 17.11 the INS process involves tunneling of electrons through the thin barrier created by the presence of the ion. This makes the INS technique somewhat similar to field-emission spectroscopy (FES). For example, the electronic states from which tunneling is easy give enhanced contributions to both kinds of spectra. A detailed analysis of the INS and FES curves requires discussion of the corresponding processes; this will be carried out in the Appendix. We anticipate that after correction for the tunneling probability both kinds of spectra yield information about the density of states in an energy region near the Fermi level. In FES this region is limited to within 1–2 eV of the Fermi level, E_F. In INS it can be extended down to 10–15 eV below E_F, and therefore the INS technique becomes comparable with uv photoemission carried out with, for example, conventional photon sources at 16.8 or 21.2 eV.

2.   Field-Emission Spectroscopy

Field emission spectroscopy or FES (41,106,140) differs from all other techniques considered in this review because no primary beam of particles is involved. Instead, an external electric field is applied to the sample. The purpose of this electric field is to remove electrons from a solid by the tunneling process illustrated in Fig. 17.12. The process requires that the vacuum level outside the solid be dropped and that the barrier be thin enough to give a reasonably large tunneling probability. Therefore a large electric field is required in the immediate neighborhood of the surface. This is obtained by shaping the sample itself as a ultrasharp needle.

In a typical FES experiment the tip of the sample has a radius of the order of 10^{-5} cm. A typical applied voltage of 10^3 V causes a field of the order of 10^7 V/cm. One should notice in Fig. 17.12 that the thickness of the barrier rapidly increases when one considers tunneling of electrons with low energies. Even at the above-mentioned large electric fields the process is in practice limited to electrons at high energies. Only electrons within a very

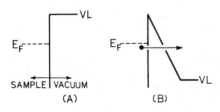

Fig 17.12.   Scheme of a field-emission process. (*A*) Ordinarily the electrons cannot leave the solid, because of the barrier between the Fermi level E_f and the vacuum level (VL). (*B*) When a strong electric field is applied, the vacuum level is lowered. If the corresponding barrier is not too thick, the electrons can go through it by a tunneling mechanism.

few electron-volts of the maximum energy—the Fermi level—can be emitted. The need for very sharp samples is a severe limitation on the range of applications of FES, which has for many years been confined to the study of tungsten. In spite of the above limits the FES technique is of high scientific interest and produces important results. The information carried by the FES spectra will be discussed together with that carried by the INS spectra.

3. Low-Energy Electron Diffraction

Low-energy electron-diffraction (LEED) experiments are confined to periodic systems, that is, solids and solid surfaces. Electrons in the primary beam are elastically scattered by the periodically distributed charge at the surface. The corresponding diffraction conditions give rise to diffracted beams that can be detected using, for example, the setup outlined in Fig. 17.13. The diffraction can be described by assuming that an incident free electron of wavefunction $\psi_i = Ae^{i\mathbf{k}\cdot\mathbf{r}}$ is scattered and detected as a free electron of wavefunction $\psi_f = Ae^{i\mathbf{k}'\cdot\mathbf{r}}$. The wavefunction ψ_f is a superposition of the waves $d\psi_f$ scattered by each fraction of the charge distribution at the surface. If $\rho(\mathbf{r})$ is the charge density at the surface, then the elementary volume $d^3\mathbf{r}$ at the position r will have a total elementary charge $\rho d^3\mathbf{r}$. The wave scattered by this elementary charge is proportional to the phase factor introduced by the scattering. Assuming that the phase factor is purely geometric, that is, originating from the change in direction from $\mathbf{k}$ to $\mathbf{k}'$, one has

$$d\psi_f = Ce^{i(\mathbf{k}'-\mathbf{k})\cdot\mathbf{r}}\rho(\mathbf{r})d^3\mathbf{r} \tag{11}$$

and therefore

$$\psi_f = C\int e^{i(\mathbf{s}\cdot\mathbf{r})}\rho(\mathbf{r})d^3\mathbf{r} \tag{12}$$

where $\mathbf{s} = \mathbf{k}' - \mathbf{k}$ is the "scattering vector," C is a constant, and the integral must be carried out over the entire charge distribution. Making use of the periodicity of the charge distribution, we can write $\mathbf{r}$ as $\mathbf{r} = \mathbf{L} + \mathbf{r}_i + \mathbf{x}$. Here $\mathbf{L}$ is a lattice vector, $\mathbf{r}_i$ defines the position of the ith atom—to which

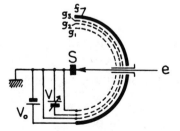

Fig. 17.13. A simple apparatus for LEED experiments. The sample (S) diffracts the electrons of the primary beam. The diffracted electrons are filtered in energy to eliminate the low-energy-electron background by the grid system (g_1, g_2, g_3). They are then detected by the fluorescent screen (f).

H. OPTICAL METHODS OF ANALYSIS

the elementary charge $d^3\mathbf{r}$ ''belongs''—in the unit cell defined by the vector $\mathbf{L}$, and $\mathbf{x}$ is the position of the elementary charge with respect to the atom. The integral in eq. 12 becomes:

$$\psi_f = \left(C \sum_{\mathbf{L}} e^{i\mathbf{s} \cdot \mathbf{L}}\right) \left(\sum_i e^{i\mathbf{s} \cdot \mathbf{r}_i} \int e^{i\mathbf{s} \cdot \mathbf{x}} \rho(\mathbf{x}) d^3\mathbf{x}\right) \tag{13}$$

The second factor in this equation—the so-called crystal structure factor—corresponds to a sum over all the atoms in the periodic cell. The first factor in equation 13 is the one that gives the diffraction conditions, as we shall now see. Since the diffraction probes only atoms at or near the surface, it is convenient to break $\mathbf{L}$ into its parallel and perpendicular components, $\mathbf{L} = \mathbf{L}_\perp + \mathbf{L}_\|$, so that the above factor becomes

$$\sum_{\mathbf{L}} e^{i\mathbf{s} \cdot \mathbf{L}} = \left(\sum_{\mathbf{L}_\perp} e^{i\mathbf{s} \cdot \mathbf{L}_\perp}\right) \left(\sum_{\mathbf{L}_\|} e^{i\mathbf{s} \cdot \mathbf{L}_\|}\right) \tag{14}$$

The first factor in equation 14 is a sum extended only over the vectors $\mathbf{L}_\perp$ near the surface, and the second factor is a sum extended over all possible $\mathbf{L}_\|$'s. That is, using the definition of $\mathbf{L}$ and therefore of $\mathbf{L}_\|$,

$$\sum_{\mathbf{L}_\|} e^{i\mathbf{s} \cdot \mathbf{L}_\|} = \left(\sum_{n_1} e^{i\mathbf{s} \cdot n_1 \mathbf{a}_1}\right) \left(\sum_{n_2} e^{i\mathbf{s} \cdot n_2 \mathbf{a}_2}\right) \tag{15}$$

where the two right-hand factors are geometric series:

$$\sum_{\mathbf{L}_\|} e^{i\mathbf{s} \cdot \mathbf{L}_\|} = \lim_{N \to \infty} \left(\frac{1 - e^{i\mathbf{s} \cdot N\mathbf{a}_1}}{1 - e^{i\mathbf{s} \cdot \mathbf{a}_1}}\right) \lim_{N \to \infty} \left(\frac{1 - e^{i\mathbf{s} \cdot N\mathbf{a}_2}}{1 - e^{i\mathbf{s} \cdot \mathbf{a}_2}}\right) \tag{16}$$

The scattered *intensity* is proportional to $|\psi_f|^2$, and therefore it will contain the *squares* of the factors in equation 16, which are

$$\lim_{N \to \infty} \frac{\sin^2\left(\dfrac{N\mathbf{s} \cdot \mathbf{a}_{1(2)}}{2}\right)}{\sin^2\left(\dfrac{\mathbf{s} \cdot \mathbf{a}_{1(2)}}{2}\right)} \propto \delta(\mathbf{s} \cdot \mathbf{a}_{1(2)} - 2\pi m_{1(2)}) \tag{17}$$

where δ is the Dirac delta function and m_1 and m_2 are integers. From equation 17 it is clear that the diffracted beams correspond to values of the vector $\mathbf{s} = \mathbf{k}' - \mathbf{k}$ with parallel components coinciding with one of the G-vectors of the surface:

$$\mathbf{s}_\| = \mathbf{G}_\| \tag{18}$$

In other words, the measured scattering vectors for the diffracted electrons directly give the vectors $\mathbf{G}_\parallel$ and therefore information on the periodicity of the surface. No special requirements for s arise from the "limited-sum" perpendicular factor in equation 14.

Besides the very valuable information about the vectors $\mathbf{G}_\parallel$ and about the surface periodicity, measurements of the diffracted-beam intensity also give information about the positions of the single atoms, through the crystal structure factor. Notice, however, that the scattered-intensity measurements do not probe ψ_f, but rather $|\psi_f|^2$. Therefore they give information not directly on the crystal structure factor, but on its square. That prevents one from obtaining directly the atomic positions, since in the above square the phase factors corresponding to the single-atomic positions are convoluted with each other. This is a severe limitation not only in LEED, but also in all other electron-diffraction techniques.

In summary, LEED yields direct information about the surface periodicity, and somewhat less direct information about the atomic positions (71). Advanced use of LEED ordinarily involves a major theoretical effort to take into account nongeometric (i.e., atomic) phase shifts, atomic scattering amplitudes, and above all to calculate the effects of multiple scattering. Multiple-scattering calculations typically require long computer times and heavy use of computer memory, and are limited by cost considerations.

III. INSTRUMENTATION AND EXPERIMENTAL METHODS

A. GENERAL CONSIDERATIONS

The common characteristics of the different electron-spectroscopy techniques are reflected by the instrumentation they require (48). These common features make it possible to use a general common scheme to describe the electron spectrometers for different techniques. In fact, the same electron spectrometer is often used for several different techniques. The general scheme of an apparatus for electron spectroscopy includes:

1. A source of particles—photons, electrons, or ions—to create the primary beam.
2. An analyzer–detector for the electrons in the state $|f\rangle$.
3. An ultrahigh-vacuum system, which is required by the extreme surface sensitivity of the electron-spectroscopy techniques.
4. The devices required for sample preparation and characterization.

The rapid expansion of electron spectroscopy in recent years has been due primarily to advances in instrumentation and in theoretical background. For example, an important factor was the complete commercialization of the ultrahigh-vacuum technology. The most spectacular advances have concerned the primary-beam sources and, more specifically, the photon sources. The advent in the 1970s of synchrotron radiation has entirely changed the conventional boundaries of photoemission spectroscopy (8,151). Other important advances have concerned the electron analyzers and the data-taking and -processing techniques. We discuss below the most significant recent advances in electron-spectroscopy instrumentation.

B. PHOTON AND ELECTRON SOURCES

The photon sources required in photoemission spectroscopy must satisfy very stringent requirements. The most severe of them involves the photon energy. In x-ray photoemission this energy is usually above 1 keV, but a large number of important results have been obtained with soft-x-ray photons at energies between 30–40 eV and 1 keV. The minimum energy in any kind of photoemission is dictated by the magnitude of the barrier—ionization potential or work function–to be overcome. Energies below 5–10 eV are typically ruled out, since the barrier is of the order of at least 4–5 eV and 3–4 more electron volts are strongly overlapping in the spectra with low-energy secondary electrons. Serious problems arise therefore from the scarcity of natural, intense sources in the energy range from 5–10 eV to 1–1.2 keV. In Table 17.II the typical sources for photoelectron spectroscopy are listed, along with the resolution, ΔE, contributed by the natural line width or monochromator response function. The natural uv sources are

1. The hydrogen-discharge continuum up to about 11 eV.
2. A small number of noble-gas emission lines: Ar at 11.2 eV, Ne at 16.8 eV, and He at 21.2 and 40.3 eV.

Above the HeII line at 40.3 eV no suitable natural photon sources are available up to the MgK_α emission at 1253 ev. Most x-ray photoemission experiments use either this emission or the AlK_α emission at 1487 eV. The natural line width of these emission lines is of the order of 1 eV, which is

TABLE 17.II
Sources for Photoelectron Spectroscopies

Source	Energy (eV)	ΔE, line width (eV)	Method acronym
H_2 discharge	7–11	~0.1	UPS
Ar, Ne, He discharge	11.7, 16.8, 21.2, 40.8	<0.1	UPS
Mg-, AlK_α	1253, 1487	1.1, 1.3	XPS
AlK_α + monochromator[a]	1487	0.5	XPS
Synchrotron radiation sources (examples)[a]			
Tantalus, NBS	7–150	0.1–0.5	UPS, SXPS
SSRL	40–1000	0.1–10	UPS, SXPS
NSLS	7–900	0.05–2	UPS, SXPS
Aladdin	7–1000	0.05–2	UPS, SXPS

[a] Additional monochromator and matching optics as well as source.

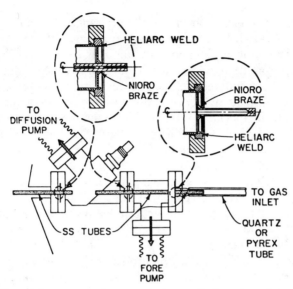

Fig. 17.14.   Differentially pumped microwave-cavity gas discharge lamp for producing uv photons for photoemission spectroscopy (119).

to say, quite large and not suitable for high-resolution photoemission spectroscopy. X-ray crystal monochromators can be used to reduce the line
width.

The uv emission sources listed above can in practice have the form of a
continuous glow discharge, of a hot-filament arc discharge, or of a microwave cavity (119). The latter typically requires less maintenance than the
other two schemes. The practical line width is of the order of 1 meV and
therefore suitable for high-resolution experiments without further monochromatization. These sources, however, have a serious compatibility problem
with the ultrahigh vacuum required in the main experimental chamber. The
operating pressure of the source is of the order of 1 torr. The main chamber
is typically at pressures below 2×10^{-10} torr, although for inert gases such
as He, Ne, and Ar, higher pressures are tolerable. No window can separate
the source from the main chamber, since no transparent materials are available in this photon-energy range—in fact, all the optics must be based on
reflections. The pressure must therefore be dropped by many orders of magnitude, using a differential pumping system. Figure 17.14 shows the scheme
of a differentially pumped microwave discharge lamp (119). Here the two
capillary tubes provide the necessary impedance to decrease the pressure.
The differential pumping is accomplished in two stages: down to the 10^{-2}-
torr range by a trapped mechanical pump, and down to the 10^{-7}-torr range
by a diffusion pump. The actual pressure in the main chamber is of the order
of 2×10^{-9} torr, and about 90% of it consists of noble-gas contributions
from the lamp itself. Therefore the equivalent pressure in a normal ultrahigh-

vacuum chamber would be in the 10^{-10}-torr range. The microwave cavity is powered by 10–100 W of radiation at a frequency of 2450 MHz.

The main limitation of the above uv lamps is their lack of spectral continuity. Except for the hydrogen-discharge lamp, which is limited to very low energies, all the sources are line sources at fixed photon energies. This necessarily limits the flexibility of photoemission spectroscopy, as we shall discuss in detail in the next section. The limitation is removed by using synchrotron radiation as a photon source (8,151). Synchrotron radiation is the electromagnetic radiation emitted by electrons that travel at relativistic energies in an accelerating machine when their trajectory is deflected by the bending magnets. The accelerating machines in which synchrotron radiation was originally detected and investigated were indeed called "synchrotrons." The machines used today as synchrotron-radiation sources are storage rings, which have the advantage of continuous operation, whereas the synchrotrons are pulsed sources.

The unique properties of the synchrotron-radiation sources can be theoretically predicted by a relativistic treatment of the motion of the stored electrons and thus of the photon-emission process. The most important properties are the following:

1.   Synchrotron radiation is a spectral continuum including infrared, visible, uv, and x-rays. In particular, it bridges completely the spectral gap between the HeII line at 40.8 eV and MgK_α line at 1253 eV. The spectral-intensity distribution is a universal function of the photon energy of the electrons circulating in the storage ring. As an example, Fig. 17.15 shows the emission spectra of the Aladdin and Tantalus storage rings at the University of Wisconsin—Madison.

2.   Synchrotron radiation is intense and collimated in the vertical plane. Thus it can provide a photon beam of reasonable intensity at any required photon energy after spectral filtering by a suitable monochromator. Several kinds of monochromators have been developed in recent years to exploit the synchrotron radiation continuum. Figure 17.16 shows the scheme of a "Grasshopper" monochromator developed by Brown, Lien, and Pruett, that has been used for photon energies up to 1 keV. A more conventional Seya–

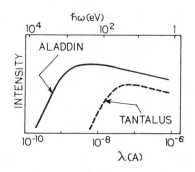

Fig. 17.15.   Spectral emissions of two different synchroton-radiation sources, the storage rings Tantalus and Aladdin at the University of Wisconsin—Madison. Notice the extremely large spectral range covered by these spectra; the cutoff depends on the energy of the electrons circulating in the storage ring.

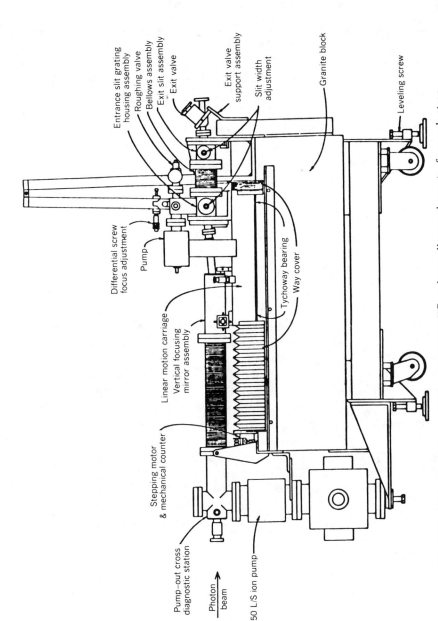

Entrance slit grating
housing assembly
Roughing valve
Bellows assembly
Exit slit assembly
Exit valve

Exit valve
support assembly

Slit width
adjustment

Granite block

Leveling screw

Differential screw
focus adjustment

Pump

Tychoway bearing

Way cover

Linear motion carriage

Vertical focusing
mirror assembly

Stepping motor
& mechanical counter

Photon
beam

Pump-out cross
diagnostic station

50 L/S ion pump

Fig. 17.16.   Scheme of a large-photon-energy-range "Grasshopper" monochromator for synchrotron-radiation experiments at energies up to 1 keV.

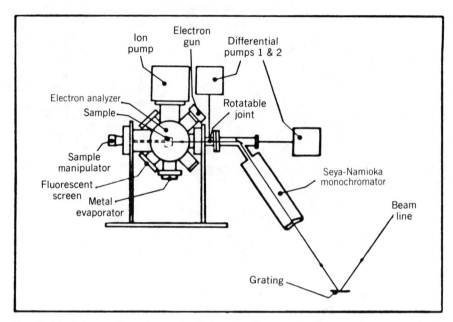

Fig. 17.17.   Scheme of synchrotron-radiation experimental beamline with a Seya–Namioka-type medium-photon-energy monochromator.

Namioka monochromator is shown in Fig. 17.17. This instrument is suitable for experiments at energies up to 40–45 eV. Other designs have recently been developed, such as the toroidal-grating monochromator based on holographic gratings.

3.   Synchrotron radiation is linearly polarized, with the electric vector in the plane of curvature of the electron-beam trajectory. In practice this natural polarization is further emphasized by several horizontal reflections along the beam line. The final polarization is better than 98%. This synchrotron-radiation property is extremely helpful in special photoemission experiments, which shall be illustrated in the next section.

The electron-source technology is more conventional than the photon-source technology. In practice, the electron guns developed for cathode ray tubes can be used after minor modifications for all the techniques such as Auger spectroscopy that require a primary electron beam. Figure 17.18 shows a typical electron-gun scheme. The cathode in this scheme is heated by a tungsten filament, although we shall see that special cathodes are required in certain techniques, such as the Auger microprobe. The energy spread of the electron beam given by the gun is 0.5–0.6 eV. This is insufficient for high-resolution experiments such as vibrational electron energy-loss spectroscopy. In those cases the beam produced by the gun is filtered by one of the electron analyzers illustrated in the next section to reduce its energy spread. For example, the monochromatized beams for vibrational

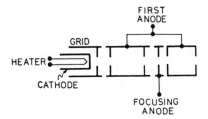

FIRST
ANODE

GRID

HEATER

CATHODE

FOCUSING
ANODE

Fig. 17.18.   Scheme of electron gun with accelerating and focusing electrodes.

energy-loss spectroscopy allow resolutions of the order of a few millielectron volts at energies of the order of a few electron volts.

C. ELECTRON ANALYZERS

The electron analyzers measure the energy and in some cases the direction of the electrons in a vacuum. The energy analysis can be accomplished by bandpass filtering or by high-pass filtering. In the first case, electrons are detected only in a small energy window centered around a value E. The energy distribution $N(E)$ is directly obtained by monitoring the output of the analyzer while scanning E. In the second case, all the electrons at energies above a lower limit E are detected, and the output signal corresponds to $\int_E^\infty N(E')dE'$ rather than to $N(E)$. A small modulation of the lower-limit energy E and a phase-sensitive detection of the corresponding modulation of the number of collected photoelectrons gives the first derivative of the above integral, that is, the energy distribution $N(E)$. Alternatively, one can simulate the modulation with a computer that numerically finds the difference between the outputs $\int_E^\infty N(E')dE'$ taken at two close values of the lower limit E.

Table 17.III presents a summary of several widely used designs of electron analyzers. The first design is the retarding-grid analyzer (101) shown in Fig. 17.19. This is a bandpass, angle-integrating analyzer. The basic configuration includes only one hemispherical grid (corresponding either to G_2 or to G_3 in Fig. 17.19) and a hemispherical collector. The limit energy E is equal to eV, where V is the retarding voltage applied between the emitting sample— which is at ground— and the grid. The total current given by the photoelectrons that reach the collector gives a (small) voltage drop superimposed on the bias in the load resistor R. A modulation in the retarding voltage V gives a modulation in E and therefore in the voltage across R. If a phase-sensitive lock-in amplifier is used to detect the first harmonic of this modulation, its reading is proportional to $N(E)$.

The above basic configuration is much improved by the four-grid design in Fig. 17.19, particularly as far as resolution is concerned. The first and fourth grids, G_1 and G_4 are both grounded, and a field-free region is created between G_1 and the sample. The retarding field is essentially confined to

TABLE 17.III

Electron Analyzers Used for Spectroscopy

Analyzer	Bandpass?	Angular range (degrees)	Energy range (eV)	Commercial suppliers?
RFA (regarding-field an., LEED grid)	No	~70–180	0–200	Yes
CMA (cylindrical-mirror an.)	Yes	12 ($\Delta\phi = 360°$)	0–2000	Yes
PMA (plane-mirror an.)	Yes	2–10	0–200	No
SDA (spherical-deflector an.)	Yes	0.5–6	0–2000	Yes
CDA (Cylindrical-deflector an.)	Yes	1.5–4	0–100	Yes

an., Analyzer.

the region between G_1 and G_2. Both G_2 and G_3 are at the same voltage—the use of two grids here instead of one has been demonstrated to improve the resolution substantially.

The main problems in the above analog-modulation scheme are the signal-to-noise ratio and the capacitive coupling between grids and collector. The first problem can be alleviated by increasing the modulation amplitude, but this in turn decreases the resolution; a compromise therefore must be found. The second problem is solved by using a computer-simulated modulation as discussed above.

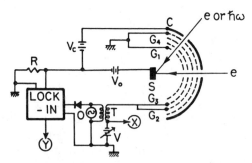

Fig. 17.19. A four-grid hemispherical retarding-grid electron analyzer (101). The retarding potential V is given by the tunable supply. A modulation is superimposed on V by means of a transformer. The corresponding photocurrent modulation in the load resistor (R) is detected by a lock-in (phase-sensitive) amplifier giving the first derivative of the photocurrent with respect to V. This derivative corresponds to the electron-energy distribution curve. The transformer and the output of the lock-in amplifier provide the X and Y inputs for an X–Y recorder. S, sample.

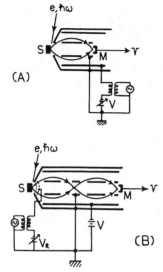

Fig. 17.20. (A) A cylindrical-mirror analyzer (CMA) (103). The voltage bias between the two cylinders, V_p, establishes the "pass energy" of the electrons that are focused at the front end of the electron multiplier (M). The electron multiplier gives an amplified negative pulse for each electron reaching its front end. A modulation can be superimposed on the bias voltage to enable first-derivative spectra to be obtained using a lock-in amplifier. (B) In the double-pass version of the CMA (102) the first stage focuses the electrons for the second stage. The electrons are preretarded or accelerated by the grid system in the front end. The electron energy is scanned by scanning the retarding voltage V rather than the bias voltage V_p.

The hemispherical-grid retarding-voltage analyzer is not *per se* an angle-resolved device. It is in fact the best angle-integrating device as far as magnitude of the acceptance angle is concerned. It can also be transformed into an angle-resolving analyzer. This is done, for example, by means of a moving aperture that substitutes for the collector, and by an electron detector beneath it (134). The detector is usually an electron multiplier. This is a device in which an input electron starts a cascade of secondary electrons if a bias of a few kiloelectron volts is applied to it. The multiplied electron current can then be detected either as an analog signal or by a digital approach.

The retarding-grid analyzer, although still widely used, has been replaced as the most used device by the cylindrical-mirror analyzer (CMA) (102,103). The geometry of this device is illustrated by Fig. 17.20. The electron-optics analysis of this geometry shows that electrons originating at the focal point are focused at a common point, provided that their trajectories form an angle of 42.3° with the axis of the CMA and that their kinetic energy is proportional to the pass voltage V_p. Here V_p is the bias voltage between the outer cylinder—which is at ground—and the inner cylinder. In commercial devices (49) the front-end aperture of the CMA accepts trajectories that form an angle with the axis of the CMA in a range of 12° centered around the "magic value" of 42.3°, and the energy/pass voltage ratio is 1.7. The signal increases with the amplitude of the accepted angle range, but the energy resolution decreases; the value of 12° for the amplitude appears to be a reasonable compromise.

The CMA is a device of excellent luminosity, and the output level is increased even more by a multiplier placed at the rear end. The focusing properties are very good, and the actual "size" of the focal point is very small. This is an advantage in those kinds of electron spectroscopy in which the emitting area is small, as when the primary beam particles are electrons—which can be focused with very high accuracy. The focusing properties of the CMA are not an advantage for spectroscopies in which the emitting area is extended, as in most photoemission experiments and, in particular, those that use conventional x-ray sources. The size of the focal point can be increased by accepting trajectories passing through a large area at the rear end (i.e., using a wide slit there), but the corresponding use of off-axis trajectories decreases the energy resolution. Another potential problem with the CMA is the way in which the energy is scanned. Since the energy window E is given by the bias voltage V_p one can sweep E by sweeping V_p. The CMA theory shows, however, that the "aperture" of the energy window, ΔE, which defines the energy resolution, has a constant ratio with respect to $E = eV_p$ ($\Delta E/eV_p$ = constant). Therefore if V_p is swept, the energy resolution ΔE changes during the sweep; in particular, it becomes very small at small V_p values (i.e., small energies), causing a drop in the signal. This feature is not a problem in certain kinds of electron spectroscopy, such as Auger spectroscopy, in which the CMA is very often used, but it is a problem with other techniques, such as photoemission with uv sources.

The problems discussed in the previous paragraph are solved by the double-pass CMA geometry (102) also illustrated in Fig. 17.20. Here the first stage of this two-stage device acts as a focusing electron lens. A large emitting area is focused by the first stage, becoming a small-area virtual source for the second stage. A small emitting area can also be used with better resolution with respect to a large area by limiting the range of accepted trajectories at the end of the first stage. The electrons are retarded or accelerated by the grid system in the front end of the analyzer *before* they enter the cylinder system. Therefore their energy can be changed by the grids to match the pass value 1.7 eV_p with minimal changes in their trajectories. The energy window position E is then given by the combination of 1.7 eV_p and of the preretarding voltage of the grids. The energy E can be scanned by scanning either V_p or the retarding voltage. In the latter mode the width ΔE of the energy window stays constant during the scanning; this removes the above-mentioned problem of changing resolution.

The excellent performance and the flexibility of the CMA explain its increasing use in different electron spectroscopies. One problem with this analyzer is that it has neither complete angular averaging nor complete angular resolution. The acceptance is limited to the angles defined by the two conical surfaces centered at the focal point. The partial angle-integrated spectra taken in this way can be affected by angular-distribution effects. However, the CMA *can* be transformed into a true angle-resolving device. This is usually done by adding between the two stages of a double-pass CMA a rotating drum with a small aperture (136). The corresponding performance is satisfactory, but the geometry is complicated and it is somewhat difficult to perform certain kinds of angular measurements.

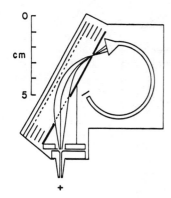

cm

0

5

Fig. 17.21. A modified plane-mirror analyzer (PMA) (139). The pass energy of the electrons is given by the bias between the front plate and the rear plate (the corresponding field is terminated by the guard plates). The electrons are focused and preretarded or accelerated by the front-end lens. The electron energy is swept by scanning the retarding voltage.

The several different kinds of angle-resolved electron analyzers we shall discuss here are the plane-mirror analyzer (PMA), the spherical-sector analyzer, the cylindrical-deflector analyzer, and the recent display analyzers.

The PMA (139) is a bandpass device with an extremely simple geometry, as shown in Fig. 17.21. The electron-optics analysis of this geometry shows that electrons originating at the focal point are focused along a line at the rear end, the exact focal point being established by their kinetic energy. The ideal focusing conditions occur for a kinetic energy of 1.8 eV_p, where V_p is the bias voltage between the two plates. For the reasons discussed above for the CMA, it is convenient to sweep the energy not by sweeping V_p, but by adding a variable retarding voltage between two grids at the front end. The advantages of the PMA scheme are its simplicity and compactness. Another potential advantage is its dispersion along a focal line of electrons with different energies, which could be exploited for parallel detection of electrons of different energies. The energy resolution is reasonably good, but it cannot match the best performances of other geometries. The most severe limiting factor on the resolution is the field termination, which is usually accomplished by means of guard plates.

The original design of the PMA was elaborated by Green and Proca (44) and improved by N. V. Smith et al. (133). A further improvement is shown in Fig. 17.21. The front-end lens added to the basic PMA design greatly improves the performance of the retarding field. This analyzer can be mounted on the compact, ultrahigh-vacuum goniometer shown in Fig. 17.22, which enables the device to be moved independently along the azimuthal and polar directions (139).

The spherical-deflector analyzer is also a bandpass device (70). Its geometry is shown in Fig. 17.23. Also shown in this figure is the geometry of another bandpass device, the cylindrical-deflector analyzer (40). In both devices the energy filtering is obtained by applying a bias voltage between the two plates. The electron-optics analysis demonstrates that good focusing is possible under certain conditions. For the spherical-deflector analyzer, the focusing conditions are optimized if the angle between the front-end plane

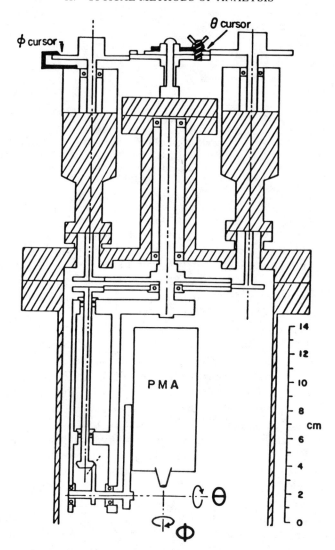

Fig. 17.22. Compact, ultrahigh-vaccum goniometer for angle-resolved electron analysis (139). The analyzer—a PMA in this case—can be moved to different positions at a pressure of 10^{-11} torr. The two angles θ and ϕ can be varied independently by using different gear trains.

and the rear-end plane is 180°. For the cylindrical-deflector analyzer, the best focusing requires that angle to be 127°. Both devices have excellent angular and energy resolution and are commercially available. The cylindrical-deflector analyzer is suitable for miniaturization, as the PMA is, and it is widely used when the analyzer must be moved under ultrahigh vacuum.

The increasing importance of angular-distribution analysis has led since 1977 to the development of new and advanced *display* analyzers (30,152). These devices reveal the angular and energy distributions of the electrons

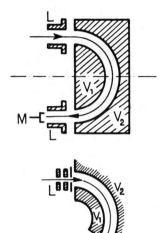

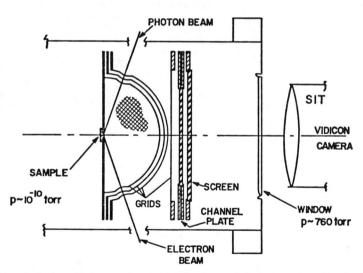

Fig. 17.23.   The schemes of two other widely used electron analyzers. Top: The 180° spherical-deflector analyzer (70). Bottom: The 127° cylindrical deflector analyzer (40). In both cases the pass energy for the focused electrons is established by the bias voltage $V_2 - V_1$. The characteristics of the analyzers are improved by the lens system (L) and by the collimater (C), and the signal level is enhanced by the electron multiplier (M).

by monitoring in parallel electrons traveling in different directions. One example (152) of a display analyzer is shown in Fig. 17.24. Here the energy analysis is performed by a retarding-grid system. The energy distribution $N(E)$ is obtained by controlling the analyzer and the signal-detection system with a minicomputer. The electrons filtered by the grids eventually reach the channel plate. This device is equivalent to an array of electron multipliers

Fig. 17.24.   A display electron analyzer detecting in parallel electrons traveling in different directions (152). The electrons are multiplied by the channel plate which is equivalent to a matrix array of electron multipliers. They are than revealed by the fluorescent screen, and the image on the screen is analyzed by a computer-controlled vidicon system. The energy analysis is provided by a standard retarding-grid system.

arranged in a plate. The distribution of electrons reaching the front face of the multiplier is reproduced by the electrons leaving the rear face, but with greatly enhanced intensity. The intensified electron distribution then reaches a high-voltage biased fluorescent screen, where it gives rise to a pattern of fluorescent emission. The screen is transparent, and the pattern can be observed from the back. For data-taking purposes the pattern is detected and stored by a minicomputer connected to a vidicon suitable for low-intensity detection. Other systems designed along with the same general principles have been developed and tested in recent years. Eastman and coworkers, for example, have developed a *bandpass* display analyzer (30). The energy bandpass filtering is obtained by electron-optics focusing in an ellipitical-mirror cavity. The measured intensity distribution at a given energy window gives directly the angle-resolved EDCs.

D.   SAMPLE PREPARATION AND VACUUM REQUIREMENTS

The stringent vacuum requirements in electron spectroscopy arise from the surface sensitivity of the experimental techniques, and from the corresponding need to avoid unwanted contamination. This is of course true when the investigated system is solid or, better, a solid surface. For free atoms or free molecules, the requirement is simply that the spurious signal from the impurities in the gas be kept below the level at which it introduces misleading spurious structure into the spectra.

The electron spectrometers *per se* and the photon and electron sources—except synchrotron radiation—could work at fairly high pressures, in the 10^{-4}-torr range. The vacuum required to keep a surface uncontaminated is much below those values. This can be estimated by a very simple calculation. The kinetic theory shows that the number of molecules striking a unit surface per unit time is of the order of $n \langle v \rangle$, where n is the density of molecules in the gas and $\langle v \rangle$ their average speed. Assuming for simplicity a unit sticking coefficient, the time required to build up a monolayer of contaminant at room temperature is of the order of 1 sec if the pressure of the gas is of the order of 10^{-6} torr. Of course, these conditions would not allow a practical experiment. A more realistic requirement is to have less than $\frac{1}{10}$ of a monolayer built over a period of the order of 1 hr. This brings down the required pressure to the 10^{-10}–10^{-11} torr range. Except for a very few chemically unreactive surfaces that can be analyzed in the 10^{-9}-torr range, all systems should be studied in a ultrahigh-vacuum environment with pressures below 4×10^{-10} torr.

The technology needed to reach and maintain pressures in the above range is now entirely available from commercial suppliers. Figure 17.25 shows a block diagram of an ultrahigh-vacuum system suitable for electron spectroscopy. The core of the system is the main pump. This is usually an ion pump; for some specialized applications, it is a cryogenic pump or a turbomolecular pump. A titanium sublimation pump is often added to the ion

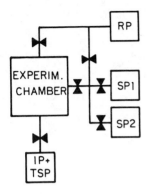

Fig. 17.25. Block diagram of a typical pumping system for the ultrahigh-vacuum chamber of an electron spectrometer. The active components of the system are a roughing pump (RP), two sorption pumps (SP 1, SP 2), and the main ion pump with titanium sublimation (IP + TSP).

pump for more efficient water-vapor pumping. The main pump can start working only at pressures below 10^{-4}–10^{-5} torr. These pressures are reached by means of a preliminary pumping system, which ordinarily consists of a oilfree mechanical pump for preliminary gas evacuation coupled to sorption pumps, to a trapped diffusion pump, or to a turbomolecular pump. The ultrahigh-vacuum system is completed by a chamber with all the necessary accessories—valves, flanges, viewports, and feedthroughs—and by pressure gauges. For the ultrahigh-vacuum pressure range, the pressure-sensing device is a nude Bayard–Alpert ionization gauge. We emphasize that only a very small number of materials are suitable for construction of ultrahigh-vacuum technology. These include primarily stainless steel, aluminium, copper, glass and nonporous ceramic. Most ultrahigh-vacuum components are made of stainless steel. The flanges must be sealed with metal gaskets made of copper or aluminum, and the valves must be metal-sealing too. Final pressures in the 10^{-10}–10^{-11} torr range can be reached only by eliminating the gas molecules (primarily water vapor) absorbed by the walls of the chamber, in particular, those trapped in small cracks or irregularities. This is done by bringing the entire system to a temperature above 130–160°C for a period of 12–24 hr during the intermediate stages of the pumpdown.

From the above discussion it is clear that the systems to be investigated by electron spectroscopy must be prepared *in situ* under ultrahigh-vacuum conditions. For free atoms or molecules the system is typically an atomic or molecular beam crossing the focal region of the analyzer or, more rarely, the background gas in the vacuum chamber. This can cause problems when the source of primary particles in the experiment is synchrotron radiation, which itself requires ultrahigh vacuum. Figure 17.26 shows schematically a beam line specially designed for gas-phase experiments. The differential pumping system brings the pressure down from 10^{-4} torr in the experimental chamber to the 10^{-10} torr pressure present in the storage ring.

For solids and solid surfaces, several preparation methods can be used for the different kinds of systems. These are cleavage, fracture, grinding, and deposition. Cleavage and fracture require a cleavage tool, and produce

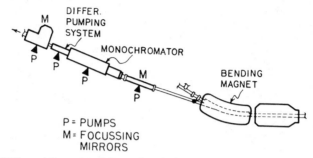

Fig. 17.26. Differentially pumped beam line for medium-vacuum, windowless experiments with a synchrotron-radiation source.

crystallographic and noncrystallographic surfaces, respectively. Cleavage is suitable only for certain faces of certain crystalline solids (e.g., the (110) face of zinc blende-structure solids.) Difficult-to-handle samples can often be cleaned *in situ* by grinding (108). This is accomplished most effectively with a small diamond grinder made of diamond imbedded in stainless steel and operated by a rotatable feedthrough. Deposition is a simple method for obtaining certain kinds of polycrystalline samples. However, difficulties arise when the deposited sample must satisfy crystallinity or stoichiometry requirements. In fact, for these samples more advanced deposition methods are required—typically expitaxial growth over another crystalline substrate. The most advanced deposition method available at present is molecular-beam epitaxy MBE (2).

Samples obtained by one of the methods described above are ordinarily characterized *in situ* before being investigated by the electron-spectroscopy technique of interest. The characterization can involve Auger spectroscopy, photoemission spectroscopy, or LEED. Most modern electron-spectroscopy systems therefore include equipment for techniques besides that of principle interest. Several techniques can usually share the same instrumentation (e.g., the electron analyzer). This makes a typical electron spectrometer an extremely versatile tool.

IV. ULTRAVIOLET PHOTOEMISSION SPECTROSCOPY

A. GENERAL FEATURES

Ultraviolet photoemission spectroscopy involves the use of photons of energies below 30–40 eV. A detailed treatment of the theory of photcemission spectroscopy is given in the Appendix. From that treatment one obtains in particular an energy-balance condition for the initial-state energy E_i, for the final-state energy E_f, and for the photon energy:

$$E_f - E_i - h\nu = 0 \qquad (19)$$

From this energy-balance condition we see that photoemission explores electronic states in an energy range $0-h\nu$ below the ionization threshold. Therefore, with uv photons its primary targets are the electrons involved in the formation of chemical bonds in molecules and solids. These states of these electrons are, for example, the bonding states in covalent systems and the occupied quasiatomic states in ionic systems. Other states can of course be present in the same energy range, for example, nonbonding states and shallow core levels. The primary information sought by uv photoemission is nevertheless about the chemical-bond states and about the bonding process itself.

We have seen in Section II that the conventional method of taking photoemission spectra yields information on the density of occupied states, that is, on the energy distribution of the electrons for the ground state of the system. This information is somewhat indirect, as we discuss in the Appendix. In particular, the conventional photoemission spectra or EDCs are proportional to the energy distribution of the joint density of states (EDJDOS)—the conventional name given to the right-hand expression in eq. A15 in the Appendix—rather than to the DOS itself. We shall see in the last part of this section that this conventional mode of photoemission originates from the technical limits of the conventional photon sources—and that it can be replaced by other modes using synchrotron radiation. Most photoemission experiments, however, are still run in the conventional of EDC mode, to which we shall dedicate most of this section.

B.   SIMPLE MOLECULES

The first photoemission experiments on molecules were carried out in 1961 by Viselov et al. (143). The experiments were limited to photon energies below 11 eV by the presence of a fluorite window between the photon source and the main chamber. The use of windowless sources and, in particular, of synchrotron radiation has enabled this branch of photoemission to make impressive advances during the past 15 years. We review in this section some representative photoemission-spectral features of simple molecules. Before starting our review, we consider some general properties of the photoemission process in molecules, such as the vibrational effects and the cross section dependence on the photon energy.

The general energy-balance for photoemission, $E_f - E_i - h\nu = 0$ (equation 19) must include for molecules the vibrational and rotational contributions to the energy of the states. Measuring the energy from the zero of the kinetic-energy scale, we have $E = E_f$ and

$$E = I_i + E_{\text{vib}}^{(i)} + E_{\text{rot}}^{(i)} + h\nu \qquad (20)$$

where $I_i + E_{\text{vib}}^{(i)} + E_{\text{rot}}^{(i)}$, the sum of the "adiabatic" electronic ionization potential and of the vibrational and rotational energies of the initial state,

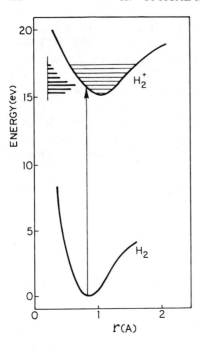

Fig. 17.27.   Energy diagram for the H_2 molecule and the corresponding H_2 ion (114). Notice the displacement of the equilibrium value of the internuclear distance for the two systems. The line shape corresponding to the ionization of H_2 is shown at the left-hand side of the H_2^+ curve.

replaces E_i. The rotational term is much smaller than the vibrational term. The role of the vibrational term $E_{vib}^{(i)}$ can be illustrated by the simple example of the H_2 molecule. On photoionization, this molecule becomes a H_2^+ ion. Figure 17.27 shows the potential-energy curves for the ground states of H_2^+ and of H_2 as a function of the internuclear distance (114). Notice that the respective minima for the two curves occur at different values of the internuclear distance. According to the Franck–Condon principle, the excitation process $H_2 \rightarrow H_2^+$ is so fast that the value of the internuclear distance stays constant during it. In Fig. 17.27 this corresponds to transitions going *vertically* from the H_2 curve to the H_2^+ curve. At least for low temperatures, the probability distribution of the internuclear distance of the H_2 molecule is Gaussian and its most probable value corresponds to the minimum of the H_2 curve. Because the transition is "vertical," the potential-energy curve for H_2^+ is reached at a vibrational level different from the lowest one. This gives in the photoemission spectrum the most probable (i.e., the most intense) peak. The total photoemission spectrum is a combination of different vibrational levels, as shown in Fig. 17.27. The position in energy of the most intense ("vertical") vibrational band, E_{max}, defines the so-called vertical ionization potential, I_{vert}:

$$I_{vert} = E_{max} - h\nu \qquad (21)$$

The photon-energy dependence of the photoionization cross section is

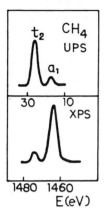

Fig. 17.28. The interplay of s-like and p-like states in the photoemission spectra of CH_4 taken at different photon energies. (adapted from Ref. 115) The s-like components are enhanced in x-ray photoemission.

discussed from an atomic point of view in the Appendix. We shall mention here one important consequence of the energy dependence of the probability of the optical transition from $|i>$ to $|f>$ (which in turn is given by the matrix element $<i\,|\,\nabla\,|\,f>$). This dependence is different for $|i>$ states of different symmetries. Conversely, it can be used to identify the relative symmetry of the states contributing to the valence-electron spectrum of a molecule. For example, it is a general rule (1,3) that s-like states have smaller transition probabilities than p-like states at low photon energies. Therefore the s-like peaks are much weaker than the p-like peaks in uv photoemission, and the intensity ratio is more favorable to s-like peaks in x-ray photoemission. The use of tunable synchrotron radiation enables one to study these changes over a continuous spectral range. An example of this s–p intensity interplay is shown in Fig. 17.28 for the CH_4 molecule (115). One particularly important process that influences both the cross section and the line shape of the photoemission spectra is *autoionization* (34,35,99). This is a two-step process in which the photon excites the molecule to a state corresponding to the increase in energy of one or more electrons *without* ionization. The total energy increase of the molecule, however, is more than that required for ionization. The molecule can rearrange its energy, expelling one electron, which appears in the photoemission spectrum. The resulting vibrational components of this spectral contribution are allowed to violate the Franck–Condon principle.

Our short review of the photoelectron spectra of simple molecules will start with those of single-bonded molecules isoelectronic with noble-gas atoms. Let us consider, for example, the H_2O and NH_3 molecules (114). The spectra taken with HeI radiation are shown in Fig. 17.29. Both molecules are isoelectronic to Ne. The Ne p^6 is degenerate. The H_2O and NH_3 molecules can be thought of as being formed by taking a Ne atom and displacing two or three protons from its nucleus. The p^6 degeneracy is thus progressively removed. On these grounds we expect to find similarities between the spectra of the above molecules and those of the H_2S and PH_3 molecules,

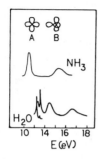

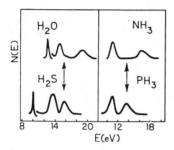

Fig. 17.29. Similarity of the photoemission spectra of the pairs of isoelectronic molecules H_2O and NH_3, H_2S and H_2O, and NH_3 and PH_3. (adapted from Ref 114).

which can be thought of as being formed in a similar way from an Ar atom. This expectation is indeed true, as we can see in Fig. 17.29. The interpretation of the spectra along these lines can include, in fact, all the hydride molecules isoelectronic with a noble-gas atom. A similar analysis can be used for molecules that have electronic structures similar to those of the above molecules plus some nonbonding electrons. This is the case, for example, for CH_3Br, the spectrum of which can be interpreted in terms of that of NH_3 with additional (sharp) bands due to nonbonding electrons.

All the above molecules are "single-bonded" systems. The next step in complexity of molecular photoelectron spectra corresponds to "multiple-bonded" systems. In these molecules the molecular orbitals are formed by combinations of s and p atomic orbitals, with the p orbitals combining not only "end-on" but also "side-on." A typical example of a multiple-bonded diatomic molecule is O_2. Figure 17.30 shows schematically the molecular orbitals for this molecule (114). The assignment of the corresponding features in the photoemission spectrum can still be based largely on qualitative considerations—that is, without full quantum-mechanical calculations. The expected energy order of the orbitals is indeed $\pi_g 2p$, $\pi_u 2p$, $\sigma_g 2p$, and $\sigma_u 2s$. This corresponds to the identification of the photoemission bands in Fig. 17.30 with the addition of two Σ bands. The analysis of the spectra for multiple-bonded diatomic molecules can often be extended to polyatomic multiple-bonded molecules. The line of reasoning is similar to that used to derive the spectra of hydrides from those of the isoelectronic noble gases. For example, the HCN molecule is isoelectronic to N_2, and its spectrum

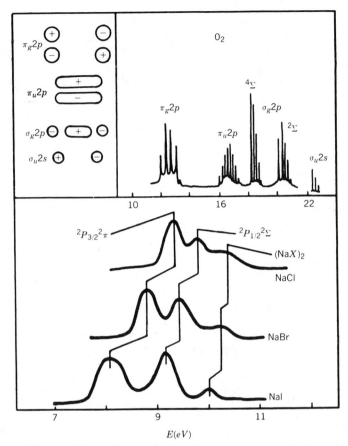

Fig. 17.30. Top: A qualitative ordering in energy of the molecular orbitals does not necessarily require full quantum-mechanical calculations. Shown here are the orbitals of the O_2 molecule (114) and the corresponding peaks in the photoemission spectrum. The identification is based largely on the expected ordering in energy. Bottom: The similarity among ionic molecules of a given series is reflected in their photoemission spectra (112). This makes it easy to identify the photoemission features once they have been identified in one molecule. The data were taken from Ref. 112 and 114.

can be "derived" from that of N_2. A similar analysis is valid for H_2C_2, which is also isoelectronic with N_2 and HCN.

The analysis becomes a little more complicated in the case of triatomic molecules. Here the linearity or nonlinearity of the molecule is a fundamental factor in establishing the energy of the different molecular orbitals. The problem is analyzed in terms of the Walsh diagrams and of the empirical rules for linearity and nonlinearity (146). In particular, molecules with less than 16 electrons are all linear in their ground states. A classic example of linearity effects is given by the comparison between the NO_2 molecule and CO_2 molecule (114). The highest-in-energy component of the CO_2 spectrum is an out-of-phase combination of p orbitals that gives a π_g molecular orbital. The spectrum of the NO_2 molecule can be analyzed in terms of that of CO_2. The NO_2 molecule has 17 electrons—one more than CO_2—and therefore it is not linear. The analysis is carried out by deriving the effects of bending on the molecular orbitals of CO_2. For example, the above π_g orbital is split by bending into $3b_2$ and $1a_2$ components.

From the above examples, it should now be clear that the analysis of molecular photoelectron spectra largely proceeds by studying in parallel large classes of similar molecules. The identification of the different molecular orbitals starts from their energy positions—or better, from their order with respect to energy—and from their vibrational structure. Once this analysis has been completed for a given molecule, the same analysis can be extended after simple changes to a homologous series of molecules, such as the systems isoelectronic to the first molecule. Another systematic approach is, of course, the classic "Aufbau" process in which orbitals of molecules with more electrons are built from those of molecules with fewer electrons. Nice examples of correlation within a certain class of molecules are offered by the ionic molecules (112). Figure 17.30 shows, for example, the spectra of three potassium halides, which are strongly correlated with one another.

C.   SOLIDS: BAND–STRUCTURE EFFECTS AND SURFACE STATES

Most experiments in uv photoemission concern the valence electrons in solids and solid surfaces. The electronic states in solid systems can be analyzed by different approaches. Broadly speaking, these approaches can be divided in two categories (51). In the first category we find those approaches that emphasize the formation of chemical bonds along lines not much different from those of molecular physics. The second category includes approaches that emphasize the periodicity of the potential "seen" by the electrons in a crystalline solid. One fundamental point in theories of this second kind is the Bloch theorem already discussed in Section I. Put in a different form, the Bloch theorem says that each electronic state in a crystalline solid can be identified by a wavevector $\mathbf{k}$. Of course this is true also in the limit case of *free* electrons, for which $\mathbf{k}$ is identified with the momentum. In particular, the energy of each Bloch state is related to $\mathbf{k}$. The energy–wavevector curves define the *band structure* of the solid. One particularly important consequence of the band structure is the presence in the energy spectrum of allowed bands separated by forbidden gaps. The position of the Fermi level in the band structure defines the general character of a solid. For metals, the Fermi level is within an allowed band, whereas for insulators and semiconductors it is in the forbidden gap separating the highest filled allowed band, or *valence band*, from the lowest empty allowed band, or *conduction band*. Band-structure calculations give the $E_i(\mathbf{k})$ curves, and from these one can calculate the density of occupied states. In turn, the density of states

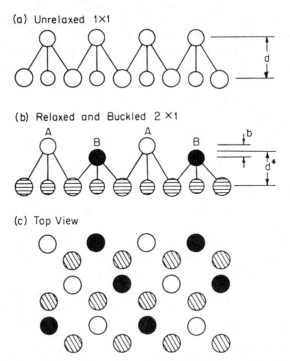

(a) Unrelaxed 1×1

(b) Relaxed and Buckled 2 ×1

(c) Top View

Fig. 17.31. Side view of the surface atoms of a SI (111) surface without (a) and with (b) a model surface reconstruction. The bottom configuration is only one of the model reconstructions that can lower the energy of this surface.

can be used to interpret the photoemission EDCs along the lines discussed in the Appendix. The interpretation is based on the assumption that the photoemission peaks primarily correspond to peaks in the density of states.

The difference between the density of states and the EDJDOS (see eq. A15 in the Appendix), together with the transition-probability effects on the peak intensities, put some limits on the correspondence between band-structure calculations and photoemission spectra. Even more severe limits are caused by the surface sensitivity of the technique. The band-structure calculations are usually based on the periodicity of the solid. The surface breaks down this periodicity in at least one direction. In many phenomena, this periodicity breakdown can be neglected. Not so in photoemission, which, as we have seen, explores a very thin region near the surface. A general consequence of the symmetry breakdown is the introduction of new localized states—the so-called *surface states*—at energies that are sometimes in a forbidden gap of the perfect-crystal energy spectrum. From a chemical-bonding point of view, the presence of these localized states on covalent surfaces can be explained in terms of nonbonding orbitals. Figure 17.31 shows, for example, the (111) surface of a silicon crystal. The presence of the surface leaves unsaturated one of the four lobes of the bonding sp^3 hybrid. The forbidden gap in silicon corresponds from a chemical-bonding point of view to the minimum distance between sp^3 bonding and antibonding states. The electrons in the unsaturated lobes at the surface are essentially nonbonding, and therefore they can be expected to be in the gap between bonding and antibonding states. Their exact position in energy depends on the rearrangement of the surface atomic positions with respect to the bulk positions—the so-called *surface reconstruction*. For Si(111) a model reconstruction (49) is also shown in Fig. 17.31. The effect of the reconstruction is to move the surface states, leaving some of them *within* the forbidden gap. In contrast, the reconstruction of the GaAs(110) surface entirely removes the local states from the forbidden gap. In "realistic" surfaces the clean-surface states are accompanied, and in most cases replaced or overwhelmed, by the chemical-bonding states for chemisorbed atoms and molecules, which we shall discuss in the next section.

The analysis of a solid-state photoemission spectrum must include consideration of the bulk band-structure states, the intrinsic clean-surface states, and the extrinsic chemisorption and physisorption states. We shall now review some representative spectra to illustrate the first two kinds of feature.

The simplest model for the electrons in a solid is the *free-electron model* based on the quantum-mechanical "particle in a box" problem. The DOS predicted by this model is a simple parabolic function, $\rho(E_i) \propto E_i^{1/2}$. The corresponding ideal photoemission spectrum would be a parabolic curve truncated at the Fermi energy, as shown in Fig. 17.32. Some of the features of this very simple model can be observed in the uv photoemission spectra of the simple metals (i.e., of the group III elements). For example, Fig. 17.32 shows the valence-band spectrum of a thick aluminum layer on silicon (93). A simple parabolic function, however, does not explain the structure observed in this spectrum. This structure is related to the details of the chem-

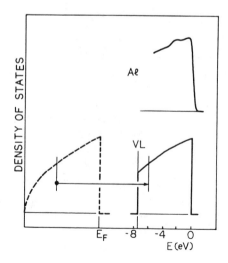

Fig. 17.32.   In a free-electron metal the density of states is a parabola up to the Fermi level and zero above the Fermi level (dashed line). In the simplistic interpretation of the photoemission spectra, one can imagine that this parabula is displaced in energy by the photo energy to give the photoelectron energy distribution (bottom solid curve). Indeed, some of the predicted features are observed in the photoemission spectrum of a thick Al layer (top solid line) (93).

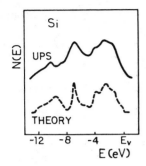

Fig. 17.33. The uv photoemission spectrum of Si can be explained almost entirely by the calculated bulk-Si density of states (63,120). The surface states (see Fig. 17.35) are an exception.

ical-bonding process in the layer. The simple metals are among the easiest solids to treat theoretically. In contrast, the transition metals present many more difficulties, due to the involvement of d electrons in the chemical-bonding process.

One of the most widely investigated materials in solid-state chemistry and physics is silicon, and photoemission spectroscopy and band-structure calculations are no exceptions to this popularity. Figure 17.33 shows the remarkably good agreement between the calculated density of states and the valence-band photoemission spectrum for this material (63,120) The structure of this spectrum is characteristic of the sp^3 hybridization and in general of all covalent semiconductors. In fact, a similar three-peaked structure can be observed for all the semiconductors with that hybridization. This qualitative structure is observed also for all semiconductors of the III–V and II–VI families (32,77). Very detailed analyses of the relation between the spectral features and the band structure for different families of semiconductors were presented in the early 1970s (32,77).

Another class of solids very widely studied by means of photoemission are the layer compounds, in particular, the III–VI compounds and the transition-metal dichalcogenides (88,94,95). These materials exhibit an extremely interesting blend of different kinds of chemical bonds. Figure 17.34 shows, for example, the structure of a layer of GaSe (94). The intralayer bonds here are either covalent (Ga—Ga) or nearly ionic (Ga—Se). furthermore, the interlayer bonds have some van der Waals components together with covalent components. These mixed characters can be found in the valence-band photoemission spectra, as we can see from the classification of the spectral features reported in Fig. 17.34.

In general, given a valence-band photoemission spectrum for a given material, a preliminary classification of the spectral features is possible on the basis of qualitative chemical-bonding considerations. A more detailed analysis requires band-structure calculations and an estimate of the density of states. These calculations are carried out using, for example, tight-binding approaches or self-consistent pseudopotential iterative schemes. The analysis of the bulk state must be preceded, however, by the identification of the nonbulk spectral features. This is typically done by contaminating the

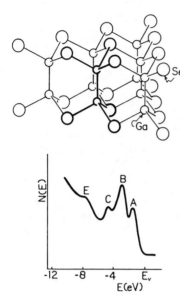

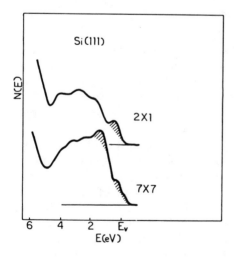

Fig. 17.34. The crystal structure of GaSe (top) corresponds to an interesting admixture of different kinds of chemical bonds. Within a layer there are covalent Ga—Ga bonds and ionic Ga—Se bonds, and the interlayer bonds have van der Waals components. The corresponding electronic states give rise to the structure observed in the photoemission spectrum of GaSe (bottom) (94). Peaks A, B, C, and E are due to cation–caton bonding p_z states, Ga—Se bonding $p_{x;y}$ states (peaks B and C), and antibonding cation–cation s states, respectively.

sample surface slightly and observing the features that are most changed in the photoemission spectra.

Figure 17.35 shows the most classic example of surface-state contributions to a photoemission spectrum—the Si(111) surface (11,31,127,145). Notice that two different spectra are shown in this figure. The two spectra correspond to two different reconstructions of the Si(111) surface. These reconstructions are detected by means of LEED patterns, and their names follow the LEED nomenclature. In our short outline of LEED in Section

Fig. 17.35. Photoemission spectra corresponding to two possible reconstructions of the Si (111) surface (127). The shaded areas emphasize the surface-state contributions.

II, we saw that electrons are diffracted along directions defined by eq. 14, or

$$\mathbf{k}_{\parallel'} = \mathbf{s}_{\parallel} + \mathbf{k}_{\parallel} = \mathbf{G}_{\parallel} + \mathbf{k}_{\parallel} \tag{22}$$

The electrons in the beam characterized by a given $\mathbf{k}_{\parallel'}$ value give a spot on the fluorescent screen used to detect them. The pattern of these spots of an *unreconstructed* surface Si(111) is called 1 × 1. When the experimental geometry used to take this pattern gives $\mathbf{k}_{\parallel} = 0$, each spot identifies directly a vector $\mathbf{G}_{\parallel}$. Shown in Fig. 17.35 are the spectra for two different recon- structions of Si(111), corresponding to the 2 × 1 and 7 × 7 patterns. Ob- viously the extra spots present in these patterns correspond to new vectors $\mathbf{G}_{\parallel}$ introduced by the surface reconstructions. According to the definition of G vectors in Section I new vectors $\mathbf{G}_{\parallel}$ correspond to a new, long-range parallel periodicity introduced by the surface reconstruction. Although the LEED patterns identify the symmetry of the reconstruction, they do not identify its physical causes. In fact, these causes have not yet been com- pletely identified for Si(111) (11,107). There is a substantial hope that the surface-state photoemission features will help clarify this point. Notice that in the two Si(111) spectra, there is more than one spectral feature due to surface states, and that the top feature is always extended above the top of the valence band of the bulk-Si spectrum. Therefore in Si there are intrinsic surface states in the forbidden gap. Also notice the substantial differences between the surface-state spectra for the two reconstructions. In particular, for the 7 × 7 reconstruction there is a small feature at the top of the spectrum that has sometimes been identified as a Fermi edge. This identification im- plies that the 7 × 7 reconstructed Si(111) surface has metallic character, as is indeed suggested by the differences between the optical-absorption spectra of the surface states of Si(111) 2 × 1 and Si(111) 7 × 7 (11,107). The whole issue of the differences and similarities between Si(111) 2 × 1 and Si(111) 7 × 7 is still quite controversial, and the spectra of Fig. 17.35 are interpreted in different ways by different authors (11,107).

D. SYNCHROTRON-RADIATION TECHNIQUES

We shall now critically analyze the conventional mode of photoemission described in this section and in Section III. The conventional photoemission mode consists of measuring the number of photoelectrons collected by the analyzer as a function of their kinetic energy, keeping constant all other parameters and, in particular, the photon energy. The corresponding spectra—the EDCs—have been related by our discussion to the distribution in energy of the electrons in the ground state of the system. We have seen that this link is quite indirect. The choice of the EDC mode of photoemission is not made to optimize the effectiveness of the technique; rather, it origi-

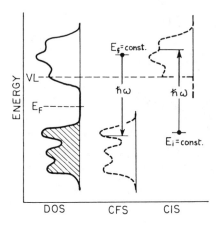

Fig. 17.36.   Scheme of the CFS and CIS modes of photoemission (72). In the CFS mode, the energy window of the electron analyzer is kept at a constant value E. The photo energy is scanned, and in this way one explores the occupied DOS (shaded area on the left) with less initial state–final state convolution than in the EDC mode. In the CIS mode, both the photon energy and the final-state energy are scanned, but their difference is kept constant. Thus the energy of the initial state of the photoionization process is kept constant. The CIS curves primarily probe the unoccupied DOS.

nates from technical limitations. In particular, the conventional photon sources *are* at constant photon energy. The limits of conventional sources are eliminated by the use of synchrotron radiation. The question then is: can one run photoemission better than in the conventional EDC mode? The work of recent years has proved that this is possible, at least for some special experiments. We shall discuss in this section the special synchrotron-radiation techniques that exploit the continuity and polarization of the synchrotron–radiation source. Another unconventional development of photoemission is the use of angular resolution, which we shall discuss in the next section.

In principle the tunability of both photon energy and the collected photoelectron kinetic energy opens the way to an infinite number of new photoemission modes (89). In practice two modes besides the EDC mode have become widely used in recent years, after having been introduced by Lapeyre and coworkers (72). These two modes are the constant-final-state (CFS) mode and the constant-initial-state (CIS) mode. Figure 17.36 schematically outlines the CFS and CIS modes.

In the CFS mode the energy window of the analyzer is kept in a constant position E while the photon energy hv is scanned by a monochromator. The purpose of this mode is to allow the investigator study the DOS in the system without having to deal with the convolution of initial and final states that cannot be avoided in the conventional mode (see eq. A15 in the Appendix). The elimination of final-state effects is not complete, since this mode fixes only the *energy* of the final state, and the energy is not sufficient to completely identify the final state itself. Furthermore, the final state plays a role in the transition probability. Nevertheless, the convolution of initial and final states is certainly much less severe than in the EDC mode. Real problems arise in the CFS mode from another factor, the interfering effects of primary and secondary electrons. In the fixed energy window in the CFS mode, a primary-electron signal is superimposed on a secondary-electron background. Changes in the photon energy can change both the primary signal and the background. The primary-electron dependence on hv will reproduce within certain limits the density of initial states at the energy $E - hv$. The secondary-electron signal will be proportional to the total number of primary electrons excited at energies above the selected window. This total number will be related to the optical-absorption coefficient of the system (45). Thus the secondary-electron effects in the CFS mode can be an interesting subject to study and not just a spurious signal. In fact, these effects retain the surface sensitivity of all other photoemission effects. Thus the secondary-electron effects in the CFS mode provide a way to study the optical-absorption coefficient of the region near the surface.

The above arguments led in recent years to the development of another synchrotron-radiation technique, called partial-yield spectroscopy (PYS) (45). A PYS experiment is in practice a CFS experiment in which the energy window is centered in the low-energy region, where the secondary-electron signal prevails over the primary-electron signal. Gudat and Kunz (45) demonstrated that the corresponding spectra taken versus hv are proportional to the surface absorption coefficient of the system. This technique has made it possible to identify a number of surface effects on the optical-absorption coefficient. The core-excitonic shifts are a typical example (116). Figure 17.37 shows the PYS curve for a Si(111) 2 × 1 surface in the spectral region of the excitation threshold of the Si $2p$ core level. Indeed, we see in the

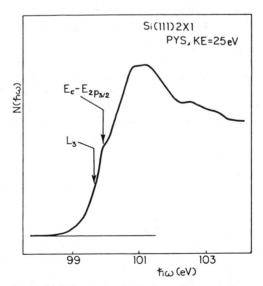

Fig. 17.37. The $L_{2,3}$ threshold of Si measured with the surface-sensitive partial-yield technique (PYS) (87). The vertical arrow labeled L_3 identifies the experimental position of the L_3 edge. The arrow labeled $E_c - E_{2p3/2}$ shows the estimated position of this threshold obtained by summing the binding energy of the Si$2p_{2/3}$ core level to the optical gap. The discrepancy is due to many-body effects, the so-called core-excitonic shifts.

spectrum the absorption threshold corresponding to that excitation (87). The threshold photon energy was estimated *a priori* by taking the distance in energy between the Si 2p peak and the top of the valence band in a conventional photoemission spectrum, and then adding to it the band gap width given by optical experiments. The estimate of the threshold photon energy obtained in this way turns out to be consistently higher than the measured PYS threshold by a few tenths of an electron volt. The discrepancy is attributed to core-excitonic effects, that is, to the interaction between the optically excited electrons at the bottom of the conduction band and the core holes they left behind. However, a complete theoretical explanation of this effect is not yet available (116).

An important by-product of the PYS technique is the surface-EXAFS experiments (19,75). The term EXAFS stands for "extended x-ray absorption fine structure." This is the small modulation present in the optical-absorption coefficients of solids and molecules above each core-level absorption threshold. Strictly speaking the surface-EXAFS technique involves photon energies in the x-ray range and does not belong in this section, but it is a natural extension of the PYS technique treated here. The mentioned modulation is due to the influence of the atoms near the excited one on the final excitation state. In an EXAFS-like approach we imagine that part of the final-state outgoing spherical wave function is backscattered by these atoms, interferes with itself, and modulates the optical transition probability. The interference can be constructive or destructive, giving maxima or minima in the modulation, depending on the interatomic distance and on the electronic wavelength. This depends in turn on the final-state energy and therefore on the energy of the exciting photon. Therefore the EXAFS modulation offers an opportunity to study *selectively* interatomic distances in the neighborhood of the atomic species corresponding to the core-level threshold. The latest development in the field of EXAFS is the surface-EXAFS technique. This corresponds to measurements of the EXAFS modulation by PYS rather than by bulk optical-absorption spectroscopy. Therefore the EXAFS modulation will correspond to the interatomic distances near the surface rather than in the bulk. This technique is very interesting in the case of chemisorbed species, for which it can be used to measure chemisorption bond lengths and, in some cases, to identify chemisorption sites. An example of a surface-EXAFS curve is presented in Fig. 17.38 (19). Here we see the surface-EXAFS of the SiK edge for a Cl-covered Si(111) surface. The observed surface-EXAFS modulation is due to the interatomic Cl–Si distance and the EXAFS modulation is used to identify an "on top" chemisorption geometry for this system.

The CIS mode of photoemission—again illustrated by Fig. 17.36—explores the density of *unoccupied* states of the system. Therefore it eliminates one of the most severe limits of conventional photoemission, which probes only occupied electronic states. This limit is eliminated in the CIS mode by minimizing the initial-state effects on the experimental spectra. In the CIS

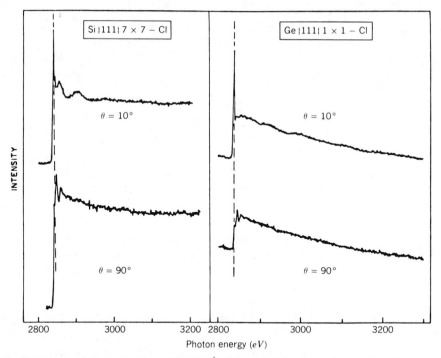

Fig. 17.38.   An example of surface-EXAFS (19): The modulation above the K optical absorption edge of chemisorbed Cl on a Si surface.

mode, spectra are taken by sweeping both the photon energy and the position of the energy window of the analyzer. The scanning is synchronized so that the difference between the two energies remains constant:

$$E - h\nu = \text{constant} \tag{23}$$

This can be accomplished either by using an analog photon-energy output from the monochromator to drive the analyzer or by controlling both energies with a minicomputer. The reasons behind the condition expressed by eq. 23 become clear if one considers that for primary photoelectrons, the kinetic energy corresponds to the final-state energy of the optical-excitation step of the photoemission process. The energy balance of eq. 19 applied to eq. 23 gives

$$E_i = \text{constant} \tag{24}$$

That is, the primary photoelectrons collected in this mode of operation are all originating from initial states at the same energy. Therefore, structure in the CIS spectra corresponds to structure in the *final* density of states, and the CIS curves are to some extent a picture of the density of unoccupied

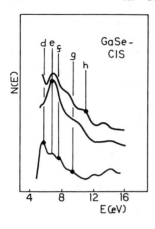

Fig. 17.39.   Constant-initial-state (CIS) spectra of GaSe (94). The vertical lines show the theoretically predicted positions of the peaks in the conduction-band DOS and emphasize the correspondence with the CIS spectral features.

states of the system. (More rigorously, if we consider eq. A15 in the Appendix, it is clear that the spectral structure observed in the CIS curves arises primarily from the condition $\nabla_k E_f = 0$, i.e., from maxima in the density of final states.) In summary, the CIS curves give a picture of the density of unoccupied states of the system.

The effectiveness of the CIS technique is illustrated by Fig. 17.39 which shows CIS curves for the layer compounds GaSe (94). The conduction-band states in this material have quasimolecular character. This gives rise to a much richer structure in the conduction-band DOS than, for example, in covalent semiconductors. The different CIS curves in Fig. 17.39 correspond to different values of the constant in eqs. 23 and 24, that is, of the constant initial-state energy. We observe a number of peaks in these curves that occur at the same *final-state* energy. These peaks are identified as conduction-band peaks in the DOS. Also shown in Fig. 17.39 are the theoretically predicted positions for these peaks (94). The correspondence between theoretical peaks and experimental peaks confirms the validity of the CIS approach. Notice, however, that the CIS curves *do not* explore the states at the bottom of the conduction band. In fact, no electron can be emitted at energies below the vacuum level. Therefore the vacuum level is the minimum energy that can be explored in the CIS mode. The states between the minimum occupied-state energy—the bottom of the conduction band for insulators and semiconductors, and the Fermi level for metals—can still be investigated using the PYS technique discussed above. This is done by taking PYS curves in the spectral region of a core-level threshold. Before the onset of the EXAFS the structure in these curves reflects the density of occupied states that can be reached by optical transitions from the core level.

The linear polarization of the synchrotron-radiation source is used in special photoemission experiments to identify the character (i.e., the symmetry) of the electronic states. This can be effectively illustrated by a practical example (94). The top of the valence band of the layered semiconductor GaSe is theoretically predicted to contain primarily *p*-like electronic states.

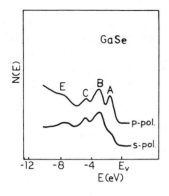

Fig. 17.40. Photon-polarization dependence of the photoemission spectrum of GaSe (94). In the s polarization, the vector potential of the photons has no components perpendicular to the surface, whereas in the p polarization, it does. Notice the dramatic effect of the change in polarization or peak A. as discussed in the text, this identifies as p_z the character of this spectral feature.

If z is the direction normal to the layers, these states could be either p_z or $p_{x,y}$. Let us now consider the optical excitation from $|i>$ to $|f>$, which is the first step of the photoemission process. As discussed in the Appendix, this transition occurs with a probability defined by the matrix element $<i|\nabla|f>$ multiplied by the photon polarization factor $\eta = \cos (\mathbf{A}, <i|\nabla|f>)$ (see eq. A3). In our case $|i>$ is a p-state. For the sake of simplicity we can consider $|f>$ to be a free-electron state, that is, plane wave [a more rigorous atomic-state analysis leads to the same conclusions (94)]. Then, since $|f> = |e^{i\mathbf{k}\cdot\mathbf{r}}>$, the direction of $<i|\nabla|f>$ is along z for $|i> = p_z$ and along the x,y plane for $|i> = p_x$ or p_y. Therefore η becomes zero for p_z states if $\mathbf{A}$ perpendicular to z. This offers a key to identifying the p_z states in a p-like manifold. In Fig. 17-40 we see the EDCs for GaSe taken for two different polarizations. In the s polarization the electric field of the synchrotron radiation, which is parallel to the $\mathbf{A}$ vector, is perpendicular to the z direction, that is, to the sample surface. A comparison between the spectra for the two polarizations shows that the states at the top of the valence band do not contribute much to the EDCs in s polarization, whereas they do in the p polarization. From this analysis we conclude that the states at the top of the GaSe valence band have p_z symmetry.

The above analysis has general validity (94). In many cases bulk bonds and chemisorption bonds involve s-like and p-like states, and their theoretical analysis is greatly facilitated by the identification of the p_z components. We shall see in the last part of this section the application of this method to chemisorption problems. Other analysis methods have been developed to exploit the polarization of the synchrotron-radiation source. Among these, one of the most powerful is the Hermanson rule (54), which can be applied to all systems that have a plane of mirror symmetry.

E. ANGLE-RESOLVED PHOTOEMISSION

In the photoemission experiments described in the previous sections, the analysis of the photoelectrons was limited to their energy. A complete characterization of the photoelectron free-electron states requires measuring the directions of the k-vector, in addition to its magnitude, which is given by the energy. Such complete characterization is the goal of angle-resolved photoemission (135). The philosophy of angle-resolved photoemission is that from a better-characterized final state it is possible to retrieve more complete information about the initial state of the system. We shall now review how this is made possible in practical experiments. As discussed in the Appendix, the photoemission process is often treated as the combination of three separate steps: optical excitation of the electron, transport to the surface, and escape into a vacuum (35). In this model one must distinguish between the final state of the photoelectron *outside* the solid—which is a free-electron state—and the final state *inside* the solid after the optical excitation. In this section we shall identify the final state *outside* the solid with the symbol $| f^* \rangle$. This state is determined once we know its k-vector $\mathbf{k}^*$. The energy measured from the vacuum level is

$$ E = \frac{\hbar^2}{2m} k^{*2} \tag{25} $$

This energy coincides with E_f for primary photoelectrons. In contrast, the k-vector $\mathbf{k}^*$ of the state $| f^* \rangle$ does *not* coincide with the k-vector $\mathbf{k}$ of the excited state $| f \rangle$—nor with that of the initial state, which is also very close to $\mathbf{k}$. Angle-resolved photoemission measures $\mathbf{k}^*$ and the analysis of its data requires one to establish a relation between the k-vector $\mathbf{k}^*$ and that of the initial state $| i \rangle$.

While making a transition from the state $| i \rangle$ to the state $| f^* \rangle$, the electron can change its k vector at each step of the photoemission process. The second step, however, does not give important contributions to the final angular distribution outside the solid. During the transport to the surface, the elastic-scattering processes can change the direction of the k-vector, but the resulting angular distribution is a smooth background that has in the measured angular distribution a role similar to that of the secondary-electron background in the EDCs. This background is usually ignored or subtracted from the angular-distribution plots. Much more important are the effects on the k-vector of the optical excitation and of the transport through the surface.

During the optical excitation, the photon, which has a k-vector of small magnitude, leaves almost unchanged the k-vector of the electron. However, the lattice *as a whole* can participate in the optical-excitation process, changing the k-vector of the electron by a G-vector without contributing to its energy. In retrieving the initial state $\mathbf{k}$ from the measured $\mathbf{k}^*$, one must therefore take into account that a G-vector may be involved.

In a similar way,, the *parallel* component of the k-vector can be changed in the third step of the photoemission process as the electron crosses the surface. Notice that the periodicity at the surface may be different from a two-dimensional projection of the bulk periodicity, due to surface reconstruction and/or ordered overlayers. Therefore the *surface* G-vectors, $\mathbf{G}_\parallel^{surf}$, do not necessarily coincide with the parallel components of bulk G-vectors. This possible new contribution must be taken into account while trying to relate $\mathbf{k}$ to $\mathbf{k}^*$. We must say in general that

$$\mathbf{k}_\parallel^* = \mathbf{k}_\parallel + \mathbf{G}_\parallel^{surf} \tag{26}$$

and only within the *surface* reduced zone does the parallel component of the k-vector not change during the process. In a real experiment a given k-vector $\mathbf{k}$ of component $\mathbf{k}_\parallel$ gives rise to a primary k-vector $\mathbf{k}_0$ with the same component, $\mathbf{k}_\parallel^* = \mathbf{k}_\parallel$, plus a number of secondary k-vectors of components given by equation 26 for $\mathbf{G}_\parallel^{surf} \neq 0$. The names given to the resulting two kinds of directions of the photoemitted electrons are "primary and secondary Mahan cones" (81).

As to the *perpendicular* component $k_\perp$, it will change in general while the electron goes through the surface, since it must overcome the potential barrier corresponding in solids to the work function. The total height of the barrier cannot be calculated easily, and must be determined empirically. Even the concept of surface barrier is somewhat difficult to define in this case. Therefore the relation between the two perpendicular components $\mathbf{k}_\perp$ and $\mathbf{k}_\perp^*$ inside and outside the sample is not easy to establish. This problem is usually solved with empirical or semiempirical approaches, as we shall discuss below.

The most spectacular application of angle-resolved photoemission is the band-mapping technique. In the previously described photoemission experiments in solids, the electronic state was characterized by measuring the DOS, that is, the distribution in energy of the electrons. This is a very limited characterization. Due to the Bloch theorem a full characterization implies a measurement of the k-vector direction as well as of the energy. In fact most physical properties of a solid are described by its band structure, that is, by the function $E_i(\mathbf{k})$. Knowledge of the band structure implies knowledge of energy and k-vector. Until recently the concept of band structure was essentially a theoretical one—no experiment had directly detected a band structure, although the experiments did detect its effects. This has been changed by angle-resolved photoemission, which allows one to *map directly* the energy–wavevector dependence (74,134).

Let us consider an angle-resolved EDC. A peak at a given energy in one of these curves reveals the presence of electrons at that energy. The k-vector of these electrons is also identified, since its magnitude is given by the energy (measured from the vacuum level) according to the equation

$$|\mathbf{k}^*| = \frac{(2mE)^{1/2}}{\hbar} \tag{27}$$

and its direction coincides with the direction in which the angle-resolved

EDC was taken. The photoelectrons so detected were originally electrons in a state $|\,i\rangle$ in the solid. The energy of the state $|\,i\rangle$ is simply $E - h\nu$. Let us now assume that we can identify the k-vector $\mathbf{k}$ of the state $|\,i\rangle$ from the measured k-vector $\mathbf{k}^*$ of the photoelectron. At this point we would have both E_i and $\mathbf{k}$. After repeating this analysis for different energies and k-vectors—by taking angle-resolved EDCs in different directions and/or at different photon energies—we would be able to build point by point the function $E_i(\mathbf{k})$. That is, we would experimentally map the band structure.

Establishing the link between $\mathbf{k}$ and $\mathbf{k}^*$ is not trival, however. Even if we neglect the possible role of bulk and surface G-vectors along certain directions, there is still the problem of the nonconservation of the perpendicular component of the k-vector as the electron crosses the surface. This last problem, however, is automatically solved for one important class of solid-state systems, the two-dimensional (or quasi-two-dimensional) systems. The layer compounds such as GaS are examples of quasi-two-dimensional systems. For these crystals the only important component of the k-vector is that parallel to the surface. For a completely two-dimensional system, that would be in fact the *only* component appearing in the band structure. Therefore the important part of the band structure is given by the relation between E_i and $\mathbf{k}_\parallel$. Except for possible—and identifiable—G-vector contributions, $\mathbf{k}_\parallel$ coincides with measured $\mathbf{k}_\perp^*$. Hence there is immediately the possibility of mapping the band structure. An example of experimentally mapped two-dimensional band structure is that of Fig. 17.41, which was deduced from the angle-resolved EDCs of GaS (124). The method of obtaining the band-

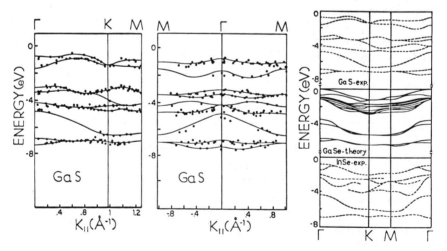

Fig. 17.41. Two-dimensional band-structure mapping with angle-resolved photoemission. Left and center: Experimental points deduced from the angle-resolved photoemission spectra of GaS (121). Right: Experimental band structure of GaS deduced from the above data, compared with the theoretical band structure of GaSe calculated by M. Schluter (see reference 96) and the experimental band structure of InSe (94).

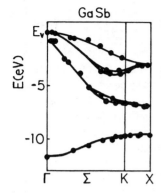

Fig. 17.42.   An example of three-dimensional band-structure mapping by angle-resolved photoemission—the band structure of GaSb (adapted from Ref. 13).

structure plot from the EDCs is the one outlined above: For each peak in the EDCs one identifies energy and $\mathbf{k}_\parallel^*$ (and therefore $\mathbf{k}_\parallel$) and this gives a point of the $E_i(\mathbf{k}_\parallel)$ plot. Also shown in Fig. 17.41 is the calculated band structure for GaSe and the mapped band structure for InSe (73), two compounds extremely similar to GaS. The similarity between theoretical and experimental curves is evidence for the validity of the band-mapping technique. This technique is one of the most spectacular results obtained in recent years in condensed-matter physics.

The problem is still left of finding a relation between $\mathbf{k}$ and $\mathbf{k}^*$, that is, between $\mathbf{k}_\perp^*$ and $\mathbf{k}_\perp$, for non-two-dimensional systems. Different authors have proposed in recent years a variety of approaches to this problem. We shall discuss here some of these. The first approach is based on the assumption that the total barrier W crossed by the electrons at the surface is measurable and defined, at least in a small range of energies. This parameter can be measured, for example, by successively comparing theoretical and experimental band structures and trying to improve the agreement between them. One can also estimate this parameter from other techniques, such as LEED. Figure 17.42 shows a three-dimensional band structure obtained by empirically determining the surface barrier. Notice that once the surface potential barrier is defined, the relation between $\mathbf{k}_\perp$ and $\mathbf{k}_\perp^*$ is given simply by the change in energy. Assuming that the excited electron inside the solid is in a nearly free state, we have

$$\frac{h^2 k^2}{2m} - W = \frac{h^2 k^{*2}}{2m} \qquad (28)$$

Since $k^2 = k_\parallel^2 + k_\perp^2$ and $k^{*2} = k_\parallel^{*2} + k_\perp^{*2}$, if we neglect G-vector effects and assume $\mathbf{k}_\parallel = \mathbf{k}_\parallel^*$, we obtain from equation 28

$$\frac{h^2 k_\perp^2}{2m} - W = \frac{h^2 k_\perp^{*2}}{2m} \qquad (29)$$

which solves the problem. A more sophisticated approach to the problem of finding the relation between $\mathbf{k}_\perp$ and $\mathbf{k}_\perp^*$ is the "triangulation method" (55). Another approach is that based on a theoretical feedback for the nearly free excited states coupled with empirical identification of zone boundaries (26).

The angle-resolved-photoemission techniques have other important applications in solid-state systems, as we shall see in the next section and in Section V. Furthermore, they have important applications in molecular and atomic physics (10,23,82). For these gas-phase systems, it is not the system itself that provides a reference for the angular distribution, as in solids. Rather, the reference frame is provided by the direction of the photon beam and by the direction of its polarization if it is linearly polarized. The angular distribution of the photoemission intensity referred to these directions can be theoretically calculated using atomic-physics approaches. Some of its features can be understood on very simple grounds. For polarized light, for example, if the initial state is an atomic s state, then the final state must be an atomic p state with its lobe pointing along the direction of the polarization vector. The intensity distribution is proportional to the square of the corresponding wavefunction:

$$N \propto \cos^2\theta \qquad (30)$$

where θ is the angle between the collection direction and the direction of the polarization vector. Equation 30 can be generalized to include initial-state orbitals of different symmetries, giving

$$N \propto 1 + \tfrac{1}{2}\beta(3\cos^2\theta - 1) \qquad (31)$$

where β is an asymmetry parameter related to the character of the initial-state orbital. For the above case, for example, $\beta = 2$ and equation 30 is obtained from equation 31. In general measuring β by detecting the angular distribution of the photoelectrons coupled to theoretical estimates of the parameter is a powerful way to study the character of the initial-state wavefunctions. Measurements with polarized photon beams are of course greatly facilitated by the use of a synchrotron-radiation source, which also allows energy scanning. For unpolarized photon beams, the parameter β can be deduced from the angular distribution around the direction of the photon beam.

F.   CHEMISORBED LAYERS

The surface sensitivity of photoelectron spectroscopy and of all other electron-spectroscopy techniques makes them ideal probes for surface chemical processes, and in particular for chemisorption processes. In this section we shall outline the application of uv photoemission spectroscopy to the study of chemisorbed species. These applications will be described using practical examples in which uv photoemission provided funamental information on chemisorption bonds, chemisorption sites, and so on.

The conventional mode of photoemission is already a powerful of information about chemisorbed species. The formation of chemical bonds on the surface changes the local distribution in energy of the electrons (i.e., the local DOS). These changes can be detected in the conventional EDCs. In turn, the character and position in energy of the EDC changes give information about the nature and the details of chemisorption-bond formation. To enhance the chemisorption-induced spectral changes, it is convenient to increase the surface sensitivity of the technique. The key to doing that is the escape-depth curve shown in Fig. 17.1. We see there that the escape depth is at a minimum or near-minimum value over the large energy interval from 10–20 to 150–200 eV. If the photon energy is selected to have kinetic energies in that range, the surface sensitivity of the photoemission probe is enhanced. As are all the states involved directly in the formation of chemical bonds, the chemisorption-bond states are located in the upper 10–20 eV from the vacuum level. Therefore the above-maximum-surface-sensitivity kinetic energies require photon energies of the order of 20–220 eV. Both the HeI and HeII sources give a reasonable surface sensitivity. The use of tunable synchrotron radiation further improves the technique—the photon energy can be selected to enhance the surface sensitivity not only for the chemisorption-bond states, but also for the deeper core levels that are also an important source of information about chemisorption processes. Furthermore, with synchrotron radiation the surface sensitivity can be emphasized or de-emphasized by tuning the kinetic energy, this makes easier the identification in the spectra of the surface contribution.

A typical example of a chemisorption system studied by conventional uv photoemission is Cl atoms chemisorbed on III–V semiconductor surfaces (92). The purpose of such a study is to identify to the best possible extent the chemisorption geometry. On cleaved (110) surfaces the anion and cation surface atoms are present in equivalent sites and in equal numbers. It is not clear *a priori* whether the Cl atoms would be bound to the cation sites or to the anion sites. Theory and simple chemical considerations indicate that the Cl p states involved in the chemisorption bonds are at higher energies for the cation-site chemisorption than for the anion-site chemisorption. This is due to the expected large transfer of electronic charge on formation of Cl–cation bonds. The corresponding theoretical DOS for GaAs–Cl is shown in Fig. 17.43 together with the DOS for the anion-site chemisorption. The difference in energy position between the two chemisorption sites is quite evident for the most intense chemisorption-induced peak in the local DOS, which indeed has Cl p character. These theoretical curves can be compared directly with the photoemission spectra for this system. The comparison is made in Fig. 17.43 and its results leave no doubt: the main spectral peak is much closer to the anion-site energy position than to the cation-site energy

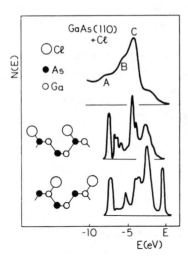

Fig. 17. 43.   The bottom and center curves show the calculated local DOS of chemisorbed Cl on GaAs for the two possible chemisorption sites shown at the left-hand side of the corresponding curve (92). The top curve is the photoemission spectrum of GaAs(110)–Cl (92). The position in energy of the main peak immediately identifies the As-site chemisorption as the correct hypothesis.

position. Therefore the experimental evidence is straightforward and in favor of the anion site.

The degree of sophistication of the photoemission technique to be applied to a given chemisorption process depends primarily on the complexity of the problems to be solved. For example, the conventional photoemission spectra of Fig. 17.43 are sufficient to identify the chemisorption site, but not to identify completely the chemisorption geometry. This can be understood if one considers that the GaAs(110) surface is relaxed and that this relaxation is likely to be changed by the Cl adatoms. The conventional EDCs are not sensitive enough to the relaxation to provide information about it. Other photoemission techniques can do better. For example, one can use the band-mapping technique illustrated in the previous section. The Cl overlayer is an ordered *two-dimensional* system for which the band structure is completely defined by the $E_i(\mathbf{k}_\parallel)$ curves in the parallel direction. The results of the band-mapping technique are shown in Fig. 17.44 for the GaAs(110)–Cl system (91). The theoretically calculated band structure is also shown in the

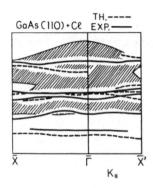

Fig. 17.44.   Two-dimensional band structure of a chemisorbed monolayer of Cl on GaAs(110) (91). The solid lines are experimental bands from angle-resolved photoemission. The dashed lines are calculated bands.

figure. The calculated curves were optimized to fit the experiment by careful selection of the surface parameters, including those that describe surface relaxation. Thus the band-mapping technique can be used indirectly to probe the surface relaxation, and in the case of Cl on III–V compounds, it shows that the adatoms do remove the relaxation, at least in part. We emphasize that the band structure is intrinsically more sensitive to the parameters of the system under investigation than the DOS is. In fact, the DOS is by definition an integral over the band structure, and therefore it averages its properties in k-space, somewhat reducing its sensitivity to the parameters of the system.

Another important source of information about chemisorption is the symmetry of the states in the chemisorption bonds. This information can be obtained using, for example, the photon-polarization techniques described above. The same techniques discussed above for identifying the p_z components in an s–p manifold can be used to study chemisorption systems. An example is Cl chemisorbed on Si(111) surfaces (89). One can *a priori* hypothesize two possible chemisorption sites for this surface. The first is a onefold symmetric "on top" site, and the second is threefold symmetric "interstitial" site. Once again the Cl p electrons play a major role in the formation of chemisorption bonds. It is easy to predict on simple chemical grounds that for the "on top" site, the $p_z - p_{x,y}$ degeneracy in energy is removed when the chemisorption bonds are formed, and that the p_z states are pushed down to energies lower than those of the $p_{x,y}$ states. In contrast, we expect the degeneracy to be reduced only slightly by the interstitial-site bond formation. The photoemission spectra for Cl on cleaved Si(111) are shown in Fig. 17.45. We see there a two-component spectrum. Furthermore, the dependence on polarization of the second peak is the same as that which was observed for the states at the top of the valence band of GaSe. We recall that our analysis linked this dependence to a p_z character of the states. Therefore, we conclude that the second peak of the Si(111)–Cl spectrum has p_z symmetry. This enables us to identify the "on top" site as the correct one for this particular system.

The study of chemisorption problems by uv photoemission is not limited to the identification of chemisorption geometries or to the investigation of the properties of the chemisorption bonds. An equally important application is the measurement of the fundamental physical parameters of the chemisorption system. One important example of this application is the formation of interfaces involving semiconductors [e.g., metal–semiconductor interfaces (93) and semiconductor–semiconductor interfaces (97)]. These interfaces are the basic components of the solid-state devices—diodes, transistors, and integrated circuits. The parameters characterizing the interface and its behavior are, for example, the Schottky barrier height, for metal-semiconductor interfaces, and the band discontinuities, for semiconductor-semiconductor interfaces. These two parameters are defined in Fig. 17.46. In most cases the measurements of the interface parameters rely on rather

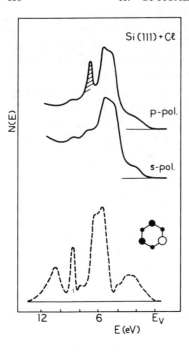

Fig. 17.45. Chemisorption-site identification by photon-polarized photoemission (89). Two chemisorption sites are proposed for Cl on cleaved Si(111), a threefold interstitial site and the onefold "on top" site shown at the bottom right. This last site removes the degeneracy of the Cl p states, creating a separated Cl p_z peak. This is the sharpest peak in the bottom spectrum which shows the calculated local DOS. The photoemission spectra indeed exhibit a separate peak and its photon-polarization dependence is that of the p_z states (see solid lines). Thus the "on top" site is the right one.

indirect methods based on the theoretical interpretation of transport experiments. The photoemission probe offers an alternative and it enables one to measure the parameters *directly*. For example Fig. 17.46 shows the spectrum of a CdS substrate covered by 10 Å of Ge (64). The valence-band region clearly exhibits two valence-band edges, one corresponding to the substrate and the other to the overlayer. From the distance in energy between the two edges, one directly obtains the valence-band discontinuity, ΔE_v. A combination of ΔE_v with the difference between the two forbidden gaps of the two semiconductors also gives the conduction-band discontinuity $\Delta E_c = \Delta E_g - \Delta E_v$. In the case of the CdS–Ge system, the result obtained from Fig. 17.46, $\Delta E_v = 1.85 \pm 0.1$ eV, is close to the theoretical estimate of 2 eV given by the LCAO (Linear Combination of Atomic Orbitals) calculations of the energy position of the top of the valence band for the two materials (50).

As to the Schottky barrier height, we see in Fig. 17.46 that this parameter is by definition the distance in energy between the bottom of the conduction band E_c and the position of the Fermi level E_F at the interfaces. The bottom of the conduction band is given by the top of the valence band E_v plus the forbidden gap: $E_c = E_v + E_g$. Thus a measurement of the Schottky barrier height $\phi_B = E_v + E_g - E_F$ requires only knowledge of E_v and measurements of E_v and E_F. These measurements can be done by photoemission spectroscopy. Figure 17.46 shows, for example, how the top of the valence band for a Si(111) surface changes its distance in energy from E_F as the surface

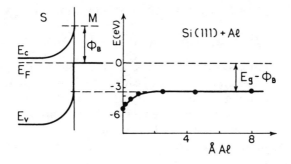

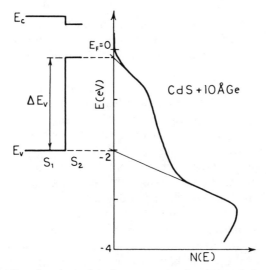

Fig. 17.46. Study of semiconductor interface parameters by photoemission. The left-hand side schemes define the valence-band discontinuity, $\triangle E_v$, for a semiconductor–semiconductor interface, and the Schottky barrier, ϕ_B, for a metal–semiconductor interface. The bottom right curve is the photoemission spectrum of CdS covered by 10 Å of Ge (64). Both valence-band edges are visible in this spectrum, and the valence-band discontinuity is directly measurable (64). The top right plot shows the position of the top of the valence band of Si with respect to the Fermi level for increasing Al coverage (93). These points were obtained from a sequence of photoemission spectra taken at each coverage step. From the final position, one obtains a direct estimate of the Schottky barrier.

is covered by Al atoms that eventually form an Al–Si(111) interface (93). The final distance of the top of the valence band from E_F gives ϕ_B, as discussed above. This approach not only enables one to measure the interface parameter ϕ_B, but also gives fundamental information on *how* the parameter is established. Notice, for example, the extreme rapidity of the shift of E_v in Fig. 17.46—a very few atoms of Al are sufficient to "create" the final situation that characterizes this interface.

V.   X-RAY PHOTOEMISSION SPECTROSCOPY

A.   GENERAL FEATURES

When the photon energy is in a range beyond 40–50 eV, the electronic states that can be studied by photoemission include not only those engaged in the formation of chemical bonds, but the atomiclike core-level states as well. The core-level peaks in photoemission spectra are an extremely valuable source of information about the chemical and physical properties of the system. On one hand, the core levels are not as deeply affected by the chemical-bond-formation process as are the electronic states directly involved in that process. On the other hand, the process does influence to a small but easily detectable extent the properties of the core levels: energy position, cross section, and so on. These properties are a vehicle of information about the chemical bond themselves. In summary, the core levels are somewhat "passive" probes of the chemical-bonding process in the system.

The investigation of core levels by x-ray photoemission spectroscopy involves the study of a number of properties. The most widely studied one is the position in energy of the core levels. This position is determined by the total potential "seen" by the electron in the deep state. The primary component of this potential is the atomic potential, since the core-level states are concentrated around the nucleus. Therefore it is not surprising that core levels for the same atomic species in different chemical environments have similar positions in energy. The positions are similar, but *not* equal. The potential "seen" by the electron is indeed influenced by the charge distribution corresponding to the chemical bonds of the system. Since the charge distribution changes with the chemical bonds from system to system, the energy position of the level changes too, by a few electron volts or more. These chemical-bonding-induced modifications of the atomic energy positions are known as *chemical shifts*.

Another source of information about the system is the *cross section* for photoionization of the core level. As we shall see, this cross section contains information not only about the atomic transition probability, but about the structural properties of the system as well. A study of cross sections ordinarily requires a tunable photon source, that is, the use of synchrotron radiation.

Finally, a third source of information is the *line shape* of core-level peaks. The line shape is influenced by many-body effects, and therefore can provide information about the electronic properties of the system beyond the one-electron approximation.

B.   CORE-LEVEL SPECTRA OF ATOMS AND SIMPLE MOLECULES

The first step in interpreting x-ray photoemission spectra is the theoretical estimate of the binding energies of the core levels. We shall now review some of the most widely used approaches to binding-energy calculations. In a first approximation, the ionization process of a core level can be regarded as a process confined entirely to a region close to the nucleus. One can assume, therefore, that the process does not involve directly—again in first approximation—the valence electrons. The effect on the valence electrons of the creation of a core hole can be considered to some extent equivalent to that of increasing by 1 the number of protons in the nucleus. This approximation is called *equivalent core* approximation (62). Let us see how it works in calculating the binding energy for **K**-shell electrons in a Li atom. The creation of a core hole in that shell can be regarded as the combination of three processes:

$$\text{Li} \rightarrow \text{Li}^+ + e^- \quad (2s \text{ electron removed}) \tag{1}$$

$$\text{Li}^+ \rightarrow \text{Li}^{2+} + e^- \quad (1s \text{ electron removed}) \tag{2}$$

$$\text{Li}^2 + e^- \rightarrow \text{Li}^+ \quad (1s \text{ hole}) \tag{3}$$

The energy for the first step is 5.39 eV, and that for the second step, 75.62 eV. The energy for the third step, that is, the negative of the energy required to remove a $2s$ electron from a Li atom with a $1s$ core hole, can be assumed equal to minus the ionization potential of Be^+, which is 18.21 eV. The sum of the three terms gives a total binding energy of 62.8 eV. By comparison, the experimental binding energy is 64.85 eV (62). The agreement is not striking, but is fair to good; a similar quality of agreement is found for all other estimates carried out in this approximation.

The real purpose of the equivalent-core method is not to calculate total binding energies, but to calculate chemical shifts in chemically bound systems using measured atomic and molecular binding energies in gas-phase systems. Its final goal is to relate the chemical shifts to measurable energies involved in chemical reactions. Let us consider, for example, the difference in binding energy between gas-phase nitrogen and molecular nitrogen in NO_2. This difference in energy can be regarded as the energy of the "reaction"

$$\text{NO}_2 + \text{N}_2^+(1s \text{ hole}) \rightarrow \text{NO}_2^+(1s \text{ hole}) + \text{N}_2$$

which, however, cannot be measured directly, because the reaction involves species not in their ground states. To transform it into a measurable reaction, one can apply the equivalent-core approximation by assuming that the N core with one $1s$ hole in $\text{NO}_2^+(1s \text{ hole})$ can be interchanged with the O core in NO^+. Therefore we can write $\text{NO}_2^+(1s \text{ hole}) = -\text{NO}^+ + \text{O}_2^+ + \text{N}_2^+(1s \text{ hole})$ and the above reaction becomes

$$\text{NO}_2 + \text{NO}^+ \rightarrow \text{O}_3^+ + \text{N}_2$$

This is a ground-state reaction, the energy of which is 3.3 eV from thermodynamic estimates. Thus we conclude that the shift in binding energy between NO_2 and N_2 is 3.3 eV for the N **K** shell. By comparison, the experimental results give a shift of 3 eV (37,62). The accuracy of this method depends, of course, on the quality of the thermodynamic data used to calculate the chemical-reaction energies.

In many cases the equivalent-core approximation method enables one to write down the chemical reaction energetically equivalent to the chemical shift, but no thermodynamic data, or too poor thermodynamic data, are available for estimation of the reaction energies. In these cases a theoretical estimate of the reaction energies can be attempted. This theoretical estimate can be either a full quantum-mechanical calculation or a simpler empirical calculation. Even full quantum-mechanical calculations exhibit several simplifying features in the case of the equivalent-core method. One of these is the involvement of closed shells only. Furthermore, the quantum-mechanical calculations ordinarily work much better in calculating the *difference* between the energies of two isolelectronic species than in calculating the absolute energy of each species. Reasonably good agreement between theoretical and experimental chemical shifts can be obtained in particular in the case of *ab initio* full quantum-mechanical calculations of the energy difference for equivalent reactions. These theoretical estimates must, of course, be corrected for all the effects not taken into account by the one-electron approaches. These effects are, for example, the relaxation effects described in the Appendix. Another important aspect of the equivalent-core approximation is its possible *reversed* used; that is, rather than using chemical-reaction energies to estimate chemical shifts, one can use chemical shifts to estimate reaction energies in cases in which thermodynamic data are not available.

Another important approach to the problem of chemical shifts is the correlation between the binding energy and the charge present in the atom (8). The physical grounds for establishing this correlation are quite simple. The energy required to remove the electron from the atom is expected to depend on the charge present in the atom. In fact, the binding energy increases with increasing positive charge and decreases with increasing negative charge. Calling q the total charge present in the atom, it is possible to relate the total binding energy to q through the equation

$$E_i = aq + eV' + E_i^R + b \qquad (32)$$

In this equation a and b are empirical constants. The term E_i^R is the relaxation-energy term discussed in the Appendix. The term eV' arises from all the atoms surrounding the one in which the photoionization process takes place. This term can be understood in the following way. When the electron is removed from the atom, it interacts not only with the charge in the atom itself, but also with the charge of all surrounding atoms. The symbol V' corresponds to the potential of the field caused by the surrounding atoms.

Equation 32 does a reasonably good job, sometimes even in the simplified form in which the correcting terms E_i^R and eV' are neglected. This is particularly true when the binding energies of *similar* compounds (i.e., compounds that have similar chemical bonds, structure, etc.) are compared with each other. The quality of the predictions of equation 32 depends, of course, on the quantitative estimates of q, the total charge of the atom. This implies that one must have a good knowledge of the chemical-bonding process and of the consequent transfer of valence-electron charge between different atoms. The methods for calculating the total valence-band charge range from crude empirical estimates to self-consistent quantum-mechanical calculations. The chemical shifts are in fact used as inputs in these calculations, since they can be used to estimate the difference in atomic charge from compound to compound, particularly in families of similar compounds. Comparison of similar compounds simplifies the use of equation 32, since the relaxation term E_i^R and the surrounding-atoms term eV' can be taken to be equal in first approximation for all such compounds. Therefore the difference in binding energy between atoms in similar compounds is related by equation 32 directly to the difference in valence charge of the atoms of interest. This direct link of the difference in binding energy to the difference in valence charge is one of the most appealing features of core-level photoemission spectroscopy. In fact, it is one of the most important reasons for the great interest of chemists in this technique. One should be careful, however, in using the link between chemical shift and atomic charge, since we have seen that this link becomes direct only after several approximations and assumptions have been made and only when reasonably similar compounds are compared with each other.

C. CORE LEVELS IN SOLIDS

The treatment of core levels in solids and, in particular, that of chemical shifts starts from the same background used for molecules, with additional complications in the estimates of the relaxation term in equation 32. Once again the power of the chemical-shift analysis relies primarily on the comparison of compounds reasonably similar to each other. A classic example of a chemical-shift trend (3,111) is shown in Fig. 17.47. Here the dots correspond to the binding energy for the Ge $3d$ level in a number of Ge compounds referred to that in pure Ge. The horizontal scale is the charge in the Ge atom. This charge was calculated from the ionicity of the chemical bonds according to Pauling:

$$I = 1 - \exp[-0.25(\Delta x)^2] \qquad (33)$$

where I is the ionicity and Δx is the difference in electronegativity between anion and cation. We see in Fig. 17.47 that the chemical shifts do *not* fall in a straight line, as equation 32 would predict if the relaxation term and the

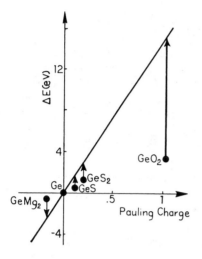

Fig. 17.47. Systematic changes of the energy position of the Ge $3d$ core level in a series of Ge compounds, plotted versus the Pauling charge (8,57). The dots are raw data from Ref. 57, and the arrows show the effects of the correction discussed in the text.

surrounding-atom terms were negligible. The discrepancy is due not to a poor estimate of the atomic charge, but to the important role of the term eV'. For a solid this term is well known: It is the so-called crystal field contribution to the binding energy (i.e., it has the character of a Madelung potential). After correcting the experimental points in Fig. 17.47 for this term, one obtains a straight line (8). The slope of this straight line is close to 14 eV per unit charge; this gives the magnitude of the valence-charge effects in the chemical shift. From the results of Fig. 17.47 it is clear that the few electron volts of chemical-shift difference observed in similar solids are the result of a delicate balance of two opposite terms—the valence (or ionic)-charge term in the atom, and the Madelung term. This balance of two large terms makes the calculation of chemical shifts a difficult task. The task is further complicated if the relaxation term must be taken into account. In particular, the chemical shift is affected by the extra-atomic relaxation, which in some cases may contribute more than 10 eV to the binding energy of core levels.

We shall now review the features of the chemical shifts in several different classes of solids. One of the simplest examples of core-level photoionization of atoms in a solid-state environment is that of noble-gas atoms in a metal matrix. The purpose of studying these systems is to isolate the binding-energy terms that are not related to charge transfer—the noble-gas atom does not form chemical bonds with the host matrix, and therefore no charge transfer occurs. In comparing binding energies for the free atoms and for the implanted atoms, it is important to use the correct zero for the energy scale in the two cases. For example, if the solid-state-environment binding energies are referred to the Fermi level, the work function must be added to these values when they are compared with the free-atom binding energies, which are referred to the vacuum level. Calculations of the experimentally observed chemical shifts have been based primarily either on the response of the electron gas (18) of the host lattice to the point perturbation caused by the creation of a core hole or on the changes occurring in the noble-gas atom itself when it is implanted in the host matrix (148). To calculate the changes occurring in the atom on implantation, we can imagine that the implanted atoms are confined to a spherical Wigner–Seitz sphere in the metal that has a radius close to the Van der Waals radius of the atom. The orbitals of the implanted atom are therefore "compressed" inside the sphere with respect to the orbitals of the free atom. The screening effect of these orbitals is enhanced by the compression, and this *decreases* the binding energy of the core levels, giving a positive contribution to their chemical shift. This effect, however, is compensated by the change in the interatomic relaxation due to the compression in the sphere. The two contributions are similar in magnitude but opposite in sign, and tend to cancel each other. The really important effects contributing to the chemical shift are the remaining relaxation effect and the possible correction to the work function. Once this conclusion is reached, the two different approaches to chemical-shift calculations for this class of systems essentially converge both conceptually and, to some extent, numerically.

The second class of solid-state systems that we shall discuss is that of ionic solids (137). Once again the analysis of the chemical shifts will be based on the analysis of the different terms in equation 32. For this kind of solid, one term is essentially negligible, that is, screening extra-atomic relaxation. In the (insulating) ionic materials, the electrons are not free to move and change their screening action. The term eV' in equation 32 is affected by a change in the electrostatic potential at the atom where the photoionization takes place due to changes in the charge distribution in the neighboring ions. The charges in these ions will respond to the creation of the core hole by becoming polarized (33). The nuclear positions will not change during the transition that creates the core hole, due to the rapidity of this transition (see Appendix). The energy term due to the changes in the neighboring ions has, therefore, the character of a polarization energy. One advantage in estimating the magnitude of this polarization energy is that it is essentially the same polarization energy that affects point-defect problems. Therefore the models developed in the past for point defects can be used for its estimation. One of these is the Mott–Gurney equation (100)

$$E_{\text{POL}} = -\frac{1}{2}\left(1 - \frac{1}{\epsilon}\right)\left(\frac{1}{R}\right) \tag{34}$$

where E_{POL} is the polarization contribution to eV', ϵ is the dielectric constant, and R is an empirical parameter that in the Mott–Gurney model corresponds to the radius of a cavity in which the central atom is supposed to reside. Other contributions to eV' arise from the usual Madelung term. The total binding energy contains, of course, the term due to the charge in the central atom in addition to the neighbor-atom term eV'. One remarkable feature of the combination of the three above terms—the polarization energy, the Madelung term, and the ionic-charge term—is that the magnitude of the polarization term is smaller than that of the other two terms, but the other two terms have opposite signs and tend to cancel each other. The value of the difference of the Madelung and ionic-charge terms is in fact similar to that of the polarization terms, which therefore end up playing an important role in the estimates of chemical shifts in ionic systems. A correct treatment should include a fourth term in the binding energy due to the repulsion of nearest neighbors (8).

Another important class of systems in chemical-shift investigations is the alloys. The problem for alloys is somewhat similar to that for noble-gas atoms implanted in solids, except for the non-charge-neutrality of the photoemitting atoms. The magnitude of the chemical shifts offers in this case an opportunity for studying the charge transfers between the constituents of the alloy. Two approaches have primarily been used to estimate the chemical shifts in alloys. The first approach (16,39,149,150) consists of making accurate calculations of all the possible terms in the binding energy of the two components; it is essentially a "first principle" approach. The use of this sophisticated method is made difficult by the fact that the contributions of different terms often cancel each other, so that only their small differences are meaningful. The second approach (66) works on more empirical, but very effective, grounds. It deals with the chemical shifts of the minority species in the alloy by comparing them with those of implanted noble-gas atoms in the same host lattice. The assumption is that for *dilute* alloys all the terms in the chemical shifts except that due to the presence of the valence charge are equivalent to those of the implanted noble-gas atom. Therefore the difference in binding energy between the noble-gas atom and the minority component of the dilute alloy is directly related to the valence charge in the latter. This method was applied by Kim and Winograd (66) to the dilute alloys of gold in silver.

Our short review of chemical shifts in different classes of solids will conclude with the recent application of chemical shifts to the detection of charge transfers due to the formation of charge-density waves (27). This problem is somewhat similar to that of alloys, but the atoms among which the charge transfer occurs are now of the same species, and the difference in charge is due not to chemical-bond formation but to a Peierls-like instability (105). It is known that one-dimensional metal crystals are unstable with respect to the creation of periodic lattice distortions with periodicity $\pi/|k_F|$ (k_F is the k-vector of the electrons with energy equal to the Fermi energy). A similar instability may occur in two-dimensional crystals (e.g., in layered transition-metal dichalcogenides) if the periodic distortion is able to affect a large enough number of electrons with energy equal to the Fermi energy (27). This gives not only lattice distortions, but also valence-charge rearrangements that take advantage of the lattice distortion to lower the total energy of the

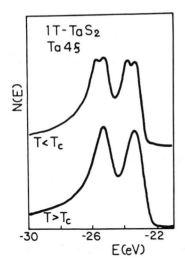

Fig. 17.48. The charge-density-wave phase transition in TaS_2 and the corresponding charge transfer between formerly equivalent atoms is revealed by the splitting of the core-level cation f peaks in photoemission spectra. (adapted from Ref. 153).

system. Figure 17.48 shows the core-level peaks in the x-ray photoemission spectra taken above and below the critical temperature for the onset of the charge-density-wave instability in TaS_2 (153). The peaks correspond to the f core levels of the transition-metal atom. The phase-transition-induced splitting of these peaks is due to the creation of two inequivalent classes of atoms, and this in turn is due to the modulation of the valence charge by the charge-density wave, which can either increase or decrease the charge in alternating transition-metal atoms. The magnitude of this extrinsic charge transfer can be estimated from the magnitude of the splitting in Fig. 17.48 that gives the difference in binding energy between the two components.

D. SPECTRAL FEATURES IN X-RAY PHOTOEMISSION SPECTROSCOPY

The analysis of x-ray photoemission spectra, and in particular the identification of binding energies and chemical shifts, is often complicated by the rich spectral structure accompanying each core-level peak and by the line shape of the peak itself. Multiplet splitting and other many-body effects and vibrational effects are the main causes of these complicating factors. The same factors are also a very interesting source of information about the properties of the system.

Multiplet splitting (129) can be understood if one considers that the final state of the optical excitation (the first step in the three-step model in the case of solids) is actually a state of the entire system and not just of the excited electron. The system includes the free electron and the ion left behind with its environment. For closed-shell systems, the state of the ion is unequivocally determined by the state of the core hole created in the process. For other systems, the ion can have different final states at different energies, and therefore the creation of a photoelectron starting from a given shell gives

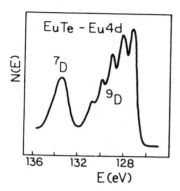

Fig. 17.49. The complicated manifold produced by multiplet splitting of the Eu 4d core level in EuTe. (data from Ref. 109).

not a single photoemission peak, but a manifold of peaks. These effects are particularly important in certain classes of open-shell systems, such as transition metals and rare earths.

The simplest possible example of multiplet splitting is the photoionization of the **K** shell of a lithium atom (129). The shell from which the electron is removed is a closed shell, but the photoionization occurs in presence of an open **L** shell. The electronic configuration of the ion left behind, Li$^+$, is ($1s^1 2s^1$). Depending on the J quantum number, this configuration can give either one of two states, ^1S or ^2S. These two states will have different energies, the energy separation being twice the Slater exchange integral. Thus the **K**-shell photoionization of Li will give a manifold of two peaks.

More complicated systems of course give more complicated multiplet splitting. The qualitative and quantitative analysis of the manifold requires a treatment of the configuration interaction, which is not necessarily difficult, but is often quite complicated. An example of a rather complicated manifold is given in Fig. 17.49 for the photoelectrons extracted from the Eu 4d states in EuTe (109). The multiplet splitting arises in this case from the configuration interaction involving the electrons in the open-shell f states. The electronic configuration for the initial state is ($4d^{10}4f^7$), corresponding, according to the Hund rules, to an ^8S state. The final state has the configuration ($4d^9 4f^7$), which gives *two* possible states, ^7D and ^9D. In fact, each one of these states has different components. This is evident in Fig. 17.49 where the manifold contains two groups of peaks, among which all five components of ^9D are resolved, whereas those of ^7D are unresolved. Particularly interesting multiplet-splitting problems are encountered in the spectra of mixed-valence compounds such as those containing Sm^{2+} and Sm^{3+} (6,110).

Even in cases in which a "main" core-level peak can be identified in the spectrum, as when multiplet splitting is absent or unresolved, this main peak may be accompanied by a number of *satellite peaks* of different nature. We shall distinguish in particular between two classes of satellites: those arising from electron correlation effects, and those arising from inelastic processes after the optical excitation, the latter requiring a solid-state environment.

The satellites arising from electron correlation are most commonly known as "shakeup" or "shakeoff" processes (9,129). The origin of these satellites is correlated with that of multiplet splitting—the final state of the optical excitation is a state of the entire system, including the photoelectron and ion left behind. In the case of multiplet splitting, we considered the possible final ion states arising from the configuration interaction involving a hole in an otherwise closed shell and electrons in an open shell. More generally, higher-energy final states of the ion may arise from electronic configurations in which one or more of the electrons left in the ions change their states to ones at higher energy. That is, in a somewhat naive picture, the photoelectron does not take all the photon energy, since part of it is given to the electrons that are "shaken up." Notice that one cannot think in terms of "one-electron" transitions for the "main" peak and "two-electron" transitions for the shakeup satellite peaks; both transitions (129) involve the ion *as a whole*, and the final states are those of the entire ion. Shakeup satellites are typically weak replicas of the main peak displaced from it by a fraction of an electron volts to a few electron volts. If the second electron goes to an ionization state rather than to another bound state, the replica of the main peak is called a "shakeoff" satellite. An important sum rule (129) says that if one considers a main peak and all its shakeup and shakeoff satellites, the center of gravity of this group of peaks is at a kinetic energy equal to the photon energy plus the one-electron orbital energy from which the photoelectron was extracted. Hence one can estimate binding energies and chemical shifts even in the case of complicated spectra with satellite peaks.

In a solid-state system, a replica of a main core-level photoemission peak may be caused also by inelastic processes during the second step of the photoemission process (transport to the surface). The inelastic processes are the same that we discussed in the case of electron energy-loss spectroscopy. The most important of these are the energy losses due to excitation of the collective modes of vibration in the "sea" of electrons (79,111). These can be thought as creating *plasmons* of energy $h\nu_p$. As a result a given main photoemission peak in a solid is accompanied by *plasmon replica* displaced to lower kinetic energies by $h\nu_p$, $2h\nu_p$, $3h\nu_p$, ... These peaks correspond to excited electrons that have lost part of their energy creating one, two, three, . . . plasmons before crossing the surface. An example of a plasmon replica (125) is shown in Fig. 17.50. Notice the several plasmon replicas accompanying the main Si $2p$ peak taken on a contaminated silicon surface.

With regard to the line shape of core-level peaks, there are several factors that can contribute to this. One of these factors—the vibrational broadening—was briefly outlined in the case of molecules (see Fig. 17.27). A similar Franck–Condon approach can be used for solid-state systems. The result of vibrational broadening in solids is a single, broad peak, rather than a series of vibrational peaks as in molecules. The equilibrium positions for the two sets of harmonic-oscillator wavefunctions corresponding to the ground and ionized states are different, so the selection rules valid for a simple

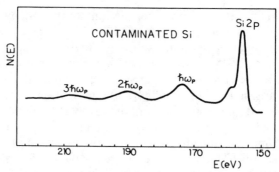

Fig. 17.50.   Plasmon replicas in the Si 2p core-level photoemission peak of a contaminated Si surface (125). The replicas correspond to the production of one, two, and three plasmons.

harmonic oscillator are not valid. The broadening of core-level peaks by vibrational effects is, of course, temperature dependent. The theory behind this phenomenon is readily available from the parent field of optical-feature broadening of point defects in the case of strong electron–phonon coupling (157). A particularly significant example is that of lattice defects with strong lattice relaxation following the optical excitation (e.g., color centers). The well-known result of the theory for these defects is a temperature-dependent broadening of the form (157)

$$W = W_0 \left(\text{ctnh} \left(\frac{a}{T} \right) \right)^{1/2} \tag{35}$$

where W is the line width, W_0 is its value at $T = 0$ K, and a is a constant that can be expressed in terms of an "equivalent" vibrational frequency $\langle \omega \rangle$ averaging all the broadening-active frequencies of the system:

$$a = \frac{\hbar \langle \omega \rangle}{2k} \tag{36}$$

Representing another core-level peak-broadening factor of importance in metals are the many-body effects (153). These effects were extensively investigated in the related problem of the line shape of x-ray optical-absorption thresholds. The line shape is expressed for the many-body-broadened threshold by the Mahan–Nozieres–De Dominicis equation (80)

$$\alpha(h\nu) \propto (h\nu - h\nu_0)^{-\epsilon} \tag{37}$$

where $\alpha(h\nu)$ is the absorption coefficient for x-ray photons of frequency ν, ν_0 is the threshold frequency, and ϵ is the so-called threshold exponent, which can be theoretically calculated. The Mahan–Nozieres–De Dominicis theory was extended by Doniach and Sunjic (28) to the case of core-level

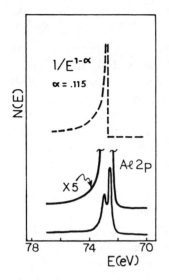

Fig. 17.51.   The solid line is the photoemission $2p$ core level of Al (153). Notice the asymmetric line shape. This asymmetry is due to many-body effects. The corresponding theoretical lineshape by Sunjic and Doniach (28) for an optimized value of the asymmetry parameter α is shown by the dashed line. A direct comparison between theory and experiment requires a convolution of the theoretical line shape with the instrumental broadening.

photoemission-peak line shapes. The physical mechanism in both theories is the coupling of the core hole with the surrounding sea of free electrons. Doniach and Sunjic predicted the resulting core-level-peak line shape $f(E)$ (E = photoelectron energy) to have a characteristic asymmetric form given by the equation

$$f(E) = \frac{\Gamma(1 - \epsilon) \cos\left[\dfrac{\pi\epsilon}{2} + (1 - \epsilon)\arctan\left(\dfrac{E}{\gamma}\right)\right]}{(E^2 + \gamma^2)^{\frac{(1 - \epsilon)}{2}}} \tag{38}$$

where ϵ is the threshold exponent, Γ is the gamma function, and γ is a parameter related to the finite core-hole lifetime. For $-E/\gamma \gg 1$, one obtains the simplified form

$$f(E) = \frac{1}{E^{1-\epsilon}} \tag{39}$$

often used in data analysis. A plot of the Doniach–Sunjic line shape is shown in Fig. 17.51.

Also shown in Fig. 17.51 is the photoemission $2p$ core-level peak for aluminum. We see in this peak the characteristic asymmetry predicted for many-body broadening. The experimental line shape, however, is affected by several effects besides the many-body asymmetric broadening of equation 39, for example the nearly Gaussian instrumental broadening.

E.   SURFACE AND CHEMISORPTION EFFECTS IN CORE-LEVEL SPECTRA

As with uv photoemission spectroscopy, the high surface sensitivity makes x-ray photoemission an ideal source of information about chemisorption problems and surface electronic properties in general. We shall outline in this section some of the applications of core-level photoemission spectroscopy to surface problems. It should once again be emphasized that the surface sensitivity of the technique depends on the kinetic energy of the photoelectron, which in turn depends on the photon energy. Photon energies can be selected to enhance the surface sensitivity, but this requires a tunable photon source, which is to say, synchrotron radiation. As we shall see, synchrotron radiation is also necessary for the use of some new techniques in chemisorption investigations, such as photoelectron diffraction.

The surface core-level shifts investigated in 1978–1979 are some of the most important examples of surface effects in core-level photoemission spectra (21,29,56). In general, we expect the binding energy of a core level in a surface atom *not* to coincide with that in a bulk atom, since the environments are different. This influences, for example, the valence charge at the atom, and through it the chemical shifts, as we have discussed above. There are two difficulties in detecting these effects, however. One difficulty is the requirement for very high surface sensitivity—surface core-level shifts can be detected by comparing bulklike spectra and surfacelike spectra. Tunable synchrotron radiation is the only answer to this requirement. A second problem is the superimposed effect of the band bending in insulators and semiconductors. Since the energy position of the core-level peaks follows the curvature of the bands, a typical photoemission experiment will sample different energy positions, giving only a single, very broadened peak. This problem can be overcome by using flat-band samples (29). Figure 17.52 gives a typical example of a surface core-level shift (29). The shift is detected by comparing the bulk core-level Ga 3d peak for GaAs with the corresponding surface core-level peak. The difference between the two spectra and the presence of a surface-shifted component in the second spectrum are evident in Fig. 17.52.

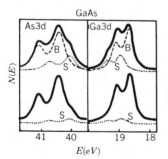

Fig. 17.52. Surface core-level shifts in the photoemission spectra of GaAs, adapted from Ref 29. The left-hand curves correspond to the As 3d level, and the right-hand, curves to the Ga 3d level. The two top curves were taken at photon energies such that the escape depth of the photoelectrons was minimized (see Fig. 17.1) and therefore the surface sensitivity was maximized. The shoulders in these spectra reveal the surface core-level components. An empirical lineshape analysis identifies the bulk components (dashed lines) as well as the surface components (dotted lines).

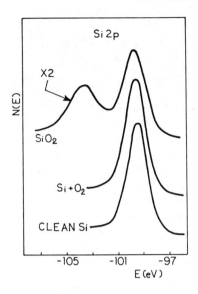

Fig. 17.53. The Si 2p core-level peak in the photoemission spectra of different Si surfaces (125). The bottom curve corresponds to ultraclean Si (111); the center curve, to the first stages of oxidation of this surface (see Fig. 17.54); the top curve, to a heavily oxidized SiO₂ surface.

Changes in the chemical and physical properties of the surface correspond to strong changes in the photoemission spectra of core levels. An example is given in Figs. 17.53 and 17.54 for the oxidation of silicon (122). We see here the O 1s and Si 2p core-level peaks for an oxygen-covered Si(111) surface produced by light exposure of a Si(111) 7 × 7 surface to O₂. We see also how the corresponding spectra appear for SiO₂. The differences affect not only the position in energy, but the line shape as well. Notice, for example, the asymmetry of the low-coverage O 1s peak. This asymmetry is explained (126) by an unresolved two-peak structure due to the presence of two inequivalent sites for oxygen atoms on this surface. Energy-loss vibrational spectroscopy (60) detected for the low-coverage Si–O system an O—O vibrational stretching mode, clearly demonstrating that the oxygen chemisorption is nondissociative. This explains the two inequivalent sites pos-

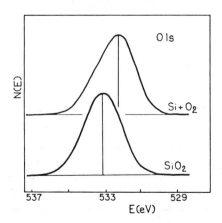

Fig. 17.54. Oxygen 1s core-level photoemission peaks for the first stage of oxidation of Si (111) and for SiO₂ (126). Notice the asymmetry of the top spectrum, attributed to an unresolved doublet, in agreement with a nondissociative chemisorption process.

tulated to justify the O $1s$ core-level line shape (126). The interpretation of the details of this chemisorption process has evolved somewhat over recent years. The original idea of an asymmetric peroxy bridge was replaced on theoretical grounds by that of a peroxy radical (43) attached to a silicon atom at one end. More recently (118), theory and experiment seemed to support the existence of an ionization state for this radical, so that the oxygen chemisorbed species must be called a "peroxy ion." The evolution in interpretation did not change the basic idea of a nondissociative process established by vibrational spectroscopy (60).

Core-level photoionization spectra can provide different kinds of information about a chemisorption system. We have just seen an example of qualitative information about chemisorption sites. More quantitative information can be provided by the chemical shifts on the valence charge involved in the formation of chemisorption bonds—along the lines discussed above in this section. In recent years a number of experimental groups have been very imaginative in finding new ways to extract information about chemisorption processes from the properties of core-level peaks in x-ray photoemission spectra. The study of surface diffusion processes and the study of photoelectron diffraction are two of these new avenues.

The idea underlying the study of diffusion processes by core-level photoemission (4,5,86) is that core-level peak intensities attenuated by the transport of the excited electrons to the surface can be used to probe the density of atoms of the corresponding species. This idea has a difficulty in common with all other quantitative-chemical-analysis uses of core-level photoemission: the peak intensity depends not only on the atomic density, but also on the photoionization cross section and on the energy-dependent attenuation during the transport to the surface. These factors combine with the linear dependence on the atomic concentration to make an *absolute* chemical analysis quite difficult. The problem is removed if *relative* atomic concentration changes are monitored. One example is the diffusion processes occurring at a GaAs(110) surface on coverage by a layer of gold (4). Figure 17.55 shows the Ga $3d$ and As $3d$ core-level peaks for stoichiometric GaAs(110). Notice the intensity difference between the two peaks, which is due not to nonstoichiometry, but to cross-section differences. Also shown in Fig. 17.55 are the same two peaks measured after the surface was covered with aluminum and gold. Aluminum is a chemically reactive species that can be added to the clean surface to "mark" its position. The intensity changes for the different core levels can be referred to the intensities of the marker-species peaks to identify the nature of the diffusion processes. The limits of this technique come from the fact that the marker sometimes influences the diffusion processes beyond one's expectations. For example, the spectra obtained after adding 2 Å of Au to the 1 Å of Al (see Fig. 17.55) reveal a preferential outward diffusion of Ga, whereas without the Al intralayer, one would observe a preferential As outward diffusion. The interpretation of the intensity changes often relies on very simple models of overlayer formation

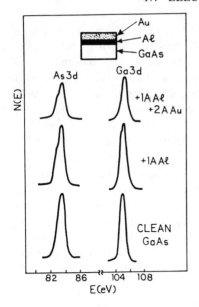

Fig. 17.55. Use of photoemission spectroscopy to study microscopic diffusion processes (4). The peaks correspond to the Ga 3d (right) and As 3d (left) core levels. The bottom spectra were taken on cleaved GaAs(110); the center spectra, on GaAs covered by 1 Å of Al; the top spectra, on the same surface after a 2-Å Au coverage. Notice the nonstoichiometric ratio between the two top peaks. This deviation is explained by a preferential Ga outdiffusion. For a layer of pure Au on GaAs one would have observed instead a preferential As outdiffusion (4). Thus the diffusion process is controlled by the ultrathin Al intralayer, and Al becomes an active ingredient rather than just a passive interface-position marker.

and on the quantitative analysis of the diffusion-induced peak-intensity changes. This method was recently applied to the study of semiconductor overlayers on semiconductor substrates (86).

Photoelectron diffraction (65,78) extends the information given by core-level photoemission to the field of structural properties of chemisorbed layers. This technique is based on the study of neighbor-atom effects on the photoionization cross section of core levels. Earlier in this review, we analyzed the atomic effects in the dependence of the cross section on the photon energy. As discussed in the Appendix, these effects are related to the properties of the matrix element $\langle i \mid \nabla \mid f \rangle$. When the atom is in a solid-state environment (e.g., chemisorbed on a surface), this environment plays a role in the cross section together with the atomic effects. One example of effects of this kind is the EXAFS modulation discussed in the previous section. The problem should be addressed in general by considering all the possible paths leading the photoelectrons from the central atom to the detector. These paths can involve single or multiple scattering events by the neighboring atoms, which in turn depend on the positions of these atoms. The effects of the scatterings can be observed in the dependence of the cross section on the photon energy. The problem of interpreting the measured intensity–photon energy curves is somewhat related to low-energy-electron diffraction, with the difference that now the electrons originate inside the sample rather than being sent into it from the outside.

Extensive theoretical and experimental studies (65,78) have demonstrated that the core-level photoionization cross section for adsorbates is heavily modulated by the above-mentioned neighboring-atom scattering effects. Prototype chemisorption systems were used to investigate dynamic calculations of the scattering paths or more simplified quasidynamic calculations. These theoretical studies have related the observed modulation of the core-level peak intensity–photon energy curves to the position of the adatoms on the surface. From these results it would appear that quantitative information on chemisorption bond lengths is obtainable by comparing theory and experiment. There is still much controversy, however, about how feasible it is to extend this method to nonprototype systems, and about how the effectiveness of this method compares with the alternative approach of measuring angular intensity distributions of the core-level peaks, which are also influenced by neighbor-atom scattering effects. A complete assessment of the powerfulness and limits of this technique must be delayed until a reasonable number of "new" systems, that is, systems with unknown structural properties, have been investigated using it. Among the difficulties in applying this technique to "new" systems, we emphasize the interference of the diffraction effects in the measured intensity curves with several other effects, such as the atomic cross-section effects previously discussed.

VI. AUGER-ELECTRON SPECTROSCOPY (AES)

A. GENERAL CONSIDERATIONS

The most attractive feature of AES is the intensity of the signal and the corresponding quality of the signal-to-noise ratio. Another important feature is the distance in energy between the spectral features of different elements. Combined, these features make AES a very advantageous technique for surface chemical analysis. The use of AES has indeed been primarily for the identification of the different chemical species, and in particular of the elements, of surfaces. Another use has been the measurement of the density of each species. The Auger process, however, contains in principle much more information than just that about the chemical composition of the surface. In this section we shall review the practical use of AES. Our review will start from the most traditional uses: qualitative and quantitative chemical analysis of surfaces. Then it will be extended to a discussion of the potential uses of AES beyond chemical analysis.

The necessary background for qualitative chemical analysis by AES is given by the energy-balance equation 4. The background for quantitative AES chemical analysis is given by the Auger-intensity equations A19 and A21 derived in the Appendix. As regards the other possible uses of AES, one must distinguish between two possible kinds of Auger processes in molecules and solids. The first kind of process is that involving only core electrons. For this kind of process equation 4 gives a well-defined energy or, better, spectral features of small intrinsic band width. The second kind of process involves both core electrons and valence electrons. Since the valence electrons are distributed in energy over a large range, equation 4 will correspond to a certain range of Auger kinetic energy rather than to just one energy. On the other hand, the distribution in energy of the valence electrons will also be reflected in the line shape of the Auger spectral features. We shall discuss later in this section how the line-shape analysis can be used to explore the DOS of valence electrons. We shall see in particular that the AES data analysis is in this case closer to that of INS than to that of photoemission spectroscopy.

B.   QUALITATIVE CHEMICAL ANALYSIS AND AUGER PEAK ENERGY

One of the strongest advantages of AES is that a qualitative chemical analysis can be performed without a detailed knowledge of the energy balance of the Auger processes in the particular system under consideration (48). In fact a comparison of the detected spectral features with tabulated Auger peaks of the elements is in general sufficient to identify the chemical components of the system. In most cases it is sufficient to consider the position in energy of the spectral features with an accuracy of the order of a few electron volts. If some question remains, it can often be solved by comparing the relative intensities of the different Auger features to those of the tabulated manifold of peaks corresponding to a given element. A complete tabulation of the Auger features for the elements is available (156).

Although remarkably simple and effective, the qualitative chemical analysis of surfaces by AES does require some attention. Two points must be emphasized: The first is the requirement for increasing the signal-to-noise ratio by using modulation techniques. The second is the possible effects of the primary excitation source on the system to be investigated. The requirement of using a modulation technique is due to the small magnitude of the Auger signal with respect to the total signal of the backscattered electrons. By definition, the excitation of the Auger process requires a primary-beam energy (photon or electron energy) larger than the kinetic energy of the Auger electrons to be observed. A rule of thumb is that the primary-beam energy must be 2.5–3 times larger than the maximum kinetic energy of the Auger electrons. The primary particles—electrons or photons—will create in their interaction with the system a large number of secondary electrons, distributed in energy from zero to the primary-beam energy. The energy distribution of these electrons is generally smooth, with weak features superimposed. The Auger peaks are among those weak features, and therefore they are superimposed on a strong background. It is difficult to detect the distribution in energy of the Auger electrons above that strong background. The Auger-electron-energy distribution can be detected, however, by analog differentiation techniques. For example, consider an Auger spectrum taken with a retarding-field analyzer such as the hemispheric-grid device. We have already seen that the energy distribution of the electrons is the *first* derivative of the total current collected at the end of the analyzer. The first derivative is obtained by modulating the retarding field and measuring the first-harmonic modulation of the analyzer output by phase-sensitive detection with a lock-in amplifier. If the retarding-field modulation is not strictly sinusoidal (e.g., if it is a square wave), then the current will have second-harmonic modulation as well as first-harmonic modulation. The second-harmonic modulation is proportional to the second derivative of the current, that is, to the first derivative of the distribution in energy of the electrons. We now consider an energy distribution consisting of a small peak superimposed on a (large) constant background. The first derivative of the background is zero and therefore does not contribute at all to the first-derivative spectrum. Instead, the peak will be detected in the first-derivative spectrum. The limits of this method are set by the interplay between the signal-to-noise ratio and the energy resolution. On one hand, the intensity of the first-derivative signal is proportional to the amplitude of the modulating signal of the retarding field. On the other hand, the modulation superimposed on the retarding field limits the energy resolution of the analyzer. The optimal modulation amplitude must be determined in each case from the required energy resolution. Similar considerations apply for bandpass energy analyzers such as CMA which is the analyzer most widely used in Auger spectroscopy. With these devices, however, the first derivative is obtained as a first-harmonic modulation of the analyzer output. The modulation is obtained in turn by modulating the bias voltage that determines the position of the energy window.

It might appear *a priori* that the effects of the primary beam on the system under investigation are essentially negligible. The first step of the process, the ionization of the core level, takes on the order of 10^{-16} sec. The core holes so created are filled by Auger processes with a lifetime of the order of 10^{-15} sec. Therefore, there is very small interaction between the initial ionization process and the Auger recombination process of the core hole. The primary ionization process can involve either electrons or photons; these give essentially the same Auger spectrum. As a matter of convenience, electrons are much more widely used than photons, due to the advantageous properties of electron beams with regard to selecting a given energy and focusing the beam. The conclusion that seems to emerge from the above considerations, that the primary beam has little importance in Auger spectroscopy, is misleading. The primary beam can in fact influence and modify the system under consideration in such a way that the AES results lose part of their reliability. This is particularly true for solid surfaces, and even more true for chemisorption systems. As a practical matter, the primary beam has an energy of the order of a few kiloelectron volts and a current of the order of a few microamperes. Empirical observations (90) show that a beam with these characteristics often induces contamination of the surface (e.g., by carbon compounds, giving monolayer coverage in a time from a few seconds to a few hours. Therefore interpretations of AES results for qualitative chemical analysis must always take into account the possibility of electron-beam-induced contamination. Even lowering the primary-beam energy is not safe, since in many cases a lower energy may actually be more effective in contaminating the surface. In chemisorption studies the primary beam can desorb some of the adatoms or stimulate changes in their chemical-bonding properties. A good rule is therefore always to follow the *evolution* of the Auger spectrum over a period of 10–12 min at low primary-beam currents, to deconvolve if possible the primary-beam effects from the "real" effects. As a general rule, primary photon beams are less troublesome than primary electron beams, but spurious results cannot be ruled out *a priori* in any case.

Although not strictly necessary for identifying the chemical components of the system, a theoretical explanation of the exact energy of the Auger peaks is extremely interesting, for reasons similar to those discussed in examining the chemical shifts of the core-level photoemission peaks. For Auger peaks the energy-balance equation 4,

$$E_k = E_3^* + E_2 - E_1 \tag{40}$$

links the kinetic energy to the binding energies (measured with respect to the zero of the kinetic energy) of the first and second core level and of the third core level in presence of another core hole. The role of the binding energies establishes at least in principle a link between AES and core-level photoemission spectroscopy (48). We shall now consider in more detail this link and its limits.

In equation 4 the term $(E_2 + E_3^*)$ can be identified as energy required to create a state with *two* holes, one in energy level 2 and the other in energy level 3. Even if one neglects relaxation effects, this energy is *not* simply the sum of the binding energies of these two levels, since it contains the interaction energy between the two holes, $I(2,3)$:

$$E_3^* + E_2 = E_3 + E_2 - I(2,3) \tag{41}$$

The most important contribution to $I(2,3)$ is the electrostatic repulsion between the two holes. Therefore the kinetic energy can be calculated only after the term $I(2,3)$ has been estimated. Equation 41, however, neglects relaxation effects. If relaxation is included, equation 41 becomes, according to Shirley (128),

$$E_3^* + E_2 = E_3 + E_2 - I(2,3) + R \tag{42}$$

where R is an "extra-atomic" relaxation-energy term (147). Calculations of the term $(E_3^* + E_2)$ using equation 42 have been very successful.

Another possible approach to the same problem is to approximate the binding energies in the presence of another core hole by the binding energies in the neutral atom to the immediate right in the periodic table. This approach is reminiscent of the equivalent-core approximation discussed earlier for photoemission. Along these lines Chung and Jenkins (17) proposed a total final-state energy calculated by taking an average between the modified-core binding energies E_2^{z+1} and E_3^{z+1} and the unmodified-core energies E_2 and E_3:

$$E_3^* + E_2 = \tfrac{1}{2}[(E_3^{Z+1} + E_3) + (E_2^{Z+1} + E_2)] \qquad (43)$$

This equation takes into account the symmetry of the states 2 and 3 in the Auger process. Its results are in very reasonable agreement with experiment in many cases. Once again an additional relaxation term must be added to improve the agreement by taking into account the reaction of the "spectator" electrons to the sudden creation of core holes (48).

C.   QUANTITATIVE AES CHEMICAL ANALYSIS

The basis for quantitative chemical analysis with AES is given by equation A19 from the Appendix:

$$I(E_k) \propto \int_0^\infty dz \, \exp\left(-\frac{z}{\lambda(E_k)\cos\phi}\right)$$

$$\times A_c(E_k)\rho(z) \int_0^{E_p} N_2(E_0)\sigma_c^e(E_0)dE_0 \quad (A19)$$

which relates the Auger intensity $I(E_k)$ at the kinetic energy E_k to the density of the Auger-emitting atomic species, $\rho(z)$, or better, to its distribution as a function of the depth z. This relation depends on the factors $A_c(E_k)$, the probability of Auger recombination after creation of a core hole; $\lambda(E_k)$, the mean free-path for the ionizing electrons; $N_z(E_0)$, the depth distribution of the ionizing electrons; and $\sigma_c^e(E_0)$, the ionization cross section. The proportionality constant in equation A19 depends on the analyzer collection geometry. The relation between I and ρ is simplified under the reasonable hypotheses introduced in the Appendix while deriving equation A21, giving

$$I(E_k) \propto \cos\phi_0 \frac{\rho A_c(E_k)\lambda(E_k)}{\cos\phi} \left[I_p\sigma_c^e(E_p) + \int_0^{E_p} N_B(E_0)\sigma_c^e(E_0)dE_0\right] \quad (44)$$

for constant density ρ along a depth of at least the order of $\lambda(E_k)$. All the factors in equation 44 can, at least in principle, be measured or calculated. The practical accuracy of these estimates is limited by a series of problems that we shall outline here. Measurements of the *backscattered*-electron distribution $N_B(E_0)$ ordinarily do not present insurmountable problems. The parameter λ, however, is a severe source of inaccuracy. The ionization cross section σ_c^e is known with relatively good accuracy for inner shells (42,144). The accuracy is not so good for the outer shells, and this again influences the overall accuracy of the AES quantitative analysis. We discuss in the Appendix the problems in determining the Auger recombination probability

$A_c(E_k)$—this parameter is close to 1 only if one considers the entire manifold of Auger spectral features caused by a given ionization process. Another important factor is that the holes to be Auger-recombined may not be caused directly by the ionization process, but rather may be the result of an indirect process. For example, they can be caused by a Coster–Kronig transition. All these factors must be taken into account when calculating the Auger recombination probability.

The proportionality constant not explicitly shown in equation 44 also plays a major role in quantitative AES analysis. We mention in the Appendix that this constant depends on the geometry of the analyzer. Each analyzer will integrate a certain range of angles and the Auger signal they provide is given by the corresponding integration of equation 44. If the Auger emission is isotropic, then the angular integration is straightforward and the geometry of the analyzer has no significant influence on the AES quantitative analysis. The Auger emissions, however, are typically *not* isotropic (98) and the analyzer geometry *can* influence the results. From this point of view, analyzers with large angular acceptance, such as the retarding-field hemispherical analyzer, offer some advantage over analyzers with limited angular acceptance, such as the CMA, since they average out angular effects. The CMA and similar devices, however, have the compensating advantage of a better signal-to-noise ratio further enhanced by the use of electron multipliers. This last factor can cause problems, since the gain of the multiplier changes with age. Some of the output pulses from the multiplier can with age become lower than the discriminator threshold in the single-electron-counting electronics, and this influences the accuracy of quantitative estimates by AES. Another important effect on the proportionality factor of equation 44 is that of the roughness of the sample surface. This factor is most critical at a glancing incidence of the primary beam.

The analog differentiation (or its digital equivalent) may become another source of inaccuracy in AES quantitative analysis. For example, the link between the amplitude of the original signal and that of the differentiated signal depends on the characteristics (e.g., the amplitude) of the modulation and, for a given amplitude, on its Fourier spectrum. A direct use of the undifferentiated Auger spectra is also problematic, since each Auger feature is superimposed on a smooth but very large background. The elimination of this background is somewhat arbitrary and another source of inaccuracy.

All the above factors limit the realistic accuracy of quantitative Auger analysis from first-principle considerations. The practical accuracy is of the order of a factor of 2. This accuracy is comparable in magnitude with that obtainable in x-ray analysis. The limits of the accuracy become better in the case of *relative* quantitative analysis by AES, which is based on direct comparison between the Auger peak intensity for a given element in an unknown sample and for the same element in a specimen of known quantitative composition. The known sample is used as a standard in the quantitative measurement. A simple analysis shows that this approach removes or bypasses many of the above problems, but not all of them. For example, the distribution N_B is not necessarily the same for the two systems, but the corresponding inaccuracy is not a very serious problem. The escape-depth parameter λ is also different in principle for the two materials. Systematic studies of λ show that these differences are not too large; in fact, one can almost talk about a "universal" $\lambda(E_k)$ curve independent of the material. The normalizaton method is very effective in eliminating the problems arising from the Auger recombination probability and from the ionization cross section. The elimination of the geometric factor is not necessarily automatic, since the angular distribution of the Auger electrons could be different for the two samples. Furthermore, the accuracy attainable in attempting to position the two samples in exactly equivalent positions is also limited. The normalization method does not remove the inaccuracies due to surface roughness, which can be a major factor in the overall inaccuracy. In contrast, instrumental factors such as the intensity of the primary beam and the multiplier gain are effectively taken care of by normalization. The effects of the analog differentiation are eliminated to the extent to which the Auger line shapes are equivalent in the two samples. The background subtraction for undifferentiated Auger spectra is still a serious problem, since the background is typically different for the two samples. In summary, after taking all the above corrections into account, the normalization method improves the accuracy of quantitative AES analysis by a factor of 2–3. One particularly delicate point must be considered, however. The accuracy given by the normalization method cannot be better than the accuracy of the "known" density of atoms in the reference specimen. This is not a completely trivial problem, since the density to be considered here is not the average bulk density, but rather the surface density, that is, the density of atoms in the thin region probed by AES. This density can be influenced by several factors, including contamination, surface segregation, and diffusion. Only after all these factors have been taken into account is it possible to estimate the surface density of the "known" specimen. In practice these factors are an important, although often neglected, source of inaccuracies.

D. SPECIAL TYPES OF AES ANALYSIS–DEPTH PROFILING AND MICROANALYSIS

The special features of Auger spectroscopy as an analytical probe can be advantageously exploited for special techniques, for example, Auger depth profiling and Auger microanalysis.

The purpose of Auger depth profiling (159) is to measure the variations of the density $\rho(z)$ as a function of the depth z. We have seen in the previous sections that ρ is obtained by assuming that it is constant over a range of z comparable to the magnitude of the escape depth. In practice, the Auger analysis measures the *average* value of ρ over a depth of the order of λ. This is the absolute limit of accuracy in using Auger as a depth-profiling

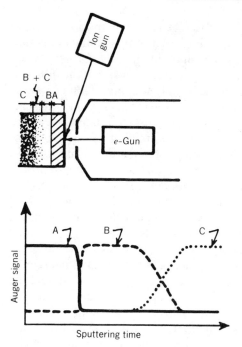

Fig. 17.56. Auger depth-profiling study (159) of a multilayer system containing the elements A, B, and C. The surface layers are progressively removed by ion sputtering while Auger electrons are created by the electron beam and analyzed by the CMA, which contains the electron gun. A plot of the intensities of the Auger signals for A, B, and C as a function of the sputtering time is shown at the bottom. From this plot, one deduces the multilayer composition A, B, B, + C, and C.

probe. Auger depth profiling requires equipment like that schematically illustrated in Fig. 17.56. This equipment is equivalent to a standard Auger spectrometer except for the presence of an ion gun. The ion gun is used to bombard the surface of the sample with ions of an inert gas, most typically argon. This bombardment removes the surface atoms of the sample at a slow rate, almost atomic layer by atomic layer. If a quantitative Auger analysis is carried out stage by stage during the ion bombardment, the results will give the composition of the sample at different depths with respect to the original surface. A hypothetical example of Auger depth profiling is also given in Fig. 17.56. Here we see that the relative strengths of the Auger signals due to the elements A, B, and C in the system ABC change with changes in z. The practical resolution in measurement of the depth profile is not very close to the absolute limit set by λ.

The Auger depth-profiling technique is affected by several problems that severely limit its applications and accuracy. The first and most important of the problems is that it is a destructive technique that requires a probe (the ion beam) strongly interacting with the system under investigation. The most serious problem is actually not so much in the destruction of the sample, since it is always local destruction limited to the area hit by the ion beam (although even this local destruction is a problem in some cases). The *real* problem is that the ion beam can very severely affect the composition depth profile itself. For example, it can selectively remove certain elements with greater efficiency and/or promote certain diffusion processes. Auger depth profiling is certainly a very "hard" probe that can give spurious and, in some cases, misleading results. Another limitation of this technique is that it is somewhat difficult to calibrate the absolute depth scale. Ordinarily the depth profiles are constructed by plotting the Auger intensity of the different elements versus the *time* during which the sample was under ion bombardment. One must assume that the ion current and the ion energy stay constant during the bombardment, and therefore that the surface atomic layers are removed at a constant rate. Even neglecting the limits of this assumption, the problem still remains of converting time of bombardment into depth. This can be done by calibrating the probe either with a known depth profile (e.g., an overlayer of known thickness) or with another depth–composition probe such as high-resolution Rutherford backscattering. In fact, Rutherford backscattering is from many points of view a far better depth–composition probe than Auger profiling, and recent developments have improved it even further (36). Rutherford backscattering, however, requires an ion accelerator, which is a permanent, large, and expensive piece of equipment. As a compromise solution, Rutherford backscattering can be used to calibrate the Auger depth-profiling probe for a certain class of systems and certain experimental conditions—ion-beam current, electron current, and so on. The constant-rate-of-removal hypothesis still limits accuracy, even after calibration, since it is not guaranteed that this rate will not depend on the chemical composition of the sample or on its structure and morphology.

Microanalysis (160) is the analysis of the chemical composition of surfaces as a function of position with a spatial resolution of better than 1 μ. The high resolution is obtained by focusing a primary beam of sufficient intensity to as small a spot as possible. This is possible in principle for the chemical analysis techniques that use electrons as primary excitation particles. In practice, focusing the beam into a small spot becomes difficult at low energies and for high primary-beam currents. Therefore Auger microanalysis typically requires high primary energies and a reasonable compromise for the primary-beam current between the requirements for space resolution and good signal-to-noise ratio. Typical primary-beam energies are a few to 10 keV. The emission source for electrons inside the electron gun must give a large enough current from a small enough area; that is, the electron source inside the gun must have high brightness. Several technical solutions are available for this problem. For example, suitable cathodes for the electron gun are the tungsten hairpin and lanthanum exaboride. The best typical performances obtained with these cathodes are currents of the order of ~0.1 μA focused into 0.01–0.1 μ.

A realistic analysis of the signal-to-noise ratio shows that with the above currents a matrix of 100 × 100 points can be probed in a time of the order of 3 hr. This shows that Auger microanalysis at very high spatial resolution is possible only when long exposure times are not a serious problem. A more typical performance is 3-μ resolution with a scan time of the order of a few minutes. The Auger microanalysis can be carried out in a scanning mode such that an "image" is obtained of the surface distribution of a given element. Special modulation techniques can make these images of good quality, and in particular, good plasticity. Another advantage is that changes in the chemical bonding of a given element can be detected from the changes in the positions in energy of the corresponding Auger peaks due to changes in the chemical shifts. Therefore Auger-microprobe scanning images taken at slightly different Auger-electron kinetic energies can give an image not only of the surface distribution of a given element, but also of its different chemical-bonding environments.

The scanning Auger microprobe is a very spectacular surface-analysis technique, but it has some limitations. In particular, although it is possible to "see" the surface distribution of the different species, it is not equally simple to convert the intensity of the Auger signal into a density of surface atoms of the corresponding species. This is due to the high primary energies required to obtain high spatial resolution. High primary energies produce, in fact, increased backscattering; this in turn will contribute more to the Auger signal. The backscattering of the primary beam can heavily depend on the properties of the substrate. Suppose that a surface consists of a mosaic of two components, A and B, covered by a smooth layer of a third component, C. In principle the Auger microprobe should give only a flat, structureless image when tuned to the Auger emission of element C of the smooth overlayer. Suppose, however, that the two components A and B have greatly different efficiencies in backscattering the primary electrons, for example, that element A is much more efficient than element B. The scanning Auger image of element C will then show a pattern reproducing the underlying mosaic. This pattern is due *not* to changes in the surface composition, but rather to changes in the substrate properties. This effect can be a serious problem in the analysis of Auger-microprobe images of unknown systems, since it could be mistaken for a real surface-composition effect. It is therefore advisable to seek independent information from other sources before trying to interpret an Auger-microprobe image. Even within these limits, the scanning Auger microprobe is an important tool in modern surface chemical analysis. This is particularly true for certain industrial problems, such as those in microelectronics.

E.   AUGER INVESTIGATIONS OF THE DENSITY OF STATES

Until now, we have discussed primarily Auger processes involving three different core levels and no valence states. Processes involving valence electrons are a very valuable source of information about the chemical and electronic properties of a given system. Some of these applications are similar to those discussed in the previous sections.

The band width of the valence states is much larger than that of core levels, and this is reflected in the band width of the Auger-spectral features. Nevertheless, the Auger features of a given element can still be distinguished

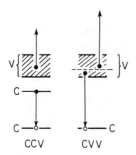

Fig. 17.57. Two possible Auger processes involving valence-band states. In the CCV process the third state is a valence-band state, whereas in the CVV process the valence band provides both the second and the third state. Notice that in the CVV process, valence states equidistant in energy from the dashed line give the same Auger-electron kinetic energy. This explains why the corresponding Auger line shapes are self-convolutions of the valence-band DOS similar to those found in INS.

from those of other elements and be used for qualitative chemical analysis. Furthermore, the processes involving valence electrons have the unique property of being probes of the *local* DOS around the atom in which the Auger recombination takes place. We shall see that the correspondence between Auger line shapes and DOS is similar to that found in INS.

Before discussing the link with the valence-electron DOS, we shall introduce the necessary nomenclature for the different kinds of Auger processes. When all the states involved in the process are core-electron states, the process is called a CCC Auger process. If valence electrons are involved, we can have either a CCV or a CVV Auger process. In this section we shall discuss the last two kinds of processes.

Figure 17.57 schematically shows the different transitions involved in the CCV and CVV processes. We shall consider the CCV processes first. From the figure it is clear that the Auger recombination of the core hole in level 1 can give rise to Auger electrons with a broad range of kinetic energies. Calling E^{*max} and E^{*min} the energies of the top and of the bottom of the valence states in the presence of the core hole in level 2, the energy-conservation law gives

$$E^{*min} + E_2 - E_1 \le E_k \le E^{*max} + E_2 - E_1 \tag{45}$$

and therefore an Auger band width of $E^{*max} - E^{*min}$. Symmetrically, if the second state of the process is a valence state and E^{max} and E^{min} are the bottom and the top of the valence-state energies, we have

$$E^{min} + E_3^* - E_1 \le E_k \le E^{max} + E_3^* - E_1 \tag{46}$$

corresponding to a band width of $E^{max} - E^{min}$. The average band width is therefore $\frac{1}{2}(E^{max} - E^{min} + E^{*max} - E^{*min})$.

We shall now consider how the Auger electrons are distributed in energy inside the above band width. In both cases it is quite clear that the distribution will be related to the DOS of the valence electrons. The formal derivation of this result is similar to that used for photoelectron spectroscopy. An important distinction must be made, however. For processes of this kind the core hole can play an important role, emphasized by the superscript asterisk in the corresponding energies in the above energy-balance equations. The question is how much the localization or delocalization of the states influences this final-state interaction. One could argue that if the two-hole final state is completely delocalized, the final-state interaction can be neglected and the Auger line shape will always be related simply to the local DOS (except for the transition probability factor, which could distort it). The localization or nonlocalization of the two-hole final state is not, unfortunately, an issue that can be resolved *a priori*. Extensive studies carried out in metals have given contrasting answers for different systems. Typical examples are the L_3VV Auger transition in copper (155) and the $L_{2,3}VV$ Auger transition in aluminum (113). In the first case, extensive investigations carried out by Kowalczyk et al. (69) have demonstrated that the final state is highly localized. In contrast, the results obtained by Powell (113) for the aluminum transition are in favor of a delocalized final state. It therefore appears necessary to study each system in detail before drawing reliable conclusions about its final state.

We shall now consider in more detail the energy balance for CVV processes. Given the position in energy of the first level, E_1, and given the kinetic energy of the Auger electron, E_k, it is clear from equation 40 that Auger electrons of that energy can be created by CVV processes in which the two valence states are at equal distances in energy above and below the level $(E_k - E_1)/2$. To find the total Auger intensity at the energy E_k we must therefore sum over all possible processes involving pairs of states equidistant from the energy $(E_k - E_1)/2$. This is reminiscent of the calculations for the INS spectral distribution (see Appendix) except for the lack of the tunneling-process-induced emphasis on the states close to the Fermi energy. Once again the distribution in energy will be a self-convolution of the valence-electron DOS (25); this sometimes makes "debugging" the data complicated. Furthermore, the final-state interaction is still present except for systems with highly delocalized two-hole states. Nevertheless, the study of the valence-electron DOS through CVV processes is an interesting topic that shares with the studies involving CCV processes the important property of emphasizing the DOS around the particular atom of the system in which Auger recombination takes place (25).

VII. LIMITATIONS OF ELECTRON SPECTROSCOPIES

A. GENERAL LIMITATIONS

In this section we shall review the techniques described in the previous sections from a critical point of view. We have seen the advantages offered by these techniques, but in only a few cases have we discussed their limits. The purpose of this section is to review realistically the limits in more detail.

We shall start from some general considerations, and then analyze each class of techniques in particular.

The most severe limit common to all electron spectroscopies arises from the vacuum requirements. We have seen that the electron-spectroscopy hardware itself requires a vacuum of the order of 10^{-4} torr or better (exceptions to this rule are the experiments involving synchrotron radiation). The extreme surface sensitivity of virtually all techniques pushes the required pressures down to the 10^{-9}-torr range or better. These pressures are not a problem if the main target of the experiment is the study of clean surfaces. The long-run aim of this entire field, is, however, the study of surface chemistry. The vacuum requirements effectively limit the surface chemistry that can be investigated to those processes that occur at very low pressures. This is unfortunate, since many important processes occur at high pressures. Attempts to reconcile high pressures and electron spectroscopy, as by means of rapid sample-introduction systems, face great technical problems.

Another kind of problem common to all types of electron spectroscopy are those problems arising from the primary-beam source. Either electrons or photons are used as primary particles in most electron-spectroscopy techniques. If photons are used, the main problem concerns the energy tunability of the source. Experiments with tunable photon sources in recent years have revealed the very severe limitations of the experiments performed without them. Thus the photon-energy tunability is an important requirement. The only suitable tunable photon sources are those based on synchrotron radiation. Experiments with these sources must be carried out at centralized facilities that can accommodate only a limited number of users. The expanding number of synchrotron-radiation facilities should alleviate this problem, which is currently very severe. Even with more sources, however, the use of synchrotron radiation does not allow as much flexibility as running experiments in one's laboratory, and in most cases is not as inexpensive, either. The tunability problem does not affect experiments in which the primary beam is an electron beam. A good electron source is not necessarily expensive but a sophisticated monochromatized source can be very expensive. The main problem with primary electron beams, however, is their effect on the system under investigation. We have already discussed this problem in the case of Auger spectroscopy; similar problems exist for energy-loss spectroscopy and other electron-beam techniques. Practical examples show that some systems, such as oxidized III–V surfaces, are completely modified by the primary electron beam for beam currents and times like those required for running an experiment with a reasonable signal level. There is no simple practical solution to this problem. The only possibility is always to monitor the evolution of the experimental results as a function of the time of exposure to the beam. Should this show very severe effects of the primary beam, it is always possible, at least in principle, to extrapolate the results to zero exposure time.

From a more fundamental point of view, there is a common problem in the interpretation of the results of all electron-spectroscopy techniques. This is the intrinsic convolution of a number of spurious effects with the experimental information of interest. For purposes of data interpretation, one would like to think in terms of one-electron transitions from an initial state $| i \rangle$ to a final state $| f \rangle$, with those states being so well defined that the information about each of them is immediately available from the experimental curves. This is not only an idealized point of view, but also a fundamentally incorrect one. The states involved in the process are always states of the entire system, for which the one-electron picture is often a bad approximation. Indeed, we have seen that many important features arise from the interaction of the collected electron with the rest of the system. If those features are completely neglected, interpretation of the data is impossible. However, the steps for going beyond the one-electron picture are in general complicated and sometimes not well defined. Furthermore, the information about the state $| i \rangle$ and that about the state $| f \rangle$ are only asymptotically separable. In all real experiments, the initial- and final-state effects are convoluted with each other. Their deconvolution is difficult and often arbitrary, as we have discussed in the case of photoemission spectroscopy.

A general consequence of the above observation is that in many cases an approximated treatment of the theory of an electron-spectroscopy technique gives unrealistically oversimplified results. This can often lead new users to have an unjustified optimism about the potential of the technique. Almost all the techniques discussed in this review could be mistaken by a superficial reader for mature techniques ready to be used routinely. On the contrary, most techniques require much experimental and theoretical work before becoming matters of routine. This will be illustrated by more detailed discussions of the peculiar limits of each technique in the following sections.

B. PHOTOEMISSION SPECTROSCOPY

It is somewhat difficult to appreciate the limits of photoemission spectroscopy, since this technique has not yet evolved to its full potential. This point can be better understood if one considers that the advents of synchrotron radiation and of angle-resolved photoemission have introduced into such experiments four new tunable scalar parameters: photon energy and polarization, and two angles. The possible combinations of tunable parameters can produce an infinite number of new techniques; the CFS and CIS modes are only two examples. New, imaginative ways to use photoemission will probably be introduced in the near future. In spite of this, photoemission spectroscopy does have limits that cannot be overcome even by imaginative experiments.

The major problems affecting photoemission experiments are the convolution problems outlined in Section VII.A. In the Appendix we show that the EDC mode of photoemission gives a complicated convolution of initial and final DOSs, the EDJDOS (or the corresponding integral over a limited portion of k-space), rather than the density of occupied states. A link between experiment and theory is made quite complicated by this convolution. The convolution between initial and final states is mitigated, but not completely removed, in the CFS and CIS modes. In fact, these modes only fix the *energy* of the initial or final state, but they still imply averages in k-space over all the possible states with that energy.

The many-body nature of the initial and final states is also a severe problem in photoemission experiments. The theoretical approaches to solving this problem are necessarily different for different classes of solids (e.g., insulators and metals) and this prevents theorists from formulating a unified theory of the photoemission process. One particularly difficult problem in calculating relaxation energies is the different time scale of different effects.

From an experimental point of view, the main problem in photoemission is energy resolution. The nature of the photoemission process implies that the overall resolution is the combination of the photon resolution with the electron resolution. In the x-ray range the resolution problems are quite severe—without some monochromatization, the band width of the natural sources is of the order of electron volts. The resolution problems have been made more dramatic in recent years by the discovery of such effects as the surface core-level shift. Correct interpretation of the data and separation of the surface from the bulk require a careful analysis of the line shapes and the identification of all the components in a given experimental peak. Even without instrumental broadening, the intrinsic line width makes this analysis quite difficult, and the instrumental resolution limits make it even worse.

The development of new photoemission techniques has raised in recent years hopes for new uses of this probe. Photoemission has indeed been successful in many experiments not strictly related to the electronic properties of the system. For example, it has been used to measure surface optical constants and to identify surface structures. The great interest of these results may generate some confusion and some overoptimistic misunderstanding. Structural investigations by photoemission spectroscopy are able to identify chemisorption sites in some cases and to give in a fraction of these cases information on chemisorption bond lengths. The number of cases in which this approach works is quite limited, however. Furthermore, each case requires a particular theoretical treatment, and even then the accuracy in determining bond lengths is not quite comparable to that of diffraction techniques. In principle, these problems could be overcome by means of photoelectron diffraction. In practice, however, the modulation caused by photoelectron diffraction interferes with a number of other effects, such as those due to the atomic cross section and Auger emissions. It is not clear whether or not photoelectron diffraction will become a routine technique for at least a limited class of chemisorption systems.

C. AUGER-ELECTRON SPECTROSCOPY

This technique is at present the electron-spectroscopy technique closest to being used routinely for analytical purposes. The only comparable technique is core-level photoemission spectroscopy, which, however, has a less favorable signal-to-noise ratio. Nevertheless, it would be premature to conclude that Auger spectroscopy has ended its evolution and become a well-established tool. In fact, not all the potential applications of Auger spectroscopy have been fully explored, and some of them are at a very early stage of development.

The main area of possible development is in the study of valence states by means of Auger processes involving valence electrons. We have seen that this relatively new field already exhibits some limits. Once again the link between experimental curves and DOS implies in many cases a self-convolution. As for INS, this makes the comparison with theory difficult.

As in photoemission spectroscopy, the exact value of the Auger-electron kinetic energy can be a valuable source of information about the coupling of the electrons involved in the process with the remainder of the system. This information is extremely important from a fundamental point of view. At the same time, however, the corresponding effects make it difficult in some cases to identify Auger peaks from their energy positions for systems with many different elements. The above considerations do not affect very much the routine use of AES for qualitative chemical analysis. The Auger peaks are so distant in energy from each other that their identification is not affected by small shifts, except for very complicated systems. When the system is very complicated, the large number of features in the Auger spectra of certain elements can have either a positive or a negative effect. On the one hand, it makes identification of that particular element more reliable, but on the other hand, it increases the probability that some component may interfere with the Auger features of other elements and hide them. In some specific cases the Auger cross section causes additional problems, since it makes it difficult to detect certain crucial elements. The most typical example is carbon, the cross section of which is very small, so that detection becomes problematic at levels below 0.05 monolayers. An *intrinsic* problem of the use of AES for qualitative analysis is that of course no Auger process can occur in a hydrogen atom. Therefore AES is completely blind to the presence of hydrogen. This is a severe limitation indeed, since hydrogen is a fundamental factor in many chemisorption problems. Furthermore, hydrogen is the typical major component of the residual-gas background in stainless-steel chambers for ultrahigh vacuum. It is usually assumed that molecular hydrogen is not highly reactive, and therefore its interaction with the sample under investigation is often neglected unless hot surfaces crack the hydrogen molecules. This cracking could in fact be caused by any hot filament in the experimental chamber, such as that of the ion gauge. Recent ion-desorption experiments (141) revealed that hydrogen can have an essential role in chemisorption processes even when no atomic residual-gas hydrogen is present. This role appears especially important in the case of semiconductor surfaces.

We have already discussed, in Section VI, the practical limits of the use of AES for quantitative chemical analysis. We recall that a number of factors lower the accuracy limit to 100% for "first principle" quantitative Auger analysis. This accuracy improves to 30–40% only if a normalization procedure is used instead of a "first principle" analysis.

The most spectacular use of Auger spectroscopy is probably the scanning Auger-microprobe analysis. This technique has very tough requirements for the primary-beam source. Even with the best currently available sources, the typical duration of a scanning confines the full use of this technique to very stable systems, that is, systems that can be probed for hours. The spectacular uses of the scanning Auger technique are therefore limited to a certain class of systems that are not necessarily the most important in practical problems.

D.   ELECTRON ENERGY-LOSS SPECTROSCOPY

Electron energy-loss spectroscopy is perhaps the most typical example of how oversimplifications can be misleading in considering the potential applications of an electron-spectroscopy technique. The oversimplification primarily arises from the apparent similarity between the loss function,

$$L(E) = \frac{\epsilon_2}{\epsilon_1^2 + \epsilon_2^2} \tag{47}$$

and the imaginary part of the dielectric function that describes the optical properties of the system. From this the idea emerges that energy loss could be an inexpensive surrogate for uv optical spectroscopy, which is made difficult by the lack of suitable photon sources. The resolution is of course much worse than in optical spectroscopy, but the experimental advantages of energy loss usually compensate for that limitation. They cannot remove, however, the fundamental reasons that make it questionable to link energy-loss spectroscopy and optical spectroscopy.

There are two such fundamental reasons: the nature of the interacting particles, and the magnitude of their k-vectors. The photons involved in optical phenomena correspond to transverse excitations. This limits the kind of effects that can cause them to exchange energy with a system. The most striking example is the creation of collective excitations of the sea of electrons in a solid, that is, the excitation of plasmons. These collective excitations are longitudinal excitations that cannot be *directly* coupled to the transverse field of the photons. Indirect coupling is possible, as through surface roughness, but direct coupling is not. Thus the plasmons do not appear in the list of phenomena directly contributing to the optical properties. They can be excited instead by electrons in electron energy-loss spectroscopy, since these electrons are not simply transverse excitations, as the photons are. The other major difference between electrons and photons is that for a given energy in the range of interest the electrons have much higher momenta than do the photons, and therefore larger k-vectors. In optical phenomena one usually assumes that the k-vector of the photon is so small as to be negligible. This is certainly not so in energy-loss phenomena. The k-vector contribution fundamentally changes the way the excitation couples the initial and final states.

As a consequence of the above differences, energy loss cannot be treated as a surface-sensitive counterpart of optical spectroscopy. A complete theory is necessary for its interpretation, and energy loss is still evolving toward being a truly routine tool.

E. FIELD-EMISSION SPECTROSCOPY AND ION-NEUTRALIZATION SPECTROSCOPY

Both the techniques treated in this section, FES and INS, have the problem of being quite limited in their applications. Neither of them has the flexibility of photoemission spectroscopy of Auger spectroscopy. Both FES and, to a minor extent, INS are limited to the study of states close to the Fermi level. Furthermore, INS has the additional problem of being linked to the DOS only through a self-convolution, and even this somewhat indirect link can be established only under a number of approximations, as discussed in the Appendix.

In spite of its limitations, INS has some potential to be a technique with more applications than those discussed herein. One aspect that could be explored in more detail is the ion-neutralization process itself, which is used at present only as a surrogate of the ionization step in Auger spectroscopy. The details of the ion-neutralization process could be very interesting in connection with the growing interest in ion–surface interactions (e.g., in desorption experiments). For example, it could be interesting to study in detail the dependence of the neutralization process on the surface parameters and on the characteristics of the ion. These possible developments make the ion-neutralization technique more interesting than FES, which is applicable only to ultrasharp needles of tungsten and a handful of other materials.

F. LOW-ENERGY ELECTRON DIFFRACTION

The most important specific limitation of LEED arises from fundamental quantum mechanics. We know that the wave function ψ_f is *not* a directly measurable quantity; only its square, $|\psi_f|^2$, is measurable, since it corresponds to the electronic-charge (i.e., probability) density. The diffraction of electrons, however, influences directly the wavefunction ψ_f itself. The situation is very similar to that of photon diffraction, in which the diffraction

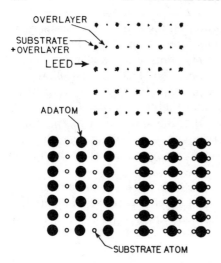

OVERLAYER

SUBSTRATE +OVERLAYER

LEED

ADATOM

SUBSTRATE ATOM

Fig. 17.58. A LEED pattern identifies the *symmetry* of a chemisorption system, but not its details, such as the chemisorption sites. For example, the pattern shown at the top of this figure could correspond either to the "on top" chemisorption site shown at the bottom left or to the "bridge" site shown at the bottom right.

itself arises from the combination of the electromagnetic fields at each point in space, but its effects are observed by detecting the intensity, which is the square of the field. As in such other diffraction techniques as x-ray crystallography, this distinction puts some severe limits on the information available from a diffraction pattern. These limits are implicit in our short review of the LEED theory in Section II.E.3. We saw there that the diffraction-pattern spots give information on the surface G-vectors $G_{\parallel}$. This in turn gives information about the periodicity of the surface according to the definition of the G-vector. The periodicity, however, does not unequivocally identify the atomic structure of the surface. In chemisorption problems, for example, a given symmetry can correspond to different chemisorption geometries, as illustrated schematically in Fig. 17.58. The complete identification of the structure requires more experimental input than just diffraction patterns. For example, the intensity of the diffracted beams as a function of the primary-electron energy is a very important source of information. Extracting this information requires complicated calculations for all the possible single- and multiple-scattering paths of the electrons in the beam. These calculations are still limited by the cost of computer time.

VIII.   CONCLUSIONS AND FUTURE DEVELOPMENTS

Our review of electron spectroscopy should have made clear at least one important point: The different electron-spectroscopy techniques are at very different stages of evolution, ranging from novel and untested ideas to well-established routine tools. It should also be clear that only a very small number of techniques are really "routine." In most other cases the application of a given technique to the study of a new class of systems requires preliminary research about the development of the technique itself. The interesting aspect of this evolution is that the next few years are likely to produce important developments in novel areas of electron spectroscopy. Some of these developments are difficult to predict, but other developments can be foreseen. We shall discuss in this section those developments that can be foreseen with reasonable confidence. Our discussion of the future developments will cover improvements in instrumentation, evolution of the theoretical background, and possible novel uses of electron spectroscopy.

A. INSTRUMENTATION

The expanding interest in the control of more parameters has emphasized in recent years the limitations of the conventional electron spectrometers. It is important to understand that each electron-spectroscopy technique is to some extent complementary to other techniques and has some instrumentation in common with them, such as the ultrahigh-vacuum hardware. It is becoming, in fact, more and more important to have *several* techniques available in parallel in the same experimental system. The most important feature required in a modern electron spectrometer is flexibility, for carrying out different kinds of experiments with basically the same instrumentation. Consider, for example, the electron analyzers. All the established electron-optics schemes have somewhat limited flexibility: Most electron analyzers are either angle integrating or angle resolving. One would like to have both features present in the same instrument. A rapid change from an angle-integrating mode to an angle-resolving mode is possible with the modified CMA, but the peculiar geometry of the CMA sometimes makes the scanning of angles complicated. The angle-resolving analyzers based on hemispherical grids and retarding fields can easily be switched to an angle-integrating mode, but these systems have the general problem of all retarding-field systems, an unfavorable signal-to-noise ratio for techniques that use primary electron beams. The new display analyzers discussed in Section III solve this particular problem. However, these analyzers are typically very expensive and the few prototypes operational at present are not available to most experimentalists. The cost of these systems is due in great part to development, so there is some hope that the display devices may become commercially available at a reasonable cost.

The display analyzers also solve the problem of shortening the time required for an experiment. This problem is connected to the typical instability, even in an ultrahigh-vacuum environment, of most surfaces investigated with electron spectroscopy. The parallel detection of electrons at many different energies and/or in many different directions is very attractive from this point of view. This area can also profit from the current spread in use of microcomputers to control the instrumentation. Other possible development areas for parallel detection are those related to the electron-optics designs that focus electrons of different energies in well-defined geometric loci. Such is the case, for example, of the PMA, which focuses electrons of different energies along a line. A pulse-delay electron system can be used in this case for parallel analysis of different kinetic energies.

The ultrahigh-vacuum technology used in electron spectroscopy is already so sophisticated that no dramatic improvements should be expected. The only exception could be an increased capability to handle low pressures and high pressures simultaneously in the same windowless systems using differential pumping. This capability would allow performing a number of important experiments that require either medium-pressure primary-beam sources and ultrahigh-vacuum sample chambers or vice versa. An already existing example of this approach is that of the gas-discharge uv lamps attached without windows to ultrahigh-vacuum sample chambers. Another possible area of development in ultrahigh-vacuum technology is that of fast pumpdown systems able to bring down the pressure after chemisorption experiments performed at intermediate pressures.

We have seen that the developments in the area of photon sources—synchrotron radiation in particular—have brought great improvements in the field of photoemission spectroscopy. Not only have substantial improvements of existing techniques been achieved, but entirely new techniques have been developed as well, such as the CIS and CFS modes of photoemission. The use of synchrotron radiation will probably bring the most important developments in the foreseeable future too. The interest of experimentalists and theorists is now moving from the "vacuum ultraviolet" photon-energy range (up to 40 eV) to the widely unexploited range between 40 and 1200–1400 eV. The late 1970s have produced important technological results from this point of view with the introduction of the Brown–Lien–Pruett "Grasshopper" monochromator and of the toroidal-grating monochromator, which in turn was made possible by the new holographic methods in grating production. Many new monochromator designs are now being developed and should become operational in the next few years. Perhaps the most challenging problem in this field is bridging the spectral gap between the range covered by grating monochromators—which have been tested up to about 1 keV—and that covered by crystal monochromators.

Further photon-source improvements will probably come from electron-accelerator technology. The development of "wigglers" and "undulators" (154), for example, has made possible the production of high-energy photons with low-energy accelerators and pushed to very high photon energies the spectral output limits of high-energy machines. There is also a reasonable possibility of using in the near future free-electron lasers in the uv or even in the x-ray range. These devices could provide an ultraintense source with easily tunable photon energy. Even in the more conventional area of storage-ring sources, the new machines scheduled to be operational in the early 1980s promise excellent performances, and in particular, very impressive levels of circulating beam current.

B. THEORETICAL BACKGROUND

All electron-spectroscopy techniques still require considerable improvement in their theoretical background before they can become routine analytic tools. We shall concentrate our attention here on the theoretical developments foreseen for the two techniques discussed in detail in this review, photoemission and Auger spectroscopy. Some of the data-interpretation problems discussed in the previous sections pose a very serious theoretical challenge, and they will most likely be substantially resolved in the near future. A typical example is the many-body contributions to the initial and final states. There is also another, more general challenge to the theory. The innovations in photoemission—and to some extent in Auger spectroscopy— are rapidly extending the information goals from the *energy* of the electrons in the system to the *state* of the electrons in the system. It is calculating states rather than just energies that produces the real new theoretical problem, since it is a task one order of magnitude less approximate than that of calculating energies. The experimental information now obtainable about electronic states includes the k-vector, the symmetry of the wavefunction, sometimes the spin, and in some cases the many-body effects. In general the states must be treated as states of the entire system rather than of the emitted electron alone, leading, for example, to multiplet splitting. In summary the evolution of electron spectroscopy has greatly changed the demands on the theory compared with the limited demands of conventional electron spectroscopy or of the optical techniques. The theory is rapidly evolving to meet these demands and to produce more efficient and reliable computational schemes.

The above discussion about the electronic states being the new target for the theory concerns not only the system under investigation, but also the process on which each electron-spectroscopy techique is based. The evolution of the different electron-spectroscopy techniques has multiplied the fundamental questions about their theoretical background. The empirical concepts that have formed until now their simplified theoretical background, concepts such as the work function, inner potential, and so on, are more and more being challenged by the experimental results. In practice, *all* the conventional theoretical descriptions of the electron-spectroscopy processes are challenged by new developments. In the field of photoemission, for example, the "constant-matrix-element" approximation, which was essentially borrowed from optical spectroscopy, is deeply questioned. The three-step model itself is being replaced by more sophisticated and more theoretically sound one-electron descriptions (155). The new results often require a comprehensive theoretical picture of the electron involved in the process and of its enviroment. The study of diffraction processes in photoemission and EXAFS emphasizes the fundamental limits of the concept of excited state in a solid—the short lifetimes of the excited states make possible a theoretical description in terms of nearest-neighbor backscattering instead of an "infinite lifetime" Bloch description. In a sense the reality of the experimental results obtained with photoemission and other electron-spectroscopy techniques is forcing chemists and physicists to adopt a more realistic view of the electronic states in complicated systems. The consequence of this evolution may very well be a different conceptual approach to electronic state problems in the long run.

C. NEW TECHNIQUES

Instrumental and theoretical improvements are stimulating scientists to adopt a very innovative approach to electron spectroscopy. This innovative approach has already produced in the late 1970s and early 1980s a number of new techniques certainly unforeseen in the early 1970s. In this section we shall not try to guess all possible future developments, but we shall try instead to identify and discuss novel techniques the introduction of which is reasonably probable and which can be used as examples of what can generally be expected.

The most rapid and substantial innovations in instrumentation concern the photon sources and, more specifically, synchrotron radiation. Accordingly, novelties can be expected from their use. For example, the use in surface-geometry analysis of photoemission and photoemission-related techniques such as surface-EXAFS is likely to expand. The better control of the soft-x-ray production will bring more and more onto the experimental scene fundamental excitation thresholds such as those of oxygen, nitrogen, and carbon. Another promising field is that of high-resolution core-level-threshold absorption spectroscopy, for which the partial-yield technique can provide surface-sensitive detection.

Some of the parameters that can be used in electron spectroscopy have been neglected in the past, but will probably be explored more extensively in the future. A typical example is spin polarization. Spin-polarized electron-spectroscopy experiments (7,147) have been very rare in the past, due to the very severe technical limitations of the electron sources and detectors that control or reveal spin polarization. Things are changing fast in this field, with the recent development of new and efficient sources and detectors (147).

Some of the electron-spectroscopy techniques are less likely to produce novelties than others, since they are approaching either routine use or the limits of their performances, or both. Even in those cases some evolution can be expected. For example, FES is severely limited by the need for ultrasharp needles. The capability to manufacture ultrasharp needles of new materials (e.g., iron or silicon) is a true innovation in that field. Auger spectroscopy is approaching a steady state in which it is primarily a routine probe for surface chemical analysis; the same thing is happening to a lesser extent to photoemission spectroscopy in the x-ray range. Innovative uses of Auger spectroscopy can be foreseen, however, in the areas of local-DOS mapping and of angular effects—a field neglected until now. X-ray photoemission is a very productive source of novel techniques—surface-EXAFS, photoelectron diffraction, and near-edge absorption spectroscopy are the recent examples, and more novelties should be expected in the future.

One technique that has some chance of joining photoemission and Auger spectroscopy among the group of leading electron spectroscopies is energy loss. The breakthrough in this field came several years ago, with the introduction of high-resolution, low-energy analyzers for detecting surface vibrational modes (12). There was hope for some time that this surface vibrational spectroscopy was going to have the same characteristics of straightforwardness and basic theoretical simplicity as do other vibrational spectroscopies. The most recent results seem to indicate that the theoretical problems in this field are more complicated than expected. Nevertheless this technique is potentially one of the most important kinds of electron spectroscopy, and stands a good chance of becoming a routine tool in the next 4–6 years.

The impressive evolution in the field of electron spectroscopy that we have witnessed in the past 10 years is still going on. From the above discussion, it should be clear that novelties are possible and indeed very probable in the next few years. Our review will probably have to be updated in the near future with a number of new results. As a concluding remark, therefore, we would like to invite the interested reader to scan the coming literature in this field carefully for more exciting news.

APPENDIX: ELECTRONIC TRANSITION PROCESSES AND EMISSION INTENSITY

All of the techniques outlined in this review involve the transition of an electronic system from a state $|i\rangle$ to a state $|f\rangle$. The link between $|i\rangle$, $|f\rangle$, and the experimental results depends on the kind of process used by each technique. This link is conceptually simple—but computationally complicated—for the elastic-scattering processes. The corresponding theory is briefly outlined in Section II.E.3. We shall outline here the treatment of the $|i\rangle \rightarrow |f\rangle$ transitions for the other techniques. Our analysis will be dedicated primarily to photoemission and Auger spectroscopy, and to a somewhat lesser extent to INS and FES.

A. PHOTOEMISSION PROCESSES

The theoretical treatment of the $|i\rangle \rightarrow |f\rangle$ transition is slightly different for atoms, molecules, and solids (22,24). In an atom or a molecule the photoemission process consists of an optical transition between the ground state of the system and a final state corresponding to an ionized atom or molecule plus a free electron. The only requirement for the electron to become a photoelectron is that it be excited above the energy threshold for photoionization. In a solid the optical transition *per se* is not sufficient to create a photoelectron. The excited electron must also be able to travel to the surface and to cross the surface without decaying in energy to below the threshold. Therefore, the photoemission process is a complex combination of several different processes. In this section we shall consider the optical excitation first and then the entire photoemission process for solids.

The cross section for the optical transition between a state $|i\rangle$ and a state $|f\rangle$ on absorption of a photon of energy $h\nu$ is proportional to the transition probability, which in turn is given by the Fermi golden rule:

$$P_{if} \propto \frac{1}{h\nu} \sum_n |\langle i | H'_n | f \rangle|^2 \qquad (A1)$$

where the right-hand term in equation A1 is the matrix element squared of the operator H'_n, which describes the energy perturbation of the nth electron of the system by the incoming photon. The perturbation arises from the interaction between the vector potential of the electromagnetic field of the photon, $\mathbf{A} = \mathbf{A}\exp(ik \cdot r)$, and the nth electron. In the dipole approximation the plane wave in the vector potential is approximated by a constant. The perturbation operator becomes, according to Dirac,

$$H'_n = \mathbf{A} \cdot \nabla_n + \nabla_n\mathbf{A} \approx 2(\mathbf{A}_0 \cdot \nabla_n) \qquad (A2)$$

where ∇_n is proportional to the momentum operator for the nth electron. From equation A2 we see that the dipole approximation, $\mathbf{A} \approx \mathbf{A}_0 = $ constant, not only changes the first term in H'_n but also makes the second term equal to the first one. If we define a polarization factor $\eta_n = \cos(\mathbf{A}_0, \langle i | \nabla_n | f \rangle)$, then the transition probability becomes

$$P_{if} \propto \frac{1}{h\nu} \sum_n \eta_n^2 |\langle i | \nabla_n | f \rangle|^2 \qquad (A3)$$

In atoms and molecules the photoionization cross section $\sigma_i(h\nu)$ for the state $|i\rangle$ and a photon of energy $h\nu$ is simply given by the sum of the probabilities P_{if} over all possible states $|f\rangle$ that satisfy the energy balance. Calling E_i and E_f the energies of the states $|i\rangle$ and $|f\rangle$, the energy balance is given by equation 19:

$$E_f = E_i + h\nu \qquad (A4)$$

Therefore

$$\sigma_i(h\nu) \propto \frac{1}{h\nu} \sum_f \sum_n \delta(E_i - E_f - h\nu)\eta_n^2 |\langle i | \nabla_n | f \rangle|^2 \qquad (A5)$$

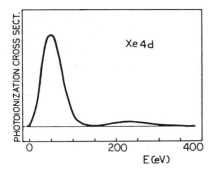

Fig. 17.59. Theoretical dependence of the photoionization cross section of a 4d core level on the kinetic energy of the resulting photoelectron, that is, on the energy of the photon (53). Notice the "Cooper minimum" 140–150 eV above the threshold.

where δ is the Dirac delta function. Calculations of the cross section require, therefore, knowledge of the matrix element $\langle i \mid \nabla_n \mid f \rangle$. In turn, this requires a good knowledge of the wavefunctions of the states $|i\rangle$ and $|f\rangle$ (83). These wavefunctions are given by the solutions of the Schrödinger equation. For the systems of interest, the solution is exact only in the case of the hyrogen atom; otherwise it is approximated. For atoms other than hydrogen, typical approximations are the central-field calculations, the Hartree–Fock calculations and the schemes that include correlation effects (83). In the central-field calculations, the true potential energy is approximated by a central potential. The approximated wavefunctions are antisymmetric products of one-electron wavefunctions containing the spherical harmonics. Different forms of central potential can be used. One of the most effective is the Hartree–Slater or Herman–Skillman potential, which includes an approximate correction for exchange effects (84). The central-field calculations correctly predict certain qualitative features of the cross section. For example, Fig. 17.59 shows the cross section for photoionization of the Xe 4d core level versus photon energy. This curve exhibits a slow rise to a maximum. This "delayed maximum" is caused by the dominating role of the $d \rightarrow f$ transitions in the photoionization process (22). The continuum f wavefunction at threshold is "kept out" of the core region by the centrifugal repulsion not overcome by the electrostatic interaction. Therefore the matrix element $\langle i \mid \nabla_n \mid f \rangle$ is small at the threshold and increases as $h\nu$ and the kinetic energy increase, making the final state more penetrating in the core region. The central-field calculations do not provide a good treatment of exchange effects or of electron–electron correlation effects. The first problem is better solved by the Hartree–Fock method (52,131), whereas the correlation effects (i.e., the many-body effects within the atom) require more sophisticated approaches (83), for example, the multiconfigurational wavefunction approximation for the ground state coupled to the continuum configuration interaction method (132), the close-coupling approximation (142), and the many-body perturbation theory in the random-phase approximation (RPA) (1).

The same problem of calculating matrix elements and therefore knowing the wavefunctions is present for molecules and solids. A variety of approximate methods have been developed to solve the Shrödinger equations for these systems (51). For solids and solid surfaces, for example, some of the most widely used methods are the tight-binding approach; the pseudopotential method, in particular, its self-consistent versions; and the cluster calculations (51). All these schemes are required for the valence states, whereas the core levels can still be treated by atomiclike schemes with corrections for the solid-state or molecular environment. These corrections depend also on the valence-charge distribution and therefore require in principle some knowledge of the wavefunctions of the valence states.

The relation between the photoionization cross section given by equation A5 and the experimental data requires a sum over all possible initial and final states contributing to the number of photoelectrons collected by the analyzer with a certain set of parameters (e.g., energy and direction). In a conventional photoemission experiment photons of fixed energy $h\nu$ are used and the analyzer accepts only photoelectrons with energies in a small range, as from E to $E + dE$. In angle-resolved photoemission the analyzer further selects the electrons that travel in a given direction, that is, those with a given k-vector **k**. In angle-integrated photoemission, therefore, the number of collected photoelectrons is given by $N(E)dE$, where the distribution function $N(E)$ is

$$N(E) \propto \frac{1}{h\nu} \sum_f \sum_i \delta(E_f - E_i - h\nu)\delta(E_f - E) \, | \langle i \, | \, \nabla \, | \, f \rangle |^2 \qquad \text{(A6)}$$

In this equation we assume that the optical transition occurs between one-electron states and that the final-state energy E_f coincides with the kinetic energy E of the photoelectron. Therefore the sum of matrix elements in equation A5 is replaced in equation A6 by a single matrix element for the one electron considered. Equation A6 shows that the angle-integrated EDCs are related to the distribution in energy of the electrons in the ground state of the system. But it also shows that this relation is somewhat indirect, contrary to what we assumed in our oversimplified analysis in Section II. We suggested there that the EDCs were simply proportional to the distribution in energy of the electrons in the ground state of the system rigidly displaced by $h\nu$. This would give $N(E) \propto (1/h\nu) \sum_i \delta(E - h\nu - E_i)$. Equation A6 differs from this expression in that it contains the matrix element $\langle i \, | \, \nabla \, | \, f \rangle$ that accounts for the transition probabilities. Furthermore, it sums a different kind of delta function and takes into account the possible role of the final states $|f\rangle$.

The above link between the EDCs, $N(E)$, and the distribution in energy of the electrons in the ground state is expressed in mathematical terms in the case of solids. The Bloch theorem assigns a k-vector to each electronic state in a solid. The sums over initial and final states in equation A6 can be replaced by sums over k-vectors. The k-vectors in a crystal are quantized, but their allowed values are so close to each other that they can be considered a continuum for integration purposes. Thus the sums over k-vectors can be approximated by integrals in k-space:

$$N^*(E) \propto \frac{1}{h\nu} \int_{k-\text{space}} \delta(E_f(\mathbf{k}) - E_i(\mathbf{k}) - h\nu)\delta(E_f(\mathbf{k}) - E) \, | \langle i \, | \, \nabla \, | \, f \rangle \, |^2 d^3\mathbf{k} \quad (A7)$$

Notice that in equation A7 we implicitly assume that $|i\rangle$ and $|f\rangle$ have the same k-vector $\mathbf{k}$. This in turn implies that neither the absorbed photon nor other causes much change the k-vector during the optical excitation. This assumption is reasonable, since the magnitude of the photon k-vector is typically small with respect to that of the electron k-vector in the states $|i\rangle$ and $|f\rangle$. The possibility of transitions without the k-conservation implied by equation A7 must be considered, however, and it gives results different from those here presented (137).

There is a particularly important class of non-k-conserving processes that must be considered. These are the processes in which the crystal lattice *as a whole* contributes a k-vector to the optical excitation process. The lattice contribution is similar to the scattering processes considered in LEED; that is, it does not change the energy of the electron. As in LEED, the corresponding changes of the k-vector must coincide with one of the G-vectors of the crystal, but unlike in LEED this condition is now in three dimensions rather than just two. An important consequence of the presence of these processes is that k-vectors differing by a G-vector are equivalent to each other in the description of the electronic properties. Indeed, the state corresponding to the first k-vector can be transformed into the state corresponding to the second k-vector by an energy-conserving lattice scattering. It is possible and convenient, therefore, to *reduce* each k-vector by combining it with a suitable G-vector so that it becomes as close as possible to the origin in k-space. These reduced k-vectors are all contained in a *reduced zone* of k-space near the origin. The reduced zone can be conveniently used for a description of the electronic properties. For example, equation A7 can be rewritten to include the lattice scattering effects:

$$N^*(h\nu) \propto \frac{1}{h\nu} \int_{\text{RZ}} \delta(E_i(\mathbf{k}) - E_f(\mathbf{k}) - h\nu)\delta(E_f(\mathbf{k}) - E) \, | \langle i \, | \, \nabla \, | \, f \rangle \, |^2 d^3\mathbf{k}$$

$$(A8)$$

where the integral is now carried out over the reduced zone (RZ) only, and each k-vector is either in the reduced zone *per se* or it has been reduced to the reduced zone. Also notice in equation A7 that the function $N(E)$, which describes the distribution in energy of the collected photoelectrons, is replaced by a different function $N^*(E)$. This emphasizes that the $N^*(E)$ describes the distribution of (excited) electrons *inside* the solid rather than in vacuum. Similarly, the final state $|f\rangle$ is now the excited state inside the solid, rather than the final state of the photoelectron. Equation A7 further assumes that $N^*(E)$ can be identified as a separate function. Thus it implies that the different contributions to the photoemission process—optical excitation, and transport to and through the surface—can be expressed in the final $N(E)$ function as a product of factors. This is an approximation, since the photoemission process *per se* is a one-step process not separable into noninteracting steps. The approximation is very widely used and is known as *three-step model* (138). It gives the EDCs as simple products:

$$N(E) = N^*(E)T(E)S(E) \qquad (A9)$$

Here the functions $N^*(E)$ (given by equation A7), $T(E)$, and $S(E)$ describe the optical excitation, the transport to the surface, and the release into a vacuum. In practice the important structure in the spectra $N(E)$ is all coming from the function $N^*(E)$. The function $T(E)$ gives the smooth secondary-electron background already shown in Fig. 17.4. The function $S(E)$ gives the low-energy cutoff in the spectra, and has an important influence on the direction of the photoelectrons in angle-resolved photoemission.

Calculating the k-space integral that appears in equation A7 requires knowing the matrix element $\langle i \mid \nabla \mid f \rangle$ for each k-vector, that is, knowing the corresponding initial- and final-state wavefunctions. A crude simplification of this almost impossible-to-meet requirement is usually applied—that of assuming a constant matrix element. This widely used approximation originates from optical spectroscopy, where it is on much firmer ground due to the much smaller range of energies usually considered. In the present case it is not really a good approximation and can give a number of spurious effects. Nevertheless, in most cases one cannot avoid it. Using this approximation, we transform equation A7 into

$$N^*(E) \propto \frac{|\langle i \mid \nabla \mid f \rangle|^2}{h\nu} \int_{\text{RZ}} \delta(E_f(\mathbf{k}) - E_i(\mathbf{k}) - h\nu)\delta(E_f(\mathbf{k}) - E)d^3\mathbf{k} \quad (A10)$$

The integral in equation A10 closely resembles two other integrals of great importance in solid-state physics: the DOS (already defined by equation 3) and the joint DOS (JDOS). The DOS at an energy $E - h\nu$ is, by definition,

$$\rho(E - h\nu) \propto \int_{\text{RZ}} \delta(E - h\nu - E_0(\mathbf{k}))d^3\mathbf{k} \qquad (A11)$$

where $E_0(\mathbf{k})$ is the energy–wavevector relation for the electronic states of the solid. If one considers occupied states, then $E_0(\mathbf{k})$ can be identified with the function $E_i(\mathbf{k})$ in equation A10. Simple mathematical steps transform equation A11 into a surface integral in k-space:

$$\rho(E - h\nu) \propto \int_{\text{RZ}} \frac{dS^*}{|\nabla_k E_i(\mathbf{k})|} \qquad (A12)$$

where S^* is the k-space surface defined by the equation $E - h\nu - E_i(\mathbf{k}) = 0$.

The JDOS is a function of the photon energy that defines the imaginary part of the dielectric function, ϵ_2, and therefore the optical-absorption coefficient. It is defined by

$$j(h\nu) \propto \int_{k-\text{space}} \delta(E_i(\mathbf{k}) - E_f(\mathbf{k}) - h\nu) d^3\mathbf{k} \qquad (A13)$$

which becomes, after some mathematical steps,

$$j(h\nu) \propto \int_{\text{RZ}} \frac{dS^*}{|\nabla_k(E_i(\mathbf{k}) - E_f(\mathbf{k}))|} \qquad (A14)$$

where now the k-space surface S^* is defined by the equation $E_f(\mathbf{k}) - E_i(\mathbf{k}) - h\nu = 0$. Notice that, in contrast to the function ρ, the function j is a *convolution* of the energy–wavevector functions of the occupied and unoccupied states of the system. Similar mathematical steps transform the function $N^*(E)$, defined by equation A10, into

$$N^*(E) \propto \int_{\text{RZ}} \frac{dL^*}{|\nabla_k E_i(\mathbf{k}) \times \nabla_k E_f(\mathbf{k})|} \qquad (A15)$$

where the integral is now a *line* integral in k-space. It is carried out along the line L^* defined by the two surfaces $E_f(\mathbf{k}) - E_i(\mathbf{k}) - h\nu = 0$ and $E_f(\mathbf{k}) - E = 0$. Equations A12, A13, and A15 once again illustrate the similarities and differences between the angle-integrated EDCs and the DOS. Maxima in the function $\delta(E - h\nu)$—the DOS rigidly displaced by $h\nu$—occur when the gradient $\nabla_k E_i(\mathbf{k})$ is zero, that is, at the so-called *critical points* of the band structure of the solid. The condition $\nabla_k E_i(\mathbf{k}) = 0$ also gives maxima in $N^*(E)$ and therefore in the EDCs. However, other maxima can occur in $N^*(E)$ if $\nabla_k E_f(\mathbf{k}) = 0$ or if the two gradients are parallel to each other. Furthermore the position in energy and the intensity of the "true" DOS maxima, those due to $\nabla_k E_i(\mathbf{k}) = 0$, can be affected by the other factors mentioned above, such as the matrix element. One should also emphasize that the comparison between equations A12 and A15 is based on the assumption that the k-space integral is carried out in all directions. In a photoemission experiment, however, photoelectrons are *not* collected in all directions. The photoelectron collection can be extended at most to a hemisphere, and the angular integration *inside* the sample is over the solid angle corresponding to that hemisphere. Many widely used electron analyzers collect photoelectrons over much less than a hemisphere. Therefore the possibility must always be considered of spurious angular effects due to the limited collection geometry affecting the structure of the EDCs.

The angle-resolved photoemission spectra of solids are in principle much easier to interpret than the angle-integrated spectra. In fact, the integral in *k*-space that appears in the equations defining the EDCs is reduced to a very small region. In the limit case, only one *k*-vector is selected. Notice, however, that this is not the *k*-vector $\mathbf{k}_0$ of the collected photoelectron. In fact, the third step of the photoemission process can strongly change the *k*-vector from its value inside the solid, $\mathbf{k}$, to $\mathbf{k}_0$. The problem of linking the measured $\mathbf{k}_0$ to $\mathbf{k}$ is discussed at length in Section V.

B. AUGER-ELECTRON EMISSION INTENSITY

We have seen that Auger spectroscopy is based on a two-step process consisting of a preliminary ionization step followed by an Auger recombination. These two steps are physically separated and therefore can be treated as independent processes without approximations. We shall first consider the ionization of an atom at a distance z from the surface. In this process a hole is created in the core state $|c\rangle$ by an electron belonging to or created by the primary electron beam. The primary-beam energy is E_p. We shall call $\sigma_c^e(E_p)$ the ionization cross section for that particular atom and for the core state $|c\rangle$ with electrons of energy E_p. The primary electrons will excite

$$dz\rho(z)I_p\sigma_c^e(E_p) \tag{A16}$$

atoms per unit time in a slab of thickness dz at a distance z from the surface (48). Here $\rho(z)$ is the density of those particular atoms and I_p is the intensity of the primary beam. Equation A16 does not give the total number of ionized atoms, since it is unrealistic to assume that only primary-beam electrons can contribute to the ionization. The primary electrons will give rise through inelastic scattering to a continuous distribution $N_z(E_0)$ of electrons in z in the energy range between 0 and E_p. Therefore equation A16 should be more realistically completed in the following way:

$$dz\rho(z) \int_0^{E_p} N_z(E_0)\sigma_c^e(E_0)dE_0 \tag{A17}$$

We now assume that $A_c(E_k)$ is the probability that the above ionized atom emits an Auger electron of kinetic energy E_k. The number of Auger electrons created inside the sample at a distance z from the surface is:

$$A_c(E_k)dz\rho(z) \int_0^{E_p} N_z(E_0)\sigma_c^e(E_0)dE_0 \tag{A18}$$

These Auger electrons will be able to travel for only a short distance before losing energy. The corresponding mean free path depends on the energy: $\lambda = \lambda(E_k)$. Calling ϕ the angle with respect to the sample normal at which the Auger electrons are emitted, the geometric attenuation factor for the slab at z is $\exp[-z/(\lambda(E_k)\cos\phi)]$. Therefore the number of Auger electrons of energy E_k collected along ϕ is

$$I(E_k) \propto \int_0^\infty dz \exp\left(-\frac{z}{\lambda(E_k)\cos\phi}\right) A_c(E_k)\rho(z) \int_0^{E_p} N_z(E_0)\sigma_c^e(E_0)dE_0$$

(A19)

where the proportionality constant depends on the solid angle of acceptance of the analyzer. If an angle-integrated analyzer is used, then the total signal will be the integral of equation A19 over an appropriate range of directions.

Equation A19 shows that the Auger intensity carries information about the density $\rho(z)$ of the atoms of the species giving that particular Auger emission. If one assumes that $\rho(z)$ is constant over a distance of the order of $\lambda(E_k)$ then it becomes a simple multiplication factor in equation A19. The evaluation of this factor from the measured Auger intensity $I(E_k)$ requires knowledge of all other factors in equation A19, that is $\lambda(E_k)$, $A_c(E_k)$, $N_z(E_0)$, and $\sigma_c^e(E_0)$. Most of these factors can be determined empirically and the normalization problems can be solved by using the known density of that atomic species in a reference system. For example, the cross section $\sigma_c^e(E_0)$ can be measured accurately from Auger experiments on gas-phase atoms and molecules. The mean free path $\lambda(E_k)$ can also be measured, as by photoemission experiments with synchrotron radiation. We shall now discuss the parameters A_c and $N_z(E_0)$, which are a little more troublesome.

The role of the Auger recombination probability A_c could appear *a priori* unrealistically simple. In fact, the filling of a core hole can occur by either an Auger recombination or by emission of x-ray photons, but the probability ratio is overwhelmingly in favor of the Auger process for the core-level energies of interest, <2–2.5 eV. One could conclude, therefore, that A_c is simply equal to 1. However, the Auger recombination of a given core hole can give rise to Auger electrons in several different final states. The energies of these states can differ by more than 100 eV. It is only the *total* probability of all processes involving all possible final states that is equal to 1, that is, independent of energy. The actual factor A_c for each Auger peak must be estimated theoretically for the corresponding process or be measured. Once again the problem is automatically solved by calibrating the Auger intensity with that of a compound with a known density of the atomic species of interest.

The problem of estimating $N_z(E_0)$ appears simple enough if one considers that the sample regions contributing to the Auger emission must be not farther from the surface than the mean free path $\lambda(E_k)$. The primary energy

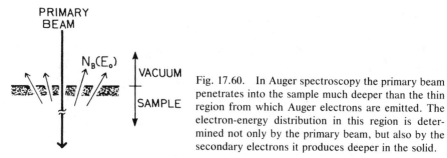

Fig. 17.60.   In Auger spectroscopy the primary beam penetrates into the sample much deeper than the thin region from which Auger electrons are emitted. The electron-energy distribution in this region is determined not only by the primary beam, but also by the secondary electrons it produces deeper in the solid.

E_p is typically much larger than the Auger kinetic energy E_k. The corresponding mean free path is larger too. In a first approximation one can assume that the primary electrons travel through all the "active" region without losing energy, and that the probability for them to be removed from the beam by elastic scattering is also negligible. This does not imply, however, that the distribution $N_z(E_0)$ is all "concentrated" at the primary energy E_p. In fact, the primary electrons will eventually lose their energy or be elastically scattered inside the sample. These events will occur a few hundred angstroms from the surface. The resulting lower-energy electrons can find their way back to the thin "active" slab and contribute to the $N_z(E_0)$ distribution in it. In practice, the contribution to $N_z(E_0)$ of these backscattered electrons is comparable in magnitude to that of the primary electrons. Since the backscattered electrons have a small probability of interacting with the sample in the thin "active" slab (see Fig. 17.60), one usually assumes that the energy distribution of the backscattered electrons is the same that one measures *outside* the sample, $N_B(E_0)$. The integral over the energy E_0 in equation A19 can be separated in two parts, one corresponding to the primary electrons and the other to the backscattered electrons. For normal incidence, for example, the integral becomes

$$\int_0^{E_p} N_z(E_0)\sigma_c^e(E_0)dE_0 = I_p\sigma_c^e(E_p) + \int_0^{E_p} N_B(E_0)\sigma_c^e(E_0)d(E_0) \quad (A20)$$

and it does not depend on z anymore. Thus if one assumes that the density $\rho(z)$ is also constant, equation A19 can be integrated, giving, again for normal incidence of the primary beam,

$$I(E_k) \propto \frac{\rho A_c(E_k)\lambda(E_k)}{\cos \phi} \left[I_p\sigma_c^e(E_p) + \int_0^{E_p} N_B(E_0)\sigma_c^e(E_0)dE_0 \right] \quad (A21)$$

In the case of nonnormal incidence at an angle ϕ_0 the above equation must simply be corrected by a multiplication factor $\cos \phi_0$.

The analysis of the Auger spectra must take into account the influence of the initial core hole on the final state of the process. We see in equation

4 that this influences the final energy of the Auger electrons, E_k. Therefore it also influences all the parameters that depend on E_k. The problem of the final states and of their energy is examined in detail in Section VII. Also important is the analysis of the Auger line shapes in cases in which valence states are involved in the Auger recombination. We shall see that this analysis yields information comparable to that of the valence-band photoemission spectra or of the INS curves.

C. FIELD-EMISSION SPECTROSCOPY AND ION-NEUTRALIZATION SPECTROSCOPY

We emphasized in the previous section that FES and INS have the common feature of probing with enhanced sensitivity the surface electronic states that have high tunneling probability. The information about electronic states, however, is provided in different forms by the two kinds of spectra. This can be illustrated by the mathematical expressions for the corresponding transition probabilities and emission intensities.

The FES process, illustrated in Fig. 17.12, has been analyzed by several authors (41,106,140). Penn and Plummer (106) demonstrated that the field-emission EDC, $N(E)$, can be approximately written

$$N(E) \propto \sum_i (D_0(E_\perp^i))^2 |\, x_i(x_i)\,|^2 \delta(E - E_i) \qquad \text{(A22)}$$

where x_i is the wavefunction of the state $|i\rangle$, which has energy E_i; $D_0(E_\perp^i)$ is the tunneling probability, and x_i is the position of the classic "turning point." Notice that the tunneling probability is a function not of E_i directly, but of the corrected energy:

$$E_\perp^i = E_i - \frac{(\hbar k_\parallel^i)^2}{2m} \qquad \text{(A23)}$$

where $k_\perp^i$ is the parallel component of the electron momentum in the state $|i\rangle$.

Equation A22 shows that the spectrum $N(E)$ probes the sum of electronic densities $|\, x\, |^2$ for all the states at the energy E, that is, the local DOS, at the point x_i. Since x_i is 1–2 Å from the surface this can be considered as the local DOS near the surface. However, the local DOS is heavily weighted in equation A22 by the tunneling probability. For example D_0 eliminates all the states deeper in energy than 1–2 eV from the Fermi level. Indeed, electrons in the deep states cannot tunnel through the barrier, due to its large width. The tunneling probability decreases so rapidly at large distances in energy from the Fermi level that a correction of the raw data is usually necessary. This is done by dividing the experimentally measured spectrum by a calculated free-electron spectrum. This data reduction eliminates the rapid intensity decrease with decreasing energy without introducing spurious

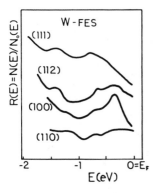

Fig. 17.61.   Field-emission spectra of different tungsten surfaces (see reference 48). The raw data were processed as described in the text to compensate for the dependence of the tunneling probability on the energy.

structure, since the free-electron spectrum is of course smooth (see Fig. 17.61). The D_0 factor also heavily favors states that have large $E_\perp^i$ (i.e., small $k_\parallel^i$) values, since they obviously have larger tunneling probabilities. The adsorbate atoms on the surface strongly influence the tunneling processes and therefore contribute to the $N(E)$ spectra in FES.

For INS the total number of emitted electrons must be calculated (46,48,53), taking into account the tunneling probability to the ion and the two-electron nature of the Auger recombination process. From the energy-level diagram of Fig. 17.11 it is clear that free electrons of energy E_k can be produced by Auger recombination processes involving any pair E_1, E_2 of energy levels that is symmetrically placed around the energy $(E_k + E_n)/2$. The total number of electrons emitted with energy E_k is given by the sum of the products of the DOSs at energies symmetric with respect to E_0, weighted by a transition probability $|M|^2$:

$$N(E_k) \propto \int_{E_0+E_F}^{E_0-E_F} |M|^2 \pi(E_0 - \Delta)\pi(E_0 + \Delta)d\Delta \qquad (A24)$$

Here the transition probability is described by the matrix element M, which includes the tunneling probability to the adsorbed ion. The analysis of this probability shows that the integrand in equation A24 can be replaced by the approximation

$$N(E_k) \propto \int_{E_0+E_F}^{E_0-E_F} \rho_I(E_0 + \Delta)\rho_S(E_0 - \Delta)d\Delta \qquad (A25)$$

where ρ_I and ρ_S are the *local* DOSs at the ion and at the outermost layer of the solid. We can define a *convolution mean* local DOS ρ by the equation

$$N(E_k) \propto \int_{E_0+E_F}^{E_0-E_F} \rho(E_0 + \Delta)\rho(E_0 - \Delta)d\Delta \qquad (A26)$$

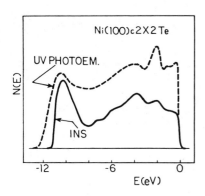

Fig. 17.62. Two different ways to measure the local DOS on a tellurium-covered nickel surface (47). The dashed line is a photoemission EDC taken with a photon energy of 21.2 eV. The solid line is the local DOS deduced from an ion-neutralization spectrum as discussed in the text.

The convolution appearing in equation A26 can be inverted; one thereby obtains the mean DOS ρ from the experimental spectra. This function will exhibit structure due to both ρ_I and ρ_S. Figure 17.62 shows, for example, the mean local DOS ρ deduced from the INS spectrum of Te-covered Ni(100) (47). This function should be compared with the local DOS ρ obtained by uv photoemission and shown in the same figure. We saw in Section VII that the formalism developed here for INS is somewhat similar to that used to discuss certain kinds of Auger line shapes for processes involving valence states.

D. FINAL-STATE POLARIZATION AND RELAXATION EFFECTS

A correct interpretation of the results of any electron-spectroscopy technique requires a correct theoretical description of the corresponding initial and final states $|i\rangle$ and $|f\rangle$. These states are states of the *entire* system (130) and only in a first approximation can they be considered one-electron states. We have already included exchange and correlation effects in calculating photoionization cross sections for atoms. In general, many-body effects play an important role in all electron-spectroscopy techniques. In this section we shall describe some of these effects and their specific consequences in core-level photoemission spectroscopy and in Auger spectroscopy. Since the most important parameter in electron spectroscopy for chemical analysis is the energy, we shall concentrate our attention on the many-body contributions to the transition energy. We emphasize, however, that many-body effects are present in all aspects of the data analysis. For example, they influence the cross sections and, as we saw in Section V, the line shapes.

The energy balance in a photoemission process is defined by equation A4:

$$E_f = E_i + h\nu \qquad (A4)$$

Since at the end of the process there is a photoelectron at (practically) infinite distance from the system, the final-state energy can be written

$$E_f = E_f^* + E \tag{A27}$$

where E is the kinetic energy of the photoelectron and E_f^* is the energy of the system minus the photoelectron. In a one-electron picture, E_f would coincide with E and $|i\rangle$ and $|f\rangle$ would be one-electron states; for atoms, for example, they would be the $1s$, $2s$, $2p$, . . . states. In the general case in which $E_f^* \neq 0$, we shall define

$$E_B = E_i - E_f^* \tag{A28}$$

the *binding energy* of the state $|i\rangle$. In a one-electron picture equation A28 would become $E_B = E_i$. A formally similar result can be obtained even after including in the description of the state the effects of the exchange interaction. The exchange interaction is included in the theoretical description of the electronic states by the Hartree–Fock method or by other self-consistent field formulations (52,83). The Hartree–Fock states can be effectively described in terms of a basis set with hydrogenic quantum-number identification. Furthermore, the Hartree–Fock method associates with each one of its states a "pseudo-energy-eigenvalue." We shall call E_i^{HF} the pseudo-energy-eigenvalue of the Hartree–Fock state $|i\rangle$. The *Koopmans theorem* (67) identifies this energy with the binding energy:

$$E_B = E_i^{HF} \tag{A29}$$

in the limit in which the other Hartree–Fock states are *not* modified by the removal of one electron from the Hartree–Fock state $|i\rangle$, and as long as the Hartree–Fock method gives a reasonably "true" description of the system. The Koopmans theorem, equation A29, is a first step beyond the crudest one-electron approximation. It includes exchange effects, but it does not include other non-one-electron contributions to the binding energy. In general the binding energy must be written including all these corrections:

$$E_B = E_i^{HF} + E_i^R + \Delta E_{corr} + \Delta E_{rel} \tag{A30}$$

The three correcting terms to the Koopmans theorem in equation A30 are the *relaxation energy* E_i^R and the differences in the correlation and relativistic energies between the initial and final states. The latter two terms are usually small, so that equation A30 can be approximated by

$$E_B = E_i^{HF} + E_i^R \tag{A31}$$

and as long as E_i^{HF} can be identified with E_i, the relaxation energy is linked to E_f^*:

$$E_f^* = -E_i^R \qquad (A32)$$

From a qualitative point of view the relaxation-energy term is due to the rearrangement of the charge distribution of the other electrons due to the removal of the emitted electron. Before the electron is emitted the ground-state charge distribution is dictated by the requirement for minimizing the energy and by the Pauli principle—which in quantum-mechanical terms influences the form of the wavefunction of the allowed states and their occupancy. After the charge of the emitted electron is removed, these conditions are no longer met, and some rearrangement of the system must be expected. The driving force for the rearrangement is a lowering of the total energy of the system. The energy balance of the photoemission process is altered, and this is described by the relaxation-energy term E_i^R in equation A31.

In the case of isolated atoms, the term E_i^R consists of two different contributions, $E_i^{R,intra}$ and $E_i^{R, outer}$. The first term arises from relaxation of the electrons in the same shell from which the photoelectron was removed (intrashell relaxation) and the second term from electrons in the outer shells—the relaxation of electrons in the inner shells can usually be neglected. The term $E_i^{R,outer}$ is the largest one, and can be calculated with good accuracy by assuming that the removed electron charge completely shielded the outer-shell electrons from one of the nuclear proton charges. In this approach the real core is substituted for by an "equivalent core," following the method discussed in Section IV.

When the photoemitting atom belongs to a molecule or to a solid, the relaxation-energy term E_i^R contains extra-atomic contributions as well as the above intra-atomic contribution. The different contributions are assumed to be linearly combined with each other:

$$E_i^R = E_i^{R,intra} + E_i^{R,outer} + E_i^{R,extra} \qquad (A33)$$

Notice that only *electronic* relaxation gives strong contributions to $E_i^{R,extra}$, since the nuclear positions do not have time to change appreciably during the photoemission process. This is equivalent to the Franck–Condon principle in optics and is a general consequence of the difference in mass between electrons and nuclei. The nature of the extra-atomic relaxation is different for different kinds of molecules and solids. For molecules, the term $E_i^{R,extra}$ can be calculated using self-consistent molecular-orbital approaches. The extra-atomic relaxation indeed involves the polarization of the charge in the chemical bonds of the molecule, and the corresponding energy can be calculated using the methods valid for the bonds themselves. If the photoemission process starts from a core level, the calculations are simplified by the space localization of the core hole that it creates. A similar simplified case is that of the *localized* molecular orbitals. The extra-atomic relaxation energy is in general small, and is practically negligible for nonlocalized molecular orbitals.

In the case of *ionic* solids, the extra-atomic relaxation term arises from a dynamic polarization of the charge in the ions near that in which the photoemission process starts. Calculations by Fadley et al. (33) and by Citrin and Thomas (20) demonstrate that the term $E_i^{R,extra}$ is in this case of the order of 0–3 eV.

In the case of *semiconducting* solids the situation is similar to that for covalent free molecules, and the molecular results can often be used by extrapolation. This is the case, for example, of the K-shell relaxation energy for graphite (24).

For *metallic* solids the term $E_i^{R,extra}$ was thought for a long time to be negligible (3). This assumption was proved wrong, and calculations of the relaxation energy now give values of up to 10 eV, or even more (76). The qualitative picture of the relaxation process is that of a quasiexcitonic state in which the charge of the core hole is shielded by electrons in states just above the Fermi level. Quantitative calculations of the relaxation energy are usually based on the Friedel sum rule (38).

In summary, the photoemission binding energy is affected by both intra-atomic and extra-atomic relaxation. Similar relaxation processes of course affect the energy balance in AES—we saw in Section VI how these effects can be taken into account. In many cases there is a simple relation between the photoemission relaxation energies and the Auger relaxation energies. For example, Kowalczyk et al. (76) found in alkali halides that the relaxation term in AES is about twice the corresponding photoemission term, which we have seen above to be due primarily to neighbor-ion electronic polarization.

REFERENCES

1. Altick, P. L., and A. E. Glassgold, *Phys. Rev.*, **133**, A61 (1964).

2. Arthur, J. R., *J. Appl. Phys.*, **39**, 4032 (1968); Cho, A. Y. *J. Appl. Phys.*, **41**, 782 (1970).

3. Bearden, T. A., and A. F. Burr, *Rev. Mod. Phys.*, **31**, 616 (1967).

4. Brillson, L. J., G. Margaritondo, and N. G. Stoffel, *Phys. Rev. Lett.*, **44**, 667 (1980).

5. Brillson, L. J., G. Margaritondo, N. G. Stoffel, R. S. Bauer, R. Z. Bachrach, and G. Hansson, *J. Vac. Sci. Technol.*, **17**, 880 (1980).

6. Campagna, M., E. Bucher, G. K. Wertheim, D. N. E. Buchanan, and L. D. Longinotti, *Phys. Rev. Lett.*, **32**, 885 (1974).

7. Campagna, M., D. T. Pierce, F. Meier, K. Sattle, and H. Siegmann, in L. Marton, Ed., *Adv. Electronics Electron Phys.*, Vol. 41, Academic, New York, 1976, p. 113.

8. Cardona, M., and L. Ley, *Photoemission in Solids*, Springer-Verlag, New York, 1978, Chapter I.

9. Carlson, T. A., *Phys. Rev.*, **156**, 142 (1967).

10. Carlson, T. A., M. O. Krause, E. A. Grimm, J. D. Allen, Jr., D. Mehaffy, P. R. Keller, and J. W. Taylor, *Phys. Rev.*, **A23**, 1316 (1981), and references therein.

11. Chadi, D. J., R. S. Bauer, R. H. Williams, G. V. Hansson, R. Z. Bachrach, J. C. Mikkelsen, Jr., F. Houzay, G. M. Guichar, P. Pinchaux, and Y. Petroff, *Phys. Rev. Lett.*, **44**, 119 (1980).

12. Cheng, C. C., in P. Kane and G. B. Larrabee, Eds., *Characterization of Solid Surfaces*, Plenum, New York, 1974, Chapter 20; Davis, L. E., A. Joshi, and P. W. Palmberg, in A. W. Czanderna, Ed., *Methods of Surface Analysis*, Elsevier, Amsterdam, 1975, Chapter 5.

13. Chiang, T.-C., aand D. E. Eastman, *Phys. Rev.*, **B22**, 2940 (1980).

14. Chiarotti, G., and S. Nannarone, *Surface Sci.*, **49**, 315 (1975).

15. Chiarotti, G., S. Nannarone, R. Pastore, and P. Chiaradia, *Phys. Rev.*, **B4**, 3396 (1971).

16. Chou, T. S., M. L. Perlman, and R. E. Watson, *Phys. Rev.*, **B14**, 3248 (1976).

17. Chung, M. F., and L. H. Jenkins, *Surface Sci.*, **22**, 479 (1970).

18. Citrin, P. H., and D. R. Hamann, *Phys. Rev.*, **B10**, 4948 (1974).

19. Citrin, P. H., and J. E. Rowe, unpublished.

20. Citrin, P. H., and T. D. Thomas, *J. Chem. Phys.*, **57**, 4446 (1972).

21. Citrin, P. H., G. K. Wertheim, and Y. Baer, *Phys. Rev. Lett.*, **41**, 1425 (1978).

22. Cooper, J. W., *Phys. Rev. Lett.*, **13**, 762 (1974).

23. Cooper, J., and R. N. Zare, in S. Geeltman, K. Mahantharpa, and W. Brittin, Eds., *Lectures in Theoretical Physics*, Gordon & Breach, New York, 1969, p. 317.

24. Davis, D. W., and D. A. Shirley, *J. Electron Spectr.*, **3**, 137 (1974).

25. Davis, G. D., and M. G. Lagally, *J. Vac. Sci. Technol.*, **15**, 1311 (1978).

26. Dietz, E., and D. E. Eastman, *Phys. Rev. Lett.*, **41**, 1674 (1978).

27. DiSalvo, F. J., D. E. Moncton, and J. Waszczak, *Phys. Rev.*, **B14**, 4321 (1976).

28. Doniach, S., and M. Sunjic, *J. Phys. C.*, **3**, 285 (1970).

29. Eastman, D. E., T.-C. Chiang, P. Heimann, and F. J. Himpsel, *Phys. Rev. Lett.*, **45**, 656 (1980).

30. Eastman, D. E., J. J. Donelon, N. C. Hien, and F. J. Himpsel, *Nucl. Instr. Meth.*, **122**, 327 (1980).

31. Eastman, D. E., and W. D. Grobman, *Phys. Rev. Lett.*, **28**, 1378 (1972).

32. Eastman, D. E., W. D. Grobman, J. L. Freeouf, and M. Eburdak, *Phys. Rev.*, **B9**, 3473 (1974).

33. Fadley, C. S., S. B. M. Hagstrom, M. P. Klein, and D. A. Shirley, *J. Chem. Phys.*, **48**, 3779 (1968).

34. Fano, U., *Phys. Rev.*, **124**, 1866 (1961).

35. Fano, U., and J. W. Cooper, *Rev. Mod. Phys.*, **40**, 441 (1968).

36. Feldman, L. C., and I. Stensgaard, *Progress in Surface Science* (in press).

37. Finn, P., R. K. Pearson, J. M. Hollander, and W. L. Jolly, *Inorg. Chem.*, **10**, 378 (1971).

38. Friedel, J., *Philos. Mag.*, **43**, 153 (1952).

39. Friedman, R. F., J. Hudis, M. L. Perlman, and R. W. Watson, *Phys. Rev.*, **B8**, 2433 (1973).

40. Froitzheim, H., and H. Ibach, *Z. Phys.*, **269**, 17 (1974).

41. Gadzuk, J. W., and E. W. Plummer, *Rev. Mod. Phys.*, **45**, 487 (1973).

42. Glupe, G., and W. Mehlhorn, *Phys. Lett.*, **25A**, 274 (1967).

43. Goddard, W. A., A. Redondo, and T. C. McGill, *Solid State Commun.*, **18**, 981 (1976).

44. Green, T. S., and G. A. Proca, *Rev. Sci. Instr.*, **41**, 1409 (1970).

45. Gudat, W., and C. Kunz, *Phys. Rev. Lett.*, **29**, 169 (1972).

46. Hagstrum, H. D., *Phys. Rev.*, **150**, 495 (1966).

47. Hagstrum, H. D., and G. E. Becker, in E. Drauglis and R. I. Jaffee, Eds., *The Physical Basis for Heterogeneous Cathalysis*, Plenum, New York, 1975, p. 173.

48. Hagstrum, H. D., J. E. Rowe, and J. C. Tracy, *Experimental Methods in Catalytic Research*, Vol. 3, Academic, New York, 1976, p. 41.

49. Haneman, D., *Phys. Rev.*, **121**, 1093 (1961).

50. Harrison, W. A., *J. Vac. Sci. Technol.*, **14**, 1016 (1977).

51. Harrison, W. A., *Electronic Structure*, Freeman, San Francisco, 1980.

52. Hartree, D. R., *The Calculation of Atomic Structures*, Wiley, New York, 1957, Chapter III.

53. Heine, V., *Phys. Rev.*, **151**, 561 (1966).

54. Hermanson, J., *Solid State Commun.*, **22**, 9 (1977).

55. Himpsel, F. J., *Appl. Optics*, **19**, 1364 (1560), and references therein.

56. Himpsel, F. J., P. Heimann, T.-C. Chiang, and D. E. Eastman, *Phys. Rev. Lett.*, **45**, 1112 (1980).

57. Hollinger, G., P. Kumurdijan, J. M. Mackowsky, P. Pertosa, L. Porte, and T. M. Duc, *J. Electron. Spectr.*, **5**, 237 (1974).

58. Houston, J. E., and R. L. Park, *Phys. Rev.*, **B5**, 3808 (1972).

59. Ibach, H., H. Froitzheim, and K. G. Mills, *Phys. Rev.*, **B11**, 4980 (1975).

60. Ibach, H., K. Horn, R. Dorn, and H. Luth, *Surface Sci.*, **38**, 433 (1973).

61. Ibach, H., and J. E. Rowe, *Phys. Rev.*, **B9**, 1951 (1974).

62. Jolly, W. L., in C. R. Brundle, and A. D. Baker, Ed., *Electron Spectroscopy, Academic*, New York, 1977, Vol. 1, Chapter 3.

63. Kane, E. O., *Phys. Rev.*, **146**, 558 (1966).

64. Katnani, A. D., R. R. Daniels, T. X. Zhao, and G. Margaritondo, *J. Vac. Sci. Technol.* **20**, 662 (1982).

65. Kevan, S. D., D. H. Rosenblatt, D. Denley, B.-C. Lu, and D. A. Shirley, *Phys. Rev. Lett.*, **41**, 1565 (1978).

66. Kim, K. S., and N. Winograd, *Chem. Phys. Lett.*, **30**, 91 (1975).

67. Koopmans, T., *Physica*, **1**, 104 (1934).

68. Kowalczyk, S. P., Ph.D. Thesis, Univ. of California, Berkeley, 1976 (*Lawrence Berkeley Lab. Rep.*, **LBL-4319**).

69. Kowalczyk, S. P., R. A. Pollak, R. A. McFeely, L. Ley, and D. A. Shirley, *Phys. Rev.*, **B8**, 2387 (1973).

70. Kuyatt, C. E., and J. A. Simpson, *Rev. Sci. Instr.*, **38**, 103 (1967), and references therein.

71. Lagally, M., and M. B. Webb, *Solid State Physics*, **28**, 301 (1973).

72. Lapeyre, G. J., A. D. Baer, J. Hermanson, J. Anderson, J. A. Knapp, and P. L. Gobby, *Phys. Rev. Lett.*, **33**, 1290 (1977).

73. Larsen, P. K., S. Chiang, and N. V. Smith, *Phys. Rev.*, **B15**, 3320 (1977).

74. Larsen, P. K., G. Margaritondo, J. E. Rowe, M. Schluter, and N. V. Smith, *Phys. Lett.*, **58A**, 623 (1976).

75. Lee, P. A., *Phys. Rev.*, **B13**, 5261 (1976); Citrin, P. H., P. Eisenberg, and B. M. Kincaid, *Phys. Rev. Lett.*, **36**, 13346 (1976); Stohr, J. *J. Vac. Sci. Technol.*, **16**, 37 (1979).

76. Ley, L., S. P. Kowalczyk, F. R. McFeely, R. A. Pollak, and D. A. Shirley, *Phys. Rev.*, **B8**, 2392 (1973).

77. Ley, L., R. A. Pollak, F. R. McFeely, S. P. Kowalczyk, and D. A. Shirley, *Phys. Rev.*, **B9**, 600 (1974).

78. Li, C. H., and S. Y. Tong, *Phys. Rev.*, **42**, 901 (1979); **43**, 526 (1979).
79. Lundqvist, B., *Phys. Kondens. Mater.*, **6**, 193 (1967).
80. Mahan, G. D., *Phys. Rev.*, **163**, 612 (1967); Nozieres, P., and C. T. DeDominicis, *Phys. Rev.*, **178**, 1097 (1969).
81. Mahan, G. D., *Phys. Rev.*, **B2**, 4334 (1970).
82. Manson, S. T., *Chem. Phys. Lett.*, **19**, 76 (1973).
83. Manson, S. T., reference 8, Chapter 3, and references therein.
84. Manson, S. T., and J. W. Cooper, *Phys. Rev.*, **165**, 182 (1968).
85. Margaritondo, G., *Surface Interf. Anal.*, **3**, 5 (1981).
86. Margaritondo, G., L. J. Brillson, and N. G. Stoffel, *Appl. Phys. Lett.*, **37**, 917 (1980).
87. Margaritondo, G., A., Franciosi, N. G. Stoffel, and H. S. Edelman, *Solid State Commun.*, **36**, 297 (1980).
88. Margaritondo, G., F. Levy, N. G. Stoffel, and A. D. Katnani, *Phys. Rev.*, **B21**, 5768 (1980).
89. Margaritondo, G., and J. E. Rowe, *J. Vac. Sci. Technol.*, **17**, 561 (1980).
90. Margaritondo, G., and J. E. Rowe, unpublished.
91. Margaritondo, G., J. E. Rowe, C. M. Bertoni, C. Calandra, and F. Manghi, *Phys. Rev.*, **B23**, 509 (1981).
92. Margaritondo, G., J. E. Rowe, C. M. Bertoni, C. Calandra, and F. Manghi, *Phys. Rev.*, **B20**, 1538 (1979).
93. Margaritondo, G., J. E. Rowe, and S. B. Christman, *Phys. Rev.*, **B14**, 5396 (1976).
94. Margaritondo, G., J. E. Rowe, and S. B. Christman, *Phys. Rev.*, **B15**, 3844 (1977).
95. Margaritondo, G., J. E. Rowe, and S. B. Christman, *Phys. Rev.*, **B19**, 3266 (1979).
96. Margaritondo, G., J. E. Rowe, M. Schluter, and H. Kasper, *Solid State Commun.*, **22**, 753 (1977).
97. Margaritondo, G., N. G. Stoffel, A. D. Katnani, H. S. Edelman, and C. M. Bertoni, *J. Vac. Sci. Technol.*, **18**, 784 (1981).
98. McDonnell, L., D. P. Woodruff, and B. W. Holland, *Surface Sci.*, **51**, 24 (1975).
99. Mies, F. H., *Phys. Rev.*, **175**, 164 (1968).
100. Mott, N. F., and R. W. Gurney, *Electronic Processes in Ionic Crystals*, Dover, New York, 1964, p. 56.
101. Palmberg, P. W., *Appl. Phys. Lett.*, **13**, 183 (1968).
102. Palmberg, P. W., *J. Electron Spectr.*, **5**, 691 (1974).
103. Palmberg, P. W., G. K. Bohn, and J. C. Tracy, *Appl. Phys. Lett.*, **15**, 254 (1969).
104. Park. R. L., and J. E. Houston, *J. Vac. Sci. Technol,*, **11**, 1 (1974).
105. Peierls, R. E., *Quantum Theory of Solids*, Oxford Univ. Press, New York, 1955, p. 108.
106. Penn, D. R., and E. W. Plummer, *Phys. Rev.*, **B9**, 1216 (1974).
107. Phillips, J. C., *Phys. Rev.*, **168**, 905 (1968).
108. Picco, P., I. Abbati, L. Braicovich, F. Cerrina, F. Levy, and G. Margaritondo, *Phys. Lett.*, **65A**, 447 (1978).
109. Pollak, R. A., in reference 129.
110. Pollak, R. A., F. Holtzberg, J. L. Freeouf, and D. E. Eastman, *Phys. Rev. Lett.*, **33**, 820 (1974).
111. Pollak, R. A., L. Ley, F. R. McFeely, S. P. Kowalczyk, and D. A. Shirley, *J. Electron Spectr.*, **3**, 381 (1974).
112. Potts, A. W., T. A. Williams, and W. C. Price, *Proc. R. Soc. London*, **A341**, 147 (1974).
113. Powell, J. C., *Phys. Rev. Lett.*, **30**, 1179 (1973).

114. Price, W. C., in C. P. Brundle, and A. D. Baker, Ed., *Electron Spectroscopy*, Academic, New York, 1977, Vol. 1, Chapter 4.

115. Price, W. C., A. W. Potts, and D. G. Streets, in D. A. Shirley, Ed., *Electron Spectroscopy*, North Holland, Amsterdam, 1972, p. 187.

116. Quattropani, A., F. Bassani, G. Margaritondo, and G. Tinivella, *Nuovo Cim.*, **51B**, 335 (1979).

117. Raether, H., *Solid State Excitations by Electrons*, Springer, Berlin, 1965.

118. Redondo, A., C. A. Swarts, W. A. Goddard, and T. C. McGill, *J. Vac. Sci. Technol.*, **19**, 498 (1981).

119. Rowe, J. E., S. B. Christman, and E. E. Chaban, *Rev. Sci. Instr.*, **44**, 1675 (1973).

120. Rowe, J. E., and H. Ibach, *Phys. Rev. Lett.*, **32**, 421 (1974).

121. Rowe, J. E., H. Ibach, and H. Froitzheim, *Surface Sci.*, **48**, 44 (1975).

122. Rowe, J. E., and G. Margaritondo, unpublished.

123. Rowe, J. E., and G. Margaritondo, unpublished.

124. Rowe, J. E., and G. Margaritondo, unpublished.

125. Rowe, J. E., G. Margaritondo, and S. B. Christman, unpublished.

126. Rowe, J. E., G. Margaritondo, H. Ibach, and H. Froitzheim, *Solid State Commun.*, **20**, 277 (1976).

127. Rowe, J. E., and J. C. Phillips, *Phys. Rev. Lett.*, **32**, 1315 (1974).

128. Shirley, D. A., *Phys. Rev.*, **A7**, 1520 (1973).

129. Shirley, D. A., in reference 8, Chapter 4.

130. Shirley, D. A., in reference 114, Chapter 2, and references therein.

131. Slater, J. C., *The Quantum Theory of Atomic Structure*, McGraw-Hill, New York, 1960, Chapter 17.

132. Slater, J. C., *ibid.*, Chapter 18, Fano, U., *Phys. Rev.*, **124**, 1866 (1961); Mies, F., *Phys. Rev.*, **175**, 164 (1968).

133. Smith, N. V., P. K. Larsen, and M. M. Traum, *Rev. Sci. Instr.*, **48**, 454 (1977).

134. Smith, N. V., and M. M. Traum, *Phys. Rev. Lett.*, **31**, 1247 (1973), and references therein.

135. Smith, N. V., and M. M. Traum, *Phys. Rev.*, **B11**, 2087 (1975).

136. Smith, N. V., M. M. Traum, J. A. Knapp, J. Anderson, and G. J. Lapeyre, *Phys. Rev.*, **B13**, 4462 (1976).

137. Spicer, W. E., *Phys. Rev. Lett.*, **11**, 243 (1963).

138. Spicer, W. E., *Phys. Rev.*, **112**, 114 (1968).

139. Stoffel, N. G., and G. Margaritondo, *Rev. Sci. Instr.* **53**, 18 (1982), and references therein.

140. Swanson, L. W., and A. E. Bell, *Adv. Electron. Electron Pys.*, **32**, 1403 (1973).

141. Traum, M. M., N. G. Stoffel, and G. Margaritondo, unpublished.

142. Truhlar, D. G., J. Abdallah, Jr., and R. L. Smith, *Adv. Chem. Phys.*, **25**, 211 (1974).

143. Viselov, F. I., B. C. Kurbatov, and A. N. Terenin, *Dokl. Akad. Nauk SSSR*, **138**, 1320 (1961).

144. Vrakking, J. J., and F. Meyer, *Phys. Rev.*, **A9**, 1932 (1974).

145. Wagner, L. F., and W. E. Spicer, *Phys. Rev. Lett.*, **28**, 1381 (1972).

146. Walsh, A. D., *J. Chem. Soc. London*, 2260 (1953).

147. Wang, G. C., B. I. Dunlas, R. J. Celotta, and D. T. Pierce, *Phys. Rev. Lett.*, **42**, 1349 (1979), and references therein.

148. Watson, R. E., J. F. Herbst, and J. W. Wilkins, *Phys. Rev.*, **B14**, 18 (1976).

149. Watson, R. E., J. Hudis, and M. L. Perlman, *Phys. Rev.*, **B4**, 4139 (1971).

150. Watson, R. E., and M. L. Perlman, *Photoelectron Spectrometry*, Vol. 24, Springer, Berlin 1974, p. 83.

151. Weaver, J. H., and G. Margaritondo, *Science*, **206**, 151 (1979).

152. Weeks, S. P., J. E. Rowe, S. B. Christman, and E. Chaban, *Rev. Sci. Instr.*, **50**, 12449 (1979).

153. Wertheim, G. K., and P. H. Citrin, in reference 8, Chapter 5.

154. Winick, H., G. Brown, K. Halbach, and J. Harris, *Physics Today*, **34**, 50 (1981).

155. Woodruff, D. P., B. W. Holland, N. V. Smith, H. H. Farrell, and M. M. Traum, *Phys. Rev.*, **B41**, 1130 (1978); Liebsch, A., *Phys. Rev.*, **B13**, 544 (1976).

156. *Handbook of Auger Electron Spectroscopy*, Physical Electronics Ind., Edina, Minnesota, 1976.

157. *Physics of Color Centers*, W. B. Fowler, Ed., Academic, New York, 1968.

158. See references 4 and 5 in reference 114.

159. See Lagally, M., in F. W. Wiffen and J. A. Spitznagel, Eds., *Advanced Techniques for the Characterization of Microstructures*, AIME, Pittsburg, in press.

160. See Powell, B. D., and D. P. Woodruff, *J. Phys. E*, **8**, 548 (1975), and references therein.

ADDENDUM

Electron spectroscopy is a rapidly developing area and the recent months have been particularly productive. A complete discussion of all the new developments would be beyond the limits of this presentation, but an update of the material presented in this chapter seems necessary. The update will be limited to a few of the recent trends in electron spectroscopy.

A. PHOTOEMISSION SPECTROSCOPY

Synchroton-radiation photoemission has definitely become the most productive area of photoemission spectroscopy. In fact, the most spectacular instrumentation developments in electron spectroscopy are the new synchrotron–radiation sources commissioned in the 1980s, which allow improved energy resolution of the order of 200 meV for core-level XPS studies. In the United States, there are now three federally funded facilities suitable for photoemission experiments: SSRL at Stanford, NLS at the Brookhaven National Laboratory, and SRC in Wisconsin. The expanded availability of photon sources has enhanced the use of synchrotron-radiation photoemission and stimulated the development of several new techniques. For example, the photon-energy tunability is now routinely used to systematically vary the surface sensitivity in microscopic diffusion experiments (1), and photoemission is routinely used as a microscopic morphology probe (5). Photo-energy tunability and polarization have been used extensively to identify the nature of molecular states in the gas phase (2). Great progress has also been made in the area of electronic-state identification for solid systems. One example of this progress is the *resonant photoemission* technique (3). This is based on the quantum interference between different processes initiated by the absorption of a photon and leading to the creation of a free electron in a vacuum. One of these processes is, for example, the direct creation of a photoelectron on excitation of a valence electron. Another possible path is the excitation of a core electron to a conduction-band state, followed by the Auger-like filling of a core hole involving a valence electron that becomes an Auger electron of energy equal to that of a direct photoelectron. The interference between these two processes gives strong modulations of the photoelectric cross section at photon energies close to the threshold for excitation of the core electron into a conduction-band state. Since the core-electron wavefunction is spatially localized near the atomic centers, the resonant photoemission processes are easily visible for localized d-derived and f-derived valence states. The corresponding modulations of the cross section are extremely helpful in identifying d-components or f-components in a complicated valence-band structure.

Among the other synchrotron-radiation photoemission methods, a rapidly expanding technique has been surface-EXAFS. New detection methods, involving, for example, ion fragmentation, have recently expanded the range of applications of this approach; however, instrumentation improvements are still necessary to extend EXAFS to the intermediate photon-energy range 500–2000 eV between the energies of maximum efficiency for crystal monochromators and for grating monochromators. Of particular importance in the near future will be the extensive application of surface-EXAFS to systems containing aluminum, whose K-edge falls in that range.

B. OTHER AREAS OF DEVELOPMENT

Perhaps the most spectacular growth in electron spectroscopy concerns the surface vibrational applications of electron energy-loss spectroscopy. Commercial spectrometers are now available that deliver the required resolution and background rejection for the performance of surface vibrational experiments. As a result, a growing number of experimentalists are now using this technique to identify the nature of surface chemical bonds. It has been suggested that surface vibrational spectroscopy may become in the second half of the 1980s the leading electron-spectroscopy technique, along with photoemission.

Another area of expansion has recently appeared in electron energy-loss spectroscopy. This is an alternate approach to surface-EXAFS based on the detection of the absorption fine structure in energy-loss spectra (4). The technique has advantages and disadvantages with respect to the normal surface-EXAFS technique. Its signal-to-noise ratio is generally better, but the data interpretation may be more complicated and there are potential severe problems of contamination by the primary electron beam. The most attractive feature of energy-loss surface-EXAFS is, of course, its low cost and high flexibility, caused by the replacement of a synchrotron-radiation source with a simple electron gun.

There is at present a renewed, strong interest in a borderline technique between electron and ion spectroscopy—electron-stimulated desorption. This technique is based on the detection of ions desorbed from a surface by a beam of electrons. Electron-stimulated desorption and the complementary technique, photon-stimulated desorption, can be used, at least in principle, to study the geometrical and even the dynamic properties of the surface chemical processes, but development of the field requires a good understanding of the physics of the desorption processes, which is not yet available. One interesting recent development (6) was the extension of electron-stimulated-desorption spectroscopy to neutral species, made possible by the detection of the secondary radiation they emit while flying away from the surface.

As a final comment, we would like to emphasize that the most important general trend that has emerged in electron spectroscopy in recent years is the tendency to use many techniques in parallel instead of using only a single technique. This is a very positive attitude, since it is becoming more and more clear that no single technique can give satisfactory answers to all the relevant questions about a system. It is from this "integrated" use of electron spectroscopy that the most important results are expected in the future.

ADDENDUM REFERENCES

1. Brillson, L. J., M. L. Slade, A. D. Katnani, M. K. Kelly, and G. Margaritondo, *Appl. Phys. Lett.*, **44**, 110 (1984), and references therein.
2. Carlson, T. A., M. A. Krause, F. A. Grimm, P. K. Keller, and J. W. Taylor, *Chem. Phys. Lett.*, **87**, 552 (1982), and references therein.
3. Guillot, C., Y. Ballu, J. Paigne, J. Lecante, K. P. Jain, P. Thiry, Y. Petroff, and L. M. Falicov, *Phys. Rev. Lett.*, **39**, 1632 (1977).
4. Rosei, R., M. De Crescenzi, F. Sette, C. Quaresima, A. Savoia, and P. Perfetti, Phys. Rev., **B28**, 1161 (1983).
5. Stoffel, N. G., M. K. Kelly and G. Margaritondo, *Phys. Rev.*, **B27**, 6571 (1983).
6. Tolk, N. H., L. C. Feldman, J. S. Kraus, R. J. Morris, M. M. Traum, and J. C. Tully, *Phys. Rev. Lett.*, **46**, 134 (1981).

Subject Index